BioNMR in Drug Research

Edited by
Oliver Zerbe

Methods and Principles in Medicinal Chemistry

Edited by
R. Mannhold, H. Kubinyi, G. Folkers

BioNMR in Drug Research

Edited by
Oliver Zerbe

WILEY-VCH

Series Editors

Prof. Dr. Raimund Mannhold
Biomedical Research Center
Molecular Drug Research Group
Heinrich-Heine-Universität
Universitätsstraße 1
40225 Düsseldorf
Germany
raimund.mannhold@uni-duesseldorf.de

Prof. Dr. Hugo Kubinyi
BASF AG Ludwigshafen
c/o Donnersbergstraße 9
67256 Weisenheim am Sand
Germany
kubinyi@t-online.de

Prof. Dr. Gerd Folkers
Department of Applied Biosciences
ETH Zürich
Winterthurer Straße 190
8057 Zürich
Switzerland
folkers@pharma.ethz.ch

Volume Editor

Dr. Oliver Zerbe
Department of Applied Biosciences
ETH Zürich
Winterthurer Straße 190
8057 Zürich
Switzerland
oliver.zerbe@pharma.anbi.ethz.ch

Library of Congress Card No.: applied for

British Library Cataloguing-in-Publication Data
A catalogue record for this book is available from the British Library.

Bibliographic information published by Die Deutsche Bibliothek
Die Deutsche Bibliothek lists this publication in the Deutsche Nationalbibliografie; detailed bibliographic data is available in the Internet at http://dnb.ddb.de.

Printed in the Federal Republic of Germany
Printed on acid-free paper

Typesetting K+V Fotosatz GmbH, Beerfelden
Printing Strauss Offsetdruck GmbH, Mörlenbach
Bookbinding Litges & Dopf Buchbinderei GmbH, Heppenheim

ISBN 3-527-30465-7

Contents

Preface

Some decades have gone since NMR spectroscopy first hit the analytical scene, and yet its capabilities and applications continue to evolve. Originally designed as a way to verify the structure of relatively small compounds, the technology of NMR boomed and became a valuable means for studying protein structure. Traditionally, X-ray crystallography has been used for solving the structure of proteins; however, it is useful only for those that can be coaxed into a crystalline state. The development of multidimensional NMR and more powerful instruments opened the door for solving the structure of proteins and peptides in an aqueous environment, as they exist in biological systems. NMR allows one to observe the physical flexibility of proteins and the dynamics of their interactions with other molecules, a huge advantage when studying the biochemical function of proteins. The structural information, achieved from NMR studies, can be used to understand the function, mechanism of action, and binding specificity of these proteins.

In addition, NMR can be used to design high affinity ligands for proteins using the SAR by NMR approach (Structure Activity Relationships by Nuclear Magnetic Resonance) introduced by Stephen Fesik. Using this method, small organic molecules that bind to proximal subsites of a protein are identified, optimized, and linked together. The method reduces the amount of chemical synthesis and time required for the discovery of high affinity ligands.

These few remarks highlight that NMR spectroscopy has adapted a central role in drug discovery and design. It is the intention of the present volume to document this role in adequate detail. Accordingly, the book is divided into four larger sections. A methodological section summarizes the technical state of the art including general aspects of spectroscopy; the following section deals in detail with spectroscopic techniques for structure determination of commonly encountered classes of biomolecules. NMR techniques for investigating drug-receptor complexes as well as strategies for drug development using NMR are the topics of the remaining two sections.

The series editors would like to thank the authors and in particular Oliver Zerbe as the volume editor that they devoted their precious time to compiling and structuring the comprehensive information on NMR techniques in drug research. Last, but not least we want to express our gratitude to Frank Weinreich and Gudrun Walter from Wiley-VCH publishers for the fruitful collaboration.

August 2002

Raimund Mannhold, Düsseldorf
Hugo Kubinyi, Ludwigshafen
Gerd Folkers, Zürich

Foreword

Two decades ago, NMR in pharmaceutical industry was mainly used as an analytical tool to validate and identify compounds that were synthesized by medicinal chemists. Following the rapid developments in biomolecular NMR that resulted in the first polypeptide structure being published by the Wüthrich group in the mid eighties, it has thereafter increasingly been used to also determine structures of biomolecules. During the late eighties/early nineties triple-resonance experiments and isotope labeling methods were established which helped to increase the molecular size limits remarkably, and, even more importantly, allowed to more rapidly assign the proton frequencies. While the development of spectroscopic tools still continued, characterized by the advent of TROSY techniques and the use of dipolar couplings, investigations into interactions between drugs or lead candidates and their corresponding receptors were initiated. Early work from Feeney and Roberts in that area was followed by the systematic use of these techniques by the Abott group lead by S. Fesik. In particular, detection of low-affinity ligands has proven to be invaluable in the early development phase of drugs. Accordingly, this book is intended to present an update both in the area of spectroscopic techniques as well as on methodology for screening.

The book is separated into five major sections: One short section on general aspects of spectroscopy, molecular biology and data evaluation is followed by an introduction into the NMR of commonly encountered classes of biomolecules. Thereafter, recent developments in spectroscopic techniques are highlighted. The next section describes experiments and practical aspects useful for the characterization of protein-ligand interactions. The final section presents an account on strategies for drug development using NMR written by experts from pharmaceutical industry.

It is also intended to present techniques which are not routinely applied nowadays but which have the potential to become very useful. Membrane-bound proteins such as G-protein coupled receptors, for example, are important pharmaceutical targets but have so far been very difficult to study by solution-state NMR or single crystal diffraction, the latter mainly due to the enormous difficulties in preparing crystal suitable for diffraction, and the more involved molecular biology to produce them. Solid-state NMR is a technique which may contribute much to that field, and the present progress both in the field of molecular biology as well as in solid-state NMR methodology will certainly stimulate groups to tackle membrane proteins in the future. To provide those readers, which lack experience in that field, with the necessary background, a more detailed introduction to the concepts and applications of solid-state NMR is provided.

The chapters are aiming at providing an overview into a particular field. For a more rigorous description of the experiments, the reader is usually referred to the original literature for which the references are included. A particular emphasis has been placed on practical aspects of work, which originated from the authors' experience with the experiments. As part of this effort, a short chapter on trouble shooting of hardware has been included, in order to make sure that no precious measuring time is wasted. In general, these practical hints are intended for those that decide to move into the field and to provide them with the knowledge necessary to successfully use the tools but should also help to recognize limitations of the techniques.

Biomolecular NMR is a complex technique, which is still rapidly evolving, and people have to learn how the strength of NMR can be exploited best. The editor hopes that the readers will be lucky in choosing the right targets together with the most efficient strategies and wishes that the chapters will help them to do so. Rational drug development is certainly strongly depending upon the input of experimental structural data and hence it is the editor's firm belief that NMR will always play an important role in the drug development process and that new methodology will help to even increase it.

I am deeply grateful to all authors who have patiently responded to my numerous requests for additions and modifications. I would like to especially thank Drs. Marcel Blommers, Wolfgang Jahnke, Werner Klaus, Alfred Ross and Hans Senn for very helpful discussions during the early phase of the book. I would also like to thank Frank Weinreich from Wiley-VCH for the competent help during this project and various publishers for giving the permission to reproduce figures from the original publications. Finally, I would like to thank my wife Katja and my sons Yannick and Benjamin who had to be without me on so many weekends and evenings while I was editing this book. I dedicate this book in love to them.

Zürich, July 2002 *Oliver Zerbe*

List of Authors

Eva de Alba
NHLBI, NIH
Bethesda, MD 20892
USA
eva@madrid.nhlbi.nih.gov

Reto Bader
Department of Chemistry
University of Cambridge
Unilever Building UB7
Lensfield Road
Cambridge CB12 1EW
United Kingdom
Rub20@cam.ac.uk

Marcel J. J. Blommers
Core Technologies
Novartis Pharma AG
P.O. Box
CH-4002 Basel
Switzerland
marcel_jj.blommers@pharma.novartis.com

Florence Cordier
Divison of Structural Biology
Biozentrum
University of Basel
CH-4056 Basel
Switzerland
florence.cordier@unibas.ch

Silvia Davalli
GlaxoSmithKline Medicines
Research Centre
Via Fleming 4
37135 Verona
Italy
sd6223@gsk.com

Andrew J. Dingley
Department of Biochemistry
and Molecular Biology
University College London
Gower Street
London, WC1E 6BT
United Kingdom
dingley@biochemistry.ucl.ac.uk

Volker Dötsch
Department of Pharmaceutical Chemistry
University of California, San Francisco
513 Parnassus Ave
San Francisco, CA 94143-0446
USA
volker@picasso.ucsf.edu

Matthias Ernst
ETH Zürich
Physical Chemistry
ETH-Hönggerberg
CH-8093 Zürich
Switzerland
maer@nmr.phys.chem.ethz.ch

Erhard Fernholz
Roche Diagnostics GmbH
Nonnenwald 2
D-82372 Penzberg
Germany
erhard.fernholz@roche.com

Radovan Fiala
National Centre for Biomolecular Research
Masaryk University
Kotlášká 2
611 37 Brno
Czech Republic
fiala@chemi.muni.cz

Andreas Flörsheimer
Oncology Research
Novartis Pharma AG
P. O. Box
CH-4002 Basel
Switzerland
andreas.floersheimer@pharma.novartis.com

David Fushman
Center of Biomolecular Structure
and Organization
Department of Chemistry and Biochemistry
University of Maryland
College Park, MD 20742
USA
fushman@wam.umd.edu

Gerd Gemmecker
TU München
Institut für Organische Chemie
und Biochemie II
Lichtenbergstr. 4
D-85747 Garching
Germany
gerd.gemmecker@ch.tum.de

Natalie Goto
Department of Molecular Biology
The Scripps Research Institute
10550 North Torrey Pines Road
La Jolla, CA 92037
USA
gonat@scripps.edu

Christian Griesinger
Max-Planck-Institut für Biophysikalische
Chemie
Am Fassberg 11
D-37077 Göttingen
Germany
cigr@mpibpc.gwdg.de

Stephan Grzesiek
Divison of Structural Biology
Biozentrum
University of Basel
CH-4056 Basel
Switzerland
stephan.grzesiek@unibas.ch

Peter Güntert
Riken Genomics Center
1-7-22 Suehiro, Tsurumi
Yokohama, 230-00-45
Japan
guentert@gsc.riken.go.jp

Markus Heller
Lehrstuhl II für organische Chemie
Technische Universität München
Lichtenbergstrasse 4
85747 Garching
Germany
markus.heller@ch.tum.de

Xuemei Huang
TRIAD Therapeutics, Inc.
5820 Nancy Ridge Drive
San Diego, CA 92121
USA

Hideo Iwai
Institute of Biochemistry
University of Zurich
Winterthurer Strasse 190
CH-8057 Zurich, Switzerland
iwai@acess.unizh.ch

Victor A. Jaravine
Divison of Structural Biology
Biozentrum
University of Basel
CH-4056 Basel
Switzerland
victor.jaravine@unibas.ch

Wolfgang Jahnke
Core Technologies
Novartis Pharma AG
P.O. Box
CH-4002 Basel
Switzerland
wolfgang.jahnke@pharma.novartis.com

Werner Klaus
Hoffmann-La Roche AG
Dep. PRBT-S
Building 65/514
4070 Basel
Switzerland
werner.klaus.wk1@roche.com

Christopher A. Lepre
Vertex Pharmaceuticals Inc.
130 Waverly Street
Cambridge, MA 02139-4242
USA
lepre@vpharm.com

Mirjam Lerch
Institute of Pharmaceutical Sciences
Department of Applied Biosciences
ETH Zürich
Winterthurer Str. 190
CH-8057 Zürich
Switzerland
mlerch@pharma.anbi.ethz.ch

Kenneth Lundstrom
Bio-Xtal,
Biopole Epalinges
Les Croisettes 22
CH-1066 Epalinges (Lausanne)
Switzerland
kenneth.lundstrom@mepnet.org

Carla Marchioro
GlaxoSmithKline Medicines
Research Centre
Via Fleming 4
I-37135 Verona
Italy
carla.marchioro@gsk.com

Beat H. Meier
ETH Zürich
Physical Chemistry
ETH Hönggerberg
CH-8093 Zürich
Switzerland
beme@nmr.phys.chem.ethz.ch

David Meininger
TRIAD Therapeutics, Inc.
5820 Nancy Ridge Dr.
San Diego, CA 92121
USA

Detlef Moskau
Bruker BioSpin AG
Industriestraße 26
CH-8117 Fällanden
Switzerland
Detlef.Moskau@bruker-biospin.ch

Heiko Patzelt
Department of Chemistry, College of Science
Sultan Qaboos University
P.O. Box 36, Al-Khod 123
Sultanate of Oman
Patzelt@squ.edu.om

Maurizio Pellecchia
The Burnham Institute
10901 North Torrey Pines Road
La Jolla, CA 92037
USA
mpellecchia@burnham.org

Stefano Provera
GlaxoSmithKline
Medicines Research Centre
Via Fleming 4
I-37135 Verona
Italy
sp15157@gsk.com

Roland Riek
Structural Biology Laboratory
The Salk Institute
La Jolla, CA 92037
USA
riek@sbl.salk.edu

Alfred Ross
Hoffman-La Roche AG
Pharmaceutical Research
CH-4002 Basel
Switzerland
alfred.ross@roche.com

Gordon C.K. Roberts
Biological NMR Centre
Department of Biochemistry
University of Leicester
P.O. Box 138
University Road
Leicester, LE1 9HN
United Kingdom
gcr@le.ac.uk

Simon Rüdisser
Core Technologies
Novartis Pharma AG
P. O. Box
CH-4002 Basel
Switzerland
simon.ruedisser@pharma.novartis.com

Daniel S. Sem
TRIAD Therapeutics, Inc.
5820 Nancy Ridge Dr.
San Diego, CA 92121
USA

Hans Senn
Hoffman-La Roche AG
Pharmaceutical Research
CH-4002 Basel
Switzerland
hans.senn@roche.com

Vladimír Sklenář
National Centre for Biomolecular Research
Masaryk University
Kotlárská 2
611 37 Brno
Czech Republic
sklenar@chemi.muni.cz

Nico Tjandra
NHLBI, NIH
Bethesda, MD 20892
USA
nico@helix.nih.gov

Philip T. F. Williamson
ETH Zürich
Physical Chemistry
ETH Hönggerberg
CH-8093 Zürich
Switzerland
Phil.Williamson@nmr.phys.chem.ethz.ch

Oliver Zerbe
Institute of Pharmaceutical Sciences
Department of Applied Biosciences
ETH Zürich
Winterthurer Str. 190
CH-8057 Zürich
Switzerland
oliver.zerbe@pharma.anbi.ethz.ch

List of Abbreviations

ADME	absorption, distribution, metabolism, excretion
BHK	baby hamster kidney
CD	circular dichroism
CHO	chinese hamster ovary
CMV	cytomegalovirus
COSY	correlation spectroscopy
CSA	chemical shift anisotropy
CSM	chemical shift modulation
DHFR	dihydrofolate reductase
DD	dipole-dipole
DFT	density functional theory
DQ	double quantum
DQC	double-quantum coherence
E.COSY	exclusive correlation spectroscopy
EGF	epidermal growth factor
EPL	expressed protein ligation
FHSQC	fast heteronuclear single-quantum coherence
FIDS	fitting of doublets from singlets
GDP	guanosine diphosphate
HEK	human embryonic kidney
HETCOR	heteronuclear correlation
HMBC	heteronuclear multiple-bond correlation
HMQC	heteronuclear multiple-quantum correlation
HMQC	heteronuclear multiple-quantum correlation
HSQC	heteronuclear single-quantum correlation
HTS	high-throughput screening
IPL	intein-mediated protein ligation
MD	molecular dynamics
MQ	multiple quantum
NOE	nuclear Overhauser effect
PK	pharmacokinetic
PCA	principal component analysis
RDC	residual dipolar coupling

RMSD	root mean square deviation
SAR	structure-activity relationship
SQ	single quantum
TOCSY	total correlation spectroscopy
TROSY	transverse relaxation-optimized spectroscopy
ZQ	zero quantum
ZQC	zero-quantum coherence

1
Modern Methods for the Expression of Proteins in Isotopically Enriched Form

Heiko Patzelt, Natalie Goto, Hideo Iwai, Kenneth Lundstrom, and Erhard Fernholz

1.1
Introduction

The introduction of stable isotopes into proteins has significantly reduced the time requirements for structure elucidation of biomolecules. Moreover, structural studies of proteins with molecular weights exceeding the 10 kDa limit are usually not possible without uniform isotope labeling because of severe resonance overlap and inefficient coherence transfer along the rather small ^3J ^{1}H-^{1}H couplings. Nowadays, efficient expression of recombinant proteins is a prerequisite for many techniques used in structural biology, but the requirement for isotope labeling in particular often precludes NMR (nuclear magnetic resonance) structure determination of proteins isolated from natural sources. Specifically, proteins that have been uniformly labeled with ^{15}N and/or ^{13}C are commonly required for NMR spectroscopy, especially for backbone chemical shift assignment procedures, which are greatly facilitated by the use of a series of rather sensitive multi-dimensional triple-resonance NMR experiments (see Chapt. 4) [1], in a process that can also be automated with good success [2]. Moreover, the random replacement of nonexchangeable protons by deuterons reduces ^{1}H-^{1}H dipolar interactions and scalar couplings, thereby reducing peak line widths considerably and allowing structure elucidation of proteins exceeding 30 kDa [3, 4]. Random fractionally deuterated protein samples also permit the use of longer mixing times in NOESY (nuclear Overhauser effect spectroscopy) experiments, since spin-diffusion pathways are largely eliminated. In addition, transverse-relaxation optimized spectroscopy (TROSY [5], see also Chapt. 10), which has been used for molecules larger than 100 kDa [6], benefits dramatically from deuteration. This stems from the fact that the TROSY component that is narrowed by the DD-CSA compensation is broadened by dipolar interactions with nearby protons.

Besides uniform labeling approaches, stable isotopes can also be introduced at specific sites in proteins in order to simplify the assignment process and to isolate spectral information from the region of interest. For example, biosynthetically-directed fractional ^{13}C labeling offers the possibility of making stereospecific assignments of all isopropyl methyl groups of Val and Leu residues [7]. In another approach often used for solid-state NMR termed *residue-specific* labeling, isotope labels are introduced at single sites in a protein, as described in another chapter of this volume (Chapt. 11). A related scheme, called amino acid-type labeling, is accomplished by expression in an amino acid-based medium where only targeted amino acids contain isotope labels [8]. Preparing a series of samples with different isotopically enriched single amino acid types provides a useful approach for the assignment of large systems not accessible by traditional methods (for an example see Ref. [9]). However, labeling specificity and yield can suffer from isotope scrambling arising from metabolic conversion of amino acids. This problem can be circumvented through the use of specially engineered amino acid-type specific auxotrophic strains. An interesting alternative that can be used to achieve residue and/or amino acid type-specific labeling is presented by the *in vitro* cell-free expression systems. These systems are additionally advantageous when expression products display cell toxicity.

Another labeling strategy geared toward the study of large proteins combines the favorable relaxation properties conferred by extensive deuteration with site-specific strategies for introducing protons. For example, methyl groups and/or aromatic amino acids can be targeted for protonation in otherwise fully deuterated proteins. An alternative approach for the study of large proteins features segmental labeling methods based on protein splicing methodology. Consequently, longer stretches of protein are isotopically enriched, leaving the remainder unlabeled. Isotope editing will remove all signals from unlabeled segments of the proteins, thereby largely reducing resonance overlap and facilitating assignment. Protein splicing methods can also be used to introduce non-natural amino acids or chemical modifications into the sequence. Therein, a protein segment containing an unnatural residue can be chemically synthesized and then ligated to a recombinantly produced (isotopically enriched) segment. Protein splicing can also be used to produce proteins which have high potential for cytotoxicity [10] or to stabilize proteins through cyclization.

This chapter will mainly focus on recent developments in protein labeling methodology. For an introduction to methods involved in the generation and yield optimization of protein samples labeled with ^{13}C and/or ^{15}N by recombinant methods in *E. coli* the interested reader is referred to excellent reviews published in the literature [8, 11–18]. Here we first describe the use of labeled algal hydrolyzates for the production of labeled proteins in *E. coli* or other organisms. We then review methods used for the introduction of isotope labels into specific sites, and this is followed by a section on segmental labeling approaches. Subsequently, we will summarize recombinant protein expression methods in hosts other than *E. coli* that have proven to be especially suitable for post-translationally modified proteins and membrane proteins, before concluding with an introduction to cell-free expression systems.

1.2
Isotope-Labeled Proteins from Hydrolyzates of the Green Alga *Scenedesmus obliquus*

Although alternative expression systems have been successfully adapted for the production of isotope-labeled proteins (see Sect. 1.5), heterologous expression in *E. coli* often remains the method of choice for NMR sample preparation. There is a fundamental difference, however, with respect to the kind of medium in which the cells are cultivated. In a so-called "chemically defined" or "minimal" medium only one or a very limited number of carbon sources is provided, e.g. glucose or glycerol. All bacterial metabolites have to be biosynthesized by the cells through the various, sometimes lengthy and energy-demanding, metabolic pathways. In a "complex" or "rich" medium, the cells grow, as the name suggests, on a complex mixture of amino acids and/or carbohydrates. Amino acid interconversions, and thus the potential for isotope scrambling in selectively labeled samples, are here reduced to a minimum. Unlabeled fermentations are usually performed in complex media (i.e. yeast extract-containing LB for *E. coli*), since protein yield and cell density are here considerably higher than in minimal media. The same is desirable for isotope-labeled fermentations, but the limited availability and/or high price of commercial amino acid/sugar mixtures in the required isotope composition often impose fermentations on a single carbon source.

Several companies are meanwhile supplying labeled amino acid/sugar mixtures of acceptable quality. However, especially if larger-scale or repeated preparations of labeled proteins are envisaged, investment of some extra time for the in-house production of labeled complex growth media for *E. coli* or other host cells clearly becomes advantageous, especially from the point of view of financial considerations.

The most frequently employed source for complex amino acid/sugar mixtures, labeled in any combination of ^2H,^{13}C and/or ^{15}N in *E. coli* continues to be a phototrophic green alga. *Scenedesmus obliquus* was introduced for that purpose in 1972 by Crespi and Katz [19]. In recent years, the original protocols have been modified by several groups, leading to improvements in the yield and purity of the algal amino acid mixtures, thereby enhancing protein labeling efficiency and expression levels in the hosts [20–22].

Depending on the desired labeling pattern, ^2H_2O, $^{13}CO_2$ and/or $^{15}NH_4Cl$ or $Na^{15}NO_3$ are used as exclusive isotope sources during the algal fermentation. All of the above have become commercially available at affordable rates. For the preparation of random partially isotope-labeled amino acid/sugar mixtures, unlabeled water, carbon dioxide or nitrogen salts are simply admixed to the labeled starting material in the appropriate proportions.

Thus, the three basic steps required for the preparation of a uniformly labeled protein for NMR experiments are

1. production of isotope-labeled algal hydrolyzates,
2. adaptation of the protein overproducing organism (usually, but not always, *E. coli*) to growth on the algal medium, and
3. preparation and purification of the isotope-labeled protein on a preparative scale.

If specific amino acid-type labeling is required, the labeled amino acid is added to the fermentation of the expression host (topic 1 above, see Sect. 1.2.3). In this case, a thorough isotope analysis of the expressed protein is advisable prior to NMR spectroscopic investigations. This is preferentially achieved by GC-MS analyses of the hydrolyzed amino acids from the protein product.

1.2.1
Production of Isotope-Labeled Algal Hydrolyzates

Cultures of *S. obliquus* can easily be grown photoautotrophically in two-tier flasks in an inorganic medium. A stepwise replacement of H_2O by ^2H_2O leads to deuterated cultures, and a replacement of CO_2 (the sole carbon source) by $^{13}CO_2$ and/or the replacement of the nitrogen-containing salts by their ^{15}N-isotopomers produces ^{13}C- and/or ^{15}N-labeled cultures. These are used as inoculi for larger fermentations in Fernbach flasks or stirred or airlifting fermenters. All fermentations are continuously illuminated with standard plant light bulbs or fluorescent tubes. In the last stage of $^{2}H_2O$ fermentations, all salts containing crystal water are repeatedly dissolved in small amounts of $^{2}H_2O$ and lyophilized before addition to the medium. If the algal cells are harvested under sterile conditions, the recovered medium can be re-inoculated for up to four further fermentations, and only phosphate is supplemented when the growth rate declines [22]. The recycling of the media cuts isotope costs by about 90%.

The algal cell mass is then purified from low-molecular-weight metabolites and hydrolyzed in HCl (^{2}HCl in the case of a deuterated fermentation). After neutralization and lyophilization, a white powder (typically 2–2.5 g L^{-1} of medium and fermentation cycle) is obtained, containing around 50% amino acids (for composition see Ref. [23]), 30% sugars (composition in Ref. [24]) and 20% NaCl. This amino acid/sugar mixture for complex microbial growth media can be produced with any combination of ^{2}H, ^{13}C and/or ^{15}N, including random fractional label distributions. Used as carbon source it enables the simple and quick preparation of isotope-labeled, complex microbial growth media for the production of labeled proteins.

1.2.2
Adaptation of the Protein Overproducer to the Algal Medium

The described algal hydrolyzate contains amino acids and sugars in a physiological composition, i.e. in a relative composition similar to that required by most host cells. Amino acid biosynthesis, interconversion, and thus the potential for isotope scrambling, are minimized. When all potentially inhibitory low-molecular-weight compounds are removed by extraction prior to hydrolysis, most organisms grow well in media containing their typical salt and trace element composition, with the exception of NaCl, which is introduced as part of the algal hydrolyzate. Only the carbon sources are substituted by the algal amino acid/sugar mixture (e.g. yeast extract is replaced by algal hydrolyzate with the required isotope composition). Examples of the production of isotope-labeled proteins in Bacteria [25], Archaea [23, 26] and Eucarya [27] can be found in the recent literature.

Since changing the carbon source may influence bacterial growth and expression characteristics, a series of unlabeled test experiments is recommended in order to establish the minimum hydrolyzate concentration required, as well as the reproducibility of protein expression levels.

While expression in ^{13}C- and/or ^{15}N-labeled media is usually straightforward, most organisms need to be adapted in 3 to 4 steps to growth in $^{2}H_2O$ (e.g. 50, 75, 90, 100%). In addition, while formulating any deuterated medium, it is important to recall that the reading on pH meters equipped with normal glass electrodes is about 0.4 units lower in $^{2}H_2O$ than in H_2O with the same hydrogen/deuterium ion concentration [28]. Because of the different physical properties of $^{2}H_2O$, growth may be slower than usual, and the timing for induction of protein expression may require adjustment. The extent of deuteration depends on the type of experiments that will be performed. For investigations of internal dynamics using ^{15}N relaxation or for backbone assignment with triple-resonance spectroscopy, 100% deuteration at the nonexchangable carbon sites maximizes signal sensitivity and resolution. On the other hand, structural information traditionally relies on distance restraints derived from $^{1}H,^{1}H$ NOEs. Nietlispach *et al.* have calculated and measured the effects of various levels of random fractional deuteration and found 50–70% deuteration most useful for larger proteins [29].

1.2.3
Preparation of Homogenously Isotope-Labeled Protein by Fermentation on Algal Media

Production of isotope-labeled proteins on a larger scale from the optimized test conditions is typically routine as long as the physical growth parameters (reactor type, aeration, etc.) are not changed significantly. However, the purification of deuterated proteins may require some adjustments, depending on the techniques utilized. For example, because of their considerably higher density, centrifugation gradients must be adapted. Also, the chromatographic properties of deuterated proteins may display differences relative to their unlabeled (^{1}H at natural abundance) counterparts, reflecting potential shifts in isoelectric point, stronger intramolecular hydrogen bonds and weaker van der Waals interactions. A final consideration is the re-introduction of protons at the exchangeable amide sites. Since the quantitative exchange of amide protons from the protein core can be extremely slow, it is sometimes necessary to expose the sample to denaturing conditions, followed by refolding in H_2O, if possible.

1.2.4
Amino Acid-Type Specific Labeling

The principal difficulty associated with the preparation of amino acid-type specific labeled proteins is the suppression of metabolic scrambling of the label into other amino acid types through the common metabolic pathways in the host cell. The use of a complex amino acid/sugar mixture, such as the one present in the algal hydrolyzates, reduces this danger greatly compared to fermentations on a single carbon source. In fact, the activity of many enzymes responsible for amino acid interconversions appears to be low or absent in bacteria grown under these conditions. For example, for the production of a fully deuterium-labeled protein containing ^{1}H-Trp, the host cell is grown on fully ^{2}H-labeled algal hydrolyzate in $^{2}H_2O$ to which unlabeled Trp is admixed. The individual amount of the differentially labeled amino acid to be added to the fermentation may vary for the different residues and depends on the biosynthetic origin of the amino acid [30] as well as on its background concentration in the algal hydrolyzate [22]. Labels in the biosynthetically central amino acids Asx and Glx show a pronounced tendency for biosynthetically-directed isotope relocation, since these molecules may be used as metabolic precursors for a number of different downstream amino acids (e. g. Met, Lys, Thr, and Ile or Pro and Arg, for Asx and Glx, respectively). The metabolically peripheral amino acids can usually be labeled specifically with much higher isotope purity.

For every new amino acid, a small series of test experiments is normally sufficient to establish a compromise between the minimum concentration required for high specific labeling (usually at least ten times the amount introduced with the algal hydrolyzate) and the toxicity limit for the respective amino acid. Before a large-scale fermentation is attempted, the isotope composition downstream from the labeled amino acid should be analyzed. The most probable sites of undesired isotope incorporation are found in the same biosynthetic group [30]. Labeled Tyr, for instance, may be found after media supplementation with labeled Phe, whereas Cys and Gly labeling may result from the addition of labeled Ser.

1.2.5
Mass Spectrometric Analysis of the Labeled Amino Acids

For amino acid analysis the labeled protein needs to be hydrolyzed and derivatized. Most commonly the hydrolysis is performed in 6 M HCl, and the amino acids are converted into their isopropyl ester and pentafluoropropanamide derivatives (Fig. 1.1) before GC/MS analysis. The molecular ion is not always visible after standard electron impact (EI) ionization, and the fragment after loss of the carboisopropoxy group is the highest observable peak. This leaves m/e=175 plus the mass of the amino acid side chain, from which the degree of labeling can be directly deduced.

Fig. 1.2 shows illustrative mass spectra for derivatized Phe with various isotope patterns. In unlabeled Phe (Fig. 1.2a), the highest observable peak at m/e=266 accounts for the aforementioned fragment of 175 plus the mass of the benzyl group (C_7H_7=91). The strongest peak is due to the tropylium cation (C_7H_7) at m/e=91. Complete deuteration (Fig. 1.2d) takes molecular ion to m/e=274 (175+$C_7{}^2H_7$), and the tropylium signal ($C_7{}^2H_7$) accordingly to m/e 98. As expected, the spectrum of ^{13}C,^{15}N-labeled Phe (Fig. 1.2e) shows the corresponding signals at m/e=275 (175+2+$^{13}C_7H_7$) and 98 ($^{13}C_7H_7$).

The signals in the spectra of the partially deuterated amino acid (Figs. 1.2b–d and 1.2f) show a statistical distribution of the masses around the calculated values for all fragments and are thus an unambiguous proof of a complete random distribution of the labels.

1.3
Selective Labeling Schemes

While general labeling strategies relying on expression of proteins using hydrolyzates of algae are useful for uniform or amino acid-specific labeling, expression can also be performed in minimal media in a cost-effective manner. These types of expression media

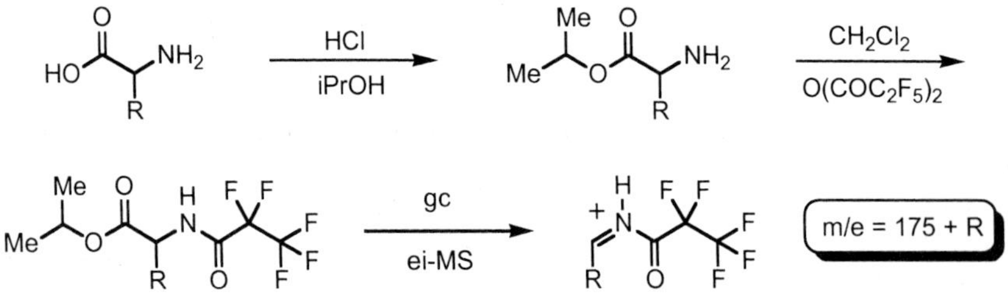

Fig. 1.1 Most common derivatization of amino acids for GC-MS analysis. The fragment without the carboisopropoxy group normally produces the highest observable peak, which is used for the determination of the isotope composition.

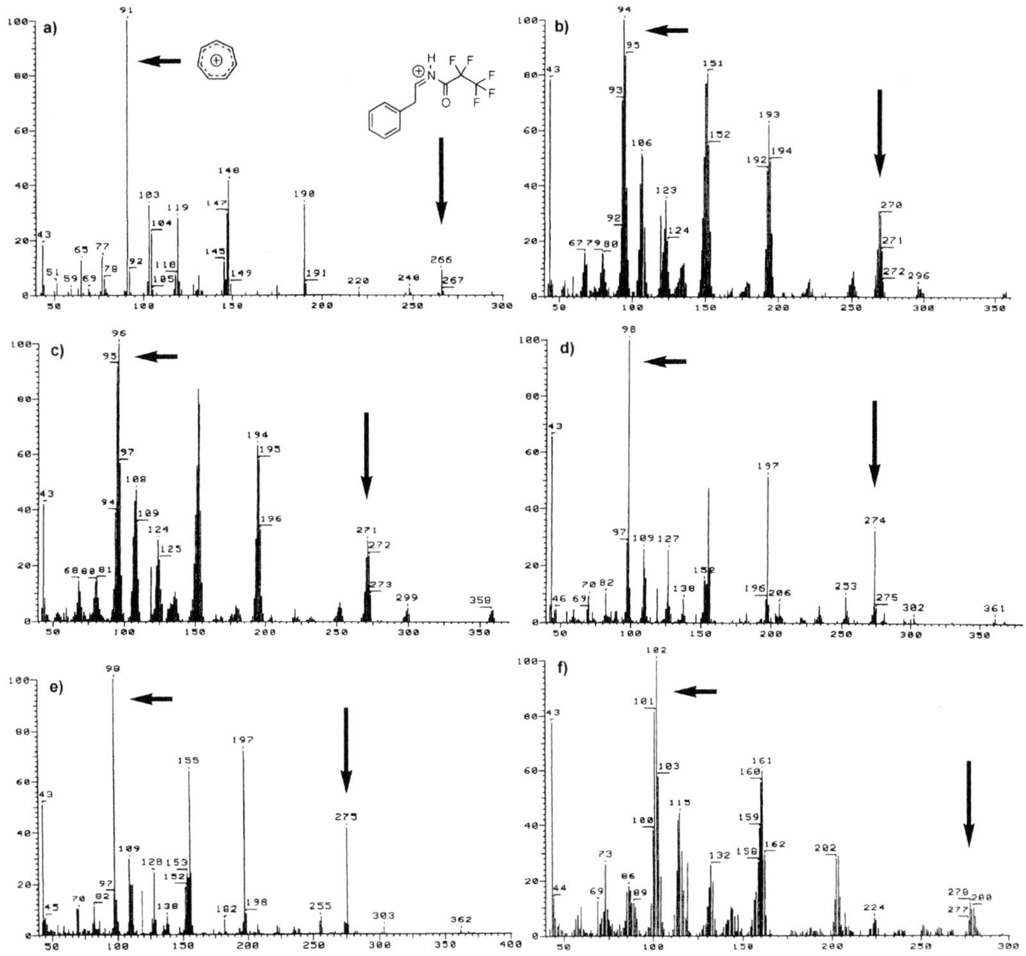

Fig. 1.2 EI mass spectra of derivatized phenylalanine, isolated from *S. obliquus*; **a** unlabeled, **b** random 50% ^{2}H-, **c** random 70% ^{2}H-, **d** 100% ^{2}H-, **e** ^{13}C,^{15}N-, and **f** ^{13}C,^{15}N/random 75% ^{2}H-labeled.

typically use glucose and ammonium salts as the sole carbon and nitrogen sources, respectively, and include D_2O as the medium base when deuteration is required. These growth conditions are easily adapted for uniform or random fractional isotope labeling. In addition, modifications to this general formula can be made to introduce isotope labels into specific sites, for example, by supplemention of the expression media with specifically labeled amino acids or amino acid precursors. In the following section we will review a number of methods used for different patterns of selective labeling in *E. coli*. Reagents used in these different strategies, with corresponding literature references, are summarized in Tab. 1.1.

Tab. 1.1 Chemical structures of metabolites involved in selective isotope labeling strategies (from Ref. [14] with kind permission)

Labeling agent	***Chemical structure***	***Incorporated as***	***Reference***
$[3\text{-}^2H]$ α-keto-*iso*valerate	$^*CH_3 - {}^*CD(^*CH_3) - CO - {}^*COO^-$	(^{1}H-δ methyl)-Leu (^{1}H-γ methyl)-Val	43
$[3,3\text{-}^2H_2]$ α-keto-butyrate	$^*CH_3 - {}^*CD_2 - {}^*CO - {}^*COO^-$	(^{1}H-δ1 methyl)-Ile	42
$[\varepsilon\text{-}^{13}C]$-L-phenylalanine	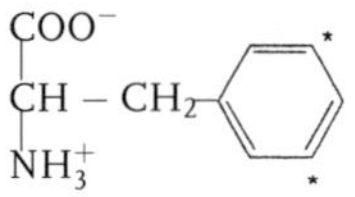	$[^{13}C]$ Phe	52
$[2\text{-}^{13}C]$ or $[1,3\text{-}^{13}C_2]$-glycerol	$CH_2 - OH$, $^*CH - OH$, $CH_2 - OH$ or $^*CH_2 - OH$, $CH - OH$, $^*CH_2 - OH$	$^{12}C\text{-}^{13}C\text{-}^{12}C$ pattern	162
^{13}C pyruvate	$^*CH_3 - {}^*CO - {}^*COO^{-\P}$	(^{1}H-δ methyl)-Leu (^{1}H-γ methyl)-Val (^{1}H-γ2 methyl)-Ile (^{1}H-β methyl)-Ala	35
$[3\text{-}^{13}C]$ pyruvate	$^*CH_3 - {}^*CO - {}^*COO^{-\Pi}$	(^{1}H-δ methyl)-Leu (^{1}H-γ methyl)-Val (^{1}H-γ2 methyl)-Ile (^{1}H-β methyl)-Ala	39, 163

* Indicates positions labeled by ^{13}C. Growth conditions optimized to maximize yields of CH_3 isotopomers in target methyl groups. Π Protocol for expression leads to production of all possible methyl isotopomers, with desired species CHD_2 for ^{13}C relaxation measurements.

1.3.1
Reverse-Labeling Schemes

1.3.1.1 Selective Protonation of Methyl Groups in ^2H-Labeled Proteins

As described in the introduction, deuteration is routinely used to reduce the rapid transverse relaxation rates characteristic of larger proteins, leading to improvements in peak line widths and experimental sensitivity. Deuteration of all the carbon-bound protons maximizes the sensitivity gains that can be obtained from this labeling strategy, and has thus proven useful for in the assignment of backbone ^1H, ^{15}N, ^{13}C and side-chain ^{13}C chemical shifts of large proteins. However, the elimination of all but the exchangeable protons significantly impedes structural studies that rely on conventional NOE approaches. Although in some cases it is possible to use only backbone $^1HN\text{-}^1HN$ NOEs to obtain a protein global fold, the accuracy of these structures is very low because of the small proportion of distance restraints between protons from nonsequential residues. These global folds tend to be less compact than high-resolution structures, with the backbone pairwise root-mean-squared deviation (rmsd) to the high-resolution structure ranging from 5 to 8 Å [31, 32].

For the purpose of increasing the number of protons in the protein core while maintaining the benefits of extensive deuteration, it is possible to re-introduce protons using a "reverse isotope" labeling approach. In some of the original approaches, side chains of target amino acid types were selectively protonated in deuterated proteins by adding protonated forms of these amino acids to the D_2O growth medium (see, for example, Refs. [33, 34]). In a variation of this theme, Rosen and coworkers developed a protocol to selectively incorporate protons at the methyl positions of Ala, Val, Leu and Ile $\gamma 2$ [35]. Methyl groups are enriched in protein hydrophobic cores and hence are attractive targets for selective protonation [36]. In addition, NMR spectroscopic properties of methyl groups are favorable owing to reasonably well-resolved ^{13}C-^{1}H correlations and rapid rotation about the methyl symmetry axis that reduces peak line widths [37].

The original protocol for the production of ^{15}N, ^{13}C, ^{2}H, ^{1}Me (protonated methyl)-proteins utilized ^{13}C,^{1}H-pyruvate as the exclusive source of carbon in 100% D_2O minimal media [35]. Pyruvate can be diverted into the tricarboxylic acid (TCA) cycle to produce many of the intermediates used in the biosynthesis of amino acids [38], but can also be directly incorporated into amino acids either by transamination (Ala), or reactions with threonine (Ile), pyruvate (Val) or both pyruvate and acetyl CoA (Leu). As was observed, pyruvate that is directly incorporated into these amino acids will largely retain the methyl protons, while those amino acids synthesized indirectly via intermediates will be highly deuterated. However, the incorporation of protons at each methyl site tends to be variable, with the result that methyl isotopomers such as $^{13}CHD_2$ and $^{13}CH_2D$ are also produced [35]. The additive deuterium isotope effect on both carbon and proton chemical shifts produces upfield shifts relative to the $^{13}CH_3$ peak by 0.02 and 0.3 ppm per ^{2}H atom in the ^{1}H and ^{13}C dimensions respectively. As a result, ^{13}C-^{1}H correlation spectroscopy of the methyl region shows three peaks for every methyl group labeled in this way, translating into resolution and sensitivity problems. Nonetheless, through the use of ^{2}H-purging pulse schemes during the acquisition of carbon chemical shift, it is possible to remove the peaks arising from methyl groups containing ^{2}H [32]. In addition, it should be noted that other pyruvate-based labeling schemes have also found great utility in the measurement of side-chain dynamics involving these methyl-containing side chains [39–41].

More recently, the yield and uniformity of methyl group protonation was enhanced through the use of α-ketoisovalerate in combination with α-ketobutyrate to produce ^{15}N, ^{13}C, ^{2}H-labeled proteins with protons introduced at the methyl positions of Leu, Val and Ile ($\delta 1$) [42, 43]. Proteins expressed in D_2O/^{13}C, ^{2}H-glucose/$^{15}NH_4Cl$ minimal media can be supplemented with ^{13}C, [3, ^{2}H] α-ketoisovalerate for the selective protonation of the Val and Leu methyl groups and [3,3-$^{2}H_2$], ^{13}C α-ketobutyrate for Ile $\delta 1$ methyl group labeling. Using this strategy, labeling efficiency of the target methyl groups was shown to exceed 90%, while high levels of deuteration were maintained at other sites without production of methyl group isotopomers containing deuterium. Since the selectively deuterated form of these amino acid precursors can be obtained by base-catalyzed proton exchange in aqueous buffer, the protonated, commercially available forms of these precursors are straightforwardly adapted to this labeling scheme.

1.3.1.2 Structure Determination of Selectively Methyl Protonated Proteins

High levels of deuteration combined with selective methyl protonation using one of the schemes outlined above permits the measurement of $^1H^N$-$^1H^N$, $^1H^N$-methyl and methyl-methyl NOEs. Global folds can be determined using this subset of NOEs, where the quality of these structures is a reflection of protein topology, secondary structure content, and the location and distribution of methyl groups in the protein [32]. In the case of a 30 kDa cell adhesion fragment from intimin, for example, intradomain backbone rmsd values of the ensemble of structures ranged between 1.5 and 1.8 Å from the mean [44]. In contrast, MBP structure quality was lower, with intradomain backbone rmsds between NMR and crystal structures of 3.1–3.8 Å [45]. Although structures produced by this methodology are often of a preliminary quality, they can nonetheless be useful in the localization of ligand or protein interaction sites and the identification of homologous proteins (e.g. Ref. [46]). Global folds can also be used as a structural stepping stone in the generation of high-resolution structures, since the assignment of additional NOEs from random fractionally deuterated samples or fully protonated molecules is facilitated by the use of a preliminary structure [47, 48]. Further improvements in the quality of structures can also be obtained through the incorporation of additional restraints such as dipolar couplings ([45]) or homology modeling (e.g. Ref. [49]).

1.3.1.3 Introducing ^1H,^{12}C Aromatic Residues into Otherwise ^{13}C Uniformly Labeled Proteins

Alternative schemes involving selective protonation have also been developed to increase the number of side-chain distance restraints that can be obtained in highly deuterated proteins. For example, ^1H, ^{12}C Phe and Tyr can be directly incorporated into an otherwise uniformly ^{13}C-labeled protein expressed in minimal media [50]. Since these amino acids are also preferentially located in the hydrophobic cores of proteins, as well as at ligand binding interfaces, distance restraints involving these residues can be very valuable. This labeling strategy was shown to be useful for proteins under 30 kDa with relatively few Phe and Tyr such as a 24 kDa Dbl homology domain [48] and the 25 kDa antiapoptotic protein Bcl-xL [51]. In cases where overlap in the aromatic spectrum becomes problematic, a synthetic strategy has been introduced to produce Phe that is ^{13}C labeled only at the epsilon position [52]. An illustration of the utlility of this approach is provided by the structure determination of a 21 kDa Dbl homology domain containing seven phenylalanine residues [47].

1.3.1.4 Backbone-Labeled Proteins

Protocols for selective isotope labeling of protein backbone atoms are also being developed, since the prevention of ^{13}C incorporation at the C^{β} site circumvents resolution problems associated with homonuclear $^1J_{C\alpha C\beta}$ coupling. Toward this end, syntheses of backbone-labeled amino acids have been described for ten different amino acids starting from ^{15}N, $^{13}C_2$-glycine [53, 54]. While original demonstrations of backbone labeling utilized a CHO cell expression system to prevent isotope scrambling [53], bacterial cell expression systems have also proven amenable to this strategy [54]. In this case, the expression medium must contain the full complement of amino acids, which are then replaced with those that are $^{13}C\alpha$, ^{13}CO, ^{15}N, and $^1H\alpha$ (or 50% $^2H\alpha$) labeled just prior to induction of protein expression. As was demonstrated for ubiquitin backbone-labeled with a subset of amino acids, sensitivity and resolution in HNCA-type experiments is enhanced, and couplings can be readily measured from IPAP ^{1}H-^{13}C HSQC (heteronuclear single-quantum correlation) spectra [54, 55].

1.3.2
Selective ^{13}C Methyl Group Labeling

To reduce the expense of producing selectively methyl-labeled proteins, it is possible to use ^{13}C-methyl iodide to synthesize α-ketobutyrate and α-ketoisovalerate containing ^{13}C only at the methyl sites [56]. A larger than 20-fold reduction in the cost of precursor molecules can be achieved using this synthetic strategy in place of the commercially available uniformly ^{13}C-labeled isotopomers. Although these compounds can be adapted to the selective methyl protonation scheme described above, they can also be used to produce proteins that only contain ^{13}C at the methyl positions of Val, Leu, and Ile $\delta 1$, with ^{12}C at all other sites. The reduced cost, spectral simplification and sensitivity enhancement of ^{13}C methyl-labeled proteins over uniformly ^{13}C-labeled samples facilitates the use of chemical shift mapping in the search for potential lead compounds in the drug discovery process. If, on the other hand, selective methyl protonation is required, the ^{1}H, ^{13}C methyl-labeled α-ketoisovalerate and α-ketobutyrate would be added to D_2O expression media containing ^{2}H, ^{13}C-labeled glucose and ^{15}N-labeled ammonium salt. However, compared to the uniformly ^{13}C-labeled selectively protonated samples described previously, structure determination is less straightforward since backbone carbon atoms for isoleucine and valine are derived from the nonmethyl portion of the α-ketobutryate and α-ketoisovalerate, respectively, and would therefore contain the ^{12}C isotope. Nonetheless, once assignments are made, NOE measurements involving these methyl groups benefit from elimination of one-bond ^{13}C-^{13}C couplings leading to narrow carbon line widths without the requirement for constant-time evolution periods [51].

1.4
Intein-Based Protein Engineering for NMR Spectroscopy

Recently, new advances in biochemistry have opened up a novel approach for protein engineering, which utilizes a protein-splicing domain. Protein splicing is a post-translational chemical modification discovered in nature, which catalyzes the excision of an intervening polypeptide (*int*ernal prote*in*, intein) while simultaneously ligating both the N- and C-terminal flanking polypeptide chains (Fig. 1.3, 1.5 A), analogous to RNA splicing [57, 58]. This unique enzymatic process has been used in various biochemical and biotechnological applications such as protein purification, protein ligation, backbone cyclization and C-terminal modifications. In particular, protein ligation using inteins has opened up a new way for the production of segmentally labeled proteins, thereby reducing the complexity of NMR spectra. For larger proteins or multimeric proteins, it may be necessary to combine selective labeling and segmental isotope-labeling approaches. Moreover, it will be useful in cases where information is desired only for a small part of a large protein, e.g. to characterize interactions with known binding sites or to detect conformational changes in a specific region. Another potentially useful application of inteins for NMR is backbone cyclization to enhance the stability of proteins.

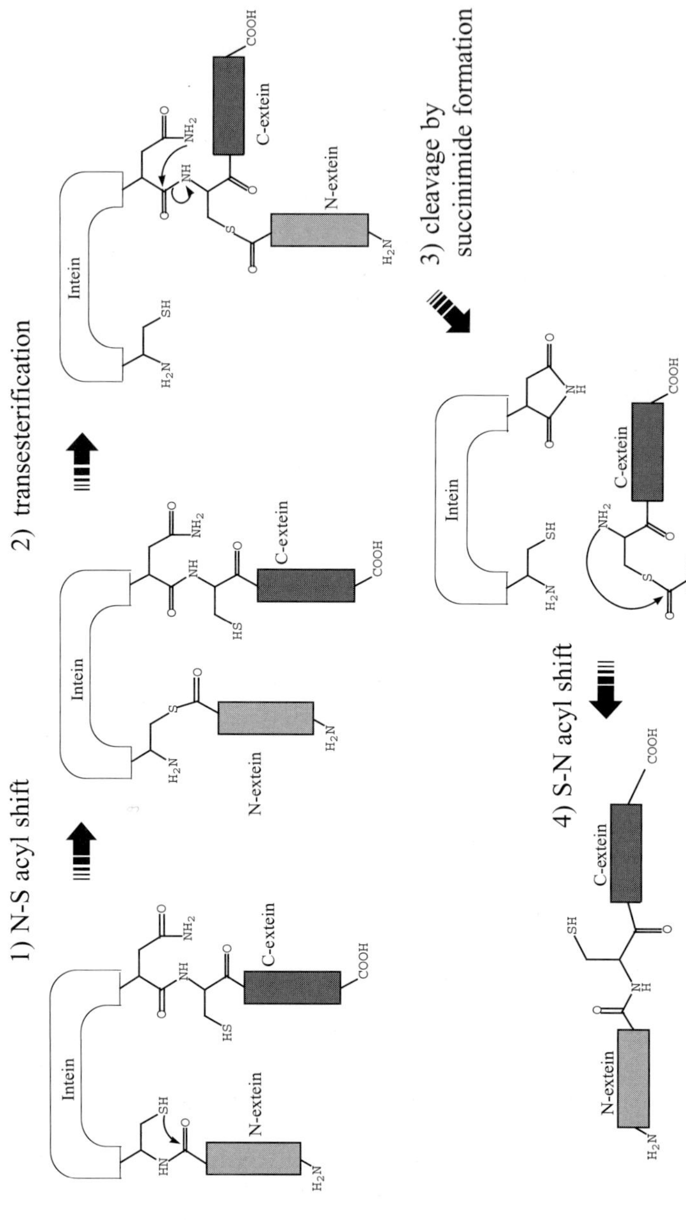

Fig. 1.3 The currently accepted chemical mechanism of protein splicing. **1** N-S(O) acyl shift, **2** transesterification, **3** cleavage by succinimide formation, **4** S(O)-N acyl shift.

1.4.1
Segmental Labeling of Proteins

Although more than 100 protein splicing domains have been found in nature [59], only a handful have been used for segmental labeling purposes, namely *Sce*VMA (PI-*Sce*I), PI-*Pfu*I and PI-*Pfu*II. The currently accepted reaction mechanism for protein splicing consists of the following four steps, namely, (i) N-S(O) acyl shift, (ii) transesterification, (iii) succinimide formation, and (iv) S(O)-N acyl rearrangement (Fig. 1.3). Either a subset of these four steps or the entire reaction process can be used for protein ligation. For example, the intein-mediated protein ligation (IPL) approach utilizes a subset of these reactions by using a modified intein as described in Sect. 1.4.1.1. On the other hand, the split intein approach requires all four reaction steps. In this case the success of the reaction depends on refolding properties of the split intein (Sect. 1.4.1.2).

1.4.1.1 Intein-Mediated Protein Ligation (IPL)/Expressed Protein Ligation (EPL) using the IMPACT System

The IMPACT (Intein-Mediated Purification with an Affinity Chitin-binding Tag) system was originally developed as a novel purification method by New England Biolab. It makes use of a modified intein, *Sce*VMA, in which the active site was mutated from His-Asn-Cys to His-Ala-Cys, so that the usual cleavage due to succinimide formation involving the side-chain of Asn is no longer possible (Figs. 1.4 A and 1.6 A). Since a chitin-binding domain (CBD) is fused to the C-terminus of the intein, this protein can be immobilized to a chitin column, providing a convenient tool for purification [60]. The desired protein segment is fused at the N-terminus of the intein, which can be liberated from the intein-CBD portion of the fusion protein by addition of nucleophiles such as dithiolthreitol (DTT), ethanethiol, 2-mercaptoethane sulfonic acids (MESNA), hydroxylamine or cysteine. The IMPACT system provides a good opportunity to expand the application of native chemical ligation, which was originally developed by the Kent group, to a variety of protein targets, because the C-terminus of the N-terminal fusion polypeptide can be converted easily into a thioester group by the intein-mediated cleavage [61]. For native chemical ligation, an N-terminal peptide segment containing a C-terminal thioester is chemoselectively ligated to a C-terminal peptide segment that has an N-terminal cysteine in aqueous solution, without protecting any functional groups in the peptides. With the intein-based *E. coli* expression system, it is now possible to produce larger protein segments with a C-terminal thioester group easily, which can subsequently be used for native chemical ligation. Using this approach, Xu et al. have demonstrated the domain-selective ^{15}N labeling of the SH2 domain of the Abl-kinase SH domain [62]. In this experiment, the ^{15}N-labeled SH2 domain containing an N-terminal cysteine capped with a specific proteolytic sequence, which can be removed, was expressed and purified in ^{15}N-labeled media. The protective N-terminal sequence was subsequently removed by proteolysis in order to create an N-terminal thionucleophile (N-terminal cysteine). The N-terminal segment of the SH3 domain was separately produced as the intein-fusion protein in unlabeled media and eluted with ethanethiol to form the C-terminal thioester. The unlabeled SH3 domain and ^{15}N-labeled SH2 domain were ligated in aqueous solution at pH 7 in the presence of thiophenol and benzyl mercaptan, achieving a yield of 70%.

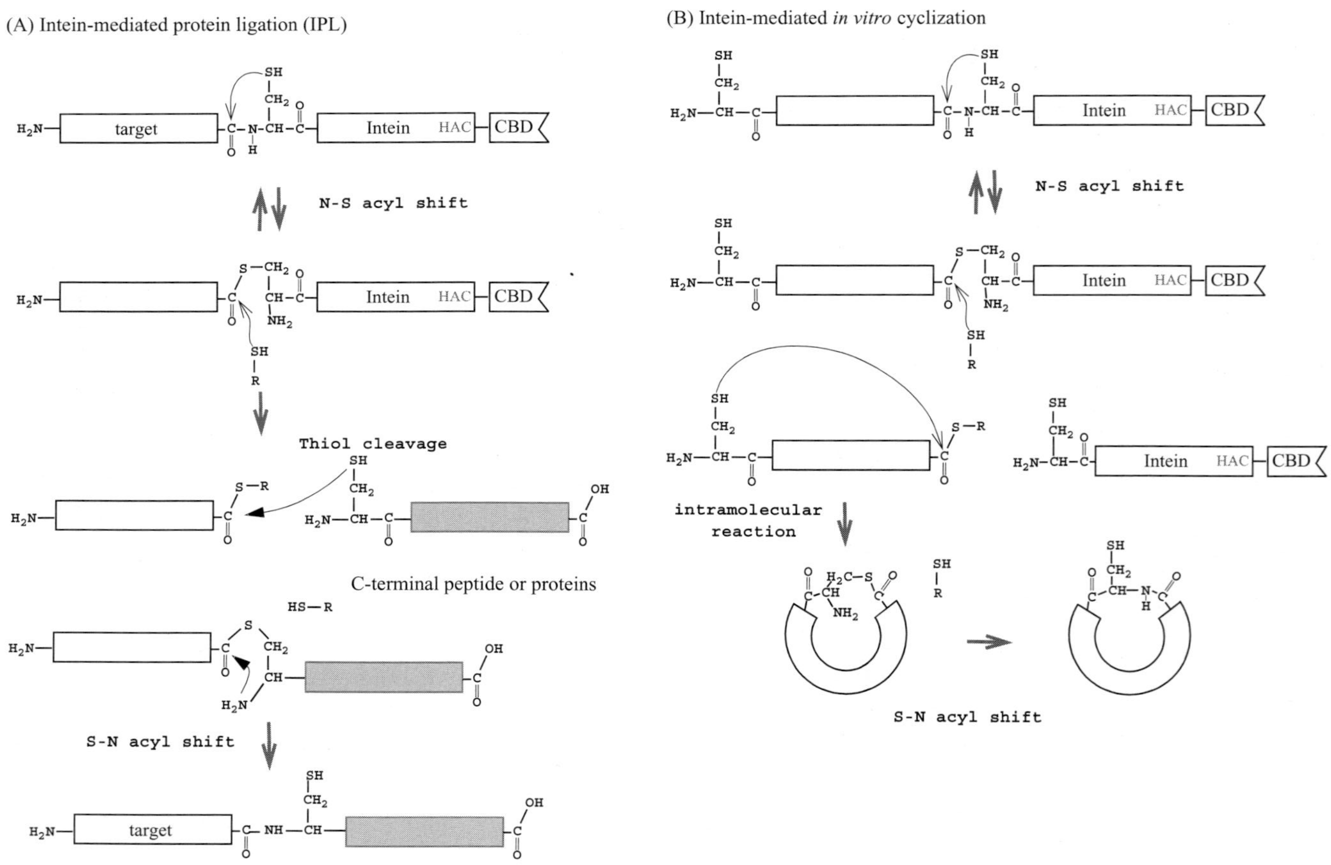

Fig. 1.4 **A** Intein-mediated protein ligation (IPL) (also called expressed protein ligation EPL). **B** The chemical mechanism of protein cyclization by the IPL/EPL approach. HAC denotes the sequence of the active site of the intein, e.g. His-Ala-Cys. CBD stands for chitin-binding domain.

It has also been demonstrated that multiple ligation steps can be performed with the IPL/EPL approach, thereby illustrating its potential use in central-segment labeling [63].

One of the intrinsic limits of the IPL/EPL system is the requirement for a cysteine residue at the site of protein ligation, which will replace the thioester group. Recently, it was shown that this requirement could be avoided by using a cleavable thiol-containing auxiliary group. Low et al. demonstrated protein ligation by introducing a cleavable thiol-containing auxiliary group, 1-phenyl-2-mercaptoethyl, at the alpha-amino group of a chemically synthesized peptide, which is removed upon protein ligation [64]. Unfortunately, this modification at the N-terminus could be difficult to introduce into proteins which are prepared from bacterial expression systems, and hence its use could be limited to situations where the C-terminal peptide can be chemically synthesized. Nevertheless, the removal auxiliary approach presents an opportunity for segmental isotope labeling regardless of the primary sequence.

A second approach that can be adopted to overcome the intrinsic requirement for cysteine at the N-terminus of C-terminal fragment utilizes the enzyme subtiligase, a double mutant of subtilisin, which is able to join two unprotected peptides. Thioester-modified proteins were shown to present good substrates of subtiligase [65]. However, although this approach could be potentially useful for general isotope labeling, the efficiency of this process remains to be proven.

It is noteworthy that there is another limiting factor in the choice of amino acid types at the junction sites which affect the enzymatic process of the intein. For example, in the case of *Sce*VMA (also called PI-*Sce*I) from the IMPACT system, proline, cysteine, asparagine, aspartic acid, and arginine cannot be at the C-terminus of the N-terminal target protein just before the intein sequence. The presence of these residues at this position would either slow down the N-S acyl shift dramatically or lead to immediate hydrolysis of the product from the N-S acyl shift [66]. The compatibility of amino acid types at the proximal sites depends on the specific inteins and needs to be carefully considered during the design of the required expression vectors. The specific amino acid requirements at a particular splicing site depends on the specific intein used and is thus a crucial point in this approach.

1.4.1.2 **Reconstitution of Split Inteins**

It has been demonstrated that an intein can be split into two fragments and reconstituted *in vitro* as well as *in vivo* to form an active intein capable of *trans* splicing [67–69]. This *trans*-splicing activity can be directly used for protein ligation as an alternative to the native chemical ligation step, which requires additional thionucleophile groups. Yamazaki et al. have applied *trans*-splicing to the segmental labeling of RNA polymerase subunit α by splitting an intein from *Pyrococcus furiosus*, PI-*Pfu*I (Fig. 1.5 B) [70]. Each half of the split intein fused to the N- (or C)-extein was produced separately, one in isotopically labeled and the other in unlabeled medium. The independently prepared protein fragments were expressed as inclusion bodies and purified under denaturing conditions. The two independently prepared fragments were reassembled and refolded in aqueous solution in order to form a functional intein, resulting in a ligated extein fragment and a spliced intein. The splicing reaction was found to be efficient at elevated temperature (70 °C), presumably because PI-*Pfu*I is a thermophilic enzyme. However, the general use

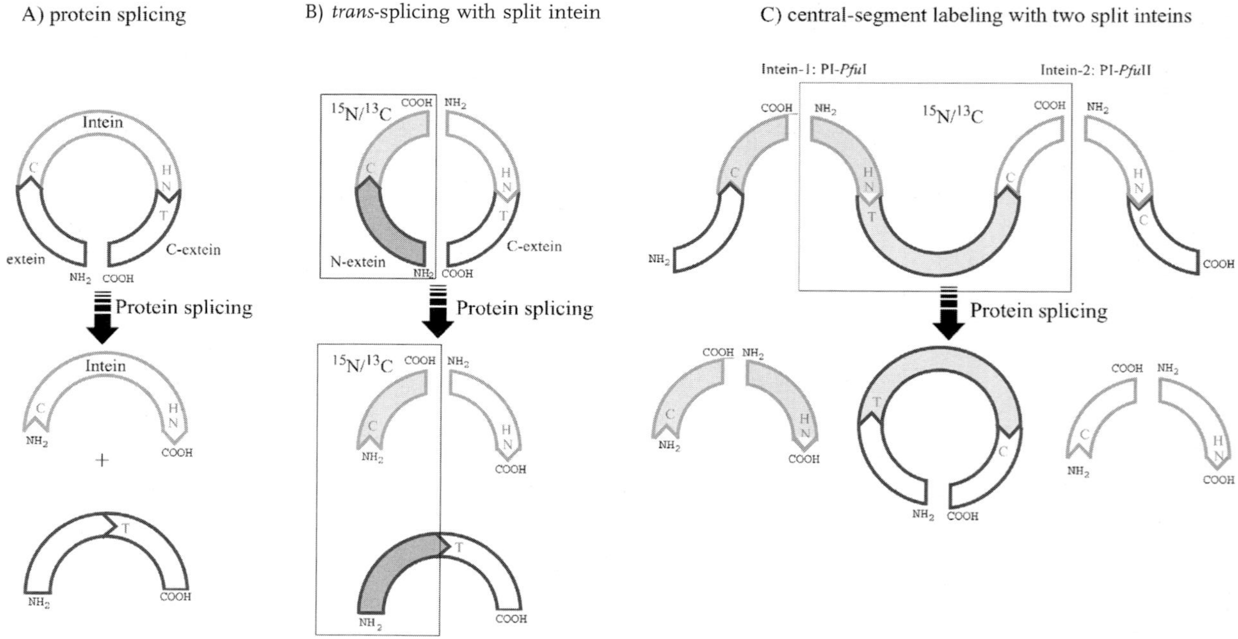

Fig. 1.5 **A** Natural protein splicing, **B** *Trans* splicing with a split intein. The two fragments can be prepared separately and reassembled *in vitro* to form an active intein domain for protein ligation. **C** Central segment labeling using two different split inteins. Residues involved in the protein splicing reaction are shown by a single character code of amino acids. N- and C-termini are indicated by NH_2 and COOH respectively.

of elevated temperatures may often be unfavorable, because many proteins denature at higher temperatures. Therefore, conditions for refolding and ligation such as temperature, pH, additives like glycerol etc. must be carefully optimized for each protein system. On the other hand, the split intein approach does not require any additional thionucleophile, in contrast to the IPL/EPL approach.

Remarkably, this method can even be extended to joining three segments by using two different inteins, PI-*Pfu*I and PI-*Pfu*II (Fig. 1.5 C). Otomo et al. have successfully demonstrated this approach by isotope labeling a central segment of the 370-residue maltose-binding protein [71]. Three protein fragments were constructed for the two ligation reactions. The first fragment was a fusion protein of the N-terminal domain of the split target protein and the N-terminal split intein-1. The second segment consisted of the C-terminal split intein-1 (PI-*Pfu*I), the central part of the split target protein fragment and the N-terminal split intein-2 (PI-*Pfu*II). The third segment contained the C-terminal split intein-2 fused to the C-terminal fragment of the split target protein. Ligation of the first and the second fragment was facilitated by intein-1, while the second and the third fragments were ligated by intein-2. These ligation reactions were highly specific because two different inteins were used. The second fragment was prepared in isotopically enriched medium, and hence the ligated protein was isotopically labeled only in the central part.

The intrinsic limitation of this approach, as in the case of the IPL/EPL approach, is the requirement for specific proximal residues near or at the ligation sites. In inteins, the residues at the junction where ligation occurs are highly conserved because of the chemical mechanism and typically require either cysteine, serine, or threonine [72]. Therefore, at least the N-terminal amino acid of the C-extein must be one of these residues, depending on which intein is used. In addition, the peptide sequence preceding the intein as well as the sequence following the ligation site play an important role for the efficiency of the protein-splicing reaction. For example, in the case of the DnaE intein, the five residues preceding the intein N-terminus and the three residues following the ligation site must seem to be native extein residues in order to achieve efficient splicing [73]. However, these sequential and structural determinants are presently not well understood for all known inteins. Otomo et al. have speculated that the flexibility of the junction region could be one of the important factors for ligation with PI-*Pfu*I. Such requirements would restrict the position of ligation sites, probably to (flexible) linker regions.

One advantage of the split intein over the IPL/EPL approach is the direct use of intein splicing activity, eliminating the requirement for additional thionucleophiles such as thiophenol. An another potential advantage is the ability to perform multiple ligations in a one-pot process, greatly simplifying the reaction procedure for the ligation of several fragments. In contrast, the IPL/EPL approach requires ligation reactions to be performed sequentially for multiple fragment ligations.

Although the use of split inteins for segmental isotope labeling has great potential, the number of inteins which have been adapted to this purpose is currently limited. Additional difficulties arise from the fact that the determinants influencing the success of an intein splicing reaction are not well understood. Moreover, the refolding requirements of split inteins could hinder its use as a general method for the ligation of protein fragments. Hence, further biochemical characterization of inteins is required for the advance of intein-mediated protein ligation methods.

1.4.2
Stabilizing Proteins by Intein-Mediated Backbone Cyclization

Limited protein stability often hampers successful structure elucidation by X-ray crystallography and/or NMR spectroscopy. Relaxation properties are usually improved at elevated temperatures, and multidimensional NMR experiments require sample lifetimes to extend over several days to weeks in order to acquire all the necessary data. In addition, the activity of contaminating proteases that are sometimes present in purified samples can be significant at the experimental temperatures. Therefore, the stability of a target protein can be a concern, in particular for expensive isotope-labeled proteins.

There have been many attempts to improve protein stability and protein properties, utilizing methods such as random mutagenesis, directed evolution, and rational protein design approaches. In general, these methods are far from straightforward and can be time-consuming. In addition, the stabilization of proteins without loss of function is not a trivial problem.

One new approach to stabilizing proteins without changing the primary sequence is to introduce backbone cyclization [74]. No mutations in the primary sequence are introduced by this method, although it might be necessary to insert a flexible linker comprising several residues to join the termini [74–76]. Polymer theory by Flory predicts an improvement in protein stability upon cyclization, because the entropy of the unfolded states should be reduced [77]. Backbone cyclization has long been used for small peptides to reduce the accessible conformational space. Recent advances in intein technology have opened up a new avenue for the cyclization of large proteins, because these proteins can be produced with recombinant techniques in bacterial expression systems [74, 76]. Cyclized proteins can be produced either *in vitro* or *in vivo*, as discussed in the following two sections. Statistical analysis of the structure database reveals that more than 30% of all known proteins might have termini in relatively close proximity, and hence the use of backbone cyclization to stabilize proteins has a good chance of success even in cases where the structure is not yet known [78].

1.4.2.1 *In vitro* Cyclization of Proteins

The IPL/EPL method described in Sect. 1.4.1.1 can be used for cyclization of the backbone polypeptide chains of proteins simply by introducing a nucleophilic thiol group at the N-terminus of the protein (Fig. 1.4 B, 1.6 A). This can be achieved either by creating an N-terminal cysteine by removing residues at the front of the cystein by specific proteolysis or by introducing a cysteine right after the methionine start codon, which is then removed enzymatically *in vivo* [74, 76]. Another method is to use the so-called TWIN system developed by New England Biolab, in which the target protein is fused into the middle of two different modified inteins (Fig. 1.6 B). The N-terminal nucleophilic cysteine is produced by an intein fused to the N-terminus of the target protein. At the same time the C-terminus can be transformed into the corresponding thioester by another intein fused to the C-terminus of the target protein [79]. This system circumvents the requirement for a specific proteolytic site in order to create the N-terminal cysteine, thereby simplifying the cyclization procedure.

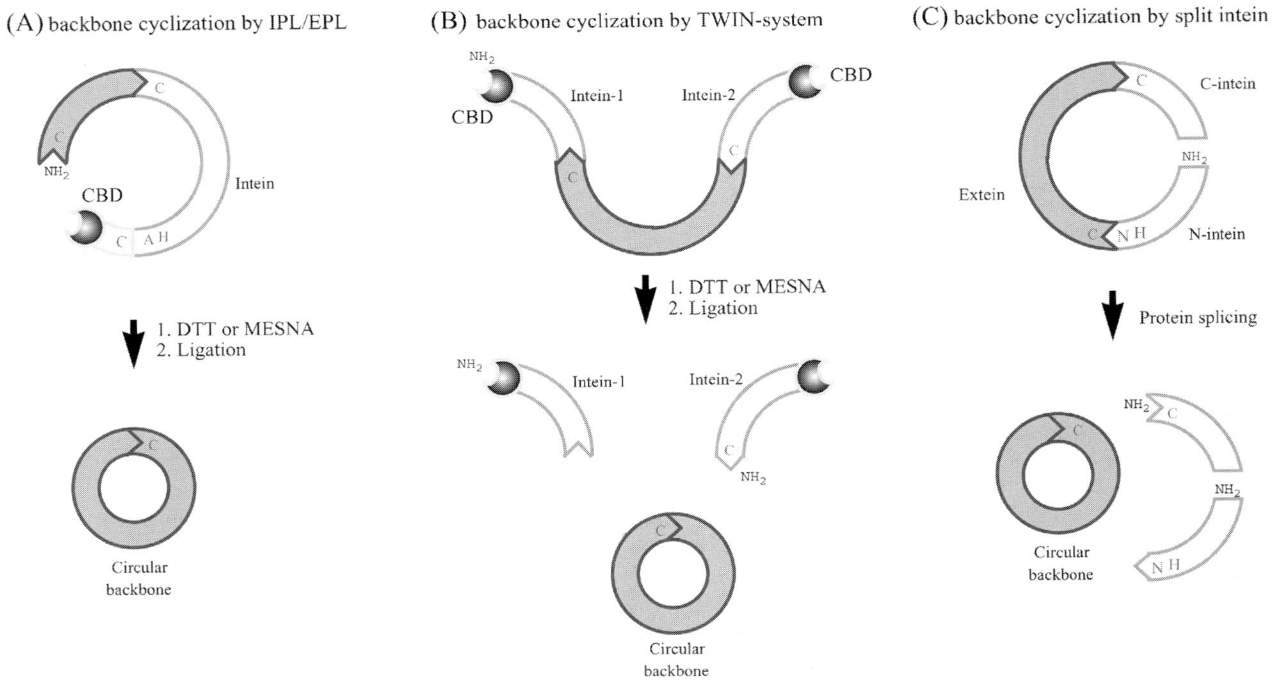

Fig. 1.6 **A** IPL/EPL cyclization with the IMPACT system, **B** backbone cyclization using the TWIN system, **C** backbone cyclization with split inteins. CBD stands for chitin-binding domain. N-termini are indicated by NH_2.

The biggest problems associated with *in vitro* cyclization methods using the IPL/EPL or the TWIN system are competing intermolecular reactions such as polymerization and hydrolysis, which complicate purification as well as reduce yields [74, 79].

1.4.2.2 *In vivo* Cyclization

The principle of *in vivo* cyclization is based on the circular permutation of precursor proteins containing an intein (Fig. 1.6C) [74, 75, 80, 81]. A naturally occurring split intein, DnaE from *Synechocystis* sp. PCC6803, was first successfully used for cyclization. However, similarly to the IPL/EPL approach, a mixture of linear and circular forms is obtained, presumably because of hydrolysis of an intermediate [73, 75]. On the other hand, artificially split inteins such as PI-*Pfu*I, DnaB, and the RecA intein have been successfully applied for *in vivo* cyclization, and only circular forms were observed [80–82], suggesting that the circular permutation approach is more suitable for cyclization. Compared to the IPL/EPL or the TWIN system, *in vivo* cyclization does not require any external thiol group for cyclization, similarly to protein ligation with split inteins. Moreover, there are no undesired products, such as linear forms or polymers, originating from intermolecular reactions.

1.4.2.3 Stability Enhancement by Backbone Cyclization

The effect of backbone cyclization was originally tested on BPTI, but no stabilization effects were observed, presumably because the three disulfide bridges reduce entropic gains [83]. Nevertheless, intein-mediated backbone cyclization has opened the way to a study of cyclization effects on protein stability (including membrane proteins) in more detail. Experimentally improved thermal and/or chemical stability has been shown to result from backbone cyclization of a range of proteins, including β-lactamase, DHFR, *E. coli* IIA^{Glc}, a destabilized mutant of SH3 domain and the N-terminal domain of DnaB [74, 75, 81, 82, 84]. An additional advantage imparted by backbone cyclization is essentially complete resistance to exopeptidases.

1.5 Alternatives to *E. coli* Expression Systems

Structural biology and the structural understanding of the genome, popularly called structural genomics, play an increasingly important part in drug discovery today. Fast and reliable protein expression tools are therefore of prime importance. To this end, the choice of protein expression systems has become increasingly important. While in the not so distant past, only *Escherichia coli*-based expression was used, today a variety of expression systems have been developed ranging from Archaebacteria to mammalian expression vectors. Needless to say, there is no universal expression system, and hence it is often necessary to balance various parameters to achieve optimal expression. For instance, considering the cost of isotopically enriched media, it can be advantageous to sacrifice some native characteristics of a recombinant protein in order to benefit from the higher yields that can be achieved in a more basic expression system. In contrast, specific modifications of the target protein (e.g. glycosylation) predominantly occur in certain cell types, which therefore require the development of special expression vectors.

Here, we describe the various alternatives to the use of *E. coli* expression hosts for heterologous gene expression. Advantages and disadvantages for the different expression systems are discussed, and practical aspects of expression technologies are also described. The feasibility of isotope labeling of recombinantly expressed proteins and their potential use for NMR is also discussed, since the costs and quantities of recombinant proteins produced depend on the system being used.

1.5.1 Expression Vectors

Traditionally, prokaryotic expression, especially employment of *E. coli*-based vectors, has been the system of choice. However, bacteria are unable to provide many vital components required for post-translational modifications including various forms of glycosylation or lipid attachment and protein processing, all of which can also be important for proper protein folding. For this reason, it is not surprising that much time and effort has been dedicated to the development of alternative systems, summarized in Tab. 1.2.

1.5.1.1 *Halobacterium salinarum*

Archaea are interesting organisms in the sense that they represent a phylogenetically distinct group of Prokarya, which is as distantly related to Eubacteria as to Eukarya [85]. *H. salinarum*, the best characterized Archaeon harbors a purple-colored plasma membrane consisting of a complex of one protein, bacterio-opsin (Bop) and its chromophore retinal in a 1:1 ratio [86]. The complex was named bacteriorhodopsin, and it forms typically highly ordered two-dimensional structures in the purple membrane, which allowed its purification and the determination of a high-resolution structure [87]. Recently, a system for heterologous gene expression was constructed for *H. salinarum* [88]. Fusion constructs between the Bop gene and heterologous sequences have been introduced into the *H. salinarum* expression vectors as follows: (i) C-terminally tagged bacteriorhodopsin [88]; C-terminal fusion with (ii) the catalytic subunit of *E. coli* aspartate transcarbamylase [88], (iii) the human muscarinic M1 receptor [88], (iv) the human serotonin 5-HT2c receptor [88], (v) the yeast α-mating factor Ste2 receptor [88]. The Bop-transcarbamylase fusion was well expressed, generating yields of 7 mg receptor per liter of culture. However, introduction of tags at the C-terminus of the Bop gene significantly reduced its expression levels. This was partly because of a decrease in Bop-fusion protein mRNA levels compared to the wild-type Bop. More dramatically, expression studies of fusion constructs between the Bop gene and mammalian GPCRs (G protein-coupled receptors, human muscarinic M1 receptor, platelet-activating receptor and angiotensin-1 receptor) failed to detect fusion protein expression detected by Western blotting [89]. In this case, coding region swaps between Bop and GPCRs improved RNA yields and resulted in detectable levels of Ste2 receptor. These results suggest that *H. salinarum* can be considered as a potentially interesting alternative. The simple and rapid large-scale culture technology is attractive; however, improvements are still required concerning heterologous gene expression. In addition, questions related to codon usage and fusion construct optimization need to be properly addressed.

Tab. 1.2 Features of expression systems

Vector	***Advantage***	***Disadvantage***
Halobacterium salinarum	Rapid expression Colorimetric expression Easy scale-up	Cloning and transformation complicated Requires fusion protein strategy Lack of post-translational modifications
E. coli	Rapid cloning procedure High expression levels Easy scale-up	Toxicity of foreign membrane proteins Lack of post-translational modifications
Saccharomyces cerevisiae	Good secretion machinery Post-translational modifications Easy scale-up	Selection procedure required Tendency to overglycosylation Thick cell wall complicates purification
Schizosacharomyces pombe	Genetics well understood Mammalian promoters applicable	Selection procedure required
Pichia pastoris	High GPCR expression levels	Selection procedure required
Baculovirus	Improved procedure Infection of insect cells High expression yields	Relatively slow virus production Different post-translational processing
Stable mammalian	High authenticity Large-scale set up	Slow procedure to generate cell lines Low recombinant protein yields Stability problems
Transient mammalian	High authenticity Relatively fast methods	Scale-up difficult Transfection methods cell line-specific
Semliki Forest virus	Rapid virus production Broad host range Extreme yields of receptors Large-scale technology established	Safety concerns

1.5.1.2 *Saccharomyces cerevisiae*

Yeast expression vectors have been among those most commonly used since the beginning of gene technology. Vectors based on baker's yeast, *Saccharomyces cerevisiae,* have been especially popular for robust expression of many types of recombinant proteins [90]. For instance, the first commercially available recombinant vaccine, the hepatitis B surface antigen vaccine, was produced from an *S. cerevisiae* vector [91]. Many other recombinant proteins have also been efficiently expressed in yeast including α1-Antitrypsin [92], insulin [93], Epstein-Barr virus envelope protein [94], superoxide dismutase [95] and interferon-α [90].

The genetics and fermentation technology of *S. cerevisiae* are well characterized. Several strong yeast promoters like alcohol dehydrogenase (ADH1), galactose (GAL1/GAL10), 3-phosphoglycerate kinase (PGK) and mating Factor-α (MFα1) have been applied as well as the selection marker genes β-isopropylmalate (LEU2) and oritidine 5′-decarboxylase (URA3) [90]. Among transmembrane proteins, the *S. cerevisiae* α-Factor receptor Ste2p has been expressed with C-terminal FLAG and His_6 tags [96]. Ste2p belongs to the family of GPCRs with a 7-transmembrane topology. Yields of up to 1 mg of almost homologous receptor were obtained, and the purified receptor was reconstituted into artificial phospholipid vesicles. However, restoration of ligand-binding activity required the addition of solubilized membranes from an Ste2p negative yeast strain. Also, the human dopamine D1A receptor was expressed with C-terminal FLAG and His_6 tags in *S. cerevisiae*, which allowed for purification and reconstitution of receptor [97].

1.5.1.3 ***Schizosaccharomyces pombe***

Another yeast strain that has received much attention as an expression host is the fission yeast *Schizosaccharomyces pombe*. In contrast to *S. cerevisiae*, no budding occurs, and the yeast only reproduces by means of fission and by spores [98]. Two types of expression vectors have been developed for *S. pombe*. The chromosomal integration type of vector maintains the foreign gene stably in the chromosome [99], and the episomal vector replicates autonomously in yeast cells [100]. Some mammalian promoters like the human chorionic gonadotropin and CMV promoters are functional in *S. pombe* [101]. The fission yeasts possess many similar features to mammalian cells. *S. pombe* has a signal transduction system similar to the mammalian G protein-coupling system [102], and the mammalian endoplasmatic reticulum retention signal KDEL is also recognized [103]. The glycosylation pattern for *S. pombe* is also different from that of *S. cerevisiae* and other yeast species.

A wide range of mammalian proteins have been expressed in *S. pombe*. In a successful example, the human lipocortin I comprised 50% of the total soluble proteins in yeast cells and showed high activity, indicating that the post-translational modifications were mammalian-like [104]. Membrane proteins including cytochrome P450 were expressed at ten times the levels of those in other yeast systems [105]. Also, GPCRs have been expressed in *S. pombe*, where the human dopamine D2 receptor was correctly inserted into the yeast cell membrane and demonstrated expression levels three times those of *S. cerevisiae* [106].

1.5.1.4 ***Pichia pastoris***

The methylotrophic yeasts including *Pichia pastoris, Hansenula polymorpha* and *Kluyveromyces lactis* have become potentially attractive expression hosts for various recombinant proteins [107]. In addition to the relative ease with which molecular biology manipulations can be carried out, *P. pastoris* has demonstrated a capacity for performing many post-translational modifications such as glycosylation, disulfide bond formation and proteolytic processing [108]. *P. pastoris* utilizes the tightly methanol-regulated alcohol oxidase 1 (AOX1) promoter, and the vector is integrated as several copies into the yeast host genome. When human insulin was expressed in *P. pastoris*, the secretion was comparable to that obtained for *S. cerevisiae*. Peptide mapping and mass spectrometry confirmed identical processing of human insulin in yeast and mammalian cells. *P. pastoris* has also been used as a host for expression of GPCRs [109]. The mouse 5-HT_{5A} receptor and the human β_2-adrenergic receptor were fused to the prepropeptide sequence of the *S. cerevisiae* α-factor, which enhanced the expression levels by a factor of three. Multiple chromosomal integrations further improved the expression twofold. In the case of the β_2-adrenergic receptor, addition of the antagonist alprenol to the culture medium increased the number of specific binding sites. A similar but weaker effect was seen for the 5-HT_{5A} receptor after addition of yohimbine. The binding activity for the β_2-adrenergic receptor and the 5-HT_{5A} receptor were 25 pmol and 40 pmol, respectively, per milligram of membrane protein. The pharmacological profiles assayed by ligand-displacement analysis were similar to those obtained from receptors expressed in mammalian cells.

1.5.1.5 **Baculovirus**

Heterologous gene expression has been studied to a great extent in insect cells with the aid of baculovirus vectors. The popularity of the baculovirus system is mainly due to the high expression levels obtained for various recombinant proteins resulting from the use of strong viral promoters [110]. Generally, heterologous genes are expressed from the polyhedrin promoter of *Autographa californica* in several insect cell lines such as *Spodoptera frugiperda* (Sf9), *Trichoplusia ni* (Tn), *Mamestra brassicae* and *Estigmene acrea* [111]. Although baculovirus vectors have been used for expression of various mammalian recombinant proteins, a limitation has been the differences in the N-glycosylation pathway between insect and mammalian cells. However, *Estigmene acrea* cells can produce a similar glycosylation pattern as occurs in mammalian cells [112]. Moreover, modifications of baculovirus vectors by replacing the polyhedrin promoter with a CMV promoter made it possible to carry out expression studies in mammalian cell lines [113]. Using this so-called BacMam system, milligram quantities of a cellular adhesion protein (SAF-3) could be produced in CHO cells [114]. Baculovirus vectors have been used extensively to express GPCRs and ligand-gated ion channels [115]. Expression levels up to 60 pmol receptor per milligram have been obtained, which has led to relatively efficient purification procedures. In attempts to further enhance the expression level of the β2-adrenergic receptor, an artificial sequence was introduced, which resulted in approximately double the receptor levels in insect cells [116].

1.5.1.6 **Transient Mammalian Expression**

Several approaches have been taken to develop efficient transient mammalian expression systems. The most straightforward process has been to engineer expression vectors with strong promoters. Relatively high expression levels for cytoplasmic and even some transmembrane proteins have been obtained in adherent cells on a small scale. However, a major problem arises in the scale-up of these growth procedures, which are also relatively expensive [117]. In spite of this difficulty, transient transfection experiments using a modified calcium-phosphate coprecipitation method have been carried out in HEK293 EBNA cells adapted to suspension cultures grown on a 100 L scale [118]. More than 0.5 g of a monoclonal antibody was produced from this system, although similar methods have yet to be developed for receptor expression.

1.5.1.7 **Stable Mammalian Expression**

Generation of various cell lines (BHK, CHO and HEK293) with the target gene inserted downstream of a strong promoter into the genome is a common approach to achieve overexpression in mammalian hosts. However, one drawback is the time-consuming procedure involved in the establishment of stable cell lines, which generally requires 6–8 weeks. Other problems associated with this approach are related to the relatively low expression levels and the instability of generated cell lines. These issues have been addressed by engineering inducible expression systems, which are usually based on tetracyclin-based regulation (Tet on-off systems) [119]. A highly interesting development has been the generation of a cold-inducible expression system based on the Sindbis virus replicon [120]. Because of a point mutation in one of the replicase genes, the viral replicase complex is totally inactive at 37 °C, whereas a shift in temperature below 34 °C results in high replication activity and high levels of heterologous gene expression. Using this approach, the serotonin transporter gene, characterized by its low expression levels in any system tested, generated reasonable yields (approximately 250,000 copies per cell) [121].

1.5.1.8 Viral Vectors

The two common features that have made viral vectors attractive for recombinant protein expression are their high infection rates for a broad range of mammalian cell lines and their strong promoters. Adenovirus vectors have shown high expression levels in, for instance, human embryonic kidney (HEK293) cells, but their use has been to some extent restricted by the fairly complicated virus generation procedure [122]. Another potentially useful class of viruses are the poxviruses. Recombinant gene expression of herpes simplex virus thymidine kinase (TK) has been established for vaccinia virus vectors [123]. Moreover, the engineering of a hybrid bacteriophage-vaccinia virus vector by applying the T7 promoter has simplified and broadened the use of pox virus-based expression systems [124]. However, vaccinia vectors are still quite complicated to use for rapid recombinant protein expression, and they have instead found applications in the field of vaccine development. Alphavirus vectors have proven to be highly efficent for heterologous gene expression. Both Semliki Forest virus- (SFV) [125] and Sindbis virus-based [126] expression vectors have been engineered to rapidly generate high-titer recombinant particles, which are susceptible to a broad range of mammalian cell lines and primary cells in culture [127]. Typically, both GPCRs and ligand-gated ion channels have been expressed at extreme levels, i.e. up to 200 pmol receptor per milligram protein [128]. Large-scale SFV technology has been established, which has generated large quantities of, for instance, mouse serotonin 5-HT3 receptor, purified to homogeneity and subject to structural characterization [129]. Moreover, several GPCRs have been expressed at levels of 5–10 mg receptor yields per liter suspension culture of mammalian host cells [130], which has provided material for large-scale purification.

1.5.2 Comparison of Expression Systems

Comparison between different systems for transmembrane protein expression are always difficult to make, and they obviously reflect individual needs and are strongly influenced by personal experience. It is, however, important to define the usefulness of each system by taking into account different aspects such as ease of handling, expression levels, time and labor requirements, safety, costs, and the quality of the produced recombinant protein (Tab. 1.3).

Obviously, prokaryotic systems are easy to use, the costs for their large-scale applications are low, and no safety risks are involved. The drawbacks are their limited capacity for post-translational modifications and generally low yields of complex mammalian

Tab. 1.3 Application of expression systems for mammalian membrane proteins

Vector	*Handling*	*Expr./Scale-up*	*Authenticity*	*Time/Labour*	*Safety*	*Costs*
Halobacterium salinarum	Easy	Low/easy	Low	Short/easy	High	Low
E.coli	Easy	Moderate/easy	Low	Short/easy	High	Low
Saccharomyces cerevisiae	Rel. easy	High/easy	Mod.	Mod./mod.	High	Mod.
Schizosacharomy-ces pombe	Rel. easy	High/easy	Mod.	Mod./mod	High	Mod.
Pichia pastoris	Rel. easy	High/easy	Acceptable	Mod./mod	High	Mod.
Baculovirus	Mod.	High/mod.	Rel. high	Mod./mod.	High	Rel. high
Stable mammalian	Difficult	Low/difficult	High	Long/intensive	High	High
Transient mammalian	Difficult	Mod./difficult	High	Mod./mod.	High	High
Semliki Forest virus	Easy	Extreme/easy	High	Short/easy	Of concern	High

transmembrane proteins. Yeast expression systems are competitive with bacterial vectors with respect to scaleability, costs and safety. Although the time required from gene construct to expressed recombinant protein is slightly longer, the yields are significantly higher. Yeast can also provide some post-translational and protein-processing capacity, although not identical to mammalian cells. The thick yeast cell wall is of some concern, because it makes the purification of intracellular and transmembrane proteins more complicated. Baculovirus vectors have the advantages of high expression levels in insect cells and the fairly simple though more expensive scale-up procedure. Needless to say, the optimal host for expression of mammalian transmembrane proteins from the viewpoint of a molecular biologist must be mammalian cell lines. Obviously, the drawbacks with the conventional transient or stable expression approaches have been the labor intensiveness and time-consuming procedures. The yields have also been disappointingly low and the costs for large-scale production high. Viral vectors are therefore potentially very attractive because of their high capacity for gene delivery and extreme expression levels of heterologous genes. Naturally, viral vectors always pose a higher safety risk of possible infection of laboratory personnel. Not only have the rapidly generated replication-deficient SFV vectors been demonstrated to be free from any contaminating replication-proficient particles, but also the amounts of residual infectious particles associated with cells or even in the medium are negligible [131]. Today the SFV system is classified as BL1 level in several European countries (Germany, Finland, Sweden, Swizerland, UK) but is currently BL2 in the United States. Moreover, the broad host range provides an additional opportunity for the study of gene expression and protein processing in several mammalian host cell lines in parallel. Evidently, large-scale cultivation of mammalian cells is more expensive than bacterial or yeast cell equivalents, but these costs are significantly reduced by using a serum-free medium for suspension cultures.

1.5.3 Isotope Labeling and NMR

Recent developments in technologies within structural biology should also play an important part for transmembrane proteins. The potential to incorporate stable isotopes would facilitate structure determination by NMR techniques. Although NMR technologies were long considered to be applicable only to smaller proteins, the development of transverse relaxation-optimized spectroscopy (TROSY) has made it possible to use NMR for larger proteins also [5], even integral membrane proteins. For example, the OmpX and OmpA integral membrane proteins of *E. coli* were labeled with $^{13}C/^{15}N/^{2}H$ isotopes and overexpressed as inclusion bodies in bacterial cells. After solubilization in 6 M Gdn-HCl and reconstitution in detergent micelles, solution NMR techniques could be used to identify regular secondary structural elements [132].

Proteins that require non-*E.coli* expression systems are generally too large in size to be used for NMR studies without uniform ^{15}N, ^{13}C and and sometimes ^{2}H labeling. Hence, when using the expression systems described in this chapter, it is important to ensure that cells can be grown on defined, isotopically enriched media. This fact at the moment excludes, for example, the use of fetal calf serum. However, special isotope-enriched defined media are available from commercial suppliers which present fully rich, serum-free (protein-free) media at reasonable costs containing labeled amino acids and carbohydrates and which can be used to effectively express heterologous proteins in insect cells applying the baculovirus vector system [133]. Similarly, rich media are also available for expression in *S. cerevisiae* [134]. Conversely, the methylotropic yeast *P. pastoris* can be grown on minimal media, facilitating its use as a potential host for isotope labeling. In fact, there have been a number of successful examples where *P. pastoris* was used to produce isotopically enriched samples for solution NMR studies, including a cysteine-rich glycosylated domain of thromobomodulin [135], a glycosylated EGF module [136], domains from rat calretinin [137], and tick anticoagulant peptide [138]. Even more importantly, it was shown that expression in yeast enables the production of heterologous proteins in deuterated form [139], which would be very difficult or even impossible to achieve in mammalian expression systems because of the cell toxicity of deuterated water. Similarly, *H. salinarum* can easily be grown in D_2O and on the above-described algae hydrolyzates. Moreover, residue type-specific labeling is possible in this host [26]. Examples of proteins expressed in this host in isotopically labeled form can be found in the literature, e.g. Refs. [25] and [140]. In order to obtain proteins with more native-like glycosylation patterns, CHO mammalian cell expression systems have also been developed for NMR sample generation [141–143]. However, the requirement for rich media in mammalian cell-based expression systems combined with difficulties associated with generating high expression levels have to date prevented a more widespread utilization for NMR. Nonetheless, considering the range of proteins that may be inaccessible to *E. coli*-based expression systems as well as the potential information that can be gained by studying the native-like post-translationally modified forms of proteins, the adaptation of heterologous protein expression systems to the purpose of isotope labeling for solution NMR clearly requires further development.

Reconstitution of GPCRs in appropriate membrane mimetics or membrane preparations is still very difficult, and success requires efficient expression systems to yield enough protein material. As large quantities of an increasing number of recombinant proteins become available, it will be possible to develop techniques for solubilization, purification and reconstitution in a high-throughput format. The first global structural genomics project for membrane proteins, MePNet (Membrane Protein Network) was recently initiated with the aim of comparing the overexpression of 100 GPCRs in three systems based on *E. coli, P. pastoris* and SFV vectors [144]. The goal of this three year program is to verify the expression levels for the 100 targets and establish platforms for solubilization, purification and crystallization technologies which should form a solid base for obtaining novel high-resolution structures of GPCRs. Technological developments arising from this initiative should also benefit the field of solution and solid-state NMR.

1.5.4
Target Proteins

The choice of expression system is to a large extent dictated by the type of target protein. In general, many soluble proteins, which are fairly easy to express and have a low molecular weight, are efficiently expressed from bacterial vectors. As a rule of thumb one can suggest that *E. coli*-based expression should be used whenever possible. However, *E. coli* vectors are not suitable for expression of many authentic mammalian transmembrane proteins. Additionally, the near-completion of the human genome sequence has revealed a multitude of genes as potential targets for structural analyses. Many of these include transmembrane proteins, whose properties quite often make them more difficult to express. Many of these transmembrane proteins, including receptors and channels, are important targets for drug discovery. GPCRs alone stand for approximately 50% of drug targets today. Moreover, a quarter of the top 200 drugs in the United States are based on GPCRs, and the annual sales in 2000 reached more than 18.5 billion US$ [145]. It is therefore understandable that so much interest has been focused on transmembrane proteins today. At present, it is possible to express complicated transmembrane receptors in several vector systems at reasonably high levels, cost-effectively in a near-to-native state. As more recombinant proteins become available, the solubilization, purification and crystallization for NMR technology can also be developed, which will contribute to the achievement of rates of structure determination similar to those common for soluble proteins today. However, the determination of high-resolution structures of membrane proteins, and particularly mammalian ones, has been modest compared to soluble proteins (Tab. 1.4). Recently, methodological advances in NMR spectroscopy have significantly raised the size limit amenable to NMR investigations, and hence membrane protein structures determined by NMR have also appeared in the literature. The outer membrane protein A from *E. coli* was reconstituted in DPC micelles and the structure determined by heteronuclear multidimensional NMR [146]. Similarly, the structure of the outer membrane protein OmpX was elucidated in DHPC micelles [147]. Both proteins were expressed in *E. coli*. However, in the case of OmpA, a series of selectively ^{15}N-labeled samples was prepared, whereas in the case of OmpX, spectroscopy was performed on a single triply ^{2}H,^{13}C,^{15}N-labeled sample. The best successes have been achieved when it was possible to isolate proteins from their natural sources instead of using recombinantly expressed forms. The results have been particularly poor for GPCRs, and the only example of a mammalian 7-transmembrane protein for which a high-resolution structure has been obtained is bovine rhodopsin [148].

1.6
The Use of Cell-Free Protein Expression for NMR Analysis

The expression of proteins usually requires optimization by trial and error, since conditions leading to production of high yields of active protein are difficult to predict. Therefore, additional time-consuming steps are often involved that may require changes in the choice of expression system, of reaction conditions, and of tests for the expression of protein sub-domains. In structural genomics programs this problem was rapidly identified as a major bottleneck, since high-throughput production of soluble proteins in milligram amounts is essential for success. Hence, systems that allow rapid, productive expression, are easy to manipulate, and can be run in a parallel format are highly desired in order to rapidly screen for the best conditions.

Until recently, cell-free protein expression (also sometimes erroneously named *in vitro* protein expression) did not exhibit the productivity required for preparation of NMR samples, especially considering the high cost of using isotopically labeled starting material. Rather, it was exclusively used as an analytical tool that served to verify correct cloning or to study promotor sites. Because of the very low yields, detection of the expressed product usually required incorporation of a radioactive label (usually via ^{35}S-methionine).

This situation changed fundamentally when Spirin published his substantial improvements in 1988 that resulted in much higher product yields [149]. He developed a set-up in which the coupled transcription/translation reaction could be continuously supplied with all the essential low-molecular-weight components (nucleotides, amino acids, energy components) while maintaining a constant reaction volume by continuously removing the product through a membrane (continuous-flow cell-free system, CFCF). Under these conditions, the system would remain active for more than 1 day, in contrast to the usual upper limit of 2 h which was observed in the conventional batch system. The prolonged expression period indicated that sufficient amounts of all factors necessary for translation were still available in the system, although the membrane clearly allowed the leakage of at least some of them. Soon, these findings were used to run the reaction in a more robust dialysis mode, keeping the reaction volume constant while feeding the system with all necessary low-molecular-weight consumables.

In the following years the productivity of the method was tremendously improved, so that milligram amounts of protein per milliliter of reaction mixture could be obtained [150–154]. This finally opened the door to cell-free protein expression to be used for the production of isotopically enriched proteins suitable for NMR analysis. However, although widely considered to be a promising method for labeling proteins at specific positions and therefore facilitating the process of chemical shift assignment as well as reducing spectral overlap [155, 156], in fact it was used in very few laboratories worldwide. This was mainly due to the fact that the preparation of the lyzate was tedious and the expression levels of each batch varied from lot to lot.

Recently, a system for cell-free protein production has became commercially available (the Rapid Translation System, RTS) [157]. In the following sections this system will be described, and advantages as well as limitations will be discussed.

1.6.1
The Cell-Free Protein Expression System RTS

The RTS system includes two different technology platforms for cell-free protein expression as well as a number of tools for finding optimal conditions (Scheme 1.1). All expression systems use the T7-polymerase for transcription and an *E. coli* lyzate with reduced nuclease and protease activity for translation. The conditions are optimized for a coupled transcription/translation reaction so that the DNA can be directly used as the template.

The first platform (RTS 100 HY) is designed as a screening tool. It uses the batch format, so that the reaction time does not exceed 3 h. In particular, in the RTS 100 HY system the exonuclease activity is reduced, so that the direct use of PCR-generated DNA templates is possible. To facilitate the generation of the PCR templates there is a special product available (linear template kit), which introduces all regulatory elements [T7-promotor, gene10 enhancer sequence and the Shine-Dalgarno (RBS) sequence]. Consequently, RTS 100 HY can be used for the rapid evaluation of the best template, without spending time with cloning, and for optimization of the reaction conditions (e.g. temperature, choice of additives like detergents, chaperones etc.). In addition, a bioinformatic tool (the program ProteoExpert) facilitates the process of designing the optimal template by analyzing and improving the secondary structure of the corresponding mRNA (without changing the amino acid sequence of the protein).

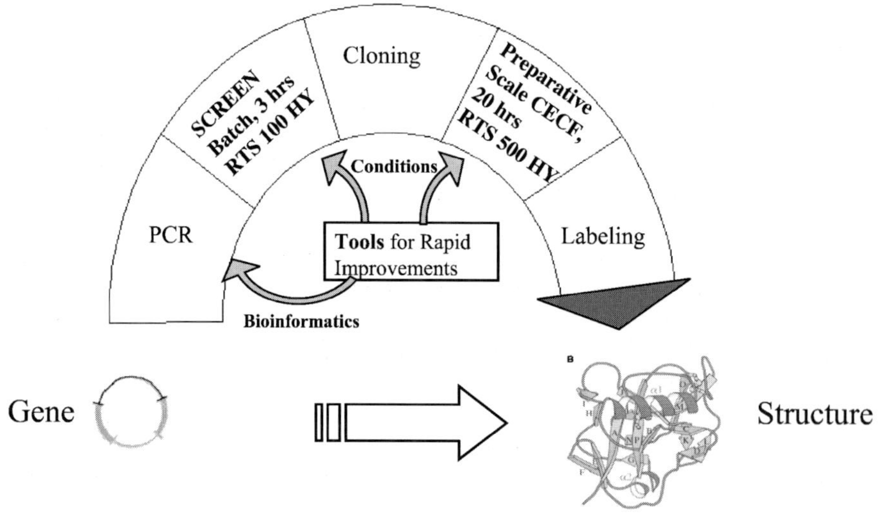

Scheme 1.1 From gene to protein structure via cell-free protein expression.

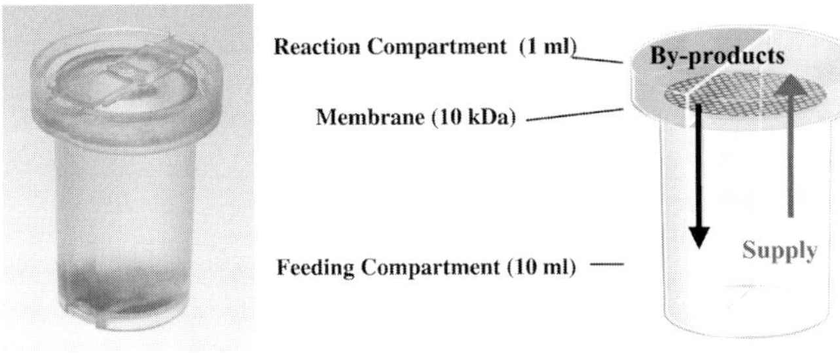

Fig. 1.7 Device of the RTS 500 format using the Continuous Exchange Cell-Free principle.

The second platform (RTS 500 HY) is designed for the production of proteins on a preparative scale. It is based on the CECF principle and utilizes a device with two chambers (Fig. 1.7). This design, with a proper choice of reaction conditions, gives a reaction time of 24 h, yielding up to 6 mg of protein per ml. A scaled-up version is also available (RTS 9000 HY), providing up to 50 mg protein per run.

In all formats, the amino acids are supplied separately, so they can be conveniently exchanged for labeled ones.

1.6.2 From PCR Product to ^{15}N-Labeled Protein

As an example of how cell-free protein expression can be used to rapidly generate a protein sample suitable for NMR analysis, an SH3-domain (8 kDa) was expressed using the RTS system [158]. Since it was initially found that the yield using a template carrying the wild-type sequence was too low, the sequence of the template was analyzed using the ProteoExpert program. The suggested sequences were subsequently evaluated by running expressions in the batch mode (RTS 100) using PCR-generated templates (linear template kit). As a result, all ten test sequences showed significantly higher yields than those with the original wild-type RNA template (data not shown). One of these was selected and ligated into the TOPO Cloning vector (Invitrogen Corp.) for expression in RTS 500 HY. Comparison with expression levels obtained with the wild-type template showed that also by using a circular template the yield could be improved more than fivefold (Fig. 1.8), resulting in approximately 3 mg product/mL reaction mix. In the next step, uniformly ^{15}N-labeled protein was produced by using a mixture of ^{15}N-labeled amino acids. Expression levels of the ^{15}N-labeled reaction were identical to the first reaction performed with unlabeled amino acids. The product was purified to homogeneity, and a [^{15}N,^{1}H]-HSQC spectrum was obtained (Fig. 1.9) to confirm the overall integrity of the protein fold.

For the residue-type assignment of the cross peaks, the SH3 domain was expressed with specifically labeled glycine or arginine residues by using amino acid mixtures where only Gly and Arg were ^{15}N-labeled. Again, yields for the labeled proteins were identical to those of the unlabeled product. Analysis of the corresponding [^{15}N,^{1}H]-HSQC spectra

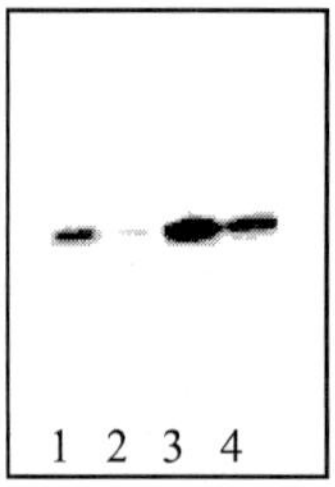

Fig. 1.8 Western blot analysis (via His_6-tag) of the expression of the SH3 domain using the wild-type DNA-sequence (lanes 1 and 2) or the optimized DNA-sequence (lanes 3 and 4). On lanes 1 and 3 0.25 μL DNA were loaded on the gel, while on lanes 2 and 4 0.125μL were applied.

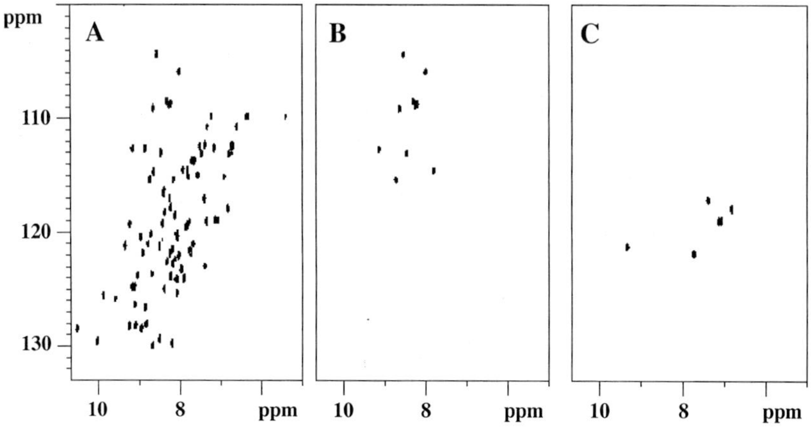

Fig. 1.9 [^{15}N,^{1}H]-HSQC spectrum of ^{15}N-uniformly labeled SH3 domain (**A**) and of a sample selectively labeled with only ^{15}N-Gly (**B**) or ^{15}N-Arg (**C**).

(Fig. 1.9) verified that the numbers of cross peaks for the ^{15}N-Gly as well as for the ^{15}N-Arg labeled protein were identical to the predicted ones (10 for the 10 Gly and 6 for the 3 Arg, respectively). Importantly, no scrambling of the ^{15}N-labels could be detected. All cross peaks were contained in the [^{15}N,^{1}H]-HSQC spectrum of the uniformly labeled protein (Fig. 1.9) and no additional signals occurred. Therefore, by simply overlaying the HSQC spectra, the cross peaks belonging to the glycine and the arginine residues could be readily assigned.

1.6.3
Discussion and Outlook

This example demonstrates how cell-free protein expression can be used to rapidly optimize the reaction conditions as well as to shorten the time for spectral assignments by producing individually labeled proteins.

To discuss in general the applicabilities of cell-free protein expression technology [159], two features are most valuable. First, only the protein of interest is produced, because the highly active T7-polymerase is used for translation. Consequently, labeled amino acids are almost exclusively incorporated into the newly produced protein. Since the labeled amino acids can be supplied as a mixture or added individually, the time for assigning the cross peaks to the particular amino acid(s) can be significantly reduced. This approach will enable partial assignments to be made in molecules that are far too large to allow spectral assignments from uniformly labeled protein by classical methods.

The second important feature is that cell-free protein expression can be considered as an "open system", meaning that no lipid membrane barriers are present. Consequently, chemicals, proteins (e.g. chaperones) as well as PCR-generated templates can be added directly to the reaction solution. Even major changes of the reaction conditions are possible (e.g. using a redox system to produce active proteins containing correctly formed disulfide bonds [160]). These features, as well as the ease of sample handling, dramatically reduce the time for optimizing the expression conditions (in fact, pipetting robots can be designed to run the reactions). Moreover, proteins can be synthesized which display cell toxicity and which therefore can hardly be expressed in classical systems.

However, certain limitations do exist that need to be considered. Although enzymes necessary for post-translational modifications can be added, in principle there is currently no productive system available for the preparation of glycosylated proteins, although some interesting results have already been obtained [161]. Also, the expression of functional membrane proteins in quantities necessary for structural analysis will be a challenging task for the future.

Nonetheless, the speed and flexibility of this emerging technology could provide the key to meeting the demands of high-throughput structure determinations.

1.7 References

1 Bax, A.; Grzesiek, S. *Acc. Chem. Res.* **1993**, *26*, 131–138.
2 Zimmerman, D. E.; Kulikowski, C. A.; Huang, Y. P.; Feng, W. Q.; Tashiro, M.; Shimotakahara, S.; Chien, C. Y.; Powers, R.; Montelione, G. T. *J. Mol. Biol.* **1997**, *269*, 592–610.
3 Yamazaki, T.; Lee, W.; Arrowsmith, C. H.; Muhandiram, D. R.; Kay, L. E. *J. Am. Chem. Soc.* **1994**, *116*, 11655–11666.
4 LeMaster, D. M.; Richards, F. M. *Biochemistry* **1988**, *27*, 142–150.
5 Pervushin, K.; Riek, R.; Wider, G.; Wüthrich, K. *Proc. Natl. Acad. Sci. USA* **1997**, *94*, 12366–12371.
6 Salzmann, M.; Pervushin, K.; Wider, G.; Senn, H.; Wüthrich, K. *J. Am. Chem. Soc.* **2000**, *122*, 7543–7548.
7 Neri, D.; Szyperski, T.; Otting, G.; Senn, H.; Wüthrich, K. *Biochemistry* **1989**, *28*, 7510–7516.
8 Muchmore, D. C.; McIntosh, L. P.; Russel, C. B.; Anderson, D. E.; Dahlquist, F. W. In *Methods Enzymol.*; Oppenheimer, N. J., James, T. L., Eds.; Academic Press: San Diego, **1989**; Vol. 177, pp 44–73.
9 Patzelt, H.; Simon, B.; Ferlaak, A.; Kessler, B.; Kühne, R.; Schmieder, P.; Oesterhelt, D.; Oschkinat, H. *Proc. Natl. Acad. Sci. USA* **2002**, *99*, 9765–9770.
10 Evans, T. C., Jr.; Benner, J.; Xu, M. Q. *Protein Sci.* **1998**, *7*, 2256–2264.
11 Pardi, A. *Curr. Opin. Struct. Biol.* **1992**, *2*, 832–835.
12 Markley, J. L.; Kainosho, M. In *NMR of Macromolecules: A Practical Approach*; Roberts, G. C. K., Ed.; Oxford University Press: Oxford, **1993**; pp 101–152.
13 LeMaster, D. M. *Progr. NMR Spectrosc.* **1994**, *26*, 371–419.
14 Goto, N. K.; Kay, L. E. *Curr. Opin. Struct. Biol.* **2000**, *10*, 585–592.
15 Bolton, P. H. *Progr. NMR Spectrosc.* **1990**, *22*, 423–452.
16 Kainosho, M. *Nature Struct. Biol.* **1997**, *4 Suppl.*, 858–861.
17 Lian, L.-Y.; Middleton, D. A. *Prog. NMR Spectrosc.* **2001**, *39*, 171–190.
18 Makrides, S. C. *Microbiol. Rev.* **1996**, *60*, 512–538.
19 Crespi, H. L.; Katz, J. J. *Methods Enzymol.* 1972, *26*, 627–637.
20 Brodin, P.; Drakenberg, T.; Thulin, E.; Forsen, S.; Grundstroem, T. *Protein Eng.* **1989**, *2*, 353–357.
21 Sorensen, P.; Poulsen, F. M. *J. Biomol. NMR* **1992**, *2*, 99–101.
22 Patzelt, H.; Kessler, B.; Oschkinat, H.; Oesterhelt, D. *Phytochemistry* **1999**, *50*, 215–217.
23 Patzelt, H.; Ulrich, A. S.; Egbringhoff, H.; Dux, P.; Ashurst, J.; Simon, B.; Oschkinat, H.; Oesterhelt, D. *J. Biomol. NMR* **1997**, *10*, 95–106.
24 Takeda, H. *Phytochemistry* **1996**, *42*, 673–675.
25 Schweimer, K.; Marg, B.-L.; Oesterhelt, D.; Rösch, P.; Sticht, H. *J. Biomol. NMR* **2000**, *16*, 347–348.
26 Reat, V.; Patzelt, H.; Ferrand, M.; Pfister, C.; Oesterhelt, D.; Zaccai, G. *Proc. Natl. Acad. Sci. U. S. A.* **1998**, *95*, 4970–4975.
27 Celeste, S. G.; Jay, G. D.; Brenda, G. S.; Engin, S. H. *Arch. Biochem. Biophys.* **1995**, *319*, 203–210.
28 Glasoe, P. K.; Long, F. A. *J. Phys. Chem.* **1960**, *64*, 188–190.
29 Nietlispach, D.; Clowes, R. T.; Broadhurst, R. W.; Ito, Y.; Keeler, J.; Kelly, M.; Ashurst, J.; Oschkinat, H.; Domaille, P. J.; Laue, E. D. *J. Am. Chem. Soc.* **1996**, *118*, 407–415.
30 For more information see: http://www.genome.ad.jp/kegg/kegg2.html
31 Venters, R. A.; Metzler, W. J.; Spicer, L. D.; Mueller, L.; Farmer, B. T. *J. Am. Chem. Soc.* **1995**, *117*, 9592–9593.
32 Gardner, K. H.; Rosen, M. K.; Kay, L. E. *Biochemistry* **1997**, *36*, 1389–1401.
33 Crespi, H. L.; Rosenberg, R. M.; Katz, J. J. *Science* **1968**, *161*, 795–796.
34 Crespi, H. L.; Katz, J. J. *Nature* **1969**, *224*, 560–562.
35 Rosen, M. K.; Gardner, K. H.; Willis, R. C.; Parris, W. E.; Pawson, T.; Kay, L. E. *J. Mol. Biol.* **1996**, *263*, 627–636.
36 Janin, J.; Miller, S.; Chothia, C. *J. Mol. Biol.* **1988**, *204*, 155–164.
37 Kay, L. E.; Bull, T. E.; Nicholson, L. K.; Griesinger, C.; Schwalbe, H.; Bax, A.; Torchia, D. A. *J. Magn. Reson.* **1992**, *100*, 538–558.

38 Gottschalk, G. *Bacterial Metabolism*; 2 ed.; Springer Verlag: New York, **1986**.

39 Lee, A. L.; Urbauer, J. L.; Wand, A. J. *J. Biomol. NMR* **1997**, *9*, 437–440.

40 Ishima, R.; Louis, J. M.; Torchia, D. A. *J. Biomol. NMR* **2001**, *21*, 167–171.

41 Mulder, F. A.; Hon, B.; Mittermaier, A.; Dahlquist, F. W.; Kay, L. E. *J. Am. Chem. Soc.* **2002**, *124*, 1443–1451.

42 Gardner, K. H.; Kay, L. E. *J. Am. Chem. Soc.* **1997**, *119*, 7599–7600.

43 Goto, N. K.; Gardner, K. H.; Mueller, G. A.; Willis, R. C.; Kay, L. E. *J. Biomol. NMR* **1999**, *13*, 369–374.

44 Kelly, G.; Prasannan, S.; Daniell, S.; Fleming, K.; Frankel, G.; Dougan, G.; Connerton, I.; Matthews, S. *Nat. Struct. Biol.* **1999**, *6*, 313–318.

45 Mueller, G. A.; Choy, W. Y.; Yang, D.; Forman-Kay, J. D.; Venters, R. A.; Kay, L. E. *J. Mol. Biol.* **2000**, *300*, 197–212.

46 Berardi, M. J.; Sun, C. H.; Zehr, M.; Abildgaard, F.; Peng, J.; Speck, N. A.; Bushweller, J. H. *Struct. Fold. Des.* **1999**, *7*, 1247–1256.

47 Liu, X.; Wang, H.; Eberstadt, M.; Schnuchel, A.; Olejniczak, E. T.; Meadows, R. P.; Schkeryantz, J. M.; Janowick, D. A.; Harlan, J. E.; Harris, E. A.; Staunton, D. E.; Fesik, S. W. *Cell* **1998**, *95*, 269–277.

48 Aghazadeh, B.; Zhu, K.; Kubiseski, T. J.; Liu, G. A.; Pawson, T.; Zheng, Y.; Rosen, M. K. *Nat. Struct. Biol.* **1998**, *5*, 1098–1107.

49 DeRose, E. F.; Li, D. W.; Darden, T.; Harvey, S.; Perrino, F. W.; Schaaper, R. M.; London, R. E. *Biochemistry* **2002**, *41*, 94–110.

50 Vuister, G. W.; Kim, S. J.; Wu, C.; Bax, A. *J. Am. Chem. Soc.* **1994**, *116*, 9206–9210.

51 Medek, A.; Olejniczak, E. T.; Meadows, R. P.; Fesik, S. W. *J. Biomol. NMR* **2000**, *18*, 229–238.

52 Wang, H.; Janowick, D. A.; Schkeryantz, J. M.; Liu, X. H.; Fesik, S. W. *J. Am. Chem. Soc.* **1999**, *121*, 1611–1612.

53 Coughlin, P. E.; Anderson, F. E.; Oliver, E. J.; Brown, J. M.; Homans, S. W.; Pollak, S.; Lustbader, J. W. *J. Am. Chem. Soc.* **1999**, *121*, 11871–11874.

54 Giesen, A. W.; Bae, L. C.; Barrett, C. L.; Chyba, J. A.; Chaykovsky, M. M.; Cheng, M. C.; Murray, J. H.; Oliver, E. J.; Sullivan, S. M.; Brown, J. M.; Dahlquist, F. W.; Homans, S. W. *J. Biomol. NMR* **2001**, *19*, 255–260.

55 Giesen, A. W.; Bae, L. C.; Barrett, C. L.; Chyba, J. A.; Chaykovsky, M. M.; Cheng, M. C.; Murray, J. H.; Oliver, E. J.; Sullivan, S. M.; Brown, J. M.; Homans, S. W. *J. Biomol. NMR* **2002**, *22*, 21–26.

56 Hajduk, P. J.; Augeri, D. J.; Mack, J.; Mendoza, R.; Yang, J. G.; Betz, S. F.; Fesik, S. W. *J. Am. Chem. Soc.* **2000**, *122*, 7898–7904.

57 Hirata, R.; Ohsumk, Y.; Nakano, A.; Kawasaki, H.; Suzuki, K.; Anraku, Y. *J. Biol. Chem.* **1990**, *265*, 6726–6733.

58 Kane, P. M.; Yamashiro, C. T.; Wolczyk, D. F.; Neff, N.; Goebl, M.; Stevens, T. H. *Science* **1990**, *250*, 651–657.

59 Perler, F. B. *Nucleic Acids Res* **2002**, *30*, 383–384.

60 Chong, S.; Montello, G. E.; Zhang, A.; Cantor, E. J.; Liao, W.; Xu, M. Q.; Benner, J. *Nucleic Acids Res* **1998**, *26*, 5109–5115.

61 Dawson, P. E.; Kent, S. B. *Annu. Rev. Biochem.* **2000**, *69*, 923–960.

62 Xu, R.; Ayers, B.; Cowburn, D.; Muir, T. W. *Proc. Natl. Acad. Sci. USA* **1999**, *96*, 388–393.

63 Cotton, G. J.; Ayers, B.; Xu, R.; Muir, T. W. *J. Am. Chem. Soc.* **1999**, *121*, 1100–1101.

64 Low, D. W.; Hill, M. G.; Carrasco, M. R.; Kent, S. B.; Botti, P. *Proc. Natl. Acad. Sci. USA* **2001**, *98*, 6554–6559.

65 Welker, E.; Scheraga, H. A. *Biochem. Biophys. Res. Commun.* **1999**, *254*, 147–151.

66 Chong, S.; Mersha, F. B.; Comb, D. G.; Scott, M. E.; Landry, D.; Vence, L. M.; Perler, F. B.; Benner, J.; Kucera, R. B.; Hirvonen, C. A.; Pelletier, J. J.; Paulus, H.; Xu, M. Q. *Gene* **1997**, *192*, 271–281.

67 Mills, K. V.; Lew, B. M.; Jiang, S.; Paulus, H. *Proc. Natl. Acad. Sci. USA* **1998**, *95*, 3543–3548.

68 Wu, H.; Xu, M. Q.; Liu, X. Q. *Biochim. Biophys. Acta* **1998**, *1387*, 422–432.

69 Southworth, M. W.; Adam, E.; Panne, D.; Byer, R.; Kautz, R.; Perler, F. B. *EMBO J.* **1998**, *17*, 918–926.

70 Yamazaki, T.; Otomo, T.; Oda, N.; Kyogoku, Y.; Uegaki, K.; Ito, N.; Ishino, Y.; Nakamura, H. *J. Am. Chem. Soc.* **1998**, *120*, 5591–5592.

71 Otomo, T.; Teruya, K.; Uegaki, K.; Yamazaki, T.; Kyogoku, Y. *J. Biomol. NMR* **1999**, *14*, 105–114.

72 Paulus, H. *Annu. Rev. Biochem.* **2000**, *69*, 447–496.

73 Evans, T.C., Jr.; Martin, D.; Kolly, R.; Panne, D.; Sun, L.; Ghosh, I.; Chen, L.; Benner, J.; Liu, X. Q.; Xu, M.Q. *J. Biol. Chem.* **2000**, *275*, 9091–9094.
74 Iwai, H.; Plückthun, A. *FEBS Lett.* **1999**, *459*, 166–172.
75 Scott, C.P.; Abel-Santos, E.; Wall, M.; Wahnon, D.C.; Benkovic, S.J. *Proc. Natl. Acad. Sc. USA* **1999**, *96*, 13638–13643.
76 Camarero, J.A.; Muir, T.W. *J. Am. Chem. Soc.* **1999**, *121*, 5597–5598.
77 Flory, P.J. *J. Am. Chem. Soc.* **1956**, *78*, 5222–5235.
78 Trabi, M.; Craik, D.J. *Trends Biochem. Sci.* **2002**, *27*, 132–138.
79 Evans, T.C., Jr.; Benner, J.; Xu, M.Q. *J. Biol. Chem.* **1999**, *274*, 3923–3926.
80 Iwai, H.; Lingel, A.; Plückthun, A. *J. Biol. Chem.* **2001**, *276*, 16548–16554.
81 Williams, N.K.; Prosselkov, P.; Liepinsh, E.; Line, I.; Shapiro, A.; Littler, D.R.; Curmi, P.M.; Otting, G.; Dixon, N.E. *J. Biol. Chem.* **2002**, *277*, 7790–7798.
82 Siebold, C.; Flukiger, K.; Beutler, R.; Erni, B. *FEBS Lett.* **2001**, *504*, 104–111.
83 Goldenberg, D.P.; Creighton, T.E. *J. Mol. Biol.* **1983**, *165*, 407–413.
84 Camarero, J.A.; Fushman, D.; Sato, S.; Giriat, I.; Cowburn, D.; Raleigh, D.P.; Muir, T. W. *J. Mol. Biol.* **2001**, *308*, 1045–1062.
85 Woese, C.R.; Fox, G.E. *Proc. Natl. Acad. Sci. USA* **1977**, *74*, 5088–5090.
86 Murata, K.; Mitsuoka, K.; Hirai, T.; Walz, T.; Agre, P.; Heymann, J.B.; Engel, A.; Fujiyoshi, Y. *Nature* **2000**, *407*, 599–605.
87 Luecke, H.; Schobert, B.; Richter, H.T.; Cartailler, J.P.; Lanyi, J.K. *J. Mol. Biol.* **1999**, *291*, 899–911.
88 Turner, G.J.; Reusch, R.; Winter-Vann, A.M.; Martinez, L.; Betlach, M.C. *Protein Expr. Purif.* **1999**, *17*, 312–323.
89 Winter-Vann, A.M.; Martinez, L.; Bartus, C.; Levay, A.; Turner, G.J. In *Perspectives on Solid State NMR in Biology*, Kühne, S. R., de Groot, H. J. M., Eds.; Kluwer: Netherlands, **2001**; pp 141–159.
90 Hinnen, A.; Buxton, F.; Chaudhuri, B.; Heim, J.; Hottiger, T.; Meyhack, B.; Pohlig, G. In *Gene expression in recombinant Microorganisms*; Smith, A., Ed.; Marcel Dekker: New York, **1994**; pp 121–193.
91 Valenzuela, P.; Medina, A.; Rutter, W.J.; Ammerer, G.; Hall, B.D. *Nature* **1982**, *298*, 347–350.
92 Cabezon, T.; De Wilde, M.; Herion, P.; Loriau, R.; Bollen, A. *Proc. Natl. Acad. Sci. USA* **1984**, *81*, 6594–6598.
93 Emini, E.A.; Ellis, R.W.; Miller, W.J.; McAleer, W.J.; Scolnick, E.M.; Gerety, R.J. *J. Infect.* **1986**, *13 Suppl. A*, 3–9.
94 Schultz, L.D.; Tanner, J.; Hofmann, K.J.; Emini, E.A.; Condra, J.H.; Jones, R.E.; Kieff, E.; Ellis, R.W. *Gene* **1987**, *54*, 113–123.
95 Hallewell, R.A.; Mills, R.; Tekamp-Olson, P.; Blacher, R.; Rosenberg, S.; Otting, F.; Masiarz, F.R.; Scandella, C.J. *Biotechnol.* **1987**, *5*, 363–366.
96 David, N.E.; Gee, M.; Andersen, B.; Naider, F.; Thorner, J.; Stevens, R.C. *J. Biol. Chem.* **1997**, *272*, 15553–15561.
97 Andersen, B.; Stevens, R.C. *Protein Expr. Purif.* **1998**, *13*, 111–119.
98 Giga-Hama, Y.; Kumagai, H. *Biotechnol. Appl. Biochem.* **1999**, *30*, 235–244.
99 Grallert, B.; Nurse, P.; Patterson, T.E. *Mol. Gen. Genet.* **1993**, *238*, 26–32.
100 Maundrell, K.; Hutchison, A.; Shall, S. *EMBO J.* **1988**, *7*, 2203–2209.
101 Toyama, R.; Okayama, H. *FEBS Lett.* **1990**, *268*, 217–221.
102 Xu, H.P.; White, M.; Marcus, S.; Wigler, M. *Mol. Cell Biol.* **1994**, *14*, 50–58.
103 Hildebrandt, V. In *Foreign Gene Expression in Fission Yeast Schizosaccharomyces pombe*; Giga-Hama, Y., Kumagai, H., Eds.; Springer Verlag: Berlin, **1997**; pp 97–110.
104 Giga-Hama, Y.; Tohda, H.; Okada, H.; Owada, M.K.; Okayama, H.; Kumagai, H. *Biotechnology* **1994**, *12*, 400–404.
105 Yasumori, T. In *Foreign Gene expression in Fission Yeast Schizosaccharomyces pombe*; Giga-Hama, Y., Kumagai, H., Eds.; Springer Verlag: Berlin, **1997**; pp 111–121.
106 Sander, P.; Grunewald, S.; Reilander, H.; Michel, H. *FEBS Lett.* **1994**, *344*, 41–46.
107 Kjeldsen, T.; Pettersson, A.F.; Hach, M. *Biotechnol. Appl. Biochem.* **1999**, *29*, 79–86.
108 Cereghino, J.L.; Cregg, J.M. *FEMS Microbiol. Rev.* **2000**, *24*, 45–66.
109 Weiss, H.M.; Haase, W.; Michel, H.; Reilander, H. *Biochem. J.* **1998**, *330*, 1137–1147.
110 Luque, T.; O'Reilly, D. R. *Mol. Biotechnol.* **1999**, *13*, 153–163.

111 Possee, R.D. *Curr. Opin. Biotechnol.* **1997**, *8*, 569–572.

112 Ogonah, O.W.; Freedman, R.B.; Jenkins, N.; Patel, K.; Rooney, B.C. *Nat. Biotechnol.* **1996**, *14*, 197–202.

113 Condreay, J.P.; Witherspoon, S.M.; Clay, W.C.; Kost, T.A. *Proc. Natl. Acad. Sci. USA* **1999**, *96*, 127–132.

114 Payne, L.A.; Fornwald, J.A.; Kane, J.F.; McNulty, D.E.; Trill, J.J.; Ramos, L. In *Animal Cell Technology: From Target to Market*; Lindner-Olsson, E., Chatissavidou, N., Lillau, E., Eds.; Kluwer: Netherlands, **2001**; pp 94–100.

115 Mazina, K.E.; Strader, C.D.; Fong, T.M. *J. Recept. Res.* **1994**, *14*, 63–73.

116 Guan, X.M.; Kobilka, T.S.; Kobilka, B.K. *J. Biol. Chem.* **1992**, *267*, 21995–21998.

117 Gu, H.; Wall, S.C.; Rudnick, G. *J. Biol. Chem.* **1994**, *269*, 7124–7130.

118 Girard, P.; Derouazi, M.; Baumgartner, G.; Bourgeois, M.; Jordan, M.; Wurm, F. M. In *Animal Cell Technology: From Target to Market*; Lindner-Olsson, E., Chatzissavidou, N., Lillau, E., Eds.; Kluwer Academic Publishers: Netherlands, **2001**; pp 37–42.

119 Fux, C.; Moser, S.; Schlatter, S.; Rimann, M.; Bailey, J.E.; Fussenegger, M. *Nucleic Acids Res.* **2001**, *29*, E19.

120 Boorsma, M.; Nieba, L.; Koller, D.; Bachmann, M.F.; Bailey, J.E.; Renner, W.A. *Nat. Biotechnol.* **2000**, *18*, 429–432.

121 Tate, C.G. *FEBS Lett.* **2001**, *504*, 94–98.

122 Graham, F.L. *Immunol. Today* **2000**, *21*, 426–428.

123 Panicali, D.; Paoletti, E. *Proc. Natl. Acad. Sci. USA* **1982**, *79*, 4927–4931.

124 Ward, G.A.; Stover, C.K.; Moss, B.; Fuerst, T.R. *Proc. Natl. Acad. Sci. USA* **1995**, *92*, 6773–6777.

125 Liljestrom, P.; Garoff, H. *Biotechnol.* **1991**, *9*, 1356–1361.

126 Xiong, C.; Levis, R.; Shen, P.; Schlesinger, S.; Rice, C.M.; Huang, H. V. *Science* **1989**, *243*, 1188–1191.

127 Lundstrom, K. *J. Recept. Signal. Transduct. Res.* **1999**, *19*, 673–686.

128 Lundstrom, K. In *Perspectives on Solid-State NMR in Biology*, Kühne, S.R., de Groot, J.J.M., Eds.; Kluwer: The Netherlands, **2001**; pp 131–139.

129 Hovius, R.; Tairi, A.-P.; Blasey, H.; Bernard, A.R.; Lundstrom, K.; Vogel, H. *J. Neurochem.* **1998**, *70*, 824–834.

130 Lundstrom, K.; Schweitzer, C.; Rotmann, D.; Hermann, D.; Schneider, E.M.; Ehrengruber, M.U. *FEBS Lett.* **2001**, *504*, 99–103.

131 Lundstrom, K.; Rotmann, D.; Herrmann, D.; Schlaeger, E.-J. *Cytotechnology* **2001**, *35*, 213–221.

132 Fernandez, C.; Hilty, C.; Bonjour, S.; Adeishvili, K.; Pervushin, K.; Wüthrich, K. *FEBS Lett.* **2001**, *504*, 173–178.

133 Bellizzi, J.J.; Widom, J.; Kemp, C.W.; Clardy, J. *Structure* **1999**, *7*, R263–267.

134 Bushnell, D.A.; Cramer, P.; Kornberg, R.D. *Structure* **2001**, *9*, R11–R14.

135 Wood, M.J.; Komives, E.A. *J. Biomol. NMR* **1999**, *13*, 149–159.

136 Morgan, W.D.; Birdsall, B.; Frenkiel, T.A.; Gradwell, M.G.; Burghaus, P.A.; Syed, S.E.H.; Uthaipibull, C.; Holder, A.A.; Feeney, J. *J. Molec. Biol.* **1999**, *289*, 113–122.

137 Palczewska, M.; Groves, P.; Kuznicki, J. *Protein Expr. Purif.* **1999**, *17*, 465–476.

138 Laroche, Y.; Storme, V.; Demeutter, J.; Messens, J.; Lauwereys, M. *Bio-Technology* **1994**, *12*, 1119–1124.

139 Morgan, W.D.; Kragt, A.; Feeney, J. *J. Biomol. NMR* **2000**, *17*, 337–347.

140 Helmle, M.; Patzelt, H.; Ockenfels, A.; Gartner, W.; Oesterhelt, D.; Bechinger, B. *Biochemistry* **2000**, *39*, 10066–10071.

141 Wyss, D.F.; Withka, J.M.; Knoppers, M.H.; Sterne, K.A.; Recny, M.A.; Wagner, G. *Biochemistry* **1993**, *32*, 10995–11006.

142 Hansen, A.P.; Petros, A.M.; Mazar, A.P.; Pederson, T.M.; Rueter, A.; Fesik, S.W. *Biochemistry* **1992**, *31*, 12713–12718.

143 Lustbader, J.W.; Birken, S.; Pollak, S.; Pound, A.; Chait, B.T.; Mirza, U.A.; Ramnarain, S.; Canfield, R.E.; Brown, J. M. *J. Biomol. NMR* **1996**, *7*, 295–304.

144 See http://www.mepnet.org for further information

145 Bailey, W.J.; Vanti, W.B.; George, S.R.; Levins, R.; Swarminathan, S.; Bonini, J.A.; Smith, K.E.; Weinshank, R.L.; O'Dowd, B.F. *Expert Opin. Ther. Patents* **2001**, *11*, 1861–1887.

146 Arora, A.; Abildgaard, F.; Bushweller, J.H.; Tamm, L.K. *Nat. Struct. Biol.* **2001**, *8*, 334–338.

147 Fernandez, C.; Adeishvili, K.; Wüthrich, K. *Proc. Natl. Acad. Sci. USA* **2001**, *98*, 2358–2363.
148 Palczewski, K.; Kumasaka, T.; Hori, T.; Behnke, C. A.; Motoshima, H.; Fox, B. A.; Le Trong, I.; Teller, D. C.; Okada, T.; Stenkamp, R. E.; Yamamoto, M.; Miyano, M. *Science* **2000**, *289*, 739–745.
149 Spirin, A. S.; Baranov, V. I.; Ryabova, L. A.; Ovodov, S. Y.; Alakhov, Y. B. *Science* **1988**, *242*, 1162–1164.
150 Kim, D. M.; Kigawa, T.; Choi, C. Y.; Yokoyama, S. *Eur. J. Biochem.* **1996**, *239*, 881–886.
151 Kigawa, T.; Yabuki, T.; Yoshida, Y.; Tsutsui, M.; Ito, Y.; Shibata, T.; Yokoyama, S. *FEBS Lett.* **1999**, *442*, 15–19.
152 Madin, K.; Sawasaki, T.; Ogasawara, T.; Endo, Y. *Proc. Natl. Acad. Sci. USA* **2000**, *97*, 559–564.
153 Kim, D. M.; Swartz, J. R. *Biotechnol. Prog.* **2000**, *16*, 385–390.
154 Kim, D. M.; Swartz, J. R. *Biotechnol. Bioeng.* **2001**, *74*, 309–316.
155 Kigawa, T.; Muto, Y.; Yokoyama, S. *J. Biomol. NMR* **1995**, *6*, 129–134.
156 Yabuki, T.; Kigawa, T.; Dohmae, N.; Takio, K.; Terada, T.; Ito, Y.; Laue, E. D.; Cooper, J. A.; Kainosho, M.; Yokoyama, S. *J. Biomol. NMR* **1998**, *11*, 295–306.
157 See www.proteinexpression.com for more information.
158 Fernholz, E.; Besir, H.; Kühlewein, A.; Mayr, D.; Schmitt, R.; Schwaiger, M. unpublished results **2002**.
159 Spirin, A. S. *Cell-Free Translation Systems*; Springer Verlag: Berlin, **2002**.
160 Fernholz, E.; Zaiss, K.; Besir, H.; Mutter, W. In *Cell–Free Translation Systems*; Spirin, A. S., Ed.; Springer Verlag: Berlin, **2002**; pp 175–179.
161 Tarui, H.; Imanishi, S.; Hara, T. *J. Biosci. Bioeng.* **2000**, *90*, 508–514.
162 LeMaster, D. M.; Kushlan, D. M. *J. Am. Chem. Soc.* **1996**, *118*, 9255–9264.
163 Ishima, R.; Louis, J. M.; Torchia, D. A. *J. Am. Chem. Soc.* **1999**, *121*, 11589–11590.
164 Toyoshima, C.; Nakasako, M.; Nomura, H.; Ogawa, H. *Nature* **2000**, *405*, 647–655.
165 Hunte, C.; Koepke, J.; Lange, C.; Rossmanith, T.; Michel, H. *Struct. Fold. Des.* **2000**, *8*, 669–684.
166 Yoshikawa, S.; Shinzawa-Itoh, K.; Nakashima, R.; Yaono, R.; Yamashita, E.; Inoue, N.; Yao, M.; Fei, M. J.; Libeu, C. P.; Mizushima, T.; Yamaguchi, H.; Tomizaki, T.; Tsukihara, T. *Science* **1998**, *280*, 1723–1729.
167 Stock, D.; Leslie, A. G.; Walker, J. E. *Science* **1999**, *286*, 1700–1705.
168 Lancaster, C. R.; Kroger, A.; Auer, M.; Michel, H. *Nature* **1999**, *402*, 377–385.
169 Unger, V. M.; Kumar, N. M.; Gilula, N. B.; Yeager, M. *Science* **1999**, *283*, 1176–1180.
170 Fu, D.; Libson, A.; Miercke, L. J.; Weitzman, C.; Nollert, P.; Krucinski, J.; Stroud, R. M. *Science* **2000**, *290*, 481–486.
171 Kolbe, M.; Besir, H.; Essen, L. O.; Oesterhelt, D. *Science* **2000**, *288*, 1390–1396.
172 Auer, M.; Scarborough, G. A.; Kühlbrandt, W. *Nature* **1998**, *392*, 840–843.
173 Chang, G.; Spencer, R. H.; Lee, A. T.; Barclay, M. T.; Rees, D. C. *Science* **1998**, *282*, 2220–2226.
174 Williams, K. A. *Nature* **2000**, *403*, 112–115.
175 Miyazawa, A.; Fujiyoshi, Y.; Stowell, M.; Unwin, N. *J. Mol. Biol.* **1999**, *288*, 765–786.
176 Lancaster, C. R.; Michel, H. *J. Mol. Biol.* **1999**, *286*, 883–898.
177 Zouni, A.; Witt, H. T.; Kern, J.; Fromme, P.; Krauss, N.; Saenger, W.; Orth, P. *Nature* **2001**, *409*, 739–743.
178 Meyer, J. E.; Hofnung, M.; Schulz, G. E. *J. Mol. Biol.* **1997**, *66*, 761–765.
179 Doyle, D. A.; Morais Cabral, J.; Pfuetzner, R. A.; Kuo, A.; Gulbis, J. M.; Cohen, S. L.; Chait, B. T.; MacKinnon, R. *Science* **1998**, *280*, 69–77.

2
Structure Calculation Using Automated Techniques

Peter Güntert

2.1
Introduction

Understanding the relationship between structure and function of biological macromolecules is one of the key elements of rational drug design. In this context, the three-dimensional structure has a pivotal role, since its knowledge is essential to understand the physical, chemical, and biological properties of a protein [1, 2]. Until 1984 structural information at atomic resolution could only be determined by X-ray diffraction techniques with protein single crystals [3]. The introduction of NMR spectroscopy [4] as a technique for protein structure determination [5] has made it possible to obtain structures with comparable accuracy also in a solution environment that is much closer to the natural situation in a living being than the single crystals required for protein crystallography.

It has been recognized that many of the time-consuming interactive tasks carried out by an expert during the process of spectral analysis could be done more efficiently by automated computational systems [6]. Over the past few years, this potential has been realized to some degree. Today automated methods for NMR structure determination are playing a more and more prominent role and can be expected to largely supersede the conventional manual approaches to solving three-dimensional protein structures in solution.

The structure of this chapter is as follows: Section 2.2 introduces the various types of conformational constraints used in NMR structure calculations. Section 2.3 is devoted to modern structure calculation algorithms. Section 2.4 gives an account of the general principles and the practice of automated NOESY assignment.

2.2
Conformational Constraints for NMR Structure Calculations

For use in a structure calculation, geometric conformational constraints have to be derived from suitable conformation-dependent NMR parameters. These geometric constraints should, on the one hand, convey to the structure calculation as much as possible of the structural information inherent in the NMR data, and, on the other hand, be simple enough to be used efficiently by the structure calculation algorithms. NMR parameters with a clearly understood physical relation to a corresponding geometric parameter generally yield more trustworthy conformational constraints than NMR data for which the conformation dependence was deduced merely from statistical analyses of known structures.

2.2.1 Constraints from Covalent Structure

NMR data alone would not be sufficient to determine the positions of all atoms in a biological macromolecule. It has to be supplemented by information about the covalent structure of the protein – the amino acid sequence, bond lengths, bond angles, chiralities, and planar groups – and the steric repulsion between nonbonded atom pairs. Depending on the degrees of freedom used in the structure calculation, the covalent parameters are maintained by different methods: in Cartesian space, where in principle each atom moves independently, the covalent structure has to be enforced by potentials in the force field, whereas in torsion angle space the covalent geometry is fixed at the ideal values and there are no degrees of freedom that affect covalent structure parameters.

Depending on the structure calculation program used, special covalent bonds such as disulfide bridges or cyclic peptide bonds have to be enforced by distance constraints. Disulfide bridges may be fixed by restraining the distance between the two sulfur atoms to 2.0–2.1 Å and the two distances between the C^b and the sulfur atoms of different residues to 3.0–3.1 Å [7].

2.2.2 Steric Repulsion

Usually a simple geometric force field is used for the structure calculation that retains only the most dominant part of the nonbonded interaction, namely the steric repulsion in the form of lower bounds for all interatomic distances between pairs of atoms separated by three or more covalent bonds from each other. Steric lower bounds are generated internally by the structure calculation programs by assigning a repulsive core radius to each atom type and imposing lower distance bounds given by the sum of the two corresponding repulsive core radii. For instance, the following repulsive core radii are used in the program Dyana [8]: 0.95 Å (1 Å=0.1 nm) for amide hydrogen, 1.0 Å for other hydrogen, 1.35 Å for aromatic carbon, 1.4 Å for other carbon, 1.3 Å for nitrogen, 1.2 Å for oxygen, and 1.6 Å for sulfur and phosphorus atoms [9]. To allow the formation of hydrogen bonds, potential hydrogen bond contacts are treated with lower bounds that are smaller than the sum of the corresponding repulsive core radii.

2.2.3 Distance Constraints from Nuclear Overhauser Effects

The principle source of experimental conformational data in an NMR structure determination is constraints on short interatomic distances between hydrogen atoms obtained from NMR measurements of the nuclear Overhauser effect (NOE). NOEs result from cross-relaxation mediated by the dipole-dipole interaction between spatially proximate nuclear spins in a molecule undergoing Brownian motion [10] and are manifested by cross peaks in NOESY spectra [11, 12]. NOEs connect pairs of hydrogen atoms separated by less than about 5 Å in amino acid residues that may be far away along the protein sequence but close together in space.

The intensity of an NOE, given by the volume V of the corresponding cross peak in a NOESY spectrum [11, 13, 14] is related to the distance r between the two interacting spins by

$$V = \langle r^{-6} \rangle f(\tau_c) \,. \tag{1}$$

The averaging indicates that in molecules with inherent flexibility the distance r may vary and thus has to be averaged appropriately. The remaining dependence of the magnetization transfer on the motion enters through the function $f(\tau_c)$, which includes effects of global and internal motions of the molecule. Since, with the exceptions of the protein surface and disordered segments of the polypeptide chain, globular proteins are relatively rigid, the structure calculation is usually based on the assumption that there exists a single rigid conformation that is compatible with all NOE data simultaneously, provided that the NOE data are interpreted in a conservative, semi-quantitative manner [5]. More sophisticated treatments that take into account the fact that the result of a NOESY experiment represents an average over time and space are, if used at all, usually deferred until the structure refinement stage [15].

In principle, all hydrogen atoms of a protein form a single network of spins, coupled by the dipole-dipole interaction. Magnetization can be transferred from one spin to another not only directly but also by "spin diffusion", that is, indirectly via other spins in the vicinity [11, 16]. The approximation of isolated spin pairs is valid only for very short mixing times in the NOESY experiment. However, in order to detect an observable NOE the mixing time cannot be made arbitrarily short. In practice, a compromise has to be made between the suppression of spin diffusion and sufficient cross-peak intensities, usually with mixing times in the range of 40–100 ms for high-quality structures. Spin diffusion effects can be included in the structure calculation by complete relaxation matrix refinement [17–19]. Because also parameters about internal and overall motions that are difficult to measure experimentally enter into the relaxation matrix refinement, care has to be taken not to bias the structure determination by overinterpretation of the data. Relaxation matrix refinement has been used mostly in situations where the conservative and robust interpretation of NOEs as upper distance limits would not be sufficient to define the three-dimensional structure, especially in the case of nucleic acids [20–22].

The quantification of an NOE amounts to determining the volume of the corresponding cross peak in the NOESY spectrum. Since the linewidths can vary appreciably for different resonances, cross-peak volumes should in principle be determined by integration over the peak area rather than by measuring peak heights. However, one should also keep in mind that, according to Eq. (1), the relative error of the distance estimate is only one sixth of the relative error of the volume determination. Furthermore, Eq. (1) involves factors that have their origin in the complex internal dynamics of the macromolecule and are beyond practical reach such that even a very accurate measurement of peak volumes will not yield equally accurate conformational constraints.

On the basis of Eq. (1), NOEs are usually treated as upper bounds on interatomic distances rather than as precise distance constraints, because the presence of internal motions and, possibly, chemical exchange may diminish the strength of an NOE [23]. In fact, much of the robustness of the NMR structure determination method is due to the use of upper distance bounds instead of exact distance constraints in conjunction with the observation that internal motions and exchange effects usually reduce rather than increase the NOEs [5]. For the same reason, the absence of an NOE is in general not interpreted as a lower bound on the distance between the two interacting spins.

Upper bounds b on the distance between two hydrogen atoms are derived from the corresponding NOESY cross peak volumes V according to "calibration curves", $V=f(b)$. Assuming a rigid molecule, the calibration curve is

$$V = \frac{k}{b^6}, \tag{2}$$

with a constant k that depends on the arbitrary scaling of the NOESY spectrum. The value b obtained from the equation may either be used directly as an upper distance bound, or NOEs may be calibrated into different classes according to their volume, using the same upper bound b for all NOEs in a given class. In this case, it is customary to set the upper bound to 2.7 Å for "strong" NOEs, 3.3 Å for "medium" NOEs, and 5.0 Å for "weak" NOEs [7]. The constant k in Eq. (2) can be determined on the basis of known distances, for example the sequential distances in regular secondary structure elements or by reference to a preliminary structure [24]. In an automatic NOESY assignment procedure it is convenient to get an estimate of the calibration constants k independently of knowledge of certain distances or preliminary structures. This can be obtained by automated structure-independent calibration or by automated structure-based calibration. Automated structure-independent calibration [25] defines the calibration constant such that the average of the upper distance bounds for all peaks involving a given combination of atom types attains a predetermined value that has been found to vary little among different structures. Structure-based automated calibration [26] sets the calibration constant such that an available preliminary structure does not violate more than a predetermined (small) percentage of the upper distance bounds.

NOEs that involve groups of protons with degenerate chemical shifts, in particular methyl groups, may be referred to pseudoatoms located in the center of the protons that they represent, and the upper bound is increased by a pseudoatom correction equal to the proton-pseudoatom distance [27, 28]. Another method that usually incurs a smaller loss of information [29] is to treat NOEs for groups of protons with degenerate chemical shifts as ambiguous distance constraints (see Eq. (13) below).

A related but not identical problem occurs because the standard method for obtaining resonance assignments in proteins [5] cannot provide stereospecific assignments, i.e. individual assignments for the two diastereotopic substituents of a prochiral center, for example in methylene groups and in the isopropyl groups of valine and leucine. In the absence of stereospecific assignments, restraints involving diastereotopic substituents can also be referred to pseudoatoms [27] or otherwise treated such that they are invariant under exchange of the two diastereotopic substituents, which inevitably results in a loss of information and less well-defined structures [30]. To minimize such loss of information, the absence of stereospecific assignments for diastereotopic groups can be treated by periodic optimal swapping of the pairs of diastereotopic atoms for minimal target function value during the simulated annealing [31]. It is also possible to determine stereospecific assignments by various methods, including biosynthetical fractional ^{13}C-labeling of valine and leucine isopropyl groups [32, 33], systematic analysis of the local conformation through grid searches [30], or comparison with preliminary three-dimensional structures [24, 28].

2.2.4 Hydrogen Bond Distance Constraints

Slow hydrogen exchange indicates that an amide proton is involved in a hydrogen bond [34]. However, hydrogen exchange measurements cannot identify the acceptor oxygen or nitrogen atom. Recently, NMR experiments have been developed that can unambiguously identify hydrogen bonds by experimental observation of scalar couplings over hydrogen bonds [35]. If the acceptor oxygen or nitrogen atom cannot be identified experimentally, one has to rely on NOEs in the vicinity of the postulated hydrogen bond or on assumptions about regular secondary structure to define the acceptor. In this way, the standard backbone-backbone hydrogen bonds in regular secondary structure can be identified with higher reliability than hydrogen bonds with side-chains. Therefore, unless based on cross-hydrogen bond scalar couplings, hydrogen bond constraints are either largely redundant with the NOE network or involve structural assumptions and should be used with care or not at all. They can, however, be useful during preliminary structure calculations of larger proteins when not enough NOE data are available yet. Hydrogen bonds constraints are introduced into the structure calculation as distance constraints, typically by confining the acceptor-hydrogen distance to the range 1.8–2.0 Å and the distance between the acceptor and the atom to which the hydrogen atom is covalently bound to 2.7–3.0 Å. The second distance constraint restricts the angle of the hydrogen bond. Being tight medium- or long-range distance constraints, their impact on the resulting structure is considerable. In regular secondary structure elements they significantly enhance their regularity. In fact, α-helices and, to a lesser extent, β-sheets become well defined by hydrogen bond constraints alone without the use of NOE constraints [36]. On the other hand, hydrogen bond constraints may lead, if assigned mechanically without clear-cut evidence, to overly regular structures in which subtle features such as a 3_{10}-helix-like final turn of an α-helix may be missed.

2.2.5 Torsion Angle Constraints from Chemical Shifts

Chemical shifts are very sensitive probes of the molecular environment of a spin. However, in many cases their dependence on the structure is complicated and either not fully understood or too intricate to allow the derivation of reliable conformational constraints [37, 38]. An exception in this respect are the deviations of $^{13}C^{\alpha}$ (and, to some extent, $^{13}C^{\beta}$) chemical shifts from their random coil values that are correlated with the local backbone conformation [39, 40]: $^{13}C^{\alpha}$ chemical shifts larger than the random coil values occur for amino acid residues in α-helical conformation, whereas deviations towards smaller values are observed for residues in β-sheet conformation. Such information can be included in a structure calculation by restricting the local conformation of a residue to the α-helical or β-sheet region of the Ramachandran plot, using torsion angle constraints in the form of allowed intervals for the backbone torsion angles ϕ and ψ [41]. Some care should be applied because the correlation between chemical shift deviation and structure is not perfect. Similarly to hydrogen bond constraints, conformational constraints based on $^{13}C^{\alpha}$ chemical shifts are therefore in general only used as auxiliary data in special situations, in particular at the beginning of a structure calculation when the NOE network is still sparse. There have also been attempts to use ^{1}H chemical shifts as direct constraints in structure refinement [42, 43]. This is more difficult than with $^{13}C^{\alpha}$ shifts because the secondary structure is not the dominant determinant of ^{1}H chemical shifts. Therefore, ^{1}H chemical shifts are more often used indirectly to delineate secondary structure elements by the "chemical shift index" [44].

2.2.6 Torsion Angle Constraints from Scalar Coupling Constants

Vicinal scalar coupling constants, ^{3}J, between atoms separated by three covalent bonds from each other are related to the enclosed torsion angle, θ, by Karplus relations [45].

$$^{3}J(\theta) = A\cos^{2}\theta + B\cos\theta + C\,. \tag{3}$$

The parameters A, B and C have been determined for various types of couplings by a best fit of the measured J values to the corresponding values calculated with Eq. (3) for known protein structures [36]. When interpreting scalar coupling constants using Eq. (3) one has to take into account not only the measurement error but also that there may be averaging due to internal mobility and that both the functional form and the parameters of the Karplus curves are approximations. In contrast to NOEs, scalar coupling constants give information only on the local conformation of a polypeptide chain. They can nevertheless be useful to accurately define the local conformation, to obtain stereospecific assignments for diastereotopic protons (usually for the β-protons) [30], and to detect torsion angles (usually χ^{1}) that occur in multiple rotamer states.

Torsion angle constraints in the form of an allowed interval are used to incorporate scalar coupling information into the structure calculation. Using Eq. (3), an allowed range for a scalar coupling constant value in general leads to several (up to four) allowed intervals for the enclosed torsion angle [36]. Restraining the torsion angle to a single interval that encloses all torsion angle values compatible with the scalar coupling constant then often results in a loss of structural information because the torsion angle constraint may encompass large regions that are forbidden by the measured coupling constant. It is therefore often advantageous to combine local data – for example all distance constraints and scalar coupling constants within the molecular fragment defined by the torsion angles ϕ, ψ, and χ^{1} – in a systematic analysis of the local conformation and to derive torsion angle constraints from the results of this grid search rather than from the individual NMR parameters [30].

Alternatively, scalar coupling constants can also be introduced into the structure calculation as direct constraints by adding a term of the type

$$V_J = k_J \sum_i ({}^3J_i^{\text{exp}} - {}^3J_i^{\text{calc}})^2 \tag{4}$$

to the target function of the structure calculation program [46, 47]. The sum in Eq. (4) extends over all measured couplings, k_J is a weighting factor, and ${}^3J_i^{\text{exp}}$ and ${}^3J_i^{\text{calc}}$ denote the experimental and calculated value of the coupling constant, respectively. The latter is obtained from the value of the corresponding torsion angle using Eq. (3).

2.2.7
Orientation Constraints

Orientation constraints originate from residual dipolar couplings in partially aligned molecules and provide information on angles between covalent bonds and globally defined axes in the molecule, namely those of the magnetic susceptibility tensor [48, 49]. In contrast to vicinal scalar couplings or ^{13}C secondary chemical shifts that probe exclusively local features of the conformation, residual dipolar couplings can provide information on long-range order that is not directly accessible from other commonly used NMR parameters.

Residual dipolar couplings arise because the strong internuclear dipolar couplings are no longer completely averaged out – as is the case in a solution of isotropically oriented molecules – if there is a small degree of molecular alignment with the static magnetic field due to an anisotropy of the magnetic susceptibility. The degree of alignment depends on the strength of the external magnetic field and results in residual dipolar couplings that are proportional to the square of the magnetic field strength [50]. They are manifested in small, field-dependent changes of the splitting normally caused by one-bond scalar couplings between directly bound nuclei and can thus be obtained from accurate measurements of ^1J couplings at different magnetic field strengths [48, 49]. The magnetic susceptibility anisotropy is relatively large in paramagnetic proteins but in general very small for diamagnetic globular proteins. It can, however, be enhanced strongly if the protein is brought into a liquid-crystalline environment [51, 52]. One obtains structural information on the angle θ between the covalent bond connecting the two scalar coupled atoms and the main axis of the magnetic susceptibility tensor, which can be incorporated into the structure calculation by adding orientation constraints that measure the deviation between the experimental residual dipolar coupling value and the corresponding value calculated from the structure. It has been shown [53] that such orientation constraints can be used in conjunction with conventional distance and angle constraints during the structure calculation, and that they can improve the quality of the resulting structure.

2.3
Structure Calculation Algorithms

The calculation of the three-dimensional structure forms a cornerstone of the NMR method for protein structure determination. Because of the complexity of the problem – a protein typically consists of more than a thousand atoms which are restrained by a similar number of experimentally determined constraints in conjunction with stereochemical and steric conditions – it is in general neither feasible to do an exhaustive search of allowed conformations nor to find solutions by interactive model building. In practice, the calculation of the three-dimensional structure is therefore usually formulated as a minimization problem for a target function that measures the agreement between a conformation and the given set of constraints. At present, the most widely used algorithms are based on the idea of simulated annealing [54]. These will be discussed in detail here. Earlier methods have been reviewed extensively already [55–58], and most of them are rarely used today. Special emphasis is thus given to the currently most efficient way of calculating NMR structures of biological macromolecules by torsion angle dynamics.

There is a fundamental difference between molecular simulation that has the aim of simulating a molecular system as realistically as possible in order to extract molecular quantities of interest and NMR structure calculation that is driven by experimental constraints. Classical molecular dynamics approaches rely on a full empirical force field to ensure proper stereochemistry and are generally run at a constant temperature close to room temperature. Substantial amounts of computation time are required because the empirical energy function includes long-range pair interactions that are time-consuming to evaluate and because conformation space is explored slowly at room temperature. When similar algorithms are used for NMR structure calculations, however, the objective is quite different. Here, such algorithms simply provide a means to efficiently optimize a target function that takes the role of the potential energy. Details of the calculation, such as the course of a trajectory, are unimportant, as long as its end point comes close to the global minimum of the target function. Therefore, the efficiency of NMR structure calculation can be enhanced by modifications of the force field or the algorithm that do not significantly alter the location of the global minimum (the correctly folded structure) but shorten (in terms of computation time needed) the way by which it can be reached from the start conformation. A typical "geometric" force field used in NMR structure calculation therefore retains only the most important part of the nonbonded interaction by a simple repulsive potential that replaces the Lennard-Jones and electrostatic interactions in the full empirical energy function. This short-range repulsive function can be calculated much faster and significantly facilitates the large-scale conformational changes required during the folding process by lowering energy barriers induced by the overlap of atoms.

2.3.1
Simulated Annealing by Molecular Dynamics Simulation in Cartesian Space

One major method for NMR structure calculation is based on numerically solving Newton's equation of motion in order to obtain a trajectory for the molecular system [59]. The degrees of freedom are the Cartesian coordinates of the atoms. In contrast to "standard" molecular dynamics simulations [60–62] that try to simulate the behavior of a real physical system as closely as possible (and do not include constraints derived from NMR), the purpose of a molecular dynamics calculation in an NMR structure determination is simply to search the conformation space of the protein for structures that fulfill the constraints, i.e. that minimize a target function which is taken as the potential energy of the system. Therefore, simulated annealing [54, 56, 63] is performed at high temperature using a simplified force field that treats the atoms as soft spheres without attractive or long-range (i.e. electrostatic) nonbonded interactions and does not include explicit consideration of the solvent. The distinctive feature of molecular dynamics simulation when compared to the straightforward minimization of a target function is the presence of kinetic energy that allows barriers of the potential surface to be crossed, thereby greatly reducing the problem of becoming trapped in local minima. Since molecular dynamics simulation cannot generate conformations from scratch, a start structure is needed, and this can be generated either by metric matrix distance geometry [63] or by the variable target function method [9, 28], but – at the expense of increased computation time – it is also possible to start from an extended structure [64] or even from a set of atoms randomly distributed in space [65]. Any general molecular dynamics program, such as Charmm [66], Amber [67], or Gromos [62], can be used for the simulated annealing of NMR structures, provided that pseudoenergy terms for distance and torsion angle constraints have been incorporated. In practice, the programs best adapted and most widely used for this purpose are Xplor [68] and its successor, CNS [69].

The classical dynamics of a system of n particles with masses m_i and positions $\boldsymbol{r}_i$ is governed by Newton's equation of motion,

$$m_i \frac{d^2 \boldsymbol{r}_i}{dt^2} = \boldsymbol{F}_i \,, \qquad (5)$$

where the forces $\boldsymbol{F}_i$ are given by the negative gradient of a potential energy function E_{pot} with respect to the Cartesian coordinates: $\boldsymbol{F}_i = -\nabla E_{\text{pot}}$. For simulated annealing, a simplified potential energy function is used that includes terms to maintain the covalent geometry of the structure by means of harmonic bond length and bond angle potentials, torsion angle potentials, terms to enforce the proper chiralities and planarities, a simple repulsive potential instead of the Lennard-Jones and electrostatic nonbonded interactions, as well as terms for distance and torsion angle constraints. For example, in the program Xplor [68],

$$\begin{aligned} E_{\text{pot}} = & \sum_{\text{bonds}} k_b (r - r_0)^2 + \sum_{\text{angles}} k_\theta (\theta - \theta_0)^2 + \sum_{\text{dihedrals}} k_\phi (1 + \cos(n\phi + \delta)) \\ & + \sum_{\text{dihedrals}} k_\varphi (\phi - \delta)^2 + \sum_{\text{nonbonded pairs}} k_{\text{repel}} (\max(0, (sR_{\min})^2 - R^2)) \\ & + \sum_{\text{distance constraints}} k_d \Delta_d^2 + \sum_{\text{angle constraints}} k_a \Delta_a^2 \end{aligned} \qquad (6)$$

k_b, k_θ, k_ϕ, k_{repel}, k_d and k_a denote the various force constants, r the actual and r_0 the correct bond length, respectively, θ the actual and θ_0 the correct bond angle, ϕ the actual torsion angle or improper angle value, n the number of minima of the torsion angle potential, δ an offset of the torsion angle and improper potentials, $R_{\min}$ the distance where the van der Waals potential has its minimum, R the actual distance between a nonbonded atom pair, s a scaling factor, and Δ_d and Δ_a the size of the distance or torsion angle constraint violation. As an alternative to the square-well potential of Eq. (6), distance constraints are often represented by a potential with linear asymptote for large violations [68]. To obtain a trajectory, the equations of motion are numerically integrated by advancing the coordinates $\boldsymbol{r}_i$ and velocities $\boldsymbol{v}_i$ of the particles by a small but finite time step Δt, for example according to the "leap-frog" integration scheme [59, 70]:

$$\begin{aligned} \boldsymbol{v}_i(t+\Delta t/2) &= \boldsymbol{v}_i(t-\Delta t/2) + \Delta t F_i(t)/m_i + O(\Delta t^3) \\ \boldsymbol{r}_i(t+\Delta t) &= \boldsymbol{r}_i(t) + \Delta t \boldsymbol{v}_i(t+\Delta t/2) + O(\Delta t^3)\,. \end{aligned} \tag{7}$$

The $O(\Delta t^3)$ terms indicate that the errors with respect to the exact solution incurred by the use of a finite time step Δt are proportional to Δt^3. The time step Δt must be small enough to sample adequately the fastest motions, i. e. of the order of 10^{-15} s. In general the highest frequency motions are bond length oscillations. Therefore, the time step can be increased if the bond lengths are constrained to their correct values by the Shake method [71]. To control the temperature the system is loosely coupled to a heat bath [73]. For the simulated annealing of a (possibly distorted) start structure, certain measures have to be taken in order to achieve sampling of the conformation space within reasonable time [63]. In a typical simulated annealing protocol [68], the simulated annealing is performed for a few picoseconds at high temperature, say 2000 K, starting with a very small weight for the steric repulsion that allows atoms to penetrate each other, and gradually increasing the strength of the steric repulsion during the calculation. Subsequently, the system is cooled down slowly for another few picoseconds and finally energy-minimized. This process is repeated for each of the start conformers. The alternative of selecting conformers that represent the solution structure at regular intervals from a single trajectory is used rarely because it is difficult to judge whether the spacing between the "snapshots" is sufficient for good sampling of conformation space. Simulated annealing by molecular dynamics requires substantially more computation time per conformer [68] than pure minimization methods such as the variable target function approach [9, 28, 72], but this potential disadvantage is in general more than compensated by a higher success rate of 40–100% of the start conformers ending up in a conformation in the vicinity of the global minimum. This effect is due to the ability of the simulated annealing algorithm to escape from local minima.

2.3.2
Torsion Angle Dynamics

Torsion angle dynamics, i. e. molecular dynamics simulation using torsion angles instead of Cartesian coordinates as degrees of freedom [8, 74–82], provides at present the most efficient way to calculate NMR structures of biomacromolecules. This is in stark contrast to a widespread but incorrect belief that dynamics in generalized coordinates is hopelessly complicated and cannot be done efficiently. In this section the torsion angle dynamics algorithm implemented in the program Dyana [8] is described in some detail. Dyana employs the fast torsion angle dynamics algorithm of Jain et al. [78] that requires a computational effort proportional the system size, as is also the case for molecular dynamics simulation in Cartesian space. "Naïve" approaches to torsion angle dynamics require a computational effort proportional to the third power of the system size (e.g. Ref. [77]), and are therefore not suitable for macromolecules. With the fast torsion angle dynamics algorithm in Dyana, the advantages of torsion angle dynamics, especially the much longer integration time steps that can be used, are effective for molecules of all sizes. There is a close analogy between molecular dynamics simulation in Cartesian and torsion angle space [36].

The key idea of the fast torsion angle dynamics algorithm in Dyana is to exploit the fact that a chain molecule such as a protein or nucleic acid can be represented in a natural way as a tree structure consisting of n+1 rigid bodies that are connected by n rotatable bonds (Fig. 2.1) [74, 83]. Each rigid body is made up of one or several mass points (atoms) with invariable relative positions. The tree structure starts from a base, typically

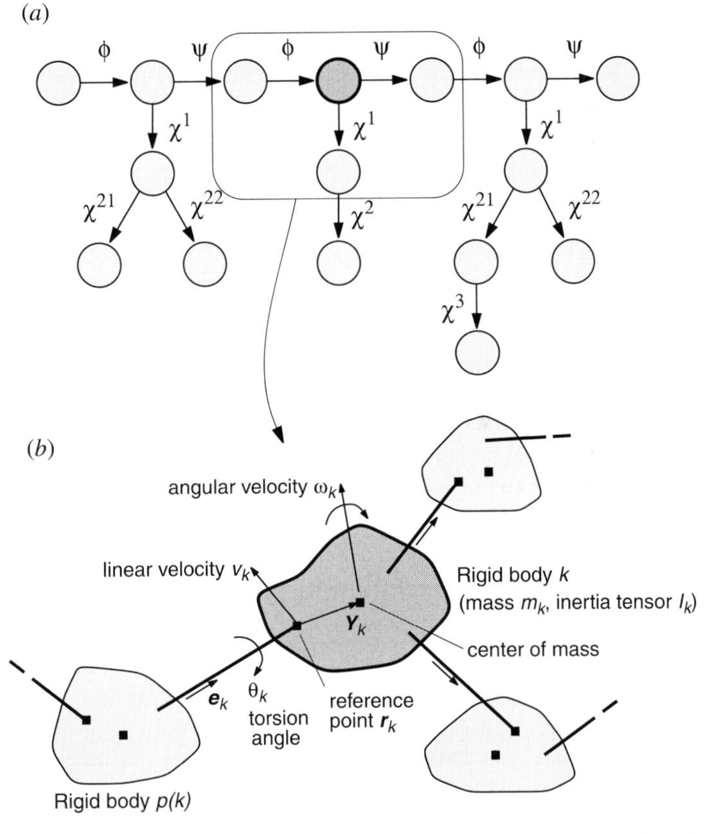

Fig. 2.1 **a** Tree structure of torsion angles for the tripeptide Val-Ser-Ile. Circles represent rigid units. Rotatable bonds are indicated by arrows that point toward the part of the structure that is rotated if the corresponding dihedral angle is changed. **b** Excerpt from the tree structure formed by the torsion angles of a molecule, and definition of quantities required by the Dyana fast torsion angle dynamics algorithm.

at the N-terminus of the polypeptide chain, and terminates with "leaves" at the ends of the side-chains and at the C-terminus. The degrees of freedom are n torsion angles, i.e. rotations about single bonds. The conformation of the molecule is thus uniquely specified by the values of all torsion angles. Covalent bonds that are incompatible with a tree structure because they would introduce closed flexible rings, for example disulfide bridges, are treated, as in Cartesian space dynamics, by distance constraints.

The role of the potential energy is taken by the Dyana target function [8, 28] that is defined such that it is zero if and only if all experimental distance constraints and torsion angle constraints are fulfilled and all nonbonded atom pairs satisfy a check for the absence of steric overlap. A conformation that satisfies the constraints more closely than another one will lead to a lower target function value. The exact definition of the Dyana target function is:

$$V = \sum_{c=u,l,v} w_c \sum_{(\alpha,\beta)\in I_c} (d_{\alpha\beta} - b_{\alpha\beta})^2 + w_a \sum_{i\in I_a} \left[1 - \frac{1}{2}\left(\frac{\Delta_i}{\Gamma_i}\right)^2\right] \Delta_i^2 \tag{8}$$

Upper and lower bounds, $b_{\alpha\beta}$, on distances $d_{\alpha\beta}$ between two atoms a and b, and constraints on individual torsion angles θ_i in the form of allowed intervals $[\theta_i^{\min}, \theta_i^{\max}]$ are considered. I_u, I_l and I_v are the sets of atom pairs (α, β) with upper, lower or van der Waals distance bounds, respectively, and I_a is the set of restrained torsion angles. w_u, w_l, w_v and w_a are weighting factors for the different types of constraints. $\Gamma_i = \pi - (\theta_i^{\max} - \theta_i^{\min})/2$ denotes the half-width of the forbidden range of torsion angle values, and Δ_i is the size of the torsion angle constraint violation. The torques about the rotatable bonds, i.e. the negative gradients of the potential energy with respect to torsion angles, are calculated by the fast recursive algorithm of Abe et al. [83].

The angular velocity vector ω_k and the linear velocity $\boldsymbol{v}_k$ of the reference point of the rigid body k (Fig. 2.1b) are calculated recursively from the corresponding quantities of the preceding rigid body $p(k)$:

$$\begin{aligned} \omega_k &= \omega_{p(k)} + \boldsymbol{e}_k \dot{\theta}_k\,, \\ \boldsymbol{v}_k &= \boldsymbol{v}_{p(k)} - (\boldsymbol{r}_k - \boldsymbol{r}_{p(k)}) \wedge \omega_{p(k)}\,. \end{aligned} \tag{9}$$

Denoting the vector from the reference point to the center of mass of the rigid body k by $\mathbf{Y}_k$, its mass by m_k, and its inertia tensor by I_k (Fig. 2.1b), the kinetic energy can be computed in a linear loop over all rigid bodies

$$E_{\text{kin}} = \frac{1}{2} \sum_{k=0}^{n} [m_k \boldsymbol{v}_k^2 + \omega_k \cdot I_k \omega_k + 2\boldsymbol{v}_k \cdot (\omega_k \wedge m_k \boldsymbol{Y}_k)]\,. \tag{10}$$

The calculation of the torsional accelerations, i.e. the second time derivatives of the torsion angles, is the crucial point of a torsion angle dynamics algorithm. The equations of motion for a classical mechanical system with generalized coordinates are the Lagrange equations

$$\frac{d}{dt}\left(\frac{\partial L}{\partial \dot{\theta}_k}\right) - \frac{\partial L}{\partial \theta_k} = 0 \qquad (k = 1, K, n) \tag{11}$$

with the Lagrange function $L = E_{\text{kin}} - E_{\text{pot}}$ [84]. They lead to equations of motion of the form

$$M(\ddot{\theta}) + C(\theta, \dot{\theta}) = 0\,. \tag{12}$$

In the case of torsion angles as degrees of freedom, the mass matrix $M(\theta)$ and the n-dimensional vector $C(\theta, \dot{\theta})$ can be calculated explicitly [76, 77]. However, to integrate the equations of motion, Eq. (12) would have to be solved in each time step for the torsional accelerations $\ddot{\theta}$). This requires the solution of a system of n linear equations and hence entails a computational effort proportional to n^3 that would become prohibitively expensive for larger systems. Therefore, in Dyana the fast recursive algorithm of Jain et al. [78] is implemented to compute the torsional accelerations, which makes explicit use of the tree structure of the molecule in order to obtain $\ddot{\theta}$ with a computational effort that is only proportional to n. The Dyana algorithm is too involved to be explained in detail here. Suffice it to say that the torsional accelerations can be obtained by executing a series of three linear loops over all rigid bodies similar to the one in Eq. (10) used to compute the kinetic energy.

The integration scheme for the equations of motion in torsion angle dynamics is a variant of the leap-frog algorithm used in Cartesian dynamics. In addition to the basic scheme of Eq. (7), the temperature is controlled by weak coupling to an external bath [73], and the length of the time step is adapted automatically based on the accuracy of energy conservation [8]. It could be shown that in practical applications with proteins, time steps of about 100, 30 and 7 fs at low (1 K), medium (400 K) and high (10000 K) temperatures, respectively, can be used in torsion angle dynamics calculations with Dyana [8], whereas time steps in Cartesian space molecular dynamics simulation generally have to be in the range of 2 ns. The concomitant fast exploration of conformation space provides the basis for the efficient Dyana structure calculation protocol.

The potential energy landscape of a protein is complex and studded with many local minima, even in the presence of experimental constraints in a simplified target function of the type of Eq. (8). Because the temperature, i. e. kinetic energy, determines the maximal height of energy barriers that can be overcome in a molecular dynamics simulation, the temperature schedule is important for the success and efficiency of a simulated annealing calculation. Consequently, elaborated protocols have been devised for structure calculations using molecular dynamics in Cartesian space [63, 68]. In addition to the temperature, other parameters such as force constants and repulsive core radii are varied in these schedules, which may involve several stages of heating and cooling. The faster exploration of conformation space with torsion angle dynamics allows for much simpler schedules. The standard simulated annealing protocol used by the program Dyana [8] will serve as an example here.

The structure calculation is started from a conformation with all torsion angles treated as independent uniformly distributed random variables and consists of five stages:

Stage 1. Short minimization to reduce high energy interactions that could otherwise disturb the torsion angle dynamics algorithm: 100 conjugate gradient minimization steps are performed, including only distance constraints between atoms up to 3 residues apart along the sequence, followed by a further 100 minimization steps including all constraints. For efficiency, until step 4 below, all hydrogen atoms are excluded from the check for steric overlap, and the repulsive core radii of heavy atoms with covalently bound hydrogens are increased by 0.15 Å with respect to their standard values. The weights in the target function of Eq. (8) are set to 1 for user-defined upper and lower distance bounds, to 0.5 for steric lower distance bounds, and to 5 $Å^2$ for torsion angle constraints.

Stage 2. Torsion angle dynamics calculation at constant high temperature: One fifth of all N torsion angle dynamics steps are performed at a constant high reference temperature of, typically, 10,000 K. The time step is initialized to 2 fs.

Stage 3. Torsion angle dynamics calculation with slow cooling close to zero temperature: The remaining $4N/5$ torsion angle dynamics steps are performed during which the reference value for the temperature approaches zero according to a fourth-power law.

Stage 4. Incorporation of all hydrogen atoms into the check for steric overlap: After resetting the repulsive core radii to their standard values and increasing the weighting factor for steric constraints to 2.0, 100 conjugate gradient minimization steps are performed, followed by 200 torsion angle dynamics steps at zero reference temperature.

Stage 5. A final minimization consisting of 1000 conjugate gradient steps.

Throughout the torsion angle dynamics calculation the list of van der Waals lower distance bounds is updated every 50 steps using a cutoff of 4.2 Å for the interatomic distance.

With the Dyana torsion angle dynamics algorithm it is possible to efficiently calculate protein structures on the basis of NMR data. Even for a system as complex as a protein the program Dyana can execute several thousand torsion angle dynamics steps within minutes of computation time. Computation times are of the order of one minute for NMR-size proteins on generally available computers. Furthermore, since an NMR structure calculation always involves the computation of a group of conformers, it is highly efficient to run calculations of multiple conformers in parallel. Nearly ideal speedup, i.e. a reduction of the computation time by a factor close to the number of processors used, can be achieved with Dyana [8].

2.4
Automated NOESY Assignment

2.4.1
The NOESY Assignment Problem

In *de novo* three-dimensional structure determinations of proteins in solution by NMR spectroscopy, the key conformational data are upper distance limits derived from nuclear Overhauser effects (NOEs) [11, 14]. In order to extract distance constraints from a NOESY spectrum, its cross peaks have to be assigned, i.e. the pairs of hydrogen atoms that give rise to cross peaks have to be identified. The basis for the NOESY assignment are previously determined ^1H chemical shift values that result from sequence-specific resonance assignment. However, because the accuracy with which NOESY cross peak positions and chemical shift values can be measured experimentally is limited, it is in general not possible to unambiguously assign all NOESY cross peaks on the basis of the known chemical shift values alone. It can be shown [25] that the number of NOESY cross peaks that can be assigned unambiguously from knowledge of the ^1H chemical shifts decreases exponentially with increasing uncertainty of the chemical shift or peak position information and drops below 10% of the total number of cross peaks for typical protein data sets. Obtaining a comprehensive set of distance constraints from a NOESY spectrum is thus by no means straightforward but becomes an iterative process in which preliminary structures, calculated from limited numbers of distance constraints, serve to reduce the ambiguity of cross peak assignments. In addition to this problem of resonance and peak overlap, considerable difficulties may arise from spectral artifacts and noise, and from the absence of expected signals because of fast relaxation. These inevitable shortcomings of NMR data collection are the main reason that until recently laborious interactive procedures have dominated three-dimensional protein structure determinations.

2.4.2
Semi-Automatic Methods

Semi-automated approaches to NOESY assignment [85–87] use the chemical shifts and a model or preliminary structure to provide the user with the list of possible assignments for each cross peak. The user decides interactively about the assignment and/or temporary removal of individual NOESY cross peaks, possibly taking into account supplementary information such as line shapes or secondary structure data, and performs a structure calculation with the resulting (usually incomplete) input. In practice, several cycles of NOESY assignment and structure calculation are required to obtain a high-quality structure.

2.4.3
General Principles of Automatic NOESY Assignment

Automated procedures follow the same general scheme but do not require manual intervention during the assignment/structure calculation cycles (Fig. 2.2). Two main obstacles have to be overcome by an automated approach starting without any prior knowledge of the structure: First, because the number of cross peaks with unique assignment based on chemical shifts is, as pointed out before, in general not sufficient to define the fold of the protein, the automated method must make use also of those NOESY cross peaks that cannot yet be assigned unambiguously. Second, the automated program must be able to deal with the amount of erroneously picked or inaccurately positioned peaks and with the incompleteness of the chemical shift assignment that is present in typical experimental data sets. An automated procedure needs devices to substitute the intuitive decisions made by an experienced spectroscopist in dealing with the imperfections of experimental NMR data.

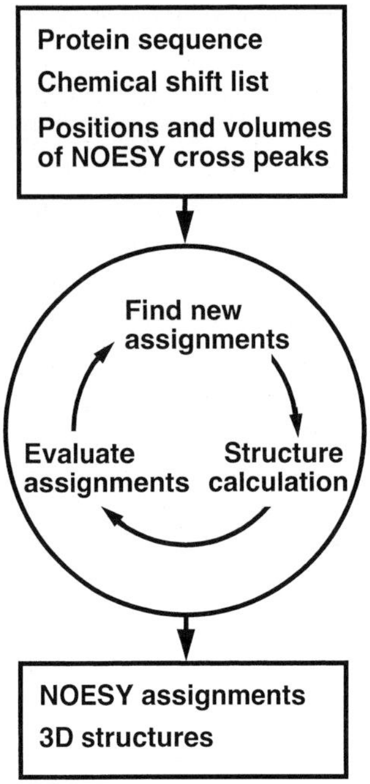

Fig. 2.2 General scheme of automated combined NOESY assignment and structure calculation.

2.4.4
Requirements on Input Data

Current automated NOESY assignment procedures do not attempt to correct or extend the sequence-specific resonance assignments and cannot normally make up for lack of chemical shift assignments. It is therefore important that the input chemical shift list includes nearly complete sequence-specific resonance assignments. Two requirements that the input data should meet in order to be a sufficient basis for a safe and successful automated *de novo* NMR structure determination of a globular protein emerged from test calculations and experience gained in *de novo* structure determinations with the automated NOESY assignment method Candid (see Sect. 2.4.6 below). Other automated NOESY assignment algorithms with fewer built-in safeguards against erroneous input data might call for more stringent requirements:

Requirement 1. Completeness of assignment: The input chemical shift list must contain more than 90% of the nonlabile and backbone amide ^{1}H chemical shifts. If three-dimensional or four-dimensional heteronuclear-resolved [^{1}H,^{1}H]-NOESY spectra are used, more than 90% of the ^{15}N and/or ^{13}C chemical shifts must also be available.

Requirement 2. Self-consistency: The peak lists must be faithful representations of the NOESY spectra, and the chemical shift positions of the NOESY cross peaks must be correctly calibrated to fit the chemical shift lists within the chemical shift tolerances. The range of allowed chemical shift variations ("tolerances") for ^{1}H should not exceed 0.02 ppm when working with homonuclear [^{1}H,^{1}H]-NOESY spectra, or 0.03 ppm when working with heteronuclear-resolved three-dimensional or four-dimensional NOESY spectra, and the tolerances for the ^{15}N and/or ^{13}C shifts should not exceed 0.6 ppm.

The requirement on the completeness of the chemical shift list is very important. A missing (or wrong) entry in the chemical shift list will make it impossible for the algorithm to correctly assign any of the NOEs of the corresponding atom. Therefore, the more NOESY cross peaks are expected for a certain atom, the more important it is to know its chemical shift. Special care should be taken to assign as extensively as possible the chemical shifts of the backbone and the hydrophobic core side-chains, whereas leniency is more tolerable for chemical shifts of flexible hydrophilic side-chains. The second requirement ensures that assignments already present in the input NOESY peak list can be reproduced by Candid. Typically, this includes many intra-residual and sequential NOEs that have been assigned in the preceding sequence-specific assignment and have been used to generate the chemical shift list(s) that are adapted to the NOESY spectra used for structure determination. Chemical shift tolerances should be chosen as small as possible but such that the second requirement is always fulfilled.

2.4.5
Overview of Algorithms

In a first approach to automated NOESY assignment, the programs Diana [28] and Dyana [8] were supplemented with the automated NOESY assignment routine Noah [25, 88]. In Noah, the multiple assignment problem is treated by temporarily ignoring cross peaks with too many (typically, more than two) assignment possibilities and instead generating independent distance constraints for all assignment possibilities of the remaining cross peaks, where one takes into account that part of these distance constraints may be incorrect. Noah requires high accuracy of the chemical shifts and peak positions in the input. It makes use of the fact that only a set of correct assignments can form a self-consistent network, and convergence towards the correct structure has been achieved for several proteins [25].

Another automated NOESY assignment procedure, Aria [89, 90], has been interfaced with the programs Xplor [68] and CNS [69], and a similar approach has been implemented by Savarin et al. [91]. Aria introduced the important concept of ambiguous distance constraints [92] for handling of ambiguities in the initial, chemical shift-based NOESY cross-peak assignments. When ambiguous distance constraints are used, each individual NOESY cross peak is treated as the superposition of the signals from each of its multiple assignments, using relative weights proportional to the inverse sixth power of the corresponding interatomic distance in a preliminary model of the molecular structure. A NOESY cross peak with a unique assignment possibility gives rise to an upper bound b on the distance $d_{\alpha\beta}$ between two hydrogen atoms, α and β. A NOESY cross peak with $n > 1$ assignment possibilities can be seen as the superposition of n degenerate signals and interpreted as an ambiguous distance constraint, $\bar{d} \leq b$, with

$$\bar{d} = \left(\sum_{k=1}^{n} d_{\alpha_k \beta_k}^{-6} \right)^{-1/6}. \qquad (13)$$

Each of the distances $d_{\alpha_k\beta_k}$ in the sum corresponds to one assignment possibility, (α_k, β_k). Because the "r^{-6}-summed distance" $\bar{d}$ is always shorter than any of the individual distances $d_{\alpha_k\beta_k}$, an ambiguous distance constraint is never falsified by including incorrect assignment possibilities, as long as the correct assignment is present. However, having more assignment possibilities decreases the information content of an ambiguous distance constraint and makes it more difficult for the structure calculation algorithm to converge to the correct structure.

It is therefore important to eliminate as far as possible incorrect assignment possibilities before the start of the structure calculation. To this end, the assignment possibilities are weighted by their generalized volume contributions, and only those with a sufficiently high contribution enter the ambiguous distance constraints used for the structure calculation. If the three-dimensional structure is known, the normalized relative contribution of the kth individual assignment possibility to the total volume of the cross peak can be estimated by $(d_{\alpha_k\beta_k}/\bar{d})^{-6}$ [89]. In this way, information from cross peaks with an arbitrary number of assignment possibilities can be used for the structure calculation, and although inclusion of erroneous assignments for a given cross peak results in a loss of information, it will not lead to inconsistencies as long as one or several correct assignments are among the initial assignments.

Both of these automated methods are quite efficient for improving and completing the NOESY assignment once a correct preliminary polypeptide fold is available, for example, based on a limited set of interactively assigned NOEs. On the other hand, unless a fair number of long-range assignments is provided by the user, obtaining a correct initial fold at the outset of a *de novo* structure determination often proves to be difficult because the structure-based filters used in both of these procedures for the elimination of erroneous cross peak assignments are then not operational. Aria has been used in the NMR structure determinations of various proteins [90].

A third approach that uses rules for assignments similar to the ones used by an expert to generate an initial protein fold has been implemented in the program AutoStructure, and applied to protein structure determination [6, 93].

The latest approach to automated NOESY assignment is the Candid algorithm [26], which will be explained in detail in the following sections.

2.4.6
The Candid Algorithm

Candid [26] combines features from Noah and Aria, such as the use of three-dimensional structure-based filters and ambiguous distance constraints, with the new concepts of network-anchoring and constraint combination that further enable an efficient and reliable search for the correct fold in the initial cycle of *de novo* NMR structure determinations. A flowchart of the Candid algorithm is given in Fig. 2.3.

The automated Candid method proceeds in iterative *cycles*, each consisting of exhaustive, in part ambiguous, NOE assignment followed by a structure calculation with the Dyana torsion angle dynamics algorithm. Between subsequent cycles, information is transferred exclusively through the intermediary three-dimensional structures, in that the protein molecular structure obtained in a given cycle is used to guide further NOE as-

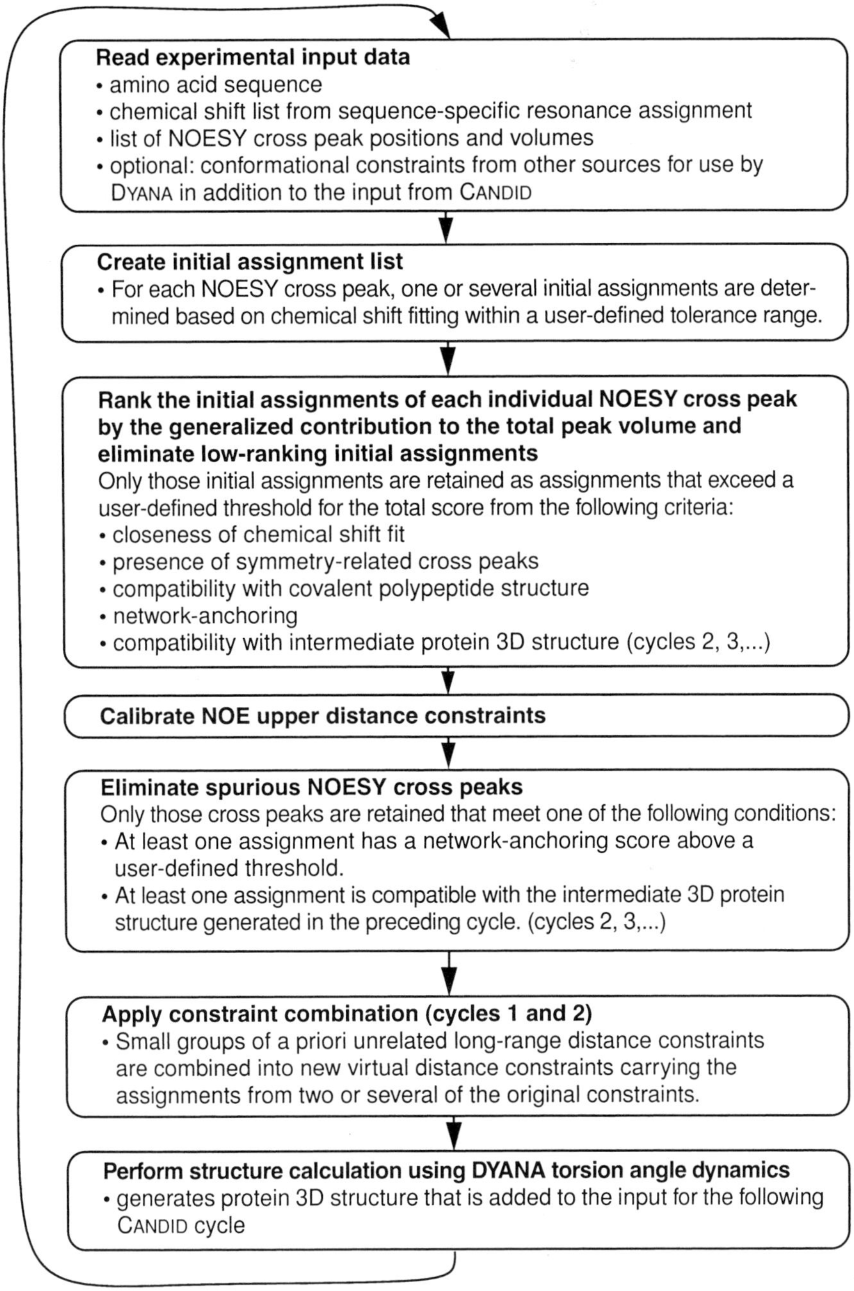

Fig. 2.3 Flowchart of NMR structure determination using the Candid method for automated NOE cross peak assignment.

signments in the following cycle. Otherwise, the same input data are used for all cycles, that is, the amino acid sequence of the protein, one or several chemical shift lists from the sequence-specific resonance assignment, and one or several lists containing the positions and volumes of cross peaks in 2D, 3D or 4D NOESY spectra. The input may further include previously assigned NOE upper distance constraints or other previously assigned conformational constraints. These will then not be changed by Candid, but used for the Dyana structure calculation.

A Candid cycle starts by generating for each NOESY cross peak an initial assignment list, i.e., hydrogen atom pairs are identified that could, from the fit of chemical shifts within the user-defined tolerance range, contribute to the peak. Subsequently, for each cross peak these initial assignments are weighted with respect to several criteria (listed in Fig. 2.3), and initial assignments with low overall scores are then discarded. In the first cycle, network anchoring has a dominant impact, since structure-based criteria cannot be applied yet. For each cross peak, the retained assignments are interpreted in the form of an upper distance limit derived from the cross peak volume. Thereby, a conventional distance constraint is obtained for cross peaks with a single retained assignment, and otherwise an ambiguous distance constraint is generated that embodies several assignments. All cross peaks with a poor score are temporarily discarded. In order to reduce deleterious effects on the resulting structure from erroneous distance constraints that may pass this filtering step, long-range distance constraints are incorporated into "combined distance constraints" (Fig. 2.3). The distance constraints are then included in the input for the structure calculation with the Dyana torsion angle dynamics algorithm.

The structure calculations typically comprise seven Candid cycles. The second and subsequent Candid cycles differ from the first cycle in the use of additional selection criteria for cross peaks and NOE assignments that are based on assessments relative to the protein three-dimensional structure from the preceding cycle. Since the precision of the structure determination normally improves with each subsequent cycle, the criteria for accepting assignments and distance constraints are tightened in more advanced cycles of the Candid calculation. The output from a Candid cycle includes a listing of NOESY cross peak assignments, a list of comments about individual assignment decisions that can help to recognize potential artifacts in the input data, and a three-dimensional protein structure in the form of a bundle of conformers.

In the final Candid cycle, an additional filtering step ensures that all NOEs have either unique assignments to a single pair of hydrogen atoms or are eliminated from the input for the structure calculation. This allows for the direct use of the Candid NOE assignments in subsequent refinement and analysis programs that do not handle ambiguous distance constraints, and in this paper enables direct comparisons of the Candid results with the corresponding data obtained by conventional interactive procedures.

The core of the current version of Candid is implemented in standard Fortran-77 and has been built upon the data structures and into the framework of the user interface of the program Dyana. The standard schedule and parameters for a complete automated structure determination with Candid and Dyana are specified in a script written in the interpreted command language Inclan that gives the user high flexibility in the way automated structure determination is performed without the need to modify the compiled core part of Candid [26].

2.4.7
Network-Anchoring of NOE Assignments

Network-anchoring exploits the observation that the correctly assigned constraints form a self-consistent subset in any network of distance constraints that is sufficiently dense for the determination of a protein three-dimensional structure. Network-anchoring thus evaluates the self-consistency of NOE assignments independently of knowledge of the three-dimensional protein structure, and in this way compensates for the absence of three-dimensional structural information at the outset of a *de novo* structure determination (Fig. 2.4). The requirement that each NOE assignment must be embedded in the network of all other assignments makes network-anchoring a sensitive approach for detecting erroneous, "lonely" constraints that might artificially constrain unstructured parts of the protein. Such constraints would not otherwise lead to systematic constraint violations during the structure calculation, and could therefore not be eliminated by three-dimensional structure-based peak filters.

The network-anchoring score $N_{\alpha\beta}$ for a given initial assignment of a NOESY cross peak to an atom pair (α,β) is calculated by searching all atoms γ in the same or in the neighboring residues of either α *or* β that are connected simultaneously to both atoms α *and* β. The connection may either be an initial assignment of another peak (in the same or in another peak list) or the fact that the covalent structure implies that the corresponding distance must be short enough to give rise to an observable NOE. Each such indirect path contributes to the total network-anchoring score for the assignment (α,β) an amount given by the product of the generalized volume contributions of its two parts, $(\alpha \rightarrow \gamma)$ and $(\gamma \rightarrow \beta)$. $N_{\alpha\beta}$ has an intuitive meaning as the number of indirect connections between the atoms α and β through a third atom γ, weighted by their respective generalized volume contributions.

The calculation of the network-anchoring score is recursive in the sense that its calculation for a given peak requires the knowledge of the generalized volume contributions

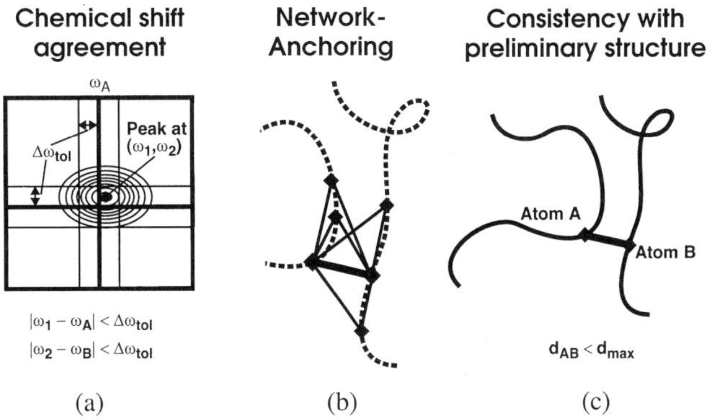

Fig. 2.4 Three conditions that must be fulfilled by valid NOESY cross peak assignments in Candid: **a** Agreement between chemical shifts and the peak position, **b** network-anchoring, and **c** spatial proximity in a (preliminary) structure.

from other peaks, which in turn involve the corresponding network-anchored assignment contributions. Therefore, the calculation of these quantities is iterated three times, or until convergence. Note that the peaks from all peak lists contribute simultaneously to network-anchored assignment.

2.4.8 Constraint-Combination

In the practice of NMR structure determination with biological macromolecules, spurious distance constraints in the input may arise from misinterpretation of stochastic noise, and similar. This situation is particularly critical at the outset of a structure determination, before the availability of a preliminary structure for three-dimensional structure-based screening of constraint assignments. Constraint-combination aims at minimizing the impact of such imperfections on the resulting structure at the expense of a temporary loss of information. Constraint combination is applied in the first two Candid cycles. It consists of generating distance constraints with combined assignments from different, in general unrelated, cross peaks (Fig. 2.5). The basic property of ambiguous distance constraints that the constraint will be fulfilled by the correct structure whenever at least one of its assignments is correct, regardless of the presence of additional, erroneous assignments, then implies that such combined constraints have a lower probability of being erroneous than the corresponding original constraints, provided that the fraction of erroneous original constraints is smaller than 50%.

Candid provides two modes of constraint combination (further combination modes can be envisaged readily): "$2 \rightarrow 1$" combination of all assignments of two long-range peaks each into a single constraint, and "$4 \rightarrow 4$" pair wise combination of the assignments of

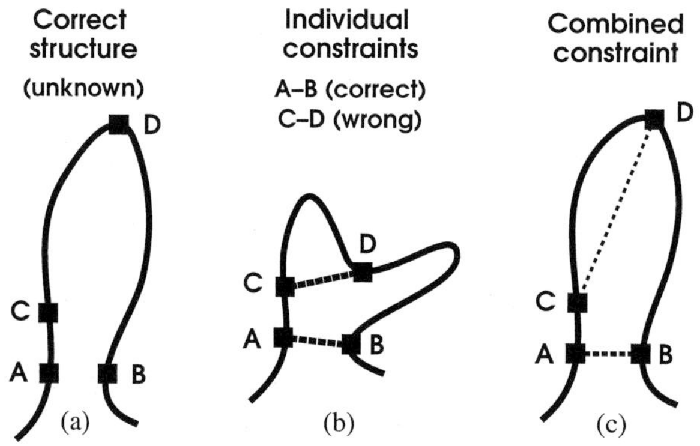

Fig. 2.5 Schematic illustration of the effect of constraint combination in the case of two distance constraints, a correct one connecting atoms A and B, and a wrong one between atoms C and D. A structure calculation that uses these two constraints as individual constraints that have to be satisfied simultaneously will, instead of finding the correct structure (**a**), result in a distorted conformation (**b**), whereas a combined constraint, that will be fulfilled already if one of the two distances is sufficiently short, leads to an almost undistorted solution (**c**).

four long-range peaks into four constraints. Let A, B, C, D denote the sets of assignments of four peaks. Then, $2 \rightarrow 1$ combination replaces two constraints with assignment sets A and B, respectively, by a single ambiguous constraint with assignment set $A \cup B$ (the union of sets A and B). $4 \rightarrow 4$ pair-wise combination replaces four constraints with assignments A, B, C, D by four combined ambiguous constraints with assignment sets $A \cup B$, $A \cup C$, $A \cup D$, and $B \cup C$, respectively. In both cases constraint combination is applied only to the long-range peaks, i.e. the peaks with all assignments to pairs of atoms separated by at least 5 residues in the sequence, because in case of error their effect on the global fold of a protein is much stronger than that of erroneous short- and medium-range constraints. The number of long-range constraints is halved by $2 \rightarrow 1$ combination but stays constant upon $4 \rightarrow 4$ pair-wise combination. The latter approach therefore preserves more of the original structural information, and can furthermore take into account the fact that certain peaks and their assignments are more reliable than others, because the peaks with assignment sets A, B, C, D are used 3, 2, 2, 1 times, respectively, to form combined constraints. To this end, the long-range peaks are sorted according to their total residue-wise network-anchoring and $4 \rightarrow 4$ combination is performed by selecting the assignments A, B, C, D from the first, second, third, and fourth quarter of the sorted list.

To estimate quantitatively the effect of constraint combination on the expected number of erroneous distance constraints in the case of $2 \rightarrow 1$ combination, assume an original data set containing N long-range peaks and a uniform probability $p \ll 1$ that a long-range peak would lead to an erroneous constraint. By $2 \rightarrow 1$ constraint combination, these are replaced by $N/2$ constraints that are erroneous with probability p^2. In the case of $4 \rightarrow 4$ combination, assume that the same N long-range peaks can be classified into four equally large classes with probabilities ap, p, p, $(2 - a)p$, respectively, that they would lead to erroneous constraints. The overall probability for an input constraint to be erroneous is again p. The parameter $a(0 \leq a \leq 1)$ expresses how much "safer" the peaks in the first class are compared to those in the two middle classes and in the fourth "unsafe" class. After $4 \rightarrow 4$ combination, there are still N long-range constraints but with an overall error probability of $(a + (1 - a^2)/4)p^2$, which is smaller than the probability p^2 obtained by simple $2 \rightarrow 1$ combination provided that the classification into more and less safe classes was successful ($a < 1$). For instance, $4 \rightarrow 4$ combination will transform an input data set of 900 correct and 100 erroneous long-range cross peaks (i.e., $N = 1000$, $p = 0.1$) that can be split into four classes with $a = 0.5$ into a new set of approximately 993 correct and 7 erroneous combined constraints. Alternatively, $2 \rightarrow 1$ combination will yield under these conditions approximately 495 correct and 5 erroneous combined constraints. In general, $4 \rightarrow 4$ combination is thus preferable over $2 \rightarrow 1$ combination in the first two Candid cycles.

The upper distance bound b for a combined constraint is formed from the two upper distance bounds b_1 and b_2 of the original constraints either as the r^{-6} sum, $b = (b_1^{-6} + b_2^{-6})^{-1/6}$, or as the maximum, $b = \max(b_1, b_2)$. The first choice minimizes the loss of information if two already correct constraints are combined, whereas the second choice avoids the introduction of too small an upper bound if a correct and an erroneous constraint are combined.

2.4.9
Has it worked?

On the basis of experience gained in *de novo* structure determinations with Candid so far, a set of four criteria for the evaluation of proper performance of combined automated NOESY assignment and structure calculation independent of the availability of an interactively determined reference structure was proposed [26]. These guidelines apply to Candid calculations using input data that fulfills the requirements for the input data presented in Sect. 2.4.4 above and are designed to ensure that the resulting structure has the correct fold if all four criteria are met simultaneously. They do not, however, automatically guarantee a high-quality structure. The four output-based criteria for "safe" Candid runs are:

Criterion 1. Target function: The average final target function value of the structures from the first Candid cycle should be below 250 Å^2, and the final target function value of the structure from the last Candid cycle should be below 10 Å^2.

Criterion 2. RMSD radius: The average backbone RMSD to the mean coordinates (excluding unstructured parts of the polypeptide chain) should be below 3 Å for the structure from Candid cycle 1.

Criterion 3. RMSD drift: The backbone RMSD between the mean structures of the first and last Candid cycles (excluding unstructured parts of the polypeptide chain) should be smaller than 3 Å and should not exceed by more than 25% the average RMSD to the mean coordinates of cycle 1.

Criterion 4. Eliminated peaks: More than 80% of the NOESY peaks should have been assigned by Candid, and less than 20% of the NOESY cross peaks with exclusively long-range assignments (spanning 5 or more residues) should have been eliminated by the peak filters of Candid.

These criteria again emphasize the crucial importance of getting good results from the first Candid cycle. For reliable automated NMR structure determination, the bundle of conformers obtained after cycle 1 should be reasonably compatible with the input data (criterion 1) and show a defined fold of the protein (criterion 2). Structural changes between the first and subsequent Candid cycles should occur essentially within the conformation space determined by the bundle of conformers obtained after cycle 1, with the implicit assumption that this conformation space contains the correct fold of the protein (criterion 3). The output criteria for target function and RMSD values might need to be slightly relaxed for proteins with more than 150 amino acid residues and tightened for small proteins of less than 80 residues.

If the output of a structure calculation based on automated NOESY assignment with Candid does not fulfill these guidelines, the structure might in many cases still be essentially correct, but it should not be accepted without further validation. Within the framework of Candid, the correct approach is to improve the quality of the input chemical shift and peak lists and to perform another Candid run until the criteria are met. Usually, this can be achieved efficiently because the output from an unsuccessful Candid run, even though the structure should not be trusted *per se*, clearly points out problems in the input, e.g., peaks that cannot be assigned and might therefore be artifacts or indications of erroneous or missing sequence-specific assignments. Candid provides for each peak informational output that greatly facilitates this task: the list of its chemical shift-based assignment possibilities, the assignment(s) finally chosen, and the reasons why an assignment is chosen or not, or why a peak is not used at all. Of course, even when the criteria are already met, a still higher precision and local accuracy of the structure might be achieved by further improving the input.

In principle, a *de novo* protein structure determination requires one round of 7 Candid cycles. This is realistic for projects where an essentially complete chemical shift list is available and much effort was made to prepare a complete high-quality input of NOESY peak lists. In practice, it turned out to be more efficient to start a first round of Candid analysis without excessive work for the preparation of the input peak list, using an slightly incomplete list of "safely identifiable" NOESY cross peaks, and then to use the result of the first round of Candid assignment and structure determination as additional information from which to prepare an improved, more complete NOESY peak list as input for a second round of 7 Candid cycles.

The Candid method has been evaluated in test calculations [26] and in various *de novo* structure determinations, including, for instance, three mutants of the human prion protein [94], the calreticulin P-domain [95] the pheromone-binding protein from *Bombyx mori* [96] (Fig. 2.6), and the class I human ubiquitin-conjugating enzyme 2b [97]. These structure determinations have confirmed that the new methods of network-anchored assignment and constraint combination enable reliable, truly automated NOESY assignment and structure calculation without prior knowledge about NOESY assignments or the three-dimensional structure. All NOESY assignments and the corresponding distance constraints for these *de novo* structure determinations were made by Candid, confining interactive work to the stage of the preparation of the input chemical shift and peak lists.

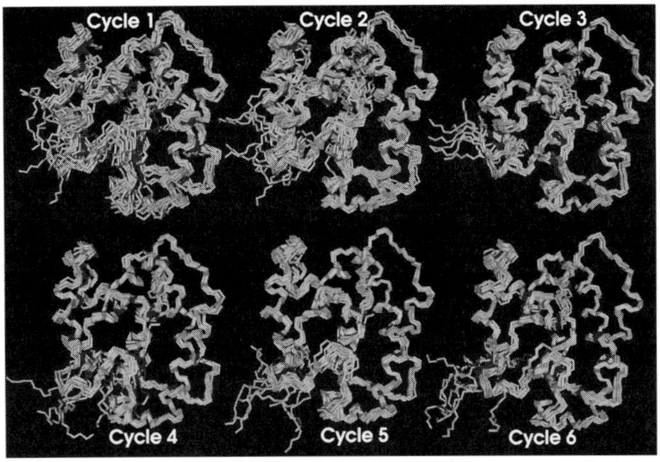

Fig. 2.6 Structures of the pheromone binding protein (form A) from the silk moth *Bombyx mori* [96] obtained in six iterative cycles of combined automated NOESY assignment with Candid and structure calculation with Dyana.

If used sensibly, automated NOESY assignment with Candid has no disadvantage over the conventional interactive approach but is a lot faster and more objective. Network-anchored assignment and constraint combination render the automated Candid method stable also in the presence of the imperfections typical of experimental NMR data sets. Simple criteria basis on the output of Candid allow one to assess the reliability of the resulting structure without cumbersome recourse to independent interactive verification of the NOESY assignments. Candid is a generally applicable, reliable method for automated NOESY assignment. With Candid, the evaluation of NOESY spectra is no longer the time-limiting step in protein structure determination by NMR.

2.5 References

1 Branden, C., Tooze, J. (1991), *Introduction to protein structure.* New York & London: Garland Publishing.
2 Creighton, T. (1993), *Proteins. Structures and molecular properties.* 2nd ed. New York: Freeman.
3 Drenth, J. (1994), *Principles of protein X-ray crystallography.* New York: Springer.
4 Abragam, A. (1961), *Principles of Nuclear Magnetism.* Oxford: Clarendon Press.
5 Wüthrich, K. (1986), *NMR of Proteins and Nucleic Acids.* New York: Wiley.
6 Moseley, H. N. B., Montelione, G. T. (1999), *Curr. Op. Struct. Biol.* **9**, 635–642.
7 Williamson, M. P., Havel, T. F., Wüthrich, K. (1985), *J. Mol. Biol.* **182**, 295–315.
8 Güntert, P., Mumenthaler, C., Wüthrich, K. (1997), *J. Mol. Biol.* **273**, 283–298.
9 Braun, W., Gô, N. (1985), *J. Mol. Biol.* **186**, 611–626.
10 Solomon, I. (1955), *Phys. Rev.* **99**, 559–565.
11 Macura, S., Ernst, R. R. (1980), *Mol. Phys.* **41**, 95–117.
12 Neuhaus, D., Williamson, M. P. (1989), *The nuclear Overhauser effect in structural and conformational analysis.* New York: VCH.
13 Jeener, J., Meier, B. H., Bachmann, P., Ernst, R. R. (1979), *J. Chem. Phys.* **71**, 4546–4553.
14 Kumar, A., Ernst, R. R., Wüthrich, K. (1980), *Biochem. Biophys. Res. Comm.* **95**, 1–6.
15 Torda, A. E., Scheek, R. M., van Gunsteren, W. F. (1989), *Chem. Phys. Lett.* **157**, 289–294.
16 Kalk, A., Berendsen, H. J. C. (1976), *J. Magn. Reson.* **24**, 343–366.
17 Keepers, J. W., James, T. L. (1984), *J. Magn. Reson.* **57**, 404–426.
18 Yip, P., Case, D. A. (1989), *J. Magn. Reson.* **83**, 643–648.
19 Mertz, J. E., Güntert, P., Wüthrich, K., Braun, W. (1991), *J. Biomol. NMR.* **1**, 257–269.
20 Pardi, A. (1995), *Meth. Enzymol.* **261**, 350–380.
21 Varani, G., Aboul-ela, F., Allain, F. H. T. (1996), *Prog. NMR Spectrosc.* **29**, 51–127.
22 Wijmenga, S. S., Mooren, M. M. W., Hilbers, C. W. (1993), In *NMR of macromolecules. A practical approach* (ed. G. C. K. Roberts), pp. 217–288, Oxford: Oxford University Press.
23 Ernst, R. R., Bodenhausen, G., Wokaun, A. (1987), *The principles of nuclear magnetic resonance in one and two dimensions.* Oxford: Clarendon Press.
24 Güntert, P., Qian, Y. Q., Otting, G., Müller, M., Gehring, W. J., Wüthrich K. (1991b), *J. Mol. Biol.* **217**, 531–540.
25 Mumenthaler, C., Güntert, P., Braun, W., Wüthrich, K. (1997), *J. Biomol. NMR* **10**, 351–362.
26 Herrmann, T., Güntert, P., Wüthrich, K. (2002), *J. Mol. Biol.*, **319**, 209–227.
27 Wüthrich, K., Billeter, M., Braun, W. (1983), *J. Mol. Biol.* **169**, 949–961.
28 Güntert, P., Braun, W., Wüthrich, K. (1991), *J. Mol. Biol.* **217**, 517–530.
29 Fletcher, C. M., Jones, D. N. M., Diamond, R., Neuhaus, D. (1996), *J. Biomol. NMR* **8**, 292–310.
30 Güntert, P., Braun, W., Billeter, M., Wüthrich, K. (1989), *J. Am. Chem. Soc.* **111**, 3997–4004.

31 Folmer, R. H. A., Hilbers, C. W., Konings, R. N. H., Nilges, M. (1997), *J. Biomol. NMR* **9**, 245–258.

32 Senn, H., Werner, B., Messerle, B. A., Weber, C., Traber, R., Wüthrich, K. (1989), *FEBS Lett.* **249**, 113–118.

33 Neri, D., Szyperski, T., Otting, G., Senn, H., Wüthrich, K. (1989), *Biochemistry*, **28**, 7510–7516.

34 Wagner, G., Wüthrich, K. (1982), *J. Mol. Biol.* **160**, 343–361.

35 Dingley, A. J., Grzesiek, S. (1998), *J. Am. Chem. Soc.* **120**, 8293–8297.

36 Güntert, P. (1998), *Q. Rev. Biophys.* **31**, 145–237.

37 Oldfield, E. (1995), *Protein Sci.* **5**, 217–225.

38 Williamson, M. P., Asakura, T. (1997), In *Protein NMR techniques* (ed. D. G. Reid), pp. 53–69, Totowa, New Jersey: Humana Press.

39 Spera, S., Bax, A. (1991), *J. Am. Chem. Soc.*, **113**, 5490–5492.

40 de Dios, A. C., Pearson, J. G., Oldfield, E. (1993), *Science* **260**, 1491–1496.

41 Luginbühl, P., Szyperski, T., Wüthrich, K. (1995), *J. Magn. Reson.* **B109**, 229–233.

42 Ösapay, K., Theriault, Y., Wright, P. E., Case, D. A. (1994), *J. Mol. Biol.* **244**, 183–197.

43 Kuszewski, J., Gronenborn, A. M., Clore, G. M. (1995a), *J. Magn. Reson.* **B107**, 293–297.

44 Wishart, D. S., Sykes, B. D., Richards, F. M. (1992), *Biochemistry* **31**, 1647–1651.

45 Karplus, M. (1963), *J. Am. Chem. Soc.* **85**, 2870–2871.

46 Kim, Y., Prestegard, J. H. (1990), *Proteins* **8**, 377–385.

47 Torda, A. E., Brunne, R. M., Huber, T., Kessler, H., van Gunsteren, W. F. (1993), *J. Biomol. NMR.* **3**, 55–66.

48 Tolman, J. R., Flanagan, J. M., Kennedy, M. A., Prestegard, J. H. (1995), *Proc. Natl. Acad. Sci. USA* **92**, 9279–9283.

49 Tjandra, N., Omichinski, J. G., Gronenborn, A. M., Clore, G. M., Bax, A. (1997), *Nature Struct. Biol.* **4**, 732–738.

50 Gayathri, C., Bothner-By, A. A., van Zijl, P. C., MacLean, C. (1982), *Chem. Phys. Lett.* **87**, 192–196.

51 Tjandra, N., Bax, A. (1997), *Science* **278**, 1111–1114.

52 Losonczi, J. A., Prestegard, J. H. (1998), *Biochemistry* **37**, 706–716.

53 Tjandra, N., Grzesiek, S., Bax, A. (1997), *J. Am. Chem. Soc.* **118**, 6264–6272.

54 Kirkpatrick, S., Gelatt Jr., C. D., Vecchi, M. P. (1983), *Science* **220**, 671–680.

55 Braun W. (1987), *Q. Rev. Biophys.* **19**, 115–157.

56 Brünger, A. T., Nilges, M. (1993), *Q. Rev. Biophys.* **26**, 49–125.

57 James, T. L. (1994), *Curr. Opin. Struct. Biol.* **4**, 275–284.

58 Nilges, M. (1996), *Curr. Opin. Struct. Biol.* **6**, 617–623.

59 Allen, M. P., Tildesley, D. J. (1987), *Computer Simulation of Liquids*. Oxford: Clarendon Press.

60 McCammon, J. A., Harvey, S. C. (1987), *Dynamics of proteins and nucleic acids*. Cambridge, UK: Cambridge University Press.

61 Brooks III, C. L., Karplus, M., Pettitt, B. M. (1988), *Proteins. A theoretical perspective of dynamics, structure, and thermodynamics*. New York: Wiley.

62 van Gunsteren, W. F., Billeter, S. R., Eising, A. A., Hünenberger, P. H., Kröger, P., Mark, A. E., Scott, W. R. P., Tironi, I. G. (1996), *Biomolecular Simulation: The Gromos 96 Manual and User Guide*. Zürich: vdf Hochschulverlag.

63 Nilges, M., Clore, G. M., Gronenborn, A. M. (1988a), *FEBS Lett.* **229**, 317–324.

64 Nilges, M., Gronenborn, A. M., Brünger, A. T., Clore, G. M. (1988), *Protein Eng.* **2**, 27–38.

65 Nilges, M., Clore, G. M., Gronenborn, A. M. (1988), *FEBS Lett.* **239**, 129–136.

66 Brooks, B. R., Bruccoleri, R. E., Olafson, B. D., States, D. J., Swaminathan, S., Karplus, M. (1983), *J. Comp. Chem.* **4**, 187–217.

67 Cornell, W. D., Cieplak, P., Bayly, C. I., Gould, I. R., Merz Jr., K. M., Ferguson, D. M., Spellmeyer, D. C., Fox, T., Caldwell, J. W., Kollman, P. A. (1995), *J. Am. Chem. Soc.* **117**, 5179–5197.

68 Brünger, A. T. (1992), *X-PLOR, version 3.1. A system for X-ray crystallography and NMR*. New Haven: Yale University Press.

69 Brünger, A. T., Adams, P. D., Clore, G. M., DeLano, W. L., Gros, P., Grosse-Kunstleve, R. W., Jiang, J. S., Kuszewski, J., Nilges. M., Pannu, N. S., Read, R. J., Rice, L. M., Simonson, T., Warren, G. L. (1998), *Acta Crystallogr.* **D 54**, 905–921.

70 Hockney, R. W. (1970), *Methods Comput. Phys.* **9**, 136–211.

71 Ryckaert, J.-P., Ciccotti, G., Berendsen, H. J. C. (1977), *J. Comput. Phys.* **23**, 327–341.

72 Güntert, P., Wüthrich, K. (1991), *J. Biomol. NMR* **1**, 446–456.

73 Berendsen, H.J.C., Postma, J.P.M., van Gunsteren, W.F., DiNola, A., Haak, J.R. (1984), *J. Chem. Phys.* **81**, 3684–3690.

74 Katz, H., Walter, R., Somorjay, R.L. (1979), *Computers, Chemistry* **3**, 25–32.

75 Bae, D.S., Haug, E.J. (1987), *Mech. Struct. Mech.* **15**, 359–382.

76 Mazur, A.K., Abagyan, R.A. (1989), *J. Biomol. Struct. Dynam.* **4**, 815–832.

77 Mazur, A.K., Dorofeev, V.E., Abagyan, R.A. (1991), *J. Comp. Phys.* **92**, 261–272.

78 Jain, A., Vaidehi, N., Rodriguez, G. (1993), *J. Comp. Phys.* **106**, 258–268.

79 Kneller, G.R., Hinsen, K. (1994), *Phys. Rev. E* **50**, 1559–1564.

80 Mathiowetz, A.M., Jain, A., Karasawa, N., Goddard III, W.A. (1994), *Proteins* **20**, 227–247.

81 Rice, L.M., Brünger, A.T. (1994), *Proteins* **19**, 277–290.

82 Stein, E.G., Rice, L.M., Brünger, A.T. (1997), *J. Magn. Reson.* **124**, 154–164.

83 Abe, H., Braun, W., Noguti, T., Gô, N. (1984), *Computers, Chemistry* **8**, 239–247.

84 Arnold, V.I. (1978), *Mathematical methods of classical mechanics.* New York: Springer.

85 Güntert, P., Berndt, K.D., Wüthrich, K. (1993), *J. Biomol. NMR* **3**, 601–606.

86 Meadows, R.P., Olejniczak, E.T., Fesik, S.W. (1994), *J. Biomol. NMR* **4**, 79–96.

87 Duggan, B.M., Legge, G.B., Dyson, H.J., Wright, P.E. (2001), *J. Biomol. NMR* **19**, 321–329.

88 Mumenthaler, C., Braun, W. (1995), *J. Mol. Biol.* **254**, 465–480.

89 Nilges, M., Macias, M., O'Donoghue, S.I., Oschkinat, H. (1997), *J. Mol. Biol.* **269**, 408–422.

90 Linge, J.P., O'Donoghue, S.I., Nilges, M. (2001), *Methods Enzymol.* 339, 71–90.

91 Savarin, P., Zinn-Justin, S., Gilquin, B. (2001), *J. Biomol. NMR* **19**, 49–62.

92 Nilges, M. (1993), *Proteins* **17**, 297–309.

93 Greenfield, N.J., Huang, Y.J., Palm, T., Swapna, G.V.T., Monleon, D., Montelione, G.T., Hitchcock-DeGregori, S.E. (2001), *J. Mol. Biol.* **312**, 833–847.

94 Calzolai, L., Lysek, D.A., Güntert, P., von Schroetter, C., Riek, R., Zahn, R., Wüthrich, K. (2000), *Proc. Natl. Acad. Sci. USA* **97**, 8340–8345.

95 Ellgaard, L., Riek, R., Herrmann, T., Güntert, P., Braun, D., Helenius, A., Wüthrich, K. (2001), *Proc. Natl. Acad. Sci. USA* **98**, 3133–3138.

96 Horst, R., Damberger, F., Luginbühl, P., Güntert, P., Peng, G., Nikonova, L., Leal, W. S., Wüthrich, K. (2001), *Proc. Natl. Acad. Sci. USA* **98**, 14374–14379.

97 Miura, T., Klaus, W., Ross, A., Güntert, P., Senn, H. (2002), *J. Biomol. NMR*, **22**, 89–92.

3
Achieving Better Sensitivity, Less Noise and Fewer Artifacts in NMR Spectra

Detlef Moskau and Oliver Zerbe

3.1
Introduction

Both structure determination of large biomolecules by multidimensional NMR and screening techniques require high sensitivity and artifact-free spectra in order to be performed reliably at the lowest possible concentrations. Moreover, automated data analysis software still has not solved the problem of distinguishing artifacts from genuine signals satisfactorily. Hence, the best way to circumvent associated problems is to have as few artifacts or noise as possible. Substantial progress has been made in the last decade in the development of spectrometer hardware and probehead design. Improvements of the hardware have aimed at increasing the signal-to-noise ratio (S/N) and resolution while reducing the amount of spurious, unwanted signals.

In principle, unwanted signals may be due to artifacts or to noise. The main source of "real" noise is Brownian motion of electrons in the receiver coil. In "white noise", frequency components are statistically distributed. In contrast, artifacts are "wrong" signals occurring at well-defined frequencies. Instabilities in hardware may lead to noise and/or artifacts. Improvements in S/N will help to distinguish genuine peaks from noise [1]. Although it is sometimes possible to remove noise peaks by symmetrization routines [2, 3] or when the peak can be recognized as false because either the F1 or F2 frequency is impossible [e.g. in the INADEQUATE experiment, where $\nu(F1) = \nu_A(F2) + \nu_B(F2)$], datasets with insufficient S/N are mostly useless.

The term noise in one-dimensional NMR spectra is related to any noise source in the receiver path which determines the sensitivity or signal-to-noise ratio in the spectrum. In multidimensional NMR spectra, a random modulation of intensities of signals will lead to a continuous band of frequencies translating into t_1 noise bands in the indirect dimension. Although, strictly speaking, t_1 noise is not real noise but rather an artifact, it behaves similarly (no well-defined frequencies of signals), and we will hence use the common term t_1 noise. Another property of (white) noise it that is occurs statistically distributed in frequency and phase and therefore will not add up coherently. Instabilities in any of the system components (either electronics, magnet system or probehead) will lead to elevated noise levels or additional artifacts. Hence, any measure that results in less disturbance of the system will improve the overall performance by reducing noise and/or artifacts. The total noise in the receiver path is a sum of thermal noise, mainly stemming from the probehead or the preamplifier, and noise introduced in the analog-to-digital converter (ADC). The latter is due to quantization errors and analog errors from the thermal noise in the ADC. Quantization errors arise because the signal is expressed in bits, and, because of the discrete steps, not all values can be presented exactly in binary format.

Artifactual peaks are even more dangerous than noise since they may not always be immediately recognized and may lead to erroneous assignments. An important source of artifacts is instability in the steady-state condition, e.g. if the relaxation delay is set too short. A commonly encountered example is presented by peaks which occur at the double quantum frequencies in DQF-COSY spectra. For detailed treatments of aspects of noise and artifacts see [4, 5].

Considerable effort has been made by the manufacturers to reduce instabilities or nonlinearities in electronics, fluctuations in temperature and other variables. Such instabilities may result in changes in phase or amplitude in both the transmitter and receiver path and will have dramatic effects on the quality of spectra. Nonlinearities introduce unexpected behavior when interpolating parameter settings. For example, low-power pulse lengths are often not determined but rather calculated from the hard pulse assuming that the amplifier possesses high linearity. Obviously, when this is not the case the pulse length settings will be wrong. Unexpected changes in phase will introduce (severe) artifacts in multiple-pulse experiments or lead to signal loss.

The sensitivity on the other hand is dictated by the spin density and the polarization (the relative population of α- and β-states). The latter in turn depends on the energy separation of α- and β-states, which increases concomitantly with field strength. Changing to a higher field will therefore not only increase spectral dispersion but also increase sensitivity because the polarization increases. The remarkable increase in resolution that is gained by going from 600 to 800 MHz is shown in Fig. 3.1.

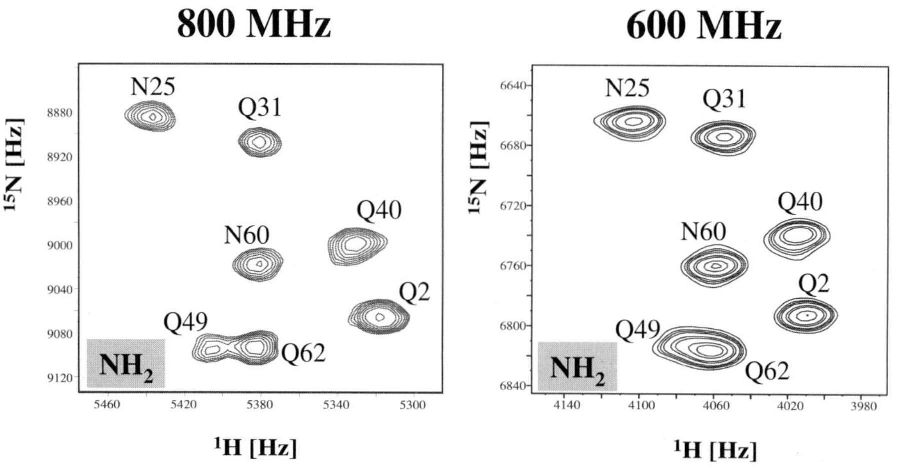

Fig. 3.1 Comparison of resolution of a [^{15}N,^{1}H]-HSQC experiment performed on ubiquitin at 800 MHz (left) and 600 MHz (right) recorded at 295 K. The gain in resolution is particularly clearly visible for the Q49/Q62 peak cluster.

However, the development of stronger magnets is technically very demanding, and the additional cost of adding another 100 MHz seem to at least double the expense for the magnet systems. The strongest magnet system which is presently commercially available (2002) corresponds to 900 MHz ^{1}H frequency. At a given field, the signal primarily depends on probe design and coil materials, and efforts to improve the sensitivity have largely been focused on their development. A major breakthrough in probe technology was achieved with the introduction of cryogenic probe technology, which dramatically reduces the thermal noise. Using this, the S/N could be typically enhanced by a factor of three compared to that of conventional probes.

Artifacts may be roughly categorized into those due to inherent limitations (e.g. pulses cannot excite unlimited bandwidths even if all hardware components work perfectly) and those that result from improper set-up of the experiment or nonideal functioning of the NMR spectrometer system. In this chapter we will mainly focus on the latter two. These artifacts are more likely to appear in multiple-pulse experiments. Quite often, they are avoided by clever programming of the experiments (e.g. interleaved acquisition of data for NOE spectra, use of pulsed-field gradients instead of phase-cycling).

The most prominent problem is presented by t_1 noise in multidimensional spectra. This stems from an additional random modulation of the signal amplitude and is manifested in spectra as continuous noise bands parallel to one of the indirect frequency dimensions. In contrast to the situation a decade ago, t_1 noise in spectra has been tremendously reduced. The achievements are primarily due to reduced sensitivity of the hardware to temperature fluctuations. Other artifacts are additional parallel diagonals. These artifacts are due to oscillations of the signal amplitude with a discrete time constant [6]. Since this modulation is not random but follows a time-dependence it will transform into a frequency component after FT. Another artifact often encountered in homonuclear two-dimensional spectra is best described as a diamond pattern. It is caused by constant phase errors on the pulse which experiences the phase cycle for ω_1-quadrature. It should be noted here that the introduction of pulse field gradients has improved many experiments in which artifacts resulted from signal subtraction [7–10]. This is due to the fact that coherence selection is performed during a single scan and is not relying on long-term stability required for phase-cycling schemes. Another advantage is that only the desired coherence is detected, and hence receiver overload or dynamic range problems for small signals that arise when large (solvent) signals have to be removed in successive scans are avoided.

In order to reveal sources of artifacts and noise we will give a brief description of spectrometer hardware and probehead technology. Of course, spectrometer manufacturers do their best to construct hardware with optimal performance. However, experience shows that all hardware components may occasionally fail. It is one of the goals of this chapter to present ways to recognize malfunction quickly and to locate the source of problem.

3.2 The Transmitter and Receiver System

The schematic transmitter and receiver pathways of a modern NMR spectrometer are displayed in Fig. 3.2.

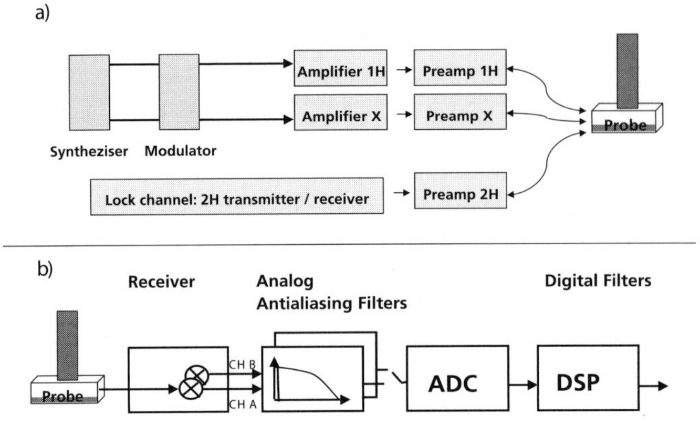

Fig. 3.2 Transmitter (**a**) and receiver (**b**) path.

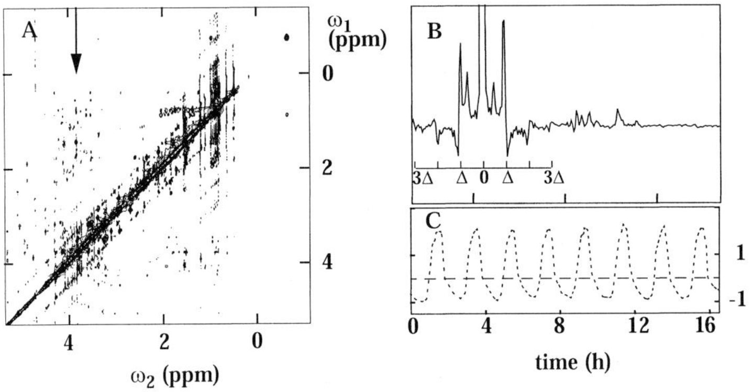

Fig. 3.3 A contour plot of a 750 MHz two-dimensional NOESY spectrum of the mating pheromone *Er*-22 adapted from Ref. [6]. **B** Cross section along ω_1 displaying strong artifacts at multiples of $\Delta = 340$ Hz corresponding to a time variation of ca. 2 h during acquisition (total acquisition time 16.5 h). **C** Plot of the room temperature, units at the y-axis correspond to degrees. The total amplitude of the temperature oscillation is about 3°.

Although the exact settings may be different for different vendors the underlying principle is the same. The transmitter channel basically consists of a unit for generating and controlling the frequency and consists of a synthesizer and corresponding control unit followed by a unit for setting the amplitudes for the pulses, and finally the amplifiers. The latter are low-noise, linear amplifiers. From there on the signal will be passed through the preamplifier and onto the probehead. In addition to the components displayed in Fig. 3.3, there is an extra element, which controls the timing of the spectrometer.

Defects in the transmitter path may cause total signal loss, increased t_1 noise or diamond patterns. Problems with the frequency control unit or the synthesizer will lead to no signal or amplitude or phase instabilities. Random variations in amplitude or phase will increase t_1 noise. A more subtle point arises when the phase presetting time is too short such that the phase is not yet correct when the pulse starts; this will lead to diamond patterns. Problems associated with the amplitude setting lead to nonlinearities or discontinuities in the amplifier output. This may result in wrong shapes, wrong excitation bandwidths or, in the case of amplitude- or phase-modulated pulses, wrong excitation offsets. Failure of the amplifiers leads to signal loss or much prolonged pulse widths (which, however, can still be calibrated in contrast to problems due to the probehead). Too low power for pulses (and hence too long pulse lengths) transform into nonuniform excitation and the problems associated with these. The amplifier linearity is particularly sensitive to changes in room temperature, which can lead to amplitude and phase errors. Problems in the logical unit that controls the timing may lead to the curious situation that some of the pulse sequences run perfectly whereas others do not. It has also been observed that in the case of malfunction of the timing generation certain increments are wrongly timed while others are timed perfectly. Furthermore, experiments may stop suddenly without obvious reason.

The receiver path is depicted in Fig. 3.2b. After the signal has been picked up in the detection coil it is amplified for the first time in the preamplifier. It is then mixed down to audio frequency by subtracting the transmitter frequency to ensure that all of the subsequent hardware components, which are frequency dependent, can be used for frequencies of all possibly interesting nuclei. Thereafter, it passes analog anti-aliasing filters to remove frequencies higher than the Nyquist frequency. If the receiver gain has been set too low, the subsequent analog-digital converter (ADC) will not use the full dynamic range. This results in higher quantization error and may also lead to incorrect intensities for very weak signals. Setting the receiver gain too high will lead to baseline problems and improper function of phase cycles. It has long been known that the analog anti-aliasing filters introduce baseline curvature and change signal intensities. These problems have now been removed by oversampling [11–14] in the directly observed dimension, in which a much larger spectral width is recorded and digital filtering used to select the appropriate region. When using oversampling (at the maximal rate allowed by the digitizer), baselines are almost perfectly flat, as the filter width of the analog anti-aliasing filters is much greater than the spectral width.

In has been mentioned in the previous paragraph that the correct choice of receiver gain is very important. A subtle point is that many multi-dimensional experiments such as the DQF-COSY do not have full signal intensity in the first increment. Adjusting the receiver gain for the first increment may result in receiver overload for later increments. Under this situation the phase cycling of the receiver channel would fail, causing a diamond pattern. Another example is that residual water intensity may be quite different for the sine- and the cosine-modulated increments. Even if this has been taken into account the poorer baseline properties for some increments (those that contain "more water" signal) leads to problems in F1 quadrature detection, resulting in anti-diagonals etc.

3.3 The Magnet, Shim and Lock System

Of course, the most precious part of a spectrometer is the magnet. Both field stability and field homogeneity are of prime importance and have a huge effect on the quality of spectra. A little calculation will show how difficult it is to achieve the required field stability. The proton resonance frequency is 42.5 MHz/T. In order to achieve a resolution of 0.1 Hz at 800 MHz, the field stability must be better than 2.3×10^{-6} Gauss, 1/100,000 of the earth's magnetic field. Field drifts of the magnet that cannot be completely avoided are compensated by the lock system. Continuous drifts, e.g. drifts fixed over time in strength in one direction, are easily followed by the lock system. However, random drifts, e.g. jumps of the field strength leading to lower and higher field strengths, may not be compensated so easily. Such random drifts are mostly caused by large changes in the air pressure (sudden weather changes), and devices are available from the vendors to reduce pressure changes in the helium dewar.

Good homogeneity of the field is achieved by the shim system. Good shimming is mandatory in order to yield proper line shapes and to enable good solvent elimination. Gradient-based shimming methods [15] have largely abolished the often tedious and time-consuming manual shimming procedure. This is particularly important when using special NMR tubes, such as the Shigemi tube, which possess much different shim settings. Continuous homogeneity drifts result from changes in the helium or nitrogen levels in the dewar and cannot be avoided. They may be compensated by methods, which continuously check the homogeneity by means of the lock signal intensity and readjust selected shims. However, this method leads to random modulation of the line shape and hence to more t_1 noise. Furthermore, autoshim is difficult when performing experiments with pulsed field gradients on older systems. Random drifts of shim values mainly result from sudden changes in air pressure. Since random field homogeneity changes cannot be compensated easily it is best to stabilize the pressure in the helium dewar (see above).

The lock system deserves some special comments. The lock usually works on the solvent deuterium line, and any factor that changes the frequency of the deuterium line will impair the performance of the lock [16]. For its optimal performance both lock phase and lock power have to be adjusted correctly. A wrong lock phase will prevent correct shimming. It will also introduce artifacts into gradient-based experiments, e.g. negative spikes at the base of signals. Last but not least, the field regulation, which is based on recognizing the zero-transition of the dispersive deuterium signal, will be poor. Setting the lock power too low results in more noise on the lock signal and hence bad performance. Too high lock power leads to saturation of the lock signal, and again deminished performance. We recommend setting the lock power during shimming to low values (values far from saturation) yielding fast response of the lock and increasing it for the actual measurement to a value just below saturation (making sure that the lock is still stable). Another feature that can be controlled by the spectroscopist is connected to the regulation properties of the lock channel. One parameter is the lock response time. When measuring in solvents with weak lock signal, e.g. 90% H_2O/D_2O, longer periods for signal averaging are required and hence a fast lock response is not possible. For such solvents, it is particularly important to apply a low-pass filter to the lock signal in order to reduce the noise in the final spectrum. Again, good S/N on the lock signal will result in less noisy spectra.

3.4
Sample Conditions and Environmental Set-up

Instabilities in temperature usually have a disastrous effect on the spectral quality [6]. Changes in room temperature will influence the electronics, thereby introducing changes in phase or amplitude of the signals. Changes in temperature in the sample will shift the resonance position of the solvent. The deuterium frequency of some solvents, e.g. water, is especially sensitive to changes in temperature. Hence, spectra recorded in water tend to contain more t_1 noise than those measured in DMSO.

Random changes in temperature are always a result of wrong spectrometer set-up or bad environmental variables. One such mis-adjustment is setting the regulation parameters of the probe air poorly. Temperature regulation can be particularly unstable when the temperature of the probe air is close to the measuring temperature. We recommend cooling the probe air for best temperature stability. The response of the temperature regulation circuit should be calibrated on the sample *at the chosen gas flow.*

Sometimes, oscillating behavior of room temperatures is observed, usually because of the regulation characteristics of the air conditioning [6]. Such oscillations have a distinct temperature dependence and hence will lead to additional (discrete) signals. Fig. 3.3 displays the disastrous effects that such instabilities may have. Quite often, artifactual diagonals result, running parallel to the true diagonal, and the frequency difference depends on the time constant of the regulation circuit of the air conditioning. Doubling or reducing the number of scans will change the amount of time required per increment and hence shift the position of those additional diagonals, thereby helping to identify the source. It is also good practice to independently monitor the room temperature [6].

Temperature gradients in the sample impair the line shape and hence will increase the line width. In order to reduce temperature gradients the sample should be thermally equilibrated before starting the shimming and the measurements. In addition, the gas flow needs to be properly adjusted. It needs to be high enough to allow quick regulation and to be able to transport heat, resulting from decoupling etc., away from the sample but low enough to avoid vibrations of the spinner.

Another serious source of additional noise is floor vibrations. Vibrations will alter the position of the main coil, thereby changing the z-component of the B_0 field and hence modulate resonance frequencies. The presence of large vibrations will strongly increase t_1 noise and will also tremendously reduce center band suppression in [^{13}C,^{1}H]-HSQC spectra performed on natural ^{13}C abundance samples. Vibration dampers are available from the vendors. It should be noted here that it is much easier to reduce high-frequency vibrations than low-frequency components. The latter will add spikes at the signal base, and proper shimming based upon observation of the line shape may become impossible. Vibrations due to increased boil-off after helium refill are common, and noise-sensitive experiments like NOESY should only be started at least half a day after refill. Depending on the location of the magnet systems, magnetic "noise" or vibrations may also be due to trolley buses, trams etc. In these cases the additional noise mostly occurs during peak times of traffic.

3.5
Probeheads

The most dramatic improvements in NMR technology have taken place in the field of probehead design during recent years. Important parameters of an NMR probe are its sensitivity, the pulse widths, radio-frequency (B_1) homogeneity, resolution, solvent suppression properties and background signal. It is well known that not the sensitivity but rather the signal-to-noise ratio (S/N) determines whether signals can be recognized unambiguously or not. The signal intensity itself is largely determined by the coil design. A shorter coil reduces the required pulse widths by reducing the resistance, and a similar benefit is achieved by increasing the filling factor. However, the former reduces the rf homogeneity in z. Larger diameter coils (e.g. going from 5 to 8 mm probes) reduce the fill factor and thereby decrease the signal content relative to the amount to be expected for the increased numbers of spins in the larger volume. Short pulse widths are required for uniform excitation at high fields, e.g. for ^{13}C excitation at 900 MHz, and to keep power levels at acceptable level in order to avoid sample heating during decoupling etc. Bad B_1 homogeneity leads to signal loss and accumulation of artifacts (imperfect 180° pulses due to bad B_1 homogeneity will cause spurious excitation, change of coherence pathways etc.) in multiple-pulse NMR experiments. Solvent suppression properties [17–19] depend of course on the shim but to a large extent also on the probe design. A fundamental challenge in probe design is to optimize one parameter, e.g. the sensitivity, without affecting the performance of others, e.g. the B_1 homogeneity and line shape. The way in which the sensitivity, pulse widths, B_1-homogeneity, resolution and solvent suppression properties interact is displayed in Fig. 3.4 for a triple-resonance (inverse-detection) probehead.

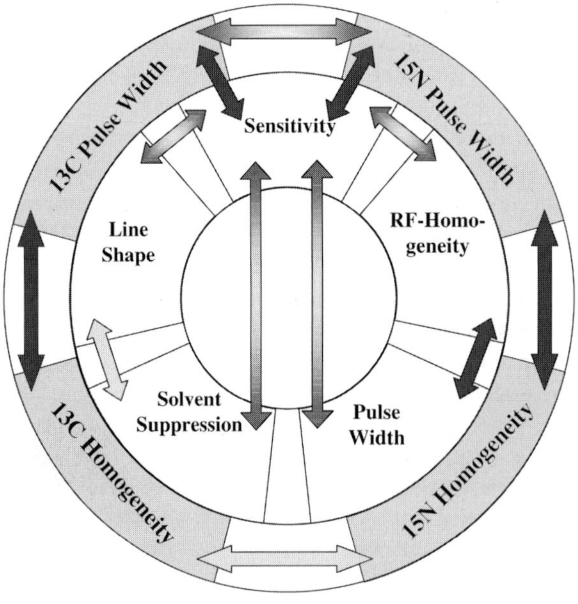

Fig. 3.4 Parameters of a $^{1}H, ^{13}C, ^{15}N$ triple resonance probehead and their mutual interactions.

The signal-to-noise ratio of a signal correlates to the ratio of induced voltage to noise voltage [1, 20]. It can be described as

$$\frac{S}{N} \propto \frac{U_I}{U_N} \propto \frac{\omega \cdot M_0 \cdot V \cdot B_1/I_{coil}}{\sqrt{4 \cdot k_B \cdot \Delta f \cdot T \cdot R}}$$

The resonance frequency ω will increase with the field strength. The equilibrium magnetization M_0 depends on the field strength, the nucleus, temperature and spin density. The volume V is determined by the sample volume and hence its diameter. For mass-limited samples it is advantageous to work in small volumes (e.g. 1 mm probes), whereas for solubility-limited cases it is better to use larger diameter probes (e.g. 8 mm). The coil current I_{coil} can be influenced through proper coil design. All the factors described above contribute to the signal. Lowering the noise may have an even more dramatic effect. The noise voltage depends on thermal noise, related to the temperature T, and on the resistance R, related to the coil material, the frequency, geometry and temperature. Two recent developments deserve special interest. Probably the most dramatic improvement in probe technology ever achieved is due to the development of cryogenic probes [21–23]. The concept behind this development is the reduction of thermal noise by cooling the probe coil and the preamplifier. Typically, the temperature of the probe coil is about 20 K; a temperature of about 70 K is sufficient for the preamplifier. The thermal noise introduced by the preamp is proportional to $1/\sqrt{T_{PA}}$. However, cold preamplifiers only achieve significant reduction in thermal noise in combination with a cryogenic receiver coil. The cooling is performed by purging preamp and probe with cold helium. In order to operate such systems at reasonable cost a closed-loop helium system had to be developed. This is nowadays achieved with a system that re-cools the helium. With cryogenic probes the S/N figures typically are threefold higher [24]. Since the S/N increases with the square of the number of scans, this improvement translates into dramatic reductions of measuring times. Alternatively, less material needs to be used. As an example, a threefold higher sensitivity of a cryogenic probe allows one to increase the number of screening experiments by a factor of 9 at a given total quantity of ^{15}N-labeled protein. Alternatively, sample concentrations may be reduced by a factor of 3, thereby often significantly reducing the problems associated with aggregation effects. Many pharmaceutical companies are nowadays using cryogenic probes for their screenings [22]. Another advantage of cryogenic probes is just emerging: triple-resonance experiments required for backbone assignment may be performed on uniformly ^{15}N-labeled protein, thereby possibly alleviating the need for ^{15}N,^{13}C doubly labeled protein. Since shift mapping is mainly performed by monitoring changes in peak positions of [^{15}N,^{1}H]-HSQC spectra, the latter usually need to be assigned even if the structure of the protein is known, e.g. from X-ray data. In such cases the possibility to perform backbone assignment on ^{15}N-labelled protein is very attractive and will largely reduce the costs of the project. An example displaying a plane from an HNCO experiment performed on 1 mM ubiquitin measured in 10 min is shown in Fig. 3.5 below.

Although ubiquitin displays very favorable relaxation properties, these spectra indicate the potential for performing these experiments at ^{13}C natural abundance for smaller proteins.

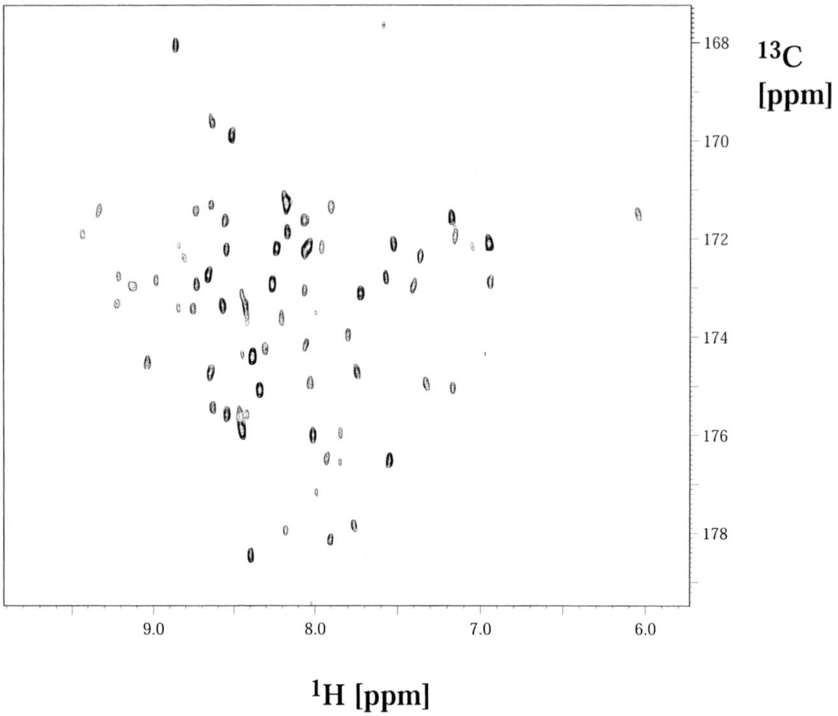

Fig. 3.5 ^{13}C plane from three-dimensional HNCO experiment on 1 mM ^{15}N-labeled ubiquitin with ^{13}C at natural abundance on a Bruker 800 MHz Triple Resonance $^{1}H\{^{13}C,^{15}N\}$ CryoProbe™. The experiment time was 10 min.

With the use of cryogenic probes, carbon detection becomes possible in protein NMR experiments [23]. This may be required for perdeuterated proteins or when signal loss in INEPT transfer steps becomes prohibitively large.

For mass-limited samples, small-diameter (conventional) probes present an attractive alternative to cryogenic probes. In collaboration with Hoffmann-La Roche, Bruker has recently developed a 1 mm probe with a sample volume of 5 μL. This probe offers a mass sensitivity that typically is fourfold higher than a conventional 5 mm probe. An example displaying a [^{13}C, ^{1}H]-HSQC of 1 μg Ibuprofen at ^{13}C natural abundance is shown in Fig. 3.6 below [25].

A number of problems with probeheads may occur. The most frequently encountered source of malfunction is due to probe arcing. This results from degradation of the electronic elements of the resonance circuit. Depending on how serious the problem is, either no pulse is applied at all or pulses from a certain power level upwards are wrong or have an unstable pulse width. The problem is easily detected during pulse calibration: the required pulse lengths at a given power level are far too long, and the zeroing of signal intensity, either for the 180° or preferably the 360° pulse, cannot be achieved. In addition, during pulse calibration, pulses may have undefined flip angle. It should be noted here that the spectroscopist is strongly encouraged to not exceed the power limits sug-

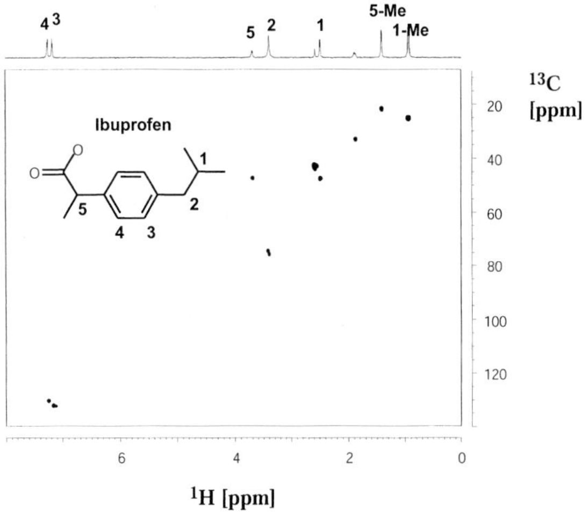

Fig. 3.6 [^{13}C,^{1}H]-HSQC of 1 μL Ibuprofen in 5 μL solvent (d_6-DMSO). The total experiment time was 14 h.

gested by the manufacturer. We have realized that using too much power can actually lead to a sound (clicking) inside the probe when the pulse is applied (however, this may also be due to a faulty gradient coil). Especially at high fields and for inverse detection probes, all channels of the probe should be tuned and matched with the actual sample inserted in the magnet before any pulse is applied. Furthermore, proton pulse lengths need to be calibrated for each sample, since they may deviate from the "standard" values significantly, especially in the presence of salt.

A frequently observed problem arises during the use of experiments with pulsed field gradients. Too short delays for gradient recovery lead to phase and amplitude errors resulting in a multitude of problems (degradation of water suppression, F1 quad images, etc.). After intense use of pulsed-field gradients, connections in the gradient circuit may also be broken. This is easily detected by measuring the resistance in the gradient circuit with an ohm-meter.

Problems in spectrometer hardware or with probes will be noticed immediately when spectra contain large artifacts or very little signal. However, a gradual decrease may not be realized. Slightly wrong pulse widths on gradient-based experiments will still deliver spectra, however, with reduced sensitivity. It is therefore of great importance to monitor the performance of the spectrometer at regular intervals. We recommend checking pulse lengths and gradient recovery times at least bi-monthly. General problems are easily recognized by routinely running a few representative experiments, always using the same sample. In addition, shimming the probes on a line-shape sample so that values within the specifications are reached should be done monthly. When these adjustments are done often, changes are small and the new values can be found very quickly. It cannot be overemphasized that, since NMR spectrometer time is very expensive, especially on very high-field instruments, running the system with components with sub-optimal performance is a waste of financial resources.

3.6 Acknowledgements

We would especially like to thank Dr. D. Nietlispach and Dr. G. Wider for very useful comments on the manuscript. The authors are indebted to Götz Schlotterbeck (F. Hoffmann-La Roche AG, Basel, Switzerland) and Till Kühn (Bruker BioSpin AG, Fällanden, Switzerland) for providing the data for Fig. 3.6.

3.7 References

1 HOULT, D.I.; RICHARDS, R.E.; *J. Magn. Reson.* **1976**, *24*, 71–85.
2 BAUMANN, R.; WIDER, G.; ERNST, R.R.; WÜTHRICH, K. *J. Magn. Reson.* **1981**, *44*, 402–406.
3 BAUMANN, R.; KUMAR, A.; ERNST, R.R.; WÜTHRICH, K. *J. Magn. Reson.* **1981**, *44*, 76–83.
4 FREEMAN, R. *Nuclear Magnetic Resonance*; Longman: Essex, 1988.
5 HULL, W.E. In *Two-dimensional NMR Spectroscopy*; CROASMUN, W.R., CARLSON, R.M.K., Eds.; VCH-Wiley: Weinheim, 1994; pp 67–456.
6 BRAUN, D.; WIDER, G.; WÜTHRICH, K. J. Magn. Reson. B **1996**, *110*, 313–315.
7 BAX, A.; DE JONG, P.G.; MEHLKOPF, A.F.; SMIDT, J. *Chem. Phys. Lett.* **1980**, *69*, 567–570.
8 BARKER, P.; FREEMAN, R. *J. Magn. Reson.*, **1985**, 334–338.
9 HURD, R.E.; JOHN, B.K.; PLANT, H.D. *J. Magn. Reson.* **1991**, *93*, 666–670.
10 KEELER, J.; CLOWES, R.T.; DAVIS, A.L.; LAUE, E.D. *Methods Enzymol.* **1994**, *239*, 145–207.
11 HALÁMEK, J.; VONDRA, V.; KASAL, M. *J. Magn. Reson. A* **1994**, *110*, 194–197.
12 HOULT, D.I.; CHEN, C.-N.; EDEN, H.; EDEN, M. *J. Magn. Reson.* **1983**, *51*, 110–117.
13 KUPCE, E.; BOYD, J.; CAMPBELL, I. *J. Magn. Reson. A* **1994**, *109*, 260–262.
14 WIDER, G. *J. Magn. Reson.* **1990**, *89*, 406–409.
15 VAN ZIJIL, P.C.M.; SUKUMAR, S.; O'NEIL JOHNSON, M.; WEBB, P.; HURD, R.E. *J. Magn. Reson. A* **1994**, *111*, 203–207.
16 BRAUN, S.; KALINOWSKI, H.-O.; BERGER, S. in *150 and More Basic NMR Experiments*; VCH-Wiley: Weinheim, 1998; pp 6–7.
17 GUÉRON, M.; PLATEAU, P.; KETTANI, A.; DECORPS, M. *J. Magn. Reson.* **1992**, *96*, 541.
18 HORE, P.J. in *Nuclear Magnetic Resonance*; OPPENHEIMER, N.J., JAMES, T.L., Eds.; Academic Press Inc.: San Diego, 1989; Vol. 176, pp 64–77.
19 SMALLCOMBE, S.H.; PATT, S.L.; KEIFER, P.A. *J. Magn. Reson. A* **1995**, *117*, 295–303.
20 HOULT, D.I. in *Encyclopedia of Nuclear Magnetic Resonance*; GRANT, D.M., HARRIS, R.K., Eds.; Wiley & Sons: New York, 1994; Vol. 7.
21 STYLES, P.; SOFFE, N.F.; SCOTT, C.A.; CRAGG, D.A.; ROW, F.; WHITE, D.J.; WHITE, P.C.J. *J Magn. Reson.* **1984**, *60*, 397–404.
22 HAJDUK, P.J.; GERFIN, T.; BOEHLEN, J.-M.; HÄBERLI, M.; MAREK, D.; FESIKE, S.W. *J. Med. Chem.* **1999**, *42*, 2315–2317.
23 SERBER, Z.; RICHTER, C.; DÖTSCH, V. *CHEMBIOCHEM* **2001**, *2*, 247–251.
24 RUSSELL, D.J.; HADDEN, C.E.; MARTIN, G.E.; GIBSON, A.A.; ZENS, A.P.; CAROLAN, J.L. *J. Nat. Prod.* **2000**, *63*, 1047–1049.
25 SCHLOTTERBECK, G.; ROSS, A.; HOCHSTRASSER, R.; SENN, H.; KÜHN, T.; MAREK, D.; SCHETT, O. *Analytical Chemistry* **2002**, in press.

4
NMR Strategies for Protein Assignments

Volker Dötsch

4.1
Introduction

Over the past two decades the methods that are used to assign the NMR resonances of proteins have dramatically changed. To date, several hundred different pulse sequences have been published, each one focusing on a particular subset of all protein NMR resonances. Since no NMR pulse sequence exists that can be used to assign all protons as well as all heteronuclear resonances from a single experiment, combinations of several different NMR experiments must be used in order to complete the assignment. Each experiment contains subsets of NMR resonances, and resonances that are common to two of these experiments can be used to link the two data sets. Usually, different experiments or different versions of one experiment are available that will provide the desired information. However, depending on the individual object, and in particular on the size and folding state of the protein, some of these experiments or versions will be more sensitive than others, and the success of the NMR investigation will depend critically on choosing the right set of experiments. In addition, optimizing the solution conditions and other parameters of the measurement, such as the length of the relaxation delay, may have a strong influence on the results. Rather than reviewing all published methods, which would be far beyond the scope of this chapter, the following article is intended as a guide to the most common approaches to obtaining the assignments of protein resonances. It is divided into four parts, starting with the optimization of solution conditions and finishing with a brief discussion of assignment strategies.

4.2
Optimization of Solution Conditions

One of the biggest challenges in structural biology is to identify suitable targets. For NMR spectroscopy the biggest limitation is the size of the protein or complex to be studied. The identification of the smallest possible structural core is, therefore, the first crucial step. The domain boundaries can be obtained either by theoretical methods, such as secondary structure prediction programs and multiple alignments, or by experimental methods. The most commonly used experimental method to identify well-folded domains is limited proteolysis, particularly with trypsin or chymotrypsin. Protease-resistant fragments of the protein can be identified by separating the proteolytic products by PAGE-gelelectrophoresis or HPLC combined with N-terminal sequencing and mass spectrometry. One interesting alternative used by several laboratories is to perform the limited trypsin digestion directly in the NMR tube and follow it with one-dimensional NMR experiments [1, 2]. The appearance of a well-defined single domain protein can be followed by changes in the line width of the proton resonances. Resonance lines that are originally broad in a larger protein become sharper, without changing their position, in the smaller protein. Taking samples from the NMR tube for further analysis by PAGE-gelelectrophoresis, and possibly N-terminal sequencing, can then identify the desired domain. This method is also suitable for removing unfolded parts that cause aggregation. Once a domain or protein construct has been defined and the protein can be expressed and purified, the folding state of the protein has to be determined. In particular, for protein domains that have been defined through sequence alignment or secondary structure prediction, it is crucial to verify that the selected sequence can fold into a well-determined three-dimensional structure. Domains that have been selected through proteolysis are more likely to be folded, although their proteolysis resistance is not a guarantee that they are structured. One quick way to investigate whether the protein is structured is by CD-spectroscopy [3, 4]. Alternatively, one-dimensional proton NMR spectroscopy can be used to investigate whether the chemical shift dispersion of the protein indicates a folded state. This, however, requires more material than CD-spectroscopy. The best way to judge the suitability of a particular protein for structure determination by NMR spectroscopy, however, is to measure a $[^{15}N,^{1}H]$-HSQC spectrum. Unfolded proteins will display most of their amide proton resonances in the proton spectral region between 7.8 and 8.4 ppm, and hence the observed chemical shift dispersion immediately helps to recognize folded constructs. Although this particular method requires ^{15}N-labeled protein, one advantage is that it can be used on crude cell extracts, because only protein resonances of the over-expressed protein are visible [5]. This method allows for rapid screening of different protein constructs to investigate their structural integrity without lengthy purification.

Once the domain boundaries of a particular protein have been defined, its oligomeric state has to be determined. This is important since relatively small domains can form higher order oligomers that could make a high resolution NMR structure determination (but not the backbone assignment, see below) very difficult or even impossible. In addition, it is important to investigate whether the protein aggregates, which also can lead to significant increases in line width. Both the oligomeric state and potential aggregation can be determined by dynamic light scattering [6] and ultra-centrifugation [7, 8] methods. In addition, gel filtration, which is often already a step during the purification procedure, can be used to assess the state of the protein.

As the next step, suitable sample conditions have to be found which maximize the protein's solubility and stability in order to prepare samples that are stable for several weeks, allowing the spectroscopist to collect all necessary spectra without having to prepare a new sample after each experiment. In particular, if aggregation has been detected, solution conditions have to be screened to produce a monodispersed protein solution. The most elegant way to screen many different solution conditions with a minimum amount of protein is the use of either the microdialysis button [9] or the microdrop screen [10, 11]. In these screens, small amounts (1–5 μL) of concentrated protein solutions are mixed with solutions in which the pH, buffer composition, salt concentration and detergents are varied systematically. Visual inspection of the sample for precipitation allows the researcher to identify favorable conditions. Details of the screens and screening strategies have been described in the literature [11].

One of the most important parameters is the pH value of the solution. A protein's solubility is lowest at its isoelectric point, where it has equal amounts of positive and negative charges. The theoretical pI value of a protein can be calculated from its amino acid composition, and several different tools are available on web pages (for example www.expasy.ch). To avoid problems with low solubility, the pH of the solution should be at least 1 unit above or below the pI of the protein. Another restriction on the pH of the solution is imposed by the pH-dependent exchange rate of the amide protons. This is at a minimum at a pH of approximately 3, increasing with both increasing and decreasing pH [12]. Although the introduction of water flip-back pulses [13] and the FHSQC [14] method have reduced the detrimental effect of fast amide proton exchange with the water, minimization of this exchange rate can significantly increase the sensitivity. Note that the pH at which the exchange rates are the lowest may depend on the environment, and in micellar solutions its value differs by 1–1.5 units (see Chapt. 5). In addition to the pH value, detergents and salt concentration can strongly affect the solubility of a protein. The interactions of salts and other compounds can become very complex and can depend on their concentration and other parameters like the pH of the solution [9, 15–18]. In principle, three classes exist. Compounds of the first class stabilize protein structures regardless of their concentration. This class contains most osmolytes and salting-out salts such as Na_2SO_4. Compounds of the second class destabilize proteins and include denaturants such as urea and guanidinium hydrochloride. The third class contains a variety of different compounds including some salts ($MgCl_2$) and some amino acids which can act as either stabilizers or destabilizers depending on their concentration. To increase the stability of proteins, the addition of osmolytes like (deuterated) glycerol, glycine or sorbitol has been suggested [19, 20]. However, addition of these stabilizers can increase the viscosity of the solution, and some stabilizers like salting out reagents also increase the risk of aggregation. The exact solution conditions have, therefore, to be determined carefully for each individual protein.

Finally, the last step of the procedure for optimizing experimental conditions is to identify the denaturation temperature of the protein. This step is important because the rotational tumbling rate of a protein increases with temperature, and faster tumbling results in sharper resonance lines. Therefore, the temperature during the NMR experiments should be as high as possible without denaturating the protein. The denaturation temperature can best be determined by either CD-spectroscopy or one-dimensional NMR.

4.3 Labeling and Overexpression

The selection of the pulse sequence to be used out of the hundreds that have been published depends on the information desired. NMR spectroscopy cannot only be used to determine high-quality three-dimensional structures, but can provide information about the global fold, interactions with other molecules or just the identification of the secondary structure [21]. These different levels of "structural characterization" by NMR spectroscopy have very different size limitations. Each of these different levels also requires a different pattern of labeling with the isotopes ^{15}N, ^{13}C and ^{2}H. Therefore, based on the molecular size of the protein and the scientific goal of the experiments, a suitable isotopic labeling pattern has to be chosen.

The first structures of biological macromolecules determined by NMR spectroscopy were solely based on [^{1}H,^{1}H]-proton correlation experiments, from which protein structures up to a size of 10 kDa can be obtained without any isotopic enrichment [22]. However, for larger proteins the increased resonance overlap makes structure determination increasingly difficult. The introduction of isotopic labeling with the NMR-active nuclei ^{15}N and ^{13}C has extended the molecular weight range to approximately 20 kDa by reducing resonance overlap through separation of the peaks along one or more heteronuclear frequency dimensions [23] (see also Chapt. 1). Since even for proteins smaller than 10 kDa isotopic labeling can significantly accelerate the structure determination process, virtually all NMR structure determinations of recombinantly expressed proteins are nowadays carried out with isotopic labeling. The only exceptions are proteins that are isolated from natural sources. Examples are snake and scorpion toxins [24, 25] and pheromones [26, 27]. Homonuclear assignment techniques are based on first identifying the proton spin systems of the individual residues and then linking them through the NOE effect, thereby following the sequential resonance assignment procedure developed by Wüthrich and coworkers [22]. However, since these homonuclear techniques are nowadays of less importance than heteronuclear experiments, the rest of this article will focus only on assignment techniques based on heteronuclear NMR experiments.

Structures of proteins up to 20 kDa can routinely be obtained with ^{15}N,^{13}C-doubly labeled NMR samples [28]. Suitable protein samples can be expressed in bacteria that are grown on minimal medium that contains (>98%) ^{15}N-labeled ammonium chloride or sulfate as the sole nitrogen source and ^{13}C-labeled glucose (or sometimes acetate) as the sole carbon source. Alternatively, $^{13}C/^{15}N$ doubly-labeled full media consisting of algae or bacterial extracts can be used. These labeled full media usually produce significantly higher yields, but they are much more expensive. Whether minimal or full medium is used for protein expression depends on the individual protein and expression construct, and the exact expression protocol has to be optimized for each individual project. For larger proteins, additional deuteration is necessary (see below) [29–40]. Different levels of deuteration can be achieved by adding D_2O to the expression medium. For highly deuterated proteins, however, media containing >99% D_2O and perdeuterated glucose or acetate have to be used. In addition, algae and bacterial extracts with different deuteration levels exist, again increasing the protein yield. Expression of highly deuterated proteins also requires stepwise adaptation of the bacteria to these high deuteration levels. Detailed protocols of these procedures are available in the literature [32, 38]. Protein production with deuterated media usually results in significantly lower yield than that with nondeuterated media, and "isotope effects" cause considerable longer expression times. Careful optimization of the expression conditions is, therefore, particularly important for deuterated samples. In general, small scale ($\sim$10 mL) expression tests in different media followed by PAGE gelelectrophoresis provide information about the expression level in an efficient way. Separating the soluble and the insoluble fractions in the bacterial samples can further help to optimize the protein production conditions. Many proteins are expressed in an insoluble form in inclusion bodies. While proteins in inclusion bodies can, in principle, be refolded, it is often desirable to find conditions that allow direct expression of soluble protein. Often decreasing the temperature from 37 °C to as low as 15 °C and reducing the amount of IPTG used to induce protein production can significantly increase the amount of soluble protein. These conditions, however, require longer expression periods. Sometimes, similar effects can be obtained by switching from minimal medium to full medium. A number of proteins, especially those belonging to the class of membrane proteins or membrane-bound receptors, may be well folded but still display a high tendency toward aggregation. In order to make these proteins amenable to NMR investigations, solubility-enhancing tags have been developed [41, 42]. These highly water-soluble tags are fused to the protein sequence of interest via flexible linkers and help to improve expression yields and solubility. The tags may also be used to quickly screen for folded constructs.

Occasionally, however, driving the overexpressed protein into inclusion bodies can be advantageous, for example if the protein is unfolded and protease-sensitive. Forcing proteins into inclusion bodies can be achieved by increasing the expression temperature (to approximately 40 °C) and by using more IPTG. Alternatively, special bacterial overexpression plasmids exist that target a protein into inclusion bodies.

4.4 NMR Experiment

4.4.1 Small and Medium-Sized Proteins

In general, the pulse experiments that are used for isotope-labeled proteins are very different in sensitivity. The two parameters that have the strongest influence on the sensitivity are the size of the coupling constants involved in the coherence transfer pathways and the transverse relaxation times of the involved nuclei. Usually, $^{13}C\alpha$ and $^{13}C\beta$ atoms display the shortest transverse relaxation times (T_2), and any sequence that transfers magnetization along small couplings involving these nuclei will be insensitive for larger proteins. In contrast, ^{13}CO and ^{15}N nuclei have much longer transverse relaxation times. A simple one-dimensional proton spin-echo experiment using two different spin echo times [43] allows one to estimate the amide proton transverse relaxation times according to the formula

$$T_2 = 2(\tau_a - \tau_b)/\ln(I_a/I_b)$$

where τ_a and τ_b are delays and I_a and I_b are the intensities of the amide protons in the experiments measured with τ_a and τ_b respectively. Note that this experiment may be used with nonlabeled protein and hence may help to decide whether a protein is amenable to NMR analysis and therefore worth producing in isotope-enriched form. To check whether a particular pulse sequence is suitable for the protein under investigation, inspection of the signal content of the first two-dimensional plane of the three-dimensional experiment is helpful, provided that the sequence is programmed such that zero chemical shift evolution has occurred during the first increment. However, the success of the experiment may also depend on whether pulse lengths are correctly calibrated, whether the pulse excitations widths are set such that only the region of interest is (completely) excited, and whether the sequence has been properly programmed. The technical aspects are very well covered in the review of G. Wider [44].

The first experiment to be recorded on isotope-labeled proteins is the [^{15}N,^{1}H]-HSQC experiment. Inspection of the [^{15}N,^{1}H]-correlation map and simple counting of cross peaks reveals whether multiple conformers exist, whether some parts of the backbone signals are broadened, possibly because of slow conformational exchange, or whether parts of the sequence are not visible at all. As mentioned above, the spectrum will also show whether the protein is well folded or not.

The first step in a structure determination by NMR spectroscopy of a doubly labeled protein is the assignment of the backbone resonances. This is achieved by combining pairs of heteronuclear NMR experiments, one of which provides intra-residual correlations between backbone resonance frequencies and the other one interresidual correlations of the same set of frequencies. The most important pair consists of the HNCA [45–47] and the HN(CO)CA [45, 48] experiment. Both pulse sequences are performed as three-dimensional experiments, and the three frequency coordinates are represented by the amide proton chemical shift, the backbone nitrogen chemical shift and the α-carbon chemical shift. In the HNCA experiment these three frequencies correspond to the intra-residual frequencies of each amino acid, i.e. the frequencies $^{1}H_i$, $^{15}N_i$ and $^{13}C\alpha_i$ of the amino acid i (Fig. 4.1). In addition, a smaller coupling between the ^{15}N spin of amino acid i and the $^{13}C\alpha$ spin of amino acid i–1 exists that creates an additional peak with the chemical shifts $^{1}H_i$ and $^{15}N_i$ of the amino acid i and $^{13}C\alpha_{i-1}$ of the preceding amino acid i–1 (Fig. 4.1).

In order to also find the amide proton frequency and the nitrogen frequency of this preceding amino acid, the HNCA spectrum is searched for a peak representing an intra-residual correlation with a $^{13}C\alpha$ frequency that matches the $^{13}C\alpha_{i-1}$ frequency of the inter-residual peak. In addition to this intra-residual peak, this residue also shows itself a sequential peak with the $^{13}C\alpha$ frequency of its preceding amino acid, and its sequential neighbor can therefore be identified by the same frequency matching procedure. In this way, the entire backbone of the protein can be assigned. An example of this procedure is shown in Fig. 4.2. One critical aspect of this method is that the interresidual and intra-re-

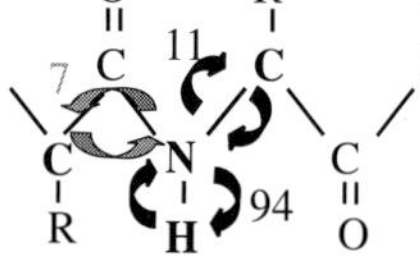

Fig. 4.1 Schematic diagram of the magnetization transfer pathway in the HNCA experiment. Two pathways exist, that transfer the magnetization from the amide proton to the amide nitrogen and then either to the intra-residual α-carbon (black arrows) or to the α-carbon of the preceding amino acid (gray arrows) and back to the amide proton for detection. The values of the coupling constants (in Hz) are indicated next to the arrows.

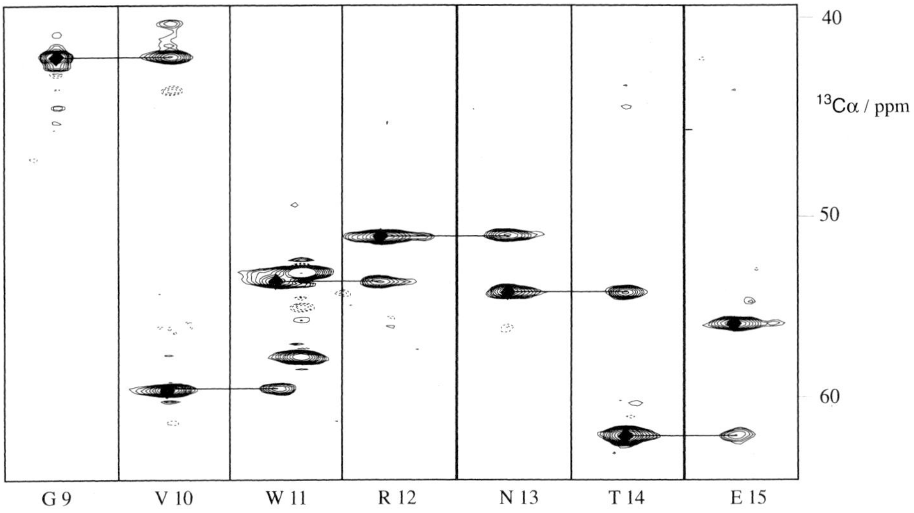

Fig. 4.2 Strips taken from an HNCA experiment of the transcription factor CDC5. The amide proton frequency is along the *x*-axis, the $^{13}C\alpha$ frequency is along the *y*-axis and the ^{15}N-frequency is perpendicular to the planes shown. Intra-residual peaks are marked with a black diamond symbol. Lines connect sequential peaks in strips of amino acid *i* with intra-residual peaks in strips of amino acid *i*–1. The amino acid sequence is indicated at the bottom.

sidual peaks have to be distinguished. Usually, the sizes of these peaks can be used for their identification: the coupling constant between the ^{15}N spin and the intra-residual $^{13}C\alpha$ spin is 11 Hz, which is greater than the coupling constant between the ^{15}N spin and the interresidual $^{13}C\alpha$ (7 Hz). Consequently, the intra-residual peak is larger than the interresidual one. However, depending on the relaxation behavior of the two $^{13}C\alpha$ coherences, this distinction might not be correct. In other cases only one peak might be visible, for example if the intra-residual and interresidual $^{13}C\alpha$ frequencies overlap. In these cases the HN(CO)CA experiment can be used to identify the inter-residual connectivity. In the HN(CO)CA experiment the coherence transfer pathway involves the carbonyl spin, which suppresses the intra-residual pathway and keeps only the inter-residual one. In this experiment the frequency of the carbonyl spin itself is not recorded but is only used to relay the coherence to the inter-residual $^{13}C\alpha$ spin. The HN(CO)CA spectrum shows, instead of two peaks, only one with the $^{1}H_i$, $^{15}N_i$ frequencies of amino acid *i* and the $^{13}C\alpha_{i-1}$ frequency of the preceding residue. The combination of HNCA and HN(CO)CA experiments, therefore, allows one to distinguish between the intra-residual and the inter-residual $^{13}C\alpha$ chemical shifts.

The larger the protein, the greater the resonance overlap. Often several different residues have degenerate $^{13}C\alpha$ frequencies which will make an unambiguous assignment difficult or even impossible. In these cases, additional information from different NMR experiments is required. One possibility is to use the carbonyl chemical shift instead of the $C\alpha$ chemical shift and measure the HNCO/HN(CA)CO pulse sequence pair [37, 45, 46, 49, 50]. As with the HNCA/HN(CO)CA combination, one of the experiments, the

HN(CA)CO, provides both intra-residual and sequential connectivities by correlating the $^{1}H_{i}$ and $^{15}N_{i}$ chemical shifts with the intra-residual $^{13}CO_{i}$ chemical shift and with the preceding $^{13}CO_{i-1}$ frequency. The other experiment, the HNCO, yields exclusively sequential connectivities. A disadvantage of the HNCO/HN(CA)CO pair is the small chemical shift dispersion of the carbonyl spins, which can cause significant resonance overlap. If chemical shift degeneracies of both the $^{13}C\alpha$ and the ^{13}CO spins prevent the complete backbone assignment, the $^{13}C\beta$ spin can be used to obtain additional information. Since the β-carbons of the individual amino acids show a much greater variety of chemical substituents than the α-carbons, their chemical shifts are also spread over a wider frequency range than those of the α-carbons. This makes experiments that correlate the $^{13}C\beta$ chemical shift with the backbone resonances very useful. Their main disadvantage is that they require one more coherence transfer step (from the β-carbon to the α-carbon), which makes the pulse sequence longer and leads to decreased sensitivity. For medium sized proteins, the most important pulse sequence, which connects the β-carbon frequencies with the backbone nuclei, is the CBCA(CO)NH experiment [51]. This correlates the $^{13}C\beta$ and $^{13}C\alpha$ chemical shifts of amino acid i–1 with the ^{1}H and ^{15}N chemical shifts of residue i. The CBCA(CO)NH spectrum, therefore, contains the same information as the HN(CO)CA with the addition of the $^{13}C\beta$ chemical shifts (Fig. 4.3). If the target protein is unstable or if the entire measurement time is limited, the HN(CO)CA can therefore be replaced with the CBCA(CO)NH experiment.

In the CBCA(CO)NH experiment the coherence pathway includes the carbonyl spin, which suppresses the intra-residual coherence transfer. However, as for the other two pairs of pulse sequences mentioned above, an experiment can be designed that allows simultaneous intra- and inter-residual coherence transfer. In the CBCANH experiment

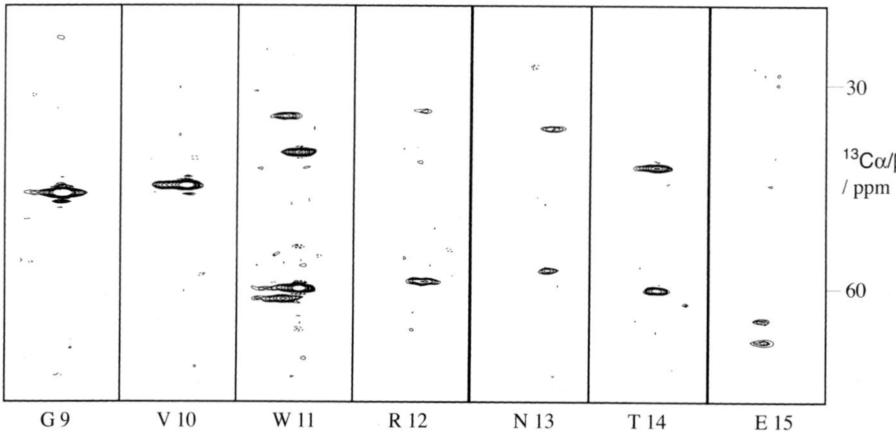

Fig. 4.3 Example of sections taken from a CBCA(CO)NH experiment of the transcription factor CDC5. Strips are shown for the same set of amino acids as in Fig. 4.2. The two peaks per strip indicate the ^{13}C-frequencies of the α- and the β-carbon of the preceding amino acid with the α-position marked with a black diamond symbol. Strips of amino acids that follow a glycine show only one peak. Note the different spectral widths in the CBCA(CO)NH and the HNCA experiments.

[52], each ^{1}H amide proton frequency is correlated with four carbon frequencies, the intra-residual ^{13}Cβ and ^{13}Cα and the inter-residual ^{13}Cβ and ^{13}Cα. Although this experiment contains more information than any of the others, it is also the least sensitive. More sensitive alternatives are described in the section about large proteins.

In addition to these triple-resonance experiments, ^{15}N-edited NOESY experiments [53] can provide further information that can be used to identify the sequential neighbor. The use of NOESY-based experiments for the sequential assignment process is based on the observation of characteristic sequential NOEs, which are, however, secondary-structure dependent [22]. For α-helices, sequential amide protons are in close contact, giving rise to amide-amide NOEs. On the other hand, in β-sheets the distance between the α-proton of amino acid i and the amide proton of residue i+1 becomes so short that the resulting sequential NOE is stronger than the NOE between the amide proton and its own α-proton. To use these αN-NOEs in β-sheets for the sequential assignment process, the chemical shifts of all α-protons have to be known. This can be achieved, for example, by measuring a HNHA experiment [54] that correlates the α-proton with its own amide proton. In addition to obtaining the α-proton assignment, this experiment also yields the amide-proton α-proton coupling constant that can be used to restrict the value of the backbone Φ-angle during structure calculations. In addition to these secondary-structure specific sequential NOEs, the overall NOE pattern of sequential neighboring residues is similar (but not identical) because they share a similar environment. Comparison of the NOE pattern of two amide protons can therefore provide additional information to help with the sequential assignment.

Once the sequential assignment is complete, the secondary structure of the protein can be determined from the secondary-structure specific chemical shifts of the ^{13}Cα nuclei. Other nuclei, including the ^{13}Cβ and the ^{13}CO, also show characteristic chemical shifts that can be used to identify the secondary structure [55–60]. In addition, the secondary structure can also be assigned with the help of the identified secondary structure-specific sequential NOEs and coupling constants obtained from the HNHA experiment.

If the goal of the project is the determination of the three-dimensional structure, the full assignment of the side-chain resonances is required. This is usually achieved by using different implementations of the TOCSY (total correlation spectroscopy) method [61, 62], which spreads the magnetization throughout the entire side chain. The simplest version is the ^{15}N-edited [^{1}H,^{1}H]- TOCSY pulse sequence, which correlates all side chain proton frequencies with the amide proton frequency. Since the experiment is performed as a three-dimensional one, with the ^{15}N-chemical shift forming the third frequency dimension, it usually achieves excellent separation of the resonances. Its main disadvantage, however, is that it relies on small proton-proton coupling constants, and this requires long mixing times for magnetization transfer. These long mixing times make [^{1}H,^{1}H]-TOCSY experiments less useful for large proteins. In addition, the efficiency of the magnetization transfer during the mixing time is dependent on the protein's secondary structure, since the amide-proton α-proton coupling constant is secondary structure dependent.

A more sensitive alternative is to use HCCH-TOCSY experiments [63–65], which transfer magnetization through the much larger ^{13}C-^{13}C coupling constants. In addition, these coupling constants are almost independent of the secondary structure and very uniform throughout the entire side chain. Their main disadvantage is resonance overlap. Another type of experiment, the HC(CCO)NH-TOCSY [66–69] combines the sensitivity of the HCCH-TOCSY with the excellent chemical shift dispersion of the ^{15}N-edited TOCSY. In the HC(CCO)NH-TOCSY the magnetization is transferred after the ^{13}C-^{13}C mixing period via the backbone carbonyl to the backbone amide nitrogen and amide proton. In the resulting spectrum, the amide proton and amide nitrogen of amino acid i are correlated with all side-chain protons of amino acid i–1. This experiment can also be run as the C(CCO)NH-TOCSY, in which the side-chain proton frequencies are replaced with the side-chain carbon chemical shifts. This experiment is useful to verify the backbone assignment as well as to obtain the side-chain carbon assignment, which is important for the interpretation of ^{13}C-edited NOESY (nuclear Overhauser effect spectroscopy) experiments.

Another useful experiment is the HBHA(CBCACO)NH pulse sequence [70]. This is based on the CBCA(CO)NH experiment but replaces the carbon dimension with the proton dimension and correlates the Hα and Hβ resonances of amino acid i–1 with the backbone amide nitrogen and amide proton of amino acid i. In combination with the TOCSY experiments, the HBHA(CBCACO)NH pulse sequence allows one to distinguish the β- and the γ-proton assignments, which are often difficult or impossible to distinguish on the basis of chemical shifts only. A special class of residues whose side chains are difficult to assign but which are often found to give rise to many important restraints for the core of the proteins are the aromatic residues. Two sequences have been developed to correlate the C-β with the aromatic carbons or protons [71, 72], connectivities that are absent in conventional HCCH TOCSY experiments because of the limited bandwidth of the isotropic mixing schemes. Furthermore, three-dimensional sequences have been designed for assignment of the aromatic ring spin systems [73]. It should be emphasized here that, although conceptually easy, complete side-chain assignments are notoriously difficult to obtain and usually require much more time than backbone assignments. Furthermore, backbone assignments are more easily adaptable to automation procedures.

The experiments described above include the most commonly used and most sensitive ones. As mentioned earlier, many more pulse sequences exist that can provide important information. Some of the experiments are also able to distinguish between certain types of amino acids based on their unique spin coupling network or chemical shifts [74–79]. Whether these experiments are used during the assignment process depends on the individual project.

4.4.2
Large Proteins

With the experiments outlined above, protein structures up to 20 kDa can routinely be assigned. For larger proteins or complexes the slower tumbling rate leads to substantial line broadening of the resonance peaks, which can result in the loss of many signals. For large proteins, often only flexible residues in the loops and at the N- and C-terminus are visible. In these cases replacing the side-chain protons with deuterons reduces the ^{13}C relaxation rates significantly [29–31]. The size of the protein and the final goal of the research decide the extent of deuteration. For proteins in the 20–40 kDa range, for which the full three-dimensional structure is the final goal, a 50–70% deuteration level has proven to be optimal. That deuteration level reduces the relaxation rates significantly and also leaves some protons for NOE-based structure determination. For even larger proteins the global fold can be determined based on a relatively small number of selected structural constraints (NOEs). However, because of the high degree of deuteration that is required in this molecular weight range, a detailed three-dimensional structure cannot, in most cases, be obtained by NMR. In contrast to the tertiary structure, the secondary structure can be obtained even for much larger proteins. The recent introduction of TROSY-type techniques [80, 81] (see Chapt. 10), which at the appropriate magnetic field strength lead to a dramatic reduction of the relaxation rates, has extended the applicability of NMR experiments beyond the 100 kDa mark [82]. These experiments, however, require very high degrees of deuteration, making NOE-based structure determination difficult. However, it has been demonstrated that protonated methyl groups can be incorporated into perdeuterated proteins, which allows the measurement of NOEs between these methyl groups and the amide protons [83]. In combination with the recently developed residual dipolar coupling methods [84, 85], the global fold of the protein can be determined. However, since this chapter focuses on protein assignment and not on the actual structure determination, the following paragraphs will describe the necessary modifications when dealing with very large perdeuterated proteins.

The only protons present in perdeuterated proteins are the amide protons. Since the sensitivity of NMR experiments is highest if the pulse sequence both begins and ends with proton spins, the lack of any side-chain carbon-bound protons forces experiments like the CBCA(CO)NH to start with ^{13}C magnetization directly, making them far less sensitive. Therefore, perdeuteration requires experiments to begin and end on the amide protons. The HNCA, HNCOCA, HNCACO and HNCO experiments fulfill this requirement and have to be modified only slightly. The only necessary modification is the introduction of deuterium decoupling during delays with carbon coherences. In addition, the slower relaxation allows the spectroscopist to change the incremented indirect α-carbon detection periods into constant-time evolution periods, resulting in increased resolution. For very large proteins, all experiments mentioned above can be performed as TROSY versions, which, however, require further changes in the pulse sequences.

To obtain information about the β-carbon chemical shifts, new pulse sequences have to be used which begin and end with amide proton magnetization. The HNCACB experiment transfers magnetization from the amide proton via the amide nitrogen to the α-carbon and further to the β-carbon spins [86]. As is the case with all other experiments that transfer magnetization from the amide nitrogen to the β-carbon spin, both intra-residual and inter-residual couplings are active, and the resulting spectrum contains correlations between the amide proton and nitrogen of amino acid i and the α- and β-carbons of amino acids i and i–1. The information obtained from the HNCACB is therefore equivalent to that from the CBCANH experiment. A detailed comparison of the two experiments, however, has revealed that the HNCACB pulse sequence is of higher sensitivity than the CBCANH experiment [86] and should therefore be used for smaller proteins also. To distinguish intra- from inter-residual correlations, the usual method of forcing the coherence transfer pathway to include the carbonyl spin can be used. The HN(CO)CACB experiment [31, 87] correlates the amide proton and amide nitrogen of amino acid i with the α-carbon and β-carbon shifts of amino acid i–1. It should be noted here that at very high fields, because of the dependence of the chemical shift anisotropy relaxation mechanism on the square of the field strength, transverse relaxation of carbonyl nuclei may become faster than that of deuterated carbons.

Since in perdeuterated proteins all side-chain protons are exchanged with deuterons, experiments to assign side-chain resonances are restricted to assignments of the carbon chemical shifts. Although these chemical shifts are very useful to identify amino acid types and can therefore support the above-mentioned backbone assignment experiments, they contain only very limited structural information. For that reason, the assignment process of perdeuterated proteins usually focuses only on the backbone resonances and uses the chemical shift index to identify the secondary structure. Expert systems like the program TALOS [60] that was developed by the Bax group may be used for that purpose. The latter tries to match chemical shifts and residue types in the sequence to those from structure/chemical shift databases. If the secondary chemical shifts coincide for a larger number of segments in the data base ϕ and ψ restraints for structure calculations are derived from the data base.

However, if side-chain carbon assignments are wanted, C(CC)(CO)NH experiments [33] that start directly with carbon magnetization and transfer it further to the amide proton for detection are available. If protonated substituents, for example methyl groups, have been introduced into the otherwise perdeuterated protein, the usual HC(C)(CO)NH-TOCSY pulse sequence can be used to obtain the proton chemical shifts. These protons can provide a small number of NOEs that, together with residual dipolar couplings and the secondary structure identification from chemical shifts, make the determination of the global fold of large proteins possible.

The experiments described in this chapter aim at assigning the complete protein. Sometimes, however, assignments are required only for a particular part of the sequence, e.g. for residues of the active site. For this case, special experiments have been designed that are aimed at assigning only a subset of resonances in the region of interest (see Chapt. 21).

4.5 Assignment Procedures

Once all necessary experiments have been recorded, the information contained in them has to be analyzed. As mentioned above, the sequential backbone assignment is achieved by searching for matching chemical shifts. For the actual search it is most efficient to reduce the three-dimensional space of a three-dimensional experiment to a two-dimensional problem. This is done by cutting out narrow strips around every amide proton position along the indirect carbon dimension (triple resonance experiments) or indirect proton dimension (^{15}N-edited NOESY-HSQC). These strips can then be compared and sorted according to their assignment. Tools to create these strips and to sort them are contained in basically every software package for NMR spectra analysis.

Usually this strip sorting is performed manually. However, frequency matching is – at least in principle – amenable to computer automation. Several different research groups have developed software packages in recent years that take several triple and double reso-

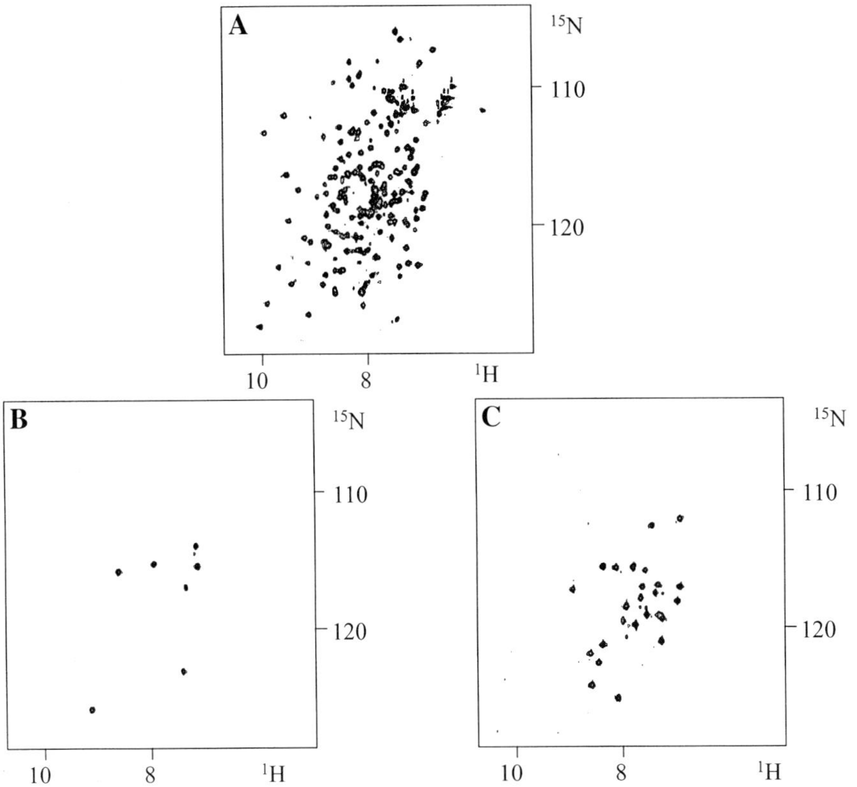

Fig. 4.4 Example of amino acid-type selective labeling. **A** [^{1}H,^{15}N]-HSQC spectrum obtained with the fully ^{15}N-labeled monomeric form of the KSHV protease. **B** and **C** [^{1}H,^{15}N]-HSQC spectra measured with the same parameters as in **A** with a sample of the KSHV-protease that was selectively labeled on phenylalanines and leucines respectively.

nance NMR spectra and the protein sequence as input and perform the backbone assignment fully automatically by searching for the correlations expected for the particular pulse experiments used [88–94]. While for small or medium sized proteins and spectra of high quality the results of these automatic procedures are very encouraging, spectral artifacts and missing or extra peaks caused by multiple conformations can cause problems, in particular with spectra of larger proteins that show more peak overlap. In these cases a trained spectroscopist is still superior to automatic assignment routines. If additional information is known, for example the amino acid type of at least some of the peaks, these automatic assignment algorithms usually become more reliable. Amino acid type selective labeling can be achieved by feeding certain auxotrophic bacterial strains with ^{15}N-labeled amino acids (Fig. 4.4) [95–98]. Although this method provides important information that can make automatic assignment procedures more reliable, it has so far not often been used. The main problem was the need to purify a separate sample of each selectively ^{15}N-labeled amino acid. However, several research groups have since demonstrated that spectra without background signals can be obtained from crude bacterial lysates containing the overexpressed protein [5, 99], which allows the researcher to identify the chemical shifts of several different amino acids without protein purification in a very efficient way [98]. In the future, further improvement of NMR technologies, including tools to use in vitro transcription/translation systems to produce NMR samples [100] and methods for segment-selective labeling [101–104] will help to make the assignment more efficient and more automated, even for larger proteins.

4.6 References

1 Freund, J.; Kellner, R.; Houthaeve, T.; Kalbitzer, H. R. *Eur. J. Biochem.* **1994**, *221*, 811–819.
2 Gitti, R. K.; Lee, B. M.; Walker, J.; Summers, M. F.; Yoo, S.; Sundquist, W. I. *Science* **1996**, *273*, 231–235.
3 Greenfield, N. J. *Anal. Biochem.* **1996**, *235*, 1–10.
4 Woody, R. W. *Methods Enzymol.* **1995**, *246*, 34–71.
5 Gronenborn, A. M.; Clore, G. M. *Protein Sci.* **1996**, *5*, 174–177.
6 Wyatt, P. J. *Anal. Chim. Acta* **1993**, *272*, 1–40.
7 Hensley, P. *Structure* **1996**, *4*, 367–373.
8 Laue, T. M.; Stafford, W. F. *Annu. Rev. Biophys. Biomol. Struct.* **1999**, *28*, 75–100.
9 Bagby, S.; Tong, K. I.; Liu, D.; Alattia, J. R.; Ikura, M. *J. Biomol. NMR* **1997**, *19*, 279–282.
10 Lepre, C. A.; Moore, J. M. *J. Biomol. NMR* **1998**, *12*, 493–499.
11 Bagby, S.; Tong, K. I.; Ikura, M. *Methods Enzymol.* **2001**, *339*, 20–41.
12 Wüthrich, K.; Wagner, G. *J. Mol. Biol.* **1979**, *130*, 1–18.
13 Grzesiek, S.; Bax, A. *J. Am. Chem. Soc.* **1993**, *115*, 12593–12594.
14 Mori, S.; Abeygunawardana, C.; Johnson, M. O. N.; Zijl, P. C. M. v. *J. Magn. Reson. B* **1995**, *108*, 94.
15 Timasheff, S. N. *Annu. Rev. Biophys. Biomol. Struct.* **1993**, *22*, 67–97.
16 Timasheff, S. N. *Biochemistry* **1992**, *31*, 9857–9864.
17 Bagby, S.; Tong, K. I.; Ikura, M. *Methods Enzymol.* **2001**, *339*, 20–41.
18 Kremer, W.; Kalbitzer, H. R. *Methods Enzymol.* **2001**, *339*, 3–19.
19 Horita, D. A.; Farnsworth, D. W.; Byrd, R. A. *J. Biomol. NMR* **2000**, *16*, 339–342.
20 Matthews, S. J.; Leatherbarrow, R. J. *J. Biomol. NMR* **1993**, *3*, 597–600.
21 Wider, G.; Wüthrich, K. *Curr. Opin. Struct. Biol.* **1999**, *9*, 594–601.
22 Wüthrich, K. *NMR of Proteins and Nucleic Acids*; John Wiley & Sons: New York, 1986.
23 Bax, A. *Curr. Opin. Struct. Biol.* **1994**, *4*, 738–744.
24 Gilquin, B.; Lecoq, A.; Desne, F.; Guenneugues, M.; Zinn-Justin, S.; Menez, A. *Proteins* **1999**, *34*, 520–532.
25 Berndt, K. D.; Güntert, P.; Wüthrich, K. *J. Mol. Biol.* **1993**, *234*, 735–750.
26 Lüginbuhl, P.; Wu, J.; Zerbe, O.; Ortenzi, C.; Luporini, P.; Wüthrich, K. *Protein Sci.* **1996**, *5*, 1512–1522.
27 Liu, A.; Lüginbuhl, P.; Zerbe, O.; Ortenzi, C.; Luporini, P.; Wüthrich, K. *J. Biomol. NMR* **2001**, *19*.
28 Ikura, M.; Kay, L. E.; Bax, A. *Biochemistry* **1990**, *29*, 4659–4667.
29 Grzesiek, S.; Anglister, J.; Ren, H.; Bax, A. *J. Am. Chem. Soc.* **1993**, *115*, 4369–4370.
30 Yamazaki, T.; Lee, W.; Revington, M.; Mattiello, D. L.; Dahlquist, F. W.; Arrowsmith, C. H.; Kay, L. E. *J. Am. Chem. Soc.* **1994**, *116*, 6464–6465.
31 Yamazaki, T.; Lee, W.; Arrowsmith, C. H.; Muhandiram, D. R.; Kay, L. E. *J. Am. Chem. Soc.* **1994**, *116*, 11655–11666.
32 Shan, X.; Gardner, K. H.; Muhandiram, D. R.; Rao, N. S.; Arrowsmith, C. H.; Kay, L. E. *J. Am. Chem. Soc.* **1996**, *118*, 6570–6579.
33 Farmer, B. T.; Venters, R. A. *J. Am. Chem. Soc.* **1995**, *117*, 4187–4188.
34 Farmer, B. T.; Venters, R. *J. Biomol. NMR* **1996**, *7*, 59–71.
35 Shirakawa, M.; Walchli, M.; Shimizu, M.; Kyogoku, Y. *J. Biomol. NMR* **1995**, *5*, 323–326.

36 Matsuo, H.; Kupce, E.; Li, H.; Wagner, G. *J. Magn. Reson. B* **1996**, *111*, 194–198.

37 Matsuo, H.; Li, H.; Wagner, G. *J. Magn. Reson. B* **1996**, *110*, 112–115.

38 Nietlispach, D.; Clowes, R.T.; Broadhurst, R.W.; Ito, Y.; Keeler, J.; Kelly, M.; Ashurst, J.; Oschkinat, H.; Domaille, P.J.; Laue, E.D. *J. Am. Chem. Soc.* **1996**, *118*, 407–415.

39 Lin, Y.; Wagner, G. *J. Biomol. NMR* **1999**, *15*, 227–239.

40 Henning, M.; Ott, D.; Schulte, P.; Lowe, R.; Krebs, J.; Vorherr, T.; Bermel, W.; Schwalbe, H.; Griesinger, C. *J. Am. Chem. Soc.* **1997**, *119*, 5055–5056.

41 Huth, J.R.; Bewley, C.A.; Jackson, B.M.; Hinnebusch, A.G.; Clore, G.M.; Gronenborn, A.M. *Protein Sci.* **1997**, *6*, 2359–2364.

42 Zhou, P.; Lugovsky, A.A.; Wagner, G. *J. Biomol. NMR* **2001**, *20*, 11–14.

43 Sklenar, V.; Bax, A. *J. Magn. Reson.* **1987**, *74*, 469–479.

44 Wider, G. *Prog. Nucl. Magn. Reson. Spectrosc.*. **1998**, *32*, 193–275.

45 Kay, L.E.; Ikura, M.; Tschudin, R.; Bax, A. *J. Magn. Reson.* **1990**, *89*, 496–514.

46 Grzesiek, S.; Bax, A. *J. Magn. Reson.* **1992**, *96*, 432–440.

47 Farmer, B.T.; Venters, L.D.; Spicer, L.D.; Wittekind, M.G.; Müller, L. *J. Biomol. NMR* **1992**, *2*, 195–202.

48 Bax, A.; Ikura, M. *J. Biomol. NMR* **1991**, *1*, 99–104.

49 Muhandiram, D.R.; Kay, L.E. *J. Magn. Reson. B* **1994**, *103*, 203–216.

50 Clubb, R.T.; Thanabal, V.; Wagner, G. *J. Magn. Reson.* **1992**, *97*, 213–217.

51 Grzesiek, S.; Bax, A. *J. Am. Chem. Soc.* **1992**, *114*, 6291–6293.

52 Grzesiek, S.; Bax, A. *J. Magn. Reson.* **1992**, *99*, 201–207.

53 Talluri, S.; Wagner, G. *J. Magn. Reson. B* **1996**, *112*, 200–205.

54 Vuister, G.W.; Bax, A. *J. Am. Chem. Soc.* **1993**, *115*, 7772–7777.

55 Wishart, D.S.; Sykes, B.D.; Richards, F. M. *Biochemistry* **1992**, *31*, 1647–1651.

56 Metzler, W.J.; Constantine, K.L.; Friedrichs, M.S.; Bell, A.J.; Ernst, E.G.; Lavoie, T.B.; Mueller, L. *Biochemistry* **1993**, *32*, 13818–13829.

57 Gronenborn, A.M.; Clore, G.M. *J. Biomol. NMR* **1994**, *4*, 455–458.

58 Ösapay, K.; Case, D.A. *J. Biomol. NMR* **1994**, *4*, 215–230.

59 Wishart, D.S.; Sykes, B.D. *Methods Enzymol.* **1994**, *239*, 363.

60 Cornilescu, G.; Delaglio, F.; Bax, A. *J. Biomol. NMR* **1999**, *13*, 289–302.

61 Braunschweiler, L.; Ernst, R.R. *J. Magn. Reson.* **1983**, *53*, 521–528.

62 Bax, A.; Davis, D.G. *J. Magn. Reson.* **1985**, *65*, 355–360.

63 Bax, A.; Clore, G.M.; Gronenborn, A.M. *J. Magn. Reson.* **1990**, *88*, 425–431.

64 Olejniczak, E.T.; Xu, R.X.; Fesik, S.W. *J. Biomol. NMR* **1992**, *2*, 655–659.

65 Kay, L.E.; Xu, G.Y.; Singer, A.U.; Muhandiram, D.R.; Forman-Kay, J.D. *J. Magn. Reson. B* **1993**, *101*, 333–337.

66 Grzesiek, S.; Anglister, J.; Bax, A. *J. Magn. Reson. B* **1993**, *101*, 114–119.

67 Logan, T.M.; Olejniczak, E.T.; Xu, R.; Fesik, S.W. *FEBS Lett.* **1992**, *314*, 413–418.

68 Montelione, G.T.; Lyons, B.A.; Emerson, S.D.; Tashiro, M. *J. Am. Chem. Soc.* **1992**, *114*, 10974–10975.

69 Gardner, K.H.; Konrat, R.; Rosen, M.K.; Kay, L.E. *J. Biomol. NMR* **1996**, *8*, 351–356.

70 Grzesiek, S.; Bax, A. *J. Biomol. NMR* **1993**, *3*, 185–204.

71 Grzesiek, S.; Bax, A. *J. Am. Chem. Soc.* **1995**, *117*, 6527–6531.

72 Yamazaki, T.; Kay, J.D.; Kay, L.E. *J. Am. Chem. Soc.* **1993**, *115*, 11054–11055.

73 Zerbe, O.; Szyperski, T.; Ottiger, M.; Wüthrich, K. *J. Biomol. NMR* **1996**, *7*, 99–106.

74 Dötsch, V.; Oswald, R.E.; Wagner, G. *J. Magn. Reson.* **1996**, *110*, 304–308.

75 Dötsch, V.; Wagner, G. *J. Magn. Reson. B* **1996b**, *111*, 310–313.

76 Feng, W.; Rios, C.B.; Montelione, G.T. *J. Biomol. NMR* **1996**, *8*, 98–104.

77 Rios, C.B.; Feng, W.; Tashiro, M.; Shang, Z.; Montelione, G.T. *J. Biomol. NMR* **1996**, *8*, 345–350.

78 Schubert, M.; Smalla, M.; Schmieder, P.; Oschkinat, H. *J. Magn. Reson.* **1999**, *141*, 34–43.

79 Schubert, M.; Oschkinat, H.; Schmieder, P. *J. Magn. Reson.* **2001**, *148*, 61–72.

80 Pervushin, K.; Riek, R.; Wider, G.; Wüthrich, K. *Proc. Natl. Acad. Sci. USA* **1997**, *94*, 12366–12371.

81 Pervushin, K.; Riek, R.; Wider, G.; Wüthrich, K. *J. Am. Chem. Soc.* **1998**, *120*, 6394–6400.

82 Pellechia, M.; Sebbel, P.; Herrmanns, U.; Wüthrich, K.; Glockshuber, R. *Nat. Struct. Biol.* **1999**, *6*, 336–339.

83 Goto, N. K.; Gardner, K. H.; Mueller, G. A.; Willis, R. C.; Kay, L. E. *J. Biomol. NMR* **1999**, *13*, 369–374.

84 Bax, A.; Kontaxis, G.; Tjandra, N. *Methods Enzymol.* **2001**, *339*, 127–174.

85 Tolman, J. R.; Flanagan, J. M.; Kennedy, M. A.; Prestegard, J. H. *Proc. Natl. Acad. Sci. USA* **1995**, *92*, 9279–9283.

86 Wittekind, M.; Mueller, L. *J. Magn. Reson. B* **1993**, *101*, 201–205.

87 Dötsch, V.; Matsuo, H.; Wagner, G. *J. Magn. Reson. B* **1996**, *112*, 95–100.

88 Buchler, N. E. G.; Zuiderweg, E. R. P.; Wang, H.; Goldstein, R. A. *J. Magn. Reson.* **1997**, *125*, 34–42.

89 Leutner, M.; Gschwind, R. M.; Liermann, J.; Schwarz, C.; Gemmecker, G.; Kessler, H. *J. Biomol. NMR* **1998**, *11*, 31–43.

90 Lukin, J. A.; Grove, A. P.; Talukdar, S. N.; Ho, C. *J. Biomol. NMR* **1997**, *9*, 151–166.

91 Moseley, H. N. B.; Monleon, D.; Montelione, G. T. *Meth. Enzymol.* **2001**, *339*, 91–108.

92 Zimmerman, D. E.; Montelione, G. T. *Curr. Opin. Struct. Biol.* **1995**, *5*, 664–673.

93 Zimmerman, D. E.; Kulikowski, C. A.; Huang, Y. P.; Feng, W. Q.; Tashiro, M.; Shimotakahara, S.; Chien, C. Y.; Powers, R.; Montelione, G. T. *J. Mol. Biol.* **1997**, *269*, 592–610.

94 Bartels, C.; Güntert, P.; Billeter, M.; Wüthrich, K. *J. Comput. Chem.* **1997**, *18*, 139–149.

95 LeMaster, D.; Richards, F. M. *Biochemistry* **1988**, *27*, 142–150.

96 Lee, K. M.; Androphy, E. J.; Baleja, J. D. *J. Biomol. NMR* **1995**, *5*, 93–96.

97 McIntosh, L. P.; Dahlquist, F. W. *Quart. Rev. Biophys.* **1990**, *23*, 1–38.

98 Ou, H. D.; Lai, H. C.; Serber, Z.; Dötsch, V. *J. Biomol. NMR* **2001**, *21*, 269–273.

99 Almeida, F. C. L.; Amorim, G. C.; Moreau, V. H.; Sousa, V. O.; Creazola, A. T.; Americo, T. A.; Pais, A. P. N.; Leite, A.; Netto, L. E. S.; Giordano, R. J.; Valente, A. P. *J. Magn. Reson.* **2001**, *148*, 142–146.

100 Yabuki, T.; Kigawa, T.; Dohmae, N.; Takio, K.; Terada, T.; Ito, Y.; Laue, E. D.; Cooper, J. A.; Kainosho, M.; Yokoyama, S. *J. Biomol. NMR* **1998**, *11*, 295–306.

101 Yamazaki, T.; Otomo, T.; Oda, N.; Kyogoku, Y.; Uegaki, K.; Ito, N.; Ishino, Y.; Nakamura, H. *J. Am. Chem. Soc.* **1998**, *120*, 5591–5592.

102 Otomo, T.; Ito, N.; Kyogoku, Y.; Yamazaki, T. *Biochemistry* **1999**, *38*, 16040–16044.

103 Xu, R.; Ayers, B.; Cowburn, D.; Muir, T. W. *Proc. Natl. Acad. Sci. USA* **1999**, *96*, 388–393.

104 Muir, T. W.; Sondhi, D.; Cole, P. A. *Proc. Natl. Acad. Sci. USA* **1998**, *95*, 6705–6710.

5
NMR of Membrane-Associated Peptides and Proteins

Reto Bader, Mirjam Lerch, and Oliver Zerbe

5.1
The Biochemistry of Membrane Interactions

5.1.1
Introduction

Membranes play a central role in all forms of life. They form parts of functional units, establish communication between the inside and outside of cells and between different cells, or are involved in the transport of molecules into and out of cells. Most fundamental biochemical functions involve membranes at some point [1, 2]. Whereas both prokaryotes and eukaryotes have a plasma membrane that separates the extracellular from the intracellular fluid, higher organisms additionally possess membranes that surround the organelles such as the nucleus, mitochondria and lysosomes. The plasma membrane additionally serves as a matrix for integral membrane proteins such as ion channels or membrane-embedded receptors, e.g. the G-protein coupled receptors (GPCRs). Many pharmacologically important drugs exhibit their biological action through binding to these.

Unfortunately, high-resolution NMR is presently incapable of elucidating structures of peptides binding to whole cells. The methods that have been developed to study this class of peptides or proteins have recently been reviewed in Refs. [3–9]. Table 5.1 contains an updated summary of structures of peptides determined in the presence of membrane mimetics by high-resolution NMR. We shall in the next sections describe models that have been successfully used in the past to determine the structure and dynamics of membrane-bound peptides. In order to understand which simplifications are inherent in the models used and what biological significance these differences have we will briefly review membrane biology and biophysics here. We shall see that most of the important differences concern the lipid composition, the functional groups and molecules presented on the surface, the curvature and charge of the membrane, and hence its membrane potential.

It is obvious why the spectroscopist wants to investigate the structure of integral membrane proteins or enzymes, whose biological action is linked to the presence of phospholipids such as phospholipase, in a membrane-mimicking environment. Why such an environment should also be used for other peptides like hormones becomes more clear when we take into account the membrane compartment theory [10–12] as postulated by R. Schwyzer. This theory states that peptides that target membrane-embedded receptors

Tab. 5.1 Summary of peptide structures determined in micelle systems

Class	*Peptide/ class*	*PDB No.*	*Size [AA]*	*Detergent*	*Result*	*Ref.*
Peptide hormones/ neuropeptides	Glucagon	n.a.	29	DPC	Two hydrophobic patches	96
	Motilin	n.a.	22	SDS	Hydrophobic N-terminus in micelle interior, amphiphilic α-helix at C-terminus parallel to surface	97
	Galanin	n.a.	30	SDS	Three turn-like regions at water-membrane interphase	98
	Calcitonin, Eel-CT, CT-M6	1BKU, 1BYV, 1BZB	32	SDS	Amphipathic helix and extended region	99, 100
	Parathyroid hormone (1-34)	n. a.	34	DPC	Two helices	101
	Peptide E	n. a.	25	SDS	Two beta-turns, intermediate order parameters	102
	Tubero-infundibular peptide	n. a.	39	DPC	Two helices	103
	Neuro-peptide Y	1F8P	36	DPC	Flexible N-terminus, amphiphilic C-terminal α-helix	83
	PACAP27	n. a.	27	DPC	Helix, flexible N-terminus	104
Membrane-perturbing toxins/ antibiotics	Melittin	n. a.	26	SDS	Bent helix	105
	Alamethicin	n.a.	20	SDS	Helix	106
	Trichorzianin		19	SDS	Two helices	107
	Tritrypticin	1D6X	13	SDS	Two turns, Tyr/Trp at membrane-water interphase	108
	Mastoparan	n. a.	14	DMPC/ DHPC	Helix	50
	Gramicidin A, B, C	1JNO, 1JO3, 1JO4	15	SDS	Helix	109
Bacteriophage coat proteins	M13 fd	2CPB, 2CPS	50	DPC, SDS (comparison)	Amphipathic helix at surface, membrane-spanning hydrophobic helix	95, 110, 111
	Ike		53	MPG	dito	112

Tab. 5.1 (continued)

Class	*Peptide/ class*	*PDB No.*	*Size [AA]*	*Detergent*	*Result*	*Ref.*
Virus coat	HIV gp41	1JAV	19	DPC	Membrane-associated helix	113
Fragment of integral membrane proteins	Bacterio-opsin (1-36), (1-71)	1BHA	36, 71	SDS	Hydrophobic helix	114
	Aβ(1-40), (1-42)	n.a.	40, 42	SDS	Membrane-associated amphiphilic helix	115
	Glycophorin A(62-101)	1AFO	80	DPC	Alpha-helical homodimer	116
	PTH receptor 1 (168-198)	1BL1	31	DPC	Three helices	101
	PTH receptor 1 (241-285)		45	DPC	Helix	117
Signal sequences	PhoE	n.a.	21	SDS	Helix-break-helix	118
Fragments of peripheral membrane proteins	N-myristoylated ARF (1-15)	n.a.	15	DPC/ bicelles	Amphipathic helix	119
Cell/organelle penetrating	Penetratin	n.a.	16	SDS	Helix	120
	TrFd chloroplast membrane transit peptide	n.a.	52	DPC, DPG (phosphoglycol)	Ill-defined helix, extended	121
Ligand-receptor-complexes	CCK-8/ CCKA-receptor (329-357)	1HZN	8/29	DPC	Asp/Arg-interactions	122
Varia	Apolipoprotein C-I	1IOJ	57	SDS	Two helices	123
	Apoligoprotein C-II	1I5J	79	SDS	Two helices	124
	β subunit of LH1	1JO5	48	Zwittergent 3:12	Two helices	125
	cPLA2-C2	n.a.	138	DPC/glycerophosphocholine	8 antiparallel beta-strands	126

associate with the membrane prior to receptor binding and that it is the membrane-bound conformation that is initially recognized by the receptor. As a consequence, parts of the ligands are optimized for membrane binding (the so-called address), other parts for receptor binding (the message). Hence, for all the classes of peptides or proteins listed above, useful structural information can only be expected when conditions in the NMR sample match the natural environment as closely as possible. Although solvent mixtures like chloroform/methanol may help to perfectly solubilize these peptides, their structure may be quite different from those of the membrane-bound form.

5.1.2
Biological Membranes

From X-ray studies it is known that many biological membranes are 5 nm thick [13, 14]. One of the first generalized and today well-accepted views of cell membranes is described by a fluid, dynamic mosaic of alternating globular proteins and phospholipid bilayer constituting a two-dimensional oriented viscous solution [15]. For comparison, the apparent effective viscosity of the membrane fluid phase is about 10^3 to 10^4 times that of water. The fluid mosaic model has been applicable to all functional membranes despite the tremendous diversity of membrane compositions and functions. Membranes usually exist in both the gel state and in the liquid crystalline state [16]. The former is characterized by a largely reduced mobility and is usually encountered at lower temperatures. The exact transition temperature depends on the lipid composition. Lateral diffusion for completely mobile molecules (such as the phospholipids themselves) is fast in the liquid crystalline state ($\sim 10^6\ s^{-1}$). Additionally, rapid rotation occurs about its long axis ($\sim 10^9\ s^{-1}$) [17–19]. In contrast, flip-flop transitions of lipids from the outer to the inner leaflet or vice versa are rare ($\sim 10^{-5}\ s^{-1}$) [18, 20, 21], depending on the protein composition in the cell membrane.

The major components of membranes comprise lipids, proteins and glycolipids or glycoproteins [1]. Biological membranes may contain between tens and hundreds of lipid species differing in the chemical nature of the head groups, chain lengths and degree of unsaturation. The lipids are long-chain aliphatic acids coupled to various moieties such as glycerol, sphingosine or ceramide. Common lipids are glycerophospholipids, glycoglycerolipids, phosphosphingolipids and sterols. They are either zwitterionic (neutral) or negatively charged. The exact composition of the membrane is tissue-dependent, and the cell maintains an asymmetry with respect to inner and outer leaflet through flipases, flopases and scramblases (see Kuypers et al. [22]). The choline-containing phospholipids sphingomyelin (SM) and phosphatidylcholine (PC) are mainly found in the outer monolayer, whereas the occurrence of most aminophospholipids [phosphatidylserine (PS) and phosphatidylethanolamine (PE)] is limited to the inner monolayer. The asymmetric distribution of charged and uncharged phospholipids contributes to the electric potential across the membrane, the so-called transmembrane potential, which varies between –20 and –200 mV in the resting state [23]. However, not only the head groups but also the fatty acid side-chain compositions are different in the two layers. Glycolipids and glycoproteins contain large heterosaccharide chains attached to the lipids or proteins and are believed to play a pivotal role in cell-cell recognition. Often, sialic acid is found at the terminus of the chain. The exact chemical nature of the membrane components as well as their concentration varies according to the cell function and organism as well as between the different organelle membranes.

Despite the weakness and short-range nature of protein-lipid and lipid-lipid interactions, cells have nevertheless evolved means of laterally assembling into membrane-microdomains. Sphingolipid-cholesterol rafts serve to recruit a specific set of membrane proteins and exclude others [24]. Caveolae are deeply invaginated raft domains that are stabilized by caveolin protein oligomers (binding cholesterol) [25].

It should be noted that plants, fungi and most bacteria have a well-defined cell wall, which covers the plasma membrane. The bacterial cell wall consists of a complex polypeptide-polysaccharide matrix referred to as the peptidoglycan. In gram-positive bacteria it is up to 20 layers thick (15–80 nm), whereas gram-negative bacteria have a much thinner peptidoglycan layer which is covered by the outer membrane, again a lipid bilayer. Because of their high osmotic pressure, plant cells are protected by a stable cell wall. It is mainly built up of carbohydrates, namely protopectin, hemicelluloses and cellulose. The fungal cell wall contains chitin instead of cellulose.

In animal cells, the integral membrane proteins and certain lipids bear short sugar chains that point outward from the plasma-membrane and form the so-called glycocalyx (or cell coat). These carbohydrates are thought to play an important role in mediating interactions with cells or the nonliving environment. Many animal cells are further surrounded by the so-called extracellular matrix (ECM) [26], which contains mainly fibrous proteins such as collagen, proteoglycans (protein-polysaccharide complexes), laminin and fibronectin. The latter is coupled to the plasma membrane through certain cell surface receptors such as integrins [27].

The cell surface additionally displays receptors responsible for cell-cell recognition [28]. Members of this class of receptors are selectins [29] that recognize specific carbohydrates from other cells in the presence of calcium. Other cell surface receptors belong to the immunoglobulin superfamily (IgSF) [30] that promote calcium-independent cell-cell adhesion. The third important class are the calcium-dependent cell adhesion molecules, the cadherins [31], which form dimers with cadherin molecules presented on the surfaces of other cells and hence promote aggregation of similar cell types.

From the information given above it is obvious that cell surfaces display an enormous complexity. A perfect model to study the interaction of a peptide with a biological membrane would require knowledge about the cell membrane composition in that particular tissue. Even if such information were available it will most probably not be possible to fully mimic the biological environment. However, some important aspects may still be studied with the available models. Whenever possible, one should try to relate the information derived from such a model to information gained from biological data taken on real cells (cell-lines) such as binding affinities etc. in order to prove the validity of the model for the study of a particular aspect.

5.1.2.1 Protein-Membrane Interactions

Peptides may bind to a membrane either by association to its surface or by insertion into its interior. The latter class comprises the integral membrane proteins whose structures are largely α-helical or β-barrel type (see Chap. 12 in Ref. [32]). The topology of interaction of helices with membranes is displayed in Fig. 5.1.

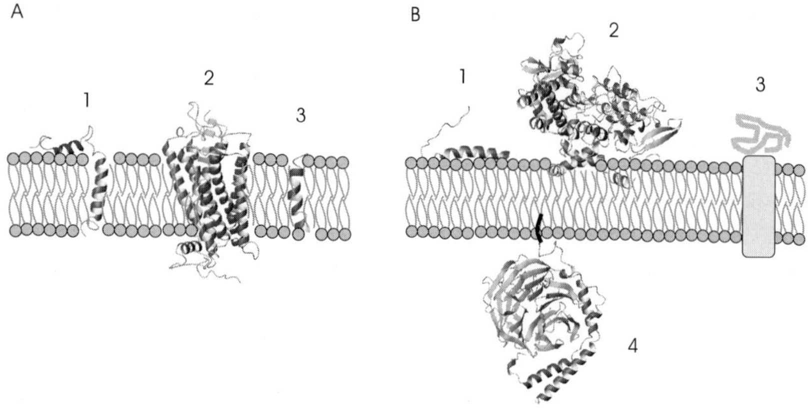

Fig. 5.1 Schematic drawing of membrane association modes of peptides: **A** Integral membrane proteins: (1) major fd coat protein gpVIII of bacteriophage M13 (pdb 1fdm), anchored by an 18-residue *trans*-membrane hydrophobic helix; (2) bovine rhodopsin, a 7 *trans*-membrane domain (G-protein-coupled) receptor (pdb 1f88); (3) ion channel peptaibol Chrysospermin C (pdb 1ee7), and **B** Peripheral membrane proteins: (1) neuropeptide Y (pdb 1f8p), associated via an amphipathic helix; (2) prostaglandin H_2 synthase-1 (pdb 1prh); membrane-anchoring domain consists of a right-handed spiral of amphipathic helices forming a planar motif; (3) schematic drawing of a protein associated to an integral membrane protein (4) transducin β-γ-subunits (pdb 1aor); the γ-subunit is farnesylated at cys-71, serving as membrane-anchor.

Binding of molecules to the membrane surface may be due to electrostatic or multiple hydrophobic interactions. Binding of proteins to exposed anionic phospholipids (mainly PS) is often characterized by low specificity and requires high anionic charge density (>10 mol%). The specific binding of proteins to individual phospholipid molecules requires low mol% and is clearly seen in the case of PI-binding PH domains [33]. For hydrophobic peptides that do not insert into the membrane interior, electrostatic attractions provided by anionic phospholipids become essential for peptide binding and insertion into membranes [34]. Interestingly, there is accumulated evidence for specific affinity of Trp and Tyr side chains for the water-membrane interface [35], a view which was recently supported by thermodynamic data [36].

An important question arises about the effects of phospholipid composition and the function of membrane-bound enzymes. The phospholipid composition and cholesterol content in cell membranes of cultured cells can be modified, either by supplementing the medium with specific lipids or by incubation with different types of liposomes. Direct effects of phospholipid structure have been observed on the activity of the Ca^{2+}-ATPase (due to changes in the phosphorylation and nucleotide binding domains) [37]. Evidence of a relationship between lipid structure and membrane functions also comes from studies with the insulin receptor [38]. Lipid alteration had no influence on insulin binding, but modified the kinetics of receptor autophosphorylation.

Protein binding to lipids also results in ordering of lipids. However, the effect is short-range, basically including only the first shell of phospholipids around the protein [39]. Interactions are mediated by electrostatic forces and by hydrogen bonding. Accordingly, the effects are selective for appropriate lipid head groups.

5.1.3
Aggregate Structures of Lipids and their Biophysics

Since detergents contain both hydrophobic tails and hydrophilic head groups, they form well-defined aggregates such that the head groups are exposed to the solvent and the hydrophobic chains are shielded from solvent access. Depending on the relative volumes of head group and aliphatic chains either spherical aggregates, micelles, or bilayers, either vesicles or bicelles, are formed. Micelles spontaneously form when head groups are more bulky than the aliphatic chains such that the overall appearance resembles a cone, e.g. dodecylphosphocholine (DPC) or sodium dodecylsulfate (SDS). In contrast, glycerol derivatives contain *two* coupled aliphatic chains and hence have a more rectangular shape (e.g. the phosphatidylcholine derivatives). The latter are most advantageously arranged as bilayers or bicelles, whereas the former rapidly assemble as micelles [1]. The chemical structures of the most prominent membrane model compounds particularly suitable for studies of protein/membrane interactions by NMR are displayed in Fig. 5.2.

5.1.3.1 **Micelles**

The classical picture of a micelle with a linear arrangement of the aliphatic chains extending radially from the center to form a sphere in which the surface is formed by the hydrophilic head groups leads to a higher atom density in the centre. Dill et al. have developed a statistical (lattice) model [40] in which the terminal carbon atoms may also be found more close to the surface. This model is more consistent with data from molecular dynamics simulation of fully solvated DPC [41, 42] or SDS [43]. In all simulations, C-12 is partially found close to the head groups and the order parameter extracted from the MD trajectories for C-H bond vectors decreases continuously on going from C-1 to C-12. The chain mobility properties are more similar to the liquid crystalline phase observed for bilayers than to the gel phase. Accordingly, Beswick et al. [44] concluded from their relaxation measurements that the dynamical behavior of phosphocholine groups from DPC micelles at 12 °C corresponds to the phosphatidyl head groups of natural lipids at 51 °C. Interestingly, also surface-associated phospholipids were encountered during the simulation. The correlation time for overall tumbling of a DPC micelle was estimated by Beswick et al. [44], based on a micelle radius of 2.3 nm [45], to be 11 ns at 25 °C and 17 ns at 12 °C.

Water molecules are absent from the hydrophobic interior, but both the choline and the phosphate headgroups are fully solvated [41]. Similarly, the first hydration shell of the sulfate headgroup of SDS is formed rather by water molecules than by sodium ions. Because of hydration the charge density due to the lipid headgroups is overcompensated by the water dipoles, thereby reducing the transmembrane potential by 50–100 mV across the lipid water interface and resulting in a negative potential at the aqueous side [42].

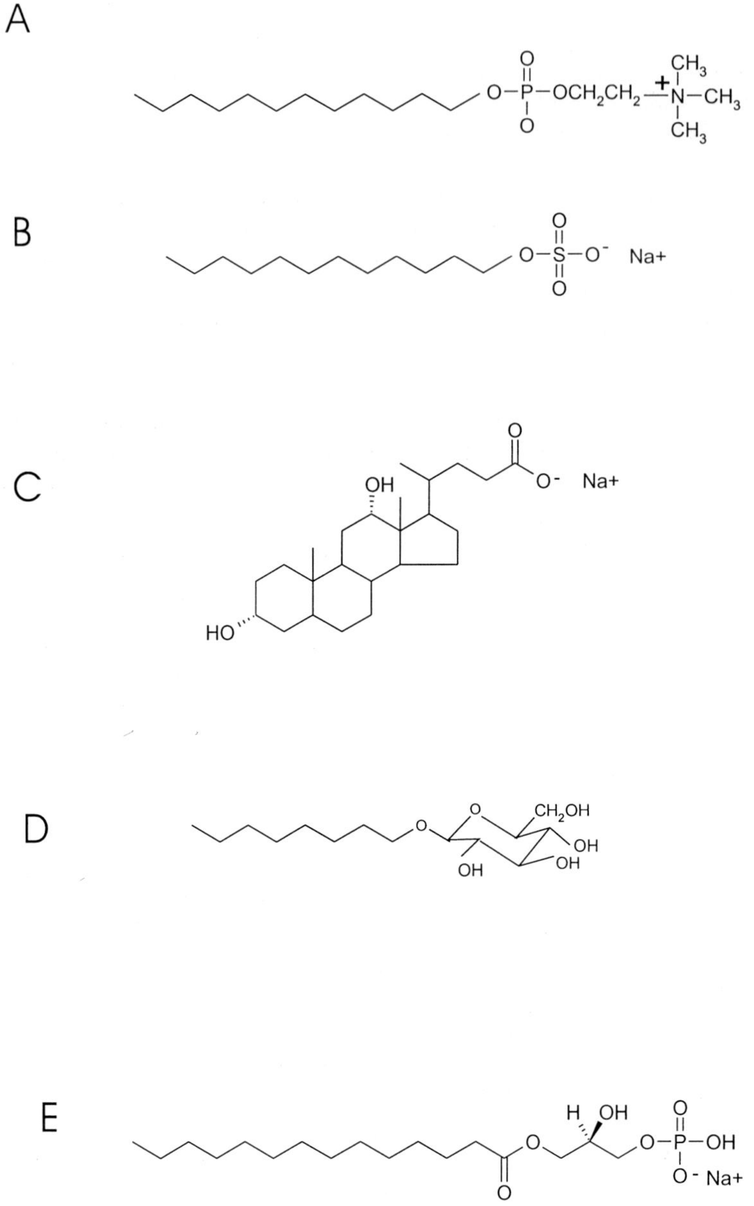

Fig. 5.2 Chemical structures of selected membrane model compounds: **A** dodecylphosphocholine (DPC), **B** sodium dodecylsulfate (SDS), **C** sodium taurodeoxylcholate (similar to CHAPS), **D** *n*-octyl-*β*-D-glucopyranoside and **E** lyso-dodecyl-phosphatidyl-glycerol.

5.1.3.2 **Bicelles**

Bicelles are disc-shaped, mixed micelles [46] composed of bilayered long-chain phospholipids such as dimyristoylphosphatidylcholine (DMPC) surrounded by a rim of short-chain detergents such as dihexanoylphosphatidylcholine (DHPC) or bile salt-like CHAPSO. Their use as a membrane model was introduced by Prestegard in 1988 [47], and new applications were reviewed by Sanders et al. [48]. The ratio of [DMPC] to [DHPC] is denoted as q and determines the diameter of the bicelles [49]. The DMPC-containing bilayer regions are conformationally and dynamically similar to DMPC in liquid crystalline phases. Recent interest in these systems was driven by the fact that solutions with $2<q<10$ containing 2–40% (w/w) phospholipids form a nematic phase over a certain temperature and pH range and the bicelles in such a phase become aligned in the presence of a magnetic field (see Chapt. 8). If peptides bind to them tightly they can only be used for solid-state NMR studies because of the large size of the disks. However, Vold [50] realized that when q is reduced to 0.5 *isotropic* solutions of bicelles [51] are generated for which bound peptides display line widths comparable to those from micelle-associated peptides in spite of its comparatively large diameter (approx. 80). This is a consequence of the high mobility of the phospholipids within the bilayer. The viscosity of the resulting solution is only twice that of water at 35 °C. In contrast to micelles, bicelles form (almost) true bilayers, and surface curvature is much less an issue in these systems. Bicelles may also be doped with up to 25% phosphatidylserine (PS) or phosphatidylglycerol (PG), resulting in negatively charged bicelles with very similar biophysical properties [52]. For the peptide mastoparan X the additional charges promoted a change in the membrane-association topology, changing from a parallel to a perpendicular orientation with respect to the bicelle surface [53].

5.1.3.3 **Vesicles or Liposomes**

Vesicles or liposomes are spherical compartments of bilayered/multilayered phospholipids enclosing a liquid (usually water). Because of their considerable size compared to micelles or bicelles the NMR study of peptides binding to them can only be performed in the solid state. However, they have been very successfully employed by Prestegard to investigate the transport of organic molecules through membranes [54]. In his elegant approach, resonances from molecules outside the vesicles were broadened through spin labels and could thereby be distinguished from those that had permeated into the interior.

For the three membrane models micelles, bicelles and vesicles (liposomes) it is of prime importance to know whether the peptides or proteins remain structurally intact and biologically active in that environment. In general, the similarity to biological membranes decreases in the order: vesicles > bicelles > mixed micelles > micelles. Functional reconstitution of membrane proteins in membrane mimetics is a well-known problem to biochemists but their successful candidate for a specific protein may not present an environment suitable for high-resolution NMR. The success also depends on the class of peptide/protein to be studied. For surface-associated peptides, micelles containing head groups similar to biological membranes will be a good mimic as long as the required surface for interaction is not too large such that the surface curvature will become a problem. The question to what extent micelle binding induces curvature was recently addressed by Chou et al. [55], who compared the structure of a peptide fragment of the HIV-1 envelope protein gp41 when bound to micelles with that of the bicelle-bound form using dipolar couplings. In their work, a clear curvature of the micelle-associated helix was observed with the hydrophobic residues pointing to the inside. In contrast, a nearly straight helix was found in the case of the bicelle-associated species. Even for membrane-integrating peptides micelles may still be useful. Arseniev et al. was able to show that peptides comprising the amino acid sequence from transmembrane parts of bacterioopsin form intact helices in SDS micelles [56]. A number of publications tested detergents for their ability to retain protein function [57–59], but the used concentrations of detergents were mostly well below those required for structural studies. A recent publication of Sanders specifically addresses that question for detergents used in high-resolution NMR studies [60]. In the case of gramicidin A, Ketchem et al. discovered by using solid-state NMR techniques that the channel-forming peptide when bound to bilayers is structurally slightly different from that of the micelle-bound form [61] and partly attributes the difference to the curvature of the micelle. For that reason bicelles are expected to better mimic biological membranes. The integral membrane protein *E.coli* diacylglycerol kinase was successfully reconstituted into bicelles [62].

5.2 The NMR Sample

5.2.1 Synthesis of Peptides and Proteins

Solid-phase peptide synthesis offers a fast and convenient route for many peptides when isotope-enriched compounds are not required. Classical synthesis additionally permits the use of non-natural amino acids and allows site-specific isotope labeling. Although Fmoc protected ^{15}N-labeled amino acids are commercially available, the cost of such synthesis is usually prohibitive, and the peptides from chemical synthesis require perdeuterated detergents and unfortunately exclude investigation of internal dynamics through measurement of ^{15}N relaxation.

In solid-phase synthesis [63, 64], peptides are mainly synthesized by applying Fmoc/tert-butyl or Boc/benzyl protection group strategy on a robot. Coupling requires activating reagents such as 1,3-diisopropylcarbodiimide (DIC) and 1-hydroxybenzotriazole (HOBt). Fmoc/tert-butyl strategy – representing the more convenient method – allows simultaneous removal of all side-chain protecting groups and cleavage of the peptide from the resin with trifluoroacetic acid. Alternatively, the peptide may be liberated from the resin under mildly acidic conditions, keeping the side chains protected. Cleavage from Wang or SASRIN resins will yield peptides with the free carboxy terminus, whereas the so-called Rink-amide resin will lead to C-terminal amides. In the last coupling step, the peptides may also be N-terminally modified, for example, by an acetyl group. The success of the coupling steps and hence the purity of the peptides will largely depend on the amino acid sequence, whereas hydrophobic peptides tend to be more difficult to synthesize and purify. Water-soluble peptides can frequently be purified with a single reversed-phase (RP)-C-18 HPLC run using an acetonitrile/water gradient.

Smaller peptides are difficult to overexpress in *E. coli* and are additionally degraded in cells. These problems are commonly circumvented if fusion peptides are used from which the peptide of interest is cleaved biochemically. These constructs, e.g. the presently popular GB1 domain of the streptococcal protein G [65], may also help to solubilize the peptide and thereby improve the expression yield. We have successfully used the system introduced by Kohno [66] in which the peptide of interest is fused to decahistidine-tagged ubiquitin. The peptide is expressed in either M9 minimal medium containing ^{15}N-NH_4Cl (and ^{13}C-glucose) or isotope-enriched rich media. Separation of the fusion peptide from others is achieved by Ni-affinity chromatography, after which the peptide is liberated from ubiquitin through the enzyme ubiquitin hydrolase. To achieve C-terminal amidation for recombinantly produced peptides an additional gly codon is added to the cDNA. After expression the Gly residue is enzymatically converted into the amide function through the peptidylglycine α-amidating enzyme [67]. Again, the resulting peptide can be purified with RP C-18 HPLC if water solubility is sufficient.

5.2.2 Choice of Detergent

Detergents commonly used to form micelles that are amenable to high-resolution NMR are summarized in Tab. 5.2, and the chemical structures of the most commonly used detergents are presented in Fig. 5.2. Unfortunately, only a few of those needed for use with nonisotopically enriched peptides are commercially available in deuterated form. Most frequently, the zwitterionic DPC or the negatively charged SDS have been used as membrane mimetics. Mixtures of DPC "doped" with small amounts of SDS may be used to modulate the charge distribution on the micelle surface. It should be emphasized here

Tab. 5.2 Commonly used detergents for work with micelles including some important biophysical properties

Detergent (MW)	***CMC [mM]***			***Aggregation number [N]***		
	0–0.05 M Na⁺	***0.1–0.2 M Na⁺***	***Ref.***	***0–0.1 M Na⁺***	***0.1–0.2 M Na⁺***	***Ref.***
SDS (288, 313[a])	3–10	0.9–2	127–129	60–75	100–150	129
DPC (351, 389[a])	1.1		45	56		45
Deoxycholate sodium (489)	2–6	1–4	127, 130–132	1.5–12	5–25	132, 133
β-D-Octylglucoside (292)	20–25	19–25	127, 128	84		134
β-D-Decylglucoside (322)	2–3	2.0	Calbiochem manual			
Zwittergent 3:12 (336)	2–4	1.4	127, 135, 136	55		136
MPG (myristoyl-lyso-phosphatidyl-glycerol)	n.a.	n.a.				
CHAPS (615)	6–10	3–5	127, 137, 138,	4–14		137, 138
CHAPSO (631)	8.0	4.1	137			

[a]) fully deuterated form

that SDS is a denaturing detergent (in contrast to CHAPS, DPC etc.) in which many membrane-bound peptides/enzymes lose their biological activity. Furthermore, the head group is not encountered in biological systems, it would hence seem to be much inferior to DPC, and is only recommended as an additive. Similarly to SDS, phosphatidylserine is negatively charged and hence presents a suitable additive for modulating the surface charge of the micelle, thereby influencing electrostatic interactions with charged peptides. CHAPSO, similarly to cholesterol, contains the rigid steroid skeleton and hence may influence the membrane rigidity substantially. It is important to note that many detergents such as β-octylglycoside or CHAPSO are not available in perdeuterated form and require ^{13}C,^{15}N doubly labeled peptides/proteins for structural studies. When working with isotopically (possibly ^{15}N,^{13}C) enriched peptides, the need for deuteration is avoided, and the above-mentioned detergents may be added.

Other alternatives are the lyso-phosphatidylglycerol derivatives in which one of the fatty acid chains coupled to glycerol have been removed. Unfortunately, the phosphatidyl lipids are not available in fully deuterated form, and hence certain regions of the proton spectra will be covered by the glycerol protons.

Opella et al. [3] have shown that the concentration of the detergent must be well above the critical micelle concentration. Ideally, the concentration should be such that each micelle is occupied on average by one peptide molecule. Otherwise, aggregation effects will lead to species with different aggregation states so that signal broadening or doubling may result. However, it has been argued that this effect is less important for smaller peptides than for larger proteins whose molecular weight is similar to that of the complete micelle. For very high concentrations of detergent (>500 mM) both viscosity and aggregation number increase rapidly. In the case of (charged) SDS, the ionic strength of the solution increases to high values so that sensitivity suffers, pulse-lengths increase, and sample heating during decoupling or spin-locking occurs.

One positive side-effect of working with detergent solutions is that peptide samples tend to be stable over longer periods at higher temperatures.

5.2.3
Choice of pH

In order to match physiological conditions, the selected pH should be close to neutral. However, because of elevated amide proton exchange at that pH for peptides with solvent-accessible amide protons, a value in the range 3–5 is usually chosen, also depending on the solubility at the pH. For globular peptides, most amide protons are involved in rather stable hydrogen bonds or are protected from solvent access and a much more basic pH may be used.

Amide proton exchange is both base- and acid-catalyzed. The mechanism for base catalysis is clear and involves abstraction of the proton from the amide nitrogen forming an imidate anion. During acid catalysis the proton may either attack the nitrogen or the oxygen and depends on the electron-withdrawing/donating capabilities of nearby groups. Titration curves were established for labile protons in peptides a long time ago [68] (for reviews see Refs. [69, 70]). It soon became obvious in studies of micelle-bound peptides that exchange kinetics are different in that environment. O'Neil et al. have determined pH titration profiles for M13 Coat protein in the presence of SDS micelles [71] and found that the minimum is shifted by 1–1.5 pH units. They also discovered that the apparent pH in the vicinity of the surface is increased by 2 pH units compared to the bulk solution. Spyrocopolous et al. determined the profiles for a set of model tripeptides [72] and concluded from their data that highly hydrophobic peptides that are likely to be placed more deeply in the hydrophobic interior display exchange kinetics reduced at least 25-fold. Perrin et al. have attributed the differences in amide proton exchange to the extent by which the transition state during catalysis, e.g. the positively or negatively charged species, are stabilized by the environment [73]. Accordingly, the pH minimum and k_{min} depend on the type of detergent used and hence will be different for positively charged, negatively charged or neutral micelles. All other effects, e.g. counterions, the head groups, the concentration of the surfactant etc., influence amide proton exchange to a much smaller extent.

The pH of course will also influence the protonation state of side chains, and hence it will possibly influence both the structure of the peptide and the affinity of the peptide toward the micelle. The latter can also determine the exchange kinetics between the free and the micelle-bound state and may thereby largely influence the line widths encountered. In our view we recommend choosing the pH from a series of spectra. The pH should then be selected such that amide protons accessible to solvent are still observable. We generally found values in the pH range 5–6 gave good spectra when working with DPC micelles, assuming reasonable affinity toward the micelle.

Figure 5.3 displays a comparison of one-dimensional spectra of a neuropeptide in the presence of either DPC or SDS micelles at various pH values. Note that the signal at approx. 8.6 ppm due to HE1 of His vanishes in SDS at a much higher pH compared to DPC. In general, spectra in SDS still yield reasonable quality at neutral pH in contrast to those recorded in DPC.

5.2.4
Choice of Temperature

It is advantageous to set the temperature for the measurement of peptide/micelle or peptide/bicelle systems at higher values consistent with stability of the peptide. We found that values in the range 30–45 °C gave spectra with reasonable line widths. At these temperatures TOCSY spectra with mixing times of 80 ms still gave good S/N for the majority of correlations on 36 amino acid peptides. Amide proton exchange is still acceptable at pH 5–6 at these temperatures, and NOESY spectra display strong cross peaks for proximate protons. In contrast, H,D exchange is very fast on lyophilized samples of peptides with μM binding affinities to the membranes and hence required ^{15}N labeled peptides in order to record [^{15}N,^{1}H]-HSQC spectra. Only with these experiments is it possible to assign the exchange to specific sites within the available time (2–20 min).

A series of one-dimensional proton spectra of a neuropeptide recorded at various temperatures between 280 and 320 K are displayed in Fig. 5.4.

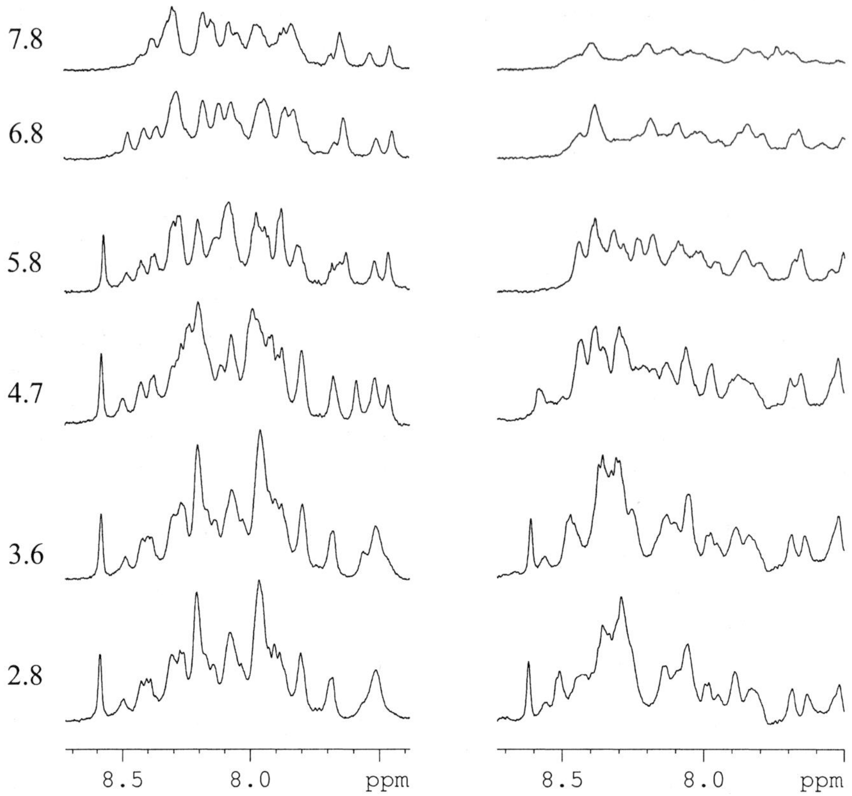

Fig. 5.3 Proton one-dimensional spectra displaying the region of amide resonances of 0.5 mM Ala31, Pro32-NPY recorded in 300 mM SDS (left) or 300 mM DPC (right) at various values of the pH.

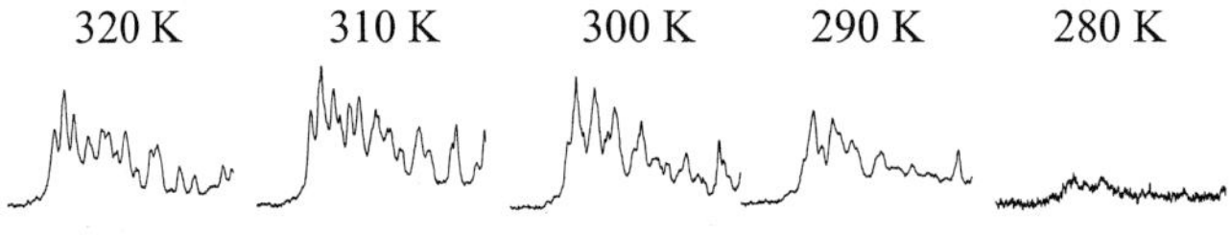

Fig. 5.4 Proton one-dimensional spectra displaying the region of amide resonances of 0.5 mM Ala31,Pro32-NPY in 300 mM DPC at various temperatures.

5.2.5
Salt Concentrations

High salt content favors higher aggregation numbers of the micelle, and hence one should try to work at low salt concentrations (<50 mM). Since peptide phospholipid interactions are often electrostatic in nature, salt may be added to modulate the binding affinity. This may be advantageous if exchange between the free and the membrane-bound state of the peptide is slow on the NMR time scale such that considerable line broadening occurs. We have found that addition of $CaCl_2$ at low concentrations (5 mM) reduced the binding affinity of neuropeptide Y toward the micelle considerably (see below) without deteriorating the physical properties of the sample significantly.

5.2.6 Practical Tips for Sample Preparation

One inherent property of peptides that interact with membranes is that self-association or even aggregation will interfere with solubilization by organic solvents or micelles. The preparation, purification and sample preparation of extremely hydrophobic (often transmembrane) peptides is nontrivial and has been addressed by only a few papers [74–79].

The first problem encountered once the peptide has been successfully synthesized is that standard purification protocols fail. Although very hydrophobic peptides are soluble in acids such as TFA, these harsh conditions are not suitable for purification, because they can reduce column life times and denature native protein structures. Hence residual acid has to be removed, and many peptides can then be redissolved in mixtures of water and *tert*-butanol. Peptides with a strong tendency to aggregate may be dissolved either in trifluoroethanol (TFE), hexafluoroisopropanol (HFIP), mixtures of 1-propanol and 1-butanol, 20% acetic acid or 70–90% formic acid.

Water-soluble peptides can easily be purified using standard C18 reverse-phase high pressure liquid chromatography running a water:acetonitrile gradient in the presence of 0.1% TFA. Again, difficulties arise with hydrophobic peptides. Firstly, it is critical to load the peptide onto the column in a deaggregated state applying one of the methods mentioned above. Secondly, standard RP-HPLC conditions may be inappropriate because of irreversible column binding or very broad peaks. However, subsititution of acetonitrile/water by either isopropanol:acetonitrile (2:3)/water (0.1% TFA) or 1-propanol:1-butanol (2:1)/water (0.1% TFA) and using a C4 or even C1 column improved the purification for a variety of hydrophobic peptides. To gain optimal yields it is important to avoid aggregation of the peptides after injection. Therefore, the starting solvent for the HPLC purification (i.e., the percentage of organic solvent/water) has to be adjusted according to the hydrophobicity of the peptide.

For hydrophobic peptides that are exclusively soluble in strong acids, Tomich [74] developed a purification scheme in which the peptide is cleaved from the resin in the presence of sodium dodecylsulfate. The precipitate is redissolved by dilute acetic acid and purified to homogeneity using a detergent-based HPLC protocol. At the same time, SDS is quantitatively exchanged by *n*-octyl-β-D-glucopyranoside. Alternatively, hydrophobic peptides may be purified by high-performace gel filtration using e.g. the Superdex Peptide HR 10/30 column (Pharmacia Biotech). With such columns isotropic elution with 70% formic acid is possible [79].

A general procedure has been reported by Killian et al. for the incorporation of hydrophobic peptides into micelles [75]. Therein, the peptides were first dissolved in TFA for deaggregation, dried under nitrogen and redissolved as 5 mM (clear) solutions in TFE or HFIP. The solution is subsequently diluted 1:1 by addition of an aqueous solution containing a varying SDS concentration (typically 500 mM, depending on the protein concentration). Water is then added to yield a 16:1 ratio of water to TFE (HFIP) by volume. Upon addition of excess water the peptide loses its solubility, but at the same time the micelles are re-formed, resulting in the homogeneous incorporation of the peptide. Vortexing and lyophilization by rapid freezing (liquid nitrogen) is followed by drying overnight under vacuum and redissolving in water to yield clear solutions.

A crucial prerequisite for the successful incorporation of peptides into micelles is that they persist in the deaggregated (mostly monomeric) form during the whole sample preparation procedure. Peptides have their lowest solubility at their pI, and therefore aggregation may be prevented by adjusting the pH of any aqueous solution (if possible by addition of buffers) such that the net charge of the peptide is at a maximum. A very nice example is reported by Zhang et al. [79] studying a 33-mer peptide derived from the extracellular loop II of the kappa opioid receptor containing 4 and 8 residues that may be positively or negatively charged, respectively. Upon purification by the Superdex-FPLC-method (see above), the peptide was redissolved in 500 mM ammonium bicarbonate (pH 8.4) and again lyophilized. The resulting powder could be directly dissolved in a 144 mM DPC solution (50 mM phosphate buffer at pH 6.8) at a concentration of 1.9 mM, and the transmembrane segments of the peptide incorporated well into the micelle interior.

Although detailed protocols for preparation of small unilamellar vesicles (SUVs) are published [80–82], their preparation is more time-demanding and complicated. In contrast, bicelles are prepared similarly to micelles with little effort.

5.3 The Structure and Dynamics of Membrane-Associated Peptides – A Case Study of Neuropeptide Y (NPY)

5.3.1 Introduction

The structural analysis of membrane-associated peptides comprises two steps: (a) the elucidation of the three-dimensional fold of the peptide and (b) the determination of the membrane-peptide interface. We will use our results gained for the 36 amino acid residue neuropeptide Y (NPY) [83] to demonstrate the approaches that can be used. NPY regulates important pharmacological functions such as blood pressure, food intake or memory retention and hence has been subject of many investigations (for a review see Ref. [84]). It targets the so-called Y receptors that belong to the class of seven transmembrane receptors coupled to G-proteins (GPCRs).

5.3.2 Structure Determination of Micelle-Bound NPY

Resonance assignment largely follows the procedure developed by the Wüthrich laboratory [85]. The repertoire of experiments that may be used depends on whether nonlabeled peptides, ^{15}N-labeled peptides or ^{13}C,^{15}N-doubly labeled peptide are available. When no labeling has been performed, the experiments for resonance assignments solely belong to the class of homonuclear experiments (mainly DQF-COSY, DQ-spectra, TOCSY and NOESY etc.) [86]. ^{15}N labeling permits the use of ^{15}N-edited three-dimensional spectra. Although signal overlap in the amide region is not so severe as to make the use of ^{15}N labeling mandatory, it was required for the determination of the internal dynamics, and it also facilitated the experiments for determination of the membrane interface. We found a (two-dimensional) NOE-relayed [^{15}N,^{1}H]-HSQC [87] particularly useful to rapidly assign the ^{15}N,^{1}H correlation map in the helical part of the peptide. In other cases, e.g. for the resonance assignment of micelle-bound bovine polypeptide (bPP) (unpublished data), we have additionally used ^{15}N-edited TOCSY and NOESY spectra for assignment. In our experience resonance assignment is speeded up so much in the case of ^{15}N-labeled peptides that the additional effort in the laboratory is quickly paid for. The HSQC spectra will also help to immediately recognize signal overlap.

The structure calculation of NPY bound to DPC micelles was based upon a 75 ms NOESY spectrum. Figure 5.5 shows expansions of spectral regions that were heavily used for backbone assignment. Sufficiently tight binding (mM range) of the peptide to the DPC micelle gives strong cross peaks and requires the use of comparably short mixing times in order to avoid spin diffusion. We have always used watergate-type solvent suppression techniques [88] to avoid saturation transfer, which rapidly takes place for residues not involved in the membrane interface at the pH 6 used. The position of helical residues is confirmed by an analysis of $^{3}J_{NH\alpha}$ couplings, which could conveniently be extracted from either the [^{1}H,^{1}H]-NOESY or [^{15}N,^{1}H]-HSQC spectra using the INFIT method [89].

The three-dimensional structure of NPY when bound to the membrane is shown in Fig. 5.6. It comprises an α-helix for residues 16 to 36 which is very well defined, and a flexible N-terminal part of the molecule. When comparing the structure of the DPC-micelle bound form to the structure in free solution, it is obvious that the α-helix is much more stable. In addition, the C-terminus of the helix comprising residues 32–36, which is flexible in solution, adopts an α-helical fold, and the Tyr36 is oriented such that it interacts with the water-membrane interface.

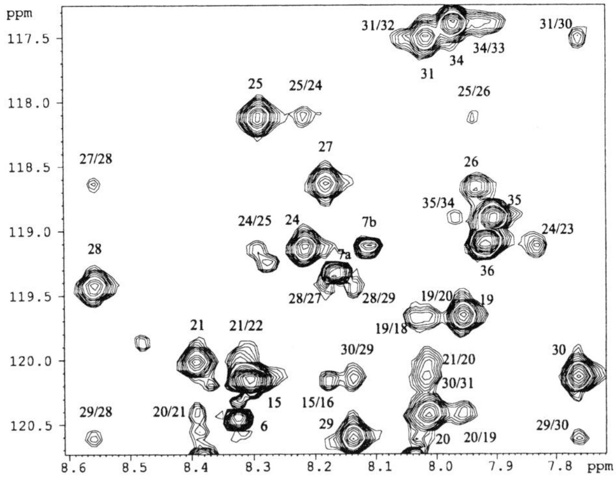

Fig. 5.5 Expansion from NOE-relayed [^{15}N,^{1}H]-HSQC spectrum of NPY.

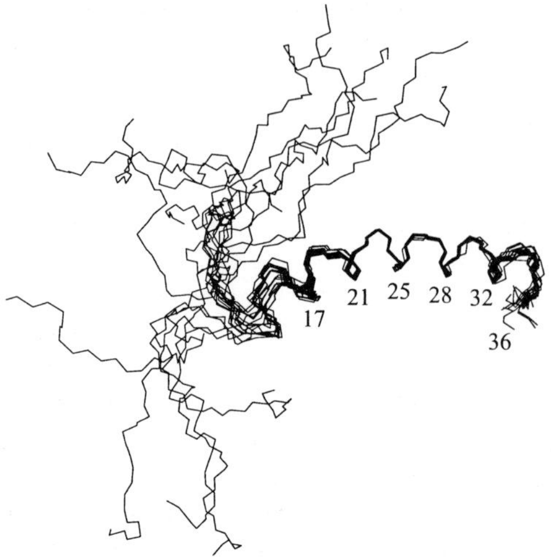

Fig. 5.6 Superposition of the backbone representation of the 20 lowest energy DYANA structures of NPY/DPC.

5.3.3 The Determination of the Topology of the Membrane-NPY Interface

5.3.3.1 Spin Labels

Spin labels contain unpaired electrons that by highly efficient electron-nuclear spin dipolar coupling lead to accelerated transverse or longitudinal relaxation. The effect is rather far-reaching (at least up to 10 Å), and its general use is described in Chapt. 15.

Micelle-integrating spin labels, such as the doxyl-substituted stearates, are commercially available. The stearates are offered with the spin label attached at different positions along the fatty acid chain: 5, 7, 12. To probe the vicinity of the membrane-water interface, the 5-doxylstearate is the most suitable. In the literature the free carboxylate has mostly been used. However, this may specifically interact e.g. with Arg or Lys side chains, and hence the methylester serves as a good control reagent. The 12-doxylstearate introduces some ambiguity into the analysis of the data because C-12 spends some time in the vicinity of the interface, as is known from MD trajectories. Beswick et al. have argued that at acidic pH the stearates are uncharged and hence their radial position is less well defined [44]. They propose the use of 1-palmitoyl-2-stearoyl-(10-doxyl)-*sn*-glycero-3-phosphocholine, which readily integrates into the micelle. Mn^{2+} ions may serve to yield a "negative print" since they will broaden all protons which are water accessible. The use of spin labels is also nicely described in the review by Damberg et al. [90].

The doxylstearates may be prepared as concentrated (e.g. 0.3 M) methanolic stock solutions. They should be added such that their final concentration in the micelle solution is about one spin label per micelle. When possible, the spin label should be added to the micelle solution before the peptide to ensure that it integrates into the micelles properly and does not bind specifically with the peptide.

In principle, spin label effects can be observed in both homonuclear and heteronuclear spectra. When no isotope-enriched peptide is available, integration of TOCSY spectra in the presence and in the absence of spin label gives information of spatial proximity of protons to the spin label. This method allows one to analyze data from both side-chain as well as from backbone protons. We found that analysis of [^{15}N,^{1}H]-HSQC spectra is very fast and reliable and superior to the use of homonuclear spectra. Because of the good signal separation and the large $^{1}J_{HN}$ coupling, cross peaks are present even in the spin label containing samples and can still be reliably integrated. Moreover, the positioning of side chains does not unambiguously enable the molecule to orient on the micelle. The so-called snorkelling effect of Arg or Lys side chains [35] has been observed frequently and may lead to erroneous interpretation of the data.

The relative signal intensities from a [^{15}N,^{1}H]-HSQC spectrum of NPY recorded in the presence of 5-doxylstearate is displayed in Fig. 5.7. It reveals two important points. Firstly, the reduction due to the spin label decreases from the start of the C-terminal α-helix up to the N-terminus continuously, proving that the N-terminus diffuses freely in solution and does not interact with the membrane. Secondly, the signal reduction along the sequence is periodic, nicely reflecting the pattern to be expected for the association of an α-helix parallel to a membrane surface. The residues affected to the greatest extent are those containing hydrophobic side chains (e.g. Leu, Ile) and are contained in the helical segment comprising residues 17–36. For a peptide inserting into the membrane interior with its backbone, a set of reduced signals from a continuous stretch of the amino acid sequence is observed.

Interestingly, upon addition of 5 mM $CaCl_2$, the affinity of NPY toward the DPC micelle is significantly reduced. Fig. 5.7b clearly displays the signal regain for the residues that have been most strongly affected by the spin label.

From the spin label studies we were able to conclude that NPY associates specifically with its hydrophobic side of the C-terminal α-helix parallel to the membrane surface.

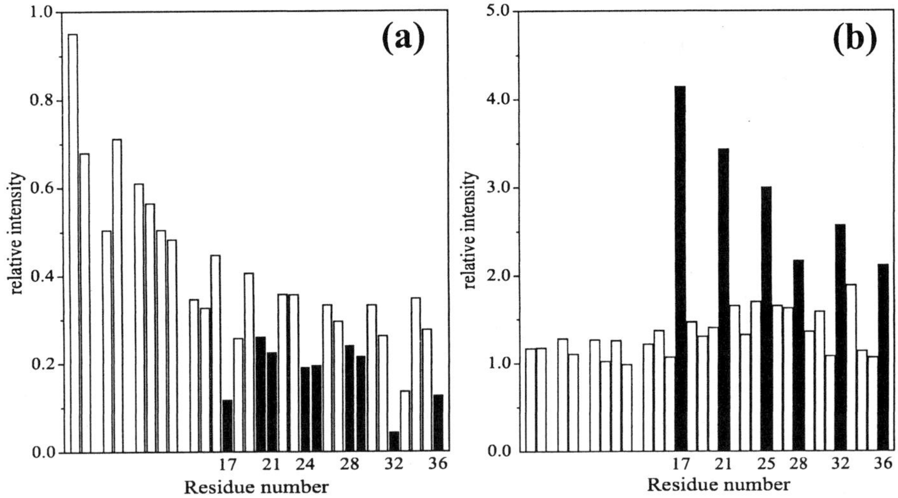

Fig. 5.7 Ratio of intensities of cross peaks in the [^{15}N,^{1}H]-HSQC spectrum in the presence of 5-doxylstearate relative to those in the absence of the spin label (**a**). Ratio of intensities in the presence of both spin label and 5 mM $CaCl_2$ to those in the presence of spin label but absence of salt (**b**).

5.3.3.2 Amide H,D Exchange

Significantly reduced rates for proton-deuterium exchange prove that the corresponding amide proton is either involved in stable hydrogen bonds, shielded from solvent access or both. Because of the ambiguity in interpretation, additional information about the persistence of hydrogen bonds stemming from structure calculations or from relaxation data should be available.

For NPY it was possible to follow the H,D exchange for certain sites by redissolving the lyophilized sample in deuterated water. Although order parameters derived from ^{15}N relaxation data indicate that the α-helix is stable between residues 17–33, residual signals in the ^{15}N,^{1}H correlation map were only visible for residues which form the hydrophobic side of the amphiphilic helix, e.g. Leu24, Ile28 Tyr27 and Ile31. This observation indicates that shielding from solvent access is the primary factor slowing down amide exchange and not formation of stable hydrogen bonds. The effect observed is much larger than the intrinsic differences of amide proton exchange rates due to neighboring atom effects. The residues mostly shielded from solvent access were identical or adjacent to those that were mostly affected by the spin label. From the spin label and H,D exchange data it is possible to position NPY on the membrane such that the hydrophobic side is pointing toward the membrane surface.

Once again, availability of ^{15}N uniformly labeled NPY has to be emphasized. In order to yield reasonably narrow lines and to avoid the need to reassign resonances of the peptide at a different temperature, the experiments were again performed at 37 °C. Only with the labeled peptide were we able to assign reduced exchange rates to specific sites in the peptide at that temperature. In our experimental set-up, lyophilized 1 mM peptide was mixed with ice-cooled deuterated water and inserted into a magnet that was preshimmed and whose probe was tuned and matched. Acquisition of the first two-dimensional data set with sufficient resolution was finished approx. 5 min after dissolving the sample in D_2O. The spectrum is displayed in Fig. 5.8. Even with μM affinity for the membrane, H,D exchange is fast enough to exclude the possibility of recording the much less sensitive ^{1}H,^{1}H homonuclear correlation spectra needed to derive the assignments on nonlabeled material.

5.3.4
Measurement of Internal Dynamics of NPY/DPC

NMR techniques are unique in their ability to resolve internal dynamics with site-specific probes. Backbone dynamics may be derived from relaxation data of ^{15}N nuclei. Relaxation data are conveniently measured in experiments that utilize [^{15}N,^{1}H]-HSQC-derived pulse sequences and hence can be performed within less than a week of total instrument time with a 1 mM sample (at one field). The underlying principle of the measurement is described in Chapt. 12 and has also been recently reviewed by Palmer [91].

Whereas ^{15}N longitudinal and transverse relaxation rates can be determined with sufficient precision, the determined values of the ^{15}N{^{1}H}-NOE differ significantly from the true values for residues involved in fast amide proton exchange. A comparison between the values for the ^{15}N{^{1}H}-NOE for NPY in solution in the presence and in the absence of DPC is displayed in the Fig. 5.9. The comparison reveals two striking differences.

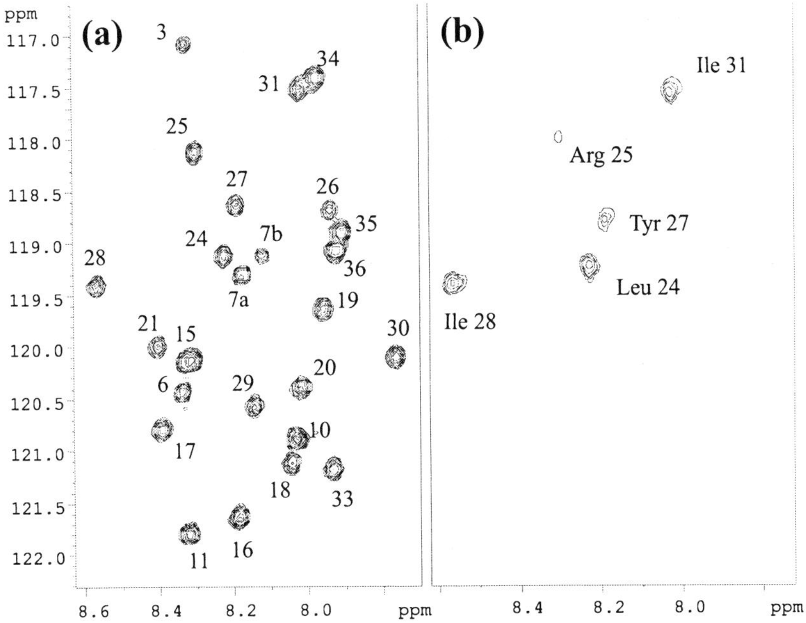

Fig. 5.8 $[^{15}N,^{1}H]$-HSQC reference spectrum of ^{15}N labeled NPY (**a**), and (**b**) same as **a** but recorded 10 min after dissolving the lyophilized sample in deuterated water.

Firstly, the (negative) values of the NOE for residues of the unstructured N-terminus that do not interact with the DPC micelle surface are larger. This result is most probably due to increased saturation transfer from the water and results from increased exchange of amide protons at the used pH of 6.0 compared to that used in the absence of DPC (pH 3.1). Secondly, the values for residues from the C-terminal pentapeptide are negative in the case of NPY free in solution whereas they are positive in the micelle-bound form. This clearly indicates that the C-terminal pentapeptide is significantly rigidified upon binding to the micelle. The result is supported by the structure calculation that displays rather low RMSD values for that part.

Physical insight into motional properties derived from the relaxation data is commonly derived from a Lipari-Szabo type analysis [92, 93]. The analysis may only be conducted when the time scale for internal and overall dynamics differs by at least one order of magnitude. In addition, as noted by Schurr et al. [94], anisotropic reorientation of the molecules may cause the analysis to erroneously reveal slow internal motions even in the absence of these. Both prerequisites may not be fulfilled for NPY in the absence of DPC, and we have therefore restricted our analysis to the above-mentioned comparison of the heteronuclear NOE. However, we feel that such an analysis is justified for NPY bound to DPC micelles. Firstly, the micelle itself reorients isotropically in solution, and binding of the hormone to the spherical micelle is not expected to change that fact dramatically. Secondly, the overall tumbling of NPY bound to a DPC micelle is slow enough (8.96 ns) to separate internal from overall dynamics. Fitting of the relaxation data was well possible within the experimental error limits with small residuals excluding residues from the N terminus.

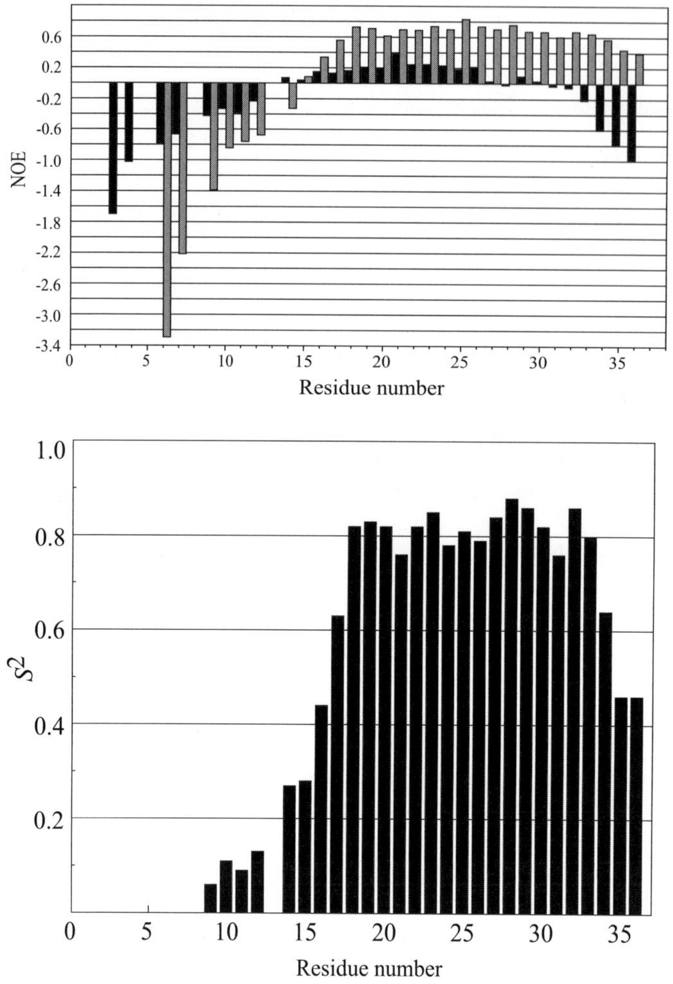

Fig. 5.9 Comparison of $^{15}N\{^{1}H\}$-NOE between NPY free in solution (black bars) and NPY bound to micelles (grey bars) (top), and (bottom) generalized order parameter S^2 for NPY bound to DPC micelles.

The generalized order parameters S^2 from a Lipari-Szabo type analysis are displayed in Fig. 5.9 (bottom). Their values are above 0.8 in the segment comprising residues 18–34. For residues 18, 19, 21, 23–28, 30 and 32 only a single variable, S^2, was required in order to yield reasonable fits to the experimental data. The data indicate that internal motions of the backbone are limited to values encountered in stable α-helices of globular proteins and show that the isolated helix upon binding to the membrane-mimicking micelle is remarkably rigidified. This result is in accordance with a comparison of the scalar $^{3}J_{NH\alpha}$ couplings that indicate rotationally averaged backbone ϕ angles in the case of free NPY and values reduced below 6 Hz for most C-terminal residues upon binding to DPC micelles. Moreover, the values drop only slightly toward the C terminus, indicating that although the backbone becomes more flexible it is still structured. Since it is this part of the molecule that is believed to be involved in forming contacts to the receptor, we believe that this result has biological significance and indicates a possible role of the membrane for recognition of NPY by its membrane-embedded receptor. The view that tight binding of helices to membranes is accompanied by a rigidification of the backbone is supported by the analysis of the backbone dynamics of M13 phage coat protein gVIIIp bound to SDS micelles as determined by Papvoine et al. [95]. They found even higher values for the micelle-spanning helix (S^2=0.96) compared to that of the membrane-associated helix (S^2=0.51).

We have recently gained confidence that our results on NPY are biologically relevant from our investigations into the Ala31,Pro32 mutant of NPY, which selectively binds to the Y5 receptor subtype. Using the strategy described above, we discovered that, although in general structural similarity for NPY and the mutant is large, significant differences exist in the C-terminal pentapeptide (the message part). In the mutant, the second last membrane anchor is removed and the backbone of the mutant forms a longer (flexible!) loop comprising residues 30–36, which is again anchored via the C-terminal Tyr36 amide. Thereby, the position of residues Arg33 and Arg35 is much less well defined and their distance to the membrane-water interface is greater on average. As a consequence, binding of these residues to an Asp receptor residue at the membrane-water interface is reduced in the Y1 receptor.

5.4 References

1 Gennis, R. B. *Biomembranes: Molecular Structure and Function,; 1st ed.; Springer: New York, 1989.*

2 Yeagle, P. L. *The membranes of cells*; Academic Press: San Diego, 1993.

3 Opella, S. J.; Kim, Y.; McDonnell, P. *Methods Enzymol.* **1994**, *239*, 536–560.

4 Henry, G. D.; Sykes, B. D. *Methods Enzymol.* **1994**, *239*, 515–535.

5 Klassen, R. B.; Opella, S. J. *Methods in Molecular Biology* **1997**, *60*, 271–297.

6 Marassi, F. M.; Opella, S. J. *Curr. Opin. Struct. Biol.* **1998**, *8*, 640–648.

7 Opella, S. J. *Nat. Struct. Biol.* **1997**, *4*, 845–848.

8 Watts, A. *Curr. Opin. Biotechnol.* **1999**, *10*, 48–53.

9 Baleja, J. D. *Anal. Biochem.* **2001**, *288*, 1–15.

10 Schwyzer, R. *J. Receptor Res.* **1991**, *11*, 45–57.

11 Schwyzer, R. *J. Mol. Recognit.* **1995**, *8*, 3–8.

12 Moroder, L.; Romano, R.; Guba, W.; Mierke, D. F.; Kessler, H.; Delporte, C.; Winand, J.; Christophe, J. *Biochemistry* **1993**, *32*, 13551–13559.

13 Edidin, M.; Finean, J. B. E.; Michell, R. H. in *Membrane Structure* **1981**, 37–82.

14 Walker, J. E.; Carne, A. F.; Schmitt, H. W. *Nature* **1979**, *278*.

15 Singer, S. J.; Nicolson, G. L. *Science* **1972**, *175*, 720–731.

16 Chapman, D.; Benga, G. in *Biological Membranes*; Chapman, D., ed.; Academic Press: New York, 1984; Vol. 5, pp 1–56.

17 Vaz, W. L. C.; Goodsaidzaldouondo, F.; Jacobson, K. *FEBS Lett.* **1984**, *1974*, 199–207.

18 Op den Kamp, J. A. F. *Annu. Rev. Biochem.* **1979**, *48*, 47–71.

19 Ganong, B. R.; Bell, R. M. *Biochemistry* **1984**, *23*, 4977–4983.

20 McNamee, M.G.; McConnel, H.M. *Biochemistry* **1973**, *12*, 2957–2958.

21 Kornberg, R.D.; McConnel, H.M. *Biochemistry* **1971**, *10*, 1111–1120.

22 Kuypers, F.A. *Curr. Opin. Hematol.* **1998**, *5*, 122–131.

23 Azzone, G.F.; Pietrobon, D.; Zoratti, M. *Curr. Top. Bioenerg.* **1984**, *13*, 1–77.

24 Harder, T.; Simons, K. *Curr. Opin. Cell Biol.* **1997**, *9*, 534–542.

25 Parton, R.G. *Science* **2001**, *293*, 2404–2405.

26 Sakai, L.Y. *Sci. Med.* **1995**, *2*, 58–67.

27 Hynes, R.O. *Cell* **1992**, *69*, 11–25.

28 Chothia, C.; Jones, E.Y. *Annu. Rev. Biochem.* **1996**, *66*, 823–862.

29 Tedder, T.F. *FASEB J.* **1995**, *9*, 866–873.

30 Uyemura, K.; Asou, H.; Yazaki, T.; Takeda, Y. *Essays Biochem.* **1996**, *31*, 37–48.

31 Geiger, B.; Ayalon, O. *Annu. Rev. Cell Biol.* **1993**, *8*, 307–332.

32 Branden, C.; Tooze, J. *Introduction to Protein Structure*; Garland Publisher: New York, 1999.

33 Backlund, B.M.; Wikander, G.; Peeters, T.L.; Graslund, A. *Biochim. Biophys. Acta* **1994**, *1190*, 337–344.

34 Liu, L.P.; Deber, C.M. *Biochemistry* **1997**, *36*, 5476–5482.

35 Killian, J.A.; von Heijne, G. *Trends Biochem. Sci.* **2000**, *25*, 429–434.

36 Wimley, W.C.; White, S.H. *Nat. Struct. Biol.* **1996**, *3*, 842–848.

37 Lee, A.G.; Dalton, K.A.; Duggleby, R.C.; East, J.M.; Starling, A.P. *Biosci. Rep.* **1995**, *15*, 289–298.

38 Meuillet, E.J.; Leray, V.; Hubert, P.; Leray, C.; Cremel, G. *Biochim. Biophys. Acta* **1999**, *1454*, 38–48.

39 Esmann, M.; Watts, A.; Marsh, D. *Biochemistry* **1985**, *24*, 1386–1393.

40 Dill, K.A.; Flory, P.J. *Proc. Natl. Acad. Sci. USA* **1981**, *78*, 676–680.

41 Wymore, T.; Gao, X.F.; Wong, T.C. *J. Mol. Struct.* **1999**, *485/486*, 195–210.

42 Tieleman, D.P.; van der Spoel, D.; Berendsen, H.J.C. *J. Phys. Chem.* **2000**, *104*, 6380–6388.

43 MacKerell, A.D. *J. Phys. Chem.* **1995**, *99*, 1846–1855.

44 Beswick, V.; Guerois, R.; Cordier-Ochsenbein, F.; Coic, Y.-M.; Huyanh-Dinh, T.; Tostain, J.; Noel, J.-P.; Sanson, A.; Neumann, J.-M. *Eur. Biophys. J.* **1998**, *28*, 48–58.

45 Lauterwein, J.; Bösch, C.; Brown, L.R.; Wüthrich, K. *Biochim. Biophys. Acta* **1979**, *556*, 244–264.

46 Small, D.M. *Gastroenterology* **1967**, *52*, 607–609.

47 Ram, P.; Prestegard, J.H. *Biochim. Biophys. Acta* **1988**, *940*, 289–294.

48 Sanders, C.R.; Prosser, R.S. *Structure* **1998**, *6*, 1227–1234.

49 Carey, M.C.; Small, M.D. *Am. J. Med.* **1970**, *49*, 590–606.

50 Vold, R.R.; Prosser, R.S.; Deese, A.J. *J. Biomol. NMR* **1997**, *9*, 329–335.

51 Glover, K.J.; Whiles, J.A.; Wu, G.; Yu, N.; Deems, R.; Struppe, J.O.; Stark, R.E.; Komives, E.A.; Vold, R.R. *Biophys. J.* **2001**, *81*, 2163–2171.

52 Struppe, J.; Whiles, J.A.; Vold, R.R. *Biophys. J.* **2000**, *78*, 281–289.

53 Whiles, J.A.; Brasseur, R.; Glover, K.J.; Melacini, G.; Komives, E.A.; Vold, R.R. *Biophys. J.* **2001**, *80*, 280–293.

54 Alger, J.R.; Prestegard, J.H. *Biophys. J.* **1979**, *28*, 1–13.

55 Chou, J.J.; Kaufman, J.D.; Stahl, S.J.; Wingfield, P.T.; Bax, A. *J. Am. Chem. Soc.* **2002**, *124*, 2450–2451.

56 Pervushin, K.V.; Orekhov, V.; Popov, A.I.; Musina, L.; Arseniev, A.S. *Eur. J. Biochem.* **1994**, *219*, 571–583.

57 Casey, J.R.; Reithmeier, R.A. *Biochemistry* **1993**, *32*, 1172–1179.

58 Kessi, J.; Poiree, J.C.; Wehrli, E.; Bachofen, R.; Semenza, G.; Hauser, H. *Biochemistry* **1994**, *33*, 10825–10836.

59 Womack, M.D.; Kendall, D.A.; MacDonald, R.C. *Biochim. Biophys. Acta* **1983**, *733*, 210–215.

60 Vinogradova, O.; Sonnichsen, F.; Sanders, C.R., 2nd *J. Biomol. NMR* **1998**, *11*, 381–386.

61 Ketchem, R.R.; Hu, W.; Cross, T.A. *Science* **1993**, *261*, 1457–1460.

62 Sanders, C.R., 2nd; Landis, G.C. *Biochemistry* **1995**, *34*, 4030–4040.

63 Merrifield, B. *Method. Enzymol.* **1997**, *289*, 3–13.

64 Fields, G.B.; Colowick, S.P. Solid-Phase Peptide Synthesis; Academic Press: London, 1997.

65 Huth, J.R.; Bewley, C.A.; Belinda, M. J.; Hinnebusch, A.G.; Clore, G.M.; Gronenborn, A.M. *Protein Sci.* **1997**, *6*, 2359–2364.

66 Kohno, T.; Kusunoki, H.; Sato, K.; Wakamatsu, K. *J. Biomol. NMR* **1998**, *12*, 109–121.

67 Eipper, B.A.; Milgram, S.L.; Husten, E.J.; Yun, H.-Y.; Mains, R.E. *Protein Sci.* **1993**, *2*, 489–497.

68 Wüthrich, K.; Wagner, G. *J. Mol. Biol.* **1979**, *130*, 1–18.

69 Englander, S.W.; Kallenbach, N.R. *Q. Rev. Biophys.* **1984**, *16*, 521–655.

70 Dempsey, C.E. *Prog. NMR Spectrosc.* **2001**, *39*, 135–170.

71 O'Neil, J.D.; Sykes, B.D. *Biochemistry* **1988**, *27*, 2753–2762.

72 Spyracopoulos, L.; O'Neil, J.D. J. *J. Am. Chem. Soc.* **1994**, *116*, 1395–1402.

73 Perrin, C.L.; Chen, J.-H.; Ohta, B.K. *J. Am. Chem. Soc.* **1999**, *121*, 2448–2455.

74 Tomich, J.M.; Carson, L.W.; Kanes, K.J.; Vogelaar, N.J.; Emerling, M.R.; Richards, J.H. *Anal. Biochem.* **1988**, *174*, 197–203.

75 Killian, J.A.; Trouard, T.P.; Greathouse, D.V.; Chupin, V.; Lindblom, G. *FEBS Lett.* **1994**, *348*, 161–165.

76 Smith, S.O.; Jonas, R.; Braiman, M.; Bormann, B.J. *Biochemistry* **1994**, *33*, 6334–6341.

77 Rigby, A.C.; Barber, K.R.; Shaw, G.S.; Grant, C.W. *Biochemistry* **1996**, *35*, 12591–12601.

78 Glover, K.J.; Martini, P.M.; Vold, R.R.; Komives, E.A. *Anal. Biochem.* **1999**, *272*, 270–274.

79 Zhang, L.; DeHaven, R.N.; Goodman, M. *Biochemistry* **2001**, *41*, 61–68.

80 Olson, F.; Hunt, C.A.; Szoka, F.C.; Vail, W.; Mayhew, E.; Paphadjopoulos, D. *Biochim. Biophys. Acta* **1980**, *601*, 559–571.

81 Bangham, A.D.; Standsih, M.M.; Watkins, J.C. *J. Mol. Biol.* **1965**, *13*, 238–252.

82 Huang, C. *Biochemistry* **1969**, *8*, 344–352.

83 Bader, R.; Bettio, A.; Beck-Sickinger, A. G.; Zerbe, O. *J. Mol. Biol.* **2001**, *305*, 307–329.

84 Beck-Sickinger, A.G.; Jung, G. *Biopolymers* **1995**, *37*, 123–142.

85 Wüthrich, K. NMR of Proteins and Nucleic Acids, 1st ed.; Wiley: New York, 1986.

86 Kessler, H.; Gehrke, M.; Griesinger, C. *Angew. Chem. Int. Ed. Engl.* **1988**, *27*, 490–536.

87 Gronenborn, A.M.; Bax, A.; Wingfield, P. T.; Clore, G.M. *FEBS Lett.* **1989**, *243*, 93–98.

88 Piotto, M.; Saudek, V.; Sklenar, V. *J. Biomol. NMR* **1992**, *2*, 661–665.

89 Szyperski, T.; Güntert, P.; Otting, G.; Wüthrich, K. *J. Magn. Reson.* **1992**, *99*, 552–560.

90 Damberg, P.; Jarvet, J.; Graslund, A. *Methods Enzymol.* **2001**, *339*, 271–285.

91 Palmer, A.G. *Annu. Rev. Biophys. Biomol. Struct.* **2001**, *30*, 129–155.

92 Lipari, G.; Szabo, A. *J. Am. Chem. Soc.* **1982**, *104*, 4546–4559.

93 Lipari, G.; Szabo, A. *J. Am. Chem. Soc.* **1982**, *104*, 4559–4570.

94 Schurr, J.M.; Babcock, H.P.; Fujimoto, B.S. *J. Magn. Reson. B* **1994**, *105*, 211–224.

95 Papavoine, C.H.; Aelen, J.M.; Konings, R.N.; Hilbers, C.W.; van de Ven, F.J. *Eur. J. Biochem.* **1995**, *232*, 490–500.

96 Braun, W.; Wider, G.; Lee, K.H.; Wüthrich, K. *J. Mol. Biol.* **1983**, *169*, 921–948.

97 Jarvet, J.; Zdunek, J.; Damberg, P.; Graslund, A. *Biochemistry* **1997**, *36*, 8153–8163.

98 Ohman, A.; Lycksell, P.O.; Jureus, A.; Langel, U.; Bartfai, T.; Graslund, A. *Biochemistry* **1998**, *37*, 9169–9178.

99 Motta, A.; Andreotti, G.; Amodeo, P.; Strazzullo, G.; Castiglione Morelli, M. A. *Proteins* **1998**, *32*, 314–323.

100 Hashimoto, Y.; Toma, K.; Nishikido, J.; Yamamoto, K.; Haneda, K.; Inazu, T.; Valentine, K.G.; Opella, S.J. *Biochemistry* **1999**, *38*, 8377–8384.

101 Pellegrini, M.; Bisello, A.; Rosenblatt, M.; Chorev, M.; Mierke, D.F. *Biochemistry* **1998**, *37*, 12737–12743.

102 Yan, C.; Digate, R.J.; Guiles, R.D. *Biopolymers* **1999**, *49*, 55–70.

103 Piserchio, A.; Usdin, T.; Mierke, D.F. *J. Biol. Chem.* **2000**, *275*, 27284–27290.

104 Inooka, H.; Ohtaki, T.; Kitahara, O.; Ikegami, T.; Endo, S.; Kitada, C.; Ogi, K.; Onda, H.; Fujino, M.; Shirakawa, M. *Nat. Struct. Biol.* **2001**, *8*, 161–165.

105 Brown, L.R.; Wüthrich, K. *Biochim. Biophys. Acta* **1981**, *647*, 95–111.

106 Franklin, J.C.; Ellena, J.F.; Jayasinghe, S.; Kelsh, L.P.; Cafiso, D.S. *Biochemistry* **1994**, *33*, 4036–4045.

107 Condamine, E.; Rebuffat, S.; Prigent, Y.; Segalas, I.; Bodo, B.; Davoust, D. *Biopolymers* **1998**, *46*, 75–88.

108 Schibli, D.J.; Hwang, P.M.; Vogel, H.J. *Biochemistry* **1999**, *38*, 16749–16755.

109 Townsley, L. E.; Tucker, W. A.; Sham, S.; Hinton, J. F. *Biochemistry* **2001**, *40*, 11676–11686.
110 Papavoine, C. H.; Christiaans, B. E.; Folmer, R. H.; Konings, R. N.; Hilbers, C. W. *J. Mol. Biol.* **1998**, *282*, 401–419.
111 Almeida, F. C.; Opella, S. J. *J. Mol. Biol.* **1997**, *270*, 481–495.
112 Williams, K. A.; Farrow, N. A.; Deber, C. M.; Kay, L. E. *Biochemistry* **1996**, *35*, 5145–5157.
113 Schibli, D. J.; Montelaro, R. C.; Vogel, H. J. *Biochemistry* **2001**, *40*, 9570–9578.
114 Pervushin, K. V.; Orekhov, V.; Popov, A. I.; Musina, L.; Arseniev, A. S. *Eur. J. Biochem.* **1994**, *219*, 571–583.
115 Shao, H.; Jao, S.; Ma, K.; Zagorski, M. G. *J. Mol. Biol.* **1999**, *285*, 755–773.
116 MacKenzie, K. R.; Prestegard, J. H.; Engelman, D. M. *Science* **1997**, *276*, 131.
117 Piserchio, A.; Bisello, A.; Rosenblatt, M.; Chorev, M.; Mierke, D. F. *Biochemistry* **2000**, *39*, 8153–8160.
118 Chupin, V.; Killian, J. A.; Breg, J.; de Jongh, H. H.; Boelens, R.; Kaptein, R.; de Kruijff, B. *Biochemistry* **1995**, *34*, 11617–11624.
119 Losonczi, J. A.; Tian, F.; Prestegard, J. H. *Biochemistry* **2000**, *39*, 3804–3816.
120 Berlose, J. P.; Convert, O.; Derossi, D.; Brunissen, A.; Chassaing, G. *Eur. J. Biochem.* **1996**, *242*, 372–386.
121 Wienk, H. L.; Wechselberger, R. W.; Czisch, M.; de Kruijff, B. *Biochemistry* **2000**, *39*, 8219–8227.
122 Pellegrini, M.; Mierke, D. F. *Biochemistry* **1999**, *38*, 14775–14783.
123 Rozek, A.; Sparrow, J. T.; Weisgraber, K. H.; Cushley, R. J. *Biochemistry* **1999**, *38*, 14475–14484.
124 MacRaild, C. A.; Hatters, D. M.; Howlett, G. J.; Gooley, P. R. *Biochemistry* **2001**, *40*, 5414–5421.
125 Sorgen, P. L.; Cahill, S. M.; Krueger-Koplin, R. D.; Krueger-Koplin, S. T.; Schenck, C. C.; Girvin, M. E. *Biochemistry* **2001**, *41*, 31–41.
126 Xu, G. Y.; McDonagh, T.; Yu, H. A.; Nalefski, E. A.; Clark, J. D.; Cumming, D. A. *J. Mol. Biol.* **1998**, *280*, 485–500.
127 Brito, R. M.; Vaz, W. L. C. *Anal. Biochem.* **1986**, *152*, 250.
128 Chattopadhyay, A.; London, E. *Anal. Biochem.* **1984**, *139*, 408.
129 Mysels, K. J.; Princen, L. H. *J. Phys. Chem.* **1959**, *63*, 1696–1700.
130 Helenius, A.; McCaslin, D. R.; Fries, E.; Tanford, C. *Methods Enzymol.* *56*, 734.
131 Small, M.; Nair, P. P.; Kritchevsky, D. in The Bile Acids: Chemistry, Physiology, and Metabolism **1971**, 249.
132 Thomas, D. C.; Christian, S. D. *J. Colloid Interf. Sci.* **1980**, *78*, 466.
133 Small, D. M. *Adv. Chem.* **1968**, *84*, 31.
134 Rosevear, P.; van Aken, T.; Baxter, J.; Ferguson-Miller, S. *Biochemistry* **1980**, *19*, 4108–4115.
135 Mast, R. C.; Haynes, L. V. *J. Colloid Interf. Sci.* **1975**, *53*, 35.
136 Herrmann, K. W. *J. Colloid Interf. Sci.* **1966**, *22*, 352.
137 Hjelmeland, L. M.; Nebert, D. W.; Osborne, J. C., Jr. *Anal. Biochem.* **1983**, *130*, 72–82.
138 Stark, R. E.; Leff, P. D.; Milheim, S. G.; Kropf, A. *J. Phys. Chem.* **1984**, *88*, 6063–6067.

6
NMR of Nucleic Acids

Radovan Fiala and Vladimír Sklenář

6.1
Introduction

During past few years NMR spectroscopy of nucleic acids underwent remarkable development, which has been fueled by the introduction of advanced and more efficient isotope-labeling techniques as well as by technical innovations in the methodology of multinuclear NMR spectroscopy. Since 1986, when Wüthrich [1] published the first extensive review of proton NMR tools for structure elucidation of proteins and nucleic acids, the capability of NMR to reveal the structural details of both DNA and RNA and to describe their dynamic behavior in solution has increased dramatically. As the limited extent of our contribution precludes covering the complete range of isotope-assisted nuclear magnetic resonance techniques, we have deliberately decided to concentrate only on the very recent developments. Attention will be paid primarily to those that have shown the potential of increasing the range of promising applications and the capability of providing new structural information. These include the ^{2}H, ^{13}C, and ^{15}N isotope-labeling techniques for RNA and DNA, sensitivity-optimized multi-dimensional NMR experiments, techniques to detect hydrogen-bond networks, applications of residual dipolar couplings for extracting structural restraints, and relaxation studies providing the information on dynamics and structure. The last few years have witnessed an explosive increase in the literature devoted to these topics. Only the basics will be reiterated here, and we refer readers to the original papers for details.

A number of reviews covering various aspects of NMR spectroscopy of nucleic acids have appeared recently. An extensive summary of both proton and multinuclear NMR of nucleic acids and the state-of-the-art of methodology up to 1997, which can be recommended as a comprehensive, introductory text, has been presented by Wijmenga and van Buuren [2]. In 2001, solid state synthesis of selectively labeled DNA [3], uniform labeling of DNA [4], and preparation of ^{2}H-labeled RNA oligonucleotides [5, 6] have been reviewed as well as investigations of unusual DNA motifs [7], resonance assignment and structure determination of RNA [8], NMR studies of cation binding to nucleic acids [9, 10], and studies of metal ion binding sites in RNA [11]. In addition, protein-DNA [12] and protein-RNA [13, 14] interactions have been described, and the methodological advances of NMR for the study of nucleic acids [15] and of computational methods for RNA structure determination [16] have been presented.

6.2
Sample Preparation

Nowadays, a typical NMR sample of nucleic acids contains a 1–2 mM solution of DNA or RNA oligonucleotide (strand concentration). The oligonucleotides are dissolved in H_2O or ^2H_2O, and buffer and salt are added to the required concentration. Phosphate buffer allows adjustment of pH in the typical range of values (pH $\sim$ 5–8). However, since the DNA/RNA oligonucleotides have a relatively large buffering capacity, adjustment of pH to the desired values without adding any buffer frequently works quite well. The pH has to be checked in any case since the pH of freshly dissolved sample can be quite low (<5). To facilitate the proper fold of the dissolved oligonucleotide the temperature annealing by heating the sample above its melting temperature and slow cooling in water bath is highly recommended.

The easiest way to make sure that the sample forms the required secondary structure is to measure a simple one-dimensional spectrum in H_2O solution in order to obtain information about the exchangeable imino and amino protons. Since these protons mostly exchange too fast with the solvent, they are usually not observable in ^2H_2O. The measurement in H_2O (typically 90–95% H_2O and 10–5% ^2H_2O) is normally performed using experiments with selective nonexcitation of the water resonance. WATERGATE with flip-back arrangement [17] or other excitation sequences which leave the water resonance unperturbed in the equilibrium state can be applied. The number of imino proton resonances and their line width indicate formation of the hydrogen-bonded base pairs. The G and T imino protons in GC and AT pairs are found betwen 12 and 15 ppm, while the imino resonance in protonated C^+s and unpaired Ts are located around 14–16 and 10–11 ppm, respectively. The 1D spectra also allow a quick assessment of sample purity and the possible presence of more than one conformation. Impurities give rise to sharp signals in the aliphatic region of the proton NMR spectrum, and extra conformations are manifested by additional sets of imino resonances. The melting stability of the sample is quickly assessed by monitoring the temperature changes of imino resonances. Although in principle all spectral information on labeled samples could be obtained from the H_2O solution, the information about sugar protons is more easily extracted from ^2H_2O spectra. The sample can be transferred from H_2O to ^2H_2O solution by lyophilization or by drying the sample in the NMR tube with a stream of filtered nitrogen. This process is repeated usually twice or three times before the sample is redissolved in 99.996% ^2H_2O. As an example, the 1D imino part of proton NMR spectrum of the ATP binding RNA aptamer [18] is shown in Fig. 6.1.

Protein-DNA complexes are usually formed by adding small portions of the protein solution to the DNA. The formation of the complex is best monitored by observing the changes of the DNA imino proton signals in a 1D spectrum after each step. A reversed approach, i.e. adding the DNA to a protein solution, is not recommended because of the formation of higher order complexes by nonselective binding of protein to DNA.

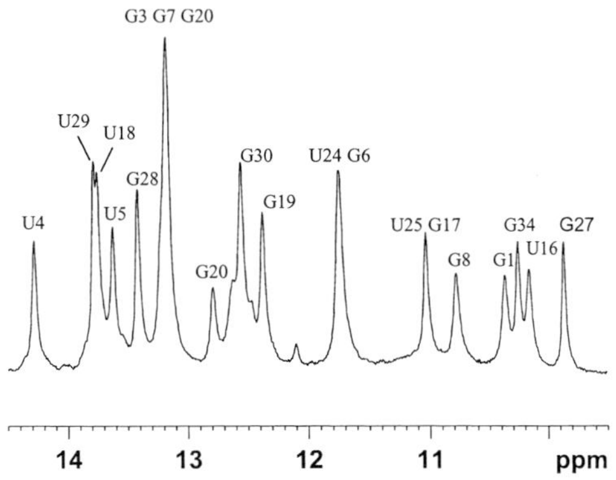

Fig. 6.1 Imino region of proton 1D spectrum of ATP-binding RNA aptamer complex consisting of 40 nucleotides and bound AMP.

6.3
Preparation of Labeled Nucleic Acids for Multinuclear NMR

Uniform labeling of RNA was introduced in the early 1990s by several groups [19–21]. The strategy is relatively simple. RNA oligomers are prepared by *in vitro* transcription with T7 RNA polymerase using synthetic DNA templates from rNTPs uniformly enriched by ^{13}C and/or ^{15}N and/or ^{2}H isotopes. rNTPs are usually produced in *E. coli* or *M. methylotropus* cells grown on enriched media ([^{13}C]-glucose or [^{13}C]-methanol, [^{15}N]-ammonium, $^{2}H_2O$). RNA is isolated from cells and cleaved to rNMPs, which are subsequently converted to rNTPs. Labeling of DNA is more complicated, and the first protocol for uniform enzymatic labeling was published in 1995 [22]. In general, dNMPs are produced from DNA fragments extracted from bacteria grown on labeled media. Subsequently, dNTPs, in a mixture or in isolated forms obtained by the conversion from dNMPs, are incorporated into oligonucleotides using a 3′–5′ exonuclease-deficient mutant of Klenow fragment of DNA polymerase and an oligonucleotide template primer. Since the first papers were published, a number of modifications have been proposed to improve the yield and efficiency of the labeling procedures for both RNA [23, 24] and DNA [4, 25–28]. In addition, new protocols for site-specific (labeling of specific atoms) and residue-specific (labeling of specific nucleotides, such as A, G, U, C or T) labeling of both RNA and DNA were developed. It has been shown that residue-specific labeling is necessary to obtain unambiguous assignments in RNAs consisting of 30–50 nucleotides [29–31].

In order to extend further the range of RNA molecular weights amenable to NMR studies, the residue-specific ^{13}C,^{15}N labeling will have to be complemented by partial deuteration. Recently, the preparation and use of ^{2}H labeled RNA and the pros and cons of such an approach have been described in detail [6]. Uniform 90–95% deuteration was obtained using the growth medium containing 90% $^{2}H_2O$ and [$^{2}H_3$]-acetate, and 75% random deuteration was reached using 80% $^{2}H_2O$ and [$^{2}H_3$]-acetate as the sole carbon source. To obtain useful information with sufficient sensitivity of NMR measurements, specific sites have to be re-protonated. Exchangeable protons in -NH and -NH_2 groups are easily recovered by dissolving the RNA in H_2O solution, while C5, C6 and C8 aromatic carbons can be re-protonated using the exchange reaction with bisulfate anion as a catalyst at the rNMP stage. This allows the retrieval of important information about H5, H6 and H8 protons, which supply details of base stacking interactions. As a consequence of reduced dipolar interactions, reduction of both longitudinal (~ 35%) and transverse (~ 10–15%) relaxation rates of exchangeable protons was observed in a 45-nucleotide RNA hairpin. The nonexchangeable protons H5, H6 and H8 showed decrease of both R_1 and R_2. Benefits have been observed in ^{31}P-^{1}H correlation spectra where substantial suppression of passive ^{1}H-^{1}H scalar interactions improved the cross-peak intensities of H5′ and H5″ signals in the case of ~ 75% ^{2}H random enriched hairpin. A modest gain in sensitivity (30–40%) was also detected in ^{1}H-^{13}C HSQC spectra. In addition to uniform ^{2}H enrichment, the protocol allowing selective introduction of ^{2}H in sugar H3′, H4′, H5′ and H5″ [32] positions was published and the advantage of this approach demonstrated using 2D CT C(CC)H TOCSY spectra of TAR RNA [33].

DNA can be isotopically labeled not only using the enzymatic approach mentioned above but also using solid-state phosphoamidite synthesis. This methodology, developed in the 1980s and modified recently to incorporate the isotope labels, is of advantage when residue- and site-specific patterns of ^{13}C and/or ^{15}N and/or ^{2}H labeling is needed [3].

6.4 Assignment Strategy – New and Sensitivity-Optimized Experiments

Resonance assignment is a crucial first step in any biomolecular NMR study. The field of both homonuclear and heteronuclear assignment methods for nucleic acids has been well researched by now and a virtually complete system of methods has been established [8]. The current development concentrates on methods for particular specialized tasks and on increasing the sensitivity and resolution of existing methods using the latest theoretical findings and improvements in the instrumentation. Here, we briefly review the basic principles of established methods of resonance assignments, and then we focus our attention on recent developments in the field. The reader interested in a more detailed treatment of the assignment strategies should refer to one of the excellent recent reviews such as Ref. [2] or directly to the original literature. Numbering of atoms in sugar, pyrimidine and purine moieties and nomenclature of torsion angles in nucleotides are shown in Fig. 6.2.

From a spectroscopist's point of view the most important question is whether an isotopically labeled sample is available. If it is not the case, then the assignment relies mainly on ^{1}H-^{1}H connectivities established through space by NOE for sequential and sugar-to-base connectivities and through scalar couplings using COSY (correlation spectroscopy) or TOCSY (total correlation spectroscopy) type experiments to detect connectivities within sugar and base moieties. These homonuclear methods are supplemented by ^{1}H-^{31}P correlation experiments to support the assignment in the sugar-phosphate backbone. Though 2D ^{1}H-^{13}C and ^{1}H-^{15}N correlation spectra can be obtained at natural abundance, they are of limited importance because of low sensitivity and the lack of other means to incorporate the carbon and nitrogen chemical shifts into a broader assignment picture.

Since the imino and amino hydrogens are subject to rapid exchange with hydrogens of water, all spectra that are intended to provide information about exchangeable protons

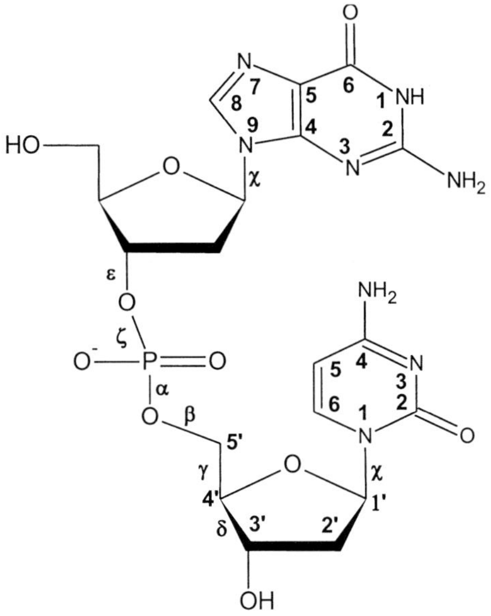

Fig. 6.2 Structure, atom numbering and labeling of torsion angles in nucleic acids.

have to be measured in H_2O solution. However, the H_2O signal falls in the region of sugar resonances, obscuring some of the desired peaks. The spectra for the assignment of nonexchangeable resonances are therefore preferably measured in ^2H_2O solution to avoid the problems associated with the strong water peak. In the following paragraph, we briefly list individual types of spectra in the order the data are typically collected and comment on information that can be extracted from them.

NOESY in H_2O First, base pairing is established through imino-imino and imino-amino sequential and intra-residual NOEs. Then, NOE contacts of the imino and amino protons with the remaining base (H6/8, H5(M), adenine H2) and H1′ protons are examined to obtain partial assignment of the nonexchangeable hydrogen atoms. After obtaining satisfactory NOESY spectra in H_2O, the sample can be lyophilized and dissolved in high quality (better than 99.99%) ^2H_2O for further experiments.

NOESY in ^2H_2O provides the assignment of nonexchangeable base and sugar proton resonances, mostly through H6/8-H1′ and H6/8-H2′/2″ sequential connectivities.

COSY and TOCSY help to identify the spin systems of sugar and pyrimidine bases. H6 and H5 or methyl (M) protons of pyrimidines always provide strong cross peaks in COSY and TOCSY spectra. In DNA, the H6-H5 and H6-M cross peaks provide an easy way to distinguish cytosine and thymine residues. In deoxyribose, TOCSY spectra provide connectivities of at least H1′ through H3′ regardless of sugar puckering thanks to the presence of H2″. In the case of 3′-endo sugar pucker, the coupling constant $^3J_{H1',H2'}$ is small, making TOCSY transfer from H1′ difficult in RNA. On the other hand, it provides an easy way of identifying the sugar puckering.

^1H-^{31}P HETCOR and/or hetero TOCSY proton-phosphorus correlation completes and/or confirms H3′, H4′ assignments made in the previous step and extend it to H5′/5″. The main limitations of the assignment strategy outlined above are spectral overlap and the ambiguity of NOE peaks. Particularly crowded spectral regions are those of base H6/8 protons and sugar H3′ through H5′/5″ (H2′ through H5′/5″ in RNA). The overlap can be in some cases resolved by using 3D homonuclear spectra such as 3D NOESY-NOESY and 3D TOCSY-NOESY. With this approach, virtually complete assignments of oligonucleotides with up to 31 residues were obtained. The ambiguity problem stems from the fact that through-space interactions are inherently conformation dependent. The approach relies on a set of distances that are well established for canonical structures but may be significantly different in regions with unusual structural motifs. Therefore there is always some degree of uncertainty in an assignment of a peak based on through-space interaction.

Availability of a sample isotopically enriched with ^{15}N and ^{13}C offers the NMR spectroscopist additional tools for overcoming the problems encountered in homonuclear experiments. In an obvious extension of the homonuclear experiments, the carbon or nitrogen dimension can be used to improve the resolution in NOESY spectra. However, recent trends clearly favor the through-bond approach where the presence of a cross peak unambiguously proves the existence of a chemical bond between the nuclei involved. Let us review briefly the experiments available for labeled oligonucleotides.

2D ^{1}H-^{15}N HSQC/HMQC (heteronuclear single-quantum correlation/heteronuclear multiple-quantum correlation) identifies imino and amino proton and nitrogen resonances. G and U(T) imino nitrogens are clearly distinguished by their chemical shifts (under and over 150 ppm when referenced to liquid ammonia, respectively). Amino nitrogens are also distinguished by their chemical shifts; however, it is usually only cytosine amino signals that are all observable. The gradient sensitivity-enhanced version of HSQC [39] works particularly well for the imino ^{1}H-^{15}N correlations.

3D ^{15}N-edited NOESY (^{15}N-NOESY-HMQC) provides imino-imino, imino-amino and aromatic-imino/amino connectivities. The approach often fails in regions such as loops and bulges where bases do not form hydrogen-bonded pairs. In that case, imino assignment can be obtained from

2D or 3D HCCNH-TOCSY experiment that correlates the imino proton with H6/8, which can be independently assigned from other experiments.

All the following experiments can be performed with the sample dissolved in $^{2}H_2O$.

2D ^{1}H-^{13}C CT-HSQC/HMQC identifies proton-carbon spin pairs in both sugar and base. The constant time evolution removes the peak splitting due to ^{13}C-^{13}C coupling. The proton and carbon resonances are subsequently connected by

2D and 3D HCCH-COSY, HCCH-RELAY and HCCH-TOCSY experiments. Large $^{1}J_{C,H}$ and $^{1}J_{C,C}$ coupling constants (approximately 140 Hz and 40 Hz, respectively, in sugars, and even larger in pyrimidine bases) remove the problem of small and conformationally dependent $^{3}J_{H,H}$ couplings in sugars. HCCH-TOCSY with a long (around 60 ms) spin-lock at low power can correlate H2 to H8 in adenine, thus helping to assign the H2 protons.

3D ^{13}C-edited NOESY (^{13}C-NOESY-HMQC) identifies sugar-to-base connectivity in a manner similar to homonuclear NOESY spectra. The intra/inter-residue ambiguity can be resolved with the help of

2D and 3D HCN and HCNCH experiments that connect H1′ to H6/8 through bond either directly (HCNCH) or based on common N1/9 (**HsCNb** and **HbCNb** from the sugar and base sides, respectively) (Figs. 6.3 and 6.4). The recent versions of the experiments (see below) provide high sensitivity and are robust enough for routine applications.

3D HCP/PCH, PCCH-TOCSY and HPHCH experiments provide assignment of the sugar-phosphate backbone in the case of HCP based on H4′-C4′-P transfers. Some of the experiments, however, do not provide enough sensitivity when applied to large molecules.

With interest continually shifting to studies of systems of higher molecular weight, the methods for resonance assignment face a double challenge, i.e. of improving the spectral resolution and sensitivity at the same time. The first part of the task has been addressed mainly by the introduction of isotopically labeled samples and multi-nuclear multi-dimensional experiments. The isotope labeling, however, is somewhat counterproductive with respect to the other side of the challenge, the sensitivity, because of an increased dipolar relaxation caused by higher density of magnetically active nuclei. The experiments that rely on polarization transfer between heteronuclei are especially vulnerable to the loss of signal due to relaxation, since the heteronuclear coupling constants tend to be small and the pulse sequences include evolution intervals of tens of milliseconds. At first sight, the relaxation rate at given conditions is a physical constant that cannot be changed by the experimenter. However, the relaxation rates of different spin states can differ dramatically, and a judicious choice of the spin state during the critical periods of the experiment can minimize the relaxation effects.

One of the possibilities is to replace the evolution of the single-quantum states by the multiple-quantum ones during extended evolution periods. As theoretically predicted by Griffey and Redfield [34] and experimentally demonstrated by Grzesiek and Bax [35], keeping a spin pair in the state of multiple-quantum coherence (MQ) eliminates most of the dipolar contribution to the spin-spin relaxation. With the X-H (X is a heteronucleus) spin pairs, this is partially offset by a higher rate of proton cross-relaxation in the transverse plane with the remote ^1H spins. Since the MQ coherences consist of transverse

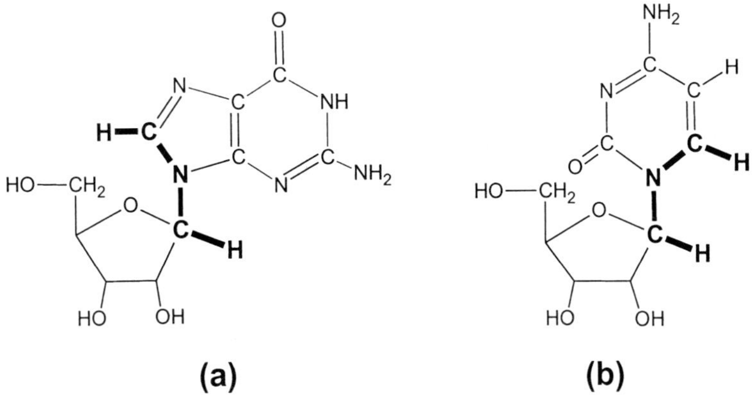

Fig. 6.3 Sugar-to-base connectivities established by through-bond HCN and HCNCH experiments. Coherence transfer pathways in purine (**a**) and pyrimidine (**b**) nucleosides are shown in bold.

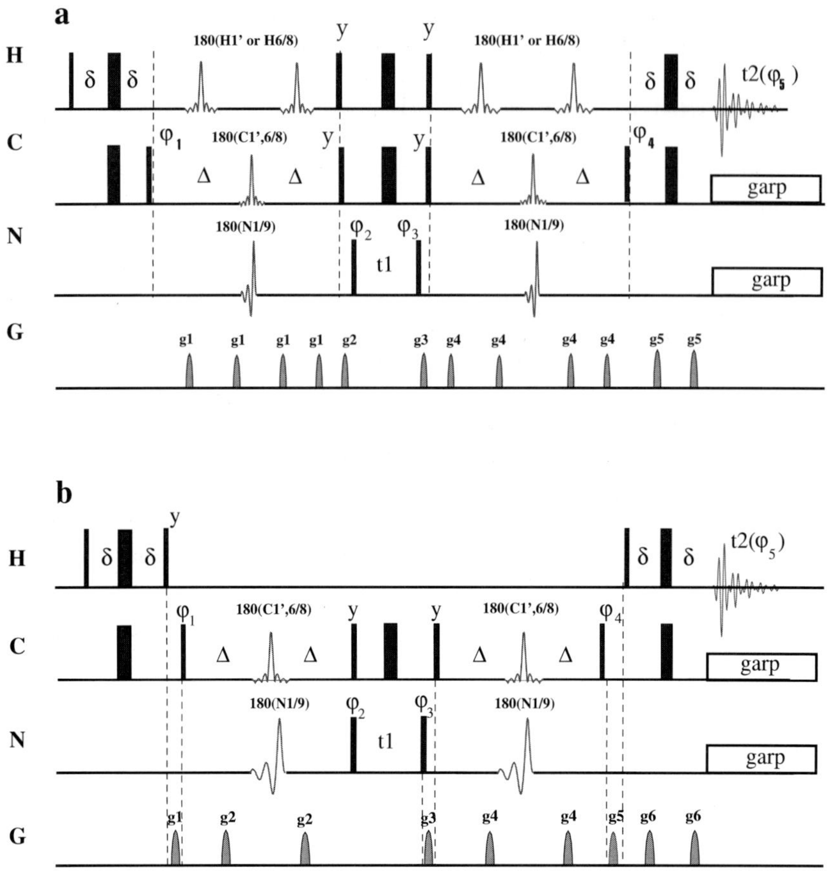

Fig. 6.4 Pulse schemes for MQ [37] (**a**) and TROSY [45] (**b**) **HCN** experiments. The thin and thick bars represent nonselective 90° and 180° pulses, respectively. $\delta = 1.60$ ms for $J_{C1'H1'}$ or 1.25 ms for $J_{C6/8H6/8}$; $\Delta = 15$ ms. The band-selective pulses are set as follows (500 MHz): proton 4 ms band-selective 180° REBURP pulses centered at 5.7 ppm and 7.8 ppm for H1' and H6/8 correlations, respectively; 3.0 ms REBURP centered at 90 ppm for **HsCNb** (**a**) or 2.5 ms REBURP centered at 140 ppm for **HbCNb** correlations in **b**; 2.0 ms IBURP-2 on ^{15}N positioned at 158 ppm. GARP decoupling of ^{13}C and ^{15}N is used during detection. The pulses are applied along the *x*-axis unless otherwise specified. Phase cycling: $\varphi_1 = x, -x$; $\varphi_2 = 2(x), 2(-x)$; $\varphi_3 = 8(x), 8(-x)$; $\varphi_4 = 4(x), 4(-x)$; $\varphi_5 = abba$, where $a = x, -x, -x, x$, and $b = -x, x, x, -x$. In addition, φ_2 is incremented in the States-TPPI manner to achieve quadrature detection in the F_1 dimension. The gradient pulses are for purging purposes only and need not to be in any specific ratios; gradients applied for 1 ms with the strengths of 2–10 G/cm are sufficient.

magnetization, they can be dephased by passive scalar interactions with neighboring nuclei, resulting in substantial decrease of attainable sensitivity. Decoupling of these interactions using band-selective pulses and constant time evolution periods is mandatory if the highest possible sensitivity is to be obtained. Because of a rather low proton density in oligonucleotides, the use of MQ coherence proved to produce a significant sensitivity improvement in the HCN experiments for the intra-residual correlation of H1′, C1′, N9/N1, C6/C8 and H8/H6 nuclei [36–38]. Sensitivity increase by a factor of 3.5–6.5 when compared to the original SQ versions were observed in the HCN correlations with incorporated band-selective ^{1}H pulses (Fig. 6.4a) on a 40-nucleotide RNA aptamer [37]. The MQ-HCN experiment was also combined with a CCH-TOCSY sequence in order to improve resolution for the ribose resonance assignment [39]. Sensitivity enhancement by incorporating the gradient scheme that preserves both coherence transfer pathways [40, 41] was further applied to the MQ-HCN-CCH-TOCSY and MQ-HCN-CCH-COSY experiments [42]. To correlate sugar-to-base resonance in the cases of severe overlap in both C1′-H1′ and C6/8-H6/8 regions, 2D and 3D bi-directional HCNCH experiments have been proposed [43] that involve as many as five different frequencies (H1′, C1′, N1/9, C6/8 and H6/8). The correlation is established by a rectangular arrangement of crossover and out-and-back peaks in the proton-carbon correlated spectrum and can be further resolved in the nitrogen dimension in a 3D experiment. The somewhat lower sensitivity of the experiment as compared to other HCN correlation schemes is compensated by the fact that information obtained is equivalent to up to three out-and-back HCN experiments.

The HMQC approach has also been employed in the measurements of ^3J(H3′P) and ^3J(C4′P) coupling constants using a 2D $\{^{31}P\}$ spin-echo difference constant-time experiment [44] to increase sensitivity by a factor of 1.5–2.4 in a 17 kDa DNA-protein complex.

TROSY-type enhancements (transverse relaxation-optimized spectroscopy, see Chapt. 10) have also been incorporated into the experiments for nucleic acids in order to achieve substantial reduction of the relaxation rates. For the TROSY to be efficient, the contributions of the CSA (chemical shift anisotropy) and DD (dipole-dipole) mechanisms to transverse relaxation must be comparable in size, and their tensors must have similar orientations. In nucleic acids, the highest reduction occurs for aromatic carbons at magnetic fields between 12 and 19 T [45], which corresponds to proton resonance frequencies of approximately 500–800 MHz. The efficiency of the TROSY approach based on the interaction of ^{13}C CSA with ^{1}H-^{13}C DD was demonstrated for the ^{1}H-^{13}C correlation experiment [46] and HCN experiments for sugar-to-base through-bond correlation [45] (Fig. 6.4b). Further sensitivity increase of about 20% (with 17 kDa *Antennapedia* homeodomain-DNA complex) in the HCN-type experiment was reported by incorporating [^{13}C-^{13}C]-TROSY of base carbon nuclei [47]. Since the CSA values of sugar carbons are low, TROSY does not provide significant sensitivity advantage for the sugar moiety, and selective multiple-quantum (MQ) experiments [37] are more sensitive for the sugar nuclei [45] even at very high magnetic fields (Fig. 6.5). The recently proposed MQ-TROSY-HCN pulse sequence [48] combines the advantages of MQ-HCN for sugar-to-base and TROSY-HCN for intra-base correlations in a single experiment.

The TROSY effect was also used in a relayed HCCH-COSY experiment to correlate adenine H2/H8 resonances in uniformly ^{13}C-labeled RNA molecules [49], and significant sensitivity over the existing HCCH-TOCSY version was reported. Magnetization is transferred simultaneously in an out-and-back manner from H2 and H8 to the three aromatic carbon spins, C4, C5 and C6, establishing thus the connectivity within the adenine base spin system.

In principle, TROSY is not limited to interference between dipole-dipole coupling and CSA. Pervushin [50] proposed the use of the differential conformational exchange-in-

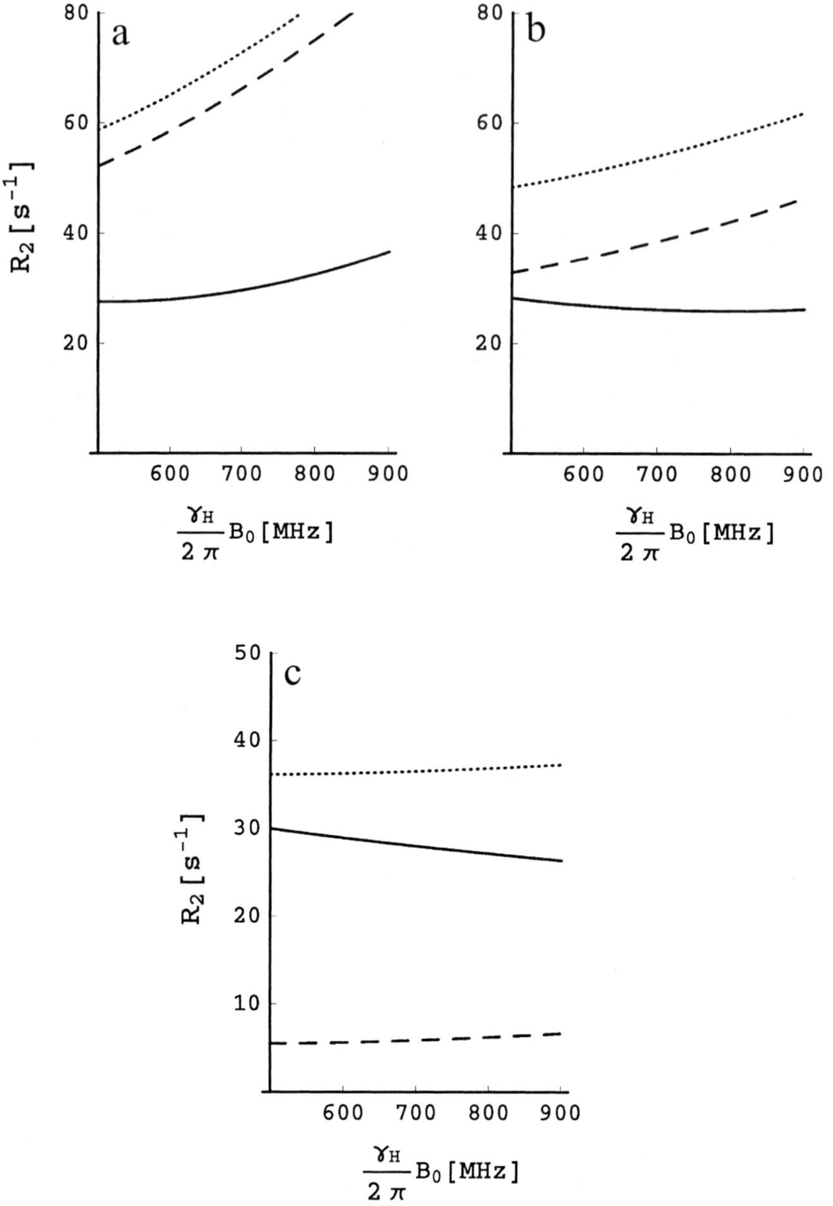

Fig. 6.5 The dependence of transverse relaxation rates on static magnetic field B_0 calculated for **a** C6 of Cytosine, **b** C8 of Guanine and **c** C1′ (average value for all nucleosides). Solid, dashed and dotted lines represent the slowly relaxing (TROSY) component of the CH doublet, multiple quantum and single quantum coherence, respectively.

duced transverse relaxation (CSX) in the ZQ (zero quantum) and DQ (double quantum) coherences to optimize transverse relaxation properties in the experiments that correlate spins and measure scalar couplings across the hydrogen bonds in nucleic acids.

Most of the early methods for assigning isotopically labeled nucleic acids were developed using RNA samples, since ^{13}C and ^{15}N isotopes can be incorporated in RNA more easily than in DNA. While most of the magnetization transfer pathways in RNA and DNA are the same, the latter contains thymine rather than uracil. To remove the ambiguity of intra- and inter-residue H6-CH3 peak assignment in thymine, the HCCCH through-bond method was proposed [51] as a more sensitive alternative to NOE or ROE.

Similarly, the assignment of exchangeable hydrogens based on imino-imino and imino-amino sequential and intrabase NOEs often fails in nonhelical regions such as bulges and loops. This problem has been addressed by through-bond TOCSY-based experiments linking the imino/amino hydrogen with the nonexchangeable protons on the same base [52–54]. In the case of pyrimidine bases, better sensitivity was reported using an experiment to correlate their imino/amino protons with H5 using consecutive INEPT steps [55].

An improved method for the ^{1}H-^{31}P correlated experiment (HP-CPMG-HSQC) was proposed [56] achieving $\sim$ 2.5-fold sensitivity increase as compared to phosphorus-excited methods. The effective correlation is achieved by the application of a CPMG pulse train in the XY-16 expansion scheme during the periods of polarization transfer. The CPMG sequence suppresses the effect of conformational exchange and cross-correlated relaxation while achieving in-phase transfer in the ^{1}H-^{1}H coupling network. The pulse sequence can be combined with an additional magnetization transfer step such as NOESY to extend the correlation to better resolved sugar or base proton resonances, e.g. H1· The method does not require an isotopically enriched sample. This approach shows several advantages over the previously published heteronuclear TOCSY experiment proposed by Kellog to obtain the identical information [57, 58].

The pulse sequences used for the resonance assignments are not discussed in this chapter. For further details the interested reader is referred to the paper of Ref. [2] and references therein.

6.5 NMR Detection of Hydrogen Bonds

Hydrogen bonds play a crucial role in the formation of secondary and tertiary structures of biological macromolecules. Recently, NMR methods have been developed to detect and measure scalar couplings via H-bonds. These methods are reviewed by Dingley et al. in Chapt. 9. The typical NOE cross-peak patterns resulting from the dipole-dipole interactions within the Watson-Crick and Hoogsteen base pairs [2] prove the hydrogen-bonding network only on the proton-donor sites. As a result, the determination of proton-acceptor sites has remained ambiguous. In nucleic acids, mainly imino –NH and amino –NH_2 serve as proton donors while aromatic nonprotonated nitrogens and –C=O groups represent the proton-acceptor sites. In addition, also –OH and –P=O groups can be involved in the hydrogen bond formation as proton donors and proton acceptors, respectively.

The detection of scalar couplings is not only a theoretically interesting direct proof of the covalent character of the hydrogen bonds. It is also of great value in studies of struc-

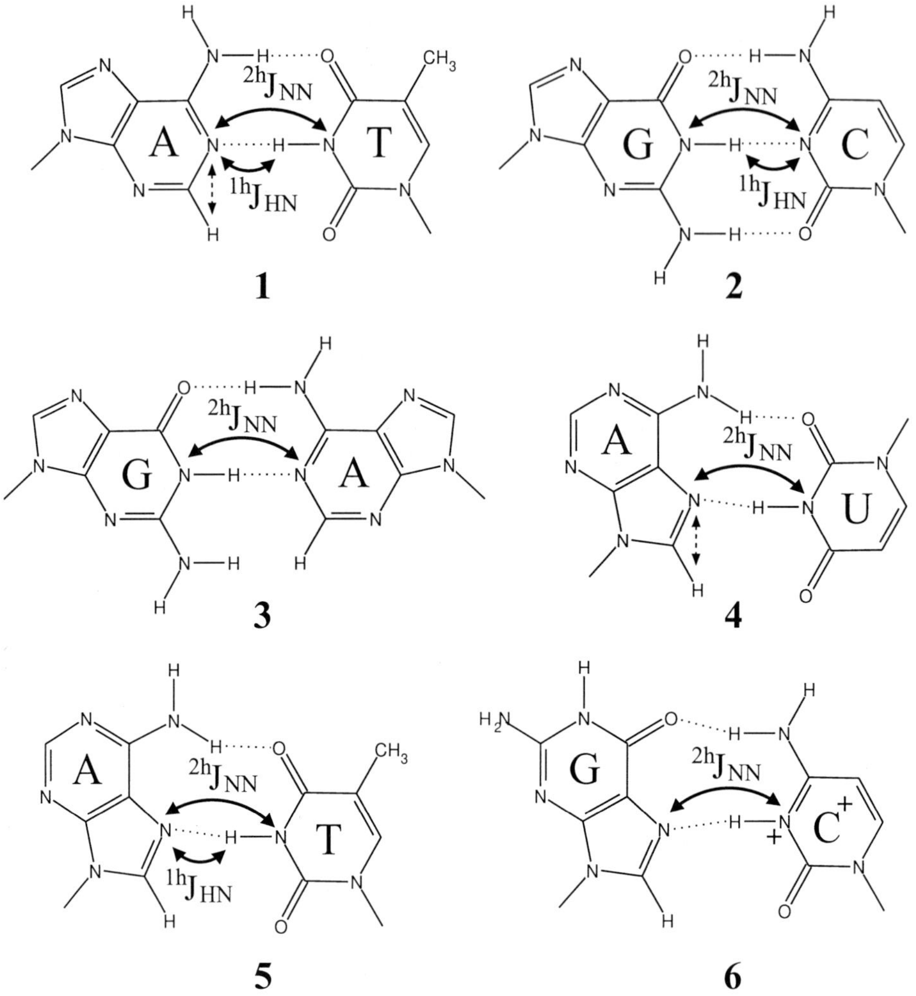

Fig. 6.6 Schematics of hydrogen bonds between purine and pyrimidine bases with indicated *trans*-hydrogen bond scalar interactions and related coupling constants, which can be measured using NMR. In addition to correlations between exchangeable protons and nitrogens, also a relayed transfer to nonexchangeable aromatic protons, shown by a dashed arrow, can be employed. The following pairs of atoms serve as proton donors and acceptors, respectively: **1**, **2** T(N3)/U(N3) – A(N1), G(N1) – C(N3) in Watson-Crick pairs [109, 110], **3** G(N1) – A(N1) in imino-hydrogen-bonded G•A pairs [112], T(N3)/U(N3) – A(N7) in (**4**) reversed Hoogsteen [112] or (**5**) Hoogsteen pairs [111], **6** C+(N3) – G(N7) in Hoogsteen pairs [111].

tures of biomacromolecules. In DNA and RNA research, the correlation via hydrogen bonds can help to assign nuclei of bases by extending the scalar coupling network to the hydrogen-bonded partners. The information about hydrogen bonds also provides very useful additional restraints for structural determination, especially in the case of noncanonical motifs. The importance of the knowledge of hydrogen bonds is even greater for

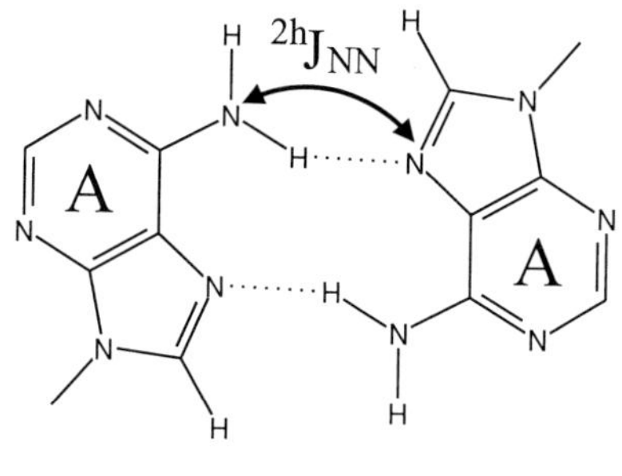

7

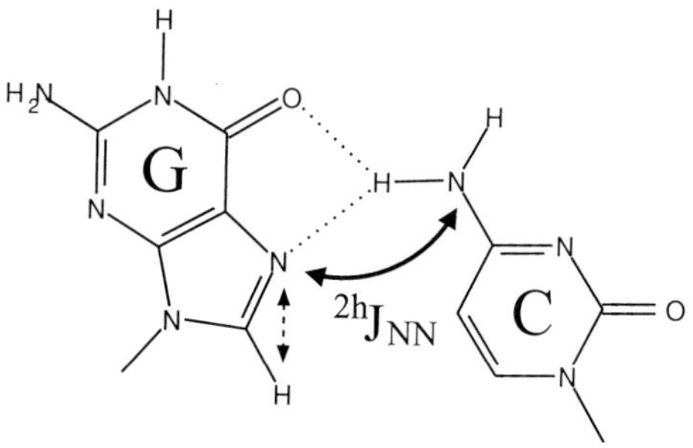

8

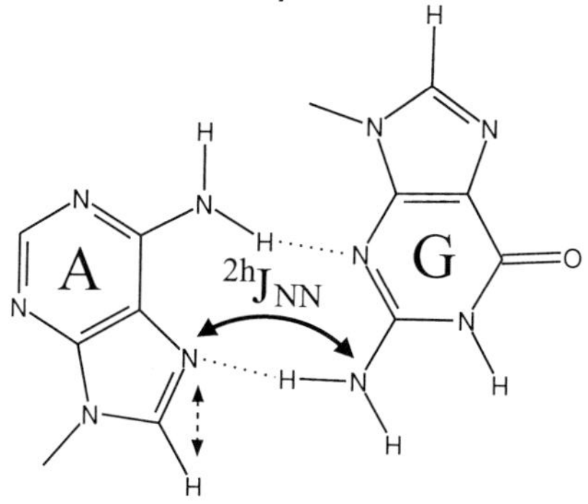

9

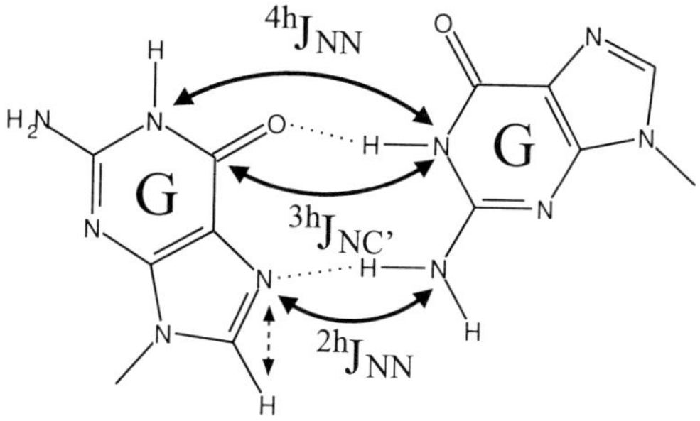

10

Fig. 6.6 (continued) **7** A(N6) – A(N7) in AA mismatch pairs [113], **8** C(N4) – G(N7) in the G•C•G•C tetrad [113], **9** A(N6) – G(N1) in the A•(G•G•G•G)•A hexad [113], and **10** G(N2) – G(N7) the G•G•G•G tetrad [114] and the A•(G•G•G•G)•A hexad [113]. G(N1) and G(O6) served as acceptor and donor, respectively, in the (G•G•G•G) tetrad [114, 115].

intermolecular complexes of DNA and RNA with proteins or small ligands. *Trans*-hydrogen bond couplings measured so far are summarized in Fig. 6.6.

Currently, the best-characterized coupling constants in nucleic acids are $^{2h}J_{NN}$ and $^{1h}J_{HN}$. Detection of hydrogen bonding via scalar coupling also helped to clarify the confusion with respect to correct assignments of N1 and N3 nitrogen chemical shifts existing in the literature. It was confirmed [59] that N1 resonates downfield from the N3 chemical shifts, typically at over 220 ppm when referenced to liquid ammonia.

Information about the H…N hydrogen bond length can also be obtained from residual dipolar couplings, as demonstrated recently by Wu et al. [60]. The N…H distance is calculated from the ratio of the inter-nucleotide dipolar coupling and the corresponding one-bond $^{15}N–^{1}H$ dipolar coupling. The data are in good agreement with those observed in crystal structures and confirm previous experience that H-bond length is shorter in A-T than in G-C base pairs.

The hydrogen bond length in Watson-Crick base pairs can be characterized using the recently developed method of measuring the cross-correlated relaxation [61] between ^{1}H chemical shift anisotropy and dipole-dipole coupling of ^{1}H and its hydrogen bond donor

^{15}N. Using the two measured cross-correlated relaxation rates, an apparent hydrogen bond length can be determined. Data for the $^{15}N3$-$^{1}H3$...$^{15}N1$ hydrogen bond in A-T base pairs of *Antennapedia* homeodomain DNA complex with a correlation time of 20 ns has been presented.

6.6
Measurement of *J*-Couplings

Measurements of scalar couplings between nonexchangeable nuclei in nucleic acids provide a large number of restraints that can be successfully used in the structure refinement. The arsenal of existing methods has recently been extended by several additions. New pulse sequences for measuring ^{1}H-^{31}P coupling constants in small, unlabeled DNA or RNA fragments with favorable relaxation properties have been proposed to obtain experimental restraints for β and ε dihedral angles [62]. The ^{1}H-^{1}H proton coupling constants in ribose and deoxyribose rings are useful sources of information concerning the sugar puckering of ribose and deoxyribose sugar rings. The prospects of specific deuteration of the deoxyribose H2″ proton and its influence on transverse cross-relaxation rate of H2′ and measurements of coupling constants to neighboring protons have recently been described [63, 64]. Substantial improvements in sensitivity and resolution in modified COSY experiments (implementation of time-shared homonuclear decoupling, application of 90° mixing pulse in P.E.COSY) have been demonstrated on the Drew-Dickerson dodecamer $d(CGCGAATTCGCG)_2$.

The conformation of the phosphodiester backbone can be assessed using the inter-residual $^{3}J(C4'(i)P(i{+}1))$ and $^{3}J(C2'(i)P(i{+}1))$, and intra-residual $^{3}J(C4'(i)P(i))$ scalar couplings, which reflect the torsion angles ε and β, respectively. Two constant-time *J*-quantitative experiments have been designed to obtain their values [65, 66]. The three-bond scalar H3′-P couplings can also be used to determine the torsion angle ε. In addition, the distance of 2.85 Å between these two nuclei in B-DNA gives rise to residual dipolar interactions in aligned media. A constant-time $^{1}H3'$-^{31}P NOE difference experiment has been proposed to obtain the H3′-P couplings and has been tested to measure the scalar couplings in an isotropic solution and the sum of the scalar and dipolar couplings in a liquid crystalline phase [64]. The torsion angle β can be estimated from H5′-P and H5″-P scalar couplings. Severe overlap in a conventional HSQC experiment usually prevents accurate measurement of these couplings. A homonuclear double-quantum experiment has been used to generate $^{1}H5'$-^{31}P and $^{1}H5''$-^{31}P cross peaks in an HMQC-type experiment. Resolution is improved as a result of efficient suppression of $^{1}H4'$-^{31}P correlations and sensitivity is enhanced by slower relaxation of the double-quantum coherence. A 3D version of the experiment has also been proposed [67].

6.7
Residual Dipolar Couplings – Use for Structure Elucidation

The lack of long-range restraints has hindered the precise and accurate structure determination of nucleic acids for a long time. The isotopic enrichment mostly improved the accuracy of local geometry without a pronounced influence on the overall shape. However, recently introduced technology for obtaining additional structural restraints using a small degree of molecular alignment with the magnetic field [68, 69] provides a direct way to obtain the desired long-range restraints. The technique is reviewed in Chapt. 8.

In structure determination of nucleic acids, mainly one-bond residual dipolar couplings have been used so far. Varani and coworkers used residual dipolar couplings for structure determination of a U1A protein-RNA complex [70]. Although they did not observe a significant improvement in the precision of the refined structures, the use of long-range orientational restraints resulted in a different mutual orientation of the two double-helical RNA stems. Tjandra, Bax and coworkers reported a structure refinement of the Drew-Dickerson dodecamer $d(CGCGAATTCGCG)_2$ using an extremely large number of residual dipolar couplings (198 ^{13}C-^{1}H, 10 ^{15}N-^{1}H, and 200 ^{1}H-^{1}H) along with a set of conventional restraints (NOEs and torsion angles) [71]. The study has unambiguously shown that inclusion of dipolar data substantially improves the structure determination. However, it has been noted that the accuracy of the determined structures is limited primarily by the uncertainty in the magnitude and rhombicity of the alignment tensor.

An essential role of long-range orientational restraints for precise and accurate determination of nucleic acid structures has been reported by Pardi and coworkers [72]. It has been shown on a theoretical study (the input restraints were derived from the target structures) that dramatic improvement in a global structure, as well as noticeable improvement in a local structure, can be obtained upon inclusion of residual dipolar couplings. The same laboratory also performed a refinement of rigid domains using a small number (27) of residual ^{15}N-^{1}H dipolar couplings to determine the relative orientation of helical stems in *E. coli* $tRNA^{val}$ [73]. A report showing that the rotational helical parameters can be determined directly from residual dipolar couplings without a prior knowledge of the refined structure has been published recently [74]. Based on the knowledge of a limited set of residual dipolar couplings for each base, the relative mutual orientation of bases can be obtained using the METHADON procedure. This concept assumes that bases in nucleic acids are nonplanar because of both intrinsic amino-group nonplanarity and out-of-plane vibrations. The degree of nonplanarity is also influenced by hydrogen bonding and stacking interactions [75]. Patel and coworkers carried out an order matrix analysis on two sets of residual dipolar couplings obtained from two different aligning environments (phage Pf1 and very high magnetic field) to determine the three-dimensional fold of C2-symmetric homodimer d(GGGTTCAGG) quadruplex [76]. The relative orientation of structural motifs within a monomeric unit, even in the presence of inter- and intra-strand NOE ambiguity, was determined.

Recently, Sibille et al. [77] proposed an interesting modification of the commonly used refinement methodology to investigate the long-range structural features of molecules in solution using the residual dipolar couplings. Structure refinement normally assumes that the molecular alignment can be characterized by a single tensor and that the molecule is essentially rigid. To obtain improved long-range order of the RNA-theophylline complex, whose long-range order was poorly defined in the NOE-based refinement, alignment tensors for two stem regions were determined independently and found to converge to similar magnitudes. In the second stage of refinement, a single alignment tensor was thus applied for the whole molecule and an average molecular conformation satisfying all experimental data was determined. This improved the long-range order and the RMSD (root mean square deviation) of all heavy atoms decreased from 3.5 Å (without RDCs) to 1.5 Å (with RDCs) in the ensemble of converged structures.

The number of one-bond ^{1}H-^{13}C and ^{1}H-^{15}N spin pairs in adenine, guanine, thymine, uracil and cytosine bases, which supply sufficiently large residual dipolar couplings, is limited to just a few. However, in addition to one-bond RDCs, two-bond interactions may also provide an important source of restraints for structure refinement in nucleic acids. In purine and pyrimidine bases, a number of two-bond and one-bond homo- and heteronuclear distances have a well-defined length and can be successfully used during the structure calculations. We have recently described a suite of spin-state-selective excitation (S3E) NMR experiments for the measurements of very small one-bond (^{13}C-^{13}C, ^{13}C-^{15}N) and two-bond (^{1}H-^{13}C, ^{1}H-^{15}N) coupling constants in order to extend the number of measurable RDCs [78]. Very small scalar and residual dipolar couplings were measured in the ^{13}C, ^{15}N-labeled DNA hairpin d(GCGAAGC) with very high precision and accuracy. The most interesting feature of minihairpins of the general formula d(GCGNAGC) (N=A, G, C, T) is their extraordinary stability, represented by their high melting temperature, high polyacrylamide gel mobility and resistance against nucleases [79, 80]. The thermodynamically most stable fragment is d(GCGAAGC) (T_m=76 °C) [81]. Although a model structure of this hairpin was proposed a few years ago [79], no accurate structure has been published so far. We used both "classical" NMR restraints and a set of residual dipolar couplings in the structure refinement of this DNA hairpin. A complete report of this research will be published elsewhere [82]. The ensembles of 13 and 14 converged structures obtained using the simulated annealing molecular dynamics protocol with NOE and dihedral angles, and NOE, dihedral angles and RDCs restraints, respectively are shown in Fig. 6.7 a, b. The average of atomic RMSD (pairwise) for these ensembles are 0.39±0.15 Å and 0.32±0.11 Å. Both structures display the same conformational features and have been determined with very high precision as documented by the RMSD values. However, a remarkable improvement in the definition of the A4 nucleotide, which sits at the top of the hairpin, has been obtained using the RDCs data (Fig. 6.7 c, d). The position of this nucleotide is not well characterized by NOEs, and just five residual dipolar couplings helped to define the structure of the A4 base with the precision close to that of other bases. The RMSD of the A4 residue improved from 0.72±0.43 Å to 0.40±0.16 Å. The experimentally obtained residual dipolar couplings can be correlated with the theoretically predicted values calculated from the resulting structure. As shown in Fig. 6.8, very good agreement has been achieved for both sugars and aromatic base, indicating that even very small RDCs can be useful in the structure refinement. The size of RDCs with various magnetogyric constants has been properly scaled to account for differences in the relative size of individual residual dipolar data.

Since NMR data may reflect the conformational averaging resulting from a substantial flexibility of DNA/RNA fragments, the parameters of the ensemble of structures obtained by MD refinement and experimental NMR restraints can display internal inconsistency. James and coworkers developed a method to determine the family of structural conformers and their populations based on NMR data [83]. This approach applies multiple-copy refinement with floating weights, which better reflects the conformational dynamics of nucleic acids.

In general, the results of NMR structure refinement should be independent of the applied force field and refinement tools and reflect primarily the experimental restraints. At

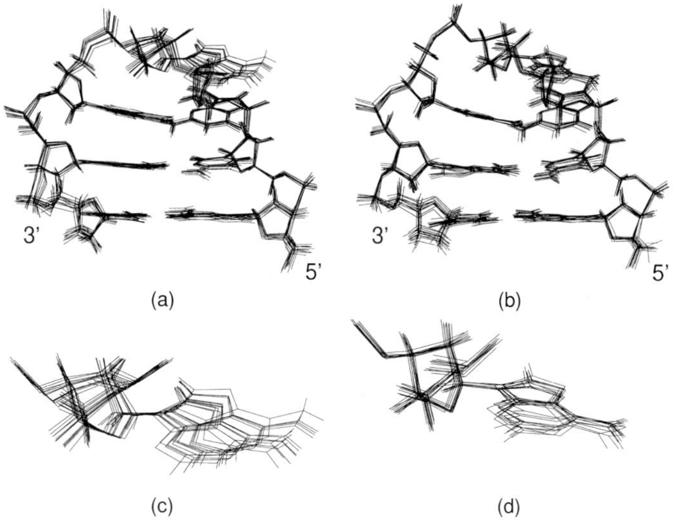

Fig. 6.7 Structure of d(GCGAAGC) hairpin calculated using the set of 99 NOEs and 37 dihedral angles restraints without (**a**) and with (**b**) inclusion of 38 residual dipolar couplings both from sugars (13) and bases (25) [82]. The ensemble **a** and **b** contains 13 and 14 converged structures with pairwise RMSD of 0.39±0.15 Å and 0.32±0.11 Å, respectively. Changes in A4 geometry after inclusion of RDCs are shown in parts **c** and **d** in detail.

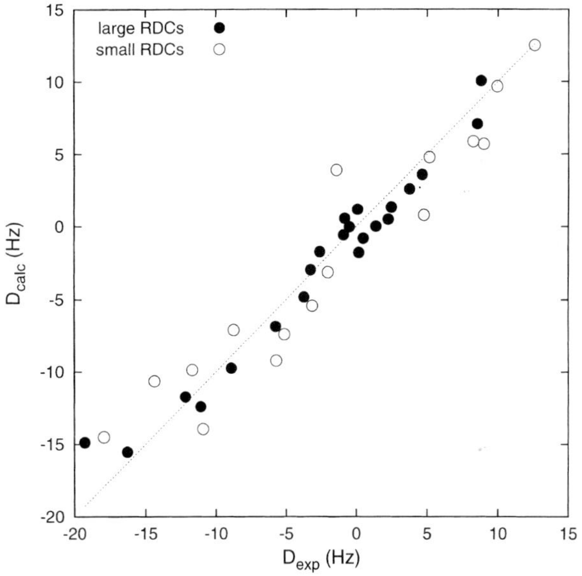

Fig. 6.8 Correlation of experimental and theoretical residual dipolar couplings calculated using the structural data obtained from restrained molecular dynamics with NOEs, dihedral angle and RDCs-restraints. The structure was refined with Amber 7.0 and recently incorporated flat-bottom RDC potential (courtesy of David Case). Filled and open circles represent large and small RDCs, respectively.

present, the resulting structures are influenced to a large extent by the computational protocols, since NMR data is mostly underdetermined. In such a situation, the impact of empirical potentials and MD simulations cannot be avoided and the use of the best available approach is essential. Application of the most accurate force fields [84, 85] with the electrostatic terms is therefore strongly recommended. Since nucleic acids are polyelectrolytes, the electrostatic interactions play an important role during a structure refinement. Recent developments of molecular dynamics simulations, which employ the explicit solvent [86] or sophisticated implicit model of solvation [87], may help to improve the correctness of the NMR structures.

6.8 Relaxation Studies of Nucleic Acids

The major cause of relaxation in NMR is the reorientation of the interacting spins with respect to each other and to the external magnetic field. The relaxation rates thus contain a wealth of information about the structure and motion of the molecule. Among the relaxation characteristics, by far the most commonly measured is the proton longitudinal cross-relaxation (NOE). To tap the information content of relaxation in greater depth, however, requires significantly more sophistication. First, a multinuclear approach is usually required, which became practical only after the isotopic labeling techniques for nucleic acids were introduced. Second, high performance hardware and new techniques are needed to measure the often subtle effects.

For the above reasons, suitable methods are only starting to emerge. In the following paragraphs we will discuss the application of the relaxation measurements to dynamics studies and structure determination separately.

Relaxation studies of protein dynamics, especially those based on proton-detected ^{15}N relaxation, have now become almost routine (for a more detailed account of ^{15}N relaxation measurements, see Chapt. 12). With nucleic acids, the situation is significantly different, and only a few relaxation-based dynamics studies have appeared so far. In close analogy with the protein studies, the relaxation of imino nitrogen can be followed [88]. However, being located only in G and U (T) bases, the imino nitrogens in nucleic acids do not occupy locations in the molecule that are as prominent as that occupied by the amide proton in the peptidic bond of proteins. The imino protons undergo fast exchange with the solvent, which complicates the measurements of the spectra, especially for those residues that do not form strong hydrogen bonds. Moreover, the chemical shift anisotropies of imino nitrogens in nucleic acid bases, which are necessary for data interpretation, were not known until recently. In the last few years the situation was rectified, and both experimental and computed values for the ^{15}N chemical shift tensors were published and are summarized in Tabs. 6.1 and 6.2. Hu et al. [89] established the principal values of the nitrogen chemical shift tensors in adenine, cytosine, guanine and thymine bases through solid-state NMR supported by quantum mechanical calculations. Several authors [90–92] studied the effect of hydrogen bonding on nitrogen chemical shielding in nucleic acids by *ab initio* methods. It seems, however, that the calculations tend to exaggerate the effect, probably as a result of neglecting the interactions with the water molecules of the solvent. Similarly, the results of investigations of ^{13}C shieldings by theoretical methods

Tab. 6.1 Calculated and experimental solid-state principle values of the chemical shifts of selected nitrogen atoms in nucleic acid bases

Calculated					*Experimental*			
Atom	δ_{11}	δ_{22}	δ_{33}	δ_{iso}	δ_{11}	δ_{22}	δ_{33}	δ_{iso}
N1(A)	396	319	14	243	365	305	2	224
N3(C)	334	247	57	213	320	242	53	205
N1(G)	227	137	98	154	222	149	79	150
N3(T)	236	156	108	167	228	150	99	159
N3(U)	233	163	112	169	223	154	102	160

δ_{iso} is isotropic chemical shift.
All values are from Ref. [116] and referenced to liquid ammonia.
δ_{nn} are components of the chemical shift tensor.

Tab. 6.2 Calculated principle values and orientations of chemical shift tensors of the imino nitrogens in nucleic acid bases and base pairs

Atom	δ_{11}	δ_{22}	δ_{33}	δ_{iso}	θ	Δ_{eff}	**Note**
N1(G)	229	120	113	151	9.4	113	a)
N1(G)	246	120	109	158	11.4	132	b)
N1(G-C)	240	153	85	159	18.4	134	a)
N1(G-C)	230	140	84	151	18.8	128	b)
N3(T)	232	140	138	170	6.8	94	a)
N3(T)	232	133	133	166	8.7	99	b)
N3(A-T)	245	195	104	181	17.3	124	a)
N3(A-T)	233	178	99	170	20.0	117	b)
N3(U)	234	142	139	172	7.5	94	a)
N3(U)	235	135	134	168	9.0	101	b)
N3(A-U)	248	197	106	184	19.1	125	a)

a) According to Czernek [69]. The values apply to base pairs with N...N distances of 2.89 Å, 2.78 Å, 2.79 Å and ∞ for G-C, A-U, A-T base pairs and individual bases, respectively.
b) According to Sitkoff and Case [117]; the bases were methylated at their N1 (pyrimidines) or N9 (purines) positions.
δ_{nn} are components of the chemical shift tensor.
δ_{iso} is isotropic chemical shift.
θ is the angle between δ_{11} and the N-H bond vector.
Δ_{eff} is defined as $\Delta_{eff}=(\delta_{11}^2+\delta_{22}^2+\delta_{33}^2 - \delta_{11}\delta_{22} - \delta_{11}\delta_{33} - \delta_{22}\delta_{33})^{1/2}$ and characterizes the autorelaxation effects of the anisotropic chemical shifts.
All values are referenced to liquid ammonia.

were described for both sugar [93–95] and base [45, 90] carbon nuclei (Tab. 6.3). Though there are some variations in the results, depending on the method of calculation used, all the chemical shifts calculated by a particular method tend to be shifted by the same value, so that the results are well usable for most purposes.

Unlike imino nitrogen, protonated carbon nuclei are abundant throughout the molecules of nucleic acids providing many sites as possible probes into the molecular motions. How-

Tab. 6.3 Calculated carbon chemical shift tensors in nucleosides.

Atom	δ_{11}	δ_{22}	δ_{33}	θ_1	θ_2	θ_3	δ_{iso}	Δ_{eff}	Note
C8(G)	211	128	75	30.4	120.4	89.1	138	119	a)
C8(G)	213	126	85	21.6	111.6	90.0	141	113	b)
C8(A)	213	138	72	29.4	119.4	89.8	141	123	a)
C8(A)	215	130	86	21.2	111.2	90.0	144	114	b)
C6(C)	245	160	34	26.9	116.9	89.3	146	184	a)
C6(C)	243	142	62	19.8	119.8	90.0	149	157	b)
C6(T)	236	120	70	19.7	119.7	90.0	142	147	b)
C6(U)	248	144	36	25.7	115.7	89.5	143	184	a)
C6(U)	248	131	57	19.3	119.3	90.0	145	167	b)
C2(A)	239	159	74	3.6	93.6	89.6	157	144	c)
C2(A)	243	160	74	3.8	93.8	90.0	159	146	b)
C5(C)	172	80	22	9.0	98.9	89.3	92	131	c)
C5(C)	175	82	22	9.6	99.6	90.0	93	134	b)
C5(U)	182	98	36	10.6	100.6	89.5	105	127	c)
C5(U)	189	101	36	10.7	100.7	90.0	109	133	b)
C1′(G)	120	103	73	22.7	109.9	78.2	99	41	a)
C1′(A)	120	104	74	21.5	107.7	78.2	99	41	a)
C1′(C)	120	105	78	28.6	116.2	79.3	101	37	a)
C1′(U)	120	105	76	23.5	111.5	81.0	100	39	a)
C1′(T)	122	103	61	23.2	111.3	81.1	95	54	c)

a) Calculated for mononucleosides [44].
b) Calculated for bases methylated at N1/9 [117].
c) J. Czernek, private communication, the same calculations as in Ref. [44].
δ_{11}, δ_{22} and δ_{33} are the principal component of the chemical shift tensor.
θ_x, θ_y and θ_z are the angles between the corresponding tensor component and the C-H bond.
δ_{iso} is the isotropic chemical shift.
The definition of Δ_{eff} is given in the footnote to Table 6.2.
The values are references to TMS, using the TMS shielding values of 183.87 ppm.

ever, if a molecule is ^{13}C-enriched to a high degree (close to 100%), scalar and dipolar couplings between carbon nuclei complicate the relaxation rate measurements. Relatively free of these effects are C8 of purine nucleotides and C2 of adenine, which are bound to two nitrogen atoms. Hall and Tang [96] used the two sites to study the dynamics of the purine bases in the iron-responsive elements RNA hairpin containing 15 residues. They measured T_1, $T_{1\rho}$ and NOE values of the base carbons as a function of temperature and analyzed the data using the Lipari-Szabo model-free formalism. To alleviate problems associated with the complex network of couplings among ^{13}C nuclei, random enrichment of about 15% is often recommended as a reasonable compromise between sensitivity and suppression of unwanted interactions. Boisbouvier et al. [97] proposed a set of experiments for measurements of ^{13}C relaxation in RNA and DNA bases, namely spin-lattice relaxation rate, spin-spin relaxation rate and CSA-dipolar cross-correlated relaxation rate. The pulse sequences achieve sensitivity and resolution enhancement through spin-state selective excitation. The authors applied the experiments to study the relaxation rates in a 15% ^{13}C-labeled 33-mer RNA-theophylline complex. Based on the measured relaxation rates, they conclude that the shielding tensors of ^{13}C are different for different atom types (C2 of adenine, C8 of purine, C6 of pyrimidine), but do not vary significantly amongst the same atom type. This is in good agreement with the data obtained by *ab initio* calculations.

The nature of the chemical shift tensor is a potential source of complications in relaxation studies. For sugar carbons, the CSAs are around 40 ppm and their contribution to relaxation of protonated carbons is nearly negligible. On the other hand, CSA values of the protonated carbons of the bases are between 120 and 180 ppm, the tensors deviate quite significantly from axial symmetry and none of their principal components is collinear with the C–H bond. This makes interpretation of the relaxation rates in terms of molecular dynamics prohibitively complicated or, if neglected, introduces an error whose magnitude has not yet been evaluated.

Other factors that potentially complicate accurate measurements of relaxation rates, e.g. dipolar or scalar couplings to other ^{13}C nuclei, short ^{1}H-^{13}C distances to nondirectly bonded protons and cross-correlation effects, were studied by Kojima et al. [98], with the conclusion that the magnitude of most of the effects is comparable to random experimental errors. The authors measured altogether nine relaxation parameters at either 500 or 600 MHz for C2 and C8 base carbons as well as for C1′, C2′ and C4′ in the sugars of a DNA decamer duplex enriched with 15% ^{13}C in purine residues. The results were interpreted complementarily by spectral density mapping and model-free analysis. Except for terminal residues, the order parameter S^2 and the corresponding fast internal motion correlation time were about 0.8 ± 0.1 and 20 ± 20 ps, respectively. A small (<3 Hz) but explicit exchange term was found. To overcome some of the complications inherent in the interpretation of R_1 and R_2, the same authors proposed new experiments to measure transverse and longitudinal cross-correlation between ^{13}C-^{1}H dipolar interactions and ^{13}C CSA [99]. The ratio of the transverse to longitudinal cross-correlation rates is independent of the structural variables such as the internuclear distance or the chemical shift tensor as well as of the effect of chemical exchange.

Cross-correlated relaxation between different relaxation mechanisms has been introduced as an additional tool for obtaining structural information in liquid-state NMR spectroscopy [100] (see also Chapt. 7). The method utilizes a principle that cross-relaxation rates depend on the relative orientations of the principal axes of the tensors responsible for relaxation. The structural interpretation of cross-correlated relaxation offers an independent method of obtaining torsion angle restraints provided the corresponding rates can be measured with sufficient accuracy. The methodology based on cross-correlated relaxation rates represents an alternative to scalar coupling constants for the conformational investigation of biomolecules without any need for parametrization. The cross-correlation rate increases linearly with the overall rotation correlation time of the molecule, which makes the method promising for large molecules where measurements of the coupling constants by E. COSY type experiments tend to be difficult. Cross-correlated relaxation rates are reflected directly in the signal intensities and integrals of the experiments, in which the single-, double- or zero-quantum coherences evolve in a constant time manner. If the cross-correlated relaxation rate evolves during a real-time evolution, the relaxation rates have to be extracted from more complicated line shapes by deconvolution.

In nucleic acids, the cross-correlation studies were applied to investigation of the sugar conformation [101–103] of the phosphodiester backbone [65] as well as to some more specialized tasks [104, 105]. The dependence of dipole-dipole cross-correlated relaxation between adjacent CH pairs was utilized to determine local conformations in ribose rings [102, 103]. The relaxation rates were measured by quantitative Γ-HCCH [102], in which the cross-relaxation serves as a means of coherence transfer. Two spectra are recorded, in which one includes effects due to cross-correlation, whereas the other is devoid of these (reference), with corresponding pulse sequences differing by the absence or presence of a specific evolution interval. The cross-relaxation rate is calculated from the ratio of the peak intensities in the cross-correlated and reference spectra. The discrimination between the C2′-endo and C3′-endo puckering modes is based on the signs of the cross-correlation rates between C1′-H1′, C2′-H2′ and/or C3′-H3′, C4′-H4′ spin pairs, or, more reliably, from the ratio of the two rates. Boisbouvier et al. [101] proposed using CSA-dipolar cross-correlated relaxation rates for the same purpose. It has been known from DFT (density functional theory) calculations [93] that the values and orientations of the principal values of the ^{13}C chemical shift tensor depend on the sugar pucker. Experimentally, this is reflected in a variation of cross-correlation rates between ^{13}C CSA and ^{13}C-^{1}H dipolar interactions that can be measured with high accuracy from the differential intensities of the doublets in a proton-coupled ^{1}H-^{13}C correlation experiment. The dependence of the relaxation rates on the dynamics parameters and the magnetic field strength can be removed in the first approximation by calculating the ratios of the cross-relaxation rates for two C-H spin pairs within the same sugar. The authors found that for the C3′/C1′ cross-relaxation rates, the ratio values were greater than 2 for C3′-endo conformation and less than 1 in the case of C2′-endo conformation.

The DD-CSA cross-correlated relaxation, namely that between ^{13}C-^{1}H dipole and ^{31}P-CSA, can also be used to determine backbone α and ζ angles in RNA [65]. The experiment requires oligonucleotides that are ^{13}C-labeled in the sugar moiety. First, ^{1}H-coupled, Γ-DQ/ZQ-HCP spectra are measured. DQ and ZQ spectra are obtained by linear combinations of four subspectra recorded for each t_1-increment. Then, the cross-relaxation rates are calculated from the peak intensity ratios of the doublets in the DQ and ZQ spectra. The observed cross-correlation rates depend on the relative orientations of CH dipoles with respect to the components of the ^{31}P chemical shift tensor. As the components of the ^{31}P chemical shift tensor in RNA are not known, the barium salt of diethyl phosphate was used as a model compound with the σ_{11}, σ_{22} and σ_{33} principal components values of –76 ppm, –16 ppm and 103 ppm, respectively [106]. Since the measured cross-correlation rates are a function of the angles β and ε as well, these angles need to be determined independently using ^{3}J(H, P) and ^{3}J(C, P) coupling constants.

Chiarparin et al. [105] used cross-correlated relaxation of multiple-quantum coherence to investigate hydrogen bonds in a 22-nucleotide RNA hairpin containing five G-C and one A-U base pairs. Since there is a resolved scalar coupling between the donor and acceptor nitrogen nuclei in a hydrogen bond, a two-spin coherence involving the two nuclei can be created. There are three main interactions responsible for the differential relaxation effect in the double- and zero-quantum doublets of the two-spin coherence, namely dipole-dipole (DD) interaction between the N-H pairs formed by the hydrogen atom and both nitrogen atoms involved in the hydrogen bond, the interaction between the chemical shift anisotropies (CSA) of the two nitrogen atoms involved, and chemical shift modulation (CSM) caused by modulations of the nitrogen chemical shifts due to conformational exchange. In a three-spin coherence involving N–H…N atoms of the hydrogen bond, the dipolar couplings between proton and nitrogen are no longer active, and only the latter two interactions are effective. Measuring the relaxation of both two-spin and three-spin coherences thus made it possible to obtain the DD cross-correlation rates as a difference between these two rates. The DD cross-correlation rate is then a function of the polar angles between the principal axes of the interaction with respect to the principal axes of the diffusion tensor. It appeared, however, that the measured cross-correlation rates were dominated by the CSM, which indicates the presence of extensive local dynamics. The accuracy of the much smaller values of dipolar cross-correlation rates was therefore rather low, which made drawing quantitative conclusions about the geometry of the hydrogen bond rather difficult. Nevertheless, the authors were able to make useful inferences on the lengths of the hydrogen bonds.

Carlomagno et al. [104] studied the extraordinary stability of a triple-stranded oligonucleotide, in which the third strand contained 2′-aminoethoxy-substituted riboses, by differential scanning calorimetry, molecular dynamics simulations and NMR relaxation, namely DD cross-correlation rates between C-H vectors of the aminoethoxy groups that depend on the dihedral angle N-C-C-O. The cross-relaxation rates were measured using a slightly modified version of the quantitative Γ-HCCH experiment [102]. Because of rather complex coherence transfer pathways, the cross-relaxation rates calculated from simple peak intensity ratios in the cross and reference spectra would be inaccurate. Instead, the rates were obtained by fitting the experimental data to simulated intensity ratios, calculated taking into account full auto- and cross-correlated relaxation.

6.9 Conclusions

This chapter has briefly described the current development of NMR in nucleic acids. The specific applications of NMR and molecular modeling to study DNA-drug interactions have recently been reviewed in two comprehensive articles [107, 108]. The main emphasis of our contribution was to assess the latest additions to the arsenal of NMR methods for structural studies of RNA and DNA. The improvement of available tools with the capability of extending the range of molecular weight that can be successfully investigated using isotope-assisted NMR is even more desirable for nucleic acids than for proteins. In contrast to polypeptides, single crystals of nucleic acids are frequently influenced by packing and lattice forces, which lead to differences between the structural data found by X-ray crystallography and NMR spectroscopy. Problems with accuracy and precision of NMR structures result mainly from the lack of long-range restraints, relatively low proton density, elongated shape of a majority of oligonucleotides and a small number of NOE restraints between distant elements of secondary and tertiary structure. For larger molecules with more than 30 residues, the uniform ^{13}C, ^{15}N isotope labeling, which, in general, is more complicated than in proteins, has to be complemented also by residue- or site-specific labeling. For oligonucleotides with more than 50 residues, partial deuteration seems to be mandatory. Despite these obstacles, the recent improvement of labeling strategies, the development of new and more sophisticated NMR methods and the progress in measurements and applications of residual dipolar couplings for structure refinement have paved the way to improving the accuracy and precision of structures obtained by NMR.

6.10 Acknowledgements

This work was supported by the Ministry of Education of the Czech Republic, grant LN00A016. The authors are grateful to Petr Padrta for help in preparation of the figures and to Lukáš Žídek and Lukáš Trantírek for critical reading of the manuscript.

6.11 References

1 K. Wüthrich *NMR of Proteins and Nucleic Acids*; Wiley: New York, 1986.
2 S. S. Wijmenga, B. N. M. van Buuren, *Prog. Nucl. Magn. Reson. Spectrosc.* **1998**, *32*, 287–387.
3 C. Kojima, A. M. Ono, A. Ono, M. Kainosho, *Methods Enzymol.* **2001**, *338*, 261–283.
4 M. H. Werner, V. Gupta, L. J. Lambert, T. Nagata, *Methods Enzymol.* **2001**, *338*, 283–304.
5 J. Cromsigt, J. Schleucher, K. Kidd-Ljunggren, S. S. Wijmenga, *J. Biomol. Struct. Dyn.* **2000**, 211–219.
6 E. P. Nikonowicz, *Methods Enzymol.* **2001**, *338*, 320–341.
7 A. T. Phan, M. Gueron, J. L. Leroy, *Methods Enzymol.* **2001**, *338*, 341–371.
8 J. Cromsigt, B. van Buuren, J. Schleucher, S. Wijmenga, *Methods Enzymol.* **2001**, *338*, 371–399.
9 J. Feigon, S. E. Butcher, L. D. Finger, N. V. Hud, *Methods Enzymol.* **2001**, *338*, 400–420.
10 N. V. Hud, M. Polak, *Curr. Opin. Struct. Biol.* **2001**, *11*, 293–301.
11 R. L. Gonzalez, I. Tinoco, *Methods Enzymol.* **2001**, *338*, 421–443.
12 V. Dotsch, *Methods Enzymol.* **2001**, *339*, 343–357.
13 M. Allen, L. Varani, G. Varani, *Methods Enzymol.* **2001**, *339*, 357–376.
14 J. M. Perez-Canadillas, G. Varani, *Curr. Opin. Struct. Biol.* **2001**, *11*, 53–58.
15 L. Zídek, R. Štefl, V. Sklenár, *Curr. Opin. Struct. Biol.* **2001**, *11*, 275–281.
16 F. Major, R. Griffey, *Curr. Opin. Struct. Biol.* **2001**, *11*, 282–286.
17 M. Piotto, V. Saudek, V. Sklenář, *J. Biomol. NMR* **1992**, *2*, 661–665.
18 F. Jiang, R. Fiala, D. Live, R. A. Kumar, D. J. Patel, *Biochemistry* **1996**, *35*, 13250–13266.
19 E. P. Nikonowicz, A. Sirr, P. Legault, F. M. Jucker, L. M. Baer, A. Pardi, *Nucleic Acids Res.* **1992**, *20*, 4507–4513.
20 R. T. Batey, M. Inada, E. Kujawinski, J. D. Puglisi, J. R. Williamson, *Nucleic Acids Res.* **1992**, *20*, 4515–4523.
21 M. J. Michnicka, J. W. Harper, G. C. King, *Biochemistry* **1993**, *32*, 395–400.
22 D. P. Zimmer, D. M. Crothers, *Proc. Natl. Acad. Sci. U. S. A.* **1995**, *92*, 3091–3095.
23 G. C. King, J. W. Harper, Z. Xi, *Methods Enzymol.* **1995**, *261*, 436–450.
24 R. T. Batey, J. L. Battiste, J. R. Williamson, *Methods Enzymol.* **1995**, *261*, 300–322.
25 J. E. Masse, P. Bortmann, T. Dieckmann, J. Feigon, *Nucleic Acids Res.* **1998**, *26*, 2618–2624.
26 G. Mer, W. J. Chazin, *J. Am. Chem. Soc.* **1998**, *120*, 607–608.
27 J. M. Louis, R. G. Martin, G. M. Clore, A. M. Gronenborn, *J. Biol. Chem.* **1998**, *273*, 2374–2378.
28 D. E. Smith, J. Y. Su, F. M. Jucker, *J. Biomol. NMR* **1997**, *10*, 245–253.
29 T. Dieckmann, J. Feigon, *J. Biomol. NMR* **1997**, *9*, 259–272.
30 M. H. Kolk, M. van der Graaf, S. S. Wijmenga, C. W. A. Pleij, H. A. Heus, C. W. Hilbers, *Science* **1998**, *280*, 434–438.
31 M. H. Kolk, S. S. Wijmenga, H. A. Heus, C. W. Hilbers, *J. Biomol. NMR* **1998**, *12*, 423–433.
32 L. G. Scott, T. J. Tolbert, J. R. Williamson, *Methods Enzymol.* **2000**, *317*, 18–38.
33 K. T. Dayie, T. J. Tolbert, J. R. Williamson, *J. Magn. Reson.* **1998**, *130*, 97–101.
34 R. H. Griffey, A. G. Redfield, *Q. Rev. Biophys.* **1987**, *19*, 51–82.

35 S. Grzesiek, A. Bax, *J. Biomol. NMR* **1995**, *6*, 335–339.

36 J.P. Marino, J.L. Diener, P.B. Moore, C. Griesinger, *J. Am. Chem. Soc.* **1997**, *119*, 7361–7366.

37 R. Fiala, F. Jiang, V. Sklenář, *J. Biomol. NMR* **1998**, *12*, 373–383.

38 V. Sklenář, T. Dieckmann, S.E. Butcher, J. Feigon, *J. Magn. Reson.* **1998**, *130*, 119–124.

39 W.D. Hu, L.C. Jiang, *J. Biomol. NMR* **1999**, *15*, 289–293.

40 A.G. Palmer, J. Cavanagh, P.E. Wright, M. Rance, *J. Magn. Reson.* **1991**, *93*, 151–170.

41 L.E. Kay, P. Keifer, T. Saarinen, *J. Am. Chem. Soc.* **1992**, *114*, 10663–10665.

42 W.D. Hu, L.C. Jiang, Y.Q. Gosser, *J. Magn. Reson.* **2000**, *145*, 147–151.

43 W.D. Hu, Y.Q. Gosser, W.J. Xu, D.J. Patel, *J. Biomol. NMR* **2001**, *20*, 167–172.

44 T. Szyperski, C. Fernandez, A. Ono, K. Wüthrich, M. Kainosho, *J. Magn. Reson.* **1999**, *140*, 491–494.

45 R. Fiala, J. Czernek, V. Sklenář, *J. Biomol. NMR* **2000**, *16*, 291–302.

46 B. Brutscher, J. Boisbouvier, A. Pardi, D. Marion, J.P. Simorre, *J. Am. Chem. Soc.* **1998**, *120*, 11845–11851.

47 R. Riek, K. Pervushin, C. Fernandez, M. Kainosho, K. Wüthrich, *J. Am. Chem. Soc.* **2001**, *123*, 658–664.

48 B. Brutscher, J.P. Simorre, *J. Biomol. NMR* **2001**, *21*, 367–372.

49 B. Simon, K. Zanier, M. Sattler, *J. Biomol. NMR* **2001**, *20*, 173–176.

50 K. Pervushin, *J. Biomol. NMR* **2001**, *20*, 275–285.

51 V. Sklenář, J. E. Masse, J. Feigon, *J. Magn. Reson.* **1999**, *137*, 345–349.

52 J.P. Simorre, G.R. Zimmermann, A. Pardi, B.T. Farmer, L. Mueller, *J. Biomol. NMR* **1995**, *6*, 427–432.

53 V. Sklenář, T. Dieckmann, S.E. Butcher, J. Feigon, *J. Biomol. NMR* **1996**, *7*, 83–87.

54 R. Fiala, F. Jiang, D.J. Patel, *J. Am. Chem. Soc.* **1996**, *118*, 689–690.

55 J. Wohnert, R. Ramachandran, M. Gorlach, L.R. Brown, *J. Magn. Reson.* **1999**, *139*, 430–433.

56 B. Luy, J.P. Marino, *J. Am. Chem. Soc.* **2001**, *ASAP*.

57 G.W. Kellogg, A.A. Szewczak, P.B. Moore, *J. Am. Chem. Soc.* **1992**, *114*, 2727–2728.

58 G.W. Kellogg, *J. Magn. Reson.* **1992**, *98*, 176–182.

59 M. Hennig, J.R. Williamson, *Nucleic Acids Res.* **2000**, *28*, 1585–1593.

60 Z.R. Wu, A. Ono, M. Kainosho, A. Bax, *J. Biomol. NMR* **2001**, *19*, 361–365.

61 R. Riek, *J. Magn. Reson.* **2001**, *149*, 149–153.

62 C.H. Gotfredsen, A. Meissner, J.O. Duus, O.W. Sorensen, *Magn. Reson. Chem.* **2000**, *38*, 692–695.

63 C. Kojima, E. Kawashima, T. Sekine, Y. Ishido, A. Ono, M. Kainosho, Y. Kyogoku, *J. Biomol. NMR* **2001**, *19*, 19–31.

64 J.P. Yang, K. McAteer, L.A. Silks, R.L. Wu, N.G. Isern, C.J. Unkefer, M.A. Kennedy, *J. Magn. Reson.* **2000**, *146*, 260–276.

65 C. Richter, B. Reif, K. Worner, S. Quant, J. P. Marino, J. W. Engels, C. Griesinger, H. Schwalbe, *J. Biomol. NMR* **1998**, *12*, 223–230.

66 C.G. Hoogstraten, A. Pardi, *J. Magn. Reson.* **1998**, *133*, 236–240.

67 A. Kaikkonen, G. Otting, *J. Biomol. NMR* **2001**, *19*, 273–277.

68 N. Tjandra, A. Bax, *Science* **1997**, *278*, 1111–1114.

69 J.H. Prestegard, *Nat. Struct. Biol.* **1998**, *Suppl. 5*, 517–522.

70 P. Bayer, L. Varani, G. Varani, *J. Biomol. NMR* **1999**, *14*, 149–155.

71 N. Tjandra, S. Tate, A. Ono, M. Kainosho, A. Bax, *J. Am. Chem. Soc.* **2000**, *122*, 6190–6200.

72 A. Vermeulen, H. Zhou, A. Pardi, *J. Am. Chem. Soc.* **2000**, *122*, 9638–9647.

73 E.T. Mollova, M.R. Hansen, A. Pardi, *J. Am. Chem. Soc.* **2000**, *122*, 11561–11562.

74 L. Trantírek, M. Urbášek, R. Štefl, J. Feigon, V. Sklenář, *J. Am. Chem. Soc.* **2000**, *122*, 10454–10455.

75 P. Hobza, J. Šponer, *Chem. Rev.* **1999**, *99*, 3247–3276.

76 H.M. Al-Hashimi, A. Majumdar, A. Gorin, A. Kettani, E. Skripkin, D.J. Patel, *J. Am. Chem. Soc.* **2001**, *123*, 633–640.

77 N. Sibille, A. Pardi, J.P. Simorre, M. Blackledge, *J. Am. Chem. Soc.* **2001**, *123*, 12135–12146.

78 L. Žídek, H.H. Wu, J. Feigon, V. Sklenář, *J. Biomol. NMR* **2001**, *21*, 153–160.

79 I. Hirao, G. Kawai, S. Yoshizawa, Y. Nishimura, Y. Ishido, K. Watanabe, K. Miura, *Nucleic Acids Res.* **1994**, *22*, 576–582.

80 I. Hirao, Y. Nishimura, Y. Tagawa, K. Watanabe, K. Miura, *Nucleic Acids Res.* **1992**, *20*, 3891–3896.

81 S. Yoshizawa, T. Ueda, Y. Ishido, K. Miura, K. Watanabe, I. Hirao, *Nucleic Acids Res.* **1994**, *22*, 2217–2221.

82 P. Padrta, R. Štefl, L. Žídek, V. Sklenář, *in press J. Biomol. NMR.*

83 A. Görler, N.B. Ulyanov, T.L. James, *J. Biomol. NMR* **2000**, *16*, 147–164.

84 W.D. Cornell, P. Cieplak, C.I. Bayly, I.R. Gould, K.M. Merz, D.M. Ferguson, D.C. Spellmeyer, T. Fox, J.W. Caldwell, P.A. Kollman, *J. Am. Chem. Soc.* **1995**, *117*, 5179–5197.

85 N. Foloppe, A.D. MacKerell, *J. Comput. Chem.* **2000**, *21*, 86–104.

86 T.E. Cheatham III, P.A. Kollman, *Annu. Rev. Phys. Chem.* **2000**, *51*, 435–471.

87 V. Tsui, I. Radhakrishnan, P.E. Wright, D. Case, *J. Mol. Biol.* **2000**, *302*, 1101–1117.

88 M. Akke, R. Fiala, F. Jiang, D. Patel, A.G. Palmer, *RNA* **1997**, *3*, 702–709.

89 J.Z. Hu, J.C. Facelli, D.W. Alderman, R.J. Pugmire, D.M. Grant, *J. Am. Chem. Soc.* **1998**, *120*, 9863–9869.

90 D. Sitkoff, D.A. Case, *Prog. Nucl. Magn. Reson. Spectrosc.* **1998**, *32*, 165–190.

91 M. Barfield, A.J. Dingley, J. Feigon, S. Grzesiek, *J. Am. Chem. Soc.* **2001**, *123*, 4014–4022.

92 J. Czernek, *J. Phys. Chem. A* **2001**, *105*, 1357–1365.

93 A.P. Dejaegere, D.A. Case, *J. Phys. Chem. A* **1998**, *102*, 5280–5289.

94 J. Czernek, V. Sklenář, *J. Phys. Chem. A* **1999**, *103*, 4089–4093.

95 P. Rossi, G.S. Harbison, *J. Magn. Reson.* **2001**, *151*, 1–8.

96 K.B. Hall, C.G. Tang, *Biochemistry* **1998**, *37*, 9323–9332.

97 J. Boisbouvier, B. Brutscher, J.P. Simorre, D. Marion, *J. Biomol. NMR* **1999**, *14*, 241–252.

98 C. Kojima, A. Ono, M. Kainosho, T.L. James, *J. Magn. Reson.* **1998**, *135*, 310–333.

99 C. Kojima, A. Ono, M. Kainosho, T.L. James, *J. Magn. Reson.* **1999**, *136*, 169–175.

100 B. Reif, M. Hennig, C. Griesinger, *Science* **1997**, *276*, 1230–1233.

101 J. Boisbouvier, B. Brutscher, A. Pardi, D. Marion, J.P. Simorre, *J. Am. Chem. Soc.* **2000**, *122*, 6779–6780.

102 I.C. Felli, C. Richter, C. Griesinger, H. Schwalbe, *J. Am. Chem. Soc.* **1999**, *121*, 1956–1957.

103 C. Richter, C. Griesinger, I. Felli, P.T. Cole, G. Varani, H. Schwalbe, *J. Biomol. NMR* **1999**, *15*, 241–250.

104 T. Carlomagno, M.J.J. Blommers, J. Meiler, B. Cuenoud, C. Griesinger, *J. Am. Chem. Soc.* **2001**, *123*, 7364–7370.

105 E. Chiarparin, S. Rudisser, G. Bodenhausen, *Chem. Phys. Chem.* **2001**, *2*, 41–45.

106 J. Herzfeld, R.G. Griffin, R.K. Haberkorn, *Biochemistry* **1978**, *17*, 2711–2718.

107 A.N. Lane, *Methods Enzymol.* **2001**, *340*, 252–281.

108 J.O. Trent, *Methods Enzymol.* **2001**, *340*, 290–326.

109 K. Pervushin, A. Ono, C. Fernandez, T. Szyperski, M. Kainosho, K. Wüthrich, *Proc. Natl. Acad. Sci. US.* **1998**, *95*, 14147–14151.

110 A.J. Dingley, S. Grzesiek, *J. Am. Chem. Soc.* **1998**, *120*, 8293–8297.

111 A.J. Dingley, J.E. Masse, R.D. Peterson, M. Barfield, J. Feigon, S. Grzesiek, *J. Am. Chem. Soc.* **1999**, *121*, 6019–6027.

112 J. Wohnert, A.J. Dingley, M. Stoldt, M. Gorlach, S. Grzesiek, L.R. Brown, *Nucleic Acids Res.* **1999**, *27*, 3104–3110.

113 A. Majumdar, A. Kettani, E. Skripkin, *J. Biomol. NMR* **1999**, *14*, 67–70.

114 A.J. Dingley, J.E. Masse, J. Feigon, S. Grzesiek, *J. Biomol. NMR* **2000**, *16*, 279–289.

115 A.Z. Liu, A. Majumdar, W.D. Hu, A. Kettani, E. Skripkin, D.J. Patel, *J. Am. Chem. Soc.* **2000**, *122*, 3206–3210.

7
Methods for the Measurement of Angle Restraints from Scalar, Dipolar Couplings and from Cross-Correlated Relaxation: Application to Biomacromolecules

Christian Griesinger

7.1
Introduction

The NMR spectra of biomolecules, especially with isotopic labeling, provide a wealth of information about interatomic distances and angular geometries that can be used as conformational restraints in structure determination. They are summarized in Fig. 7.1. In this review I will focus on the determination of scalar couplings and the use of dipolar couplings as structural restraints and cross-correlated relaxation rates to obtain structural information. The high potential of these parameters for complementing information derived from NOEs is nowadays generally recognized. In the context of biomolecular NMR applied in drug research, these techniques are of particular value for systems that display low proton density. Such systems are exemplified especially by DNA/RNA or by drug-like small organic molecules, for which NOE information is often insufficient to establish their stereochemistry or conformation. Dynamical aspects of the parameters listed above will not be part of this review. However, we refer the interested reader to recent articles on dynamics from cross-correlated relaxation [1] as well as from dipolar couplings [2b]. The reader's attention is also drawn to some other reviews that present more details and examples than the rather short review here [2a]. The use of chemical shifts and NOEs for structure determination will also not be covered.

In the rest of this chapter, the angular dependence of NMR observables will be discussed.

7.2
Coupling Constants

The magnitude and sign of homo- and heteronuclear ^nJ coupling constants are mainly determined by the local bonding geometry [3]. To a good approximation, the magnitude of ^3J coupling constants depends only on the torsion angle γ between two spins separated by the three bonds. The functional dependence can be expressed using Karplus equations [4] of the form $^3J(\gamma) = A \cdot \cos^2(\gamma) + B \cdot \cos(\gamma) + C$, for which A, B, and C are dimensionless parameters that are either calculated from quantum mechanical calculations or are parameterized empirically. In the last few years, the majority of pulse sequences designed to measure coupling constants have focused on the determination of vicinal cou-

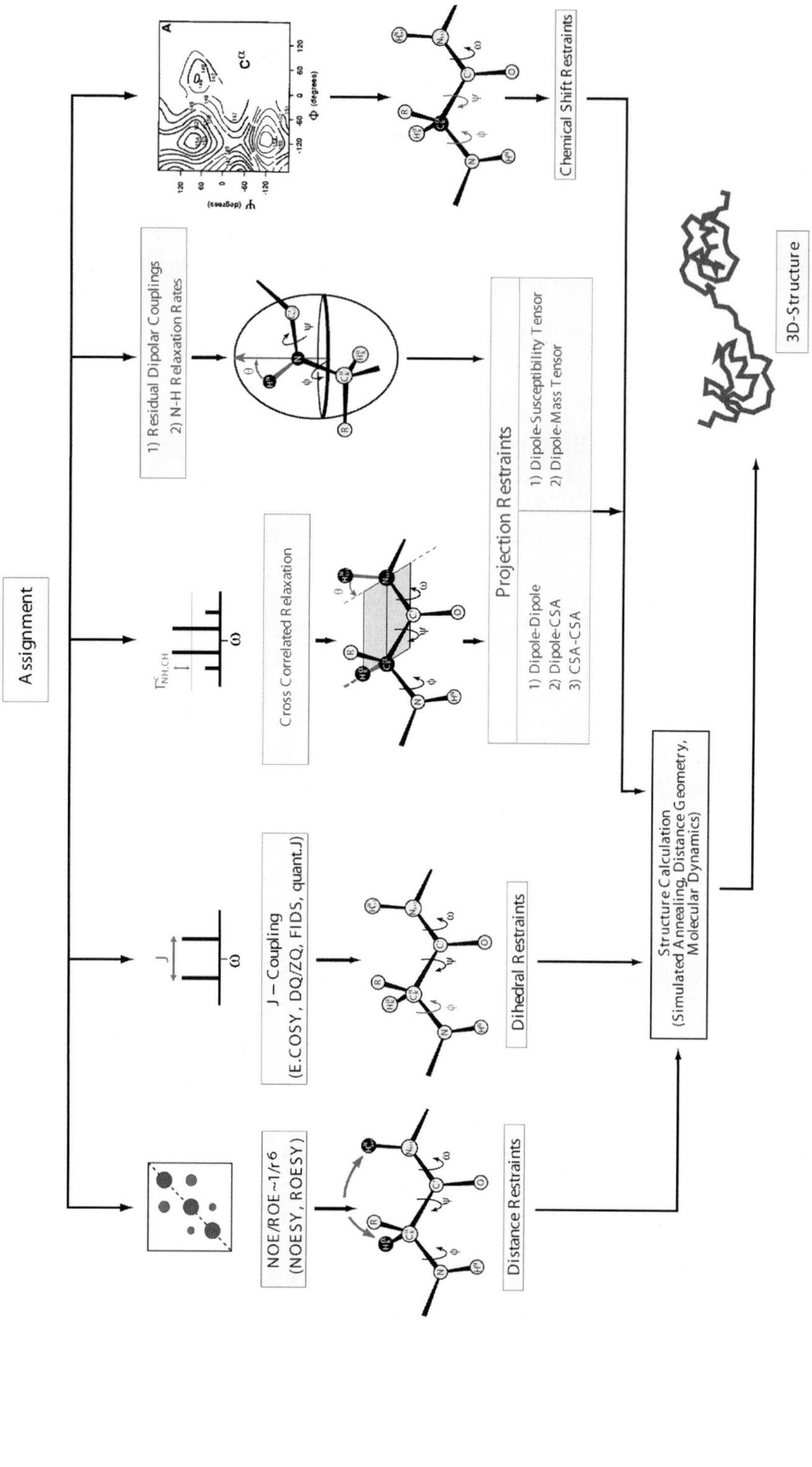

Fig. 7.1 Flow chart for structure determination with NMR, focusing on the most useful structural NMR parameters: NOE, *J* coupling, dipolar coupling, cross-correlated relaxation rate, and chemical shifts.

pling constants because their torsion angle dependence is best understood. Recently, however, geminal (^2J), direct (^1J) coupling constants and cross hydrogen bond coupling constants (see Chapt. 9) have become interesting for the following reasons: (i) Heteronuclear ^2J coupling constants can provide valuable information for making stereospecific assignments [5]. (ii) Both ^2J and ^1J couplings relate nuclei with fixed geometry such that dipolar couplings between these nuclei measured in aligning media are only orientation dependent [6, 7]. *Trans* hydrogen bond couplings allow us to quantify geometry and strength of hydrogen bonds in a better way. There are five basic principles that underlie NMR methodologies currently available for the determination of torsion angle restraints in biomolecules: E.COSY (exclusive correlation spectroscopy), DQ/ZQ, FIDS (fitting of doublets from singlets)/*J*-modulation, quantitative-*J* and cross-correlated relaxation. These principles will be introduced, and then recent applications to ^{13}C,^{15}N and ^{13}C,^{15}N,^2H labeled proteins and ^{13}C,^{15}N labeled RNA will be discussed.

7.2.1
The E. COSY Principle [8]

E. COSY (Exclusive COrrelation SpectroscopY) type measurements require three mutually coupled spins A, B, and C. If A and B are correlated such that the polarization of the third spin C remains unaffected (or is inverted), then the AB cross peak consists of only two of four possible submultiplets, for which C is either in the α-state $C^{\alpha}=1/2(1+2C_z)$ or in the β-state $C^{\beta}=1/2(1-2C_z)$. In the two-dimensional experiment, the two pairs of frequencies $\Omega_A+\pi J(A,C)$ in ω_1, $\Omega_B + \pi J(B,C)$ in ω_2, and $\Omega_A-\pi J(A,C)$ in ω_1, $\Omega_B-\pi J(B,C)$ in ω_2, are correlated. The two-dimensional frequency shift of the submultiplets is given by the displacement vector J_C [C for the passive spin C with unchanged (or inverted) spin states] with the components $J(A,C)$ in ω_1 and $J(B,C)$ in ω_2. As can be inferred from Fig. 7.2 and neglecting for the moment relaxation effects, the size of a small coupling $J(B,C)$ can be determined accurately provided that the *associated* coupling constant $J(A,C)$ is larger than the resolution in ω_1. Furthermore, the relative signs of the two couplings can be determined from the orientation of the displacement vector J_C. The J_C vector will point to the upper right (upper left) or lower left (lower right), if the signs of $J(A,C)$ and $J(B,C)$ couplings are the same (opposite).

As an example we take the measurement of the H^N,C′ coupling constant in perdeuterated ^{13}C- and ^{15}N-labeled calmodulin. The pulse sequence of the constant-time (ct) HNCA-COSY experiment is shown in Fig. 7.3. The parent experiment is here a conventional HNCA experiment that is modified in the following way: ^2H decoupling is introduced during the evolution of the aliphatic carbons and the C_{α},C′ coupling evolves during t_1. By that, an E. COSY pattern is generated with the large (approx. 55 Hz) $^1J(C_{\alpha},C')$ in ω_1 and the desired $^3J(H^N,C')$ coupling in ω_3. This experiment thus provides information about the backbone angle ϕ for deuterated proteins. Similar methods are of course available for protonated proteins as well. The resulting spectrum is shown in Fig. 7.4. The signal-to-noise ratio of this experiment is reduced by a factor of 2 as compared to the parent experiment because two peaks are observed for each correlation.

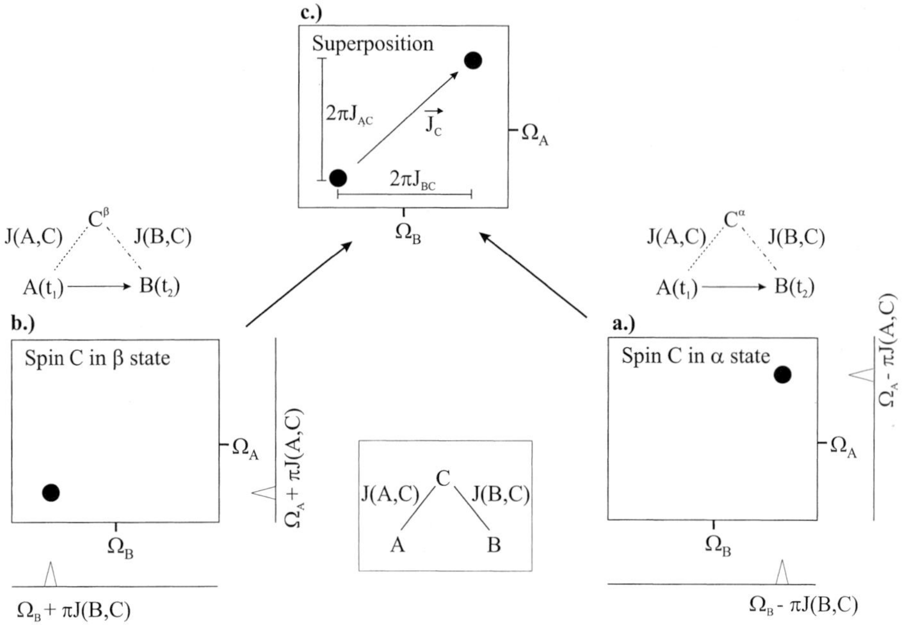

Fig. 7.2 Explanation of the E. COSY principle. The E. COSY spectrum can be conceived as a superposition of spectra originating from two sorts of molecules which differ only in the spin state of the passive spin C.

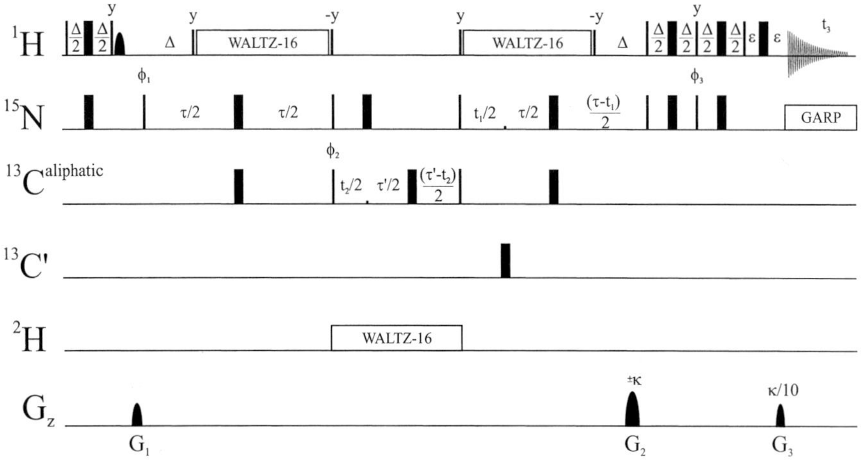

Fig. 7.3 Pulse sequence of the constant-time (ct) HNCA-COSY experiment. The passive spin is C′. Obviously, C′ is affected by only one 180° pulse, which leaves the spin states untouched except for permutation.

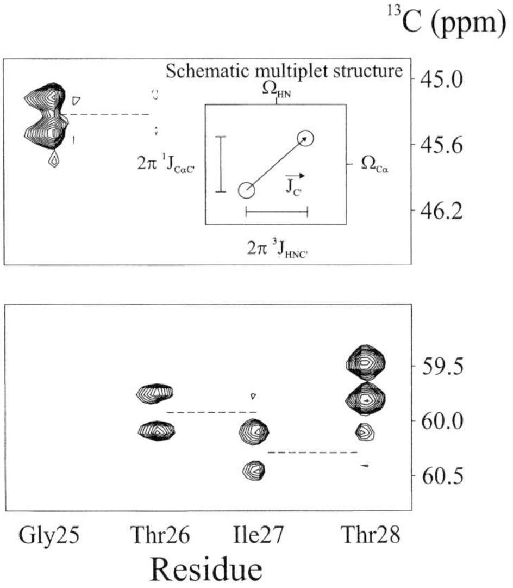

Fig. 7.4 Schematic multiplet pattern of a C_α,H^N correlation in the constant-time (ct) HNCA-COSY experiment leaving the C′ untouched. The $^3J(H^N,C')$ coupling can be derived from this experiment.

7.2.2
The DQ/ZQ Principle [9]

Differential relaxation of in-phase and anti-phase operators involving a spin C [10], which are due to additional T_1 relaxation effects active only for the anti-phase components and which depend on the geometry of the spin system, can lead to systematic errors of the coupling constant derived from cross-peak multiplets observed in an E. COSY-type experiment [11]. Since these errors depend for a given differential relaxation rate $\Delta\rho$ on the frequency difference of the coherences with C in the α or β state, according to Eq. (1) a remedy to the problem is to maximize the relevant J such that the condition $J \gg \Delta\rho/2\pi$ is fulfilled:

$$J^{\text{eff}} = \sqrt{J^2 - \left(\frac{\Delta\rho}{2\pi}\right)^2} = J - \frac{\Delta\rho^2}{(2\pi)^2(2J)} \,. \tag{1}$$

Excitation of double and zero quantum coherence (called DQ/ZQ coherence) (Fig. 7.5) between A and B in a spin system A,B,C with a large coupling $J(\text{A, C})$ and a desired coupling $J(\text{B, C})$ will yield a doublet with a splitting by $J(\text{A, C})+J(\text{B, C})$ for the DQ and $J(\text{A, C})-J(\text{B, C})$ for the ZQ coherence. Adding or subtracting the splittings observed in the DQ and ZQ spectra yields $2\cdot J(\text{A, C})$ and $2\cdot J(\text{B, C})$, respectively. Alternatively, DQ or ZQ spectra can be compared with an SQ spectrum that reflects $J(\text{A, C})$. Information about the relative sign of the two couplings $J(\text{A, C})$ and $J(\text{B, C})$ can be extracted in the same way as from the E. COSY experiment.

Here, we present the example of the *trans* hydrogen bond coupling between the C′ of the acceptor and the N of the donor $^hJ(\text{N, C}')$ that is measured by excitation of double-quantum and zero-quantum coherence between the H^N and the C′ nuclei [12] in a protein. Thus, the double-quantum coherence is split by $^hJ(\text{N, C}')+\ ^1J(\text{N, H})$ while the zero-quantum coher-

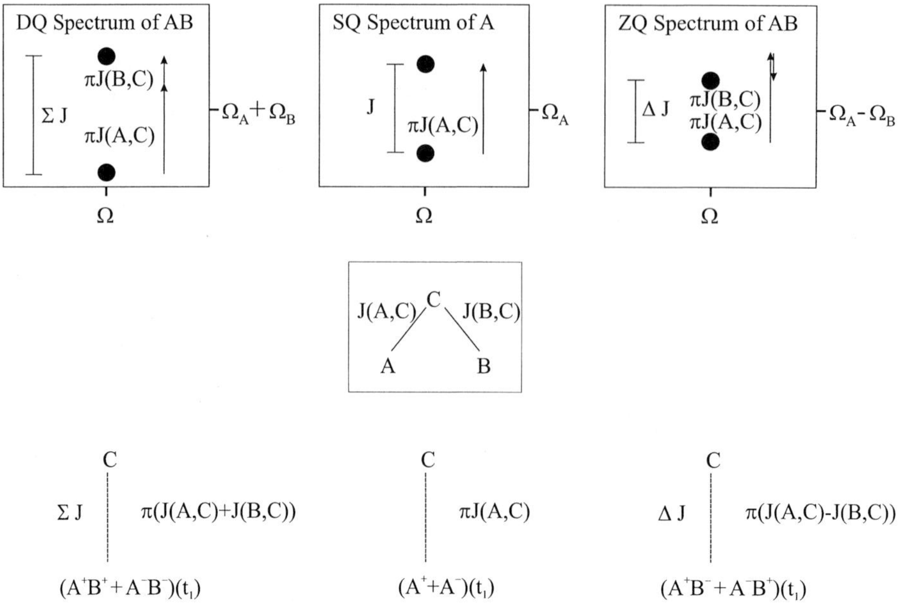

Fig. 7.5 Schematic multiplet patterns for the DQ/ZQ experiment. If either of the two coherences (DQ or ZQ) cannot be resolved, a single quantum (SQ) experiment can also be used as a reference. As in the E. COSY type experiments, the passive spin C must couple to both active spins A and B.

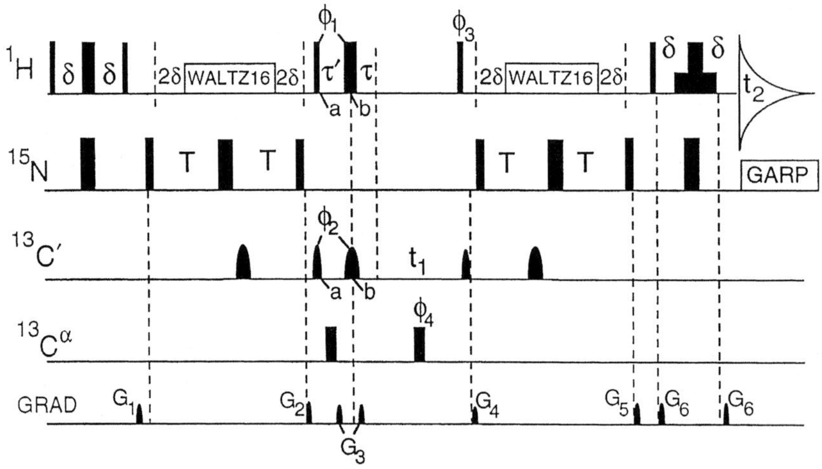

Fig. 7.6 Pulse sequence of a 2D ZQ/DQ HN(CO) experiment. In this experiment double-quantum and zero-quantum coherence between H^N and the C′ bound to the proton via a hydrogen bond is excited. N is the passive spin which induces a large splitting due to the HN coupling, on top of which the small coupling $^{h}J(N,C')$ via the hydrogen bond is added or subtracted.

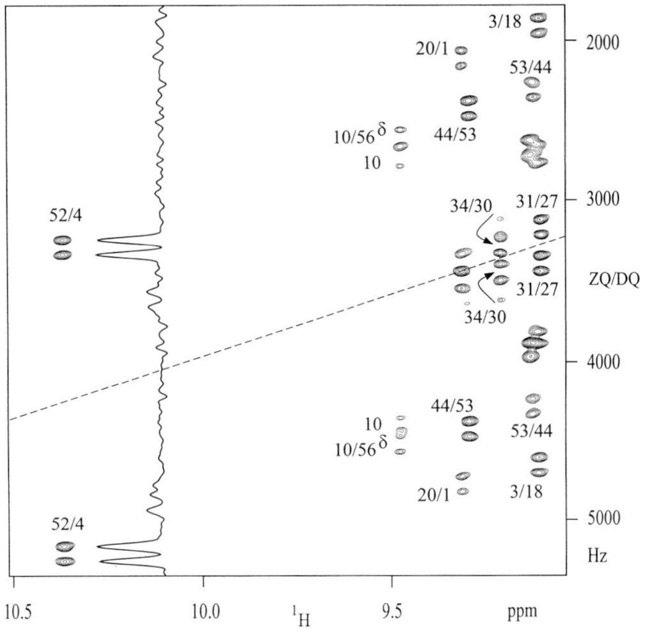

Fig. 7.7 Spectrum derived from the 2D ZQ/DQ HN(CO). The differences of the splittings in the indirect dimension is the ^{h}J(N,C′) via the hydrogen bond. Referencing to the large ^{1}J(N,H) allows one to measure the relative signs of the ^{1}J(N,H) and the ^{h}J(N,C′) couplings.

ence is split by ^{h}J(N, C′)–^{1}J(N, H). The pulse sequence used is shown in Fig. 7.6. Therein, defocusing via both the ^{h}J(N, C′) and the ^{1}J(H, N) coupling during the delay 2T leads to the operator $H_z^N N_x C'_z$. By simultaneous application of three 90° pulses on H^N, C′ and N the DQ and ZQ operators are formed. After sufficient evolution of the DQ and ZQ coherences during t_1 the transfer scheme is applied in reverse order and proton magnetization is finally detected. The spectrum shown in Fig. 7.7 exhibits doublets both due to double and zero quantum coherences and by comparison of their splittings the sign of the ^{h}J(N, C′) coupling was found to be identical to that of the ^{1}J(N, H) coupling, namely negative.

7.2.3
The FIDS Principle [13]

In the absence of a large associated coupling J(A, C), which is a requirement for the successful application of E. COSY or DQ/ZQ methodology, the FIDS (fitting of doublets from singlets) procedure can be applied. The basic principle of this experiment is outlined in Fig. 7.8.

In a B, C spin system, the multiplet of spin B contains the J(B, C) coupling constant among other couplings. This coupling can be determined by comparison of two spectra: a coupled spectrum in which the J(B, C) coupling evolves and a decoupled reference experiment in which the J(B, C) is removed.

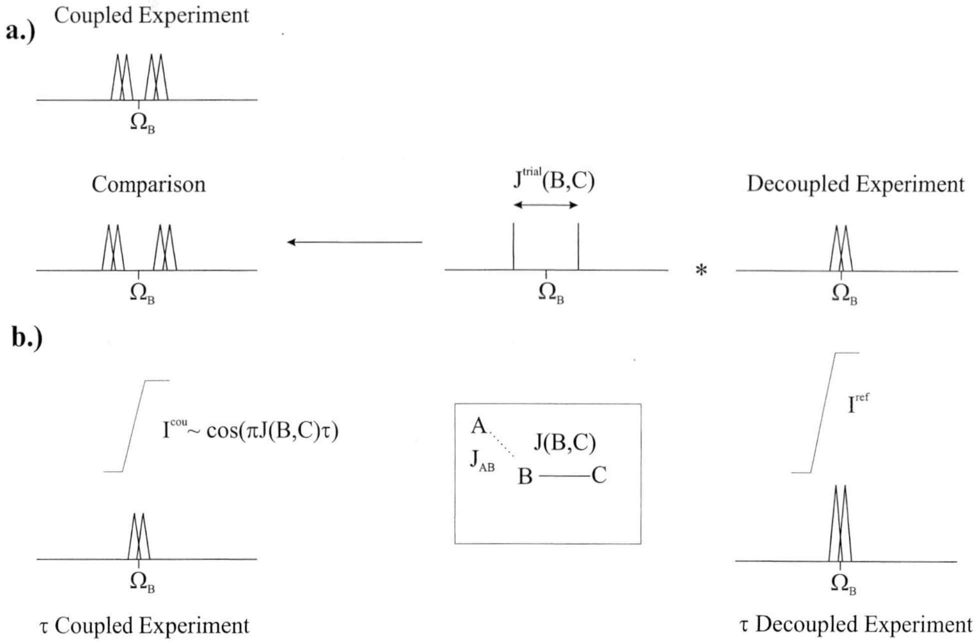

Fig. 7.8 **a** Schematics for the FIDS experiment as explained in the text. A coupled and a decoupled experiment are recorded. The decoupled experiment is convoluted with a stick multiplet in order to match the coupled experiment. **b** Schematics for a CT-HSQC experiment (see text for details).

The procedure works when each detected spin B can be selectively decoupled from a spin C, although other couplings may be present. If this is the case, the spectra obtained in the coupled experiment and the reference experiment are identical except for the additional splitting due to the coupling of interest $J(B,C)$. This splitting can be determined by convolution (* in Fig. 7.8) of the decoupled spectrum with a stick doublet to yield the coupled spectrum.

A sequence suitable for measurement of $J(H,P)$ and $J(C,P)$ couplings is shown in Fig. 7.9a. The pulse sequence is a constant-time [^{13}C, ^{1}H]-HSQC (heteronuclear single-quantum correlation), in which ^{31}P decoupling is applied in ω_1 in the first experiment and in ω_2 in the second.

Traces through the spectrum along ω_2 are shown in Fig. 7.10 together with the fitting of the coupled to the decoupled spectrum after convolution by an in-phase stick doublet. The fit delivers the coupling constant with high precision. The sensitivity of this experiment is practically identical to that of the HSQC experiment since the splitting is normally smaller or in the order of the line widths.

Conceptually similar to FIDS is the so-called *J*-modulated CT-HSQC experiment [14] (Fig. 7.8b). The coupling evolves in one experiment during τ and in a second experiment not at all. The intensity ratio between first and second experiment is $\cos(\pi J\tau)$. As an example we show the measurement of an NH dipolar coupling by this method. The pulse sequence of the constant time HSQC and the oscillatory behavior of the cross-peak intensities are shown in Fig. 7.11.

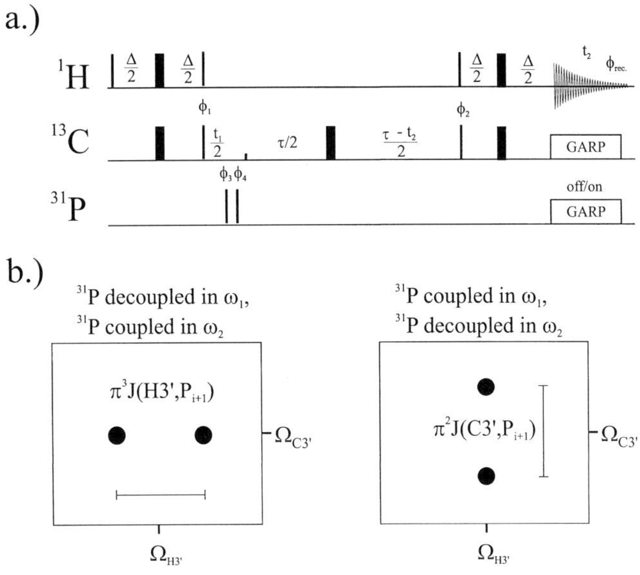

Fig. 7.9 Pulse sequence and schematic multiplet patterns expected for the FIDS-HSQC experiment to measure both the J(H,P) and J(C,P) coupling constants.

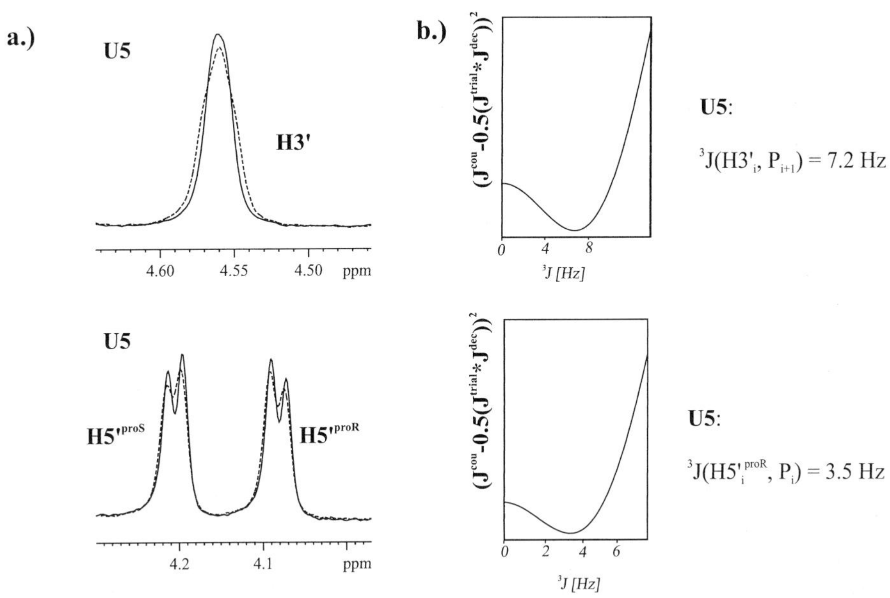

Fig. 7.10 **a** Experimental traces through the FIDS HSQCs of Fig. 7.9. The convolution is shown in **b**, indicating the very clean minimum that defines the coupling constant quite precisely.

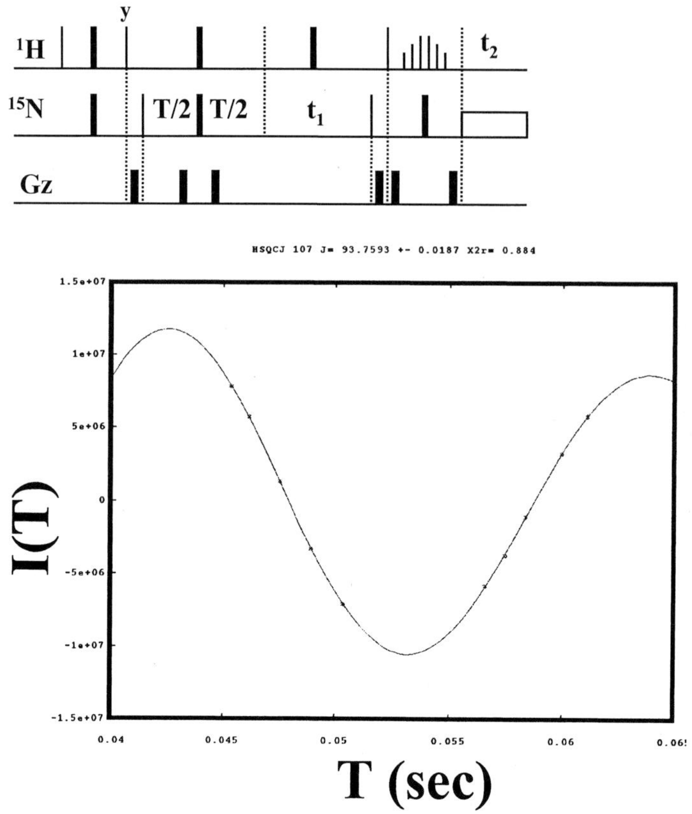

Fig. 7.11 *J* modulation in a constant time HSQC in order to measure dipolar couplings. A "Watergate" pulse is introduced in order to optimize water suppression. The intensity of cross peaks is given by $I(T) = C\cos(\pi J T)\exp(-T/T_2)$. The right hand side displays a fit of the data to the theoretical curve, resulting in $J = 93.76 \pm 0.019$ Hz ($c^2 = 0.88$, 600 MHz data) for residue 107 of the trigger factor.

Fitting of these curves yields the sum of the dipolar coupling plus the *J*-coupling. Referencing to a spectrum without alignment allows extraction of the dipolar coupling constant.

7.2.4
Quantitative *J*-correlation spectroscopy [15]

In quantitative *J*-correlations (Fig. 7.12), a coupling of interest is extracted from quantitative evaluation of the ratio of cross-peak to diagonal-peak intensities in an out-and-back correlation experiment between spins B and C.

In the preparation period of the out-and-back correlation experiment, in-phase coherence B_x evolves into anti-phase coherence $2B_yC_z$ during τ, a subsequent 90°_x(B,C) pulse converts these two operators to B_x and $2B_zC_y$, which evolve and give rise to a reference

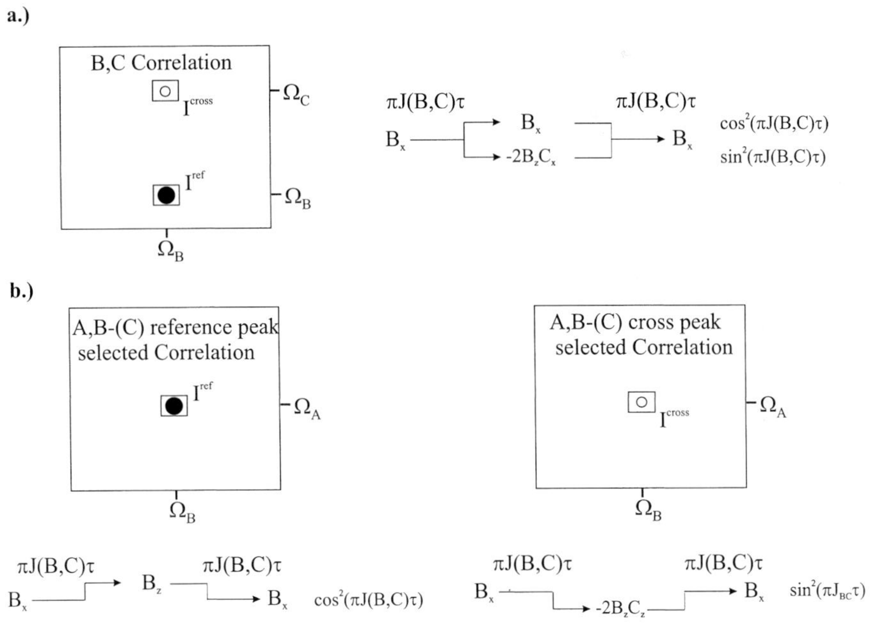

Fig. 7.12 Schematic spectra for quantitative *J*-correlation experiments. A cross peak is compared to a reference peak occurring in the same experiment.

peak derived from B_x and a cross peak ($2B_zC_y$) in ω_1. After the evolution period, coherences are transferred back to B and are finally detected on B or a spin coupled to it. The transfer amplitude of the reference peak depends on $\cos^2[\pi J(\mathrm{B,C})\,\tau]$, while the transfer amplitude of the cross peak depends on $\sin^2[\pi J(\mathrm{B,C})\,\tau]$. Further passive couplings to B contribute equally to both the reference peak and the cross peak. The coupling of interest J(B,C) can then be determined from the ratio of the reference (I_{ref}) and cross peak (I_{CP}) intensity. Quantitative *J* experiments have been developed which are slightly different depending on whether B and C are homo- (Fig. 7.12a) or heteronuclear (Fig. 7.12b).

As an example, the measurement of $J(\mathrm{C}_\alpha,\mathrm{C}^{\mathrm{ali}})$ using a quantitative-*J* HN(CO)CAC^{ali} experiment is shown. The pulse sequence is a CT–HN(CO)CA experiment (Fig. 7.13) to which a spin-echo period has been added to defocus the $J(\mathrm{C}_\alpha,\mathrm{C}^{\mathrm{ali}})$ coupling. Cross peaks that exhibit a C_α chemical shift in ω_2 are the reference peaks modulated with $\cos^2[\pi J(\mathrm{C}_\alpha,\mathrm{C}^{\mathrm{ali}})\tau'']$, whereas the cross peaks that exhibit a $\mathrm{C}^{\mathrm{ali}}$ chemical shift in ω_2 are the cross peaks modulated with $\sin^2[\pi J(\mathrm{C}_\alpha,\mathrm{C}^{\mathrm{ali}})\tau''$. The intensity ratio yields the desired *J*-coupling. Spectra of perdeuterated calmodulin are shown in Fig. 7.14 together with strips from a CC-TOCSY-(CO)NH experiment for comparison. In order to assess the sensitivity of this experiment, the parent experiment would be the HN(CO)CACB experiment, which is rather sensitive. However, the CACB dephasing rephasing step is replaced for a CAC^{ali} step involving long-range couplings. The delays involved are therefore much longer such that the fast C_α transverse relaxation comes into play. Therefore the experiment is only pratical for deuterated proteins in which the C_α transverse relaxation is

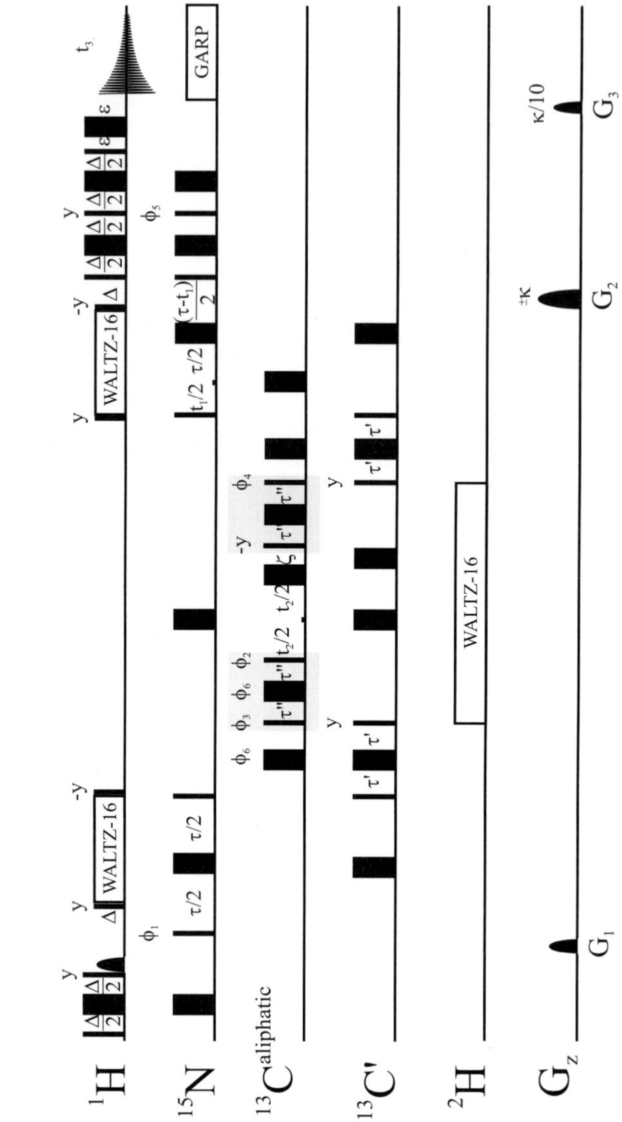

Fig. 7.13 Pulse sequence of the quantitative-J-HN(CO)CAC^{ali} experiment. The parent experiment is the constant-time HN(CO)CA experiment, to which a spin-echo period (see shaded part) has been added to defocus and refocus the $J(C_{\alpha},C^{ali})$ coupling. The cross peak will evolve chemical shift of C^{ali} in t_2 while the reference peak evolves that of C_{α}.

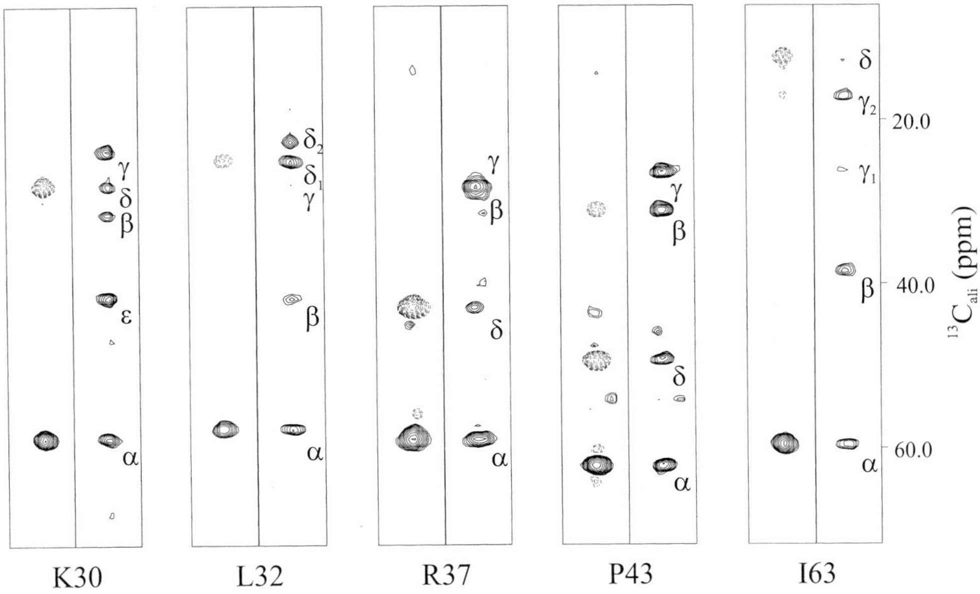

Fig. 7.14 (left) Strips from the quantitative-*J*-HN(CO)CAC^{ali} experiment showing reference and cross peaks as explained in the caption of Fig. 7.13. The assignment of side-chain protons is indicated in the strips from the CC-TOCSY (CO)NH experiment (right).

slow enough. If one has a value for this rate, then the added sensitivity penalty of the spin echo step can be assessed to be

$$e^{-T_2^{C\alpha}2\tau''} \sin^2(\pi^n J_{C_\alpha C^{ali}}\tau'') \prod_i \cos^2(\pi^n J_{C\alpha C_i}\tau'')$$

where τ'' is the dephasing and rephasing delay for the long range coupling.

7.3 Incorporation of Dipolar Couplings into Simulated Annealing Protocols

Dipolar couplings are very powerful restraints for structure determination of biomolecules and the determination of protein-protein or protein-ligand interactions [16]. They depend on the orientation of an internuclear vector, and its distance and the angular dependence is given by the following formula:

$$D^i(\beta^i a^i) = D_{\|}(3\cos^2\beta^i - 1) + \frac{3}{2}D_{\perp}(\cos 2a^i \sin^2\beta^i) \tag{2}$$

where $D_{\|}$ is the size of the main axis of the alignment tensor and $D_{\perp}$ is the rhombic part. β^i and a^i are the angles between the vector and the z-axis and the azimuthal angle, respectively (Fig. 7.15).

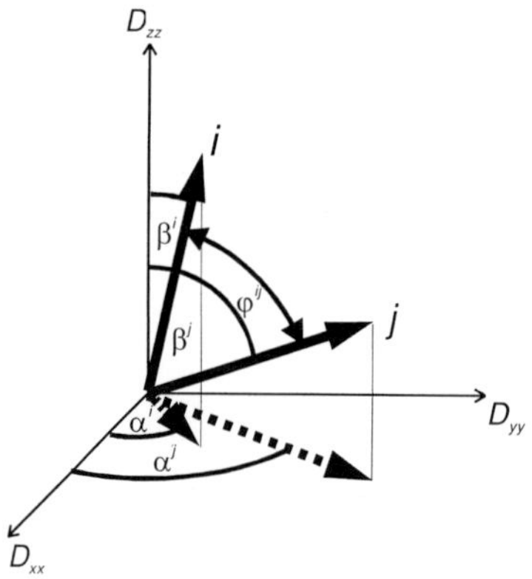

Fig. 7.15 Definition of angles for two vectors in the alignment frame. Note the relative angle φ^{ij} between the two vectors.

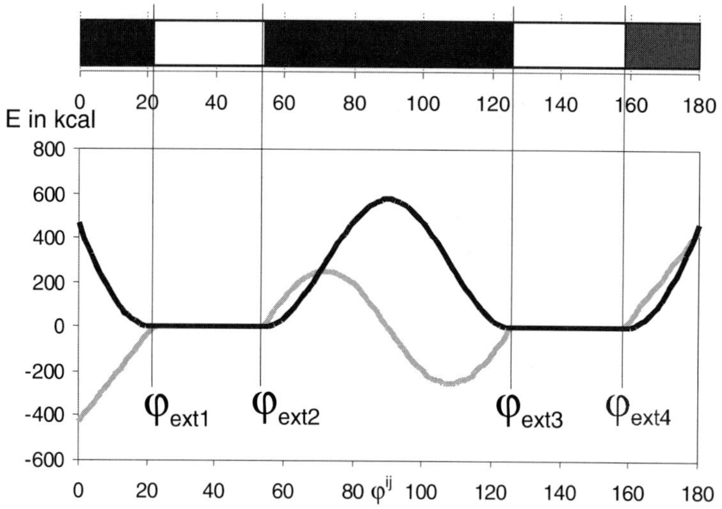

Fig. 7.16 Allowed range of intervector angles based on the measurement of dipolar couplings for each vector *i* and *j*.

It is obvious that there are two continua of orientations generated by inversion at the center of the sphere in which the dipolar couplings assume the same values. The incorporation of dipolar couplings into structure determination protocols is hampered by this multivaluedness, which causes a complicated energy hypersurface with a large number of local minima. However, translation of pairs of dipolar couplings, which may for example be derived from multiple measurements in differently orienting media, into relative angles of two vectors φ^{ij} largely solves this problem.

Using this approach [17], from n dipolar couplings $n(n-1)/2$ ranges for angles φ^{ij} can be calculated. These ranges do not include any information about tensor orientation; however, the size of the tensor is required for the protocol. A representative angle range is presented in Fig. 7.16. The angle range has two minima symmetrical about $\varphi^{ij} = 90°$. The angular restraints are constructed using the following potentials:

$$
\begin{aligned}
E^{ij}_{0\rightarrow\varphi_{\text{ext1}}} &= k_1(\varphi^{ij} - \varphi^{ij}_{\text{ext1}})^2 \\
E^{ij}_{\varphi_{\text{ext1}}\rightarrow\varphi_{\text{ext2}}} &= 0 \\
E^{ij}_{\varphi_{\text{ext2}}\rightarrow\varphi_{\text{ext3}}} &= k_2 \cos^2\left(\pi\left(\frac{\varphi^{ij} - \varphi^{ij}_{\text{ext2}}}{\varphi^{ij}_{\text{ext3}} - \varphi^{ij}_{\text{ext2}}} - \frac{1}{2}\right)\right) \\
E^{ij}_{\varphi_{\text{ext3}}\rightarrow\varphi_{\text{ext4}}} &= 0 \\
E^{ij}_{\text{ext4}\rightarrow 180^\circ} &= k_1(\varphi^{ij} - \varphi^{ij}_{\text{ext4}})^2 \,.
\end{aligned} \tag{3}
$$

During the protocol it is essential to switch on first k_1 during the high-temperature phase to generate a smooth energy hypersurface. Only during the cooling phase does the inversion symmetry of the dipolar couplings come into play, when each of the ϕ^{ij} angles has to settle to a value that is larger or smaller than 90°, and hence k_2 is then set to non-vanishing values. The structure of the protein trigger factor that has been calculated with this protocol is shown in Fig. 7.17.

7.4 Cross-Correlated Relaxation for the Measurement of Projection Angles between Tensors [18]

Angular information on molecules is not only encoded in through-bond scalar or through-space dipolar couplings, but can also be derived from measurements of relaxation rates. Relaxation is caused by the correlation of two tensorial interactions. If these interactions are different, then relaxation caused by the cross correlation of the two interactions is called cross-correlated relaxation. The cross-correlated relaxation rate depends on the strengths of the tensorial interactions as well as on the projection angle of the two tensors onto each other. To be efficient, the time-dependence of the tensorial interactions must be similar. Such a situation exists e.g. between dipolar interactions of two proximate spin pairs within well-structured parts of the molecule, which display the same time dependence because of overall tumbling. The most important tensorial interactions are the dipolar coupling and the chemical shift anisotropy. The dipole tensor between two spins A^1 and A^2 (B^1 and B^2) is axially symmetric, and the main axis is collinear with the internuclear bond vector $\mathbf{A^1A^2}$ ($\mathbf{B^1B^2}$).

The cross-correlated relaxation rate observed for double- or zero-quantum coherence involving A^1 or A^2 and B^1 or B^2 for two dipolar interactions therefore takes the following form:

$$
\Gamma^c_{A^1A^2,B^1B^2} = \frac{\gamma_{A^1}\gamma_{A^2}}{(r_{A^1A^2})^3}\frac{\gamma_{B^1}\gamma_{B^2}}{(r_{B^1B^2})^3}\left(\frac{\mu_0}{4\pi}\hbar\right)^2 \frac{1}{5}(3\cos^2\theta_{A^1A^2,B^1B^2} - 1)^* \tau_c \,. \tag{4}
$$

The characterisation of the angular dependence of the interaction of two dipole tensors $\mathbf{A^1A^2}$ and $\mathbf{B^1B^2}$ is therefore straightforward, namely it depends on the projection angle of the two bonds between A^1 and A^2 and between B^1 and B^2. The orientation and magnitude of the chemical shift anisotropy (CSA) tensor, which also can cause cross-correlated relaxation, is not know *a priori* and therefore needs to be determined experimentally or

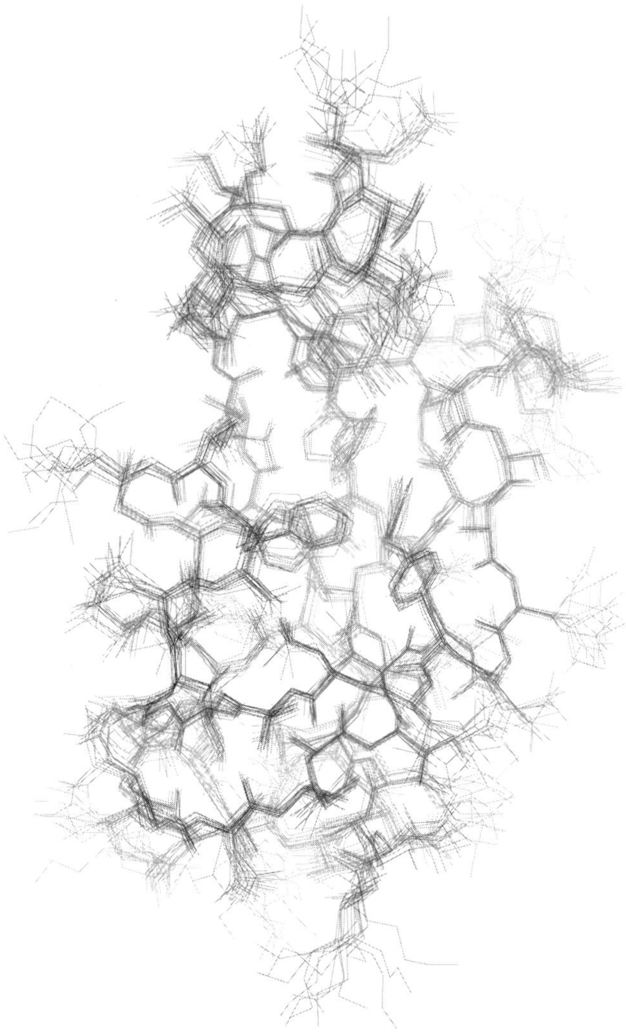

Fig. 7.17 Structure of trigger factor that has been obtained using dipolar coupling data as explained in the two previous figures.

by quantum chemical calculations. The projections of tensors have been determined and interpreted in structural terms in solid-state *local field separated* [19], spin diffusion [20] and multiple-quantum NMR spectroscopy [21]. From the side band pattern in solid-state spectra, the orientation of the two tensors with respect to each other can be derived. The cross-correlated relaxation rate between the CSA of spin A^1 and the B^1-B^2 dipole is given by the following formula:

$$\Gamma^{c}_{A^1,B^1B^2} = -\frac{2}{15}\gamma_{A^1} B_0 \tau_c \hbar \frac{\mu_0}{4\pi}\frac{\gamma_{B^1}\gamma_{B^2}}{(r_{B^1B^2})^3}\{(\sigma_{xx}-\sigma_{zz})(3\cos^2\theta_{B^1B^2,\sigma_{xx}}-1) + (\sigma_{yy}-\sigma_{zz})(3\cos^2\theta_{B^1B^2,\sigma_{yy}}-1)\}\,. \quad (5)$$

In two pairs of nuclei A^1–A^2 and B^1–B^2, projection angle-dependent cross-correlated relaxation rates due to two dipolar couplings $\Gamma^{c}_{A^1A^2,B^1B^2}$ of double and zero quantum coherences between nuclei A^1 and B^1 can be measured provided the following three requirements are fulfilled: (i) The desired double and zero quantum coherence between nuclei A^1 and B^1 can be excited. (ii) There is a resolvable scalar coupling between A^1 and A^2 as well as between B^1 and B^2. For CSA/DD cross-correlated relaxation there is no need for the A^1, A^2 coupling. (iii) The main relaxation source for single-quantum coherence of A^1 (SQC) is the dipolar coupling to A^2, and the main relaxation source of the SQC of B^1 is the dipolar coupling to B^2. Again, for CSA/DD, the requirement for the A^1 and A^2 relaxation is void.

To illustrate how cross-correlated relaxation can be used to measure the angle between two bond vectors, we will use the example of the generation of double and zero quantum coherence between spins A^1 and B^1 and call the angle between the A^1-A^2 and B^1-B^2 vectors θ (Fig. 7.18).

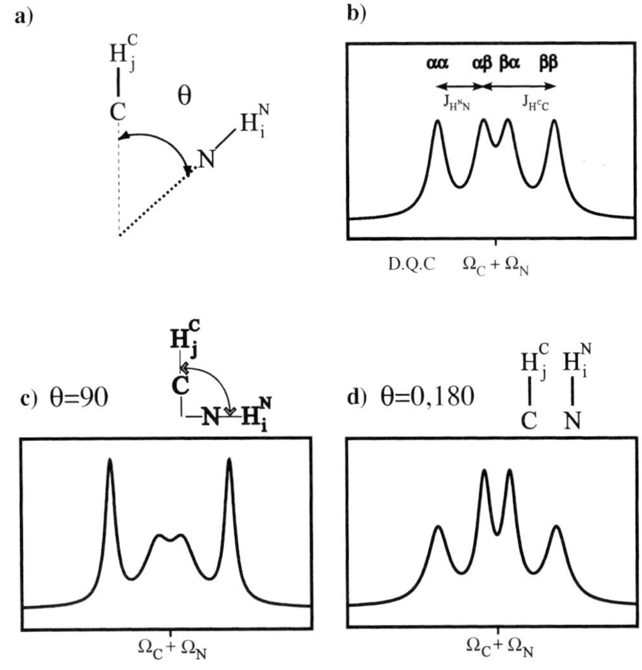

Fig. 7.18 Schematic representation of cross-correlated relaxation of double and zero quantum coherences. Depending on the relative angle of the two involved internuclear vectors, the differential relaxation affects the multiplet in the given way.

In the experiment, double- and zero-quantum coherence between A^1 and B^1 evolve chemical shifts at $\Omega_{A^1} + \Omega_{B^1}$ and $\Omega_{A^1} - \Omega_{B^1}$ in a indirect detected period t_1 or t_2. A doublet of doublet of lines is generated with splittings due to scalar coupling of $^1J(A^1,A^2)$ and $^1J(B^1,B^2)$ if neither A^2 nor B^2 are decoupled during this indirect evolution period. In the absence of cross-correlated relaxation, all four multiplet components would have the same line width and intensities (Fig. 7.18). α and β denote the polarization of A^2 and B^2, respectively. However, if one considers the effects of cross-correlated relaxation, the relative intensities of the lines will differ depending on the relative orientation of the two vectors. If the two vectors A^1–A^2 and B^1–B^2 are oriented orthogonal to each other, the inner two lines will be broader than the outer lines, and the opposite is true for parallel orientation of the two vectors (Fig. 7.18). Equal intensity for all four lines is also obtained if the two vectors span the *magic angle* $\theta = \arctan(\sqrt{2}) \approx 54.6°$ (Fig. 7.18).

For the measurement of cross-correlated relaxation rates, there are mainly three methods that have been used in practice. In the *J*-resolved constant time experiment, the multiplet lines exhibiting differential relaxation are resolved by the *J* couplings, and the line width is translated into intensity in a constant time experiment (Fig. 7.19 a,d). In the *J*-resolved real time experiment the line width of each multiplet line is measured instead (Fig. 7.19 b, d). This experiment has been applied so far only for the measurement of

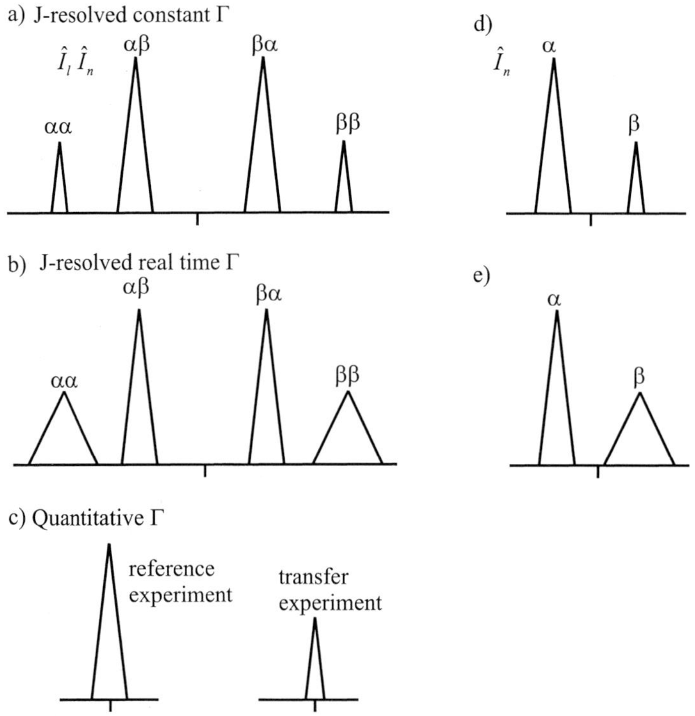

Fig. 7.19 Three approaches to the measurement of dipole/dipole and dipole/CSA cross-correlated relaxation rates.

cross-correlated relaxation rates in organometallic compounds [22]. Both experiments have the disadvantage that there are multiplets in the spectrum, which lead to lower resolution compared to a completely decoupled spectrum. This problem is removed in the third approach, the so-called quantitative Γ approach. Here, in the so called cross experiment, operators are generated based on cross-correlated relaxation, and in a second experiment a similar operator is created based on e.g. a known coupling constant. The intensity ratio of the cross peaks in the cross and reference experiment can then be used for quantitative measurement of the respective cross-correlated relaxation rate (Fig. 7.19c).

As can be seen, the cross-correlated relaxation rate depends on the correlation time of the molecule. Therefore, cross-correlated relaxation rates will increase with increasing size of the biomolecule under investigation, as opposed to *J*-coupling constants whose size does not depend on the molecular weight of the molecule. A second application lies in the fact that, in analogy to the transferred NOESY experiment, in which distances are measured for ligands when bound to large target molecules, a transferred cross-correlated relaxation experiment allows us to measure projection angles of weakly bound ligands when bound to large molecules.

7.4.1
J-Resolved Constant Time Measurement of Cross-Correlated Relaxation Rates

The method relies on the measurement of cross-correlated relaxation rates in a constant time period such that the cross-correlated relaxation rate evolves during a fixed time τ. In order to resolve the cross-correlated relaxation rate, however, the couplings need to evolve during an evolution time, e.g. t_1. The first pulse sequence published for the measurement of the cross-correlated relaxation rate between the HN_{i-1} and the $C_{a,i}$,$H_{a,i}$ vector relied on an HN(CO)CA experiment, in which the C_a chemical shift evolution period was replaced by evolution of ^{15}N,$^{13}C^a$ double and zero quantum coherences (Fig. 7.20).

Then, each of the correlation peaks exhibits a doublet of doublet structure along ω_2 because of the evolution of the ^1J(H,N) and the $^1J(C_a,H_a)$ coupling during t_2. The intensity of the individual multiplet lines is affected by the dipole, dipole cross-correlated relaxation rate Γ^c_{HN,C_aH_a} as well as the sums of two dipole/CSA cross-correlated relaxation rates: $\Gamma^c_{HN,C_a} + \Gamma^c_{HN,N}$ and $\Gamma^c_{C_aH_a,C_a} + \Gamma^c_{C_aH_a,N}$. By forming the ratios

$$\Gamma^c_{NH,CH} = \frac{1}{4T} * \ln\left(\frac{I(\alpha\beta) * I(\beta\alpha)}{I(\alpha\alpha) * I(\beta\beta)}\right) - \frac{1}{2}(\Gamma^{NOE}_{DQ} - \Gamma^{NOE}_{ZQ})$$

$$\Gamma^c_{C+N,NH} = \frac{1}{4T} * \ln\left(\frac{I(\alpha\beta) * I(\beta\beta)}{I(\alpha\alpha) * I(\beta\alpha)}\right) \tag{6}$$

$$\Gamma^c_{C+N,CH} = \frac{1}{4T} * \ln\left(\frac{I(\beta\beta) * I(\beta\alpha)}{I(\alpha\alpha) * I(\alpha\beta)}\right)$$

the different cross-correlated relaxation rates can be extracted. The term $-\frac{1}{2}(\Gamma^{NOE}_{DQ} - \Gamma^{NOE}_{ZQ})$ is the conventional NOE between the two protons, H^N_{i-1} and $H_{a,i}$. Normally, it is much smal-

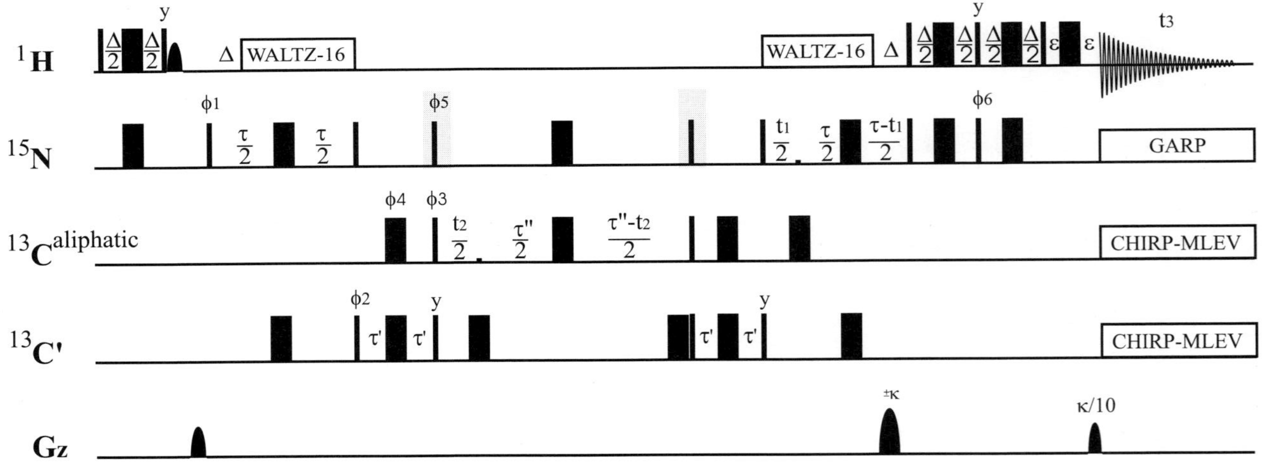

Fig. 7.20 *J*-resolved constant Γ HN(CO)CA experiment for the measurement of cross-correlated relaxation rates, especially $\Gamma^{c}_{NH,CH}$. The experiment has two 90° (^{15}N) pulses (shaded part) simultaneously with the C_{α} pulses exciting the double and zero quantum coherences.

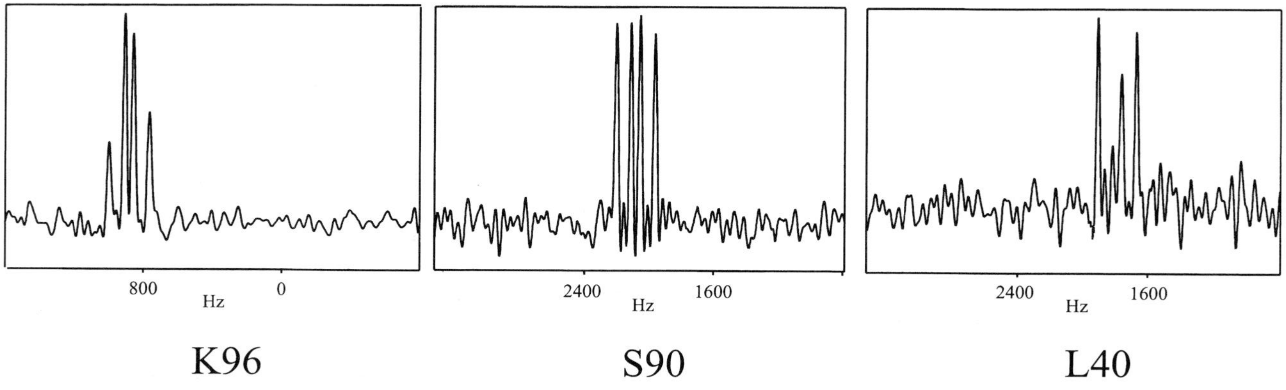

Fig. 7.21 Traces through ω_2 of the *J*-resolved constant Γ HN(CO)CA of rhodniin. The different multiplet patterns obviously indicate different geometries of the HN and H_α,C_α vectors for these residues.

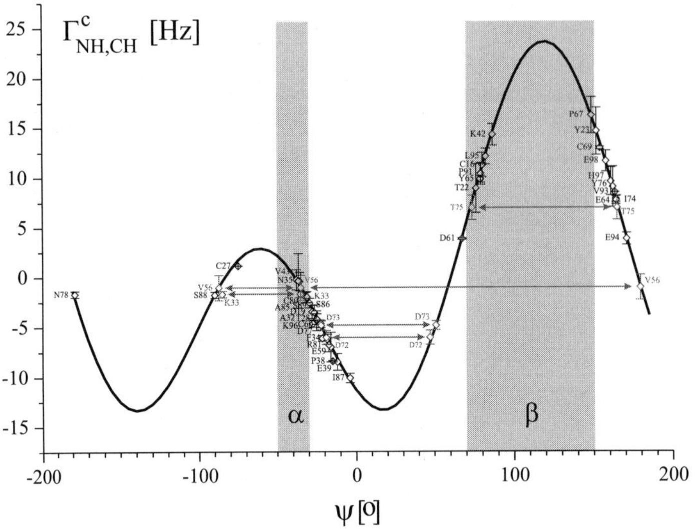

Fig. 7.22 Correlation between experimental and theoretical rates back-calculated from the structure that was refined with the $\Gamma^{c}_{NH,CH}$ values.

ler than the cross-correlated relaxation rate and can therefore safely be ignored. The application of the pulse sequence to the protein rhodniin yielded a spectrum from which traces along ω_2 (Fig. 7.21) were taken. The fit of the cross-correlated relaxation rates to a structure calculated with these cross-correlated relaxation rates as restraints is shown in Fig. 7.22.

7.4.2
Quantitative Γ Measurement of Cross-Correlated Relaxation Rates

Felli et al. [23] and Pelupessy et al. [24] have developed a method that measures dipole, dipole cross-correlated relaxation by transferring an initial operator to another operator by the evolution of the desired cross-correlated relaxation rate.

Fig. 7.23 shows the quantitative Γ sequence that measures the same $\Gamma^{c}_{NH,CH}$ cross-correlated relaxation rate as the sequence of Fig. 7.20. Again N-C$^{\alpha}$ DQ/ZQ coherence is excited using an HN(CO)CA experiment. Because of cross-correlated relaxation, N-C$^{\alpha}$ DQ/ZQ coherence evolves as described in Eq. (7):

$$4N_xC_{ax}C'_z \rightarrow 4N_xC_{ax}C'_z\begin{bmatrix}\cosh(\Gamma^{DD,DD}_{N,H^N,C_aH_a}T)\cos(\pi^1J(N,H^N)\Delta_1)\cos(\pi^1J(C_a,H_a)\Delta_2)\\-\sinh(\Gamma^{DD,DD}_{N,H^N,C_aH_a}T)\sin(\pi^1J(N,H^N)\Delta_1)\sin(\pi^1J(C_a,H_a)\Delta_2)\end{bmatrix}$$

$$+16N_yC_{ay}C'_zH^N_zH_{az}\begin{bmatrix}\sinh(\Gamma^{DD,DD}_{N,H^N,C_aH_a}T)\cos(\pi^1J(N,H^N)\Delta_1)\cos(\pi^1J(C_a,H_a)\Delta_2)\\+\cosh(\Gamma^{DD,DD}_{N,H^N,C_aH_a}T)\sin(\pi^1J(N,H^N)\Delta_1)\sin(\pi^1J(C_a,H_a)\Delta_2)\end{bmatrix}. \quad (7)$$

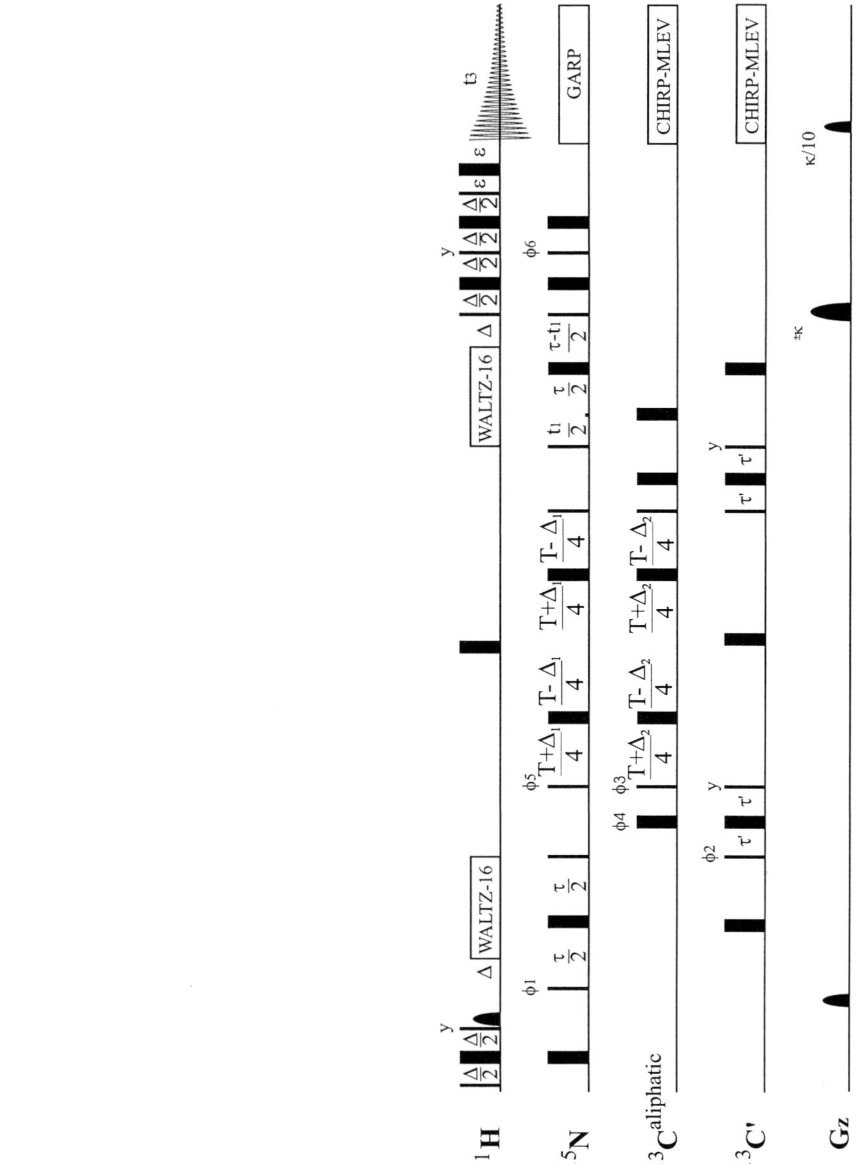

Fig. 7.23 Quantitative *Γ* version of the experiment discussed in Fig. 7.20.

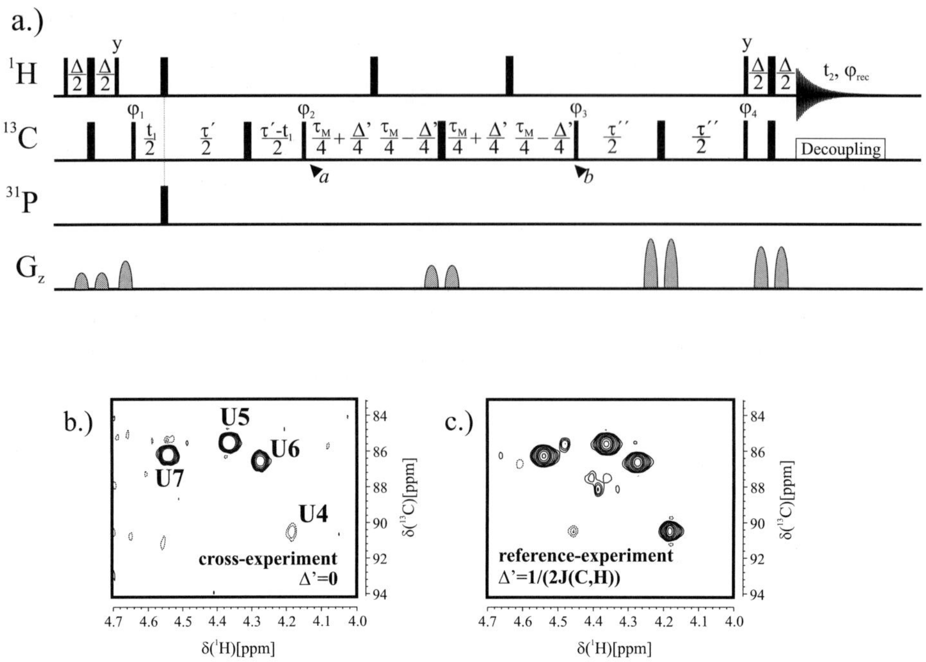

Fig. 7.24 Quantitative $\Gamma^{c}_{CH,CH}$HCCH experiment (**a**) and spectra from cross (**b**) and reference experiments (**c**).

Setting $\Delta_1 = 0$ and $\Delta_2 = 0$ yields

$$4N_x C_{\alpha x} C'_z \xrightarrow{\Delta_1=\Delta_2=0} -4N_x C_{\alpha x} C'_z \cosh(\Gamma^{DD,DD}_{N,H^N,C_\alpha H_\alpha} T) + 16 N_y C_{\alpha y} C'_z H^N_z H_{\alpha z} \sinh(\Gamma^{DD,DD}_{N,H^N,C_\alpha H_\alpha} T) \quad (8)$$

while setting $\Delta_1 = 1/2^1J(N,H^N)$ and $\Delta_2 = 1/2^1J(C_\alpha, H_\alpha)$ yields

$$4N_x C_{\alpha x} C'_z \longrightarrow -4N_x C_{\alpha x} C'_z \sinh(\Gamma^{DD,DD}_{N,H^N,C_\alpha H_\alpha} T) + 16 N_y C_{\alpha y} C'_z H^N_z H_{\alpha z} \cosh(\Gamma^{DD,DD}_{N,H^N,C_\alpha H_\alpha} T). \quad (9)$$

Out of the remaining two operators, the term is selected by adjusting the phases of the $90°(^{13}C^\alpha)$ and $90°(^{15}N)$ pulses properly, and signal intensity is modulated by $\cosh(\Gamma^{DD,DD}_{N,H^N,C_\alpha H_\alpha} T)$ for the reference experiment and by $\sinh(\Gamma^{DD,DD}_{N,H^N,C_\alpha H_\alpha} T)$ in the cross experiment.

The same idea applies in the quantitative Γ-HCCH experiment, which can be applied to aliphatic side chains or to sugars to measure the pseudorotation phase (Fig. 7.24).

Therein, cross-correlated relaxation $\Gamma^{c}_{C_iH_i,C_jH_j}$ of the double and zero quantum coherence (DQ/ZQ) $4H_{iz}C_{ix}C_{jy}$ generated at time point a creates the DQ/ZQ operator $4H_{jz}C_{jx}C_{iy}$. In the second part of the experiment, the operator $4H_{jz}C_{jx}C_{iy}$ is transferred via a 90° y-pulse applied to ^{13}C nuclei to give rise to a cross peak at $\omega_{H_j}, \omega_{C_j}$. $\Gamma^{CSA,DD}_{C_i,C_jH_j}$ and

$\Gamma^{CSA,DD}_{H_i,C_jH_j}$ cross-correlated relaxation are refocused by application of the 180° carbon and proton pulses during the mixing time τ_M. The following transfers are achieved in the sequence of Fig. 7.24:

$$4H_{iz}C_{ix}C_{jy} \rightarrow 4H_{iz}C_{ix}C_{jy}\lfloor\cosh(\Gamma^c_{C_iH_i,C_jH_j}\tau_M)\cos^2(\pi J_{CH}\Delta') - \sinh(\Gamma^c_{C_iH_i,C_jH_j}\tau_M)\sin^2(\pi J_{CH}\Delta')\rfloor$$
$$- 4H_{jz}C_{jx}C_{iy}\lfloor\sinh(\Gamma^c_{C_iH_i,C_jH_j}\tau_M)\cos^2(\pi J_{CH}\Delta') - \cosh(\Gamma^c_{C_iH_i,C_jH_j}\tau_M)\sin^2(\pi J_{CH}\Delta')\rfloor \qquad (10)$$

The last term produces a cross peak at $\omega_{C_i}, \omega_{H_j}$ due to coherence transfer between $4H_{iz}C_{ix}C_{jy}$ and $4H_{jz}C_{jx}C_{iy}$. In the experiment with $\Delta' = 0$, the intensity of the cross peak (I^{cross}) is proportional to $\sinh(\Gamma^c_{C_iH_i,C_jH_j}\tau_M)$, whereas for $\Delta' = 1/2J_{CH}$ the intensity of the cross peak (I^{ref}) is proportional to $\cosh(\Gamma^c_{C_iH_i,C_jH_j}\tau_M)$. By comparing the intensity of the cross peak measured in the two experiments one can determine

$$\frac{I^{\text{cross}}}{I^{\text{ref}}} = \tanh(\Gamma^c_{C_iH_i,C_jH_j}\tau_M)\,. \qquad (11)$$

The dependence of cross-correlated relaxation on the projection angle between the two dipole vectors can be exploited to determine local conformations in ribose rings.

Fig. 7.25 shows the cross-correlation rates $\Gamma^c_{C1'H1',C2'H2'}$, $\Gamma^c_{C2'H2',C3'H3'}$ and $\Gamma^c_{C3'H3',C4'H4'}$ as a function of pseudo rotation phase (P) and amplitude ($\nu^{\max}$) [25, 26]. Discrimination between sugar pucker modes C2′-*endo* and C3′-*endo* [25], which are shaded in gray in Fig. 7.25, can be achieved from observation of opposite signs of the two rates $\Gamma^c_{C1'H1',C2'H2'}$ and $\Gamma^c_{C3'H3',C4'H4'}$. Analysis of the relative signs and magnitudes of the two rates provides a method to distinguish between the two main sugar pucker conformations. However, as can be observed from the various plots with different $\nu^{\max}$, the absolute values of the rates are affected by variations of the sugar pucker amplitude. Fig. 7.25 shows the ratio of cross-correlated relaxation rates,

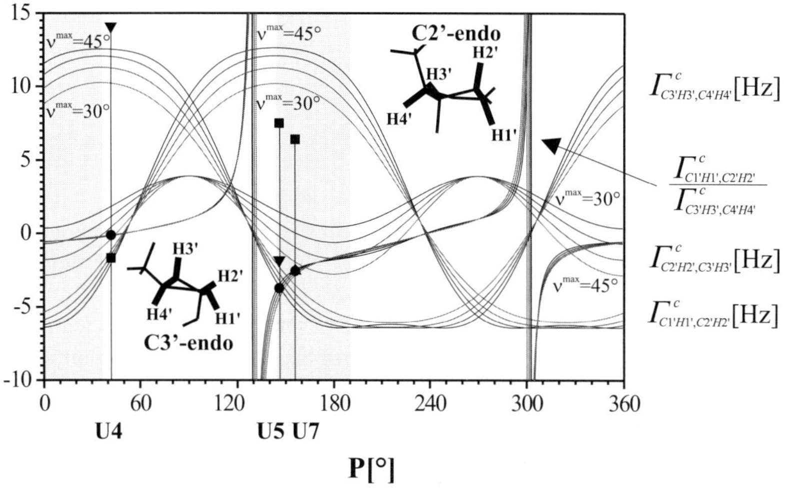

Fig. 7.25 Theoretical and experimental $\Gamma^c_{CH,CH}$ values to define the sugar pucker.

$$\frac{\Gamma^{c}_{C1'H1',C2'H2'}}{\Gamma^{c}_{C3'H3',C4'H4'}} = \frac{\langle S^{c}_{1',2'} \rangle \, 3\cos^2\theta_{1',2'} - 1}{\langle S^{c}_{3',4'} \rangle \, 3\cos^2\theta_{3',4'} - 1} \tag{12}$$

which is less sensitive to variations of ν^{max}, does not depend on τ_c and, if the fluctuations of the respective dipole tensors are comparable, is also less sensitive to $S^{c}_{i,j}$. The ratio $\Gamma^{c}_{C1'H1',C2'H2'} : \Gamma^{c}_{C3'H3',C4'H4'}$ is a measure of the sugar puckering mode, while closer analysis of individual rates reveals information on fluctuations of the sugar pucker on various time scales.

By combining the quantitative approach [23] to extract cross-correlated relaxation with resolution enhancement methods using restricted coherence transfer in a so-called forward directed TOCSY [27], Richter et al. could determine the ribose sugar conformation for all but two residues in a uniformly ^{13}C,^{15}N labeled 25mer RNA [28] and compare them to $^{3}J(H, H)$ values determined using a *forward-directed* HCC-TOCSY-CCH-E.COSY experiment [29].

7.4.3
J-Resolved Constant Time Γ Experiment for the Determination of the Phosphodiester Backbone Angles α and ζ

As an example of the measurement of cross-correlated relaxation between CSA and dipolar couplings, we choose the *J*-resolved constant time experiment [30] (Fig. 7.26 a) that measures the cross-correlated relaxation of ^{1}H,^{13}C-dipolar coupling and ^{31}P-chemical shift anisotropy to determine the phosphodiester backbone angles α and ζ in RNA. Since ^{31}P is not bound to NMR-active nuclei, NOE information for the backbone of RNA is sparse, and vicinal scalar coupling constants cannot be exploited. The cross-correlated relaxation rates can be obtained from the relative scaling (shown schematically in Fig. 7.19 d) of the two submultiplet intensities derived from an ^{1}H-coupled constant time spectrum of ^{13}C,^{31}P double- and zero-quantum coherence [DQC (double-quantum coherence) and ZQC (zero-quantum coherence), respectively]. These traces are shown in Fig. 7.26 c. The desired cross-correlated relaxation rate can be extracted from the intensities of the cross peaks according to:

$$\Gamma^{c}_{P,CH} = \frac{1}{4T} * \ln\left(\frac{I^{DQ}(\alpha) I^{ZQ}(\beta)}{I^{ZQ}(\alpha) I^{DQ}(\beta)}\right) .$$

The orientation of the ^{31}P CSA is shown in Fig. 7.26 b. The experiment determines the projection angles of a dipole tensor, e.g. the C2′H2′ dipole tensor, onto the components of the ^{31}P CSA tensors as shown in Fig. 7.26 b.

The size and orientation of the ^{31}P CSA tensor have been calibrated from single-crystal solid state NMR data of barium diethylphosphate [31]. The projection angles $\theta_{CH,\sigma 22}$ and $\theta_{CH,\sigma 33}$ between the ^{1}H,^{13}C-dipolar coupling and the components of the ^{31}P-CSA-tensor depend on a number of torsion angles. The dependence of $\Gamma^{DD,CSA}_{(C3'_i,H3'_i),(P_{i+1})}$ on the backbone angles ε and ζ is shown dark-shaded in Fig. 7.26 d and $\Gamma^{DD,CSA}_{(C2'_i,H2'_i),(P_{i+1})}$ of on ε and ζ (for the ribofuranoside ring in C3′-*endo* conformation) is shown light-shaded. For τ_c=2.5±0.2 ns, the rates vary from –10 to 20 Hz. With an experimental uncertainty of 1.5 Hz deter-

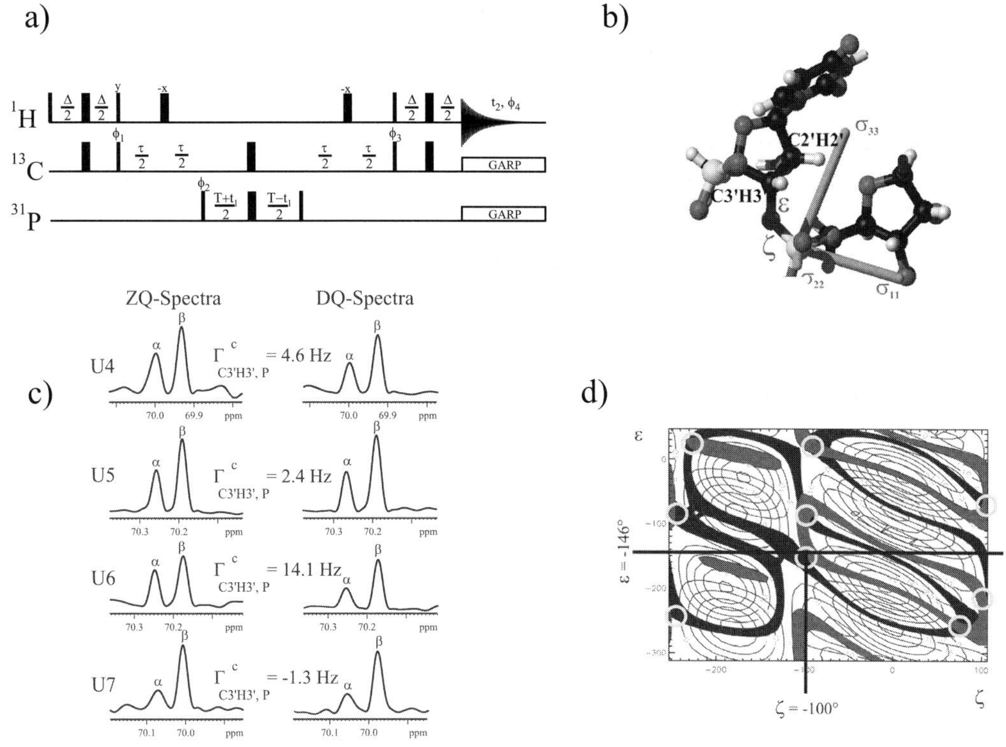

Fig. 7.26 *J*-resolved constant Γ C,H-HSQC experiment (**a**) to measure the cross-correlated relaxation rates in RNA with a geometry given in **b**. Traces through the spectrum (**c**) and the evaluation of the desired angle ζ based on known angles ε (**d**) are shown as well.

mined from measurement of the $\Gamma^{DD,CSA}_{(C',H'),(C')}$ in a nondecoupled [^{13}C,^{1}H]-HSQC, the circled areas in Fig. 7.26d can be defined for the angular pairs ε, ζ that are in agreement with the observed cross-correlated relaxation rates $\Gamma^{DD,CSA}_{(C3'_i,H3'_i),CSA(P_{i+1})}=4.6$ Hz and $\Gamma^{DD,CSA}_{(C2'_i,H2'_i),(P_{i+1})}=9.5$ Hz for U4. For U4, ε is found to be $-146°$ [32]. The angle ζ must therefore be close to $-100°$. For the angle α, the sum $\Gamma^{DD}_{(C5'_i,H5'_i),(P_i)} + \Gamma^{DD}_{(C5'_i,H5''_i),(P_i)}$, representing the arithmetic average of the two dipole tensors, can be measured. Together with scalar coupling constant information for β, the backbone angle α can be defined.

7.4.4
Transferred Cross-Correlated Relaxation

Due to the fact that cross-correlated relaxation depends linearly on the correlation time, it can be used to determine the conformation of ligands when bound to target molecules, provided that the off rate is fast enough to enable detection of the cross-correlated relaxation rate via the free ligand [33, 34]. The conditions under which such an experiment can be performed are similar to those found for transferred NOEs [35], and, for K_d values normally exceeding 10^{-6} M, ligands are in an equilibrium between a protein-bound and a free form. The period during which the molecule is in the bound conformation will contribute to relaxation to the largest extent and hence will heavily weight experiments that are based on relaxation. The population-averaged cross-correlated relaxation rate for dipole,dipole cross-correlated relaxation between two sites with fractions p_L of the free ligand and $p_{ML}=(1-p_{ML})$ of the bound ligand is given in Eq. (13):

$$\langle \Gamma^{DD,DD}_{C_mH_m,C_nH_n} \rangle = \frac{2}{5} \frac{\gamma_C^2 \gamma_H^2 \mu_0^2 \hbar^2}{(4\pi)^2 r^3_{C_mH_m} r^3_{C_nH_n}} \times \left[S^2_{mn,L} p_L \frac{3\cos^2\theta_{mn,L} - 1}{2} \tau_{c,L} + S^2_{mn,ML} p_{ML} \frac{3\cos^2\theta_{mn,ML} - 1}{2} \tau_{c,ML} \right] \quad (13)$$

If the correlation time of the free ligand $\tau_{c,L}$ is much smaller than the correlation time of the bound ligand $\tau_{c,ML}$, the observed averaged cross-correlated relaxation rate $\langle \Gamma^{DD,DD}_{C_mH_m,C_nH_n} \rangle$ is dominated by the bound conformation, and therefore precise information on the conformation of a bound ligand can be obtained. This concept has been used to derive the conformation of a phosphotyrosine peptide weakly bound to STAT-6 [34] and to determine the conformation of a tRNA analog bound to EF-Tu·GDP complex [33]. The technique should hold potential for the determination of weakly bound ligands in SAR-by-NMR approaches and for the characterization of transient species in enzyme-catalyzed reactions. Note that the technique does not suffer from spin-diffusion type artifacts inherent in transfer-NOE experiments performed on large targets. An application is shown in Fig. 7.27 for the ester of anthranilic acid with the 3′ alcohol function of adenosin. The conformation of the sugar moiety could be determined to be the 2′ *endo*.

7.5
Applicability of the Methods

Methods for the determination of coupling constants impose different requirements for labeling, amounts of sample etc. Methods for the determination of ^1J heteronuclear coupling constants are among the most sensitive measurements and have sensitivity similar to HSQC experiments. Methods for the determination of ^2J and ^3J coupling constants or cross-correlated relaxation are less sensitive. However, their sensitivity can be evaluated based on the experiments they are derived from and whose sensitivity is normally known because these parent experiments are used e.g. for assignment purposes etc. This has been discussed with each of the presented sequences exemplarily. For the cross-correlated relaxation experiments similar arguments apply. For example, the cross-correlated relaxation experiment discussed in Sect. 7.4.1. relies on the HN(CO)CA experiment, and the transferred cross-correlated relaxation experiment discussed in Sect. 7.4.3. relies on the HCCH-COSY experiment. Let us assume that the sensitivity of the HN(CO)CA (HCCH-COSY) experiment is 1. Then the sensitivity of the reference experiments for both cases is given by: $e^{-T_2^{DQ/ZQ}\tau}$, where $T_2^{DQ/ZQ}$ is the relaxation time of the carbon/nitrogen (carbon/carbon) double quantum and zero quantum coherences and τ is the time during which the double and zero quantum coherences exist in the pulse sequence. In the cross experiments, cross peaks arise because of the evolution of the cross-correlated relaxation

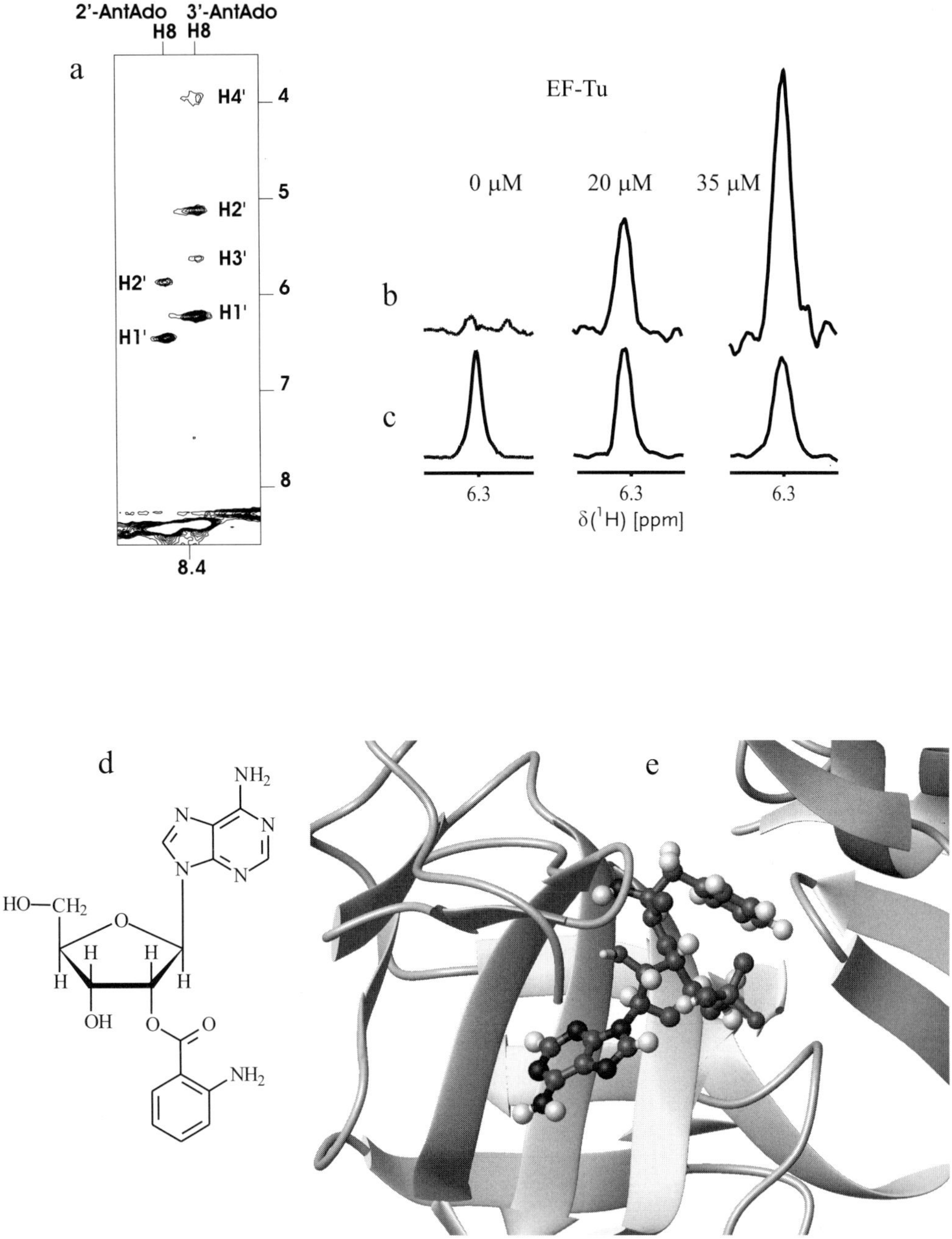

Fig. 7.27 **a** Transferred NOESY, **b** cross and **c** reference traces from the transferred $\Gamma^{c}_{CH,CH}$-HCCH experiment that display the dependence of the rate of the transferred cross-correlated relaxation rate on the EF-tu concentration. **d** Formula of Ant-Ado that was weakly bound to EF-tu. **e** Structure of the NMR-derived conformation modeled into the X-ray structure of EF-tu.

rate during the delay τ. Thus the sensitivity penalty factor for the cross experiments becomes: $e^{-T_2^{DQ/ZQ}\tau}\sinh(\Gamma_{HN,HC}\tau)(e^{-T_2^{DQ/ZQ}\tau}\sinh(\Gamma_{HC,HC}\tau))$ which reiterates the $T_2^{DQ/ZQ}$ relaxation of the double and zero quantum coherences and introduces newly the cross-correlated relaxation rate. Normally, one would try to reach values between 10 and 50% for $\sinh(\Gamma_{HN,HC}\tau)(\sinh(\Gamma_{HC,HC}\tau))$ in order to have sufficient signal-to-noise ratio. With respect to the labeling patterns required for the reviewed experiments, it is obvious that the labeling pattern required for the experiments to measure couplings or cross-correlated relaxation is the same as that required for those parent experiments. Because of the normally lower signal-to-noise ratio, increased amounts of sample or measurement time are required.

7.6 References

1 T. Bremi, R. Brüschweiler, *J. Am. Chem. Soc.* **1997**, *119*, 6672–6673; S.F. Lienin, T. Bremi, B. Brutscher, R. Brüschweiler, R.R. Ernst, *J. Am. Chem. Soc.* **1998**, *120*, 9870–9879; M. Buck, M. Karplus, *J. Am. Chem. Soc.* **1999**, *121*, 9645–9658; M. Pellecchia, Y. Pang, L. Wang, A.V. Kurochkin, A. Kumar, E.R.P. Zuiderweg, *J. Am. Chem. Soc.* **1999**, 121, 9165–9170; T. Carlomagno, M. Maurer, M. Hennig, C. Griesinger, *J. Am. Chem. Soc.* **2000**, *122*, 5105–5113; X. Pang, E.R.P. Zuiderweg, *J. Am. Chem. Soc.* **2000**, *122*, 4841–4842.

2a J.P. Marino, H. Schwalbe, C. Griesinger, *Acc. Chem. Res.* **1999**, *32*, 614–632; C. Griesinger, M. Hennig, J.P. Marino, B. Reif, C. Richter, H. Schwalbe, in "Modern Techniques in Protein NMR", Eds. N.R. Krishna, L.J. Berliner, Vol 16 of Biological Magnetic Resonance, Plenum Press, 1999; H. Schwalbe, T. Carlomagno, J. Junker, M. Hennig, B. Reif, C. Richter, C. Griesinger, *Methods Enzymol.* **2001**, *338*, 35–81; C. Griesinger, "NMR Spectroscopy as a Tool for the Determination of Structure and Dynamics of Molecules" in "Essays in Comtemporary Chemistry: From Molecular Structure to Biology", Eds. G. Quinkert, M.V. Kisakürek, Verlag Helv. Chim. Acta, Wiley-VCH **2001**, pp. 35–105; C. Griesinger, W. Peti, J. Meiler (2002) "Angular Restraints from Residual Dipolar Couplings for Structure Refinement" in "NMR for a New Millennium", Kluwer Academic/Plenum Press

2b J.R.Tolman, H.M. Al-Hashimi, L.E. Kay, J.H. Prestegard, *J. Am. Chem. Soc.* **2001**, *123*, 1416–1424; J. Meiler, W. Peti, J. Prompers, C. Griesinger, R. Brüschweiler, *J. Am. Chem. Soc.* **2001**, *123*, (25), 6098–6107; W. Peti, J. Meiler, R. Brüschweiler, C. Griesinger, *J. Am. Chem. Soc.* **2002**, *124*, 5822–5833.

3 V.F. Bystrov, *Prog. NMR. Spectrosc.* **1976**, *10*, 41.

4 M. Karplus, *J. Chem. Phys.* **1959**, *30*, 11; M. Karplus, *J. Am. Chem Soc.* **1963**, *85*, 2870.

5 N. Cyr, G.K. Hamer, A.S. Perlin (1978) *Can. J. Chem.* **56**, 297; J.A. Schwarcz, N. Cyr, and A.S. Perlin, *Can. J. Chem.* **1975**, *53*, 1872; J.V. Hines, S.M. Laundry, G. Varani, I. Tinoco, Jr., *J. Am. Chem. Soc.* **1993**, *115*, 11002–11003; J. V. Hines, S.M. Laundry, G. Varani, I. Tinoco, Jr., *J. Am. Chem. Soc.* **1994**, *116*, 5823–5831; J.P. Marino, H. Schwalbe, S.J. Glaser, C. Griesinger, *J. Am. Chem. Soc.* **1996**, *118*, 4388–4395.

6 J.A.B. Lohman, and C. MacLean, *Chem. Phys.* **1978**, *35*, 269; J.A.B. Lohman, C. MacLean, *Chem. Phys.* **1978**, *43*, 144; J.A.B. Lohman, C. MacLean, *Mol. Phys.* **1997**, *38*, 1255–1261; E.W. Bastiaan, C. MacLean, P.C.M. van Zijl, A.A. Bothner-By, *Ann. Rep. NMR Spectrosc.* **1987**, *19*, 35–37; J.R. Tolman, J.M. Flanagan, M.A. Kennedy, J.H. Prestegard, *Proc. Natl. Acad. Sci. USA* **1995**, *92*, 9279; H.C. Kung, K.Y. Wang, I. Goljer, P.H. Bolton, *J. Magn. Reson. Ser. B* **1995**, *109*, 323; N.

Tjandra, S. Grzesiek, and A. Bax, *J. Am. Chem. Soc.* **1996**, *118*, 6264; J. R. Tolman, J. H. Prestegard, *J. Magn. Reson. Ser. B* **1996**, *112*, 245; N. Tjandra, A. Bax, *J. Magn. Reson.* **1997**, *124*, 512. N. Tjandra, J. B. Omichinski, A. M. Gronenborn, G. M. Clore, A. Bax, *Nat. Struct. Biol.* **1997**, *4*, 732.

7 N. Tjandra, A. Bax, *Science* **1997**, *278*, 1111–1114.

8 C. Griesinger, O. W. Sørensen, R. R. Ernst, *J. Am. Chem. Soc.* **1985**, *107*, 6394; C. Griesinger, O. W. Sørensen, R. R. Ernst, *J. Chem. Phys.* **1986**, *85*, 6837; C. Griesinger, O. W. Sørensen, R. R. Ernst, *J. Magn. Reson.* **1987**, *75*, 474; H. Schwalbe, P. Schmidt, C. Griesinger, Encyclopedia of NMR, Eds. D. M. Grant, R. K. Harris, John Wiley and Sons, Chichester, **1995**.

9 A. Rexroth, P. Schmidt, S. Szalma, T. Geppert, H. Schwalbe, C. Griesinger, *J. Am. Chem. Soc.* **1995**, *117*, 10389–10390.

10 A. Abragam, *Principles of Nuclear Magnetism*; Clarendon Press, Oxford, **1961**; G. S. Harbison, *J. Am. Chem. Soc.* **1993**, *115*, 3026; T. J. Norwood, *J. Magn. Reson. Ser. A* **1993**, *104*, 106.

11 T. Carlomagno, H. Schwalbe, A. Rexroth, O. W. Sørensen, C. Griesinger, *J. Magn. Reson.* **1998**, *135*, 216–226.

12 G. Cornilescu, B. E. Ramirez, M. K. Frank, G. Marius Clore, A. M. Gronenborn, A. Bax, *J. Am. Chem. Soc.* **1999**, *121*, 6275–6279.

13 H. Schwalbe, W. Samstag, J. W. Engels, W. Bermel, C. Griesinger, *J. Biomol. NMR* **1993**, *3*, 479–486.

14 G. W. Vuister, A. C. Wang, A. Bax, *J. Am. Chem. Soc.* **1993**, *115*, 5334–5335; G. W. Vuister, S. Grzesiek, S. Delaglio, A. C. Wang, R. Tschudin, A. Bax, *Methods Enzymol.* **1994**, *239*, 79–105 and references cited therein.

15 A. Bax, D. Max, and D. Zax, *J. Am. Chem. Soc.* **1992**, *114*, 6924; P. R. Blake, M. F. Summers, M. W. W. Adams, J.-B. Park, Z. H. Zhou, A. Bax, *J. Biomol. NMR* **1992**, *2*, 527.

16 J. R. Tolman, J. M. Flanagan, M. A. Kennedy, J. H. Prestegard, *Proc. Natl. Acad. Sci.* **1995**, *92*, 9279–9283; Prestegard, J. H., *Nat. Struct. Biol.* **1998**, *5 (Suppl. S)*, 517–522; Tjandra, N., Bax, A., *Science* **1997**, *278*, 1111–1114; Tjandra, N., *Structure* **1999**, *7*, R205–R211; Tjandra, N.; Omichinski, J. G.; Gronenborn, A. M.; Clore, G. M.; Bax, A., *Nat. Struct. Biol.* **1997**, *4*, 732.

17 J. Meiler, N. Blomberg, M. Nilges, C. Griesinger, *J. Biomol. NMR* **2000**, *16*, 245–252.

18 B. Reif, M. Hennig, C. Griesinger, *Science* **1997**, *276*, 1230–1233.

19 R. K. Hester, J. L. Ackermann, B. L. Neff, J. S. Waugh, *Phys. Rev. Lett.* **1976**, *36*, 1081; M. Linder, A. Hšhener, R. R. Ernst, *J. Chem. Phys.* **1980**, *73*, 4959.

20 G. Dabbagh, D. P. Weli, R. Tycko, *Macromolecules* **1994**, *27*, 6183–6191.

21 K. Schmidt-Rohr, *Macromolecules* **1996**, *29*, 3975–3981; K. Schmidt-Rohr, *J. Am. Chem. Soc.* **1996**, *118*, 7601–7603; X. Feng, Y. K. Lee, D. Sandstrøm, M. Edén, H. Maisel, A. Sebald, M. H. Levitt, *Chem. Phys. Lett.* **1996**, *257*, 314–320; X. Feng, P. J. E. Verdegem, Y. K. Lee, D. Sandstrøm, M. Edén, P. Bovee-Geurts, J. W. de Grip, J. Lugtenburg, H. J. M. de Groot, M. H. Levitt, *J. Am. Chem. Soc.* **1997**, *119*, 6853–6857; M. Hong, J. D. Gross, R. G. Griffin, *J. Phys. Chem. B* **1997**, *101*, 5869– 5874.

22 H. Steinhagen, M. Reggelin, G. Helmchen, *Angew. Chem.* **1997**, *109*, 2199–2202/ *Angew. Chem. Int. Ed. Engl.* **1997**, *36*, 2108–2110; J. Junker, B. Reif, H. Steinhagen, B. Junker, I. C. Felli, M. Reggelin, C. Griesinger, *Chem. Eur. J.* **2000**, *6*, 3281–3286.

23 I. C. Felli, C. Richter, C. Griesinger, H. Schwalbe, *J. Am. Chem. Soc.* **1999**, *121*, 1956–1957.

24 P. Pelupessy, E. Chiarparin, R. Ghose, G. Bodenhausen, *J. Biomol. NMR* **1999**, *13*, 375–380.

25 C. Altona and M. Sundaralingam, *J. Am. Chem. Soc.* **1972**, *94*, 8205–8212.

26 J. L. Markley, A. Bax, Y. Arata, C. W. Hilbers, R. Kaptein, B. D. Sykes, P. E. Wright, K. Wüthrich, *J. Biomol. NMR* **1998**, *12*, 1–23.

27 H. Schwalbe, J. P. Marino, S. J. Glaser, C. Griesinger, *J. Am. Chem. Soc.* **1995**, *117*, 7251–7252; S. J. Glaser, H. Schwalbe, J. P. Marino, C. Griesinger, *J. Magn. Reson. Ser. B* **1996**, *112*, 160–180; J. P. Marino, H. Schwalbe, S. J. Glaser, C. Griesinger, *J. Am. Chem. Soc.* **1996**, *118*, 7251–7256.

28 C. Richter, C. Griesinger, I. Felli, P. T. Cole, G. Varani, H. Schwalbe, *J. Biomol. NMR* **1999**, *15*, 241–250.

29 H. Schwalbe, J. P. Marino, S. J. Glaser, C. Griesinger, *J. Am. Chem. Soc.* **1995**, *117*,

7251–7252; S. J. Glaser, H. Schwalbe, J. P. Marino, C. Griesinger, *J. Magn. Reson. Ser. B* **1996**, *112*, 160–180; J. P. Marino, H. Schwalbe, S. J. Glaser, C. Griesinger, *J. Am. Chem. Soc.* **1996**, *118*, 7251–56.

30 C. Richter, B. Reif, C. Griesinger, H. Schwalbe, *J. Am. Chem. Soc.* **2000**, *122*, 12728–12731.

31 H. Shindo, *Biopolymers* **1980**, *19*, 509–522; J. W. Keepers, T. L. James, *J. Am. Chem. Soc.* **1982**, *104*, 929–939; J. Herzfeld, R. G. Griffin, R. A. Haberkorn *Biochemistry* **1978**, *17*, 2711–2718.

32 C. Richter, B. Reif, K. Wörner, S. Quant, J. W. Engels, C. Griesinger, H. Schwalbe, *J. Biomol. NMR* **1998**, *12*, 223–230.

33 T. Carlomagno, I. C. Felli, M. Czech, R. Fischer, M. Sprinzl, C. Griesinger, *J. Am. Chem. Soc.* **1999**, *121*, 1945–1948.

34 M. J. J. Blommers, W. Stark, C. E. Jones, D. Head, C. E. Owen, W. Jahnke, *J. Am. Chem. Soc.* **1999**, *121*, 1949–1953.

35 G. M. Clore, A. M. Gronenborn, *J. Magn. Reson.* **1982**, *48*, 402–417; G. M. Clore, A. M. Gronenborn, *J. Magn. Reson.* **1983**, *53*, 423–442 ; F. Ni, *Prog. NMR Spectrosc.* **1994**, *26*, 517; L. Y. Lian, I. L. Barsukov, M. J. Sutcliffe, K. H. Sze, G. C. K. Roberts, *Methods Enzymol.* **1994**, *239*, 657–707.

8
Orientational Restraints

Eva de Alba and Nico Tjandra

8.1
General Considerations

Dipolar couplings have only recently been introduced into the structure determination of biological biopolymers by NMR [1–3]. So far, structural restraints have been obtained from scalar coupling constants (*J*), chemical shifts and Nuclear Overhauser Effect data (NOE). The first two contain local rather than long-range information, while the third contains both. Long-range structural information relates atoms or molecular fragments that may or may not be close in space but are far away in the series of chemical bonds that constitute the biopolymer. In contrast, short-range information relates atoms or parts of the molecule that are separated by a few chemical bonds. *J* coupling constants yield data on dihedral angles [4], while conformational shifts provide secondary structure information [5, 6]. Additionally, indirect evidence of the existence of hydrogen bonds has been traditionally derived from hydrogen exchange experiments. In recent work it has been demonstrated that it is possible to measure *J* coupling interactions across hydrogen bonds, therefore, providing direct evidence of their presence [7–9]. In addition, cross-correlation rates can be used to obtain information on dihedral angles as discussed in Chapt. 7. Nevertheless, the most important NMR parameter for structure determination is the Nuclear Overhauser Effect. Typically, biopolymer structure calculation requires a large amount of NOE data. Although the redundancy in these data usually leads to sufficiently precise structures, there are some cases where all the traditional NMR parameters fail to determine the three-dimensional structure accurately. This occurs in multidomain proteins, whose domains are structurally independent, or in elongated structures. In the first case the small number of NOE cross-peaks connecting the domains precludes the determination of their relative orientation. A similar problem is also present in the structure determination of nucleic acids. These biomolecules not only suffer from a low proton density but also they usually lack compact tertiary structure, and therefore only medium- and short-range NOE contacts are observed. Mainly for this reason nucleic acid structures are often poorly defined. Along with other advantages, residual dipolar couplings have proven to be very useful in obtaining long-range structural information, not available from NOE data, of nucleic acids and multidomain proteins. Additionally, dipolar couplings improve structure quality of globular proteins.

The dipolar coupling is the physical response of the interaction between the magnetic dipole of two atoms and the external magnetic field. Since it depends on the angle that the vector connecting the dipoles makes with the direction of the static field, it typically averages to zero in solution because of isotropic tumbling of the molecule. However, when rotational motion is restricted to a preferred direction (the molecule under study is anisotropically oriented), dipolar couplings are reintroduced and can easily be observed. If the two atoms are magnetically related through scalar coupling, the dipolar coupling is manifested as an apparent addition to the *J* coupling constant.

Solution NMR dipolar couplings have been measured in the past using liquid crystals [10, 11] and magnetic alignment [12]. The degree of ordering achieved with the liquid crystalline media is large, resulting in NMR spectra too complex to analyze and impairing the utilization of dipolar couplings in biopolymer structure calculation. It was shown recently that the use of dilute liquid crystals can create the required low degree of order where residual dipolar couplings are observable and the simplicity of the spectra is still conserved [13, 14]. In dilute liquid crystals the degree of order is such that the value of the static dipolar coupling is scaled down by a factor of at least 10^3, and hence only dipolar couplings of chemically bonded nuclei can easily be observed.

Biomolecules of sufficiently large anisotropy of their magnetic susceptibility adopt a preferred orientation with respect to the external magnetic field [15]. Such behavior has been shown for both paramagnetic and diamagnetic biomolecules. For paramagnetic proteins [16], the natural alignment in high magnetic fields is usually one order of magnitude smaller than that achieved by the use of orienting media. A similar degree of ordering is achieved for nucleic acids [17], whose magnetic alignment is dominated by the susceptibility anisotropy generated by the almost perfect stacking of the bases. The net alignment of diamagnetic biomolecules will, in general, be at least another order of magnitude smaller than that of their paramagnetic counterparts [18]. For this reason, field alignment tends to be impractical for common NMR biomolecular structure application.

The mathematical relationship between dipolar couplings and structural information can be explained as follows:

In the principal axis frame of the alignment tensor, **A**, the dipolar coupling between two nuclei, P and Q, as a function of the polar coordinates, θ and φ, is given by

$$D_{\mathrm{P,Q}(\theta,\varphi)} = D_{\mathrm{a}}^{\mathrm{PQ}}\{(3\cos^2\theta - 1) + 3/2R\sin^2\theta\cos 2\varphi\} \tag{1}$$

where

$$D_{\mathrm{a}}^{\mathrm{PQ}} = -(\mu_{\mathrm{o}}h/16\pi^3)S\gamma_{\mathrm{P}}\gamma_{\mathrm{Q}}\langle r_{PQ}^{-3}\rangle A_{\mathrm{a}}\,. \tag{2}$$

A_a is the dimensionless axial component of the alignment tensor, i.e., $A_{zz} - (A_{xx} + A_{yy})/2$, and $R = (A_{xx} - A_{yy})/A_{zz}$. The axial component refers to the largest component of the tensor, which coincides with the z axis according to the definition used herein. R is the rhombicity of the tensor and provides a measure of the deviation from axial symmetry. S is the generalized order parameter, which accounts for the effect of internal dynamics [19]. γ_P and γ_Q are the gyromagnetic ratios of nuclei P and Q. μ_o is the magnetic susceptibility. The polar angles θ and φ describe the orientation of the vector in the coordinate system of the alignment tensor (Fig. 8.1). The brackets indicate time averaging of the inverse cube of the P-Q distance (r_{PQ}).

Equation (1) establishes the dependence of the dipolar coupling value on the orientation of the vector connecting the two atoms with respect to the alignment tensor. Therefore, once the orientation of the alignment tensor, the dipolar coupling, and the distance between the two atoms are known, the orientation of each vector with respect to a reference frame can be derived (Fig. 8.1). Thus, dipolar couplings, in contrast to NOE data, are not typically used to obtain distance information. This applies to the general case where the two atoms are chemically bonded and the distance between them is known. The situation is different for ^{1}H-^{1}H residual dipolar couplings, since the distance separating the protons is usually unknown and structure refinement can be done using the distance and the angles as variables. Therefore, distance information can be obtained from ^{1}H-^{1}H residual dipolar couplings.

The position of the vector that connects the two atoms can be restricted to a cone with a semiangle θ about the principal axis of the alignment tensor assuming axial symmetry. In the more general case of an asymmetric tensor, the cone is distorted [20]. In reality, the dipolar coupling defines two cones of possible vector orientations, since the alignment tensor is of second rank and the direction of a second rank interaction cannot be distinguished from its inverse. This ambiguity complicates the determination of local geometry in the absence of other structural information. Fortunately, it is possible to combine residual dipolar data obtained under anisotropic conditions where the orientation and/or the rhombicity of the alignment tensor are different [20]. Dipolar couplings obtained from the two different orienting conditions can be translated into the shape of two cones. The angle at which they intersect will indicate the orientation of the vector (Fig. 8.2). Some of the procedures used to modify the alignment tensor will be discussed in Sect. 8.2.

The dependence of the residual dipolar coupling on the angle that the vector forms with a reference axis explains why the use of dipolar couplings makes possible the determination of the relative orientation of different domains in a multidomain protein and facilitates nucleic acid structure determination. Dipolar couplings can constitute up to 50% of the total structural data available for nucleic acids, while this number drops to 10–15% in proteins. Thus, the impact of the use of dipolar couplings on the structure determination of nucleic acids is generally more substantial than in the case of proteins. Furthermore, the presence or absence of tertiary structure in a protein or nucleic acid does not have a major influence on the number of dipolar couplings that can be measured, in contrast to the case of the NOE.

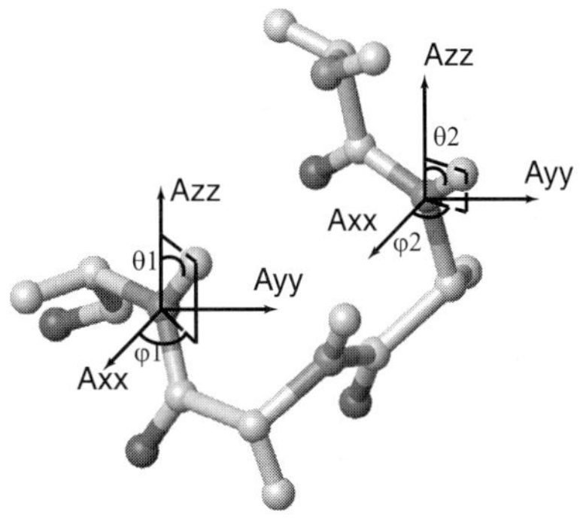

Fig. 8.1 Orientation of two dipolar coupling vectors in a protein segment. The vectors connect the amide $^1H^N$ and ^{15}N atoms. In this case the interaction vector coincides with the chemical bond. The axis system of the alignment tensor is designated as A_{xx}, A_{yy}, A_{zz}. The angles θ_1, φ_1, and θ_2, φ_2 define the orientation of the two dipolar vectors with respect to the alignment tensor. (Reproduced with permission from N. Tjandra, *Structure* **1999**, *7*, R205–R211.)

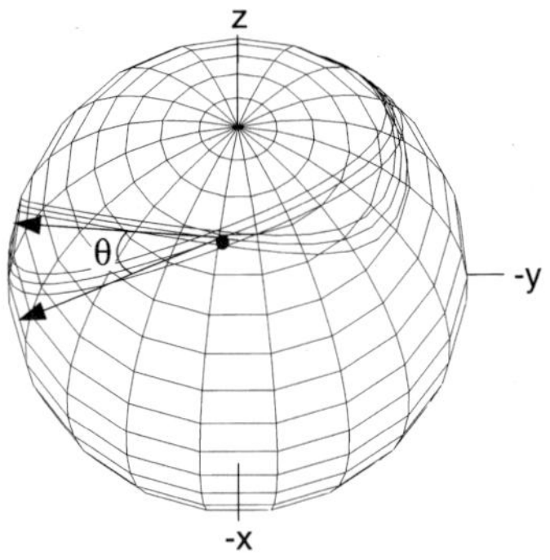

Fig. 8.2 Orientations of an amide NH dipolar coupling bond-vector of the protein ubiquitin. Each cone of orientations is compatible with two different alignment directions adopted by the protein in two different alignment media. The central lines defining each cone correspond to the orientations obtained from the measured dipolar couplings. The outer lines include orientations that are possible if the dipolar coupling values are either increased or decreased by 1 Hz. The angle at which the two cones intersect is defined by θ. The solid dot at the cone intersection determines the orientation of the dipolar coupling vector. (Reproduced with permission from B. E. Ramirez and A. Bax, *J. Am. Chem. Soc.* **1998**, *120*, 9106–9107.)

8.2
Commonly Used Systems to Orient Biopolymers

8.2.1
Bicelle Systems

If the magnetic susceptibility anisotropy is not sufficiently large to orient the biomolecule itself, such ordering can be achieved by applying electric fields [21] or using media that exist as liquid crystalline phases [10, 11]. The first system used to study protein structure in weakly aligned states was a mixture of dihexanoyl phosphatidylcholine (DHPC) and dimyristoyl phosphatidylcholine (DMPC), which undergoes a gel-to-lyotropic liquid crystal phase transition at or above room temperature in solution [13]. This type of liquid crystal is formed by disc-shaped particles (bicelles) that are diamagnetic and orient in the magnetic field. Bicelle orientation induces a slight alignment of the solute present in the medium. The interaction between the solute and the bicelle can be predominantly steric, as in the case of neutral bicelles [14], or electrostatic, as in the case of charged bicelles [20]. In this way, the necessary anisotropy to measure residual dipolar couplings is created (Fig. 8.3). DHPC and DMPC are commercially available. Because of the hygroscopic character of DHPC, the two compounds should be mixed under an inert atmosphere. The typical DMPC/DHPC molar ratio used is 3.5 : 1. This mixture is present in the macromolecule solution at a concentration of 3–5% (w/v). Under these conditions, water molecules become ordered on the surface of the bicelles, resulting in a residual ^{2}H splitting of the HOD resonance that ranges from 5 to 10 Hz at 37 °C [22]. More details concerning the preparation of these mixtures, as well as the pH and temperature conditions that should be used, can be found in the work reported by Ottiger and Bax [22].

The ester linkages in DHPC and DMPC can be replaced by ether linkages, thereby preventing ester hydrolysis [23] and resulting in chemically much more inert media. The original bicelle method (DMPC/DHPC) has a temperature range of roughly 35 to 45 °C in which the liquid crystalline state is observed [22]. For proteins that are unstable at moderately high temperatures, ternary mixtures of DMPC- and DHPC-related compounds can be used, since they allow the gel-to-liquid-crystal phase transition to occur at lower temperatures [22, 24].

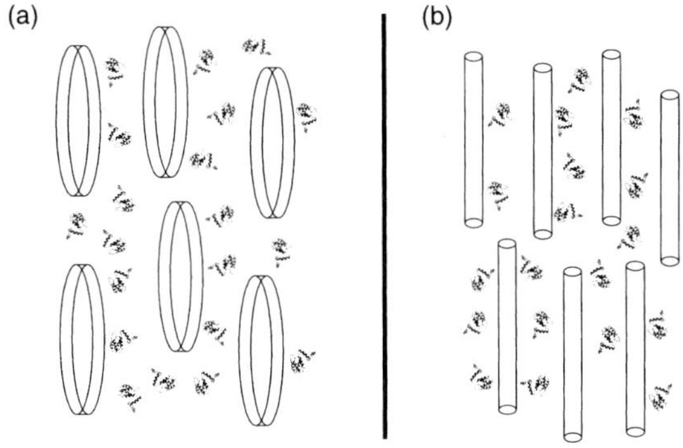

Fig. 8.3 Cartoon illustrating the alignment of the particles of two different orienting media. **a** Disc particles represent lipid bicelles. **b** Rods represent viral particles. Bicelles orient with their normal orthogonal to the magnetic field and viruses with their long axis parallel to it. (Reproduced with permission from N. Tjandra, *Structure* **1999**, *7*, R205–R211.)

Other novel bicelle systems composed of a mixture of 1,2-di-*O*-dodecyl-*sn*-glycero-3-phosphocholine and 3-(chloramidopropyl)dimethylammonio-2-hydroxyl-1-propane sulfonate have been designed to orient proteins at low pH values and over wide temperature ranges [25].

8.2.2
Other Orienting Systems

Orienting systems of quasi-ternary mixtures composed of cetylpyridinium chloride/hexanol/NaCl [26] and cetylpyridinium bromide/hexanol/NaBr [27] have been reported to form lamellar liquid crystalline phases that allow a large temperature range over which dipolar couplings can be measured. The optimum condition for protein alignment consists of a 1:1.33 (w/w) ratio of cetylpyridinium bromide/hexanol. The residual ^{2}H quadrupolar splitting of the HOD resonance increases from 5 Hz to 20 Hz as the concentration of the mixture is varied from 30 to 65 mg/mL. These quasi-ternary mixtures are positively charged.

Mixtures of *n*-alkyl-poly(ethylene glycol)/*n*-alkyl alcohol and glucopone/*n*-hexanol have been shown to form lyotropic systems for biological macromolecule orientation in solution [28]. These components are less expensive than other alignment systems and are commercially available. Since they are uncharged, they are not affected by pH changes and are only slightly affected by changes in the salt concentration. The main characteristic of these systems is that they have a very weak binding capacity for macromolecules.

Filamentous viruses have been commonly utilized to obtain a moderate degree of biomolecular alignment [29, 30]. Once the virus particles reach a certain concentration in solution (i.e. ~ 15 mg/mL) a nematic phase is formed. The viruses adopt a preferred orientation in the presence of the magnetic field, which is transferred to the solute. The phage particles align with their long axis parallel to the magnetic field. The most commonly used viruses are *fd* and *Pf1*. The viruses' coat proteins are negatively charged and electrostatic interactions with the macromolecule may be possible. One important advantage of these systems is that the macromolecule can be recovered from the phage solution by centrifugation. The viruses will form a pellet and the macromolecule will remain in the supernatant.

Biological macromolecules have also been aligned using purple membrane [31, 32]. The purple membrane is formed by two-dimensional crystals of the membrane protein bacteriorhodopsin. In contrast to other systems that have been previously mentioned, the purple membrane induces alignment in the absence of a liquid crystalline phase. Purple membranes are negatively charged, thus allowing for possible electrostatic interactions with the macromolecule. The surface of the purple membrane aligns orthogonal to the magnetic field, in contrast to bicelles and bacteriophages whose orientation is parallel.

Suspensions of cellulose crystallites have proven to be a stable liquid crystalline medium for the measurement of residual dipolar couplings of biological macromolecules [33]. The preparation contains sulfate groups; therefore the crystallites are negatively charged and electrostatic interactions with the biopolymer can be present. The long axis of the cellulose crystallites aligns orthogonal to the magnetic field.

Cross-linked polyacrylamide gel is another medium to orient biopolymers [34]. The gel is highly hydrated, and the protein solution can enter easily by diffusion from an external solution. The gel contains pores whose shape can be modified by straining. To induce anisotropy the gel is stretched or compressed. The pore shape will depend on the direction of the applied pressure. The alignment, in this case, is independent of the magnetic field. This system of orientation is very inert, and interactions other than steric-type are not expected. Unfortunately, it is still unknown whether it is possible to make the system sufficiently homogeneous to produce high resolution NMR spectra. A similar approach has been reported by Sass et al. [35], in which the alignment can also be achieved by embedding oriented purple membrane fragments into the polyacrylamide gel.

Since the alignment of a protein is governed by steric and/or electrostatic interactions, it is possible to modify the orientation of the alignment by changing the type of protein-bicelle interaction or by using different types of orienting media. The alignment tensor has been modulated by changing the net charge of the bicelle or the sample pH, or by creating a substantial increase in the net charge of the protein [20]. Additionally, two alignment tensors have been used in protein structure refinement by analyzing protein dipolar couplings under two different orienting media, such as bicelles and bacteriophages [36].

It is important to note that the alignment induced by most orienting systems is due to the existence of very weak interactions between the macromolecule and the media itself. Because of the weak character of these interactions, the rotational tumbling of the molecule is not impaired or restricted, and therefore there is no additional broadening of resonance lines due to relaxation phenomena [37].

8.3 NMR Experiments Designed to Measure Dipolar Couplings

When two atoms that give rise to a dipolar splitting are scalar coupled, the dipolar coupling translates into an addition to the isotropic J splitting. Therefore, the difference between the resonance splitting observed under anisotropic conditions and that observed under isotropic conditions provides the magnitude and sign of the dipolar coupling. Typically, chemical shifts are almost unmodified by the presence of the orienting media or by the existence of anisotropic conditions; thus no reassignment needs to be done to measure dipolar couplings. If significant changes of chemical shifts are observed under orienting conditions, it is possible that the orienting particles interact with the solute. The structural information contained in dipolar couplings of atoms that are connected by a chemical bond is easier to obtain, since the separating distance is usually known. In the case of proteins, usually five types of dipolar couplings are measured. These are the one-bond amide NH, $C_\alpha H_\alpha$, $C_\alpha C'$, C'N and the two-bond $H^N C'$. In order to obtain these dipolar couplings, the protein under study has to be enriched in magnetically active ^{13}C and ^{15}N isotopes. Although an enrichment close to 100% in both isotopes is desirable, ^{13}C-HSQC (heteronuclear single-quantum correlation) experiments could be performed at ^{13}C natural abundance in a reasonable period of time. In contrast, triple resonance experiments and experiments that depend on ^{15}N magnetization would be hard, if not impossible, to perform at natural abundance within a reasonable experimental time. Nevertheless, for certain situations where technological improvements such as the availability of very high magnetic fields (i.e. corresponding to 800 and 900 MHz) and the use of cryoprobes are combined with target proteins that are small (i.e. $\sim$ 100 amino acids or less) and can be dissolved at high concentrations, dipolar couplings could be measured on samples enriched only in ^{15}N. Simple modifications of the ^{15}N-HSQC, ^{13}C-HSQC and HNCO NMR experiments allow the measurement of the mentioned dipolar couplings. In these three NMR experiments, J coupling between the nuclei that give rise to the dipolar coupling evolves at one stage of the experiment, thus allowing this NMR parameter to be measured in the resulting NMR spectrum. In the two-dimensional ^{15}N-HSQC experiment the resonance splitting due to scalar and dipolar coupling will be observed in the ^{15}N dimension. It is preferable to record J coupling evolution in the indirect dimension (^{15}N), since transverse relaxation is slower for ^{15}N than for ^1H. Additionally, if C'N coupling is active as well during the ^{15}N chemical shift evolution [38], this dipolar coupling can also be obtained.

The two-bond $H^N C'$ dipolar coupling is observable in a ^{15}N-HSQC experiment in which the J coupling between the carbonyl atom C' and the ^{15}N amide is active. The doublet components in the ^{15}N dimension that represent the C'N coupling are displaced with respect to one other in the ^1H dimension as in an E.COSY [39] because of this two-bond coupling.

The presence of the ^{15}N-^1H splitting implies that the number of signals is twice that of a conventional, fully ^1H-decoupled, ^{15}N-HSQC spectrum. In the case of proteins of medium to large molecular weights, this will result in severely crowded spectra. In order to alleviate this problem, experiments of the ^{15}N-HSQC-type have been made spin-state selective. This selection can be done by separating the in-phase and anti-phase terms of the ^1H-coupled ^{15}N magnetization such that two spectra are acquired, each containing signal intensities that are modulated with a sine and a cosine function, respectively [40]. Alternatively, spectra containing anti-phase and in-phase doublets may be stored separately (IPAP method, Fig. 8.4) [41]. These subspectra are then added or subtracted, thereby generating spectra that contain only the upfield or the downfield components of the doublet (Fig. 8.5). Furthermore, since the contribution to the signal is derived from both subspectra, the loss in signal-to-noise ratio due to splitting is compensated. Another method that provides accurate J_{NH} values measures signal intensity variations as a function of a dephasing delay in an other-

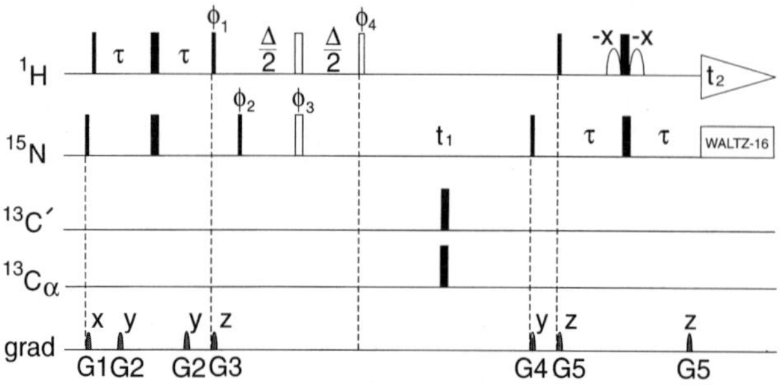

Fig. 8.4 Schematic representation of the pulse program IPAP-[^{1}H-^{15}N]-HSQC. Narrow and wide bars represent 90° and 180° pulses, with phase *x* unless indicated. The white bars represent pulses that are applied only when the anti-phase spectrum is acquired. The anti-phase and in-phase spectra are recorded in an interleaved manner. Technical details may be found in the original publication. (Reproduced with permission from M. Ottiger, F. Delaglio and A. Bax, *J. Magn. Reson.* **1998**, *131*, 373–378.)

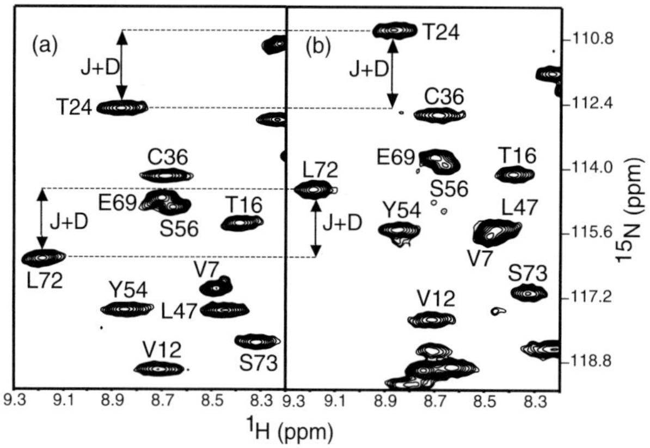

Fig. 8.5 Downfield (**a**) and upfield (**b**) components of ^{15}N doublets of a IPAP-[^{1}H-^{15}N]-HSQC for the protein saposin in Pf1 viruses. **a** results from the subtraction and **b** from the addition of the spectra containing the in-phase and anti-phase components, respectively. The sum of the residual dipolar coupling (*D*) and the *J* values can be obtained from the difference in chemical shift of the two components (arrows). The broad aspect of the signals indicates the presence of long-range ^{1}H-^{1}H dipolar couplings.

wise regular HSQC. With this experiment, changes in the J_{NH} coupling smaller than 0.2 Hz, due to the residual dipolar splitting, have been detected [18].

When the protein rotational correlation time exceeds 15 ns, the upfield component of the J_{NH} doublet broadens. In such cases the amide J_{NH} coupling can be obtained from the displacement of the ^{15}N chemical shift of the TROSY component (see Chapt. 10) relative to the one from a ^{1}H-decoupled HSQC experiment.

In those cases where spectral crowding is very severe, HNCO-based experiments allow the measurement of all or some of the four dipolar couplings (Fig. 8.6): NH, $C_\alpha C'$, C′N

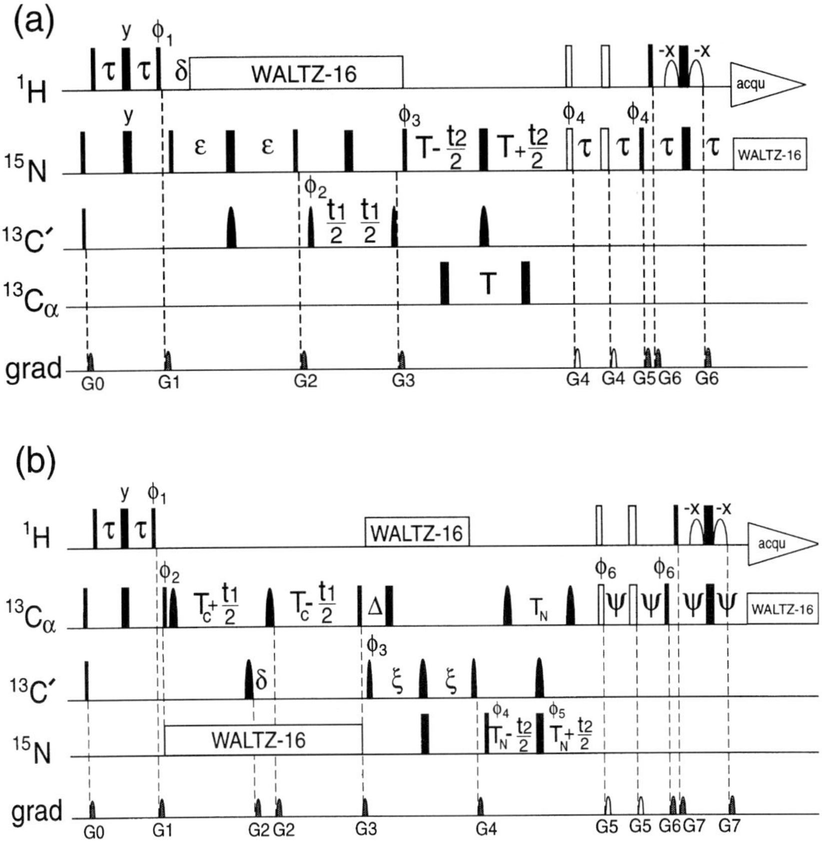

Fig. 8.6 Schematic representation of the modified experiments HNCO (**a**) and (HA)CA(CO)NH (**b**) to measure residual dipolar couplings. The white bars represent the extra pulses that are applied in an interleaved manner to collect the in-phase ^{15}N magnetization. The anti-phase ^{15}N magnetization is collected without the application of the white pulses. For further details see the original paper. (Reproduced with permission from E. de Alba, M. Suzuki and N. Tjandra, *J. Biomol. NMR* **2001**, *19*, 63–67.)

and the two-bond H^NC' [42–46]. For example, a method has been described to obtain accurate measurements of $J_{C'N}$ using the relative intensity of two TROSY-HNCO spectra. In both experiments the signal intensity is a sine function that depends on $J_{C'N}$. One of the experiments has a dephasing time of $1/(2\,J_{C'N})$, and the resulting spectrum is used as reference. The other experiment has a dephasing interval of $1/J_{C'N}$. Both experiments have to be performed under isotropic and anisotropic conditions in order to obtain the dipolar coupling value [47].

One of the NMR experiments most commonly used to measure the $C_\alpha H_\alpha$ dipolar coupling is a *J*-modulated pseudo-three-dimensional version of the constant time ^{13}C-HSQC [48, 49]. The *J* coupling values are extracted by using time-domain fitting in the third dimension. Since spectral overlap is very common in ^{13}C-HSQC of proteins, sometimes only a few

$C_\alpha H_\alpha$ dipolar couplings can be obtained with sufficient accuracy. In these cases, three-dimensional experiments such as HN(CO)CA or (HA)CA(CO)NH can be applied (Fig. 8.6). The *J*-modulated constant time ^{13}C-HSQC [49] is also applied for the measurement of dipolar couplings of methyl and methylene groups. The dipolar couplings of side chains nuclei have to be applied cautiously in structure calculations, since side chains can undergo conformational averaging. A modified version of the CBCA(CO)NH experiment [50] has been designed to measure $C_\beta H_\beta$ dipolar couplings [51]. The comparison of this coupling to the dipolar couplings of $C_\alpha H_\alpha$, $C_\alpha N$ and $C_\alpha C'$ bond-vectors of the same residue indicates its side chain rotamer, provided the rotamer is staggered.

Since the value of the dipolar coupling depends on the gyromagnetic ratios and the inter-atomic distance, the intrinsic magnitudes of the $C_\alpha C'$ and the C′N dipolar couplings are ~5 times and ~9 times smaller than the NH dipolar coupling, respectively. The magnitude of the $C_\alpha H_\alpha$ dipolar coupling is approximately twice that of the one-bond amide dipolar coupling. These differences have to be considered to estimate the degree of accuracy with which to measure the different types of dipolar couplings and therefore which NMR experiment to use for this purpose.

Homonuclear ^{1}H-^{1}H dipolar couplings can provide very important structural information for the accurate determination of protein side chain conformations. From the experiments designed to measure ^{1}H-^{1}H *J* couplings which have been reviewed in the work by Kessler et al. [52], only a few have been applied to measure ^{1}H-^{1}H dipolar couplings. This is because, under anisotropic conditions, ^{1}H-^{1}H multiplets tend to be very complex. This problem is alleviated by the use of COSY-based heteronuclear experiments, which have been utilized to determine the sign and magnitude of ^{1}H-^{1}H dipolar couplings [53–60]. Recently, an NMR experiment has been designed to measure the sign and magnitude of ^{1}H-^{1}H dipolar couplings of protons belonging to the same methyl group [61]. The structural information gained by measuring this dipolar coupling is similar to that obtained from the heteronuclear ^{13}C-^{1}H dipolar couplings of methyl groups [49]. However, the former dipolar coupling is intrinsically larger because of the gyromagnetic ratios of the nuclei involved. The advantages and technical issues of NMR experiments designed to measure ^{1}H-^{1}H dipolar couplings can be found in the review by Bax et al. [3].

8.4
Application of Dipolar Couplings to Structure Calculation

8.4.1
Protein Structure Determination and Refinement

When dipolar couplings are the only source of structural information available, it is necessary to know *a priori* the magnitude and orientation of the alignment tensor in order to use these data as conformational restraints in structure calculation. However, if a set of NOE data is available as well, only the magnitude of the alignment tensor needs to be known in advance. In this situation the orientation of the alignment tensor can be determined during the structure calculation process by treating it as a freely floating variable [62].

Clore et al. [63] have noticed that for an isotropic distribution of dipolar bond-vectors the probabilities resemble the shape of a chemical shift anisotropy powder pattern

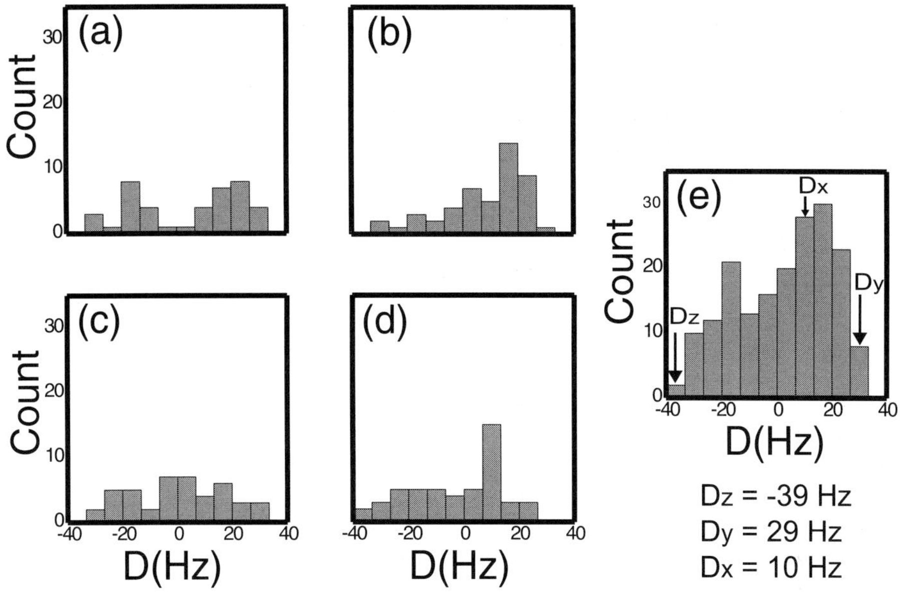

Fig. 8.7 Histograms of residual dipolar couplings observed for the C-terminal KH domain of the ribonucleoprotein K in 3.2% w/v bicelle solution [J. L. Baber, D. Libutti, D. Levens and N. Tjandra, J. Mol. Biol. **1999**, 289, 949–962]. **a** $^1D_{NH}$, **b** $^1D_{C\alpha H\alpha}$, **c** $^1D_{C'N}$, **d** $^1D_{C\alpha C'}$. None of each type of dipolar couplings properly represents a powder pattern. In contrast, a histogram that contains the addition of all normalized dipolar couplings provides a good representation of a powder pattern distribution (**e**). Values of D_{zz}, D_{yy} and D_{xx} (Hz) are indicated.

(Fig. 8.7). On the basis of this similarity they have developed a method to obtain good estimations of D_a and R. The use of dipolar couplings of all possible backbone bond-vectors will approximate to an isotropic distribution, provided that they are normalized by taking into account the differences in the gyromagnetic ratio and bond distance.

Using this procedure the values of D_a and R can be obtained from Eqs. (3) to (6) [63];

$$A_{zz} = 2D_a \tag{3}$$

$$A_{yy} = -D_a(1 + 1.5R) \tag{4}$$

$$A_{xx} = -D_a(1 - 1.5R) \tag{5}$$

where

$$|A_{zz}| > |A_{yy}| \geq |A_{xx}| \,. \tag{6}$$

Warren and Moore [64] have introduced a method that uses a grid search to find the pair of D_a and R compatible with the observed set of dipolar couplings. Their approach also estimates the errors associated with the pair of D_a and R values on the basis of the quality of the dipolar coupling data. This method has the advantage over the previous one [63] that it takes into account the intrinsically low probability of finding vectors that are parallel to the z axis of the alignment tensor.

A different approach to determining the magnitude and direction of the alignment tensor is reported in the work by Zweckstetter and Bax [65]. This method consists of a systematic search of solute orientations that have the highest tendency to clash with the orienting media, and is based on the molecular shape of the solute. It works well, provided that solute alignment is driven only by steric interactions. The method is being modified to also consider electrostatic interactions.

One of the most popular methods to incorporate dipolar coupling information into structure determination has been developed by Tjandra et al. [62] using the structure calculation program XPLOR [66]. Dipolar couplings are included as an empirical energy term added to the other conventional terms of the target function (distances from NOE data, torsion angles, covalent geometry, etc.) used in a simulated annealing protocol. This equation takes the form

$$E_{\text{dip}} = k_{\text{dip}}(D^{\text{calc}} - D^{\text{obs}})^2 \tag{7}$$

k_{dip} is a force constant and D^{calc} and D^{obs} are the calculated and observed values of the dipolar coupling, respectively.

The alignment tensor is considered as a reference axis that is represented by a pseudomolecule located far away from the biopolymer. This pseudomolecule consists of four equidistant atoms (OXYZ). One represents the axis origin (O). The other three represent the three axes, OX, OY and OZ, which are orthogonal to one another. E_{dip} is evaluated by calculating the angles that the dipolar vectors form with the reference axis. During the calculation the pseudomolecule OXYZ is continually reoriented in order to yield the best fit between the calculated and the observed dipolar couplings. The force constant in Eq. (7) is chosen such that the errors between the predicted dipolar couplings from the calculated structure and the observed ones match the experimental errors.

Since the energy landscape of a simulated annealing procedure in which only dipolar couplings are included tends to have many local minima, these orientational restraints are usually introduced once a preliminary structure has been calculated from NOE data. This procedure has also the advantage of identifying misassigned NOEs and possible conflicts among the various types of input data.

There are alternative methods to apply dipolar couplings as restraints in structure calculation. One of them calculates the projection angles between all pairs of inter-nuclear vectors for which dipolar couplings are known [67]. The calculation of these angles is done with respect to the molecular frame, and therefore does not require the knowledge of the alignment tensor orientation. This differs from the previous method mainly in the mathematical expression that relates dipolar couplings to the chosen reference frame. Apparently, the later method avoids convergence problems in the structure calculation that are present when using dipolar couplings as structural restraints.

Another method to determine the magnitude and rhombicity of the alignment tensor is based on the determination of the Saupe order matrix. The anisotropic parameter of motional averaging is represented by this order matrix, which is diagonalized by a transformation matrix that relates the principal frame, in which the order matrix is diagonal, to the initial molecular frame over which the dipolar couplings are defined. The Saupe matrix can be determined by singular value decomposition [68]. Using this method the values of D_a and R can be obtained provided that a good local structure of a molecular fragment is known, even when only a small number of dipolar couplings are available.

Among the mentioned methods it is important to distinguish between those that are used to fit a known structure to the measured residual dipolar couplings [65, 68] and those that allow structure calculation and structure refinement using dipolar coupling data [62, 69].

Equation (2) gives the dependence of the dipolar coupling value on the generalized order parameter that accounts for fast internal motion. The order parameter varies in the range of 0.85 to 0.95 for a protein backbone that adopts a regular secondary structure. In more mobile regions of a protein, residual dipolar couplings will display values which are smaller than expected for a fixed conformation. Dipolar couplings of such regions can also be used in structure refinement, provided that the obtained dipolar coupling value is considered as a lower limit.

Dipolar couplings of side chains can provide valuable structural information. Dipolar couplings for methylene and methyl sites have been applied to protein structure refinement [49]. In a side chain with a C-CH_3 moiety, rapid rotation of the methyl group results in an averaged C-H dipolar vector aligned in the direction of the C-C bond. This rapid rotation scales the real value of the C-H dipolar by a factor of $\sim -1/3$. Therefore, when including methyl dipolar couplings in protein structure refinement, the orientation of the C-CH_3 bond is the one that is restricted, rather than the C-H bond.

The NMR experiment most commonly used to measure dipolar couplings of atoms that belong to side chains [49], provides the sum of the J couplings of both C-H_1 and C-H_2 in the case of methylenes, so that the sum of both dipolar couplings is obtained. In order to use the data in the structure calculation the energy function that is minimized has the form;

$$E_{\mathrm{dip}} = k[(D^{\mathrm{cal}}_{\mathrm{CH1}} + D^{\mathrm{cal}}_{\mathrm{CH2}} - D^{\mathrm{obs}}_{\mathrm{CH1}} - D^{\mathrm{obs}}_{\mathrm{CH2}})]^2 \tag{8}$$

where D^{cal} and D^{obs} are the calculated and observed dipolar couplings, respectively.

The incorporation of side chain dipolar couplings into structure calculation must be done carefully. In the absence of quantitative information on the dynamics, the measured dipolar couplings are assumed to be lower limits of the magnitude of the real couplings.

Except for methylene groups, the distance between protons is an unknown parameter in ^{1}H-^{1}H dipolar couplings. This variable, which is fixed for one-bond heteronuclear dipolar couplings, is included in the energy term represented by Eq. (7) [70]. In the situation where the sign of the ^{1}H-^{1}H dipolar coupling cannot be determined, the calculation is carried out using only the magnitude. The energy term then takes the form:

$$E_{\mathrm{dip}} = k(|D^{\mathrm{cal}}| - |D^{\mathrm{obs}}|)^2 . \tag{9}$$

An alternative approach to obtaining a starting model for the refinement that does not need NOE data is called molecular fragment replacement [71] and treats the protein of interest as overlapping fragments of 7 to 10 residues in length. A protein database is used to find those fragments for which there is a best fit of the calculated dipolar couplings to the measured dipolar couplings of the target protein (see Sect. 8.5).

8.4.2 Nucleic Acid Structure Calculation

For molecular sizes that are amenable by NMR techniques, nucleic acids usually lack a tertiary fold. This fact, together with the characteristic low proton density, complicates NMR structural analysis of nucleic acids. As a result, local geometries and overall shapes of nucleic acids, whose structures have been determined by NMR, usually are poorly defined. Dipolar couplings provide the necessary long-range information to improve the quality of nucleic acid structures substantially [72]. Some examples can be found already in the literature where the successful application of dipolar couplings into structure calculation and structure refinement of DNA and RNA are reported [73–77].

Bacteriophages and bicelles have been used as orienting media for the measurement of residual dipolar couplings for nucleic acids. Bacteriophages are thought to be more suited for this purpose, since the similarity in the charges (negative for the bacteriophages and for the nucleic acids) are supposed to reduce nonspecific interactions. In contrast, neutral bicelles will orient primarily by steric interactions.

The types of dipolar couplings that can be measured for nucleic acids are ^{1}H-^{15}N, ^{1}H-^{13}C, ^{13}C-^{13}C, ^{1}H-^{1}H, ^{1}H-^{31}P. Some of them can be difficult to obtain or to use as structural restraints. For example, imino protons exchange fast with water, so that the number of measured ^{1}H-^{15}N dipolar couplings is usually small. Additionally, ^{1}H-^{13}C dipolar couplings can be difficult to obtain because of spectral overlap in the sugar spectral region, although this problem can be alleviated using selective labeling. As already mentioned, ^{1}H-^{1}H distances are not known *a priori* unless a previously determined structure is available. Thus, in order to use ^{1}H-^{1}H dipolar couplings as structural restraints, the relevant angles and the interproton distances have to be considered as variables during the refinement process. However, it must be emphasized that the assumption of a uniform distribution of bond-vector orientations is far from real in nucleic acids. Therefore, it is more complicated than in the case of proteins to estimate the magnitude and rhombicity of the alignment tensor from the distribution of the measured dipolar couplings.

Nucleic acid dipolar couplings have also been measured in the absence of orienting media by using paramagnetic ions to drive the magnetic alignment. One example is reported in the work by Berger et al. [78]. The authors use paramagnetic europium ions that bind to the narrow grooves of a DNA pentadecamer to increase the alignment relative to that observed for free DNA. This increase in the degree of orientation allowed the authors to measure ^{1}H-^{13}C residual dipolar couplings in the 70 Hz range.

The first RNA structure that has been refined using dipolar coupling information is part of a protein-RNA complex [74]. Starting structures of the complex that were calculated without dipolar couplings were refined using dipolar couplings measured for the protein and the RNA moiety. The main contribution of the dipolar coupling data is a better definition of the relative position of the two double helical RNA stems. Apparently, this structural feature was poorly defined before dipolar coupling refinement because of the lack of NOE data.

Dipolar couplings have been used as the main source of conformational restraints in the structure determination of a DNA dodecamer [79]. In this work Tjandra et al. measure ^{1}H-^{13}C, ^{1}H-^{15}N and ^{1}H-^{1}H residual dipolar couplings using samples in which complementary halves of the oligomer are isotopically labeled. ^{1}H-^{13}C and ^{1}H-^{15}N couplings were measured using HSQC experiments, and ^{1}H-^{1}H couplings were obtained from phase-sensitive COSY spectra. Initially, starting structures are calculated from completely random structures considering only energy terms for bonds, angles and improper torsions. In a second step, new structures are determined including NOE and hydrogen bond restraints. These last structures are refined against residual dipolar couplings in a series of molecular dynamic runs in which the dipolar force constants are increased slowly (Fig. 8.8). Once again, it is very difficult to obtain good estimates of the magnitude and rhombicity of the alignment tensor on the basis of the dipolar coupling distribution. The authors establish lower limits of the magnitude of the alignment, and derive the optimum values from a grid search in which the dipolar energy term in the structure calculation is evaluated as a function of the magnitude and the rhombicity of the alignment. The calculated ensemble of structures are similar to the structure previously determined by X-Ray crystallography [80], but are substantially different from the ones calculated without dipolar couplings and from previous NMR structures (Fig. 8.9).

Nucleic acid backbones are usually poorly defined because of the difficulty in obtaining experimental restraints that depend on conformation. For this reason ^{1}H-^{31}P dipolar couplings can have a special importance in the structure calculation of nucleic acids. Recently, experiments designed to measure ^{1}H-^{31}P dipolar couplings have been reported [81, 82]. ^{1}H-^{31}P dipolar couplings, as well as ^{31}P chemical shift anisotropy (Fig. 8.10) have been applied in the refinement of DNA, demonstrating its usefulness in constraining the orientation of phosphodiester groups [83].

8.4.3
Oligosaccharide Structure Calculation

Residual dipolar couplings were applied for the first time in the structure determination of oligosaccharides. One of the original works, carried out by Aubin and Prestegard [84], was the determination of the conformation of a sialic acid attached to the surface of a magnetically oriented membrane-mimicking surface. The authors used ^{1}H-^{13}C and ^{13}C-^{13}C dipolar couplings together with a single anisotropy-induced chemical shift. A more recent example involves the application of residual ^{1}H-^{13}C dipolar couplings of a trisaccharide in a restrained dynamic simulated annealing protocol [85]. The value of the axial component of the alignment tensor is derived from the minimum dipolar coupling value, considering that the bond-vector having this value is aligned perpendicular to the principal axis of the tensor. The rhombic component is instead derived from the usual minimization of the difference between the measured and the predicted dipolar couplings. The conformation of each monomer is fixed on the NMR time scale, and motion in the molecule is a result of torsional oscillations about the glycosidic linkages. In this case dipolar couplings restrain the conformation of the oligosaccharide better than regular short-range NOE data. This work serves to demonstrate that it is possible to partially orient small molecules such as a trisaccharide in a dilute liquid crystalline medium. Sim-

Restrained molecular dynamics protocol

1) **Randomization**
   Total time: 16 ps
   Temperature: 1000 K
   Energy terms: bonds, angles, improper torsions
   Number of calculated structures: 40

2) **Global fold**
   a) 500 cycles of energy minimization
      Energy terms: bonds, angles, improper torsions
   b) 15 ps of restrained molecular dynamics
      Energy terms: bonds, angles, improper torsions, NOE
      Soft-square NOE potential
      $k_{NOE}=50$ kcal/mol Å^2
      Step size: 1 fs
      Temperature: 1000 K
   c) 200 steps of energy minimization
   d) Restrained molecular dynamics
      20 cycles of 1 ps each
      Energy terms: bonds, angles, improper torsions, NOE, van der Waals (Lennard-Jones)
      van der Waals repulsion energy is increased by a factor of 1.23, from 0.0015 to 0.1
      Step size: 1 fs
      Temperature: 1000 K
   e) 200 steps of energy minimization
   f) Restrained molecular dynamics while cooling to 300 K
      28 cycles of 0.5 ps each
      Temperature increment: 25 K per cycle
      Step size: 1 fs
      Energy terms: bonds, angles, improper torsions, NOE, van der Waals (Lennard-Jones)
   g) 1000 cycles of energy minimization

3) **Refinement**
   a) 500 cycles of energy minimization
      Energy terms: bonds, angles, improper torsions, NOE, van der Waals (Lennard-Jones)
      Soft-square NOE potential
      $k_{NOE}=50$ kcal/mol Å^2
      $k_{DIH}=5$ kcal/mol rad^2
   b) 2.5 ps of restrained molecular dynamics while increasing the force constants of the β, γ, δ, ε torsion angles
      k_{NOE}: from 1 to 50 kcal/mol Å^2, increased by a factor of 1.22 per cycle
      k_{DIH}: from 1 to 200 kcal/mol rad^2, increased by a factor of 1.30 per cycle
      Step size: 0.5 fs
      Temperature: 1000 K

Fig. 8.8 Schematic representation of the protocol used in the structure calculation of a DNA dodecamer using residual dipolar couplings. (Reproduced with permission from N. Tjandra, S.-I. Tate, A. Ono, M. Kainosho and A. Bax, *J. Am. Chem. Soc.* **2000**, *122*, 6190-6200.)

c) 2.5 ps of restrained molecular dynamics while increasing the force constants of all torsion angles
k_{NOE}: 50 kcal/mol Å^2
k_{DIH}: from 1 to 200 kcal/mol rad^2, increased by a factor of 1.30 per cycle
Step size: 0.5 fs
Temperature: 1000 K
d) 200 steps of minimization
e) Restrained molecular dynamics while cooling to 300 K
28 cycles of 0.25 ps each
Temperature increment: 25 K per cycle
Step size: 0.5 fs
f) 1000 steps of minimization

4) Refinement with dipolar couplings
Energy terms: bonds, angles improper torsions, NOE,
van der Waals (Lennard-Jones), electrostatic, dipolar
Switched van der Waals and electrostatic functions, with switching distances of 9.5 and 10.5 Å, were used to truncate the number of possible interactions
a) 200 steps of minimization
k_{dipo}=0.001 kcal/mol Å^2
b) 15 ps of restrained molecular dynamics
k_{dipo}: from 0.001 to 0.2 kcal/mol Å^2, increased by a factor of 1.11 per cycle
Step size: 0.5 fs
Temperature: 300 K
c) 25 ps of restrained molecular dynamics
Step size: 0.5 fs
d) Analyze structures each 200 steps from 15 ps to 25 ps
e) 200 steps of minimization

Fig. 8.8 (continued)

ilar work was reported by Martin-Pastor and Bush [86]. The authors use ^{1}H-^{13}C dipolar couplings in the structure calculation of a pentasaccharide. The molecule is considered as a rigid trisaccharide moiety, whose conformation has been proposed previously, and a more flexible disaccharide. The dipolar coupling data provide information concerning the conformation of the disaccharide and also agree with the proposed model for the trisaccharide.

^{1}H-^{1}H dipolar couplings have been measured from COSY experiments in oriented oligosaccharides [87]. Only recently these experimental restraints have been applied to the conformational study of oligosaccharides. In the work reported by Martin-Pastor and Bush [88], long-range ^{1}H-^{1}H and one-bond ^{13}C-^{1}H residual dipolar couplings are used in the structure determination of an heptasaccharide. The conformational flexibility gives rise to incompatible NMR restraints, and the residual dipolar couplings are incompatible with a unique alignment tensor. However, dipolar couplings still help in refining the more rigid moieties of the oligosaccharide. Correspondingly, Tian et al. [89] have used residual dipolar couplings to gain insight into the structure and flexibility of the oligosaccharide. The Saupe order matrices are determined for individual fragments of the molecule, which are assumed to be rigid. The structural information comes from the fact that, if the whole molecule is rigid, the fragments must share a common order tensor.

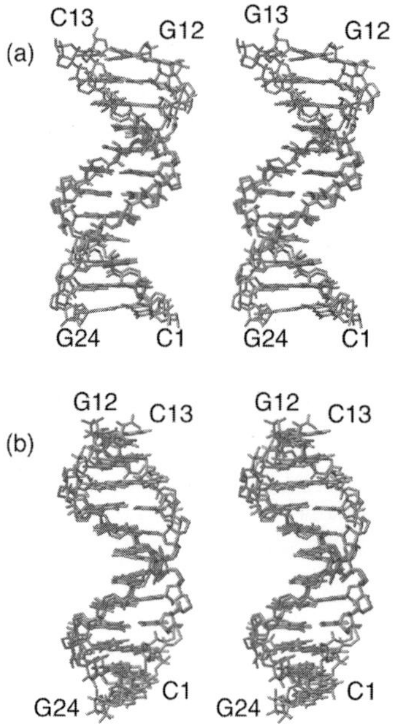

Fig. 8.9 Stereoviews of the NMR structure of a DNA dodecamer calculated using residual dipolar couplings. **a** and **b** differ in a 90° rotation around the helix axis to highlight the non-cylindrical symmetry of the molecule. These structures differ considerably from solution structures calculated without dipolar couplings. (Reproduced with permission from N. Tjandra, S.-I. Tate, A. Ono, M. Kainosho and A. Bax, *J. Am. Chem. Soc.* **2000**, *122*, 6190-6200.)

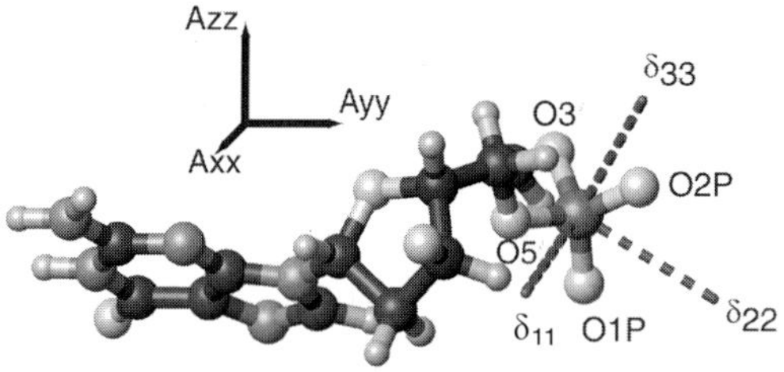

Fig. 8.10 ^{31}P chemical shift anisotropy (CSA) tensor orientation of a nucleotide from a DNA dodecamer with its helix axis pointing vertically. The DNA structure has been calculated using CSA data obtained under anisotropic conditions (Z. Wu, N. Tjandra and A. Bax, *J. Am. Chem. Soc.* **2001**, *123*, 3617-3618). The CSA tensor is defined by δ_{11}, δ_{22}, δ_{33}, δ_{11} is orthogonal to the O1P-P-O2P plane. δ_{22} bisects the O1P-P-O2P angle. A_{xx}, A_{yy}, A_{zz} define the alignment tensor.

The presence of flexibility and motion in oligosaccharides, as in the case of proteins or nucleic acids, complicates the utilization of dipolar coupling data in structure refinement. Dipolar couplings are averaged because of conformational flexibility, but it is also possible that the various existing conformations orient differently with respect to the magnetic field.

8.5 Other Applications of Dipolar Couplings

8.5.1 Protein Structure Validation Factors

Residual dipolar couplings applied to macromolecule structure refinement help in improving both precision and accuracy of the calculated structures. Structure quality factors that describe the agreement between the calculated structures and the experimental data such as the R factor used in X-Ray crystallography [90] and in NMR [91] have been widely used. The incorporation of dipolar couplings into the field of macromolecule structure determination prompted the definition of a structure quality factor (Q factor) that depends on dipolar coupling data [23]. The Q factor describes the agreement between the predicted dipolar couplings for a determined structure and the ones experimentally measured.

This Q factor is defined as follows:

$$Q = [\textstyle\sum_{i=1,\ldots,N}(D^{\mathrm{meas}} - D^{\mathrm{pred}})^2/N]^{1/2}/D^{\mathrm{rms}} \tag{10}$$

where D^{meas} and D^{pred} are the experimentally measured and the predicted dipolar couplings, respectively, for N residues of known dipolar couplings. D^{rms} is the root-mean-square value of dipolar couplings when these are randomly distributed and using an alignment tensor obtained from the best fit between the experimental data and the existing structure. If the distribution cannot be considered uniform, D^{rms} may be calculated as [92, 93]

$$D^{\mathrm{rms}} = \{D_{\mathrm{a}}^2[4 + 3(D_{\mathrm{r}}/D_{\mathrm{a}})^2]/5\}^{1/2} \tag{11}$$

where D_{a} and D_{r} are the axial and rhombic components of a traceless second rank diagonal tensor related to the alignment tensor.

In analogy to the free R factor [90, 94], Q has to be calculated for a set of restraints not used in the structure refinement.

The chemical shift is another NMR parameter that has been used to define structure validation factors. The chemical shift is slightly modified under anisotropic conditions, and the change is related to the orientation of the chemical shift anisotropy tensor with respect to the alignment tensor. The change in carbonyl chemical shift upon alignment is larger and more easily measured than for other nuclei [95]. The quality factor compares the experimentally measured carbonyl chemical shift changes and the predicted changes on the basis of a determined structure. The characteristics of the alignment tensor and the carbonyl chemical shift anisotropy tensor, which is considered uniform, have to be known.

A different way to assess structure quality is reported in the work by Nikolai et al. [96]. The authors translate dipolar couplings into angular restraints and use them to build consistency maps. These maps are utilized to examine the quality of protein structure.

8.5.2
Protein Domain Orientation

In a multidomain protein whose domains have fixed orientations relative to each other, a unique alignment tensor will represent the preferred orientation of all the domains in the anisotropic environment. Therefore, structure refinement with dipolar couplings is performed as in one-domain proteins (Sect. 8.4). Several examples are reported in the literature of cases with conformational ambiguity due to the lack of NOE contacts between the domains. One example is the determination of subdomain orientation of the ribosomal protein S4 Δ41 [97]. In this work the lack of NOE contacts between the domains produces an ambiguity in interdomain orientation. The authors use two different anisotropic media to obtain dipolar couplings (DMPC/DHPC bicelles and Pf1 filamentous bacteriophages). They conclude that subdomain orientation in solution is similar to the one present in the crystal structure.

The determination of the relative orientation of protein domains by X-Ray crystallography can be biased by the presence of crystal packing forces, and therefore may differ from the physiologically relevant structure in solution. Dipolar couplings measured under the influence of a weakly aligned medium allowed the identification of a difference between the X-ray and solution structures of the maltodextrin-binding protein (MBP) loaded with β-cyclodextrin [98]. The solution conformation of this 370-residue protein has been obtained from the X-ray structure using a large set of dipolar couplings. Hinge-rotations have been applied to various crystal structures, searching for the minimum difference between the experimentally measured and calculated dipolar couplings.

This same methodology based on crystal structures has been applied in the analysis of the dipolar coupling data of the two-domain T4 lysozyme [99]. The average solution conformation of T4 lysozyme differs from the crystal structure in the relative orientation of the domains. A similar result was found for a two-domain fragment of the barley lectin protein (BLP) [100] on comparing the solution structure to the crystal structure of a highly (approx. 95%) homologous protein. These are examples of the effect that crystal packing interactions may have in protein domain orientation and the possibility of identifying it by the use of dipolar couplings.

8.5.3
Protein-Ligand Conformation and Orientation

The function of numerous proteins is intimately related to ligand conformation in protein-ligand complexes. In addition, knowledge of ligand conformation and orientation can be used as a basis for rational drug design. X-ray crystallography and NOE-based NMR techniques are the most common methods used to obtain this type of information. Nevertheless, determining the conformation of ligands using NOE data can be difficult because of the small number of NOE contacts between the protein and its ligand, which is usually a consequence of large internuclear distances. Dipolar couplings have proved to be very useful in extracting information about ligand structure in protein-ligand complexes. One example is the study by Bolon et al. [101] of a 149-residue fragment of the mannose-binding protein-A (MBPA) bound to α-methyl mannoside (AMM). The measured one-bond ^1H-^{13}C dipolar couplings of AMM bound to MBPA present a weighted time-average of the dipolar couplings of AMM in the free and bound states. The dipolar couplings of AMM in the absence of MBPA are used as a reference for the free state. In this study the orientational tensor of the protein is defined by the protein symmetry, and the ligand orientational tensor is determined on the basis of the dipolar couplings. The superimposition of the two tensors indicates the orientation of AMM with respect to the protein.

Residual dipolar couplings, as well as other NMR parameters, have also been measured for a transducin peptide that transiently binds to rhodopsin in field-oriented rhodopsin-containing disks [102]. The binding is triggered by rhodopsin light-activation. After activation, ^{15}N relaxation rates, proton amide line width, residual dipolar couplings and ^{15}N chemical shift anisotropy all indicate that order is transferred from the oriented disks to the bound peptide. Dipolar coupling values provide the angles that some of the peptide NH bond vectors form with the magnetic field. This work exemplifies the use of dipolar couplings in extracting conformational information of ligands bound to integral membrane receptor proteins.

It is important to note that even in the presence of sufficient proton density and in the case of differential isotope labeling, efficient spin-diffusion may complicate the interpretation of the NOE data and result in structures or relative orientations that are erroneous. A similar problem does not occur when using residual dipolar couplings.

8.5.4
Structure Building using Dipolar Couplings

Based on the structural information that can be obtained from residual dipolar couplings, it is possible to define atom positions with respect to a molecular frame and hence determine the molecular structure. Attempts are being made to define protein global folds using dipolar couplings as a sole data source or with a small number of NOE restraints in order to avoid the time-consuming task of side-chain resonance assignment. Clore et al. have improved the precision and accuracy of NMR structures of the B1 domain of streptococcal protein G (GB1), the monomer of the barrier-to-autoregulation factor (BAF) and cyanovirin-N (CVN) by including dipolar couplings into protein structure calculation and using a small number of NOE distance restraints [92]. In the case of GB1, only distance information derived from hydrogen bonds has been used together with dipolar couplings to successfully determine the overall fold. The accuracy, evaluated as resemblance to the crystal structure, was improved. In contrast, in the case of BAF and CVN, NOEs connecting backbone to side chain and side chain to side chain were also included in the calculation.

As another example, the three-dimensional structure of Cytochrome c′ has been determined on the basis of structural information from pseudocontact paramagnetic chemical shifts, Curie-Dipolar cross-correlation, secondary structure constraints, dipolar couplings and ^{15}N relaxation data [103]. This protein has a paramagnetic center, and therefore the above-mentioned conformational restraints can be derived from this feature. Dipolar couplings do not average to zero because of the susceptibility tensor anisotropy of the protein. The structure determination of this protein without NOE data gives an RMSD (root mean square deviation) of 0.73 (±0.18) Å for the backbone atoms of the helical residues within the ensemble of calculated structures and 1.7 Å RMSD to the X-ray coordinates. The authors suggest that this method can be applied to a large number of metal-binding proteins or to proteins capable of binding peptides containing lanthanide ions.

The global fold of the maltodextrin-binding protein MBP (42 KD) complexed with β-cyclodextrin has also been determined using dipolar couplings and a limited set of NOE restraints [104]. Amide-to-amide, as well as amide-to-methyl and methyl-to-methyl distance restraints from NOE data are used. Structural information from hydrogen bonds and dihedral angles is also included in the calculation. The authors developed a new protocol for the use of dipolar couplings in the calculation of large protein structures with a limited set of NOE data. Their approach consists of orienting peptide planes using dipolar couplings. Five dipolar couplings for each peptide plane are necessary to overdetermine the three degrees of freedom related to the three Euler angles that describe the transformation from the coordinate to the alignment frame, assuming that both the axial and the rhombic components of the alignment tensor are known. The three Euler angles are determined by minimizing the difference between the predicted and the measured dipolar couplings. Eight possible orientations are obtained through this procedure. The appropriate orientation is chosen based on the preliminary structures that were calculated with the NOE, dihedral and hydrogen bond data. Global pair-wise RMSD of the MBP backbone is reduced from 5.5 Å to 2.2 Å and the RMSD for the existing crystal structure decreases from 5.1 Å to 3.3 Å by including the dipolar data.

Delaglio et al. have developed another approach to define global folds, in the absence of NOE data, that uses dipolar couplings, chemical shifts and a protein database with crystal structures of high resolution. The method is called molecular fragment replacement (MFR) and has been applied to the protein ubiquitin [71]. MFR selects from the data base those fragments of seven to ten residues whose predicted dipolar couplings best fit the set of measured ones, as well as the fragments with the smaller differences between the measured and calculated chemical shifts (Fig. 8.11). The preliminary structures can be subsequently refined by minimizing the difference between the measured and the calculated dipolar couplings, and analogously minimizing the differences in the chemical shifts of the target protein and the selected fragments. The final refined structures have an RMSD of 0.88 Å with respect to the crystal structure.

The ubiquitin backbone structure has been determined using dipolar couplings as the sole source of structural restraints in the work reported by Hus et al. [105]. The resulting structure is practically identical to the NMR structure calculated from NOE and other

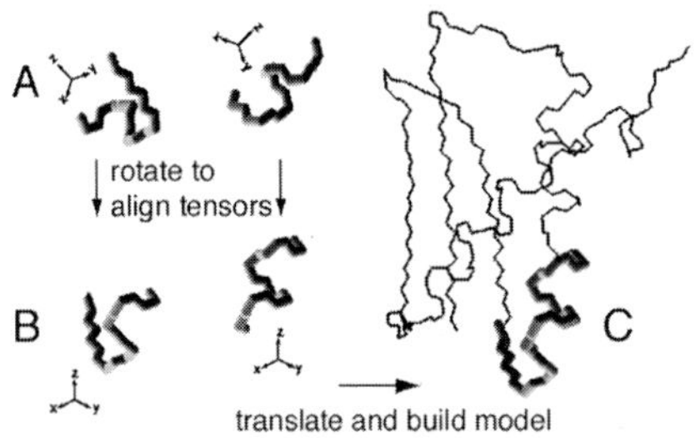

Fig. 8.11 Illustration of the basic steps followed by the Molecular Fragment Replacement method to build a model of the structure of the protein DinI.

NMR data [95]. This new method requires two sets of dipolar couplings from two different protein orientations. Together with the backbone dipolar couplings that are typically used (i.e., amide NH, C′N, $C_\alpha C'$, $C_\alpha H_\alpha$ and the two-bond $H^N C'$), $C_\alpha C_\beta$ dipolar couplings are also needed. Provided that the orientation of one peptide plane is known independently, the dipolar coupling data give rise to two possible orientations for the subsequent peptide plane, where the conformations about the alpha carbon in these two orientations are mirror images. One of the conformations can be ruled out because of chirality.

Additionally, dipolar couplings together with intermolecular NOEs have been applied in docking to form protein-protein complexes [106]. High-resolution structures of the free proteins have to be known *a priori*, and no significant backbone conformational changes must occur upon complexation. The docking is achieved by rigid body minimization using a target function that takes into account intermolecular NOEs, amide residual dipolar couplings and a purely repulsive intermolecular potential. If no NOE data are included, the method finds several possible orientations of one molecule relative to the other. The orientational information contained in the NOE data is essential to discriminate between the different possibilities.

8.5.5
Dipolar Couplings in Protein Family Search

The sequencing of genomic DNA is providing an enormous amount of data concerning new proteins, most of which are of unknown function. Tools are being developed to facilitate the identification of these functions and to overcome the time-consuming task of determining the three-dimensional structure of the proteins at high resolution. In this sense, residual dipolar couplings have proven to be very useful in protein fold recognition. Annila et al. have been pioneers in applying NMR dipolar couplings for the purpose mentioned [107]. Their idea is based on the secondary structure composition of proteins and its arrangement in a limited number of relative orientations. Backbone dipolar couplings measured for a protein of unknown structure are compared to predicted dipolar couplings of proteins of known structure that compose a data base. Such a comparison may provide critical information about the global fold of proteins of unknown structure. In this work, amide residual dipolar couplings measured for a protein of undetermined structure named calerythrin, an EF-hand containing protein, were compared to the calculated values of proteins that are members of the EF-hand family. The comparison indicated that calerythrin has a similar fold to other proteins belonging to this family. The method worked adequately even though calerythrin shares less than 30% sequence homology with any of the proteins included in the database. The comparison between the experimental and the calculated dipolar couplings gets complicated by the presence of protein loops and variability in the length of secondary structure elements. For this reason the authors recommend aligning the primary structure based on amino acid sequence homology and the secondary structure based on C_α secondary shifts before comparing dipolar couplings. Proteins with close structures will have dipolar couplings that are very similar for the majority of the residues, with only a few showing large deviations. Since the agreement is concentrated in the secondary structure fragments and dissimilarities are found in turns, it will be easier to recognize common folds of proteins composed mainly of regular secondary structure elements. Related examples of the application of these ideas have been reported by Meiler et al. [108] and Fowler et al. [109] for the proteins rhodniin and NodF, respectively.

The wide structural application of dipolar couplings is demonstrated by its use to validate models built by sequence homology methods. Additionally, dipolar couplings have been shown to reduce the RMSD between these models and the target structure. One example is the work reported by Chou et al., in which the RMSD of sequence homology models of the protein calmodulin, built from the structure of recoverin and parvalbumin, is reduced using heteronuclear dipolar couplings [110].

Although it does not come within the scope of this chapter, dipolar coupling can also provide dynamic information. Some work related to this issue can be found in Refs. [111–113].

8.6 Acknowledgements

We would like to thank Dusty Baber, Ad Bax, Frank Delaglio, Georg Kontaxis and Ben E. Ramirez for kindly providing some of the figures shown in this manuscript. We are very grateful to James A. Ferretti for the careful reading of the manuscript and very useful suggestions.

8.7 References

1 Tjandra, N. *Structure* **1999**, *7*, R205–R211.
2 Prestegard, J. H.; Al-Hashimi, H. M.; Tolman, J. R. *Q. Rev. Biophys.* **2000**, *33*, 371–424.
3 Bax, A.; Kontaxis, G.; Tjandra, N. *Methods Enzymol.* **2001**, *339*, 127–174.
4 Karplus, M. *J. Phys. Chem.* **1959**, *30*, 11–15.
5 Spera, S.; Bax, A. *J. Am. Chem. Soc.* **1991**, *113*, 5490–5492.
6 Wishart, D. S.; Sykes, B. D.; Richards, F. M. *J. Mol. Biol.* **1991**, *222*, 311–333.
7 Dingley, A. J.; Grzesiek, S. *J. Am. Chem. Soc.* **1998**, *120*, 8293–8297.
8 Pervushin, K.; Ono, A.; Fernandez, C.; Szyperski, T.; Kainosho, M.; Wüthrich, K. *Proc. Natl. Acad. Sci. USA.* **1998**, *95*, 14147–14151.
9 Wang, Y. X.; Jacob, J.; Cordier, F.; Wingfield, P.; Stahl, S. J.; Lee-Huang, S.; Torchia, D.; Grzesiek, S.; Bax, A. *J. Biomol. NMR* **1999**, *14*, 181–184.
10 Saupe, A. *Angew. Chem. Int. Ed. Engl.* **1968**, *7*, 97–112.
11 Emsley, J. W. Liquid crystals: general considerations, vol. 4, Wiley, Chichester **1996**.
12 Gayathra, C.; Bothner-By, A. A.; van Zijl, P. C. M.; MacLean, C. *Chem. Phys. Lett.* **1982**, *87*, 192–196.
13 Sanders, C. R.; Schwonek, J. P. *Biochemistry* **1992**, *31*, 8898–8905.
14 Tjandra, N.; Bax, A. *Science* **1997**, *278*, 1111–1114.
15 Tolman, J. R.; Flanagan, J. M.; Kennedy, M. A.; Prestegard, J. H. *Proc. Natl. Acad. Sci. USA.* **1995**, *92*, 9279–9283.
16 Contreras, M. A.; Ubach, J.; Millet, O.; Rizo, J.; Pons, M. *J. Am. Chem. Soc.* **1999**, *121*, 8947–8948.
17 Kung, H. C.; Wang, K. Y.; Goljer, I.; Bolton, P. H. *J. Magn. Reson. B* **1995**, *109*, 323–325.
18 Tjandra, N.; Grzesiek, S.; Bax, A. *J. Am. Chem. Soc.* **1996**, *118*, 6264–6272.
19 Lipari, G.; Szabo, A. *J. Am. Chem. Soc.* **1982**, *104*, 4546–4559.
20 Ramirez, B. E.; Bax, A. *J. Am. Chem. Soc.* **1998**, *120*, 9106–9107.
21 Hilbers, C. W.; MacLean, C. *Chem. Phys. Lett.* **1968**, *2*, 445–447.

22 OTTIGER, M.; BAX, A. *J. Biomol. NMR* **1998**, *12*, 361–372.

23 OTTIGER, M.; BAX, A. *J. Biomol. NMR* **1999**, *13*, 187–191.

24 LOSONCZI, J. A.; PRESTEGARD, J. H. *J. Biomol. NMR* **1998**, *12*, 447–451.

25 CAVAGNERO, S.; DYSON, H. J.; WRIGHT, P. E. *J. Biomol. NMR* **1999**, *13*, 387–391.

26 PROSSER, R. S.; LOSONCZI, J. A.; SHIYANOVSKAYA, I. V. *J. Am. Chem. Soc.* **1998**, *120*, 11010–11011.

27 BARRIENTOS, L. G.; DOLAN, C.; GRONENBORN, A. M. *J. Biomol. NMR* **2000**, *16*, 329–337.

28 RUCKERT, M.; OTTING, G. *J. Am. Chem. Soc.* **2000**, *122*, 7793–7797.

29 CLORE, G. M.; STARICH, M. R.; GRONENBORN, A. M. *J. Am. Chem. Soc.* **1998**, *120*, 10571–10572.

30 HANSEN, M. R.; MUELLER, L.; PARDI, A. *Nature Struct. Biol.* **1998**, *5*, 1065–1074.

31 SASS, H. J.; CORDIER, F.; HOFFMANN, A.; ROGOWSKI, M.; COUSIN, A.; OMICHINSKI, J. G.; LOWEN, H.; GRZESIEK, S. *J. Am. Chem. Soc.* **1999**, *121*, 2047–2055.

32 KOENIG, B. W.; HU, J.-S.; OTTIGER, M.; BOSE, S.; HENDLER, R. W.; BAX, A. *J. Am. Chem. Soc.* **1999**, *121*, 1385–1386.

33 FLEMING, K.; GRAY, D.; PRASANNAN, S.; MATTHEWS, S. *J. Am. Chem. Soc.* **2000**, *122*, 5224–5225.

34 TYCKO, R.; BLANCO, F. J.; ISHII, Y. *J. Am. Chem. Soc.* **2000**, *122*, 9340–9341.

35 SASS, H. J.; MUSCO, G.; STAHL, S. J.; WINGFIELD, P. T.; GRZESIEK, S. *J. Biomol. NMR* **2000**, *18*, 303–309.

36 DE ALBA, E.; DE VRIES, L.; FARQUHAR, M.; TJANDRA, N. *J. Mol. Biol.* **1999**, *291*, 927–939.

37 BAX, A.; TJANDRA, N. *J. Biomol. NMR* **1997**, *10*, 289–292.

38 DELAGLIO, F.; TORCHIA, D.; BAX, A. *J. Biomol. NMR* **1991**, *1*, 439–446.

39 GRIESINGER, C.; SORENSEN, O. W.; ERNST, R. R. *J. Magn. Reson.* **1987**, *75*, 474–492.

40 TOLMAN, J. R.; PRESTEGARD, J. H. *J. Magn. Reson. B* **1996**, *112*, 245–252.

41 OTTIGER, M.; DELAGLIO, F.; BAX, A. *J. Magn. Reson.* **1998**, *131*, 373–378.

42 YANG, D. W.; VENTERS, R. A.; MUELLER, G. A.; CHOY, W. Y.; KAY, L. E. *J. Biomol. NMR* **1999**, *14*, 333–343.

43 KONTAXIS, G.; CLORE, G. M.; BAX, A. *J. Magn. Reson.* **2000**, *143*, 184–196.

44 PERMI, P.; ANNILA, A. *J. Biomol. NMR* **2000**, *16*, 221–227.

45 PERMI, P.; ROSEVEAR, P. R.; ANNILA, A. *J. Biomol. NMR* **2000**, *17*, 43–54.

46 DE ALBA, E.; SUZUKI, M.; TJANDRA, N. *J. Biomol. NMR* **2001**, *19*, 63–67.

47 CHOU, J. J.; DELAGLIO, F.; BAX, A. *J. Biomol. NMR* **2000**, *18*, 101–105.

48 SANTORO, J.; KING, G. C. *J. Magn. Reson.* **1992**, *97*, 202–207.

49 OTTIGER, M.; DELAGLIO, F.; MARQUARDT, J. L.; TJANDRA, N.; BAX, A. *J. Magn. Reson.* **1998**, *134*, 365–369.

50 GRZESIEK, S.; BAX, A. *J. Am. Chem. Soc.* **1992**, *114*, 6291–6293.

51 CHOU, J. J.; BAX, A. *J. Am. Chem. Soc.* **2001**, *123*, 3844–3845.

52 KESSLER, H.; GEHRKE, M.; GRIESINGER, C. *Angew. Chem. Int. Ed. Engl.* **1988**, *27*, 490–536.

53 CAI, M.; WANG, H.; OLEJNICZAK, E. T.; MEADOWS, R. P.; GUNASEKERA, A.; XU, N.; FESIK, S. W. *J. Magn. Reson.* **1999**, *139*, 451–453.

54 TIAN, F.; LOSONCZI, J. A.; FISCHER, M. W. F.; PRESTEGARD, J. H. *J. Biomol. NMR* **1999**, *15*, 145–150.

55 TIAN, F.; FOWLER, C. A.; ZARTLER, E. R.; JENNEY, F. A. J.; ADAMS, M. W.; PRESTEGARD, J. H. *J. Biomol. NMR* **2000**, *18*, 23–31.

56 OTTING, G.; RUCKERT, M.; LEVITT, M. H.; MOSHREF, A. *J. Biomol. NMR* **2000**, *16*, 343–346.

57 PELLECCHIA, M.; VANDER KOOI, C. W.; KELIIKULI, K.; ZUIDERWEG, E. R. *J. Magn. Reson.* **2000**, *143*, 435–439.

58 CARLOMAGNO, T.; PETI, W.; GRIESINGER, C. *J. Biomol. NMR* **2000**, *17*, 99–109.

59 PETI, W.; GRIESINGER, C. *J. Am. Chem. Soc.* **2000**, *122*, 343–346.

60 WU, Z.; BAX, A. *J. Magn. Reson.* **2001**, *151*, 242–252.

61 KAIKKONEN, A.; OTTING, G. *J. Am. Chem. Soc.* **2001**, *123*, 1770–1771.

62 TJANDRA, N.; OMICHINSKI, J. G.; GRONENBORN, A. M.; CLORE, G. M.; BAX, A. *Nat. Struct. Biol.* **1997**, *4*, 732–738.

63 CLORE, G. M.; GRONENBORN, A. M.; BAX, A. *J. Magn. Reson.* **1998**, *131*, 159–162.

64 WARREN, J. J.; MOORE, P. B. *J. Magn. Reson.* **2001**, *149*, 271–275.

65 ZWECKSTETTER, M.; BAX, A. *J. Am. Chem. Soc.* **2000**, *122*, 3791–3792.

66 Brunger, A.T. *XPLOR:* A system for X–Ray crystallography and NMR, ed. 3.1, Yale University Press, New Haven, **1993**.

67 Meiler, J.; Blomberg, N.; Nilges, M.; Griesinger, C. *J. Biomol. NMR* **2000**, *16*, 245–252.

68 Losonczi, J.A.; Andrec, M.; Fischer, M.W. F.; Prestegard, J. H. *J. Magn. Reson.* **1999**, *138*, 334–342.

69 Guntert, P.; Mumenthaler, C.; Wüthrich, K. *J. Mol. Biol.* **1997**, *273*, 283–298.

70 Tjandra, N.; Marquardt, J.L.; Clore, G.M. *J. Magn. Reson.* **2000**, *142*, 393–396.

71 Delaglio, F.; Kontaxis, G.; Bax, A. *J. Am. Chem. Soc.* **2000**, *122*, 2142–2143.

72 Zhou, H.; Vermeulen, A.; Jucker, F.M.; Pardi, A. *Biopolymers* **1999**, *52*, 168–180.

73 Mollova, E.T.; Pardi, A. *Curr. Opin. Struct. Biol.* **2000**, *10*, 298–302.

74 Bayer, P.; Varani, L.; Varani, G. *J. Biomol. NMR* **1999**, *14*, 149–155.

75 Ramos, A.; Grunert, S.; Adams, J.; Micklem, D.R.; Proctor, M.R.; Freund, S.; Bycroft, M.; St Johnson, D.; Varani, G. *EMBO J.* **2000**, *19*, 997–1009.

76 Lynch, S.R.; Puglisi, J.D. *J. Am. Chem. Soc.* **2000**, *122*, 7853–7854.

77 MacDonald, D.; Herbert, K.; Zhang, X.; Polgruto, T.; Lu, P. *J. Mol. Biol.* **2001**, *306*, 1081–1098.

78 Beger, R.D.; Marathias, V.M.; Volkman, B.F.; Bolton, P.H. *J. Magn. Reson.* **1998**, *135*, 256–259.

79 Tjandra, N.; Tate, S.–I.; Ono, A.; Kainosho, M.; Bax, A. *J. Am. Chem. Soc.* **2000**, *122*, 6190–6200.

80 Minasov, G.; Tereshko, V.; Egli, M. *J. Mol. Biol.* **1999**, *291*, 83–99.

81 Hennig, M.; Carlomagno, T.; Williamson, J.R. *J. Am. Chem. Soc.* **2001**, *123*, 3395–3396.

82 Wu, Z.; Tjandra, N.; Bax, A. *J. Biomol. NMR* **2001**, *19*, 367–370.

83 Wu, Z.; Tjandra, N.; Bax, A. *J. Am. Chem. Soc.* **2001**, *123*, 3617–3618.

84 Aubin, Y.; Prestegard, J.H. *Biochemistry* **1993**, *32*, 3422–3428.

85 Kiddle, G.R.; Homans, S.W. *FEBS Lett.* **1998**, *436*, 128–130.

86 Martin-Pastor, M.; Bush, C.A. *Carbohydr. Res.* **2000**, *323*, 147–155.

87 Bolon, P.J.; Prestegard, J.H. *J. Am. Chem. Soc.* **1998**, *120*, 9366–9367.

88 Martin-Pastor, M.; Bush, C.A. *J. Biomol. NMR* **2001**, *19*, 125–139.

89 Tian, F.; Al-Hashimi, H.M.; Craighead, J.L.; Prestegard, J. H. *J. Am. Chem. Soc.* **2001**, *123*, 485–492.

90 Brunger, A.T. *Nature* **1992**, *355*, 472–475.

91 Gonzalez, C.; Rullmann, J.A.C.; Bonvin, A.; Boelens, R.; Kaptein, R. *J. Magn. Reson.* **1991**, *91*, 659–664.

92 Clore, G.M.; Starich, M.R.; Bewley, C.A.; Cai, M.L.; Kuszewski, J. *J. Am. Chem. Soc.* **1999**, *121*, 6513–6514.

93 Clore, G.M.; Garret, D.S. *J. Am. Chem. Soc.* **1999**, *121*, 9008–9012.

94 Brunger, A.T.; Clore, G.M.; Gronenborn, A.M.; Saffrich, R.; Nilges, M. *Science* **1993**, *261*, 328–331.

95 Cornilescu, G.; Marquardt, J.L.; Ottiger, M.; Bax, A. *J. Am. Chem. Soc.* **1998**, *120*, 6836–6837.

96 Nikolai, R.; Skrynnikov, N.R.; Kay, L.E. *J. Biomol. NMR* **2000**, *18*, 239–252.

97 Markus, M.A.; Gerstner, R.B.; Draper, D.E.; Torchia, D.A. *J. Mol. Biol.* **1999**, *292*, 375–387.

98 Skrynnikov, N.R.; Goto, N.K.; Yang, D.; Choy, W.-Y.; Tolman, J.R.; Mueller, G.A.; Kay, L.E. *J. Mol. Biol.* **2000**, *295*, 1265–1273.

99 Goto, N.K.; Skrynnikov, N.R.; Dahlquist, F.W.; Kay, L.E. *J. Mol. Biol.* **2001**, *308*, 745–764.

100 Fischer, M.W.F.; Losonczi, J.A.; Lim Weaver, J.; Prestegard, J.H. *Biochemistry* **1999**, *38*, 9013–9022.

101 Bolon, P..; Al-Hashimi, H.M.; Prestegard, J.H. *J. Mol. Biol.* **1999**, *293*, 107–115.

102 Koenig, B.W.; Mitchell, D.C.; Konig, S.; Grzesiek, S.; Litman, B.J.; Bax, A. *J. Biomol. NMR* **2000**, *16*, 121–125.

103 Hus, J.-C.; Marion, D.; Blackledge, M. *J. Mol. Biol.* **2000**, *298*, 921–936.

104 Mueller, G.A.; Choy, W.Y.; Yang, D.; Forman-Kay, J.D.; Venters, R.A.; Kay, L.E. *J. Mol. Biol.* **2000**, *300*, 197–212.

105 Hus, J.-C.; Marion, D.; Blackledge, M. *J. Am. Chem. Soc.* **2001**, *123*, 1541–1542.

106 Clore, G. M. *Proc. Natl. Acad. Sci. USA.* **2000**, *97*, 9021–9025.

107 Annila, A.; Aitio, H.; Thulin, E.; Drakenberg, T. *J. Biomol. NMR* **1999**, *14*, 223–230.

108 Meiler, J.; Peti, W.; Griesinger, C. *J. Biomol. NMR* **2000**, *17*, 283–294.

109 Fowler, C. A.; Tian, F.; Al–Hashimi, H. M.; Prestegard, J. H. *J. Mol. Biol.* **2000**, *304*, 447–460.

110 Chou, J. J.; Li, S.; Bax, A. *J. Biomol. NMR* **2000**, *18*, 217–227.

111 Tolman, J. R.; Al-Hashimi, H. M.; Kay, L. E.; Prestegard, J. H. *J. Am. Chem. Soc.* **2001**, *123*, 1416–1424.

112 Meiler, J.; Prompers, J. J.; Peti, W.; Griesinger, C.; Bruschweiler, R. *J. Am. Chem. Soc.* **2001**, *123*, 6098–6107.

113 Mittermaier, A.; Kay, L. E. *J. Am. Chem. Soc.* **2001**, *123*, 6892–6903.

9
Scalar Couplings Across Hydrogen Bonds [1]

Andrew J. Dingley, Florence Cordier, Victor A. Jaravine, and Stephan Grzesiek

Abstract

This chapter gives a summary of the recent findings on electron-mediated scalar couplings across hydrogen bonds. The main emphasis is upon a review of the various types and sizes of such couplings that have been detected in biological macromolecules. Various experimental schemes for the detection of H-bond couplings are summarized and recent applications that characterize subtle changes in H-bond geometries of biomacromolecules are presented.

9.1
Introduction

The concept of a hydrogen bond (H-bond) has long been recognized as a very intuitive explanation for the weak attractive forces between an electronegative acceptor atom and a hydrogen atom attached to a second electronegative atom [1–4]. Compared to covalent bonds, the activation energies for the formation of an H-bond and the bond energies themselves are small, so that H-bonds are readily formed and broken between various partners under conditions where donor and acceptor groups undergo diffusive processes under ambient temperatures in many common solvents. This property explains the key role that H-bonds play in the formation and stabilization of biomacromolecular structures and as participants in many chemical reactions [5, 6].

In biological systems, H-bond donors and acceptors are predominantly nitrogen and oxygen atoms. However, the π electrons of aromatic systems can also act as acceptors, and H-bonds involving sulfur groups or metallic cofactors are also known. The presence of individual H-bonds in biomacromolecular structures is usually derived from the spatial arrangement of the donor and acceptor groups once the structure of a molecule has been solved by diffractive or NMR techniques. More detailed information about H-bonds

1) This chapter is an updated version of an article previously published as "Scalar couplings across hydrogen bonds" by S. Grzesiek, F. Cordier, and A. J. Dingley, *Methods in Enzymology* **2001**, *338*, 111–133. Reprinted by permission of the publisher.

from X-ray diffraction is particularly hard to obtain since the weak scattering density of the hydrogen atom can only be detected in structures solved at highest resolutions (i.e. < ~1 Å resolution). Such structures (year 2000) comprise about 0.6% of all crystallographic structures entered in the Brookhaven protein data bank. Only in these cases is it possible to ascribe individual spatial positions to the hydrogen atoms that are independent of the use of standard covalent geometries. Recent technical innovations in neutron diffractometry have also yielded high-resolution information on the position of either proton or deuterium nuclei in H-bonds for a small number of biomacromolecules [7].

Numerous NMR observables for individual nuclei in hydrogen-bonded moieties have been shown to correlate with the stability and relative geometry of individual H-bonds [8]. Such parameters include: the reduced hydrogen exchange rates with the solvent [9, 10], fractionation factors [9, 11–13], isotope shifts for the substitution of the hydrogen-bonded proton by ^2H and ^3H [14, 15], the isotropic and anisotropic chemical shifts of the hydrogen-bonded proton and of other nuclei within the donor and acceptor groups [16–23], the size of the electric field gradient at the position of the proton within the H-bond as observed by the deuterium quadrupolar coupling constant [24, 25], and the sequential one-bond $^1J_{C'N}$ coupling constants in proteins [26, 27].

Recently, electron-mediated, scalar couplings which are active between magnetic nuclei on both sides of the hydrogen bridge have been discovered in nucleic acids [28–41], proteins [42–54] and their complexes (Tabs. 9.1–9.3) [54–56]. These couplings are closely related to similar inter- and intramolecular couplings across H-bonds in smaller chemical compounds [57–60]. It is well established [31, 58, 61–74] that such *trans* H-bond scalar couplings follow the same electron-mediated polarization mechanism as any covalent

Tab. 9.1 Observed *trans* hydrogen bond couplings in nucleic acids

Coupling	*Size [Hz]*	*Donor*	*Acceptor*	*Base pair type*	*References*
$^{h2}J_{NN}$ 1)	5–11	G–N1 U/T–N3 C–N3	C–N3 A–N1 A–N7 G–N7	Watson-Crick, T–A and C^+–G Hoogsteen, U–A Reverse Hoogsteen, G–A and G–G mismatches	28–32, 36, 38, 39
$^{h2}J_{NN}$ 2)	2.5–8	A–N6 G–N2	A–N7 G–N7 G–N3	A–A mismatch, G_4-tetrad, GCGC-tetrad, A_2G_4-hexad G–A sheared mismatch	33–35, 41
$^{h1}J_{HN}$	0.5–4	G–H1 U/T–H3 C–H3	C–N3 A–N1 A–N7 G–N7	Watson-Crick, T–A and C^+–G Hoogsteen	29–31, 39, 40
$^{h3}J_{NC'}$	~0.2 3)	G–N1	G–C6	G_4-tetrad	35
$^{h4}J_{NN}$	~0.14 3)	G–N1	G–N1	G_4-tetrad	37

1) $^{h2}J_{NN}$ couplings involving an imino donor group.
2) $^{h2}J_{NN}$ couplings involving an amino donor group.
3) Sign of coupling not determined experimentally.

Tab. 9.2 Observed *trans* hydrogen bond couplings in proteins

Coupling	Size [Hz]	Donor	Acceptor	References
${}^{h3}J_{NC'}$	–0.2 to –0.9	Backbone amide	Backbone carbonyl or side chain carboxylate	44, 46, 47, 49, 51, 81
${}^{h2}J_{HC'}$	–0.6 to –1.3	Backbone amide	Backbone carbonyl	45, 53
${}^{h3}J_{HC_\alpha}$	0 to 1.4	Backbone amide	Backbone carbonyl	52
${}^{h2}J_{NN}$	8 to 11	Histidine-Nε2	Histidine-Nε2	48
${}^{h2}J_{HMe}$	0.3 to 4 [1)]	Backbone amide	Cysteine-S coordinating ^{113}Cd or ^{199}Hg	42, 43

1) Sign of coupling not determined experimentally.

Tab. 9.3 Observed *trans* hydrogen bond couplings in protein-nucleic acid complexes

Coupling	Size [Hz]	Donor	Acceptor	References
${}^{h2}J_{NN}$	6	Arginine ${}^{\eta}$N–H_2	G-N7	55
${}^{h3}J_{NP}$	1.7–4.6 [1)]	Backbone amide	GDP-phosphate FMN-phosphate	54, 56
${}^{h2}J_{HP}$	0.5–3.4 [1)]	Backbone amide	GDP-phosphate FMN-phosphate	54, 56
${}^{h2}J_{HP}$	0.6–1.7 [1)]	Serine/threonine OH	FMN-phosphate	56

1) Sign of coupling not determined experimentally.

spin-spin coupling, and thus behave experimentally in exactly the same way as their covalent counterparts. The scalar coupling effect can be used to directly identify individual H-bonds within a biological macromolecule. This means that the frequencies of all three H-bond partners (i.e. the donor, the acceptor and the proton itself) can be correlated by a single two- or three-dimensional NMR experiment. Thus, in favorable cases, the H-bond network in nucleic acids and proteins (and therefore the secondary structure) can be established directly via a COSY experiment. In addition to this direct structural use, the size of the scalar coupling is influenced by the relative geometry of the H-bond partners. Therefore valuable information about the "strength" of individual H-bonds can be derived from an analysis of the size of the coupling constants.

It is the aim of this chapter (1) to discuss the basic experiments for the detection of *trans* H-bond couplings and to indicate experimental problems, (2) to briefly describe the range of groups for which such couplings have been observed and to summarize their sizes, and (3) to present a number of recent applications which characterize changes in H-bond geometries of biomacromolecules.

9.2 H-Bond Scalar Couplings in Biomacromolecules

9.2.1 Nucleic Acids

9.2.1.1 $^{h2}J_{NN}$-Couplings

Scalar couplings across H-bonds are expected to be strongest when the number of bonds between the two interacting nuclei is small. It is therefore no surprise that the strongest H-bond couplings in biomacromolecules are observed within nucleic acid base pairs when both donor and acceptor are ^{15}N-labeled nitrogen atoms. In such a situation, *trans* H-bond couplings only need to bridge one ($^{h1}J_{HN}$) or two ($^{h2}J_{NN}$) bonds (including the H-bond) in order to be observable (see insert to Fig. 9.2). The symbol $^{hn}J_{AB}$ was introduced by Wüthrich and coworkers [30] as a notation for *trans* H-bond scalar couplings between nuclei A and B in order to emphasize that one of the *n* bonds connecting the two nuclei in the chemical structure is actually an H-bond.

The conceptually simplest scheme for the detection and quantification of $^{h2}J_{NN}$ correlations is the non-selective quantitative J_{NN} HNN-COSY experiment [28, 30]. In this scheme, the ^{15}N-evolution period of a ^{1}H–^{15}N-TROSY has been replaced by a homonuclear ^{15}N–^{15}N COSY scheme (Fig. 9.1). As the rate of magnetization transfer in COSY experiments is given by the magnitude of the coupling constant, the intensity of the cross peaks presents a measure of the size of the coupling. As previously described in detail [28, 75], this scheme transfers magnetization from the proton (H) within the hydrogen bridge via the covalent $^{1}J_{NH}$ coupling to the ^{15}N nucleus of the H-bond donor (N_d) during the first INEPT period (time points *a* to *b*). During the first ^{15}N–^{15}N COSY delay from time points *b* to *c*, part of this magnetization is transferred via the $^{h2}J_{NN}$ coupling to the ^{15}N acceptor nucleus (N_a) while another part remains on the imino ^{15}N nucleus. This interval is followed by the ^{15}N frequency labeling period t_1, between time points *d*

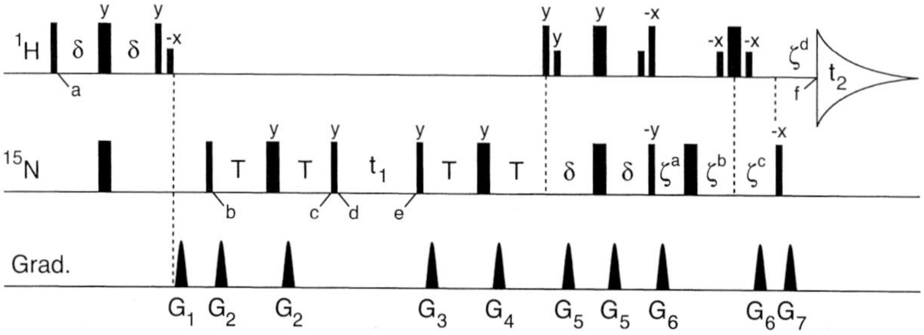

Fig. 9.1 Basic HNN COSY pulse sequence. Narrow and wide pulses correspond to flip angles of 90° and 180°, respectively, whereas low-power (water flip-back) 90° ^{1}H pulses are illustrated as smaller narrow pulses. Delays: δ=2.25 ms; T=15 ms (can be shorter or longer); ζ^a=2.5 ms; ζ^b=0.25 ms; ζ^c=2.25 ms; ζ^d=0.5 ms. Unless indicated, the phase of all pulses are applied along the *x* axis. Quadrature detection in the t_1 dimension is achieved by simultaneously incrementing both 90° ^{15}N-pulses at positions *b* and *c* in the States-TPPI manner. Gradients are sine-bell shaped, with an absolute amplitude at their center and durations (polarities) of G1, 2, 3, 4, 5, 6, 7=2.5 (+), 2.1 (–), 1.35 (+), 2.35 (+), 0.2 (+), 0.4 (+), and 0.101 ms (+). Adapted from [28].

and e, in which both ^{15}N and $^{15}N'$ nuclei precess according to their chemical shifts. Note that during the t_1-evolution the heteronuclear $^{1}J_{NH}$ coupling is also active, and is subsequently used in the TROSY-type detection [76] at the end of the pulse scheme. After this period, the second ^{15}N–^{15}N COSY interval and the TROSY part of the pulse sequence (time points e to f) transfer magnetization back to the original proton, and its oscillations are detected in the receiver during the t_2-period. As in similar quantitative J-correlation schemes, this leads to resonances in the two-dimensional spectrum at frequency positions ω_{Na}, ω_H (cross peaks) and ω_{Nd}, ω_H (diagonal peaks). The value of the $^{h2}J_{NN}$ coupling constant is derived from the intensity ratio of the cross and diagonal peaks by the formula $|^{h2}J_{NN}| = \text{atan}[(-I_{Na}/I_{Nd})^{1/2}]/(\pi T)$ where T is the COSY transfer time.

Fig. 9.2 shows a region of the Hoogsteen-Watson-Crick T•A–T triplets of an intramolecular DNA-triplex [31]. Panel A shows the diagonal resonances corresponding to the thymidine $^{1}H3$–$^{15}N3$ correlations, whereas panel B depicts the cross resonances corresponding to the Watson-Crick $^{1}H3(T)$–$^{15}N1(A)$ and the Hoogsteen $^{1}H3(T)^{15}$•N7(A) correlations. Because of the considerable size of the $^{h2}J_{NN}$ couplings, the sensitivity of the HNN-COSY experiment at field strengths of ≤ 14 T is sufficient for all H–N•••N type H-bonds to

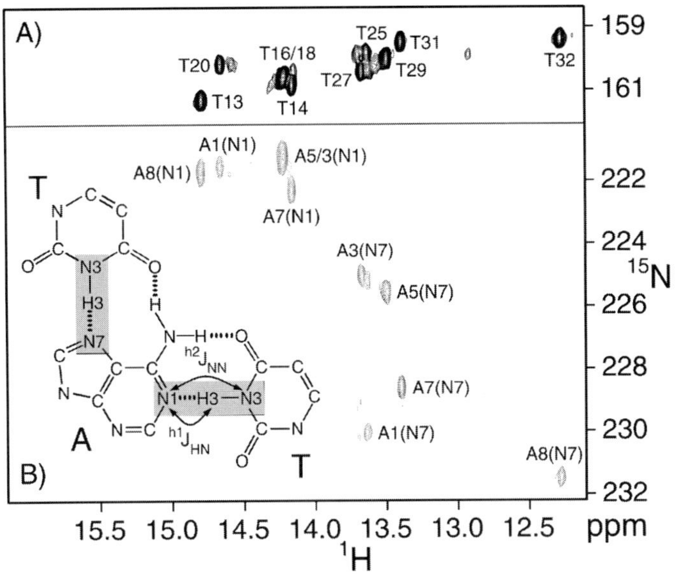

Fig. 9.2 Part of a quantitative-J_{NN} HNN-COSY spectrum of a 1.5 mM uniformly $^{13}C/^{15}N$-labeled intramolecular DNA triplex. This triplex consists of five Hoogsteen-Watson-Crick T•A–T and three Hoogsteen C$^+$•G–C base triplets. The spectral region corresponds to the 10 imino resonances of the Hoogsteen-Watson-Crick T•A–T triplets. The data matrix consisted of 250* (t_1) ×1024* (t_2) data points (where n* refers to complex points) with acquisition times of 45 ms (t_1) and 85 ms (t_2). The total measuring time was 20.1 h on a Bruker DMX500 spectrometer. Positive contours depict diagonal resonances (**A**); negative contours (**B**) correspond to cross peaks resulting from H-bond scalar ^{15}N–^{15}N magnetization transfer. Resonances are labeled with assignment information. The insert in panel **B** shows the chemical structure of the T•A–T triplet and the definition of the scalar $^{h2}J_{NN}$ and $^{h1}J_{HN}$ correlation via the H-bond.

be observed for this 32 nucleotide intramolecular triplex [31] and even for an RNA domain of 69 nucleotides [28] in overnight experiments at ~ 1.5 mM concentrations. Note that for large molecules with short transverse relaxation times of the ^{15}N-donor ($\pi T_2{}^{h2}J_{NN} \ll 1$), the sensitivity for the observation of cross-peaks is highest when the total COSY transfer time is set to a value of approximately T_2. It is envisaged that the introduction of cryoprobe technology and the measurement of $^{h2}J_{NN}$ couplings at higher magnetic field strengths than 14 T will enhance the possibility of observing such couplings at substantially lower sample concentrations and in larger nucleic acid molecules.

Several modifications have been proposed for the basic HNN-COSY experiment. For example, frequency separations between amino and aromatic ^{15}N resonances are typically in the range 100–130 ppm and therefore much larger than between imino ^{15}N donor and aromatic ^{15}N acceptor resonances. As has been pointed out by Majumdar and coworkers [33], such ^{15}N frequency separations are too large to be covered effectively by the non-selective ^{15}N pulses of the homonuclear HNN-COSY. They therefore designed a pseudo-heteronuclear H(N)N-COSY experiment, where selective ^{15}N pulses excite the amino and aromatic ^{15}N resonances separately to yield excellent sensitivity [33]. An inconvenience of this experiment is that the resonances corresponding to the amino ^{15}N nuclei are not detected, and a separate spin-echo difference experiment was used to quantify the $^{h2}J_{NN}$ values. A slightly improved version of this pseudo-heteronuclear H(N)N-COSY [35] remedies this problem by the use of phase-coherent ^{15}N pulses such that both amino and aromatic ^{15}N resonances can be detected in a single experiment.

A further modification of the HNN-COSY scheme has been proposed for the observation of $^{h2}J_{NN}$ correlations, where the hydrogen nucleus in the H-bond is unobservable [34, 36, 38, 41]. This situation is often due to intermediate-fast rotations around the exocyclic C–N bond of hydrogen-bonding amino groups or due to exchange of the imino/amino proton with the solvent. For this condition, the HNN-COSY can then be started and detected on a carbon-bound proton in the vicinity of the acceptor. This is possible for adenosine and guanosine ^{15}N7 acceptors as well as for adenosine ^{15}N1 acceptors, which can be connected to the H8 or H2 proton by means of the $^{2}J_{H8N7}$ or $^{2}J_{H2N1}$ couplings of $\sim$ 11–15 Hz [77], respectively. This concept has been recently extended to carbon-bound protons located more than two bonds away from the H-bond moiety [41]. Here, $^{h2}J_{NN}$ couplings in sheared G–A mismatches are observed via the use of adenosine H2 protons, which are separated by *four* bonds from the H–N•••N moiety. In addition to solving the exchange broadening issue, these experiments may also be performed at higher temperature as well as in D_2O, and are therefore free of potential water suppression- and saturation transfer-related problems.

Another ingenious modification of the HNN-COSY scheme involves the replacement of the homonuclear ^{15}N-COSY transfer by a ^{15}N-TOCSY transfer [37]. As the homonuclear TOCSY transfer is twice as fast as the COSY transfer, a significant sensitivity increase is achieved. The application is, however, limited to cases where the ^{15}N donor and acceptor resonances are at similar frequencies, so that the power of the ^{15}N-TOCSY radio frequency pulses needed is not too strong. A very interesting application of this scheme was presented for the sensitive detection of very small (0.14 Hz, Tab. 9.1) $^{h4}J_{N1N1}$ couplings in N1–H1•••O6=C6–N1 H-bonds of guanosine tetrads.

In general, the size of $^{h2}J_{NN}$ couplings between imino or amino ^{15}N donor and aromatic ^{15}N acceptor nuclei in nucleic acid base pairs ranges between 5 and 11 Hz

(Tab. 9.1). The quantitative HNN-COSY experiment only yields the absolute value of ${}^{h2}J_{NN}$. The sign of the ${}^{h2}J_{NN}$ coupling was determined [74, 78] as positive by an E. COSY TROSY technique, which is in agreement with theoretical predictions [31, 62]. Several density functional theory simulations show that, with the exception of charged Hoogsteen C^{+}•G base pairs [31, 74], the size of the ${}^{h2}J_{NN}$ couplings does not depend strongly on the base pair type. The range of ${}^{h2}J_{NN}$ values is rather the result of slight variations of the H-bond geometry in the different base pair types.

9.2.1.2 ${}^{h1}J_{HN}$-Couplings

${}^{h1}J_{HN}$ couplings between the imino proton and aromatic ${}^{15}N$ acceptor nuclei [30, 31, 39, 40] and between the H-bonded amino proton and its nitrogen acceptor in G_4-quartets (A. Dingley, unpublished results) have been observed in the range 0.5–4 Hz (Tab. 9.1) and have been determined by an E. COSY method as positive in sign [30]. Because of their smaller size as compared to ${}^{h2}J_{NN}$ and the faster relaxation times of the proton resonances, the measurement of ${}^{h1}J_{HN}$ couplings is more challenging. A number of methods have been described for the quantification of ${}^{h1}J_{HN}$ couplings: an E. COSY experiment [30], a quantitative *J*-correlation TROSY scheme based on the more favorable relaxation properties of ${}^{1}H$–${}^{15}N$ zero-quantum coherences [39], a quantitative HN-COSY [31], a modified HNN-TROSY [40], and a quantitative TROSY scheme based on the constructive use of conformational exchange-induced relaxation to improve relaxation properties of either the ${}^{1}H$–${}^{15}N$ zero- or double-quantum coherences for nucleic acid base moieties where conformational exchange is the dominant relaxation mechanism [29]. This latter approach also derives the ${}^{h2}J_{NN}$ coupling values.

9.2.1.3 ${}^{h3}J_{NC'}$-Couplings

Besides the aromatic nitrogen acceptor atoms, many nucleic acid base pairs involve the oxygen atoms of carbonyl groups as H-bond acceptors. In a similar approach for the detection of ${}^{h3}J_{NC'}$ couplings in protein (see below), N–H•••O=C H-bonds in nucleic acids can be detected by the use of a slightly modified version of the long-range HNCO experiment [35]. The observed ${}^{h3}J_{NC'}$ couplings for G–G base pairs of G_4-quartets are only 0.2 Hz in size and thus smaller in absolute size than the average of the analogous couplings in proteins (Tab. 9.1). Nevertheless, because of the favorable relaxation properties of the G-quartet system, all the expected ${}^{h3}J_{NC'}$ correlations could be detected. The information obtained from this experiment is sufficient to establish the G-quartet structure and the number of non-equivalent G-quartets within the entire molecule as well as to derive all the H-bond partners for the single guanine bases within each quartet.

9.2.2 Proteins

9.2.2.1 ${}^{h3}J_{NC'}$-Couplings

The predominant H-bond in proteins is the bridge between the backbone amide proton of one amino acid and the backbone carbonyl oxygen atom of a second amino acid (see insert to Fig. 9.3). Although scalar couplings across H-bonds to the magnetic isotope ${}^{17}O$ are conceivable, the fast relaxation of this quadrupolar nucleus would prevent such observations in

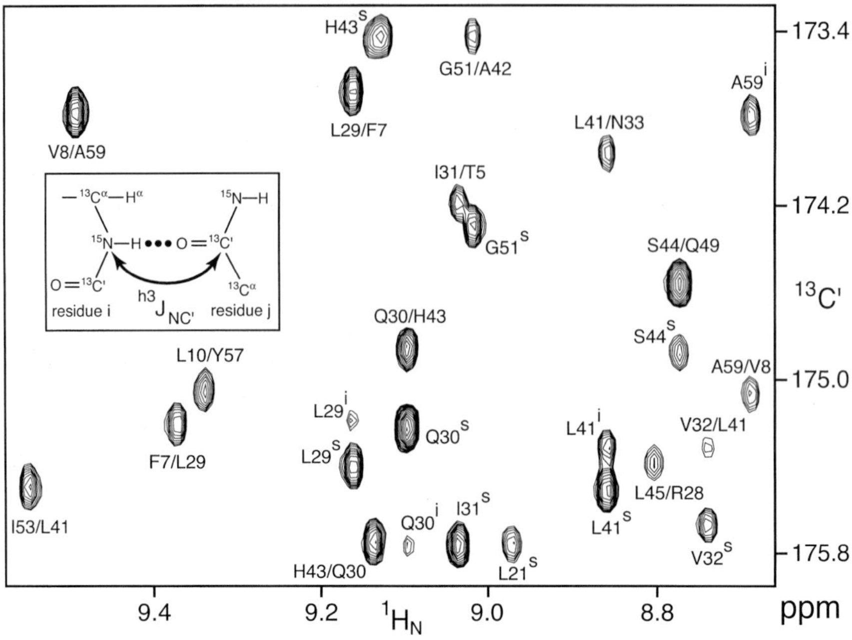

Fig. 9.3 Selected region of the two-dimensional, long-range, quantitative-$J_{NC'}$ H(N)CO-TROSY spectrum recorded on a 3.5 mM sample of $^2H/^{15}N/^{13}C$ labeled c-Src SH3 in complex with the polyproline peptide RLP2 at 95% H_2O/5% D_2O, 25 °C. The data matrix consisted of $65^*(t_1)\times 512^*(t_2)$ data points (where n^* refers to complex points) with acquisition times of 39 ms (t_1) and 53 ms (t_2). Total measuring time was 17 h on a Bruker DMX600 instrument. Cross peaks marked as Res$_i$/Res$_j$ are due to $^{h3}J_{NiC'j}$ *trans* hydrogen bond scalar couplings between the ^{15}N nucleus of residue *i* and $^{13}C'$ nucleus of residue *j*. Residue names marked by the superscript 's' denote not completely suppressed, sequential, one-bond correlations between the ^{15}N nucleus of residue *i* and $^{13}C'$ nucleus of residue *i*–1. The superscript '*i*' demarks intraresidue two-bond $^{15}N_i$-$^{13}C'_i$ correlations. The insert illustrates the definition of the scalar $^{h3}J_{NC'}$ correlation via the H-bond.

solution. Instead, H-bond couplings to the next possible nucleus of the carbonyl acceptor group, the carbonyl carbon, have been observed in proteins [44–47, 49, 52, 53].

The detection of these weak three-bond ($|^{h3}J_{NC'}|<0.9$ Hz, Tab. 1.2) correlations in proteins can be achieved with a conventional HNCO experiment where the nitrogen to carbonyl dephasing and rephasing delays are set to longer values than for the detection of the sequential one-bond J_{NiCi-1} correlations. In a standard HNCO experiment, the time for the two INEPT transfers between in-phase $N_y(i)$ and anti-phase $2N_x(i)C'_z(i\text{–}1)$ magnetization is usually set to values of about 25–30 ms. In an HNCO experiment suitable for the transfer by $^{h3}J_{NC'}$, this time is set to 133 ms $\approx 2/^1J_{NC'}$ such that the one-bond transfer from $N_y(i)$ to $2N_x(i)C'_z(i\text{–}1)$ is approximately refocused [44, 46]. In such a situation, the resulting HNCO spectrum contains mostly correlations which result from smaller, long-range $J_{NC'}$ couplings.

Fig. 9.3 shows the results of this experiment on the c-Src SH3 domain carried out as a two-dimensional H(N)CO where only amide proton and carbonyl ^{13}C frequencies were detected [79]. Clearly visible, for example, are the correlations from the amide $^{1}H-(^{15}N)$ of residue F7 to the carbonyl ^{13}C nucleus of residue L29 and from the amide $^{1}H-(^{15}N)$ of residue L29 to the carbonyl ^{13}C of residue F7 that correspond to the anti-parallel beta sheet arrangement of these two amino acids. Quantification of the long-range $J_{NC'}$ coupling constants can be achieved by comparison of the cross-peak intensities with the intensities of sequential amide-carbonyl correlations measured in a second reference experiment [44, 46]. In the latter experiment, transfer is tuned to the sequential $^{1}J_{NC'}$ coupling while the relaxation losses are kept identical to those in the long-range experiment. The quantitative long-range HNCO experiment only yields the absolute value of $^{h3}J_{NC'}$. In agreement with density functional theory simulations [62], the sign of the $^{h3}J_{NC'}$ coupling in proteins was determined as negative by a zero-quantum/double-quantum technique [47].

Because of the small size of the $^{h3}J_{NC'}$ couplings, the sensitivity of the long-range HNCO experiment is rather low. Whereas about 83% of all expected backbone-to-backbone H-bonds could be observed in a 12 h experiment on a 1.6 mM sample of uniformly $^{13}C/^{15}N$ labeled human ubiquitin (MWT 8.6 kDa), for larger systems, the shorter ^{15}N transverse relaxation times together with the long $^{15}N-^{13}C$ dephasing and rephasing times of 266 ms clearly limit the detection of $^{h3}J_{NC'}$ correlations. Several sensitivity improvements have been proposed.

An obvious enhancement in sensitivity can be obtained by the use of an HNCO-TROSY at higher magnetic field strengths (>14 T), which makes use of the cancellation of dipolar and CSA relaxation mechanisms for the downfield component of the amide ^{15}N doublet. Such an approach was successful to detect $^{h3}J_{NC'}$ correlations in the 30 kDa ribosome-inactivating protein MAP30 [49] using uniform ^{2}H, ^{13}C, ^{15}N enrichment and a protein concentration of only 0.7 mM. Experience from this experiment shows that the number of detected $^{h3}J_{NC'}$ correlations was not limited by the transverse ^{15}N relaxation time, but by the incomplete back-exchange of amide deuterons against protons and by the long amide proton T_1 relaxation times. Deuteration is essential for sensitivity gains by the HNCO-TROSY approach, because in protonated samples proton-proton spin-flips strongly reduce the line-narrowing gain of TROSY [80]. For example, for protonated ubiquitin at 25 °C and 14 T magnetic field strength, the "TROSY-T_2" and the T_2 of the amide ^{15}N singlet (using ^{1}H-decoupling) are almost identical (170–180 ms). In contrast, in perdeuterated (>85%, non-exchangeable hydrogen positions), but amide protonated ^{13}C, ^{15}N-labeled ubiquitin, the "TROSY-^{15}N-T_2" increases to 340–400 ms. Clearly larger enhancements can be achieved at higher magnetic field strengths than 14 T [76, 80].

A second sensitivity improvement of the long-range HNCO has been proposed for non-deuterated protein samples [81]. In such samples, the transverse relaxation time of the amide ^{15}N nucleus is decreased to some extent by the scalar coupling to the $^{13}C_{\alpha}$ nucleus and its relatively short longitudinal relaxation time. This relaxation mechanism is commonly referred to as "scalar relaxation of the second kind" [82]. A similar decrease in imino ^{15}N transverse relaxation times due to the scalar relaxation effect has been observed in nucleic acids [35]. The effect is particularly pronounced for smaller macromolecules where ^{13}C T_1 relaxation times are short. In the case of proteins, the scalar contribution to the ^{15}N relaxation can be removed with relative ease by continuous decoupling of the $^{13}C_{\alpha}$ nucleus. Using this approach for the 12 kDa FK506 binding protein, a sensitivity increase of about 50% was observed as compared to the conventional long-range HNCO experiment [81].

H-bond $^{h3}J_{NC'}$ couplings of –0.1 to –0.4 Hz (Tab. 9.2) have also been detected for backbone amides H-bonded to the carboxylate side chains of glutamic and aspartic acids [47, 51]. In addition, an H-bond $^{h3}J_{NC'}$ coupling involving side chain-side chain interactions has been observed between a donor arginine guanidinium nitrogen $^{15}N^{\varepsilon}$ and an aspartic acid acceptor carboxylate carbon $^{13}CO_{2}\gamma$ [50]. Apparently, the magnetization transfer across these H-bonds is very similar to that of the backbone amide to carbonyl H-bonds.

9.2.2.2 $^{h2}J_{HC'}$- and $^{h3}J_{HC\alpha}$-Couplings

Currently, two schemes have been proposed to detect $^{h2}J_{HC'}$ correlations in proteins (Tab. 9.2). The first uses very selective carbonyl pulses to separate the $^{h2}J_{HC'}$ correlations from interfering covalent two- and three-bond H_N-^{13}C-carbonyl scalar couplings [45]. The second scheme utilizes a non-selective E. COSY-HNCO-TROSY approach [53] that also yields the sign of the coupling constants. A very similar experiment was used to detect $^{h3}J_{HC\alpha}$ couplings [52]. Ranges from –0.6 to 1.3 Hz have been observed for $^{h2}J_{HiC'k}$ couplings. These values are similar to the size of $^{h3}J_{NiC'k}$ couplings, whereas the $^{h3}J_{HiC\alpha k}$ couplings have values up to 1.4 Hz (Tab. 9.2).

9.2.2.3 $^{h2}J_{NN}$-Couplings

For a pair of H-bonded histidine side chains, $^{h2}J_{NN}$ couplings in the range 8–11 Hz (Tab. 9.2) have been observed between a protonated $^{15}N\varepsilon 2$-nucleus and an unprotonated $^{15}N\varepsilon 2$ using an HNN-COSY-based technique [48]. These values are similar to the $^{h2}J_{NN}$ coupling values observed for the imino and amino N–H•••$N_{aromatic}$ H-bonds of nucleic acid base pairs (Tab. 9.1).

9.2.2.4 $^{h2}J_{HMe}$-Couplings

Probably the earliest observations of H-bond couplings in proteins were made by Summers and coworkers [42, 43] in ^{113}Cd- or ^{199}Hg-substituted rubredoxin (Tab. 9.2). In this protein, J interactions of up to 4 Hz connected backbone amide protons and the cysteine-coordinated metal atom ($^{h2}J_{HMe}$). Apparently these $^{h2}J_{HMe}$ couplings were mediated via the H-bonds from the backbone amides to the S atoms of the cysteine residues.

9.2.3 Protein-Nucleic Acid Complexes

9.2.3.1 $^{h2}J_{NN}$-Couplings

Another important development is the direct detection of H-bonds between amino acid side chains and nucleic acids by *trans* H-bond scalar couplings (Tab. 9.3). Recently, $^{h2}J_{NN}$ couplings between arginine side-chain guanidinium moieties and guanine base $^{15}N7$ nuclei have been observed in a complex of the HTLV-1 Rex peptide and an RNA aptamer [55]. With values of ~ 6 Hz, these couplings are again an indication that the chemical nature of the donor or acceptor group has a very limited influence on the size of the $^{h2}J_{NN}$ couplings.

9.2.3.2 $^{h3}J_{NP}$- and $^{h2}J_{HP}$-Couplings

Recently, *trans* H-bond scalar couplings involving protein amide or hydroxyl groups as donors and phosphate groups as acceptors have been described in molecular complexes of Ras p21 and GDP [54] as well as flavodoxin and riboflavin 5′-monophosphate [56] (Tab. 9.3). The measurement of these H-bond correlations was obtained by an HNPO or HNPO-TROSY experiment [54] that is analogous to the HNCO or HNCO-TROSY experiment, the only difference being that the ^{13}C carbonyl pulses are replaced by ^{31}P pulses. For the quantification of the $^{h3}J_{NP}$ and $^{h2}J_{HP}$ couplings, appropriate spin-echo difference HSQC schemes were proposed [54]. It was found that, depending on the geometry, both $^{h2}J_{HP}$ and $^{h3}J_{NP}$ can be as large as 3–5 Hz. Given the larger extent of phosphorus d-orbitals participating in the phosphate oxygen bonds, as compared to the more restricted carbon p-orbitals participating in the carbonylic oxygen bonds, it is perhaps not surprising that the H-bond couplings to ^{31}P in phosphates are larger than the analogous couplings to ^{13}C in the carbonyl groups. The detection of such couplings is particularly important because long-range information on the position of phosphorus in biomacromolecular complexes is very hard to obtain by traditional high-resolution NMR methods.

9.3 Relation to Chemical Shift

Strong correlations of the isotropic chemical shift of the H-bonded protons and the *trans* H-bond scalar coupling constants have been observed for both nucleic acids and proteins [78]. In nucleic acids N–H•••N H-bonds, both $^{h2}J_{NN}$ and $^{1}J_{HN}$ increase linearly with increasing imino proton chemical shifts [31]. Similarly in protein N–H•••O=C H-bonds, $|^{h3}J_{NC'}|$ [44] and $|^{h2}J_{HC'}|$ [78] values increase linearly with increasing amide proton isotropic chemical shifts. For proteins containing a large number of aromatic side chains, these correlations are considerably improved when the amide proton chemical shifts are corrected for ring current effects [F. Cordier, unpublished results].

The isotropic [10, 18, 83–85] and anisotropic [21, 22] chemical shifts of amide protons have been used as an indicator for the strength of the H-bond in proteins. In particular, there is a strong correlation between the H-bond length and the isotropic shift of the amide proton [10, 18, 83–85]. Short H-bond lengths correspond to larger values for the isotropic chemical shift. Thus, the observed correlations between H-bond coupling constants and isotropic proton chemical shifts indicate that both parameters are mainly determined by the spatial arrangement of the donor and acceptor groups.

9.4 Dependence on Geometry

9.4.1 H-Bond Lengths

Comparisons of the measured coupling constants to the geometry of the H-bond are hampered by the limited availability of very high-resolution diffraction data. Especially, no crystallographic data are available for most of the nucleic acids for which H-bond couplings have been measured. In the available high-resolution structures of nucleic acids, the variation of N1–N3 distances in Watson-Crick base pairs is clearly very limited. A survey of a number of such structures [31] showed that typical N1–N3 distances are 2.92 ± 0.05 Å and 2.82 ± 0.05 Å for G–C and A–T (A–U) base pairs in DNA and RNA. The shorter crystallographic donor-acceptor distances in A–T (A–U) base pairs as compared to G–C base pairs coincide with an increase in the value of the ${}^{h2}J_{NN}$ coupling constants from about 6–7 Hz (G–C) to about 7–8 Hz (A–T, A–U) in the NMR investigations [2 8, 30, 31]. Density functional theory simulations [31, 74] indicate that the different chemical nature of the donor and acceptor groups in Watson-Crick G–C and A–T (A–U) and in Hoogsteen A•T base pairs have a very limited influence on the ${}^{h2}J_{NN}$ coupling constants (≤ 0.2 Hz, for N–N distances from 2.7 to 4.0 Å). Therefore, the observed differences in ${}^{h2}J_{NN}$ values for G–C, A–T, A–U, and A•T base pairs should be largely due to the differences in donor-acceptor distances. Neglecting angular variations, it follows that at N–N distances of 2.8 to 2.9 Å, a change in the value of ${}^{h2}J_{NN}$ by 1 Hz corresponds to a change in donor-acceptor distance of 0.07 ± 0.01 Å. Decreases of ${}^{h2}J_{NN}$ coupling constants by about 1 Hz have been observed at the ends of helical stems and have been interpreted as a corresponding increase in the ensemble average of the donor-acceptor distance [31]. An alternative method to measure apparent H-bond lengths in A–T base pairs has recently been reported [86]. In this method, by measuring cross-correlated relaxation rates across the N–H•••N H-bond, an apparent H-bond length is determined which characterizes the amplitude of inter-base motions and the length of the H-bond.

In contrast to N–H•••N H-bonds in Watson-Crick base pairs, the geometry of N–H•••O=C H-bonds in proteins is more varied. A survey of a number of crystallographic structures [87] showed that typical values for N•••O and H•••O distances and for N–H•••O and H•••O=C angles are 2.99 ± 0.14 Å, 2.06 ± 0.16 Å, 157 ± 11°, 147 ± 9° for α-helices and 2.91 ± 0.14 Å, 1.96 ± 0.16 Å, 160 ± 10°, 151 ± 12° for β-sheet conformations, respectively. Thus, on average the N•••O and H•••O distances are 0.08–0.10 Å shorter in β-sheets than in α-helices. This coincides with an increase in the average strength of the observed ${}^{h3}J_{NC'}$ coupling from –0.38 ± 0.12 Hz for α-helical conformations to –0.65 ± 0.14 Hz for β-sheets in ubiquitin [44]. An exponential correlation between the coupling constant and the N•••O distance was described for protein G by the Bax group [47], where ${}^{h3}J_{NC'}$ is given as -59×10^3 Hz × exp[$-4 \times R_{NO}$/Å]. Neglecting again angular dependencies, it follows from this relation that for a typical α-helical or β-sheet conformation, a variation in the value of ${}^{h3}J_{NC'}$ by 0.1 Hz corresponds to a change in the N•••O distance of 0.05 or 0.07 Å, respectively.

9.4.2
H-Bond Angles

At present, knowledge about the angular dependencies of the H-bond coupling constants is limited and systematic experimental data are missing. Density functional theory simulations [62] indicate that the $|{}^{h3}J_{NC'}|$ value has a maximum for a straight N–H•••O angle of 180° and it drops by about 20–30% for a decrease in this angle to 140°. Density functional theory calculations of the H•••O=C angle dependency in the formamide dimer [63] yield similar results, i.e. that a straighter conformation yields larger values for

$|^{h3}J_{NC'}|$. Depending on other parameters of the H-bond geometry, the maximum of the $|^{h3}J_{NC'}|$ was found at an H•••O=C angle of 150–170° [63]. This seems to be in line with experimental findings that the $^{h3}J_{NC'}$ couplings in the nucleic acid G-tetrad are weaker than the average of $^{h3}J_{NC'}$ couplings observed in proteins [31]. The principal difference in the geometry of the two systems is that the H•••O=C angle in the G-tetrad is only 125±5°, whereas this angle has typical values of around 150° in proteins (see above). Similar observations were made for $^{h3}J_{NP}$ couplings in N–H•••O–P H-bonds [54]. In this study, an $^{h3}J_{NP}$ coupling of 4.6 Hz was observed for an H•••O–P angle of 173°, but the values of $^{h3}J_{NP}$ dropped below 0.35 Hz for conformations where the H•••O–P angle was smaller than 126°. This observation is supported by a very recent DFT study that examined the geometric dependence of the $^{h3}J_{NP}$ and $^{h2}J_{HP}$ scalar couplings in protein-nucleotide complexes [88].

9.5 Applications

9.5.1 Establishment of Secondary and Tertiary Structure Information

H-bond couplings will be very useful in the process of resonance assignment and for secondary or tertiary structure information. Although indirect information about H-bonds in the backbone of proteins can often be derived from NOE data, only the direct observation of an H-bond coupling unambiguously proves the existence of the H-bond. The exponential dependence of the coupling constant on the H-bond distance [47] offers very tight distance constraints in structure calculations. H-bond coupling constants along with secondary chemical shifts have recently been used as a potential approach to the rapid generation of low-resolution protein folds [89]. Instead of using the exponential correlation described above, this study uniformly converted the identified H-bonds into distant restraints with upper and lower limits of 2.5–1.7 Å between proton and carbonyl oxygen acceptor atoms and 3.5–2.3 Å between nitrogen donor and carbonyl oxygen acceptor atoms. A soft-square restraining potential was used for the H-bond restraints with a force constant of 50 kcal mol^{-1} $Å^{-2}$.

In contrast to the backbone H-bonds, the observation of H-bonds involving the side chains of glutamic and aspartic acids [47, 51], histidines [48] or arginines [50] by means of long-range NOEs is considerably more difficult. In all cases, the good spectral dispersion of $^{1}H_{N}$, ^{15}N, ^{13}C-carbonyl or ^{13}C-carboxylate resonances makes the identification of the donor and acceptor groups by H-bond *J*-couplings rather straightforward. Particularly important applications may be the unambiguous establishment of H-bond networks in the active sites of enzymes or ribozymes and the detection of intermolecular H-bonds in macromolecular interactions. The latter application was recently demonstrated for the HTLV-1 Rex peptide in complex with an RNA aptamer in which $^{h2}J_{NN}$ couplings between arginine side chain guanidinium moieties and the major groove edge of guanine bases were observed [55].

Compared to proteins, nucleic acids have a lower abundance of protons. As a consequence, structural information from NOEs is often rather limited. Since the few long-range interactions in folded nucleic acids generally involve H-bonds, the detection of the H-bond couplings not only for Watson-Crick base pairs but also for a wide variety of non-canonical base pairs and couplings to phosphates yields valuable tertiary structure information that might not be obtainable by any other NMR parameter. In this context, it is remarkable that the ${}^{h2}J_{NN}$, ${}^{h2}J_{HP}$, and ${}^{h3}J_{NP}$ couplings are large and that sensitivity issues are only a minor problem. In particular, easily obtainable information from ${}^{h2}J_{NN}$-correlations has been used in recent nucleic acid structure determinations of DNA/RNA [90] and RNA/RNA [91] kissing complexes, a GAAA tetraloop [92], and for the detection of novel structural elements in DNA triads, tetrads, and hexads [93–95].

9.5.2
Physicochemically-Induced Changes in H-Bond Geometry

Physicochemically-induced changes in H-bond geometry in nucleic acids and proteins have been traditionally investigated by indirect techniques such as monitoring changes in the chemical shift for both donor and acceptor groups [96, 97]. A direct approach to characterize the subtle changes in H-bond geometry when biomacromolecules are subject to different physicochemical conditions can be obtained by observing changes in the size of the *trans* H-bond couplings. Such changes have been recently reported for protein G [98], where small pressure-induced changes in backbone $|{}^{h3}J_{NC'}|$ values between –0.12 and 0.08 Hz have been observed and, using the established exponential relationship [47], translate into changes in the H-bond length between –0.037 and 0.045 Å. At high pressure (200 MPa), decreases in $|{}^{h3}J_{NC'}|$ values were observed for H-bonds located at the center of the α-helix, whereas increases in $|{}^{h3}J_{NC'}|$ were found for H-bonds positioned at the α-helix ends. These changes in $|{}^{h3}J_{NC'}|$ couplings were interpreted as a compaction of the α-helix ends and an increase in the helix-pitch at its center. Small reproducible decreases in backbone $|{}^{h3}J_{NC'}|$ values in ubiquitin have recently been observed when increasing the temperature from 5 to 65 °C, and these correspond to very small increases in H-bond lengths of $\leq$0.15 Å [F. Cordier, manuscript in preparation].

In DNA, an increase in temperature from 10 to 50 °C resulted in decreases in ${}^{h2}J_{NN}$ of ~ 0.7 Hz for Watson-Crick A–T and G–C base pairs [99]. The decreases in ${}^{h2}J_{NN}$ couplings were not due to hydrogen exchange, but most likely reflect an increase in the ensemble average of the N–N distance. Small decreases (<0.4 Hz) in ${}^{h2}J_{NN}$ coupling values due to isotopic substitution effects upon going from H_2O to D_2O solution have also been observed in DNA base pairs [99]. Isotopic effects have also been observed in sheared G–A mismatches [41]. The observed decreases in ${}^{h2}J_{NN}$ couplings are explained by the shorter N–D bond length compared to the N–H bond length and by a consequent increase of the N–N distance [100, 101].

9.5.3
Ligand-Induced Changes in H-Bond Geometry

Characterization of ligand-induced conformational changes in atomic detail has primarily been studied through comparison of high-resolution X-ray structures for macromolecules crystallized in both free and ligand-bound states. The possibility of monitoring such structural changes via changes in *trans* H-bond coupling values offers an efficient approach to quantitate subtle conformational changes in an environment free of crystal packing artifacts. This approach has recently been applied to monitor changes in H-bond geometry in the chicken c-Src SH3 domain upon binding the high-affinity class ligand RLP2 [79]. Fig. 9.4 A shows that SH3/RLP2 binding interaction induces changes in $|^{h3}J_{NC'}|$ values for a number of H-bonds and translates, according to the established exponential relationship [47], into changes in H-bond length ranging between 0.02 and 0.12 Å. The observed structural changes were rationalized as an induced fit mechanism in which H-bonds throughout the protein participate in a compensatory response to forces imparted at the protein-ligand interface (Fig. 9.4 B).

9.5.4 Protein Folding

NMR spectroscopy has been an invaluable tool for examining the process of biomacromolecular folding. Although NMR parameters such as chemical shift [102–104], NOEs [105–107], and hydrogen/deuterium exchange [108–110] have provided information on various biomacromolecular folding reactions, *trans* H-bond couplings offer a more quantitative approach to the characterization of individual H-bonds during the transition from the unfolded to the folded state. Thus, important details about the cooperativity and energetics of secondary and tertiary structure formation can be revealed. Recently an application was presented for the helix formation of an S-peptide analog of RNase A as a function of 2,2,2-trifluoroethanol (TFE) concentration [111]. In this study, the $|^{h3}J_{NC'}|$ couplings gave direct evidence of the closing of individual backbone N–H•••O=C H-bonds during the TFE-induced formation of the α-helix, and could be quantitatively compared to the predictions of coil-to-helix transition theories.

9.6 Conclusions

The most widely accepted application of *trans* H-bond couplings will be directed to the establishment of H-bond networks in the secondary and tertiary structures of biomacromolecules. Such observations offer valuable assignment information and should yield tight constraints in structure calculations. Recent studies which exploit the exponential dependence of the coupling constant on the H-bond distance in proteins have been able to characterize subtle structural rearrangements caused by ligand binding [79] or physicochemical inducement [98, 99]. In addition, H-bond couplings offer an approach to the quantitative characterization of macromolecular folding. These studies and continued efforts represent new steps toward the understanding of how H-bonds contribute to the stabilization and function of biological macromolecules. Further applications that can be envisaged from a quantitative analysis of the *trans* H-bond couplings include the following.

(1) The coupling constants could be used to characterize H-bond networks in the active centers of enzymes and ribozymes. In particular, the proposed low-barrier H-bonds or short strong H-bonds should give rise to very strong coupling values.

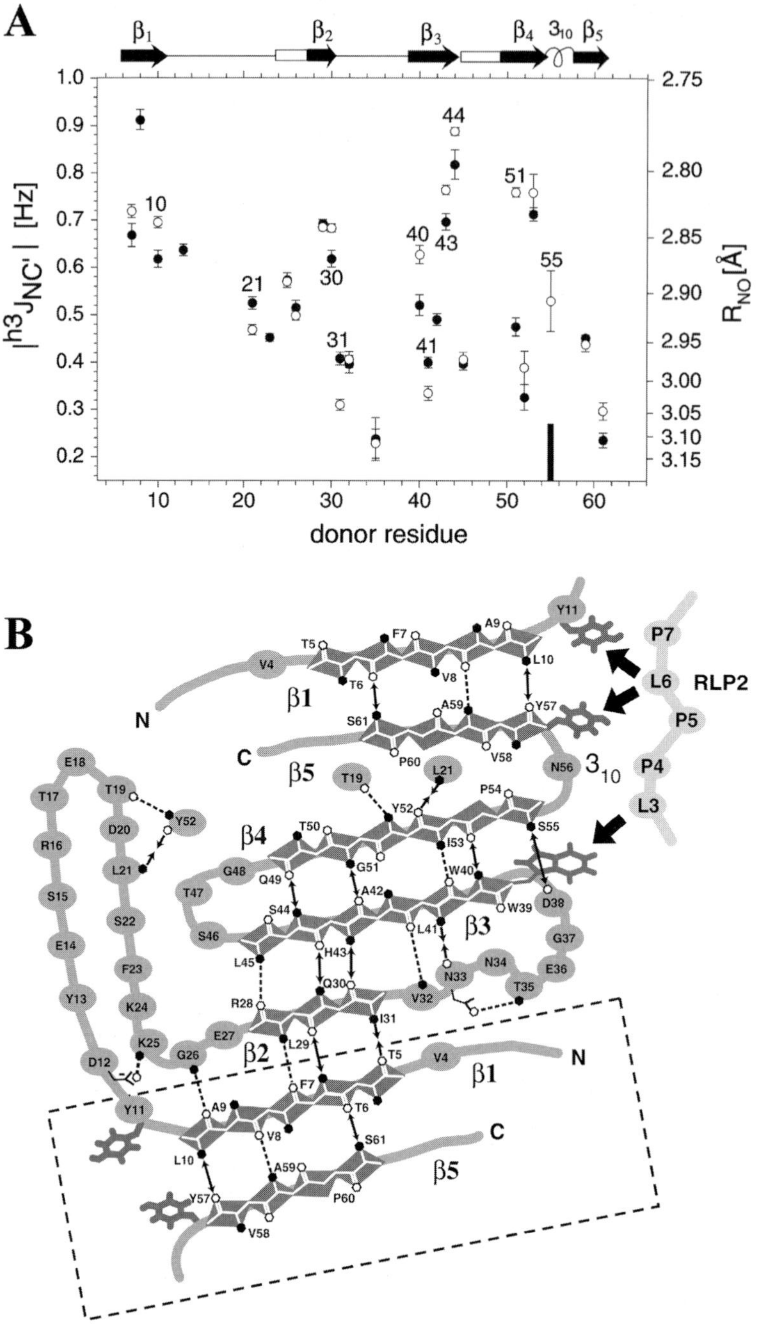
A
β1
β2
β3
β4
3₁₀
β5
|h3JNC'| [Hz]
RNO[Å]
donor residue
10
21
30
31
40
41
43
44
51
55
B
N
C
β1
β2
β3
β4
β5
3₁₀
RLP2
P7
L6
P5
P4
L3

(2) The good correlation with the proton chemical shift is particularly striking. A quantitative understanding of this phenomenon might lead to an understanding of the proton shift itself.
(3) At present, it is unclear how the size of the couplings relates to the energy of the hydrogen bond. If a simple correlation exists, the *J*-values could be used as a direct indicator of the H-bond stability.
(4) The correlation between the size of the H-bond couplings and the donor-acceptor distances is well established. A good understanding of the angular dependencies is lacking at present. Such an analysis will lead to a better understanding of the impact the geometry has on the *J*-values.

9.7 Acknowledgements

We thank Profs. Juli Feigon, Michael Barfield, Linda K. Nicholson and Masatsune Kainosho for many stimulating discussions as well as Chunyu Wang, James Masse and Dr. Robert D. Peterson for the preparation of the c-Src SH3 and DNA triplex samples. A. J. D. acknowledges funding by the Australian National Health and Medical Research Council C. J. Martin Fellowship (Regkey 987074) and by DFG grant RI 252/17-1. F. C. is a recipient of an A. v. Humboldt fellowship. This work was supported by SNF grants 31-61'757.00 and 31-43'091.95 to S.G.

Fig. 9.4 Changes in H-bond length induced by ligand binding. **A** Changes in $|^{h3}J_{NC'}|$ values and corresponding N•••O distances in the chicken c-Src SH3 domain induced by the binding of the ligand RLP2 peptide. The $|^{h3}J_{NC'}|$ values measured for free (open circles) and RLP2-bound (filled circles) SH3 are plotted as a function of donor group residue. Error bars reflect uncertainty in repeated measurements. Using the exponential relationship $R_{NO}=2.75-0.25\ \ln(-^{h3}J_{NC'})$ Å, a scale reflecting the corresponding distance between the N and O atoms of the N–H•••O=C H-bond is shown on the right side of the plot. Secondary structure elements are illustrated across the top. The black bar for residue 55 represents an upper limit for the possible value of $|^{h3}J_{NC'}|$, and is derived as previously described. **B** The two-dimensional schematic diagram illustrates the changes in H-bond length as a result of the propagation of strain across the 5-stranded β-sheet upon ligand binding. In order to visualize the H-bond network and protein ligand-interactions more clearly, the area within the broken box is repeated in the upper part of the figure. Donor amide and acceptor carbonyl groups are denoted as filled and open circles, respectively. Double-sided arrows indicate an increase, inward-pointing arrows a decrease, and dashed lines no change in H-bond length [within the error limits of (A)]. The two Leu-Pro segments of the RLP2 ligand intercalate between three aromatic side-chains (Y11, Y57 and W39) protruding from the SH3 surface. This intercalation wedges secondary structural elements within the SH3 domain apart, and the disruption is transmitted in a domino-like effect through networks of H-bonded peptide planes.

9.8 References

1 Huggins, M. L. *Thesis*, University of California, 1919.

2 Latimer, W. M.; Rodebush, W. H. *J. Am. Chem. Soc.* **1920**, *42*, 1419–1433.

3 Huggins, M. L. *J. Phys. Chem.* **1922**, *26*, 601–625.

4 Pauling, L. *The Nature of the Chemical Bond*, Cornell University Press, Ithaca, NY, **1960**.

5 Fersht, A. *Enzyme Structure and Mechanism*, W. H. Freeman, New York, **1985**.

6 Jeffrey, G. A.; Saenger, W. *Hydrogen bonding in biological structures*, Springer, New York, **1991**.

7 Niimura, N. *Curr. Opin. Struct. Biol.* **1999**, *9*, 602–608.

8 Becker, E. D. Hydrogen bonding in *Encyclopedia of Nuclear Magnetic Resonance.*, Grant, D. M., Harris, R. K. (eds.), John Wiley: New York, **1996**, pp 2409–2415.

9 Hvidt, A.; Nielsen, S. O. *Adv. Prot. Chem.* **1966**, *21*, 287–385.

10 Wagner, G. *Q. Rev. Biophys.* **1983**, *16*, 1–57.

11 Loh, S. N.; Markley, J. L. *Biochemistry* **1994**, *33*, 1029–1036.

12 Liwang, A. C.; Bax, A. *J. Am. Chem. Soc.* **1996**, *118*, 12864–12865.

13 Bowers, P. M.; Klevitt, R. *J. Am. Chem. Soc.* **2000**, *122*, 1030–1033.

14 Gunnarsson, G.; Wennerström, H.; Ega, W.; Forsen, S. *Chem. Phys. Lett.* **1976**, *38*, 96–99.

15 Altman, L. J.; Laungani, D.; Gunnarsson, G.; Wennerström, H.; Forsen, S. *J. Am. Chem. Soc.* **1978**, *100*, 8264–8266.

16 Shoup, R. R.; Miles, H. T.; Becker, E. D. *Biochem. Biophys. Res. Commun.* **1966**, *23*, 194–201.

17 Markowski, V.; Sullivan, G. R.; Roberts, J. D. *J. Am. Chem. Soc.* **1977**, *99*, 714–718.

18 Wagner, G.; Pardi, A.; Wüthrich, K. *J. Am. Chem. Soc.* **1983**, *105*, 5948–5949.

19 Asakawa, N.; Kuroki, S.; Kuroso, H.; Ando, I.; Shoji, A.; Ozaki, T. *J. Am. Chem. Soc.* **1992**, *114*, 3261–3265.

20 McDermott, A.; Ridenour, C. F. Proton chemical shift measurements in biological solids in *Encyclopedia of Nuclear Magnetic Resonance*, Grant, D. M., Harris, R. K. (eds.), **1996**, pp 3820–3825.

21 Tjandra, N.; Bax, A. *J. Am. Chem. Soc.* **1997**, *119*, 8076–8082.

22 Tessari, M.; Vis, H.; Boelens, R.; Kaptein, R.; Vuister, G. W. *J. Am. Chem. Soc.* **1997**, *119*, 8985–8990.

23 Takahashi, A.; Kuroki, S.; Ando, I.; Okaki, T.; Shoji, A. *J. Mol. Struct.* **1998**, *442*, 195–199.

24 Boyd, J.; Mal, T. K.; Soffe, N.; Campbell, I. D. *J. Magn. Reson.* **1997**, *124*, 61–71.

25 Liwang, A. C.; Bax, A. *J. Magn Reson.* **1997**, *127*, 54–64.

26 Juranic, N.; Ilich, P. K.; Macura, S. *J. Am. Chem. Soc.* **1995**, *117*, 405–410.

27 Juranic, N.; Likic, V. A.; Prendergast, F. G.; Macura, S. *J. Am. Chem. Soc.* **1996**, *118*, 7859–7860.

28 Dingley, A. J.; Grzesiek, S. *J. Am. Chem. Soc.* **1998**, *120*, 8293–8297.

29 Pervushin, K. *J. Biomol. NMR* **2001**, *20*, 275–285.

30 Pervushin, K.; Ono, A.; Fernandez, C.; Szyperski, T.; Kainosho, M.; Wüthrich, K. *Proc. Natl. Acad. Sci. USA* **1998**, *95*, 14147–14151.

31 Dingley, A. J.; Masse, J. E.; Peterson, R. D.; Barfield, M.; Feigon, J.; Grzesiek, S. *J. Am. Chem. Soc.* **1999**, *121*, 6019–6027.

32 Wöhnert, J.; Dingley, A. J.; Stoldt, M.; Gorlach, M.; Grzesiek, S.; Brown, L. R. *Nucleic Acids Res.* **1999**, *27*, 3104–3110.

33 Majumdar, A.; Kettani, A.; Skripkin, E. *J. Biomol. NMR* **1999**, *14*, 67–70.

34 Majumdar, A.; Kettani, A.; Skripkin, E.; Patel, D. J. *J. Biomol. NMR* **1999**, *15*, 207–211.

35 Dingley, A. J.; Masse, J. E.; Feigon, J.; Grzesiek, S. *J. Biomol. NMR* **2000**, *16*, 279–289.

36 Hennig, M.; Williamson, J. R. *Nucleic Acids Res.* **2000**, *28*, 1585–1593.

37 Liu, A.; Majumdar, A.; Hu, W.; Kettani, A.; Skripkin, E.; Patel, D. J. *J. Am. Chem. Soc.* **2000**, *122*, 3206–3210.

38 Luy, B.; Marino, J. P. *J. Am. Chem. Soc.* **2000**, *122*, 8095–8096.

39 Pervushin, K.; Fernandez, C.; Riek, R.; Ono, A.; Kainosho, M.; Wüthrich, K. *J. Biomol. NMR* **2000**, *16*, 39–46.

40 Yan, X.; Kong, X.; Xia, Y.; Sze, K.H.; Zhu, G. *J. Magn. Reson.* **2000**, *147*, 357–360.
41 Majumdar, A.; Kettani, A.; Skripkin, E.; Patel, D.J. *J. Biomol. NMR* **2001**, *19*, 103–113.
42 Blake, P.R.; Lee, B.; Summers, M.F.; Adams, M.W.; Park, J.B.; Zhou, Z.H.; Bax, A. *J. Biomol. NMR* **1992**, *2*, 527–533.
43 Blake, P.R.; Park, J.B.; Adams, M.W.; Summers, M.F. *J. Am. Chem. Soc.* **1992**, *114*, 4933.
44 Cordier, F.; Grzesiek, S. *J. Am. Chem. Soc.* **1999**, *121*, 1601–1602.
45 Cordier, F.; Rogowski, M.; Grzesiek, S.; Bax, A. *J. Magn. Reson.* **1999**, *140*, 510–512.
46 Cornilescu, G.; Hu, J.-S.; Bax, A. *J. Am. Chem. Soc.* **1999**, *121*, 2949–2950.
47 Cornilescu, G.; Ramirez, B.E.; Frank, M.K.; Clore, G.M.; Gronenborn, A. M.; Bax, A. *J. Am. Chem. Soc.* **1999**, *121*, 6275–6279.
48 Hennig, M.; Geierstanger, B. *J. Am. Chem. Soc.* **1999**, *121*, 5123–5126.
49 Wang, Y.X.; Jacob, J.; Cordier, F.; Wingfield, P.; Stahl, S. J.; Lee-Huang, S.; Torchia, D.; Grzesiek, S.; Bax, A. *J. Biomol. NMR* **1999**, *14*, 181–184.
50 Liu, A.; Hu, W.; Majumdar, A.; Rosen, M.K.; Patel, D. J. *J. Biomol. NMR* **2000**, *17*, 305–310.
51 Liu, A.; Hu, W.; Majumdar, A.; Rosen, M.K.; Patel, D.J. *J. Biomol. NMR* **2000**, *17*, 79–82.
52 Meissner, A.; Sørensen, O.W. *J. Magn Reson.* **2000**, *143*, 431–434.
53 Meissner, A.; Sørensen, O.W. *J. Magn Reson.* **2000**, *143*, 387–390.
54 Mishima, M.; Hatanaka, M.; Yokoyama, S.; Ikegami, T.; Wälchli, M.; Ito, Y.; Shirakawa, M. *J. Am. Chem. Soc.* **2000**, *122*, 5883–5884.
55 Liu, A.; Majumdar, A.; Jiang, F.; Chernichenko, N.; Skripkin, E.; Patel, D.J. *J. Am. Chem. Soc.* **2000**, *122*, 11226–11227.
56 Löhr, F.; Mayhew, S.G.; Rüterjans, H. *J. Am. Chem. Soc.* **2000**, *122*, 9289–9295.
57 Crabtree, R.; Siegbahn, P.; Eisenstein, O.; Rheingold, A.; Koetzle, T. *Acc. Chem. Res.* **1996**, *29*, 348–354.
58 Shenderovich, I.G.; Smirnov, S.N.; Denisov, G.S.; Gindin, V.A.; Golubev, N.S.; Dunger, A.; Reibke, R.; Kirpekar, S.; Malkina, O.L.; Limbach, H.-H. *Ber. Bunsenges. Phys. Chem.* **1998**, *102*, 422–428.
59 Kwon, O.; Danishefsky, S. *J. Am. Chem. Soc.* **1998**, *120*, 1588–1599.
60 Golubev, N.S.; Shenderovich, I.G.; Smirnov, S.N.; Denisov, G.S.; Limbach, H.-H. *Chem. Eur. J.* **1999**, *5*, 492–497.
61 Dannenberg, J.J.; Haskamp, L.; Masunov, A. *J. Phys. Chem.* **1999**, *103*, 7083–7086.
62 Scheurer, C.; Brüschweiler, R. *J. Am. Chem. Soc.* **1999**, *121*, 8661–8662.
63 Bagno, A. *Chem. Eur. J.* **2000**, *6*, 2925–2930.
64 Benedict, H.; Shenderovich, I.G.; Malkina, O.L.; Malkin, V.G.; Denisov, G.S.; Golubev, N.S.; Limbach, H.-H. *J. Am. Chem. Soc.* **2000**, *122*, 1979–1988.
65 Perera, S.; Bartlett, R.J. *J. Am. Chem. Soc.* **2000**, *122*, 1231–1232.
66 Del Bene, J.E.; Bartlett, R.J. *J. Am. Chem. Soc.* **2000**, *122*, 10480–10481.
67 Del Bene, J.E.; Jordan, M.J.T. *J. Am. Chem. Soc.* **2000**, *122*, 4794–4797.
68 Del Bene, J.E. *J. Am. Chem. Soc.* **2000**, *122*, 3560–3561.
69 Del Bene, J.E.; Perera, S.A.; Bartlett, R.J.; Alkorta, I.; Elguero, J. *J. Phys. Chem.* **2000**, *A104*, 7165–7166.
70 Pecul, M.; Leszczynski, J.; Sadlej, J. *J. Phys. Chem. A* **2000**, *104*, 8105–8113.
71 Case, D.A. *Curr. Opin. Struct. Biol.* **2000**, *10*, 197–203.
72 Ghanty, T.K.; Staroverov, V.N.; Koren, P.R.; Davidson, E.R. *J. Am. Chem. Soc.* **2000**, *122*, 1210–1214.
73 Arnold, W.D.; Oldfield, E. *J. Am. Chem. Soc.* **2000**, *122*, 12835–12841.
74 Barfield, M.; Dingley, A.J.; Feigon, J.; Grzesiek, S. *J. Am. Chem. Soc.* **2001**, *121*, 4014–4022.
75 Dingley, A.J.; Cordier, F.; Grzesiek, S. *Concept. Magnetic Res.* **2001**, *13*, 103–127.
76 Pervushin, K.; Riek, R.; Wider, G.; Wüthrich, K. *Proc. Natl. Acad. Sci. USA.* **1997**, *94*, 12366–12371.
77 Ippel, J.H.; Wijmenga, S.S.; de Jong, R.; Heus, H.A.; Hilbers, C.W.; de Vroom, E.; van der Marcel, G.A.; van Boom, J.H. *Magn. Reson. Chem.* **1996**, *34*, S156–S176.
78 Grzesiek, S.; Cordier, F.; Dingley, A.J. *Methods Enzymol.* **2001**, *338*, 111–133.
79 Cordier, F.; Wang, C.; Grzesiek, S.; Nicholson, L.K. *J. Mol. Biol.* **2000**, *304*, 497–505.
80 Kontaxis, G.; Clore, G.M.; Bax, A. *J. Magn. Reson.* **2000**, *143*, 184–196.

81 Liu, A.; Hu, W.; Qamar, S.; Majumdar, A. *J. Biomol. NMR* **2000**, *17*, 55–61.

82 Abragam, A. *The principles of nuclear magnetism*, Clarendon Press, Oxford, 1961.

83 Wishart, D. S.; Sykes, B. D.; Richards, F. M. *J. Mol. Biol.* **1991**, *222*, 311–333.

84 Kuntz, I. D.; Kosen, P. A.; Craig, E. C. *J. Am. Chem. Soc.* **1991**, *113*, 1406–1408.

85 Zhou, N. E.; Zhu, B.-Y.; Sykes, B. D.; Hodges, R. S. *J. Am. Chem. Soc.* **1992**, *114*, 4320–4326.

86 Riek, R. *J. Magn. Reson.* **2001**, *149*, 149–153.

87 Baker, E. N.; Hubbard, R. E. *Prog. Biophys. Molec. Biol.* **1984**, *44*, 97–179.

88 Czernek, J.; Brüschweiler, R. *J. Am. Chem. Soc.* **2001**, *123*, 11079–11080.

89 Bonvin, A. M. J. J.; Houben, K.; Guenneugues, M.; Kaptein, R.; Boelens, R. *J. Biomol. NMR* **2001** , *21*, 221–233.

90 Collin, D.; van Heijenoort, C.; Boiziau, C.; Toulme, J. J.; Guittet, E. *Nucleic Acids Res.* **2000**, *28*, 3386–3391.

91 Kim, C. H.; Tinoco, I., Jr. *Proc. Natl. Acad. Sci. USA* **2000**, *97*, 9396–9401.

92 Rüdisser, S.; Tinoco, I., Jr. *J. Mol. Biol.* **2000**, *295*, 1211–1223.

93 Kettani, A.; Basu, G.; Gorin, A.; Majumdar, A.; Skripkin, E.; Patel, D. J. *J. Mol. Biol.* **2000**, *301*, 129–146.

94 Kettani, A.; Gorin, A.; Majumdar, A.; Hermann, T.; Skripkin, E.; Zhao, H.; Jones, R.; Patel, D. J. *J. Mol. Biol.* **2000**, *297*, 627–644.

95 Kuryavyi, V.; Majumdar, A.; Shallop, A.; Chernichenko, N.; Skripkin, E.; Jones, R.; Patel, D. J. *J. Mol. Biol.* **2001**, *310*, 181–194.

96 Goswami, B.; Gaffney, B. L.; Jones, R. A. *J. Am. Chem. Soc.* **1993**, *115*, 3832–3833.

97 Gaffney, B. L.; Goswami, B.; Jones, R. A. *J. Am. Chem. Soc.* **1993**, *115*, 12607–12608.

98 Li, H.; Yamada, H.; Akasaka, K.; Gronenborn, A. M. *J. Biomol. NMR* **2000**, *18*, 207–216.

99 Kojima, C.; Ono, A.; Kainosho, M. *J. Biomol. NMR* **2000**, *18*, 269–277.

100 Hibbert, F.; Emsley, J. *Adv. Phys. Org. Chem.* **1990**, *26*, 255–379.

101 Benedict, H.; Limbach, H.-H.; Wehlan, M.; Fehlhammer, W.-P.; Golubev, N. S.; Janoschek, R. *J. Am. Chem. Soc.* **1998**, *120*, 2939–2950.

102 Epstein, H. F.; Schechter, A. N.; Chen, R. F.; Anfinsen, C. B. *J. Mol. Biol.* **1971**, *60*, 499–508.

103 Kiefhaber, T.; Labhardt, A.; Baldwin, R. *Nature* **1995**, *375*, 513–515.

104 Balbach, J.; Forge, V.; Lau, W. S.; van Nuland, N. A.; Brew, K.; Dobson, C. M. *Science* **1996**, *274*, 1161–1163.

105 Neri, D.; Billeter, M.; Wider, G.; Wüthrich, K. *Science* **1992**, *257*, 1559–1563.

106 Balbach, J.; Forge, V.; Lau, W. S.; Jones, J. A.; van Nuland, N. A.; Dobson, C. M. *Proc. Natl. Acad. Sci. U.S.A* **1997**, *94*, 7182–7185.

107 Mok, Y. K.; Kay, C. M.; Kay, L. E.; Forman-Kay, J. *J. Mol. Biol.* **1999**, *289*, 619–638.

108 Roder, H.; Wüthrich, K. *Proteins* **1986**, *1*, 34–42.

109 Roder, H.; Elove, G. A.; Englander, S. W. *Nature* **1988**, *335*, 700–704.

110 Udgaonkar, J. B.; Baldwin, R. L. *Nature* **1988**, *335*, 694–699.

111 Jaravine, V. A.; Alexandrescu, A. T.; Grzesiek, S. *Protein Sci.* **2001**, *10*, 943–950.

10
TROSY: Transverse Relaxation-Optimized Spectroscopy

Roland Riek

10.1
Introduction

Nuclear magnetic resonance spectroscopy (NMR) is one of the principal experimental techniques in structural biology, with abilities to determine atomic resolution structures as well as to investigate dynamic features and intermolecular interactions of biological macromolecules at near-physiological solution conditions. However, conventional NMR spectra of large molecular structures in solution typically display broad lines due to efficient transverse relaxation. It is therefore of keen interest to search for novel techniques which reduce transverse relaxation rates, opening avenues for NMR spectroscopy with larger structures [1]. The presently reviewed transverse relaxation-optimized spectroscopy (TROSY) [2, 3] now enables the recording of high-resolution NMR spectra of macromolecules and multimolecular assemblies with masses up to 100 kDa and beyond (Fig. 10.1). Immediate applications include NMR studies of structure-activity relationship (SAR by NMR [4]) with large receptor molecules, structure determination of membrane proteins reconstituted in soluble micelles or lipid vesicles [5, 6], and quite generally NMR studies of individual macromolecular components in supramolecular structures, in particular the mapping of interactions at protein-protein or protein-nucleic acid interfaces [7, 8]. In this review, the concept of TROSY is described together with experimental applications.

10.2
The Concept of TROSY

Transverse nuclear spin relaxation is dominated by dipole-dipole (DD) coupling and chemical shift anisotropy (CSA). TROSY exploits constructive interference between these two relaxation mechanisms to reduce the transverse relaxation rates during the frequency labeling period and acquisition. This interference between DD coupling and CSA has long been known and is termed cross-correlated relaxation between DD coupling and CSA [9]. It is the aim of this section to describe the concept of TROSY in a physical picture, and this is followed in a second section by TROSY requirements in pulse sequences. A detailed description of the principle of TROSY using relaxation theory is given in the Appendix.

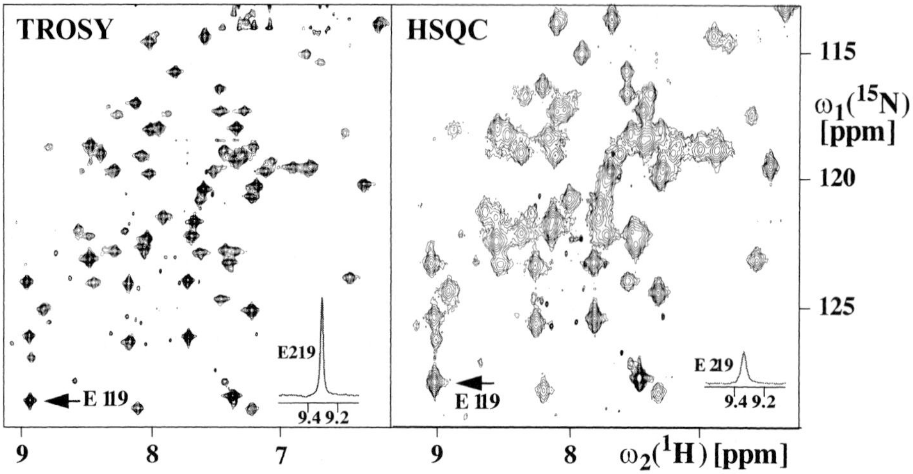

Fig. 10.1 TROSY-type and conventional $[^{15}N,^{1}H]$-correlation spectra of the $^{2}H,^{15}N$- labeled 110 kDa protein DHNA. $[^{15}N,^{1}H]$-correlation experiments afford a "fingerprint" of the protein, which is highly sensitive to changes in the protein environment and thus presents a many-parameter NMR probe for studies of intermolecular interactions and ensuing conformational changes. In the inserts cross-sections along $\omega_2(^{1}H)$ are shown for the cross peak of residue 219. The spectra were measured at 750 MHz ^{1}H frequency and at 20 °C.

10.2.1
A Physical Picture of TROSY

The relaxation theory used in the Appendix to describe the principle of TROSY clearly tells us what to expect, but it is always a little more satisfying if one can obtain a simple physical picture of what is happening. We consider a system of two isolated scalar coupled spins of magnitude ½, ^{1}H (I) and ^{15}N (S), with a scalar coupling constant J_{HN}. Transverse relaxation of this spin system is dominated by the DD coupling between spins ^{1}H and ^{15}N and by the CSA of each individual spin. The relaxation rates of the individual multiplet components of spin ^{15}N are now discussed assuming an axially symmetric ^{15}N CSA tensor with the axial principal component parallel to the ^{15}N-^{1}H vector as shown in Fig. 10.2.

The CSA of ^{15}N induces a motion-influenced time-dependent magnetic field $B^{CSA}(t)$ on spin ^{15}N [10].

$$B^{CSA}(t) \propto \gamma_N B_0 \Delta\sigma_N [3\cos^2\theta(t) - 1] \tag{1}$$

where $\theta(t)$ is the angle between the magnetic field B_0 and the axial principal component of the CSA tensor. The angle $\theta(t)$ and concomitantly $B^{CSA}(t)$ are modulated with time, since the molecule tumbles in the solution due to Brownian motion. $B^{CSA}(t)$ is further dependent on the strength of the magnetic field B_0. This motion-influenced magnetic field $B^{CSA}(t)$ couples to the precessional motion of the nuclear spin and leads to transverse relaxation and line broadening. Similarly, the DD coupling between ^{1}H and ^{15}N spins induces a motion-influenced time-dependent magnetic field [10].

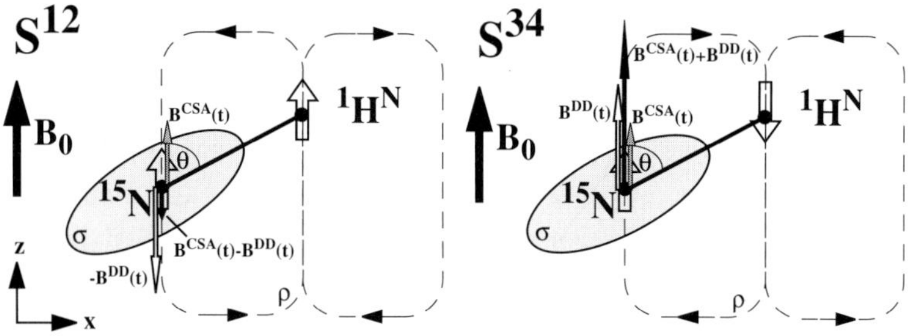

Fig. 10.2 Interactions of the local magnetic fields $B^{DD}(t)$ with $B^{CSA}(t)$ (see text). B_0 is the static magnetic field. The CSA tensor σ is displayed by an ellipse. θ designates the angle between the ^{15}N-^{1}H bond and B_0. ρ designates the dipole–dipole interaction between the spin ^{1}H and ^{15}N.

$$B^{DD}(t) \propto \gamma_H \gamma_N \hbar / r_{HN}^3 [3\cos^2\theta(t) - 1]. \quad (2)$$

It is the modulation of the motion-influenced $B^{DD}(t)$ that leads to transverse relaxation and line broadening. $B^{DD}(t)$ is B_0 independent, but the sign of $B^{DD}(t)$ depends on whether the two spins are parallel or antiparallel as indicated in Fig. 10.2.

Both time-dependent magnetic fields $B^{CSA}(t)$ and $B^{DD}(t)$ influence simultaneously the spin ^{15}N and show the same angular dependence and concomitantly the same time dependence [11]. Thus, depending on whether the ^{1}H spin is parallel or antiparallel to spin ^{15}N, the two fields either add or subtract, as demonstrated in Fig. 10.2. For the multiplet component S^{12}, $B^{DD}(t)$ compensates $B^{CSA}(t)$ and yields the favorable narrow multiplet component shown in Fig. 10.3. Correspondingly, the multiplet component S^{34} does relax fast since the two time-dependent magnetic fields $B^{DD}(t)$ and $B^{CSA}(t)$ are added together. An optimal compensation of $B^{CSA}(t)$ with $B^{DD}(t)$ and thus minimal transverse relaxation rates can be achieved adjusting the size of $B^{CSA}(t)$ by choosing an optimal magnetic field B_0 of about 950–1050 MHz (see Appendix). The principle of TROSY is the selection of the favorable multiplet component S^{12}. The half difference between the two relaxations of S^{34} and S^{12} is termed cross-correlated relaxation between DD coupling and CSA. The described effect between DD coupling and CSA is termed cross-correlation between CSA and DD coupling, because the two mechansims seem to interfere.

In conventional NMR experiments the multiplet pattern (Fig. 10.3) has routinely been collapsed by decoupling on ^{1}H. The ^{1}H decoupling flips the spin ^{1}H and concomitantly the sign of $B^{DD}(t)$. Thus, during ^{15}N-evolution each ^{15}N spin is perturbed during half of the evolution time by $B^{CSA}(t)+B^{DD}(t)$ and during the other half of the evolution time by $B^{CSA}(t)-B^{DD}(t)$, which leads to a less favorable relaxation when compared with the relaxation of the component S^{12} selected by TROSY.

For completeness, transverse relaxation-optimization is not limited to the constructive interference between DD coupling and CSA. Pervushin *et al.* [12] introduced the ZQ-TROSY, which is based on cross-correlated relaxation between ^{1}H CSA and ^{15}N CSA during the multiple quantum ^{15}N-evolution period.

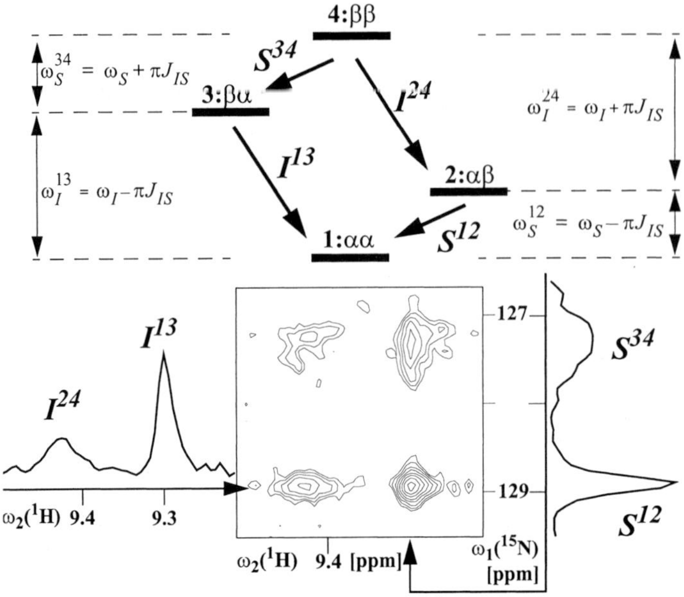

Fig. 10.3 Energy level diagram of a two spin ½ system *I* and *S* showing the identification of components of the 2D multiplet expressed via single transition basis operators, I^{13}, I^{24}, S^{12}, S^{34}. Contour plots of a ^{15}N-^1H backbone moiety with cross-sections of a conventional [^{15}N,^1H] correlation spectrum without decoupling during evolutions. The spectra were measured with ^{15}N,^2H-labeled 110 kDa DHNA at 750 MHz ^1H frequency and at 20 °C.

10.2.2
Technical Aspects of TROSY

Technically, the TROSY approach is based on the following: in heteronuclear two-spin systems, such as ^{15}N-^1H and aromatic ^{13}C-^1H, the NMR signal of each nucleus is split into two components by the scalar spin-spin coupling. In two-dimensional correlation experiments one therefore observes a four-line fine structure (Fig. 10.3). With the advent of modern multidimensional NMR, this four-line pattern has routinely been collapsed into a single, centrally located line by application of broad-band decoupling during the evolution and detection periods, with the expectation of obtaining a simplified spectrum and improved sensitivity. However, as has long been known (see Ref. [2] for a literature survey), that individual multiplet components have different transverse relaxation times and hence different line widths, which are mixed by the aforementioned decoupling (Fig. 10.3). In the TROSY technique, the multiplet structure is not decoupled, and only the narrowest, most slowly relaxing line of each multiplet is retained. Experimentally, the most slowly relaxing multiplet component is selected using pulse sequence elements, as for example the ST2-PT element in the [^{15}N,^1H]-TROSY experiment of Fig. 10.4. Mixing the multiplet components during evolution periods is prevented by the lack of any radiofrequency pulses on the attached spin. For example, during the t_1-evolution of the

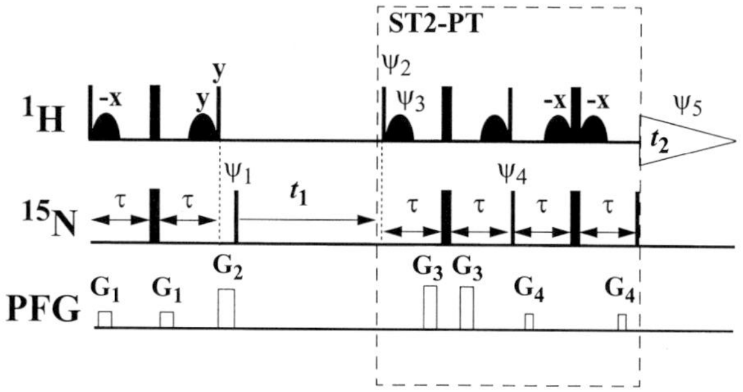

Fig. 10.4 Experimental scheme for the 2D [^{15}N,^{1}H]-TROSY using single transition to single transition polarization transfer (box labeled ST2-PT). On the lines marked ^{1}H and ^{15}N, narrow and wide bars stand for nonselective 90° and 180° radio-frequency pulses, respectively. The delay $\tau = 2.7$ ms (see text). The line marked PFG indicates the pulsed magnetic field gradients applied along the *z*-axis: G_1, amplitude 30 G/cm, duration 1 ms; G_2, 40 G/cm, 1 ms; G_3, 40 G/cm, 1 ms; G_4; 48 G/cm, 1 ms. The following two-step phase cycling scheme was used: $\Psi_1 = \{y, -x\}$, $\Psi_2 = \{-y\}$, $\Psi_3 = \{y\}$, $\Psi_4 = \{-y\}$, $\Psi_5 = \{y, -x\}$; *x* on all other pulses. To obtain a complex interferogram a second FID is recorded for each t_1 delay, with $\Psi_1 = \{y, x\}$, $\Psi_2 = \{y\}$, $\Psi_3 = \{-y\}$, $\Psi_4 = \{y\}$. The use of ST2-PT thus results in a 2D [^{15}N,^{1}H]-correlation spectrum that contains only the most slowly relaxing component of the 2D ^{15}N-^{1}H multiplet. The data are processed as described by Kay et al. [44] in an echo/antiecho manner. Water saturation is minimized by keeping the water magnetization along the *z*-axis during the entire experiment, which is achieved by the application of the water-selective 90° rf pulses indicated by curved shapes on the line ^{1}H. It was reported that on some NMR instruments the phase cycle mentioned above does select the desired multiplet component. On these instruments, the replacements of Ψ_1 with $\Psi_1 = \{y, x\}$ for the first FID and $\Psi_1 = \{y, -x\}$ for the second FID select the desired multiplet component.

[^{15}N,^{1}H]-TROSY (Fig. 10.4), ^{15}N evolves and no pulses on ^{1}H are applied (for completeness, radio-frequency pulses on ^{15}N or a short 360° pulse on ^{1}H during the ^{15}N-evolution would not destroy the TROSY-effect).

As an example, we will discuss in the following the pulse sequence of the [^{15}N,^{1}H]-TROSY experiment (Fig. 10.4) in detail. The evolution of the density operator [2, 13–16] can be schematically represented as:

$$\frac{1}{2}(uI_z + vS_z) \rightarrow \frac{u+v}{2} S_{12}^{+} \exp(-i\omega_S^{12} t_1) \rightarrow \frac{u+v}{2} I_{13}^{-} \exp(-i\omega_s^{12} t_1) \tag{3}$$

and

$$\frac{1}{2}(uI_z + vS_z) \rightarrow \frac{u+v}{2} S_{34}^{+} \exp(-i\omega_S^{34} t_1) \rightarrow \frac{u+v}{2} I_{24}^{-} \exp(-i\omega_s^{34} t_1) \tag{4}$$

The first arrow of Eqs. (3) and (4) designates coherence transfer starting from the steady-state magnetizations ^{1}H (*I*) and ^{15}N (*S*) to ^{15}N followed by the ^{15}N chemical shift evolution during the delay t_1, and the second arrow represents the coherence transfer from ^{15}N to ^{1}H using a train of pulses called ST2-PT element (single transition to single transition polarization transfer). The positive constant factors *u* and *v* reflect the relative magnitude of the steady- state ^{1}H and ^{15}N magnetization. Only in TROSY-experiments is it possible to merge synergistically the ^{1}H and ^{15}N steady-state magnetizations (Eqs. (3) and (4)) [13, 17, 18], gaining a signal increase of 10% for ^{15}N-^{1}H moieties [13] and 50–100% for ^{13}C-^{1}H aromatic moieties [17].

When both pathways indicated in Eqs. (3) and (4) are retained, two diagonally shifted signals per ^{15}N-^{1}H moiety are observed representing two out of the four ^{15}N-^{1}H multiplet components in the resulting [^{15}N,^{1}H]-correlation spectrum. The undesired polarization transfer pathway, $S_{34}^{\mp} \rightarrow I_{24}^{\mp}$, is suppressed by two-step cycling of the phases Ψ_1 and Ψ_5 (Fig. 10.4); it has been noticed that opposite phase shifts during phase cycling have to be implemented on Varian instruments when compared to Bruker, since phase shifts have been implemented differently by the spectrometer manufacturers). The remaining anti-echo polarization transfer pathway, $S_{12}^{+} \rightarrow I_{13}^{-}$, connects a single transition of spin S with a single transition of spin I, and in alternate scans with inversion of the phases Ψ_2 and Ψ_4, the corresponding echo transfer, $S_{12}^{+} \rightarrow I_{13}^{-}$, is recorded. The sensitivity loss of a factor 2 due to the use of only one out of the four multiplet components is in practice readily recovered when working with large structures. Rance et al. [14] have demonstrated, on the basis of experimental data and theoretical calculations, that relaxation-induced imbalances between the coherence transfer pathways utilized in the ST2-PT element (Fig. 10.4) give rise to additional unanticipated signals at the positions of the broader multiplet components. The intensities of these non-perfectly suppressed multiplet components are only several percent of the intensities of the TROSY component, but it would still be desirable to suppress them. Recently, the Clean TROSY [19] was proposed, which largely suppresses these artifacts by modifying the ST2-PT element of the [^{15}N,^{1}H]-TROSY.

10.3 TROSY Applications

TROSY is especially powerful for ^{15}N-^{1}H backbone moieties, ^{13}C-^{1}H aromatic moieties of proteins [2, 17], and ^{15}N-^{1}H and ^{13}C-^{1}H base moieties of oligonucleotides [18, 20–23]. Thus, applications of multidimensional TROSY-based [^{15}N,^{1}H]- and [^{13}C,^{1}H]-experiments are now discussed and compared with the conventional experiments. Some special characteristics of the TROSY-based experiments are mentioned, but for a detailed description of the experiments we refer to the referenced literature.

10.3.1 [^{15}N,^{1}H]-TROSY

Figure 10.1 shows a two-dimensional [^{15}N,^{1}H]-TROSY correlation spectrum of the ^{15}N,^{2}H- labeled 110 kDa homo-octameric protein 7,8-dihydroneopterin aldolase from *Staphylococcus aureus* (DHNA) measured with the pulse sequence of Fig. 10.4 [13]. The gain in spectral resolution and sensitivity is readily apparent from comparison with the corresponding conventional experiment. The optimal sensitivity is achieved by adjusting the polarization transfer τ in Fig. 10.4 (3 ms $<2\tau<$5.4 ms [3]). For an optimal suppression of the non-TROSY components, the so-called Clean TROSY might be used [19]. Similar signal and spectral resolution enhancements are achieved for ^{15}N,^{2}H-labeled or ^{13}C,^{15}N,^{2}H-

Tab. 10.1 Transverse $^1H^N$ and ^{15}N relaxation rates predicted for a 23 kDa protein at 750 MHz in [^{15}N,^1H]-TROSY and in [^{15}N,^1H]-HSQC [29] [a)]

^{15}N-^1H moiety	Transverse relaxation of ^{15}N [s^{-1}]		Transverse relaxation of $^1H^N$ [s^{-1}]	
	TROSY	HSQC	TROSY	HSQC
Isolated	3.0	20.9	3.2	20.3
β-sheet of a ^{13}C,^{15}N labeled protein [b)]	10.6	28.5	41.1	58.2
α-helix of a ^{13}C,^{15}N labeled protein [c)]	8.7	26.6	31.5	48.6
β-sheet of a ^{13}C,^{15}N,^2H labeled protein [d)]	3.7	21.6	6.3	23.5
α-helix of a ^{13}C,^{15}N,^2H labeled protein [e)]	5.0	22.9	13.2	30.3

a) ^{15}N and $^1H^N$ relaxation rates were calculated using Eq. (5) as is described in the text. The values listed for the [^{15}N,^1H]-HSQC are the average relaxation rates of both components of the ^{15}N and the $^1H^N$ doublets. The following parameters were used: $r_{HN}=1.04$ Å, $\Delta\sigma_N=-155$ ppm, $\Delta\sigma_H=-15$ ppm, $\vartheta_{p\delta_N}=15°$, $\vartheta_{p\delta_H}=10°$ and $\tau_c=15$ ns.

b) Remote protons considered are $^1H^N(i-1)$, $^1H^N(i+1)$, $^1H^N(j)$, $^1H^\alpha(i)$, $^1H^\alpha(i-1)$, $^1H^\alpha(j)$, $^1H^\beta(i)$, and $^1H^\beta(i-1)$ at distances of 4.3, 4.3, 3.3, 2.8, 2.2, 3.2, 2.5, and 3.2 Å, respectively, which are typical for an antiparallel β-sheet [i is the observed residue, (i–1) and (i+1) the sequential neighbours, j indicates a long-range contact across the β-sheet].

c) Remote protons considered are $^1H^N(i-1)$, $^1H^N(i+1)$, $^1H^N(i-2)$, $H^N(i+2)$, $^1H^\alpha(i)$, $^1H^\alpha(i-1)$, $^1H^\alpha(i-2)$, $^1H^\alpha(i-3)$, $^1H^\alpha(i-4)$, and $^1H^\beta(i)$ at distances of 2.8, 2.8, 4.2, 4.2, 2.6, 3.5, 4.4, 3.4, 4.2, and 2.5 Å.

d) Remote protons considered are $^1H^N(i-1)$, $^1H^N(i+1)$, $^1H^N(j)$ at distances of 4.3, 4.3, and 3.3 Å.

e) Remote protons are $^1H^N(i-1)$, $^1H^N(i+1)$, $^1H^N(-2)$, $H^N(i+2)$ at distances 2.8, 2.8, 4.2 and 4.2 Å.

labeled proteins and protein complexes with molecular weights of 20–70 kDa [1, 7, 25, 26] and at magnetic fields between 500 and 900 MHz. The enhancements in resolution and sensitivity of nondeuterated large proteins are less pronounced because of the transverse relaxation induced by the DD coupling with remote hydrogens (see Tab. 10.1). For ^{15}N-labeled and ^{15}N,^{13}C-labeled proteins with a molecular weight <15 kDa the [^{15}N,^1H]-TROSY might be somewhat less sensitive than the conventional [^{15}N,^1H]-correlation experiment, but TROSY has still the advantage of higher resolution. Using the pulse scheme of Fig. 10.4, the NH_2-moieties of Gln and Asn are suppressed (Fig. 10.1). They can be retained using $\tau=1.7$ ms, with the drawback that unwanted multiplet components are no longer perfectly suppressed [27].

10.3.2
[^{15}N,^1H]-TROSY – Triple Resonance Spectroscopy for Sequential Assignment

Obtaining good quality NMR spectra of large molecular structures is an essential first step. However, in addition, the assignment of the chemical shifts to individual nuclei is indispensable as a basis for detailed structural studies. Sequential assignment is achieved with "triple resonance experiments" [28] (chapter 4). In these experiments, magnetization is transferred between $^1H^N$, ^{15}N and ^{13}C, and they are routinely applied with molecular sizes up to about 30 kDa. The use of the [^{15}N,^1H]-TROSY in triple resonance experi-

Tab. 10.2 TROSY-triple resonance experiments for proteins

Experiment	*Reference*
For $^2H,^{13}C,^{15}N$-labeled proteins:	
3D TROSY-HNCA	Salzmann et al., 1999 [45]
3D TROSY-HNCO	Loria et al., 1999 [46]
3D TROSY-HNCACB [a)]	Salzmann et al., 1999 [47]
3D TROSY-CT-HNCA [a)]	Salzmann et al., 1999 [48]
3D TROSY-HN(CA)CB [b)]	Yang and Kay, 1999 [49]
3D TROSY-HN(CO)CA [a)]	Salzmann et al., 1999 [47]
3D TROSY-HN(CO)CACB [a)]	Salzmann et al., 1999 [47]
3D TROSY-HN(COCA)CB [b)]	Yang and Kay, 1999 [49]
3D TROSY-HN(CA)CO	Loria et al., 1999 [46]
3D MP-CT-HNCA/	Permi and Annila, 2001 [50]
sequential HNCA [c)]	Meissner and Sorensen, 2001 [51]
3D HNCAN [d)]	Löhr et al., 2000 [52]
4D TROSY-HNCACO [b)]	Yang and Kay, 1999 [49]
4D TROSY-HNCOCA [b)]	Yang and Kay, 1999 [49]
4D TROSY-HNCO$_{i-1}$CA [b)]	Konrat et al., 1999 [53]
3D SEA-TROSY [e)]	Pellecchia et. al., 2001 [54]
3D TROSY-XY-HNCA [f)]	Pervushin et. al., 2001 [55]
for $^{13}C,^{15}N$-labeled proteins:	
3D TROSY-HNCA	Eletsky et al., 2001 [56]
3D TROSY-HNCACB	Eletsky et al., 2001 [56]

a) Improved sensitivity can be gained by concatenating the ST2-PT element with the ^{15}N constant-time period as demonstrated for the TROSY-HNCA experiment [45, 46].

b) A different TROSY-component selection is used [57] with similar properties to those of the ST2-PT element (Fig. 10.4 [13]). The author prefers the ST2-PT element.

c) A TROSY-HNCA experiment detecting only the sequential cross peaks.

d) For sequential backbone resonance assignment across proline residues.

e) TROSY-type triple resonance experiments for the detection of solvent-accessible loops in large proteins.

f) TROSY-HNCA experiment with suppression of conformational exchange-induced relaxation.

ments now enables backbone assignments for larger proteins, such as the 67 kDa $^2H,^{13}C,^{15}N$-labeled p53 dimer [25] and the homo-oligomeric 110 kDa $^2H,^{13}C,^{15}N$-labeled DHNA [24]. TROSY yielded 20- to 50-fold signal enhancements for individual residues in the structured segments of the polypeptide chain of DHNA (Fig. 10.5), whereas the highly flexible C-terminal residue Lys 121 gave comparable results with the two different experiments [24]. Thus, TROSY-based triple-resonance experiments with $^2H,^{15}N,^{13}C$-labeled small and large proteins are superior to conventional triple-resonance experiments [29]. Only for small $^{13}C,^{15}N$-labeled proteins (<15 kDa) TROSY- based triple-resonance experiments might be somewhat less sensitive than the conventional experiments. A series of TROSY-based 3D and 4D triple-resonance experiments for $^2H,^{13}C,^{15}N$-labeled and $^{13}C,^{15}N$-labeled proteins are listed in Tab. 10.2. We refer to the given corresponding references for the pulse sequence schemes and technical details. In principle, the implementation of TROSY into conventional triple-resonance experiments is straightforward. The [^{15}N,^{1}H]-HSQC module is replaced by the [^{15}N,^{1}H]-TROSY of Fig. 10.4 (with $\Psi_1=\{y, x\}$ for the first FID and $\Psi_1=\{y, -x\}$ for the second FID), and the ^{1}H- and ^{15}N-decoupling is removed throughout the whole pulse sequence.

10.3.3
[^{13}C,^{1}H]-TROSY

The TROSY principle is not limited to ^{15}N-^{1}H groups, but is also very effective for ^{13}C-^{1}H groups in aromatic rings of proteins and in DNA and RNA nucleotides [17, 20–22, 30]. In 2D ct-[$^{13}C^{aromatic}$,^{1}H]-TROSY spectra of ^{13}C,^{15}N-labeled cyclophilin, a 4- to 10-fold signal enhancement was achieved when compared with conventional ct-[$^{13}C^{aromatic}$,^{1}H]-HSQC [17], and similar signal enhancements resulted for RNA and DNA molecules [20–22]. In the [^{13}C,^{1}H]-TROSY experiments only the carbon chemical shift evolution is recorded with the TROSY-method, since only the orientation of the ^{13}C CSA tensor is favorable and provides efficient compensation of relaxation due to ^{13}C-^{1}H dipolar coupling (the small ^{1}H CSA advise against the use of the TROSY-method during proton chemical shift evolution periods). The merge of the ^{1}H steady-state magnetization with the ^{13}C steady-state magnetization yielded a signal gain of up to a factor of 2 [17, 20].

10.3.4
TROSY-Based NOESY Experiments

For the collection of structural constraints in large molecules, the TROSY principle was further applied to 3D ^{15}N-resolved [^{1}H,^{1}H]-NOESY experiments [12, 16, 31, 32]. In these experiments the TROSY principle is used in all three frequency labeling periods, resulting in the extraordinary possibility of suppressing the diagonal. Furthermore, the resolution for NH-NH cross peaks is enhanced in all dimensions. For the collection of CH-NH NOEs in large proteins we recommend the 3D ^{15}N-resolved [^{1}H,^{1}H]-NOESY experiment [16] with TROSY during the ^{15}N-evolution period and the acquisition. However, the benefit of TROSY during the ^{15}N-evolution period is marginal because of the small maximal chemical shift evolution.

10.3.5
Transverse Relaxation-Optimization in the Polarization Transfers

In heteronuclear NMR experiments, magnetization is transferred between the different types of nuclei via scalar spin-spin couplings in so-called INEPT transfers (insensitive nuclei-enhanced polarization transfer [33]) (Fig. 10.4). In TROSY-type experiments, transverse relaxation-optimization is usually applied only during the evolution and detection periods [2]; transverse relaxation during the INEPT transfers therefore tends to become a limiting factor for molecular weights beyond about 100,000. CRINEPT (Cross-correlated enhanced polarization transfer) and CRIPT (cross-correlated induced polarization transfer) overcome this limitation by using cross-correlated relaxation between DD coupling and CSA, which becomes with the intrinsic use of TROSY a highly efficient transfer

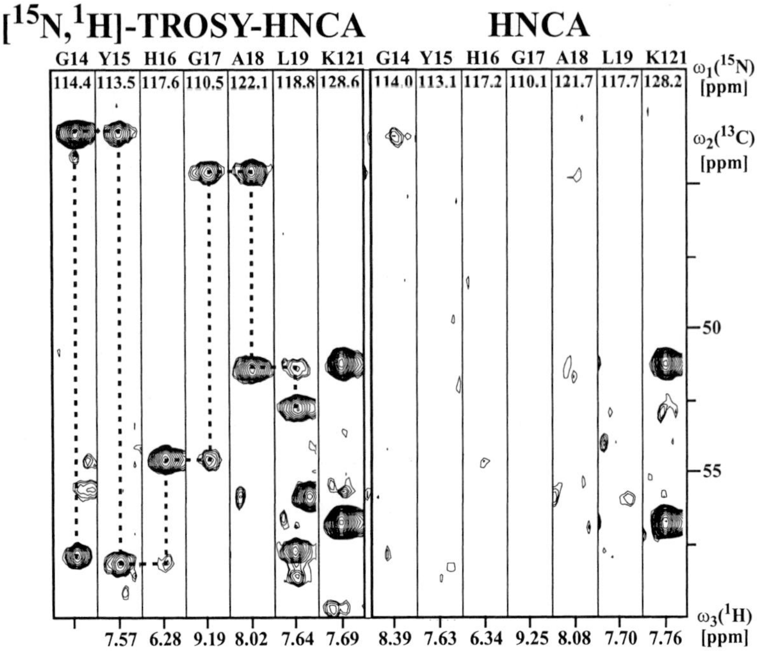

Fig. 10.5 Sequential resonance assignment of the polypeptide backbone of $^{2}H,^{13}C,^{15}N$-labeled DHNA using the HNCA triple resonance experiment, which connects the $^{1}H^{N}$ and ^{15}N resonances of the amide groups with the sequential and intraresidual $^{13}C^{\alpha}$chemical shifts. The dotted line indicates the assignment pathway based on the sequential and intraresidual HNCA cross peaks for the polypeptide segment 14–19. The highly flexible C-terminal residue Lys121 served as an internal reference in this experiment.

mechanism for molecular sizes above 200 kDa in aqueous solution at ambient temperature [3, 34]. The combination of TROSY and CRIPT or CRINEPT in so-called $[^{15}N,^{1}H]$-CRIPT-TROSY or $[^{15}N,^{1}H]$-CRINEPT-TROSY experiments results in fully transverse relaxation-optimized experiments and allows the detection of ^{15}N-^{1}H-fingerprints of molecules or supramolecular ensembles with a molecular weight up to 1 MDa. Fig. 10.6 shows a $[^{15}N,^{1}H]$-CRIPT-TROSY spectrum of the heptameric $^{15}N,^{2}H$-labeled GroES in complex with a ^{2}H-labeled single ring variant of GroEL. The molecular weight of the complex is 480 kDa. 85% of all expected resonances of GroES are observed [26, 35].

10.4 Conclusions

With TROSY, CRIPT and CRINEPT, high-resolution NMR spectra of macromolecules with masses of one to several hundred thousand Daltons can be recorded [3, 26, 35] (Figs. 10.1 and 10.6). Sequence-specific NMR assignments for large structures have already been obtained (Fig. 10.5), and TROSY-based NOESY experiments [12, 16, 31, 32] for the collection of structural constraints are also available. Practical applications for

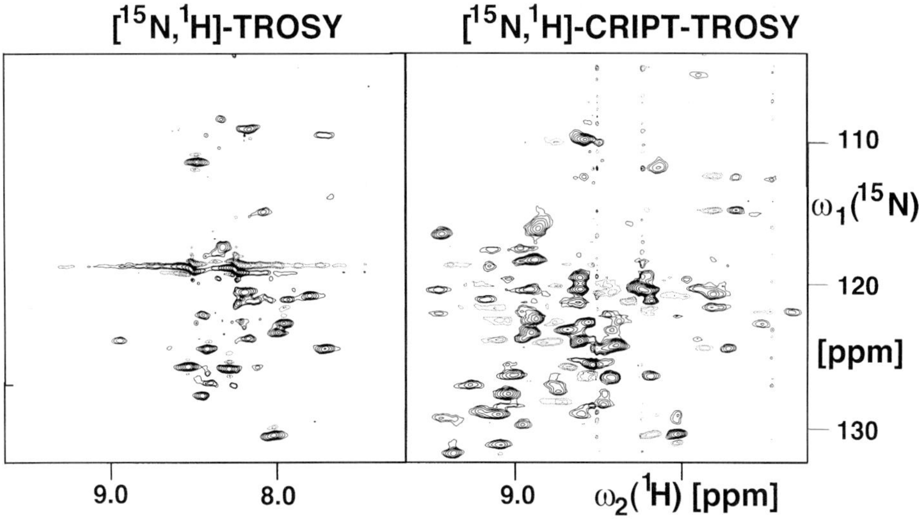

Fig. 10.6 [^{15}N,^{1}H]-TROSY spectrum and [^{15}N,^{1}H]-CRIPT-TROSY spectrum of uniformly ^{15}N,^{2}H-labeled GroES bound to SR1 with a molecular weight of 480 kDa. GroES is a homo-heptameric protein with a molecular weight 78 kDa, and SR1 is a homoheptameric single ring variant of GroEL with a molecular weight 400 kDa. The cross peaks observed in the [^{15}N,^{1}H]-TROSY spectrum are residuals of free GroES. The acquired data size was 100×1024 complex points with $t_{1,max}$=10 ms, $t_{2,max}$=100 ms. In the t_1-dimension a sine bell shifted by 30° was applied prior to Fourier transformation, and in the t_2-dimension an exponential function was applied prior to Fourier transformation.

studies of intermolecular interactions in supramolecular assemblies have already been described [7, 8], and the way is open for acquiring structure-activity relationship data [4] with large receptor molecules. Furthermore, the folds of small membrane proteins reconstituted in water-soluble lipid micelles were determined [5, 6]. The detection of small scalar couplings across hydrogen bonds in oligonucleotides [23, 36] and proteins [37, 38] and dynamical studies on hydrogen bonds [39] depend on using transverse relaxation-optimization.

10.5 Appendix: TROSY-Theory

On the basic of relaxation theory the concept of TROSY is described. We consider a system of two scalar coupled spins ½, *I* and *S*, with a scalar coupling constant J_{IS}, which is located in a protein molecule. Usually, *I* represents ^{1}H and *S* represents ^{15}N in a ^{15}N-^{1}H moiety. Transverse relaxation of this spin system is dominated by the DD coupling between *I* and *S* and by CSA of each individual spin. An additional relaxation mechanism is the DD coupling with a small number of remote protons, I_k. The relaxation rates of the individual multiplet components in a single quantum spectrum may then be widely different (Fig. 10.3) [2, 9]. They can be described using the single-transition basis operators, $I_{13}^{\pm}$, $I_{24}^{\pm}$, $S_{12}^{\pm}$, and $S_{34}^{\pm}$, which refer to the transitions 1→2, 1→3, 2→4, and 3→4 in the standard energy-level diagram for a system of two spins ½ (Fig. 10.3) [2, 9, 29]:

$$\begin{aligned} I_{13}^{\pm} &= \tfrac{1}{2} I^{+}(1-2S_z) \\ I_{24}^{\pm} &= \tfrac{1}{2} S^{\pm}(1+2S_z) \\ S_{12}^{\pm} &= \tfrac{1}{2} S^{\pm}(1-2I_z) \\ S_{34}^{\pm} &= \tfrac{1}{2} S^{\pm}(1+2I_z) \end{aligned} \quad \text{with the corresponding precession frequency} \quad \begin{aligned} \omega_I^{13} &= \omega_I - \pi J S_{IS} \\ \omega_I^{24} &= \omega_I + \pi J_{IS} \\ \omega_S^{12} &= \omega_S - \pi J_{IS} \\ \omega_S^{34} &= \omega_S + \pi J_{IS} \end{aligned}$$

Then, the first order relaxation matrix results in an uncoupled system of differential equations with the diagonal form:

$$\frac{d}{dt}\begin{bmatrix} \langle I_{13}^{\pm}\rangle \\ \langle I_{24}^{\pm}\rangle \\ \langle S_{12}^{\pm}\rangle \\ \langle S_{34}^{\pm}\rangle \end{bmatrix} = -\text{diag}\left(\begin{bmatrix} \pm i\omega_I^{13} \\ \pm i\omega_I^{24} \\ \pm i\omega_S^{12} \\ \pm i\omega_S^{34} \end{bmatrix} + 4J(0)\left(\begin{bmatrix} p^2 - 2C_{p\delta_I} p\delta_I + \delta_I^2 \\ p^2 + 2C_{p\delta_I} p\delta_I + \delta_I^2 \\ p^2 - 2C_{p\delta_S} p\delta_S + \delta_S^2 \\ p^2 + 2C_{p\delta_S} p\delta_S + \delta_S^2 \end{bmatrix} + \begin{bmatrix} \frac{1}{T_{2I}} + \frac{1}{2T_{1S}} \\ \frac{1}{T_{2I}} + \frac{1}{2T_{1S}} \\ \frac{1}{T_{2S}} + \frac{1}{2T_{1I}} \\ \frac{1}{T_{2S}} + \frac{1}{2T_{1I}} \end{bmatrix}\right)\right) \cdot \begin{bmatrix} \langle I_{13}^{\pm}\rangle \\ \langle I_{24}^{\pm}\rangle \\ \langle S_{12}^{\pm}\rangle \\ \langle S_{34}^{\pm}\rangle \end{bmatrix} \tag{5}$$

In the slow tumbling limit in the absence of radio frequency pulses only terms with $J(\omega = 0) = \frac{2\tau_c}{5}$ need to be retained. The contribution of the DD coupling is

$$p = \frac{1}{2\sqrt{2}} \gamma_I / \gamma_S \hbar / r_{IS}^3 , \tag{6}$$

and the contributions of the CSAs of spin I and S are

$$\delta_S = \frac{1}{3\sqrt{2}} \gamma_S B_0 \Delta\sigma_S \quad \text{and} \tag{7}$$

$$\delta_I = \frac{1}{3\sqrt{2}} \gamma_I B_0 \Delta\sigma_I , \tag{8}$$

where γ_I and γ_S are the gyromagnetic ratios of I and S, $\hbar$ is the Planck constant divided by 2π, r_{IS} the distance between S and I, B_0 the polarizing magnetic field, and $\Delta\sigma_S$ and $\Delta\sigma_I$ are the differences between the axial and the perpendicular principal components of the axially symmetric chemical shift tensors of spins S and I, respectively. $C_{p\delta_S} = 0.5\ (\cos(\vartheta_{p\delta_S})^2 - 1)$ where $\vartheta_{p\delta_S}$ is the angle between the tensor axes of the CSA of spin S and the I–S vector. Correspondingly, $C_{p\delta_I} = 0.5\ (\cos(\vartheta_{p\delta_I})^2 - 1)$ where $\vartheta_{p\delta_I}$ is the angle between the tensor axes of the CSA of spin I and the I–S vector. $1/T_{2I}$, $1/T_{2S}$ and $1/T_{1I}$, $1/T_{1S}$ account for the transverse relaxation and the longitudinal relaxation of spin I and spin S due to DD coupling with remote protons I_k.

In the following we will discuss the transverse relaxation of the different single transitions on the basis of Eq. (5). The two brackets multiplied by $J(0)$ in Eq. (5) are the terms of interest. The dominant relaxation mechanisms of spin I and S are the DD coupling and the CSA which are listed in the first bracket. The relaxation due to DD coupling with remote protons I_k is encountered in the second bracket and will be discussed below. Because of the terms containing $C_{p\delta_I}$ and $C_{p\delta_S}$ the relaxation rates of the individual multiplet components of spin S and spin I are different, as can be inferred also by the different linewidths of the multiplet components in the $[^{15}N,^{1}H]$-HSQC spectrum of Fig. 10.3. These terms are due to cross-correlated relaxation between DD coupling and CSA. The name cross-correlated relaxation comes from calculations using a different set of eigenvectors $(I_x, S_x, 2I_xS_z, 2S_xI_z)$. On the basis of these eigenvectors the relaxation matrix of Eq. (5) is containing off-diagonal elements with the terms $C_{p\delta_I}$ and $C_{p\delta_S}$. Thus, the term cross-correlation is anticipated.

Whenever CSA and DD coupling are comparable, *i.e.*, $p \approx \delta_S$ or/and $p \approx \delta_I$, and the angles $\vartheta_{p\delta_N} \approx 0°$ and $\vartheta_{p\delta_H} \approx 0°$, the first and third row of the bracket of interest would be close to zero, and concomitantly the transverse relaxation due to DD coupling and CSA would be small for resonances at frequencies ω_S^{12} and ω_I^{13}. TROSY is just selecting out of the four multiplet components the multiplet with frequencies ω_S^{12} and ω_I^{13}(Fig. 10.3).

Since dipole-dipole coupling is field-independent, whereas CSA increases proportionally to the field strength [Eqs. (5–8], there is actually a "magic field" at which for a specific multiplet component, the so-called TROSY component, the transverse relaxation due to DD and CSA will be near zero. For ^{15}N-^{1}H groups one approaches this situation at the highest presently available ^{1}H frequencies of 800 or 900 MHz, and a minimum of transverse relaxation is expected in the ^{1}H frequency range from about 950 to 1050 MHz [1, 2]. The ideal situation, where transverse relaxation would be completely quenched, will foreseeably not be attained in practice, for the following reasons: (i) It appears that the CSA is slightly variable depending on the residue type, sequence and possibly three-dimensional structure, so that there is no common "magic field" for all residues in a protein. The $\Delta\sigma_{HN}$ of backbone amide protons are in the range from 3 to 15 ppm [40, 41] and the $\Delta\sigma_N$ is between –125 and –216 ppm, respectively. (ii) For complete cancellation of transverse relaxation the CSA tensor would need to be collinear with the ^{15}N-^{1}H bond, which is not strictly valid. For example, the $\vartheta_{p\delta N}$ range from 6–26° [42, 43]. (iii) The remaining relaxation of the single transitions $I_{13}^{\pm}$ and $S_{12}^{\pm}$ is dominated by the DD coupling to remote protons, represented in Eq. (5) by $1/T_{1I}$, $1/T_{2I}$, $1/T_{1S}$ and $1/T_{2S}$. To evaluate these contributions we identify I and S as the ^{1}H and ^{15}N spins in a ^{15}N-^{1}H moiety. The relaxation of ^{15}N is then mainly determined by the CSA of ^{15}N and the DD interactions with the directly attached proton [29], so that the contributions with remote protons I_k, $1/T_{1S}$ and $1/T_{2S}$, can to a good approximation be neglected. However, the DD interactions between spin I and remote protons I_k have to be accounted for [29]:

$$1/T_{1I} = \sum_k (\gamma_I^2 \hbar / 2r_k^3)^2 J(0)\,, \tag{9}$$

$$1/T_{2I} = \sum_k (\gamma_I^2 \hbar / 2r_k^3)^2 \frac{5}{2} J(0)\,. \tag{10}$$

This relaxation pathways cannot be influenced by TROSY. Only with the replacement of nonlabile protons with deuterons is the transverse relaxation significantly reduced further, as can be inferred from Tab. 10.1. In Tab. 10.1 the transverse relaxation rates of ^{1}H and ^{15}N are predicted for a 23 kDa protein. In a conventional $[^{15}N,^{1}H]$-HSQC experiment with a 23 kDa protein one expects that deuteration reduces the ^1H relaxation rates 2.5-fold and 1.6-fold for β-sheets and -α-helices, respectively, and that deuteration yields only a small reduction in the ^{15}N relaxation rate by less then a factor of 1.3 [29]. For [^{15}N,^1H]-TROSY, deuteration has approximately the same absolute effects on the transverse relaxation, but because of the much smaller relaxation rates the relative improvement is larger, up to 6.5 for ^1H and up to 2.9 for ^{15}N. Conclusively, the combination of TROSY with deuteration dramatically reduces the transverse relaxation.

10.6 Acknowledgements

We thank D. Riek-Loher for careful reading and helpful discussions. Financial support was obtained from the F. Berger foundation.

10.7 References

1 K. Wüthrich, *Nat. Struct. Biol.* **1998**, *5*, 492–4952.

2 K. Pervushin, R. Riek, G. Wider, K. Wüthrich, *Proc. Natl. Acad. Sci. USA* **1997**, *94*, 12366–12371.

3 R. Riek, G. Wider, K. Pervushin, K. Wüthrich, *Proc. Natl. Acad. Sci. USA* **1999**, *96*, 4918– 4923.

4 S.B. Shuker, P.J. Hajduk, R.P. Meadows, S.W. Fesik, *Science* **1996**, *274*, 1531–1534.

5 C. Fernandez, K. Adeishvili, K. Wüthrich. *Proc. Natl. Acad. Sci. USA* **2001**, *98*, 2358–2363.

6 A. Arora, F. Abildgaard, J.H. Bushweller, L.K. Tamm, *Nat. Struct. Biol.* **2001**, *8*, 334–338.

7 M. Pellecchia, P. Sebbel, U. Hermanns, K. Wüthrich, R. Glockshuber, *Nat. Struct. Biol.* **1999**, *6*, 336–339.

8 H. Takahashi, T. Nakanishi, K. Kami, Y. Arata, I. Shimada, *Nat. Struct. Biol.* **2000**, *7*, 220–223.

9 M. Goldman, *J. Magn. Reson.* **1984**, *60*, 437–452.

10 A. Abragam, *The Principles of Nuclear Magnetism*, Clarendon, Oxford **1961**.

11 T.C. Farrar, T.C. Stringfellow, in *Encyclopedia of NMR*, D.M. Grand, R.K. Harris (Eds.), Wiley, New York **1996**, Vol. 6, pp. 4101–4107.

12 K. Pervushin, G. Wider, R. Riek, K. Wüthrich, *Proc. Natl. Acad. Sci. USA* **1999**, *196*, 9607–9612.

13 K. Pervushin, G. Wider, K. Wüthrich, *J. Biomol. NMR* **1998**, *12*, 345–348.

14 M. Rance, J.P. Loria, A.G. Palmer, *J. Magn. Reson.* **1999**, *136*, 92–101.

15 A. Meissner, T. Schulte-Herbrüggen, J. Briand, O.W. Sorensen, *Molec. Phys.* **1998**, *95*, 1137–1142.

16 G. Zhu, X.M., Kong, K.H. Sze, *J. Biomol. NMR* **1999**, *13*, 77–81.

17 K. Pervushin, R. Riek, G. Wider, K. Wüthrich, *J. Am. Chem. Soc.* **1998**, *120*, 6394–6400.

18 R. Riek, *J. Biomol. NMR* **2001**, *21*, 99–105.

19 T. Schulte-Herbrüggen, O.W. Sorensen, *J. Magn. Reson.* **1999**, *144*, 123–128.

20 B. Brutscher, J. Boisbouvier, A. Pardi, D. Marion, J.P. Simorre. *J. Am. Chem. Soc.* **1998**, *120*, 11845–11851.

21 R. Fiala, J. Czernek, V. Sklenar, *J. Biomol. NMR* **2000**, *16*, 291–302.

22 Riek, R., K. Pervushin, C. Fernandez, M. Kainosho, K. Wüthrich *J. Am. Chem. Soc.* **2001**, *123*, 658–664.

23 K. Pervushin, A. Ono, C. Fernandez, T. Szyperski, M., Kainosho, K. Wüthich *Proc. Natl. Acad. Sci. USA* **1998**, *95*, 14147–14151.

24 M. Salzmann, K. Pervushin, G. Wider, H. Senn, K. Wüthrich, *J. Am. Chem. Soc.* **2000**, *122*, 7543–7548.

25 A.F. Mulder, A. Ayed, D. Yang, C.H. Arrowsmith, L.E. Kay, *J. Biomol. NMR* **2000**, *18*, 173–176.

26 J. Fiaux, R. Riek, E. Bertelsen, A. Horwich, K Wüthrich, *in preparation.*

27 K. Pervushin, D. Braun, C. Fernandez, K. Wüthrich, *J. Biomol. NMR* **2000**, *17*, 195–202.

28 A. Bax, S. Grzesiek, *Acc. Chem. Res.* **1993**, *26*, 131–138.

29 M. Salzmann, K. Pervushin, G. Wider, H. Senn, K. Wüthrich, *Proc. Natl. Acad. Sci. USA* **1998**, *95*, 13585–13590.

30 A. Meissner, O.W. Sorensen, *J. Magn. Reson.* **1999**, *139*, 439–442.

31 Y. Xia, K. Sye, G. Zhu, *J. Biomol. NMR* **2000**, 18, 261–268.

32 A. Meissner, O.W. Sorensen, *J. Magn. Reson.* **2000**, *142*, 195–198.

33 G.A. Morris, R. Freeman, *J. Am. Chem. Soc.* **1979**, *101*, 760–762.

34 R. Brüschweiler, R.R. Ernst, *J. Chem. Phys.* **1992**, *96*, 1758–1766.

35 R. Riek, J. Fiaux, E. Bertelsen, A. Horwich, K Würthrich, *J. Am. Chem. Soc.*, in press

36 A.J. Dingley, S. Grzesiek, *J. Am. Chem. Soc.* **1998**, *120*, 8293–8297.

37 F. Cordier, S. Grzesiek, *J. Am. Chem. Soc.* **1999**, *121*, 1601–1602.

38 Y.X. Wang, J. Jacob, F. Cordier, P. Wingfield, S.J. Stahl, S. Lee-Huang, D. Torchia, S. Grzesiek, A. Bax, *J. Biomol. NMR.* **1999**, *14*, 181–184.

39 R. Riek, *J. Magn. Reson.* **2001**, *149*, 149–153.

40 N. Tjandra, A. Bax, *J. Am. Chem. Soc.* **1997**, 119, 8076–8082.

41 M. Tessari, G.W. Vuister, *J. Biomol. NMR* **2000**, *16*, 171–174.

42 D. Fushman, D. Cowburn, *J. Am. Chem. Soc.* **1998**, *120*, 7109–7110.

43 C.D. Kroenke, M. Rance, A.G. Palmer, *J. Am. Chem. Soc.* **1999**, *121*, 10119–10125.

44 L.E. Kay, P. Keifer, T. Saarinen, *J. Am. Chem. Soc.* **1992**, *114*, 10663–10665.

45 M. Salzmann, G. Wider, K. Pervushin, K. Wüthrich, *J. Biomol. NMR* **1999**, *15*, 181–184.

46 J.P. Loria, M. Rance, A.G. Palmer, *J. Magn. Reson.* **1999**, *141*, 180–184.

47 M. Salzmann, G. Wider, K. Pervushin, H. Senn, K. Wüthrich, *J. Am. Chem. Soc.* **1999**, *121*, 844–848.

48 M. Salzmann, K. Pervushin, G. Wider, K. Wüthrich, *J. Biomol. NMR.* **1999**, *14*, 85–88.

49 D. Yang, L.E. Kay, *J. Am. Chem. Soc.* **1999**, *121*, 2571–2575.

50 P. Permi, A. Annila, *J. Biomol. NMR* **2001**, *20*, 127–133.

51 A. Meissner, O.W. Sørensen, *J. Magn. Reson.* **2001**, *150*, 100–104.

52 F. Löhr, S. Pfeiffer, Y.-J. Lin, J. Hartleib, O. Klimmek, H. Rüterjans, *J. Biomol. NMR* **2000**, *18*, 337–346.

53 R. Konrat, D.W. Yang, L.E. Kay, *J. Biomol. NMR.* **1999**, *15*, 309–313.

54 M. Pellecchia, D. Meininger, A.L. Shen, R. Jack, C.B. Kaspar, D.S. Sem, *J. Am. Chem. Soc.* **2001**, *123*, 4633–4634.

55 K. Pervushin, V. Gallius, C. Ritter, *J. Biomol. NMR* **2001**, *21*, 161–166.

56 A. Eletsky, A. Kienhofer, K. Pervushin, *J. Biomol. NMR* **2001**, *20*, 177–180.

57 D. Yang, L.E. Kay, *J. Biomol. NMR* **1999**, *13*, 3–10.

11
MAS Solid-State NMR of Isotopically Enriched Biological Samples

Philip T. F. Williamson, Matthias Ernst, and Beat H. Meier

11.1
Introduction

The study of the structure and dynamics of biological systems by various experimental techniques forms the basis of understanding biological processes including enzyme mechanisms, cellular energetics, and molecular recognition. Much of the structural information acquired so far is based on either X-ray diffraction [1] or solution-state NMR [2, 3]. The application of these two techniques requires certain physical characteristics of the molecules studied. X-ray diffraction relies on the formation of crystals as an essential prerequisite to structural studies, while solution-state NMR relies on the solubility of the molecule in a suitable solvent and requires rapid tumbling of the molecules in solution, effectively limiting the size of the systems studied to approximately 100 kDa presently. Although the methods mentioned above have proved invaluable in the study of many systems, these limitations have hindered the studies of entire classes of biological molecules including integral membrane proteins, protein aggregates, and fibrous proteins. In such cases the physical nature of the molecules studied has often precluded successful crystallization, whilst the relatively large size or insufficient solubility of many of these complexes has hindered the application of solution-state NMR.

More recently, solid-state NMR has emerged as a tool for the study of biological molecules that were intractable to high-resolution liquid-state NMR techniques. The absence of molecular-weight constraints and the ability to study amorphous, nanocrystalline, or microcrystalline materials (as opposed to crystalline solids for X-ray diffraction) has permitted the development of methods for the characterization of the structure and dynamics of large membrane complexes, protein aggregates, and fibrous proteins. Technically, solid-state NMR experiments can be divided into experiments performed on static samples and experiments performed under magic-angle sample spinning (MAS) conditions.

The first approach has successfully been applied to the study of amorphous as well as to macroscopically ordered solids. Examples of applications include the determination of backbone geometries in fibrous proteins [4] or the determination of protein-backbone, side-chain, and bound-ligand orientation with respect to the membrane normal in membrane-bound proteins [5–8]. Membranes, bilayers, bicelles, or liposomes are neither solid nor liquid systems but have aspects of both and are sometimes liquid crystalline. In most of these systems, time-independent anisotropic interactions play an important role, and, with respect to the NMR techniques that have to be applied, these systems resemble proper solids.

The second method, magic-angle spinning, relies on rapid sample rotation about an axis inclined at an angle of 54.74° with the static magnetic field, to average the anisotropic interactions, thereby allowing the resolution of sites on the basis of their isotropic chemical shift. The application of MAS in combination with specifically designed pulse sequences that selectively reintroduce some desired anisotropic interactions typically allows the determination of a range of structural parameters including internuclear distances and torsion angles.

This chapter focuses on the recent developments that have been made in the application of magic-angle spinning to biological systems. Methods based on static and macroscopically aligned samples have been recently discussed elsewhere [7, 8]. It is not the aim of this contribution to present a balanced and complete review of the field but rather to introduce the basic principles and illustrate their applicability to systems of biological relevance using selected examples. The chapter is separated into five parts. Section 11.2 deals with the physical principles underlying many of the solid-state NMR experiments together with a brief overview of the basic techniques. Section 11.3 discusses the different polarization-transfer techniques which form the basis of many applications to biological systems. Section 11.4 deals with the experimental aspects of sample preparation for solid-state NMR studies, including labeling schemes to introduce NMR-sensitive isotopes into biological molecules. Section 11.5 discusses how the experimental schemes introduced can be combined to elucidate structural information, and provides some examples of how some more recent developments in the field have been applied to the analysis of biological systems.

11.2 Basic Concepts in Solid-State NMR

11.2.1 Spin Interactions

Although the same nuclear spin interactions are present in solid-state as in solution-state NMR, the manifestations of these effects are different because, in the solid, the anisotropic contribution to the spin interactions contributes large time-independent terms to the Hamiltonian that are absent in the liquid phase. Therefore, the experimental methods employed in solids differ from the ones in the liquid state. The spin Hamiltonian for organic or biological solids can be described in the usual rotating frame as the sum of the following interactions:

$$\mathcal{H} = \sum_{k} \mathcal{H}_k^{\text{CS}} + \sum_{k \neq j} \mathcal{H}_{k,j}^{\text{D}} + \sum_{k} \mathcal{H}_k^{\text{Q}} + \sum_{k \neq j} \mathcal{H}_{k,j}^{\text{J}} \tag{1}$$

Here, $\mathcal{H}_k^{\text{CS}}$ describes the isotropic and anisotropic chemical shift of spin k; $\mathcal{H}_k^{\text{D}}$ describes the anisotropic dipolar coupling between spins k and j; $\mathcal{H}_k^{\text{Q}}$ is the anisotropic quadrupolar coupling of spin k; and $\mathcal{H}_{k,j}^{\text{J}}$ describes the isotropic and anisotropic J-coupling be-

Tab. 11.1 Typical magnitude of important interactions

Interaction	*Nuclei*	*Typical magnitude*
Chemical shift range	^{1}H	~ 15 ppm
	^{13}C	~ 200 ppm
	^{15}N	~ 200 ppm
Anisotropy of CSA	^{1}H	< 10 ppm
	^{13}C	< 140 ppm
	^{15}N	< 200 ppm
Anisotropy of one-bond dipolar coupling	^{1}H-^{13}C	~ 22 kHz
	^{1}H-^{15}N	~ 20 kHz
	^{13}C-^{13}C	4.5 kHz
	^{13}C-^{15}N	2 kHz
	^{15}N-^{15}N	< 1 kHz
^{1}J coupling constant	^{1}H-^{13}C	120–250 Hz
	^{1}H-^{15}N	~ 90 Hz
	^{13}C-^{13}C	30–60 Hz
	^{13}C-^{15}N	~ 15 Hz

tween spins k and j. All four terms of Eq. (1) contain anisotropic contributions, i.e. the size of the interaction depends on the orientation of the crystallite in the static magnetic field. In a micro-crystalline powder, all possible orientations will be present, and one observes a broad line with a typical line shape in a static sample [9–11]. Table 11.1 shows some typical sizes of the different important interactions in NMR.

11.2.1.1 **The Chemical-Shift Hamiltonian**

The magnetic field at the position of a nucleus can deviate from the applied field $\vec{B}_0$. This "shielding" of the nuclear spins is governed by the distribution of electrons around the nucleus. It reflects the chemical functionality and the conformation of the molecule. The shielding of the nuclear spin by the surrounding electrons gives rise to a small (parts per million) deviation of the magnetic field $\vec{B}$ from the applied field $\vec{B}_0$. The shielding of the nuclear spin is not an isotropic quantity, and therefore the chemical shielding shows an orientation dependence with respect to the static magnetic field $\vec{B}_0$. The chemical-shielding Hamiltonian has the form:

$$\mathcal{H}_k^{\mathrm{CS}} = \omega_0 \left(\bar{\sigma}_k + \frac{\delta_k^{CSA}}{2} \left[(3 \cos^2 \beta_k - 1) + \eta_k^{\mathrm{CSA}} \sin^2 \beta_k \cos 2a_k \right] \right) \hat{I}_{kz} \, . \tag{2}$$

Here, $\bar{\sigma}_k$ is the isotropic chemical shift referenced in ppm from the carrier frequency ω_0, δ_k^{CSA} is the anisotropy and η_k^{CSA} the asymmetry of the chemical-shielding tensor, here also expressed in ppm. Note that for heteronuclear cases different reference frequencies ω_0 are chosen for different nuclei (doubly rotating frame of reference). The two Euler angles a_k and β_k describe the orientation of the chemical-shielding tensor with respect to the laboratory-fixed frame of reference. The anisotropy δ_k^{CSA} defines the width and the asymmetry η_k^{CSA} the shape of the powder line shape (see Fig. 11.1 a).

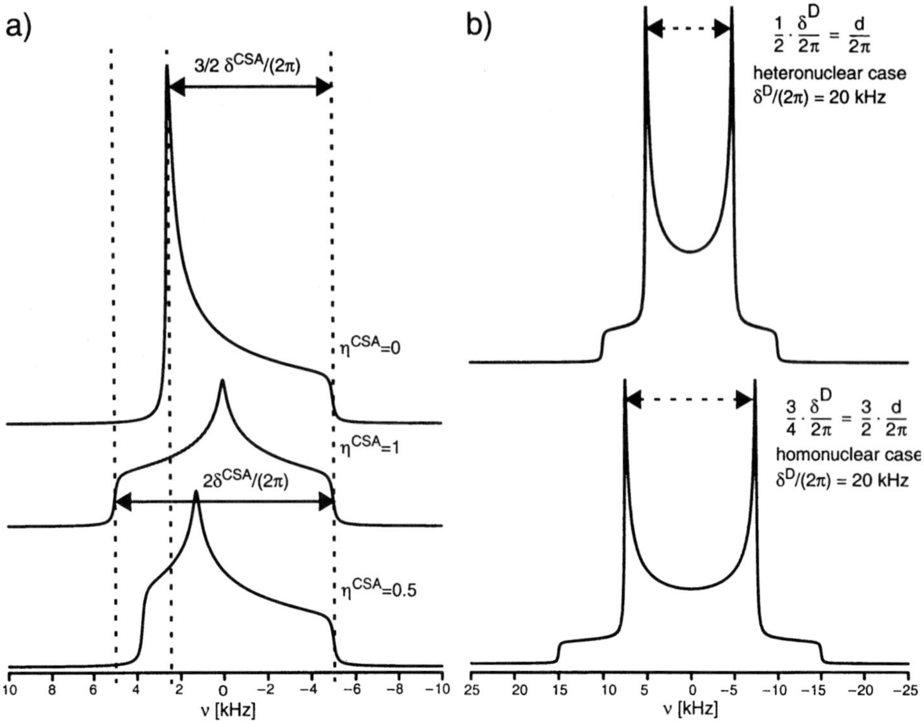

Fig. 11.1 (**a**) Powder line shapes of chemical-shielding tensors, $\delta^{\mathrm{CSA}}/(2\pi) = 5$ kHz, for three different values of the asymmetry parameter η^{CSA}. (**b**) Powder line shape for a dipolar-coupling tensor ("Pake" pattern).

11.2.1.2 **The Dipolar-Coupling Hamiltonian**

The dipolar-coupling Hamiltonian ($\mathcal{H}_{kj}^{\mathrm{D}}$) describes the through-space coupling between two nuclear spins $\vec{I}_k$ and $\vec{I}_j$. The dipolar coupling has an r_{kj}^{-3} dependence, and is key to the determination of internuclear distances in both solid-state and solution-state NMR. The high-field truncated form of the dipolar Hamiltonian is given by

$$\mathcal{H}_{kj}^{\mathrm{D}} = \frac{\delta_{kj}^{\mathrm{D}}}{2} \cdot \frac{(3\cos^2\beta_{kj} - 1)}{2} \left[2\hat{I}_{kz}\hat{I}_{jz} - \frac{1}{2}(\hat{I}_k^+\hat{I}_j^- + \hat{I}_k^-\hat{I}_j^+) \right] \tag{3}$$

for homonuclear dipolar couplings and by

$$\mathcal{H}_{kj}^{\mathrm{D}} = \frac{\delta_{kj}^{\mathrm{D}}}{2} \cdot \frac{(3\cos^2\beta_{kj} - 1)}{2} 2\hat{I}_{kz}\hat{S}_{jz} \tag{4}$$

for heteronuclear dipolar couplings. In both equations, the anisotropy of the dipolar coupling is given by

$$\delta_{kj}^{\mathrm{D}} = -2 \cdot \frac{\mu_0}{4\pi} \frac{\gamma_k \gamma_j}{r_{kj}^3} \tag{5}$$

and β_{kj} is the angle between the internuclear vector and the static magnetic field. The gyromagnetic ratios of the two spins are γ_k, γ_j and r_{kj} is the internuclear distance. The typical powder line shape generated by two dipolar coupled spins is called a Pake pattern (Fig. 11.1 b) [12]. The distance between the two "horns" of the Pake pattern is proportional to the anisotropy of the dipolar coupling (see Fig. 11.1 b). In the literature one often finds another parameter, $d_{kj} = \delta_{kj}^{D}/2$, which is sometimes called the dipolar-coupling constant.

11.2.1.3 The Quadrupolar Hamiltonian

If the spin-quantum number, I_k, of a spin k is larger than 1/2, we have an additional term in the Hamiltonian, the quadrupolar coupling, $\mathcal{H}_k^{Q}$. The quadrupolar Hamiltonian arises from the interaction between the electric-field gradient and the nuclear spin. The first-order quadrupolar Hamiltonian is given by:

$$\mathcal{H}_k^{Q_1} = \frac{\delta_k^{Q}}{2}[(3\cos^2\beta_k - 1) + \eta_k^{Q}\sin^2\beta_k\cos 2\alpha_k](3\hat{I}_{kz}^2 - I_k(I_k+1)) \tag{6}$$

Here, the two Euler angles α_k and β_k define the orientation of the quadrupolar tensor with respect to the static magnetic field;

$$\delta_k^{Q} = \frac{e^2 q_k Q_k}{2I_k(2I_k - 1)^{\hbar}} \tag{7}$$

is the anisotropy η_k^{Q} and the asymmetry of the quadrupolar coupling; I_k is the spin-quantum number of spin k; e is the elementary charge; Q_k is the quadrupole moment of the nucleus; and eq_k equals the largest principal value of the external electric-field gradient. The tensor patterns for the first-order quadrupolar interaction are, similar to the dipolar interaction, symmetric around the isotropic frequency but they consist of two non axially symmetric tensors. Because of the size of the quadrupolar coupling tensors, the first-order Hamiltonian is often not sufficient to describe the quadrupolar interaction. The functional form of the second-order quadrupolar coupling can be found in the literature [13].

11.2.1.4 The J-Coupling Hamiltonian

The J-coupling Hamiltonian ($\mathcal{H}_{kj}^{J}$) arises through the indirect interactions of two nuclear spins via the binding electrons. In principle, the J-coupling Hamiltonian contains an anisotropic component. Because the anisotropic part of the J-coupling is not discernible from the dipolar interaction, it is usually neglected and included into the dipolar interaction. The isotropic part of the J-coupling is given by

$$\mathcal{H}_{kj}^{J} = 2\pi J_{kj}\,\vec{\hat{I}}_k \cdot \vec{\hat{I}}_j \tag{8}$$

in the case of strong coupling and

$$\mathcal{H}_{kj}^{J} = 2\pi J_{kj}\,\hat{I}_{kz}\hat{S}_{jz} \tag{9}$$

in the weak-coupling limit. Here, J_{kj} is the isotropic J-coupling constant between spins k and j. Strong coupling refers to the case where the J-coupling constant is larger than or in the order of the isotropic chemical-shift difference of the two spins $\hat{I}_k$ and $\hat{I}_j$, while weak coupling refers to the case where the J-coupling constant is much smaller than the isotropic chemical-shift difference. The heteronuclear J-coupling is always a weak coupling. Because of the spatial independence and the presence of significant couplings only between adjacent spins, $(\mathcal{H}^J_{kj})$ plays an important role in the determination of through-bond connectivities, which are frequently useful in the assignment of resonances in both solid-state and solution-state NMR.

11.2.2 Basic Building Blocks for Solid-State NMR Experiments

In the absence of rapid molecular reorientation, the nuclear spin Hamiltonian of Eq. (1) is often dominated by the large anisotropic contributions from the chemical-shift interactions, the dipolar couplings, and the quadrupolar interactions. One of the major challenges for solid-state NMR methodology is to simplify these spectra such that structural and dynamic information can be obtained for specific sites in the molecule. In the case of static NMR experiments this is achieved through either the selective introduction of NMR-sensitive isotopes at particular sites or through the macroscopic orientation of the sample with respect to the magnetic field, resulting in simplified spectra.

11.2.2.1 Magic-Angle Spinning

The observation of well-resolved spectra usually requires spatial averaging of the anisotropic interactions by mechanical rotation of the sample. This has been achieved experimentally through the rapid rotation of the sample about an axis which is inclined at an angle of 54.74° ("magic" angle) to the static magnetic field (see insert in Fig. 11.2). For this angle, the second-order Legendre polynomial $(3\cos^2\theta - 1)/2$ vanishes [14, 15].

If the time-dependent Hamiltonian under MAS commutes with itself at all points in time, i.e., $[\mathcal{H}(t_1), \mathcal{H}(t_2)] = 0$ for all values of t_1 and t_2, magic-angle spinning will lead to a side-band spectrum consisting of a central resonance line at the isotropic value together with a family of side bands spaced at intervals of the rotor frequency (Fig. 11.2). In theory, the side bands are infinitely sharp. Such a Hamiltonian is sometimes called "inhomogeneous" under sample rotation. This situation is always encountered if we have only chemical-shielding and heteronuclear dipolar-coupling interactions. Detailed analysis of the side-band intensities from low-speed MAS spectra can be used to extract the principal values of the chemical-shielding or dipolar-coupling tensors [16, 17]. In many cases, chemical-shift information can be used to extract structural information about the system [18–25]. In the limit of infinitely fast spinning, the anisotropic part of the Hamiltonian is averaged out, and one obtains a spectrum which is fully characterized by the isotropic interactions in the Hamiltonian. The first-order quadrupolar Hamiltonian behaves inhomogeneously, while the second-order quadrupolar Hamiltonian has a fourth-rank tensor component which is not averaged out by MAS [13]. Such higher-order powder patterns can be removed by rotating the sample about more than one axis, as was experimentally implemented in double-rotation (DOR) and dynamic-angle spinning (DAS) [26–28].

If the time-dependent Hamiltonian does not commute with itself at all times, then one does not necessarily observe a sharp side-band spectrum under MAS. Only for spinning

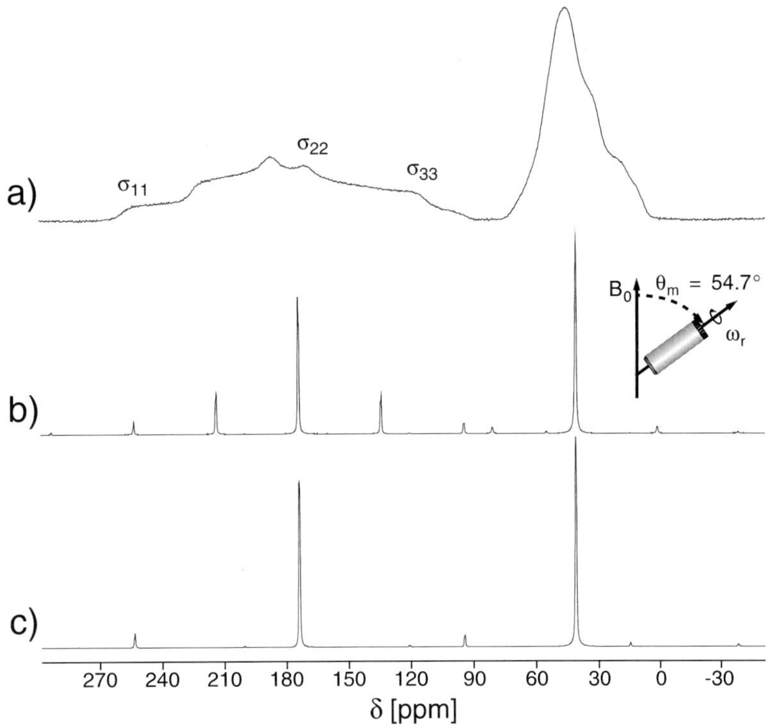

Fig. 11.2 Solid-state ^{13}C NMR spectra of a powder sample of U-^{13}C-^{15}N-glycine illustrating the broad NMR resonances in the static sample (**a**) and the effects of magic-angle spinning at 5 kHz (**b**) and 10 kHz (**c**). Both the chemicalshielding tensors and the homonuclear dipolar coupling between the adjacent ^{13}C atoms are represented in the static spectrum. MAS collapses the broad line into a central line close to the isotropic value, with rotational side bands spaced at intervals equal to the spinning frequency.

frequencies exceeding the width of the static line can one observe a significant narrowing of the line and the forming of side bands. Such a Hamiltonian is sometimes called "homogeneous". A Hamiltonian containing more than one homonuclear dipolar interaction or a homonuclear and a heteronuclear dipolar interaction can fall into this category [9–11].

11.2.2.2 **Sensitivity-Enhancement Techniques**

Solution-state NMR studies of biological molecules benefit from the fact that excitation and detection usually takes place on proton nuclei with their relatively high gyromagnetic ratio and favorable relaxation properties. In solid-state NMR the strong proton-proton dipolar coupling network results in relatively broad proton spectra with little fine structure even at high MAS frequencies [29]. Therefore, the majority of experiments performed currently rely on the detection of low-γ nuclei (e.g. ^{13}C, ^{15}N). The lower gyromagnetic ratio and the frequently lower natural abundance lead to a reduction of the strength of the homonuclear dipolar-coupling networks compared to protons (see Tab. 11.1), and one can

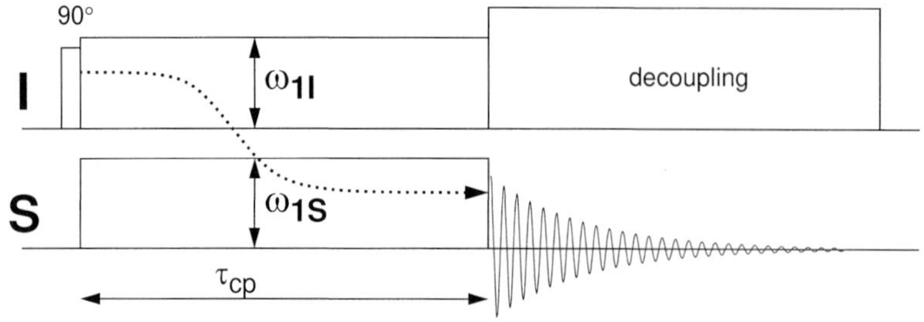

Fig. 11.3 Schematic pulse sequence of Hartmann-Hahn cross polarization to transfer polarization from the *I* spins to the *S* spins by matching the rf-field amplitudes such that the condition $|\omega_{1I}| = |\omega_{1S}|$ is fulfilled.

often obtain narrow lines under fast MAS. On the other hand, the detection of low-γ nuclei leads to a reduction in sensitivity compared to proton-detection. In some cases, however, in particular with fast MAS, proton detected solid-state NMR experiments have been shown to offer advantages in signal-to-noise ratio over low-γ detected spectra [30, 31].

In many biological solids, there is a large reservoir of protons with a higher equilibrium polarization and more favorable relaxation characteristics. It is, therefore, common to benefit from the large proton equilibrium polarization. Instead of directly exciting the low-γ nuclei, polarization is spin-locked and transferred to the low-γ nuclei using Hartmann-Hahn cross polarization [32]. The transfer is achieved through the simultaneous irradiation of the low-γ nuclei (Fig. 11.3). Transfer is optimal when the rf fields applied to the two spins fulfill the Hartmann-Hahn condition, $|\gamma_I B_{1I}| = |\gamma_S B_{1S}|$.

Using the proton polarization, the sensitivity of the low-γ nuclei can be enhanced by a factor up to the ratio of the gyromagnetic ratio of the two nuclei γ_I/γ_S, which equals 4 for I = ^{1}H, S = ^{13}C and 10 for I = ^{1}H, S = ^{15}N. The T_1-relaxation of protons is typically considerably shorter than that of low-γ nuclei, allowing a significant reduction in the recycle time. This further enhances the relative gain in signal-to-noise ratio that can be achieved through the application of cross polarization.

One can further increase the amount of transferred polarization if one carries out the cross polarization in an adiabatic fashion. In this experiment, the amplitude of one of the spin-lock fields is usually varied in a tangential shape [33–35]. In addition to the compensation of instabilities in the amplitude and rf field inhomogeneities, one can also obtain a gain in signal by a up to a factor of two. The concept of adiabatic polarization transfer will be discussed in more detail in Sect. 11.3.1.

11.2.2.3 **Heteronuclear Decoupling**

In the presence of a strong homonuclear dipolar-coupling network, the heteronuclear dipolar coupling is not spun out by MAS (Fig. 11.4a). Even at the highest MAS frequencies available today (50 kHz) the ^{13}C resonances are still significantly broadened by the presence of neighboring protons. Therefore, one has to apply proton rf irradiation in order to fully average out the residual heteronuclear dipolar couplings as well as the heteronuclear isotropic J couplings. Until recently, high-power continuous wave (CW) proton decoupling was the method of choice for heteronuclear spin decoupling under MAS.

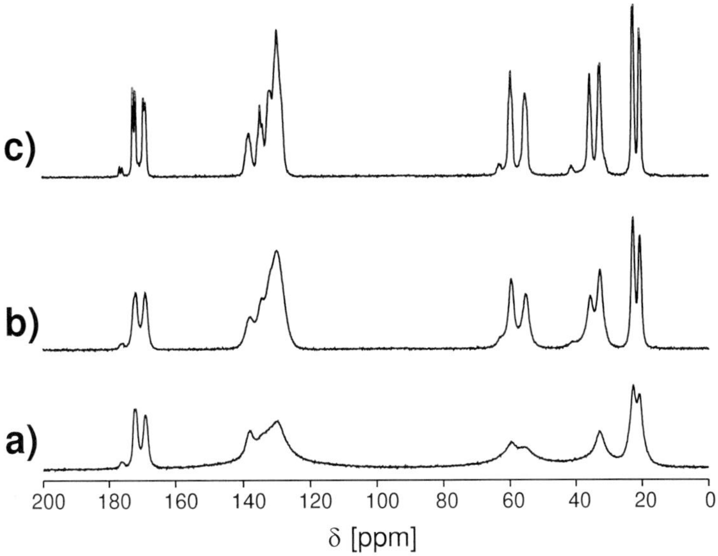

Fig. 11.4 ^{13}C spectrum of the fully ^{13}C- and ^{15}N-labeled dipeptide L-Val-L-Phe (**a**) without proton decoupling, (**b**) with CW decoupling, and (**c**) with optimized TPPM decoupling. The MAS frequency was $\nu_r = 28$ kHz, the decoupler field strength $\nu_1 = 150$ kHz, and the proton resonance frequency 300 MHz.

Composite-pulse decoupling schemes like WALTZ [36, 37], DIPSI [38], or GARP [39], which are used in solution-state NMR, have failed to offer any significant improvements in the solid state compared to CW decoupling. The residual line width in CW-decoupled spectra is dominated by a cross term between the chemical-shielding tensor of the protons and the heteronuclear dipolar-coupling tensor [40, 41].

A few years ago, a new decoupling scheme called two-pulse phase-modulated decoupling (TPPM) [42] was introduced. It consists of two pulses with a flip angle of about 180° and a phase shift of 20–60°. Both parameters depend on the sample, the MAS frequency, and the probe, and have to be optimized experimentally. TPPM decoupling significantly reduces the residual line width at a given rf field strength compared to CW decoupling. The biggest gains over CW decoupling can be achieved at high MAS frequencies.

Very recently, a rotor-synchronized decoupling sequence called X-inverse-X (XiX) [43], which consists of two pulses of equal length with a phase difference of 180°, has been proposed. The XiX decoupling sequence has only one adjustable parameter, namely the pulse length, and the optimum pulse length is quite well predefined by the spinning frequency. Under many experimental conditions, XiX decoupling seems to lead to significant improvements in line intensity compared to TPPM decoupling. The XiX sequence should be easy to optimize and is in addition less sensitive to rf field inhomogeneities than TPPM.

At MAS frequencies exceeding 40 kHz, one can significantly reduce the applied rf field amplitude by using low-power CW decoupling with a decoupling field strength of ca. $\omega_1 \approx \omega_r/4$. Low-power CW decoupling relies on the reversal of the order of the averaging processes [44]. The obtained line widths are about 10–20% larger than the ones obtained with high-power decoupling, but the power dissipated in the sample is about two orders of magnitude lower.

Special care has to be taken for the decoupling of protons during the application of pulse sequences on the low-γ nuclei. During many recoupling sequences, strong rf fields are applied to the low-γ nuclei in order to reintroduce dipolar couplings and suppress anisotropic chemical shifts. In these experimental schemes, incomplete decoupling of the protons from the irradiated low-γ nuclei leads to a rapid decay of the polarization. Therefore, additional proton irradiation is required. Experimentally it was found that for CW irradiation the proton rf field strength should be at least three times the rf field strength on the low-γ nucleus [45, 46], which is often impossible to achieve. In such cases an alternative strategy has been applied, relying on the suppression of the strong proton-proton dipolar couplings using homonuclear decoupling sequences, like Lee-Goldburg decoupling [47].

11.3
Polarization-Transfer Techniques

Polarization-transfer techniques form the basis of most NMR experiments which are currently used for the assignment of resonances and for the determination of structural parameters (distances or torsion angles) in solid-state NMR. Polarization transfer can take place via the J couplings ("through-bond") or via the dipolar couplings ("through-space"). The dipolar couplings are usually significantly larger than the J couplings (see Tab. 11.1). They are, however, averaged by MAS and must be recoupled or driven by the proton bath to be useful as a polarization-transfer mechanism. The isotropic J coupling is, of course, unaffected by MAS.

The techniques discussed in this chapter can be divided into two general classes: homonuclear and heteronuclear polarization-transfer techniques. In each of the two classes, further distinctions can be made and are used to structure the following discussion of the different techniques. We do not want to discuss the details of the experiments, which can be found in the original literature or in several reviews about dipolar recoupling techniques [48–50].

11.3.1
Adiabatic Versus Sudden Polarization Transfer

Polarization-transfer experiments which are based on a resonance condition, i.e. where a variable quantity in the experiment is matched to a parameter of the investigated spin system, can be carried out as a transient experiment or as an adiabatic experiment. Figure 11.5 illustrates the differences between these two types of experiments. In a transient or "sudden" experiment, the density operator is prepared in a state orthogonal to the effective polarization-transfer Hamiltonian (Fig. 11.5 a). When the polarization-transfer Hamiltonian is switched on, the density operator starts precessing around the effective Hamiltonian, and usually maximum polarization transfer is reached after a 180° rotation. Since often the size of the effective Hamiltonian at the matching condition depends on the orientation of the crystallite relative to the static magnetic field, it is impossible to find one time which corresponds to a 180° rotation for all crystallites. Therefore, the efficiency of such sudden experiments is limited to 73% if the size of the effective Hamiltonian depends on a single Euler angle (β) and to 56% if the size of the effective Hamiltonian depends on two Euler angles (β and γ). The first type of sequences are sometimes called "γ encoded".

In the adiabatic experiment, the density operator is prepared such that it is initially oriented along the starting effective Hamiltonian prepared to be far away from the resonance condition (Fig. 11.5 b). The direction of the effective Hamiltonian is then slowly changed (e.g. by a change in rf amplitude or frequency) to pass through the resonance condition to the final position, again far away from the resonance condition. If the change of the effective Hamiltonian is carried out adiabatically, the density operator will follow the trajectory of the Hamiltonian, and a (complete) polarization transfer can occur. The variation in the size of the effective Hamiltonian at the resonance condition only influences the condition for adiabaticity. Therefore, it is in principle possible to obtain si-

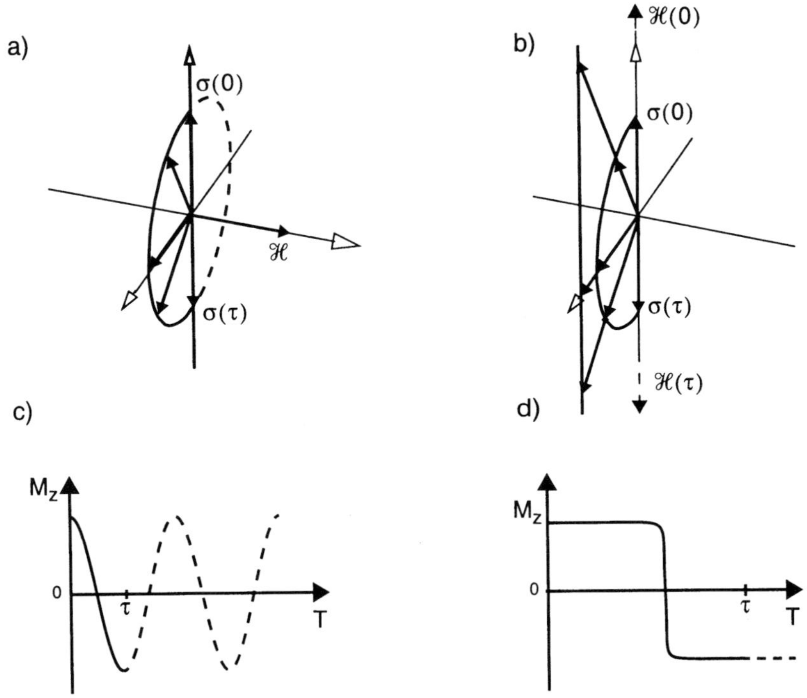

Fig. 11.5 Schematic comparison of (**a**) sudden and (**b**) adiabatic inversion of the *z*-component of the polarization vector. In the sudden case a π-pulse is applied while in the adiabatic case a frequency sweep is shown. The time evolution of the *z*-polarization as a function of the pulse duration τ is shown in (**c**) and (**d**) for sudden and adiabatic schemes, respectively. For the sudden experiment, the precise choice of τ is crucial to obtain efficient inversion; for adiabatic experiments a variation within rather wide limits does not lead to significant changes in inversion efficiency.

multaneously full polarization transfer for all possible orientations of the crystallites, and a theoretical efficiency of 100% can be reached in adiabatic experiments.

11.3.2
Homonuclear Polarization Transfer

Homonuclear polarization-transfer experiments have, so far, mostly relied on dipolar couplings for establishing connectivity as well as for measuring structural constraints, i.e. distances and dihedral angles. For distance measurements, the dipolar coupling is the method of choice, since the coupling constant and the internuclear distance are directly correlated. There are many different schemes available which allow either the selective recoupling of isolated spin pairs or the broad-band recoupling of all spins in the molecule. For the assignment of resonances by tracing out homonuclear coupling networks, there are some differences between the J and dipolar polarization-transfer mechanisms. Dipolar couplings are through-space couplings and, therefore, not directly correlated with connectivity between atoms except for the fact that often the one-bond dipolar-coupling constants are the largest ones. On the other hand, J-couplings are through-bond couplings and are directly linked to connectivity between atoms, which forms the basis of many resonance assignment strategies. For identifying one-bond connectivities, both coupling mechanisms can be employed, but in more general cases the J-couplings may be the method of choice for resonance assignment.

11.3.2.1 Dipolar Recoupling Techniques

Under magic-angle spinning conditions, the homonuclear dipolar couplings between low-γ nuclei are significantly averaged, but, through simultaneous application of MAS and synchronized rf pulses, the dipolar interaction can be recovered. The development of such experiments has permitted the reintroduction of homonuclear dipolar couplings under MAS between both isolated spin pairs and in uniformly labeled molecules. A wide variety of experiments (see Tab. 11.2) have been suggested in the literature to achieve dipolar recoupling. There are many ways of classifying the various recoupling techniques: transient versus adiabatic, selective versus broad-band, or double-quantum versus single-quantum transfer. We will use the type of transfer mechanism, either double-quantum (DQ) transfer or zero-quantum (ZQ) transfer, used for classifying the different techniques.

Zero-Quantum Recoupling

Zero-quantum recoupling schemes lead to a polarization transfer from $\hat{I}_{kz}$ to $\hat{I}_{jz}$, where the sign of the polarization and the sum polarization of the sample are preserved. Among the homonuclear zero-quantum recoupling sequences are: rotational resonance (R^2) [51–53], its variant rotational-resonance tickling (R2TR) [54–56], and its adiabatic implementation adiabatic-passage rotational resonance (APRR) [57]; radio-frequency driven dipolar recoupling (RFDR) [58] (Fig. 11.6a); simple excitation for the dephasing of the rotational echo amplitudes (SEDRA) [59]; unified spin echo and magic echo (USEME) [60]; and rotating/laboratory frame (RIL) [61, 62] (Fig. 11.6b).

Rotational resonance [51–53] is mostly used as a selective recoupling method between isolated spin pairs generated chemically by selective labeling. It reintroduces the homo-

Tab. 11.2 Characteristics of heteronuclear and homonuclear dipolar recoupling schemes

Sudden recoupling method	*Adiabatic recoupling method*	*ZQ or DQ method*	*γ-encoding*	*References*
CP	APHHCP	ZQ or DQ	yes	32–34, 95
REDOR	–	ZQ	no	102
R^2	APRR	ZQ	yes	51–53, 57
R2TR	R2TR	ZQ or DQ	yes	54–56
HORROR	DREAM	DQ	yes	67–69
C7	APC7	DQ	yes	75–77
DRAMA	–	DQ	no	70, 71
MELODRAMA	–	DQ	no	72
USEME	–	ZQ	no	60
RIL	–	ZQ	no	61, 62
DRAWS	–	ZQ and DQ	no	78
SEDRA	–	ZQ	no	59
RFDR	–	ZQ	no	58
BABA	–	DQ	no	73, 74

nuclear dipolar coupling through the matching of the spinning frequency to some submultiple of the isotropic chemical-shift difference, i.e. the condition $n\omega_r = \Omega_k^{\mathrm{iso}} - \Omega_j^{\mathrm{iso}}$ has to be fulfilled with $n = \pm 1, \pm 2, \ldots$ It is convenient to measure weak dipolar couplings in the form of a polarization-exchange experiment, which allows quite small couplings to be determined. Polarization at one of the labeled sites is selectively inverted and then the exchange between the two sites monitored as a function of the mixing time (Fig. 11.7). These polarization-exchange curves can be simulated in order to extract the strength of the dipolar coupling between the two labeled sites.

The polarization-exchange trajectories do not depend solely on the dipolar coupling. Additional contributions appear from the zero-quantum relaxation time constant, T_{2ZQ}, the size and orientation of the two chemical-shielding tensors, and from an inhomogeneous broadening of the isotropic chemical-shift difference. At the $n = 1$ rotational-resonance condition ($\omega_r = |\Omega_k^{\mathrm{iso}} - \Omega_j^{\mathrm{iso}}|$), the influence of the chemical-shielding tensors is smaller and is usually neglected. Uncorrelated inhomogeneous broadening of the two zero-quantum lines [63] leads to an incomplete decay of the difference polarization. Efficient heteronuclear spin decoupling during the mixing time is critical for the quality of the data one can obtain [64, 65]. Insufficient decoupling will lead to an additional decay of the polarization due to residual couplings.

Radio-frequency driven recoupling (RFDR) [58] uses rotor-synchronized 180°-pulses to prevent the averaging of the homonuclear dipolar coupling by the MAS rotation. A single 180°-pulse is placed in the middle of each rotor period (Fig. 11.6a), often using an XY-8 phase cycle [66]. The efficiency of the recoupling depends on the isotropic chemical-shift difference of the two spins and the size and relative orientation of their CSA tensors.

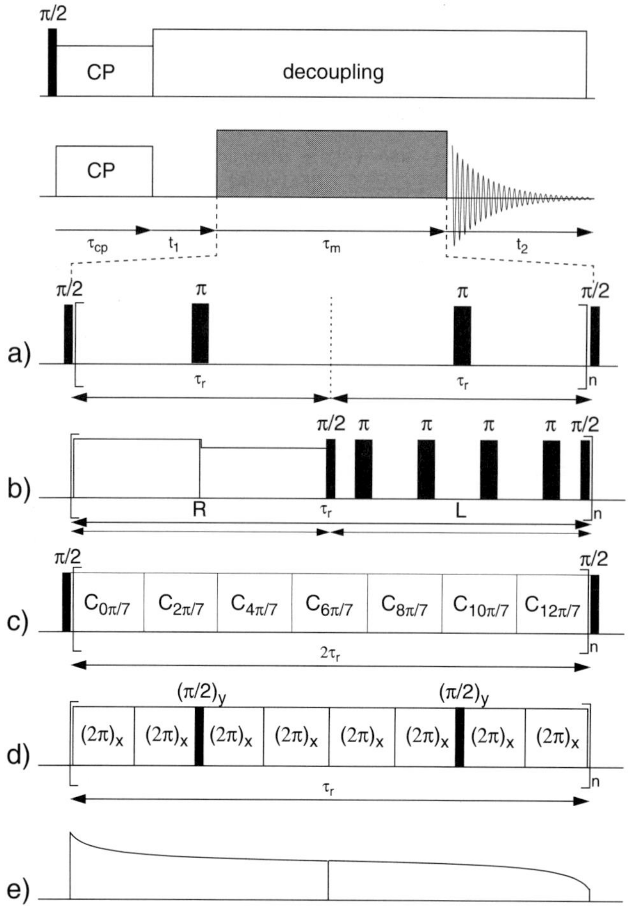

Fig. 11.6 Pulse sequence for two-dimensional homonuclear chemical-shift correlation experiments. The gray box indicates the mixing sequence, with some examples shown in more detail: **a** RFDR, **b** RIL, **c** C7, **d** DRAWS, **e** DREAM.

Double-Quantum Recoupling

Double-quantum recoupling schemes lead to a polarization transfer from $\hat{I}_{kz}$ to $-\hat{I}_{jz}$, where the sign of the polarization changes in each transfer step. For the two-spin system, the difference polarization but not the sum polarization is a preserved quantity. If one uses a double-quantum recoupling sequence as the mixing step in a two-dimensional chemical-shift correlation experiment (see Fig. 11.6), then the sign of the cross peaks will alternate between negative and positive depending on the number of polarization-transfer steps involved. Among the homonuclear double-quantum recoupling sequences are: homonuclear rotary resonance (HORROR) [67] and its adiabatic implementation dipolar recoupling enhanced by amplitude modulation (DREAM) [68, 69] (Fig.

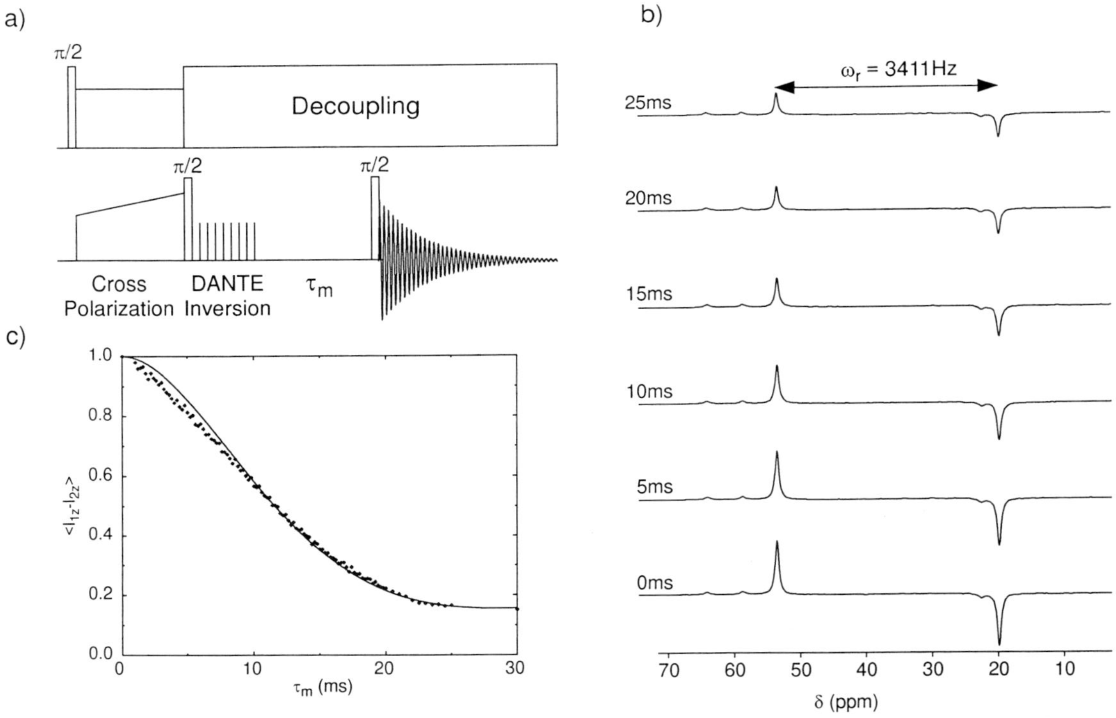

Fig. 11.7 **a** Pulse sequence for rotational-resonance recoupling of homonuclear spin pairs. **b** The spinning frequency is matched to the isotropic chemical-shift difference, and one of the resonances is selectively inverted and the polarization exchange measured as a function of the mixing time. **c** The difference polarization as a function of the mixing can be evaluated to give the dipolar coupling constant.

11.6e), dipolar recovery at the magic angle (DRAMA) [70, 71], melding of spin-locking and DRAMA (MELODRAMA) [72], back-to-back sequence (BABA) [73, 74], C7-type sequences [75, 76] (Fig. 11.6c) and their adiabatic variant APC7 [77], and dipolar recovery with a windowless sequence (DRAWS) [78] (Fig. 11.6d), which is not a pure double-quantum recoupling sequence but also reintroduces the zero-quantum part of the dipolar Hamiltonian.

A double-quantum recoupling experiment frequently employed is C7 [75, 76]. The C7 sequence reintroduces the homonuclear dipolar interaction through the application of a rotor-synchronized radio-frequency pulse cycle combined in a seven-fold symmetric phase-shift scheme. The basic element consists of a $(2\pi)_\varphi(2\pi)_{\varphi+\pi}$ unit which is phase shifted by $2\pi/7$. The C7 experiment has found application in the excitation and reconversion of double-quantum coherence in rotating solids, and has been frequently used in experiments for the determination of torsion angles. One drawback of this experiment is the relatively high rf field requirement of $\omega_1 = 7 \cdot \omega_r$. This results in high field strengths even for moderate spinning frequencies, and thus shows greater applicability at lower spinning frequencies.

Some modifications of the C7 sequence are based on a variation of the basic pulse unit. The POST-C7 sequence uses a basic element of $(\pi/2)_\varphi(2\pi)_{\varphi+\pi}(3\pi/2)_\varphi$ [76]. The CMR7 sequence uses two different basic elements which are combined in a super cycle [79]. Alternative schemes have been proposed that reduce these rf field requirements, matching the rf field amplitude to five times the spinning frequency, which may in some instances be preferable [80]. There are also newer multiple-pulse sequences based on the same symmetry principles as the C7 sequence which have significantly lower rf field requirements [81–84].

11.3.2.2 J-Coupling Polarization-Transfer Techniques

J-couplings are the basis of most homonuclear polarization-transfer techniques in liquid-state NMR. Apart from technical details, J-based transfer schemes in solids work similarly to their liquid counterparts. One has, however, to consider that pulse schemes that lead to differential polarization transfer (e.g. COSY) are usually not very helpful, because they lead to antiphase J-patterns in the spectral domain. Because the spectral resolution in solids is not always good enough to resolve J splittings between two low-γ nuclei, the antiphase lines can cancel each other out. Therefore, it is beneficial to use polarization transfer schemes with a net transfer (e.g. TOCSY). If one wants to exploit the J-coupling for polarization-transfer experiments with TOCSY-type transfer, the J-coupling Hamiltonian must be the dominant term in the effective Hamiltonian that governs the mixing time of the experiment. To achieve this, one has to employ additional rf irradiation schemes in order to compensate for isotropic chemical-shift differences and to improve the averaging of the large anisotropic interactions (dipolar couplings and chemical-shielding tensors) [85, 86]. Furthermore, because of the smallness of the couplings, longer polarization-transfer times than those for dipolar transfer are typically needed that can lead to enhanced relaxation and problems with high-power rf irradiation over extended periods of time. Because of these technical difficulties, the use of J-couplings for homonuclear polarization transfer is quite recent. It is mainly motivated by the fact that J-couplings are directly correlated with chemical bonds and allow tracing out the chemical connectivity between atoms. In addition, the isotropic nature of the J-coupling interaction allows, in theory, transfer efficiencies of up to 100%. Experiments analogous to the TOCSY experiment in solution have been developed in recent years [86–89]. These experimental schemes are used in two-dimensional polarization-transfer experiments and do not depend on the J-coupling being resolved. The efficiency of the TOBSY sequences depends on the effective suppression of both the anisotropic and isotropic chemical shifts. In addition, the pulse scheme has to be designed such that it does not reintroduce dipolar couplings which, in the absence of the multiple-pulse irradiation, would be effectively suppressed by the magic-angle spinning. These schemes are now proving useful in the assignment of sites in uniformly labeled proteins [90]. Techniques analogous to the INADEQUATE [91] experiment have been proposed and experimentally carried out [92–94]. These experiments have been employed to facilitate the assignment of sites in structurally heterogeneous materials such as wood [93].

11.3.3 Heteronuclear Polarization Transfer

In the following, we will discuss heteronuclear polarization-transfer techniques in four different contexts. They can be used as a polarization-transfer method to increase the sensitivity of a nucleus and to shorten the recycle delay of an experiment as it is widely used in ^{1}H-^{13}C or ^{1}H-^{15}N cross polarization. Heteronuclear polarization-transfer methods can also be used as the correlation mechanism in a multi-dimensional NMR experiment where, for example, the chemical shifts of two different spins are correlated. The third application is in measuring dipolar coupling constants in order to obtain distance information between selected nuclei as is often done in the REDOR experiment. Finally, heteronuclear polarization transfer also plays a role in measuring dihedral angles by generating heteronuclear double-quantum coherences.

11.3.3.1 Dipolar-Recoupling Techniques

As has been stated earlier, fast MAS will average out heteronuclear dipolar couplings. In order to efficiently use dipolar couplings for polarization transfer one has to reintroduce the dipolar couplings under MAS by applying rf irradiation schemes. Hartmann-Hahn cross polarization [32] is based on the matching of the rf field strength of two nuclei such that the Hartmann-Hahn condition $|\gamma_I B_{1I}| = |\gamma_S B_{1S}|$ is fulfilled (Fig 11.3). In the presence of sample spinning, the Hartmann-Hahn condition is split up into a series of matching sidebands $|\gamma_I B_{1I}| = |\gamma_S B_{1S}| + f \cdot \omega_r$ that are separated by the MAS frequency [95]. At these sideband conditions, the dipolar interaction is recoupled. In a zeroth-order approximation, only the $f = \pm 1, \pm 2$ matching conditions are allowed. The sidebands of the Hartmann-Hahn condition are narrower than the full matching condition in a static solid and decrease in width with increasing spinning frequency until they reach their final width determined by the heteronuclear dipolar-coupling constant. In NMR of biological molecules, cross polarization between protons and low-γ nuclei is routinely used as a way of increasing the sensitivity and shortening relaxation delays (see also Sect. 11.2.2.2). Proton spectroscopy is rarely used in biological solid-state NMR because the achievable

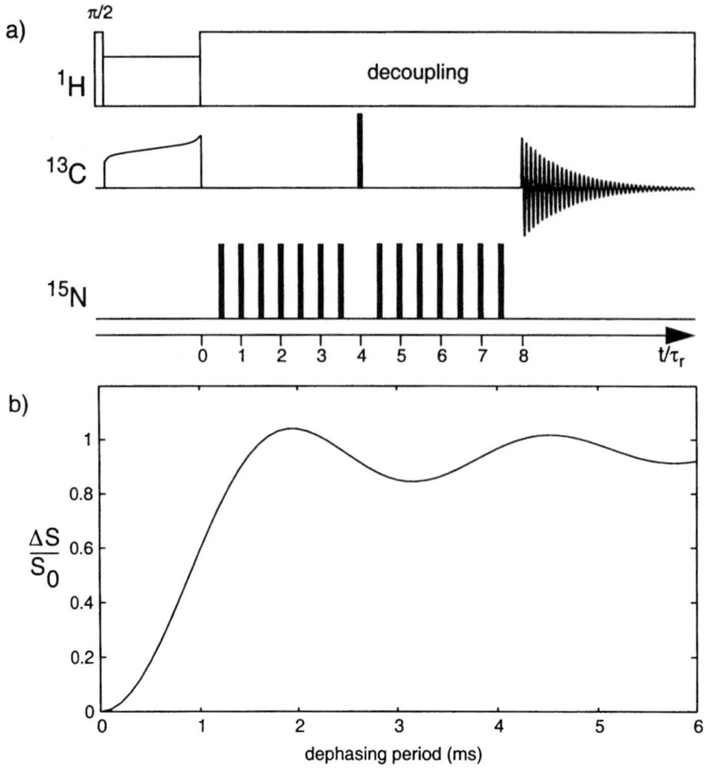

Fig. 11.8 (**a**) REDOR pulse sequence for the determination of dipolar couplings between ^{13}C and ^{15}N. Initially ^{13}C polarization is generated by cross polarization from the protons. During the following evolution period π pulses are used to prevent the averaging of the heteronuclear dipolar coupling. (**b**) Example of the dephasing curve obtained for adjacent $^{13}C/^{15}N$ spin (d=840 Hz). Typically, the degree of dephasing (ΔS) is normalized to the non-dephased data (S) collected in the absence of τ pulses applied to the ^{15}N spins for each point in the dephasing curve.

resolution is quite low unless special multiple-pulse sequences, like frequency-switched Lee-Goldburg decoupling [96–98], are used during the evolution period [99, 100].

Hartmann-Hahn cross polarization between two low-γ nuclei has been successfully used to record chemical-shift correlation spectra between ^{13}C and ^{15}N nuclei. Cross polarization between two low-γ nuclei suffers from a high sensitivity to the exact matching condition at one of the side bands of the Hartmann-Hahn condition [101]. Adiabatic methods (APHH-CP) can eliminate most of this sensitivity and lead to high transfer efficiencies [34, 62, 90].

Rotational-echo double resonance (REDOR), originally introduced by Gullion and Schaefer [102], is a method to recouple heteronuclear spin pairs. The sequence relies on a train of rotor-synchronized π pulses applied to the I spin to interrupt the spatial averaging of the heteronuclear dipolar coupling under MAS to give a nonvanishing dipolar Hamiltonian over a full rotor cycle (Fig. 11.8). Typically, REDOR data are collected by collecting the dephased signal (with I-spin π pulses), often called S, and the nondephased signal (without I-spin π pulses), often called S_0 in alternating experiments. In the absence of I-spin π pulses the decay of the signal as a function of the number of rotor cycles reflects an inherent T_2 process within the sample. In the presence of I-spin π pulses the decay will be enhanced because of the reintroduction of the I-S dipolar couplings. The analysis of this experiment is aided by the weak dependence of the dephasing curves on the relative orientation of the chemical shift tensors. For the analysis of weaker coupling where extended dephasing periods are required, efficient proton decoupling is required to prevent an overall decay of the S_0 polarization.

11.3.3.2 J-Coupling Polarization-Transfer Techniques

Solid-state analogs of the HMQC [103, 104] and HSQC [105] experiment, MAS-J-HMQC and MAS-J-HSQC [106, 107], have been proposed. They rely on the suppression of the large homonuclear dipolar couplings by FSLG irradiation of the protons. In contrast to the liquid-state implementations, both experiments use low-γ detection in the solid state.

11.3.4
A Comparison with Liquid-State NMR Methods

For readers mainly familiar with liquid-state spectroscopy, it may be interesting to make a comparison of a typical solid-state application, as illustrated in Fig. 11.6, with a corresponding solution-state experiment. Initially, polarization is transferred from protons to carbons via the dipolar couplings using heteronuclear cross polarization. In the corresponding solution-state experiment, a corresponding block could be an INEPT transfer mediated by the heteronuclear scalar couplings. During t_1, the scalar and dipolar coupling together with the chemical shift anisotropy are removed through the application of MAS and high-power proton decoupling to obtain an evolution under the isotropic chemical shift and the homonuclear J-couplings only. In isotropic solution-state NMR, only the heteronuclear J coupling has to be removed in a corresponding experiment, e.g. through the application of a 180° proton pulse in the center of t_1. During the following mixing period, a mixing scheme is applied to generate a suitable Hamiltonian for the transfer of polarization between spins. The mixing sequences in solids can employ either coherent through-bond transfer or coherent through-space transfer with isotropic mixing sequences (analogous to those employed in TOCSY experiments) or dipolar recoupling schemes for the dipolar-mediated coherent transfer of polarization between spins. In an isotropic phase, in contrast, coherent transfer is always mediated by J couplings. The incoherent cross-relaxation transfer, which is a key element for liquid-state structure determination by NOESY spectroscopy, has only then a strict analog in solids if considerable internal motion is present ("exchange-driven spin diffusion"). Otherwise, the closest analog is probably the proton-driven spin-diffusion experiment where the complicated influence of the proton bath on carbon polarization transfer can be approximately modeled by a correlation function similar to the one used for the description of relaxation. In contrast to cross-relaxation, the spin-diffusion process is not automatically accompanied by T_1 relaxation. Detection during t_2 is typically on the low γ-nuclei in solids, whereas in liquids a reverse INEPT transfer is usually inserted to detect the magnetization on the more sensitive proton nuclei. As mentioned above, indirect detection can, under certain circumstances, also be favorable in solids.

11.4 Experimental Considerations

11.4.1 Labeling Strategies

The reliance of current solid-state NMR methods on low-γ nuclei makes it necessary to introduce NMR-sensitive isotopes into a range of biological systems. Solid-state NMR experiments typically fall into two classes: (i) those which aim to obtain high-resolution data in a well-defined (small) spin system, and (ii) those which aim at obtaining assignment and structural data from the entire system or a significant part of it. Because of the differing nature of these two types of experiments, isotope-labeling procedures have been developed to permit molecules to be labeled both site-specifically and uniformly.

The reliance of several spectroscopic methods, including NMR, on isotopic labeling has driven the development of new synthetic strategies aimed at the introduction of isotopes into a wide range of natural products. The successful development of synthetic routes for such molecules should meet several criteria: (i) the synthesis should be based on simple compounds that can be purchased in specifically and highly enriched form (>99%); (ii) because of the expensive nature of the reagents, an efficient and highly-optimized synthetic scheme is required; (iii) care should be taken to ensure that no dilution or scrambling of the isotope occurs at any stage. With these design principles, synthetic strategies are now available that permit the site-selective introduction of ^{13}C and ^{15}N labels into a wide range of natural products including amino acids [108] and prosthetic groups [109] (see also Chapt. 1). An attractive alternative to the *de novo* synthesis of natural products is chemical modification of the raw material. These labeling strategies based on selective removal and reintroduction of labeled molecules into large natural products, so called retrosynthesis, permit the selective labeling of complex biomolecules in relatively high yield and with little isotopic scrambling [110].

11.4.1.1 Specific Labeling Strategies for Small Peptides

Many structural studies of small peptides, either as protein aggregates or reconstituted into bilayer systems, rely on the selective introduction of NMR-sensitive isotopes into locations suitable for the determination of dipolar couplings. The synthesis of small peptides (fewer than 50 amino acid residues) is routinely achieved by solid-phase synthesis based on either FMOC (9-fluorenylmethoxycarbonyl) [111, 112] or BOC (*tert*-butoxycarbonyl) [111] protection chemistry. Through the incorporation of the protected amino acid labeled at the appropriate position, the incorporation of NMR-sensitive isotopes can be achieved without scrambling and with reasonable yield [111, 112]. These strategies have been used in the synthesis of a range of small peptide systems which have been employed in solid-state NMR studies [113]. Many labeling schemes have been developed to probe a variety of structural problems. The conformation of the peptide backbone, for example, has been determined with backbone labeling of two times two residues, namely

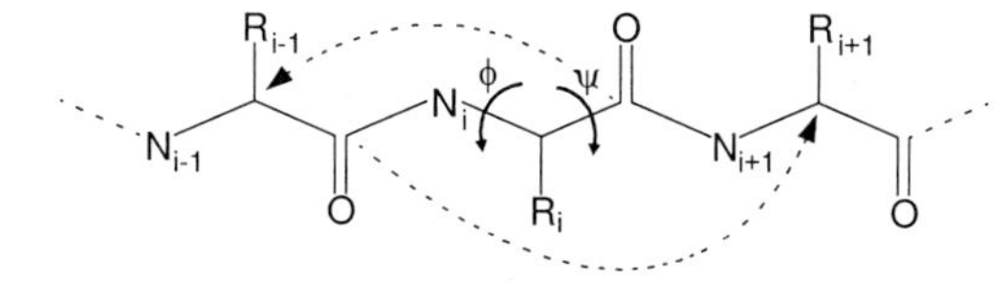

Res(i-1)	Res(i)	Res(i+1)
Cα	C'	
C'		Cα

Fig. 11.9 Diagram showing the homonuclear distance measurements necessary to define the torsion angle ϕ and Ψ of the amino acid of an arbitary amino acid, i.

at the C^{α}_{1-i} and C'_{i} positions in one selectively labeled compound and at the C'_{i-1} and C^{α}_{i+1} positions in the other. Such a scheme yields two internuclear distances, which allows the unambiguous determination of the torsion angles surrounding the amino-acid residue *i* (Fig. 11.9). Such a labeling scheme has been employed in the determination of the backbone conformation of several peptides including β-amyloid(1–42) fibrils [114] and the M13 coat protein reconstituted into lipid vesicles [113]. As mentioned above, it is also possible to directly measure torsion angles without measuring distances.

11.4.1.2 Specific Labeling of Proteins

The synthesis of larger, more complex proteins with site-selective labels still remains a real challenge. However, the expression of the proteins in suitable multi-auxotrophic media allows the incorporation of isotopes at high levels for particular types of amino acids without scrambling of the label. Although this results in the labeling of all amino-acid residues of a particular type, with a limited amount of structural knowledge it is often possible to create sufficiently well-defied systems to permit the application of solid-state NMR experiments to probe detailed local geometries.

Expression of proteins labeled site-specifically has been employed to good effect with the 120 kDa bacterial serine chemotaxis receptor (Tsr) [115, 116]. Predictions based on modeling studies have been used to identify amino acids which are located near the binding sites and are suitable for solid-state NMR measurements. Expression of the chemotactic receptor was subsequently performed in *E. coli* growing on chemically defined media containing 1-^{13}C-phenylalanine, thus introducing carbon labels into a region surrounding the ligand-binding site. Subsequent REDOR experiments carried out between ^{13}C labels in the phenylalanine groups within the ligand-binding site and ^{15}N labels in the ligand helped refine the geometry of the ligand within the intact receptor [115]. Site-directed mutagenesis can now be employed to generate proteins containing only a single occurrence of a specific amino acid in large biomolecules, which can then easily be labeled using such technology [116].

11.4.1.3 Chemical Labeling/Modification of Biomolecules

The selective modification of reactive side chains within proteins with compounds containing NMR-sensitive isotopes offers an alternative route to the labeling of proteins. The chemical modification of side chains with reagents labeled with NMR-sensitive isotopes may permit detailed structural and dynamic questions to be addressed, although perturbations of the system caused by the labeling may be of concern in cases where the chemical identity of the compound is changed. A early example of a chemical modification in solid-state NMR of proteins is the carbon labeling of the surface methionine residues in cytochrome c using methyl iodide (Fig. 11.10a) [117]. Similar methods have been employed for the labeling of cysteines with ^{19}F-containing reagents. Using the trifluoroethylthio (TET) group, CF_3–CH_2–S–, Khorana and co-worker have introduced ^{19}F labels into both native and cysteine mutants of the visual chromophore rhodopsin (Fig. 11.10b) [118]. In systems where efficient expression occurs, the ability to introduce amino acids with reactive side chains at unique locations within the protein offers, in conjunction with chemical labeling, the ability to introduce labels at predetermined sites. Such labeling methods have been used to probe distances between prosthetic groups or ligands bound to membrane proteins and unique sites in the protein [119, 120] and to probe specific tertiary contacts in membrane-bound receptors [118, 121, 122].

11.4.1.4 **Uniform Labeling of Peptides and Proteins**

For all but the smallest peptides, the most economic route to the production of uniformly labeled peptides and proteins is their expression in a host growing on labeled media. The expression of protein for NMR studies typically relies on (i) a well-optimized expression system, with relatively high yields of protein (typically several 10 mg per liter of culture); (ii) growth on well-defined chemical media which are available enriched with NMR-sensitive isotopes. Although expression of proteins in a range of hosts is now possible, because of the ease of genetic manipulation, scalability of growth and ease with which bacterial growth is attained on minimal media containing only small organic molecules and other salts, *E. coli* represents the preferred route to the expression of isotopically labeled proteins. Expression and purification of a range of systems from small peptides [123] to integral membrane proteins [124] have been demonstrated using such methodology. For proteins derived from higher organisms, post-translational processing is a vital step in the correct expression, folding, and function of the protein. The absence of many of the post-translational processing pathways in bacterial systems in some cases hinders the expression of functional active protein [125]. The expression of proteins in cell lines derived from higher organisms has been shown to be a viable alternative to bacterial expression. Successes include the expression of several members of the G-protein coupled receptor family of proteins in a range of systems including those based on the Semliki Forest virus [126], baculo virus [127] and stable expression in HEK cells [128]. Importantly, the scalability of these systems has been demonstrated with yields between 4 and 10 mg per liter of culture. Although the composition of the media is typically more complex than that required for bacterial expression, labeling is possible through the introduction of labeled sugars and amino acids. Putative problems with isotope dilution associated with the presence of unlabeled animal serum in the media have largely been prevented either through the adaptation to growth on serum-free media or the removal of unlabeled small molecules from the serum by dialysis.

11.4.1.5 **Isotopic Dilution**

At moderate MAS frequencies, solid-state NMR spectra of uniformly labeled material have somewhat broader lines than their unlabeled counterparts. This has been attributed to only partial averaging of the dipolar-coupling between the low-γ nuclei by MAS and

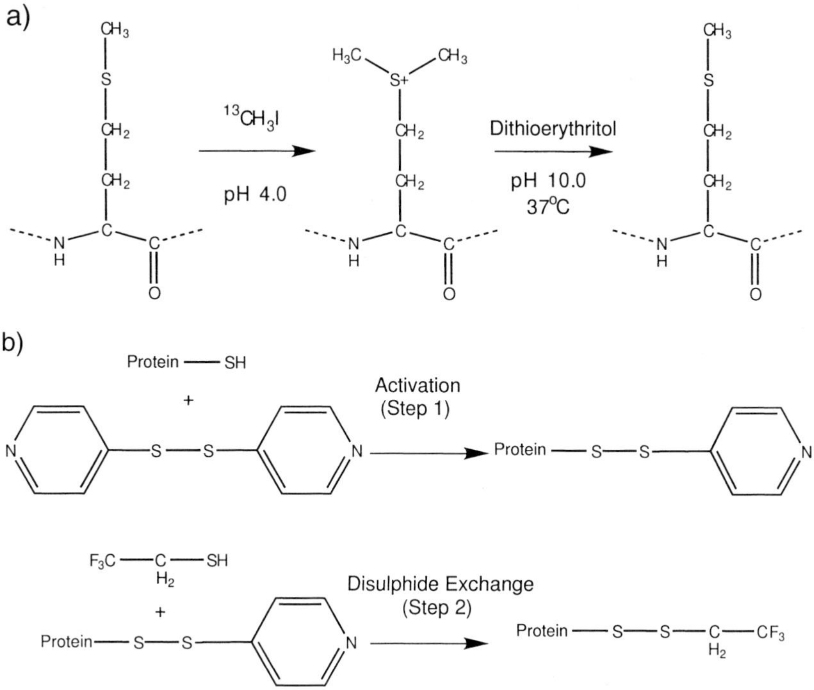

Fig. 11.10 Diagram showing the reaction scheme for the exchange of the methionine side chain using carbon-13 methyl iodide (**a**) and the labeling of cysteine side chains using the reagent trifluoroethylthio such that fluorine-19 labels can be introduced into the system (**b**). Both methods permit the introduction of NMR-sensitive isotopes into the protein following its purification.

the presence of one-bond J couplings between adjacent nuclei [129, 130]. These effects can be minimized through the incorporation of labeled nuclei into defined sites in all amino acids whilst minimizing the number of labels that are separated by a single bond only. Expressing the protein in bacteria growing anaerobically on either 1-^{13}C-glucose or 2-^{13}C-glycerol, labels are selectively accumulated in the side chains and C^{α} positions, respectively, whilst under aerobic conditions more complex labeling occurs. This results in "dilute" labeling of the protein at particular sites, reducing the size of dipolar couplings and eliminating many homonuclear J-couplings [130].

Isotope dilution has also been proposed as a mechanism to exploit the high sensitivity and favorable relaxation properties of protons in solid-state NMR. As mentioned earlier, the strong dipolar interactions between the abundant protons causes a homogeneous broadening of the proton resonances. Even experiments carried out at the highest magnetic fields (900 MHz) and using the fastest MAS frequencies (50 kHz) currently available fail to remove these [29, 131]. Isotope dilution with deuterium has been proposed as a mechanism for reducing these strong homonuclear dipolar couplings by increasing the distance between neighboring proton spins [132, 133]. Such an approach also offers the potential of determining medium- and long-range dipolar couplings between protons in solids. Typically, in a fully protonated molecule, the strong dipolar couplings between protons in close contact leads to an attenuation of the weaker, yet structurally more interesting, dipolar couplings between more distant sites [133]. Through the back exchange of amide protons in an otherwise perdeuterated molecule, isotope dilution has permitted not only the observation of resolved proton resonances (proton line width of about 0.5 ppm) but also the determination of weak dipolar couplings between amide protons [132, 133].

11.4.2 Sample Preparation

11.4.2.1 Soluble Proteins

The sensitivity and resolution of solid-state NMR experiments has been shown in part to depend on the sample preparation. Effects such as the structural homogeneity of the sample, hydration, and dynamics play an important role in defining the attainable sensitivity and resolution. Initial studies on enzyme-substrate complexes indicated that, for lyophilized samples, the line width could be reduced through the use of strong buffering conditions and the presence of cryoprotectants such as PEG-3350 or trehalose prior to rapid freezing in either liquid nitrogen or liquid propane [134]. Recent success in the assignment of proteins in the solid state have been made possible by the improved resolution achieved through the crystallization of the proteins to reduce line broadening arising from structural heterogeneity within the sample. Conditions for crystallization are similar to those that have been used previously for X-ray diffraction studies, but solid-state NMR studies do not require diffraction-quality crystals. The critical point is not the presence of long-range order within the sample – rather the uniform conformation for all molecules of the ensemble studied. Techniques successfully employed include crystallization of small polypeptides from methanol/water solution [90] and the salting out of larger less-stable proteins from aqueous solution [135]. Such approaches have led to carbon line widths of 50–60 Hz, offering the potential for the complete assignment of such systems (Fig. 11.11).

11.4.2.2 Membrane Proteins

Solid-state NMR studies have been performed on a variety of membrane-associated systems, either in the native membranes, in a detergent-solubilized form, or reconstituted into synthetic bilayers. The extraction of the protein of interest into a micellar system is often an attractive alternative to solid-state NMR. Frequently, when the micellar systems containing the protein are sufficiently small, they can readily be studied using solution-state NMR methods (see Chapt. 5). Although MAS methods have been successfully employed for the study of large detergent receptor complexes in the "viscous" state [136], typically immobilization by rapid freezing of the sample is employed, which, although increasing structural heterogeneity, permits the successful application of solid-state MAS experiments [137, 138].

Most attractive are solid-state NMR studies of membrane proteins, in their "native" membranes, or membrane proteins reconstituted into synthetic lipid vesicles with defined lipid composition. Several integral membrane proteins are present in sufficiently high concentrations in the native membranes (>25% of total membrane protein) to offer samples concentrated enough to permit the application of solid-state NMR without

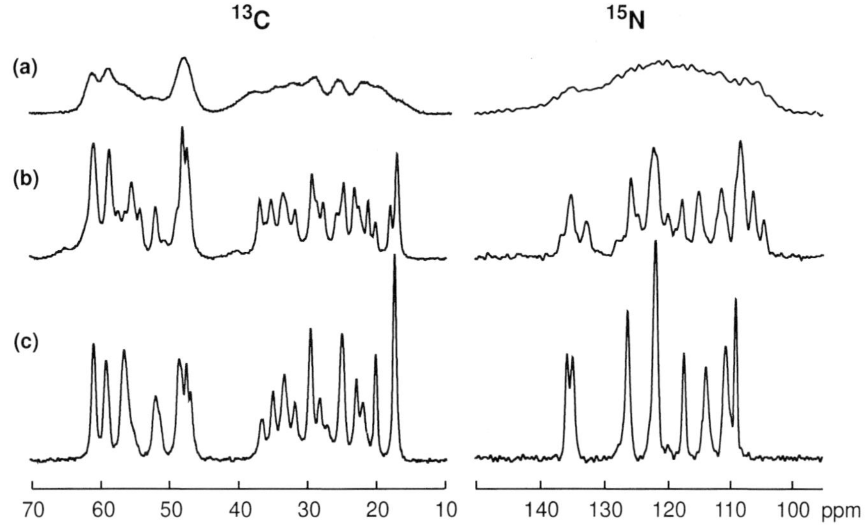

Fig. 11.11 ^{13}C and ^{15}N MAS spectra of antamanide for different sample preparations. **a** Lyophilized powder. **b** and **c** Micro-crystalline powder obtained by evaporation of the solvent from a solution of antamanide in a 7:3 methanol/water mixture. In the ^{13}C spectra, only the aliphatic region is shown. The sample leading to spectra **b** was obtained by fast evaporation of the solvent at room temperature in the presence of dry silica gel as a drying agent. The sample leading to spectra **c** was obtained by slow evaporation during several days in a controlled humidity of 76%. (Reproduced from Ref. [90] with permission)

further purification [139]. Although the study of membrane proteins in their native membrane precludes the labeling of the protein (unless bacterial in origin), labeling of reactive side chains, ligands and other prosthetic groups enables NMR-sensitive isotopes to be introduced into the system. The introduction of NMR-sensitive labels at these sites, which are frequently of functional importance, offers the possibility of obtaining structural and dynamic information from key sites in biological systems by MAS methods [110, 139]. For other integral-membrane proteins, where detergent solubilization is required to purify and concentrate the sample, reconstitution into lipid vesicles frequently permits the study of the protein in a native-like environment and typically permits higher protein concentrations than are attainable in detergent micelles, aiding sensitivity whilst suppressing dynamics [140].

11.5 Application of Polarization-Transfer Techniques to Biological Systems

11.5.1 Assignment of Resonances

The desire to obtain structural and dynamical information from multiply labeled systems stems from the fact that such an approach alleviates the need to produce multiple selectively-labeled molecules. While the use of uniformly labeled samples simplifies the sample preparation stage, problems associated with resolution and resonance assignment in the NMR experiments are often severe. Higher B_0 fields and improved sample preparation techniques, however, are areas where significant improvements have been (and will be) made with respect to resolution and sensitivity. The assignment of multiply labeled proteins in the solution state has been dominated by techniques which permit the assignment of spin systems to particular amino acids and the transfer of polarization along the peptide chain to permit sequential assignment. Recently, similar methods have been developed that permit sequential assignments of the backbone resonances in the solid state. Analogous liquid-state triple-resonance experiments e.g., HNCO, HNCA, CBCANH, CBCA(CO)NH, HBHA(CBCA)NH, and HBHA(CBCACO)NH [3], which permit the assignment of resonances in doubly $^{13}C,^{15}N$-labeled proteins, are particularly suited for application to solids. Strategies for the assignment of spin systems for particular amino acids have relied primarily on homonuclear correlation spectroscopy. In contrast to solution-state NMR, where homonuclear correlation spectroscopy makes use of J-couplings, to date many intra-residue spin-system assignments in solid-state NMR have been made using broad-band homonuclear dipolar recoupling sequences such as CMR7 [140], proton-driven spin diffusion [141], RFDR [135, 141], RIL [136], and DREAM [90]. Using such techniques with relatively short mixing times gives homonuclear correlation spectra that are dominated by the presence of cross peaks from carbon atoms belonging to the same amino acid.

More recently, homonuclear correlation techniques relying on J-couplings have also been developed [86, 87, 89] and applied to the assignment of spin systems of amino-acid residues in uniformly labeled proteins and peptides [90]. These have, in some cases, a higher information content than the comparable dipolar-mediated experiments, as relayed correlations throughout the continuous ^{13}C–^{13}C network are more easily realized at high B_0 fields [90].

For the sequential assignment of resonance lines along the peptide backbone, homonuclear dipolar correlation experiments have been used. Allowing long periods for mixing to occur through either proton-driven spin diffusion or radio frequency-driven transfer, correlations between nondirectly bonded nuclei can be observed that permit sequential assignments across the peptide bond [140–142]. However, these techniques tend to be low in sensitivity because of the long mixing times. Therefore, several triple-resonance techniques have been developed for inter-residue assignments by transferring the polarization via the ^{15}N nucleus. Examples include schemes that correlate NCA, NCO, NCACX [141], N(CO)CA [140], N(CA)CO, N(CA)CB [90, 140] (Fig. 11.12). Many of these experiments rely on the transfer mechanisms already mentioned. However, the sensitivity of these methods has been improved considerably through directed transfer of polarization between labeled sites with the aim of reducing the number of peaks in the 2D spectrum, thereby improving the signal-to-noise ratio. Methods include cross polarization optimized for transfer to occur selectively between the amide nitrogen and either the C^{α} or C'; carbons. The application of adiabatic cross polarization between carbon and nitrogen [143] at moderate radio-frequency fields and sweeping the field through the spectral region of interest was also shown to lead to the targeted transfer of polarization [141].

Similar selective homonuclear polarization-transfer schemes have been used for the subsequent homonuclear directed transfer of polarization along the polypeptide chain.

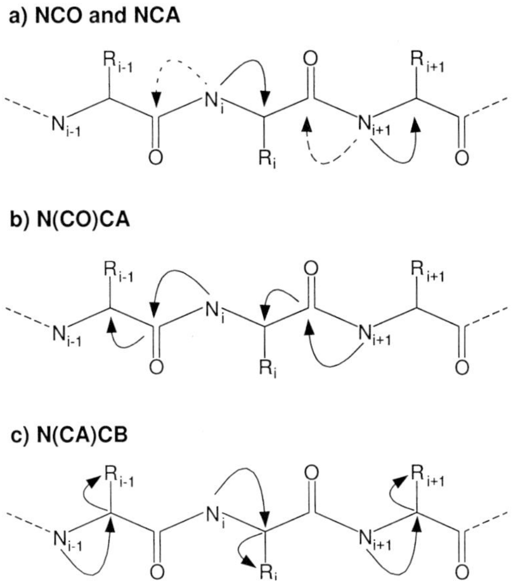

Fig. 11.12 Diagram showing the transfer of polarization for an NCO/NCA experiment (**a**), an N(CO)CA experiment (**b**), and an N(CA)CB experiment (**c**).

These selective transfers are based on dipolar recoupling sequences. For relatively large chemical-shift differences, e.g., for selective transfer between C′ and C^{α} resonances, methods such as rotational-resonance tickling [54, 56] are effective for relaying polarization. This transfer can be executed adiabatically with the corresponding gain in efficiency [90]. The methods described so far are often sufficient for the assignment of smaller peptides. For application to larger systems, a further gain in resolution can be realized by a 3D correlation experiment (NCACB) where the resonances in a NCA experiment are further dispersed with the C^{β} carbon resonance frequency, which has a larger chemical-shift dispersion [140]. In such cases, specific transfer of polarization from the amide nitrogen to either the C′ or the C^{α} carbon can then be directed toward the C^{β} carbon by means of a band-selective dipolar recoupling step. For small chemical-shift differences, e.g., transfer from the C^{α} to the C^{β}, optimum transfer has been obtained through the use of the DREAM experiment [90, 141]. The application of these techniques to the complete assignment of peptides was demonstrated by studies of the cyclic antitoxic decapeptide, antamanide [90], where polarization transfer techniques have been optimized for experimental efficiency. The full assignment of the peptide was obtained. The spin systems were assigned to residues using a J-coupled homonuclear correlation (TOBSY) experiment [86], and the sequential assignment was achieved by using dipolar NCA, NCO and N(CO)CA experiments (Fig. 11.16). Cross polarization between nitrogen and carbon nuclei was optimized through the appropriate choice of carrier frequency, rf field and sweep, such that the transfer was optimized from NH to C'. Subsequently, transfer from the C′ to the C^{α} was performed in a band-selective fashion using a rotational resonance tickling experiment, resulting in highly efficient transfer from the C' to the C^{α}. As can be seen from Fig. 11.13, the sequential assignment for this clearly resolved system allows a complete sequential assignment of the peptide to be performed.

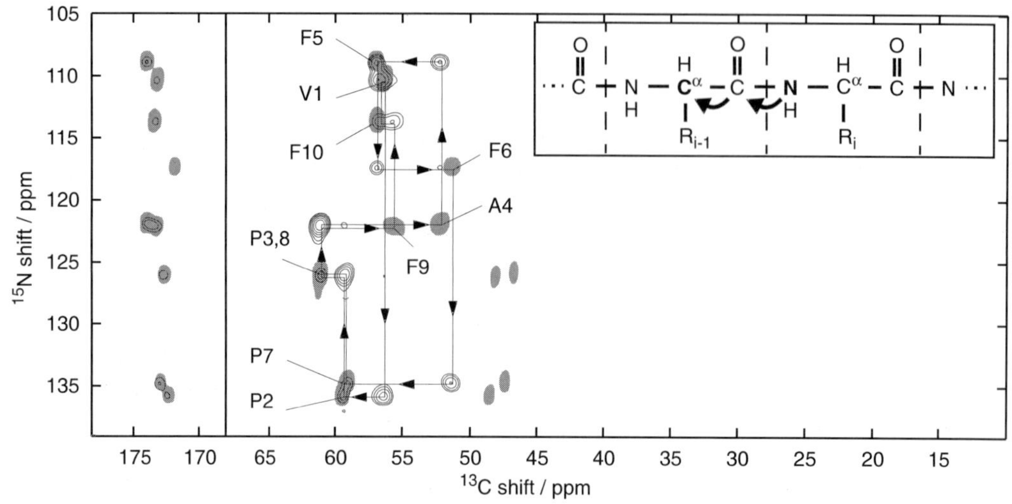

Fig. 11.13 An N(CO)CA spectrum of the decapeptide antamanide (contours) together with the CO(N)CA spectra (solid). By "walking" through these two spectra in a stepwise fashion as indicated by the arrows, all ^{15}N and ^{13}C resonances can be assigned. (Adapted from Ref. [90]).

Analogous experiments have been performed for a variety of other more complex systems including SH3 [141], LHC-I [144] and ubiquitin [140]. These experiments rely on similar principles to those outlined in the studies of the antamanide, and vary in the nature of the procedures used to perform the transfer between adjacent sites within the peptide. Although the assignment of resonances in uniformly labeled molecules provides no direct information regarding the conformation of the molecule, detailed analysis of the chemical shifts has been exploited to provide information regarding both the conformation and the environment at individual sites [145–147]. This analysis can provide information ranging from the conformation of the peptide backbone [147] to the environment of ligands upon binding to membrane receptors [146].

Although significant advances have been made in resolution and for the assignment of uniformly labeled proteins in the solid state, few experiments are currently available which could lead to the determination of the global structure of the protein or peptide. The problem is the insufficient number of long-range structural constraints. Broad-band dipolar recoupling methods, which potentially may allow long-range coupling to be probed for structural information, have provided few long-range constraints in uniformly labeled systems, because of the difficulties associated with the observation of weaker dipolar couplings in the presence of strong nearest-neighbor dipolar couplings. Experiments with extended proton-driven spin-diffusion periods can yield long-range cross peaks, but the spread of the magnetization over many nuclei leads to signal-to-noise problems and reduced spectral resolution.

11.5.2
Conformational Constraints

Applications of solid-state NMR to structural problems in biological systems have focused either on the determination of dipolar couplings between selectively introduced non-bonded homonuclear and heteronuclear spin pairs or on torsion angle measurements between neighboring spins. This enables the determination, with a relatively high degree of accuracy, of a limited number of structural constraints that can be used in the refinement of our understanding of biological systems.

11.5.2.1 Homonuclear Distance Measurements

The introduction of two NMR-sensitive isotopes site-specifically into a range of biological systems permits the determination of single distance constraints between the two labels. Homonuclear recoupling sequences have been applied to a range of biological systems, including the study of prosthetic groups and side-chain conformations in membrane-bound receptors [148, 149], the conformation of integral membrane peptides [150, 151], the conformation of ligands within enzyme binding sites, and the processes associated with tissue mineralization [152, 153].

An early example of the successful application of these methods is in the analysis of the conformation of insoluble protein aggregates derived from amyloidogenic proteins [154]. These insoluble protein aggregates arise in the brain tissue of patients suffering from Alzheimer's disease, in which the misprocessing of amyloid precursor protein results in the production of β-amyloid peptide (1–42/43). Griffin and coworkers prepared the multiple samples of the fragment 34–42 by solid-phase synthesis with ^{13}C labels introduced according to the principles shown in Fig. 11.9 [154]. Following dilution of the peptide to reduce the probability of observing intermolecular transfer, polarization exchange experiments were performed for each of the peptides containing a pair of ^{13}C labels (see Tab. 11.3 and Fig. 11.14), allowing the determination of the internuclear dis-

Tab. 11.3 Rotational resonance constraints obtained for β-amyloid protein

Measurement (constrained residue)	*Intramolecular distance*	*Intermolecular effect*
^{a}34,35(Met 35)	≥4.65 Å	–
34,^{a}36(Met 35)	4.7–56 Å	–
35,^{a}37(Val36)	4.9–5.6 Å	–
^{a}36,37(Gly37)	≥4.6–5 Å	–
36,^{a}38 (Gly37)	5.35–6.05 Å	–
^{a}37,38 (Gly38)	4.1–4.8 Å	+
37,^{a}39 (Gly38)	4.95–5.65 Å	+
^{a}38,39 (Val39)	≥4.65 Å	–
38,^{a}40 (Val39)	≥5.5 Å	–
^{a}39,40 (Val40)	≥4.35 Å	–
^{a}40,41 (Ile41)	≥4.45 Å	–
40,^{a}42 (Ile41)	4.65–5.35 Å	–

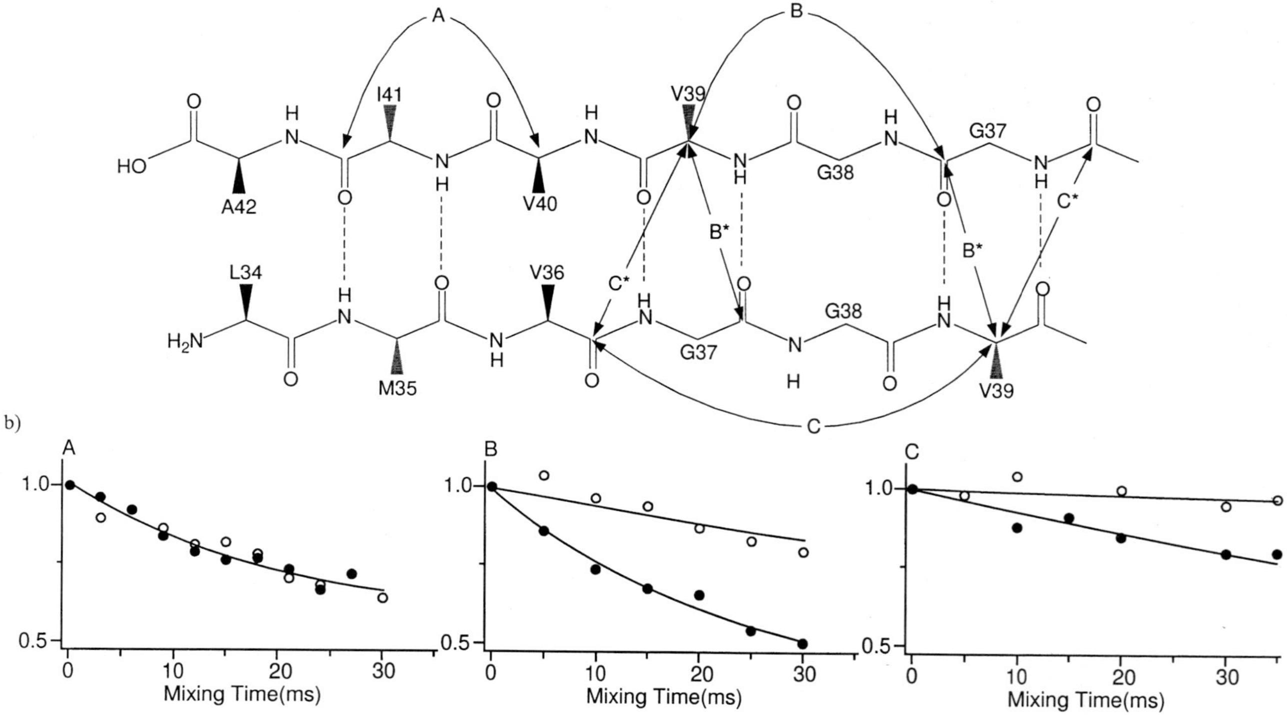

Fig. 11.14 Diagram showing the polarization exchange curves obtained between $^{a}40,41$ (**A**), $37,^{a}39$ (**B**) and $36,^{a}39$ (**C**) in β-amyloid peptide [34–42]. Data obtained from samples diluted with unlabeled peptide (open circles) and undiluted (closed circles) indicate the presence of inter-strand contacts used in subsequent modeling (**a**). The model proposed on the basis of these measurements and others given in Tab. 11.3, is shown (**b**). (Adapted with permission from Ref. [154]).

tances and the calculation of the ϕ and Ψ angles along the peptide backbone. The constraints obtained in addition to those based on backbone C^{α} and C' chemical shifts and isotope-filtered FT-IR experiments were used to screen a library of 2000 energetically favorable structures [154]. This search resulted in a family of structures consistent with the applied constraints that were predominantly β-sheet in structure, differing primarily in the *cis-trans* nature of a single GlyGly bond.

Measurements repeated on samples prepared without dilution in natural abundance peptide showed that, in some cases, significant changes in polarization exchange curves occurred, consistent with intermolecular polarization transfer. These measurements were complemented with additional rotational resonance studies on other labeled peptides to confirm the nature of this structure. On the basis of the intermolecular transfers observed in the polarization exchange curves, an antiparallel β-sheet arrangement has been proposed (Fig. 11.14). A parallel β-sheet conformation has been found for the native amyloid beta peptide(1-42/43) using multiple-quantum solid-state NMR experiments [155]. These studies have formed the basis for a number of experiments aimed at elucidating the role of a range of plaque-forming peptides in diseased states [156, 157].

11.5.2.2 **Heteronuclear Distance Measurements**

In analogy to the experiments for the determination of weak coupling between homonuclear spin pairs, the introduction of simple heteronuclear spin systems into biological molecules has allowed for the determination of weak heteronuclear couplings. Although cross polarization between low-γ nuclei is particularly effective for polarization transfer experiments, for the determination of weak heteronuclear dipolar couplings REDOR type experiments have more frequently been employed in biological systems. Using this technique, a variety of heteronuclear couplings have been analyzed in a range of systems including the characterization of novel cross-linking regimes in mussel byssus [158, 159], the measurement of metabolic flux in bacterial cultures [160], the determination of sclerotization in insect cuticular exoskeletons [161], ribosomal elongation factors [162], and conformational changes in membrane-bound receptors upon ligand binding [116, 115]. REDOR solid-state NMR studies have also found application in the study of enzyme-substrate complexes, having been applied to a wide variety of systems including tryptophan synthase [163] and both metallo/serine proteases [164, 165]. Perhaps the most extensive studies made of enzyme-substrate complexes have been performed by Schaefer and coworkers on the enzyme 5-enolpyruvylshikimate-3-phosphate (EPSP) synthase. The 46 kDa enzyme EPSP synthase catalyzes the reversible condensation reaction between shikimate-3-phosphate (S3P) and phosphoenolpyruvate (PEP), a key reaction in biosynthesis of aromatic amino acids in plants and micro-organisms. The reaction is inhibited by the commercial herbicide *N*-phosphonomethyl glycine (glyphosate, Glp), which in the presence of S3P binds the enzyme to form a stable ternary complex. Although crystal structures of the unliganded form of EPSP synthase exist [166], current attempts to form crystals of the ternary complex have not proved successful. In an attempt to characterize the structure of this ternary complex, Schaefer and coworkers have carried out a three-phase study to (i) determine the relative orientation of the two ligands, (ii) characterize the interaction between the enzyme and the two substrate molecules, and (iii) obtain long-range structural constraints capable of determining how the global conformation of

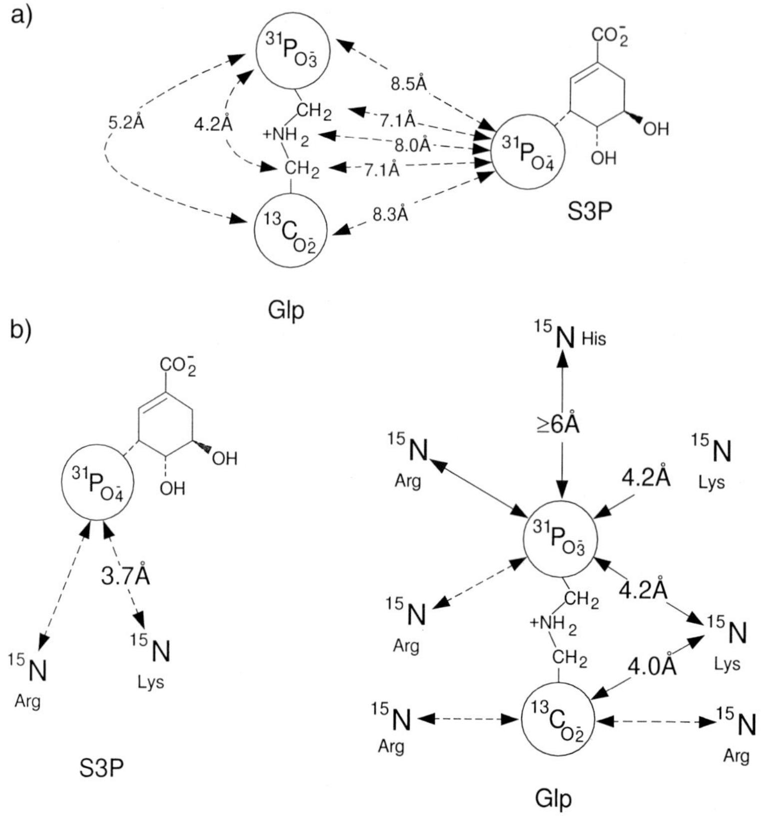

Fig. 11.15 Diagram showing the relative proximity of the two ligands Glp and S3P, which form a stable ternary complex with the enzyme EPSP synthase. The distance constraints were obtained from both homonuclear and heteronuclear dipolar couplings obtained using the REDOR and DRAMA pulse sequence, together with a model showing a selection of allowed orientations of S3P with respect to the Glp molecule (**a**) (reproduced with permission from Ref. [167]). The proximity of the basic side chains involved in charge stabilization in the ternary complex have been obtained by REDOR measurements between the basic side chains and labeled sites within the ligands (**b**).

the protein changes upon substrate binding [167]. For solid-state NMR studies, a ternary complex of EPSP synthase, S3P and Glp was prepared from a dilute solution of the ternary complex in a buffer, which was flash frozen and lyophilized, resulting in a protein immobilized in an ionic glass formed from the buffer. Preliminary ^{31}P-observed ^{13}C-dephased REDOR curves confirmed the proximity of the ^{13}C-Glp to the ^{31}P of the S3P in the enzyme binding site [168], and the geometries were subsequently refined using Glp labeled specifically at each of its ^{13}C and ^{15}N atoms [163] (Fig. 11.15 b). The proximity of the basic side chain involved in charge stabilization was subsequently probed using an elegant series of experiments measuring both ^{31}P-observed ^{15}N-dephased and ^{13}C-observed ^{15}N-dephased REDOR experiments between the protein and labeled S3P and Glp [167]. With the aid of TEDOR polarization transfer measurements which permitted the assignment of resonances to particular side chains in the protein (the model of the ternary complex as depicted in Fig. 11.15 b), an understanding of the interactions involved in the stabilization of the ternary complex was achieved. These measurements have been complemented with ^{31}P-dephased 19F REDOR data obtained from fluorinated tryptophans within the protein to the ligands in the binding site, which have indicated a folding of the two lobes of the protein around the ligands upon the formation of the ternary complex [169].

11.5.2.3 Measurement of Torsion Angles

A complementary method to the determination of multiple distance constraints in proteins is the direct determination of torsion angles. These techniques rely on the correlation of anisotropic spin interactions such as dipolar-coupling tensors and chemical-shielding tensors to obtain local structural information. Dipolar coupling tensors are axially symmetric, and, by correlation of two tensors, one angle can be obtained. The long axis of the dipolar coupling tensor is always aligned with the bond direction. In contrast, the CSA is not in general axially symmetric, and, by correlation, three angles can be obtained. However the relationship between the principle axis system of the CSA and the molecular coordinate system is more complicated. Sometimes, for example, for ^{13}C carbonyl tensors it is relatively well characterized (within 5°) from model compounds. Typically it is now possible to calculate the orientation of the chemical shift anisotropy using density-functional theory.

One of the earliest examples of this class of experiments was developed for the determination of the torsion angle in an H–C–C–H system using the properties of the ^{13}C double-quantum coherences [170]. In these experiments, ^{13}C double-quantum coherence was excited using a double-quantum dipolar recoupling scheme. The evolution of the double-quantum coherence was then followed for one rotor period. The rotor period is divided into two portions, t_1, containing a homonuclear decoupling scheme e.g. MREV8 such that the double quantum coherence evolves solely under the influence of the ^{1}H-^{13}C heteronuclear dipolar coupling, and the second containing high-power proton decoupling to effectively suppress the 1H-^{13}C dipolar interaction (Fig. 11.16 a). The double-quantum intensity was subsequently determined indirectly by reconversion to single-quantum coherence. An analysis of the evolution of the double-quantum intensity as a function of t_1 allows the determination of the relative orientation between the two ^{1}H-^{13}C bonds.

Such experiments were applied to the proton pump bacteriorhodopsin, from *Halobacterium salinarum*. Torsion-angle measurements were used to supplement data from a range of other static [171] and MAS [148] solid-state NMR results to provide information regarding the conformational changes that are undergone by the retinylidene chromophore during the photocycle responsible for the pumping of protons across the bacterial membrane. Through the introduction of ^{13}C labels at the C14 and C15 positions in the retinylidene chromophore, the torsion angles around the C14–C15 bond have been probed. The recoupling scheme CMR7 (a derivative of the C7) was employed for the excitation and reconversion of double quantum coherence between the C14 and C15 carbon atoms. The evolution of the double-quantum coherence under these heteronuclear couplings was analyzed for the ground state (bR_{568}) (Fig. 11.16 b), M_0, and the M_n states of bacteriorhodopsin by trapping each of the photointermediates at low temperature whilst

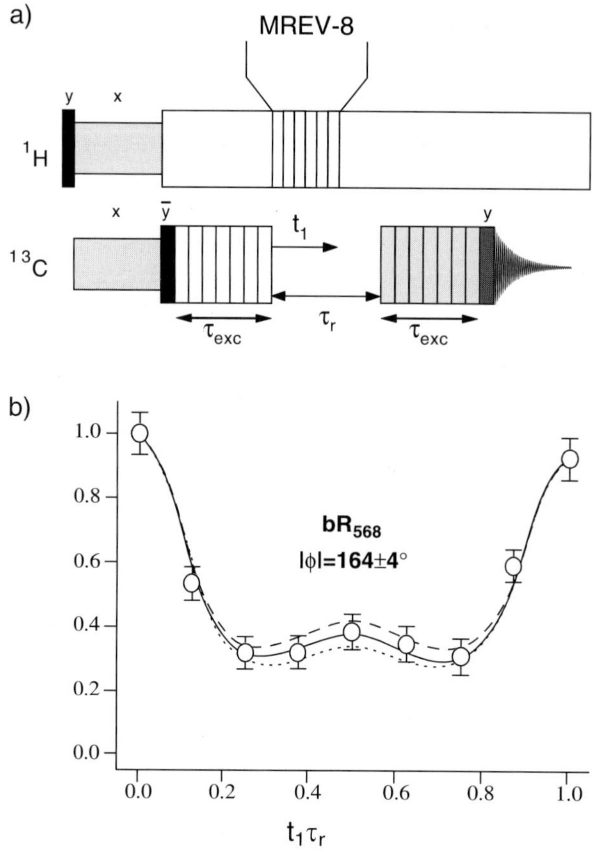

Fig. 11.16 The pulse sequence used to monitor the evolution of carboncarbon double-quantum coherence over a single rotor period in the presence of the proton-carbon heteronuclear dipolar coupling (**a**). The evolution of the double-quantum coherence between the C14 and C15 carbons in the retinal of bacteriorhodopsin in the ground state (**b**). The observed evolution is consistent with a C14–C15 torsion angle of 164° (reproduced with permission from Ref. [172]).

illuminating the sample. Analysis of data acquired in the ground state indicated a dihedral angle of 164°, indicating that even in the ground state the chromophore is distorted from its ideal planar geometry. Following activation and conversion to the M_n state, this distortion increases to a dihedral angle of 150°, which may in part explain how the Schiff base remains oriented toward the extracellular side of the protein during the first half of the photo cycle [172]. Similar experiments have been proposed that allow the determination of the torsion angle N–C–C–N through a correlation of the two N-C dipolar interactions [173]. In a manner analogous to the experiments described above, ^{13}-C^{13}C double-quantum coherence is initially excited. However, in contrast to the H–C–C–H experiment, the ^{13}C-^{15}N dipolar couplings are small and averaged by MAS and must be reintroduced, e.g. by rotary-resonance recoupling with rotor synchronous phase inversions to suppress the effects of chemical-shift anisotropy (SPI-R3) [173] or REDOR style pulse schemes [174]. These techniques have been applied to determine the backbone ϕ and ψ angles in proteins and have been successfully applied to extensively labeled systems [174]. The differential response of these experiments to characteristic secondary structural motifs in proteins has led to the proposal to use these sequences as secondary structure filters in multidimensional solid-state NMR experiments, permitting the assignment of particular resonances to particular secondary structural motifs [174, 175].

11.6
The Future of Applications/Developments of Solid-State NMR in Biology

Currently the application of solid-state NMR by MAS methods to biological systems has proved most fruitful upon the incorporation of isotopes into well-defined sites within the system under study. Although these studies provide relatively limited amounts of information, they have permitted significant advances in our understanding of biological systems.

Methods offering the potential to obtain multiple structural and dynamic constraints from single uniformly labeled samples will significantly enhance the application of solid-state NMR methods. Significant advances have recently been made in both the preparation and the assignment of uniformly labeled systems. With current developments in MAS-NMR methodology, the elucidation of structural parameters by MAS-NMR may offer an alternative route to established methods for the structural analysis of biological systems. Significant progress toward this aim has been made in the last few years, and we are optimistic about the future prospects.

11.7
References

1 R. Huber, *Angew. Chem. Int. Ed. Engl.* **1989**, *28*, 848–869.

2 K. Wüthrich, 1986, *NMR of Proteins and Nucleic Acids.* Wiley Interscience, New York.

3 J. Cavanagh, W. J. Fairbrother, A. G. Palmer, and N. J. Skelton, 1996, *Protein NMR Spectroscopy.* Academic Press, San Diego.

4 J. D. van Beek, L. Beaulieu, H. Schaefer, M. Demura, T. Asakura, and B. H. Meier, *Nature* **2000**, *405*, 1077–1079.

5 P. T. F. Williamson, J. A. Watts, G. H. Addona, K. W. Miller, and A. Watts, *Proc. Natl. Acad. Sci. U.S.A.* **2001**, *98*, 2346–2351.

6 G. Grobner, I. J. Burnett, C. Glaubitz, G. Choi, J. Mason, and A. Watts, *Nature* **2000**, *405*, 810–813.

7 X. Fu and T. A. Cross, *A. Rev. Biophys. Biomol. Struct.* **1999**, *28*, 235–268.

8 S. J. Opella, C. Ma, and F. M. Marassi, *Methods Enzymol.* **2001**, *339*, 285–313.

9 M. Mehring, 1983, *Principles of High Resolution NMR in Solids, 2nd edition.* Springer, Berlin.

10 U. Haeberlen, 1976, *High Resolution NMR in Solids: Selective Averaging.* Academic Press, New York.

11 K. Schmidt-Rohr and H. W. Spiess, 1994, *Multidimensional Solid-State NMR and Polymers.* Academic Press, London.

12 G. E. Pake and E. M. Purcell, *Phys. Rev.* **1948**, *74*, 1184.

13 D. Freude and J. Haase, 1993, *Quadrupole Effects in Solid-State Magnetic resonance,* vol. 29. Springer-Verlag, Berlin.

14 E. R. Andrew, A. Bradbury, and R. G. Eades, *Nature* **1958**, *182*, 1659.

15 I. J. Lowe, *Phys. Rev. Lett.* **1959**, *2*, 285.

16 M. M. Maricq and J. S. Waugh, *J. Chem. Phys.* **1979**, *70*, 3300.

17 J. Herzfeld and A. Berger, *J. Chem. Phys.* **1980**, *73*, 6021.

18 A. C. de Dios, J. G. Pearson, and E. Oldfield, *J. Am. Chem. Soc.* **1993**, *115*, 9768.

19 A. C. de Dios, J. G. Pearson, and E. Oldfield, *Science* **1993**, *260*, 1491.

20 A. C. DeDios and E. Oldfield, *J. Am. Chem. Soc.* **1994**, *116*, 5307.

21 A.C. Dedios and E. Oldfield, *Solid State NMR* **1996**, *6*, 101–125.

22 J. Heller, A.C. Kolbert, R. Larsen, M. Ernst, T. Bekker, M. Baldwin, S.B. Prusiner, A. Pines, and D.E. Wemmer, *Protein Science* **1996**, *5*, 1655–1661.

23 M.R. Farrar, K.V. Lakshmi, S.O. Smith, R.S. Brown, J. Raap, J. Lugtenburg, R.G. Griffin, and J. Herzfeld, *Biophys. J.* **1993**, *65*, 310–315.

24 R.H. Havlin, H.B. Le, D.D. Laws, A.C. Dedios, and E. Oldfield, *J. Am. Chem. Soc.* **1997**, *119*, 11951–11958.

25 P.T.F. Williamson, J. Watts, G. Grobner, K.W. Miller, and A. Watts, *Biochem. Soc. Trans.* **1998**, *26*, S297–S297.

26 K.T. Mueller, B.Q. Sun, G.C. Chingas, J.W. Zwanziger, T. Terao, and A. Pines, *J. Magn. Reson.* **1990**, *86*, 470.

27 A. Llor and J. Virlet *Chem. Phys. Lett.* **1988**, *152*, 248–253.

28 A. Samoson, E. Lippmaa, and A. Pines, *Mol. Phys.* **1988**, *65*, 1013.

29 A. Samoson, T. Tuherm, and Z. Gan, *Solid State NMR* **2001**, *20*, 130–136.

30 Y. Ishii and R. Tycko, *J. Magn. Reson.* **2000**, *142*, 199–204.

31 M. Hong and S. Yamaguchi, *J. Magn. Reson.* **2001**, *150*, 4348.

32 S.R. Hartmann and E.L. Hahn, *Phys. Rev.* **1962**, *128*, 2042.

33 S. Hediger, B.H. Meier, N.D. Kurur, G. Bodenhausen, and R R. Ernst, *Chem. Phys. Lett.* **1994**, *223*, 283–288.

34 S. Hediger, B.H. Meier, and R.R. Ernst, *Chem. Phys. Lett.* **1995**, *240*, 449.

35 S. Hediger, P. Signer, M. Tomaselli, R.R. Ernst, and B.H. Meier, *J. Magn. Reson.* **1997**, *125*, 291–301.

36 A.J. Shaka, J. Keeler, T. Frenkiel, and R. Freeman, *J. Magn. Reson.* **1983**, *52*, 335–338.

37 A.J. Shaka, J. Keeler, and R. Freeman, *J. Magn. Reson.* **1983**, *53*, 313.

38 A.J. Shaka, C.J. Lee, and A. Pines, *J. Magn. Reson.* **1988**, *77*, 274.

39 A.J. Shaka, P.B. Barker, and R. Freeman, *J. Magn. Reson.* **1985**, *64*, 547.

40 M. Ernst, S. Bush, A.C. Kolbert, and A. Pines, *J. Chem. Phys.* **1996**, *105*, 3387–3397.

41 M. Ernst, H. Zimmermann, and B.H. Meier, *Chem. Phys. Lett.* **2000**, *317*, 581–588.

42 A.E. Bennett, L.R. Becerra, and R.G. Griffin, *J. Chem. Phys.* **1994**, *100*, 812–814.

43 A. Detken, E. Hardy, M. Ernst, and B.H. Meier, *Chem. Phys. Lett.* **2002**, 298–304.

44 M. Ernst, A. Samoson, and B.H. Meier, *Chem. Phys. Lett.* **2001**, *348*, 293–302.

45 Y. Ishii, J. Ashida, and T. Terao, *Chem. Phys. Lett.* **1995**, *246*, 439–445.

46 A.E. Bennett, C.M. Rienstra, J.M. Griffiths, Weiguo-Zhen, Lansbury PTz Jr, and R.G. Griffin, *J. Chem. Phys.* **1998**, *108*, 9463.

47 M. Lee and W.I. Goldburg, *Phys. Rev.* **1965**, *140*, A1261–1271.

48 A.E. Bennett, R.G. Griffin, and S. Vega. Recoupling of Homo- and Heteronuclear Dipolar Interactions in Rotating Solids. In *NMR Basic Principles and Progress*, vol. 33 of *NMR Basic principles and progress, Solid-State NMR IV*, pp. 1–77. Springer Verlag Berlin, Heidelberg, 1994.

49 R.G. Griffin, *Nat. Struct. Biol.* **1998**, *5 supplement*, 508–512.

50 S. Dusold and A. Sebald, *Ann. Rep. NMR Spectrosc.*, **2000**, *41*, 185–264.

51 D.P. Raleigh, M.H. Levitt, and R.G. Griffin, *Chem. Phys. Lett.* **1988**, *146*, 71.

52 M.G. Colombo, B.H. Meier, and R.R. Ernst, *Chem. Phys. Lett.* **1988**, *146*, 189.

53 M.H. Levitt, D.P. Raleigh, F. Creuzet, and R.G. Griffin, *J. Chem. Phys.* **1990**, *92*, 6347.

54 K. Takegoshi, K. Nomura, and T. Terao, *Chem. Phys. Lett.* **1995**, *232*, 424–428.

55 K. Takegoshi, K. Nomura, and T. Terao, *J. Magn. Reson.* **1997**, *127*, 206–216.

56 P.R. Costa, B.Q. Sun, and R.G. Griffin, *J. Am. Chem. Soc.* **1997**, *119*, 10821–10830.

57 R. Verel, M. Baldus, M. Nijman, J.W.M. Vanos, and B.H. Meier, *Chem. Phys. Lett.* **1997**, *280*, 3139.

58 A.E. Bennett, J.H. Ok, R.G. Griffin, and S. Vega, *J. Chem. Phys.* **1992**, *96*, 8624.

59 T. Gullion and S. Vega, *Chem. Phys. Lett.* **1992**, *194*, 423.

60 T. Fujiwara, A. Ramamoorthy, K. Nagayama, K. Hioka, and T. Fujito, *Chem. Phys. Lett.* **1993**, *212*, 8184.

61 M. Baldus, M. Tomaselli, B.H. Meier, and R.R. Ernst, *Chem. Phys. Lett.* **1994**, *230*, 329–336.

62 M. Baldus, D.G. Geurts, and B.H. Meier, *Solid State NMR* **1998**, *11*, 157–168.

63 J. Heller, R. Larsen, M. Ernst, A.C. Kolbert, M. Baldwin, S.B. Prusiner, D.E. Wemmer, and A. Pines, *Chem. Phys. Lett.* **1996**, *251*, 223–229.

64 T. Karlsson and M.H. Levitt, *J. Chem. Phys.* **1998**, *109*, 5493–5507.

65 M. Helmle, Y.K. Lee, P.J.E. Verdegem, X. Feng, T. Karlsson, J. Lugtenburg, H.J.M. de Groot, and M.H. Levitt, *J. Magn. Reson.* **1999**, *140*, 379–403.

66 T. Gullion, D.B. Baker, and M.S. Conradi, *J. Magn. Reson.* **1990**, *89*, 479–484.

67 N.C. Nielsen, H. Bildsoe, H.J. Jakobsen, and M.H. Levitt, *J. Chem. Phys.* **1994**, *101*, 1805.

68 R. Verel, M. Baldus, M. Ernst, and B.H. Meier, *Chem. Phys. Lett.* **1998**, *287*, 421–428.

69 R. Verel, M. Ernst, and B.H. Meier, *J. Magn. Reson.* **2001**, *150*, 8199.

70 R. Tycko and G. Dabbagh, *Chem. Phys. Lett.* **1990**, *173*, 461.

71 R. Tycko and S.O. Smith, *J. Chem. Phys.* **1993**, *98*, 932–943.

72 B.Q. Sun, P.R. Costa, and R.G. Griffin, *J. Magn. Reson. A* **1995**, *112*, 1918.

73 B.H. Meier and W.L. Earl, *J. Chem. Phys.* **1986**, *85*, 4905.

74 W. Sommer, J. Gottwald, D.E. Demco, and H.W. Spiess, *J. Magn. Reson. A* **1995**, *113*, 131–134.

75 P.L. Lee, C.D. Xiao, J.H. Wu, A.F. Yee, and J. Schaefer, *Macromolecules* **1995**, *28*, 6477–6480.

76 M. Hohwy, H.J. Jakobsen, M. Eden, M.H. Levitt, and N.C. Nielsen, *J. Chem. Phys.* **1998**, *108*, 2686.

77 R. Verel. *Adiabatic methods for Homonuclear Dipolar Recoupling in Magic Angle Spinning Solid-State NMR.* PhD thesis, ETH Zurich, Diss. Nr. 14152, 2001.

78 D.M. Gregory, D.J. Mitchell, J.A. Stringer, S. Kiihne, J.C. Shiels, J. Callahan, M.A. Mehta, and G.P. Drobny, *Chem. Phys. Lett.* **1995**, *246*, 654.

79 C.M. Rienstra, M.E. Hatcher, L.J. Mueller, B.Q. Sun, S.W. Fesik, and R.G. Griffin, *J. Am. Chem. Soc.* **1998**, *120*, 10602–10612.

80 M. Hohwy, C.M. Rienstra, C.P. Jaroniec, and R.G. Griffin, *J. Chem. Phys.* **1999**, *110*, 7983–7992.

81 A. Brinkmann, M. Eden, and M.H. Levitt, *J. Chem. Phys.* **2000**, *112*, 8539–8554.

82 A. Brinkmann and M.H. Levitt, *J. Chem. Phys.* **2001**, *115*, 357–384.

83 M. Carravetta, M. Eden, X. Zhao, A. Brinkmann, and M.H. Levitt, *Chem. Phys. Lett.* **2000**, *321*, 205–215.

84 M. Eden and M.H. Levitt, *J. Chem. Phys.* **1999**, *111*, 1511–1519.

85 M. Baldus and B.H. Meier, *J. Magn. Reson. A* **1996**, *121*, 65–69.

86 E.H. Hardy, R. Verel, and B.H. Meier, *J. Magn. Reson.* **2001**, *148*, 459–464.

87 M. Baldus, R.J. Iuliucci, and B.H. Meier, *J. Am. Chem. Soc.* **1997**, *119*, 1121–1124.

88 R.J. Iuliucci and B.H. Meier, *J. Am. Chem. Soc.* **1998**, *120*, 9059–9062.

89 A.S.D. Heindrichs, H. Geen, C. Giordani, and J.J. Titman *Chem. Phys. Lett.* **2001**, *335*, 89–96.

90 A. Detken, E.H. Hardy, M. Ernst, M. Kainosho, T. Kawakami, S. Aimoto, and B.H. Meier, *J. Biomol. NMR* **2001**, *20*, 203–221.

91 A. Bax, R. Freeman, and S.P. Kempsell **1980**, *102*, 4849–4851.

92 A. Lesage, C. Auger, S. Caldarelli, and L. Emsley, *J. Am. Chem. Soc.* **1997**, *119*, 7867–7868.

93 A. Lesage, M. Bardet, and L. Emsley, *J. Am. Chem. Soc.* **1999**, *121*, 10987–10993.

94 R. Verel, J.D. van Beek, and B.H Meier, *J. Magn. Reson.* **1999**, *140*, 300–303.

95 E.O. Stejskal, J. Schaefer, and J.S. Waugh, *J. Magn. Reson.* **1977**, *28*, 105.

96 A. Bielecki, A.C. Kolbert, and M.H. Levitt, *Chem. Phys. Lett.* **1989**, *155*, 341.

97 A. Bielecki, A.C. Kolbert, H.J.M. de Groot, R.G. Griffin, and M.H. Levitt. Frequency-Switched Lee-Goldburg Sequences in Solids. In: W.S. Warren, editor, *Adv. Magn. Reson.*, volume 14, pages 111–124. Academic Press, New York, 1990.

98 E. Vinogradov, P.K. Madhu, and S. Vega, *Chem. Phys. Lett.* **2000**, *329*, 207–214.

99 B.J. van Rossum, H. Foerster, and H.J.M. de Groot, *J. Magn. Reson.* **1997**, *124*, 516–519.

100 E. Vinogradov, P.K. Madhu, and S. Vega, *J. Chem. Phys.* **2001**, *115*, 8983–9000.

101 M. Baldus, D.G. Geurts, S. Hediger, and B.H. Meier., *J. Magn. Reson.*, **1996**, *123*, 140–144.

102 T. Gullion and J. Schaefer, *J. Magn. Reson.* **1989**, *81*, 196.

103 L. Muller, *J. Am. Chem. Soc.*, **1979**, *101*, 4481–4484.

104 A. Bax, R.H. Griffey, and B.H. Hawkins, *J. Magn. Reson.* **1983**, *55*, 301–315.

105 G. Bodenhausen and D.J. Ruben, *Chem. Phys. Lett.* **1980**, *69*, 185–189.

106 A. Lesage, P. Charmont, S. Steuernagel, and L. Emsley, *J. Am. Chem. Soc.* **2000**, *122*, 9739–9744.

107 A. Lesage and L. Emsley, *J. Magn. Reson.* **2001**, *148*, 449–454.

108 J. Raap, S. Nieuwenhuis, A. Creemers, S. Hexspoor, U. Kragl, and J. Lugtenburg, *Eur. J. Org. Chem.* **1999**, *10*, 2609–2621.

109 J. Lugtenburg, A.F.L. Creemers, M.A. Verhoeven, A.A. C. van Wijk, P.J.E. Verdegem, M.C.F. Monnee, and F.J.H.M. Jansen, *Pure Appl. Chem.* **1999**, *71*, 2245–2251.

110 A. Watts, I.J. Burnett, C. Glaubitz, G. Grobner, D.A. Middleton, P.J.R. Spooner, J.A. Watts, and P.T.F. Williamson, *Nat. Prod. Rep.* **1999**, *16*, 419–423.

111 J. Jones, 1997, *Amino Acid and Peptide Synthesis.* Oxford Chemistry Primers. Oxford University Press.

112 G.B. Fields and R.L. Noble, *Int. J. Pep. Prot. Res.* **1990**, *35*, 161–214.

113 C. Glaubitz, A. Groger, K. Gottschalk, P. Spooner, A. Watts, S. Schuldiner, and H. Kessler, *FEBS Lett.* **2000**, *480*, 127–131.

114 P.R. Costa, D.A. Kocisko, B.Q. Sun, P.T. Lansbury, and R.G. Griffin, *J. Am. Chem. Soc.* **1997**, *119*, 10487–10493.

115 J.X. Wang, Y.S. Balazs, and L.K. Thompson, *Biochemistry* **1997**, *36*, 1699–1703.

116 O.J. Murphy, F.A. Kovacs, E.L. Sicard, and L.K. Thompson, *Biochemistry* **2001**, *40*, 1358–1366.

117 P.J.R. Spooner and A. Watts, *Biochemistry* **1992**, *31*, 10129–10138.

118 M.C. Loewen, J. Klein-Seetharaman, E.V. Getmanova, P.J. Reeves, H. Schwalbe, and H.G. Khorana, *Proc. Natl. Acad. Sci. U.S.A.* **2001**, *98*, 4888–4892.

119 A.D. Albert, A. Watts, P.J.R. Spooner, G. Groebner, J. Young, and Yeagle P.L., *Biochim. Biophys. Acta* **1997**, *1328*, 74–82.

120 P.J.R. Spooner, L.M. Veenhoff, A. Watts, and B. Poolman, *Biochemistry* **2000**, *38*, 9634–9639.

121 J. Klein-Seetharaman, E.V. Getmanova, M.C. Loewen, P.J. Reeves, and H.G. Khorana, *Proc. Natl. Acad. Sci. U.S.A.* **1999**, *96*, 13744–13749.

122 E.V. Getmanova, J. Klein-Seetharaman, P.J. Loewen, M.C. and Reeves, S.O. Smith, and H.G. Khorana, *Biophys. J.* **2001**, *80*, 2716.

123 P.T.F. Williamson, J.F. Roth, T. Haddingham, and A. Watts, *Prot. Exp. Purif.* **2000**, *19*, 271–275.

124 A. Ward, J. O'Reilly, N.G. Rutherford, S.M. Ferguson, C.K. Hoyle, S.L. Palmer, J.L. Clough, H. Venter, H. Xie, G.J. Litherland, G.E.M. Martin, J.M. Wood, M.A.T. Roberts, P.E. Groves, W.J. Liang, A. Steel, B.J. McKeown, and P.J.F. Henderson, *Biochem. Soc. Trans.* **1999**, *27*, 893–899.

125 R. Grisshammer and T. Tate, *Q. Rev. Biophys.* **1995**, *28*, 315–422.

126 H.D. Blasey, B. Brethon, R. Hovious, H. Vogel, A.P. Tairi, K. Lundstrom, L. Rey, and A.R. Bernard, *Cytotechnology* **2000**, *32*, 199–208.

127 C.H.W. Klaassen, P.H.M. Bovee-Geurts, G.L.J. DeCaluwem, and W.J. deGripp, *Biochem. J.* **1999**, *342*, 293–300.

128 M. Eilers, W. Ying, P.J. Reeves, H.G. Khorana, and S.O. Smith, *Methods Enzymol.* **2002**, *343*, 212–222.

129 S.K. Straus, T. Bremi, and R.R. Ernst, *Chem. Phys. Lett.* **1996**, *262*, 709–715.

130 M. Hong, *J. Magn. Reson.* **1999**, *139*, 389–401.

131 S. Ray, E. Vinogradov, G. Boender, and S. Vega, *J. Magn. Reson.* **1998**, *135*, 418–426.

132 L. Zheng, W. Fishbein K, R.G. Griffin, and J. Herzfeld, *J. Am. Chem. Soc.* **1993**, *115*, 6254.

133 B. Reif, C.P. Jaroniec, C.M. Rienstra, M. Hohwy, and R.G. Griffin, *J. Magn. Reson.* **2001**, *151*, 320–327.

134 D.L. Jakeman, D.J. Mitchell, W.A. Shuttleworth, and J.N.S. Evans, *J. Biomol. NMR* **1998**, *12*, 417–421.

135 J. Pauli, B. van Rossum, H. Forster, H.J.M. de Groot, and H. Oschkinat, *J. Magn. Reson.* **2000**, *143*, 411–416.

136 P.T.F. Williamson. *The application of solid state NMR to the study of ligand protein interaction.* PhD thesis, University of Oxford, 1999.

137 P.T.F. Williamson, S. Bains, C. Chung, R. Cooke, B.H. Meier, and A. Watts. Solid State NMR in Biology. Leiden, 2001.

138 Alia, J. Matysik, C. Soede-Huijbregts, M. Baldus, J. Raap, J. Lugtenburg, P. Gast, H.J.M. Gorkon, A.J. Hoff, and H.J.M. deGroot, *J. Am. Chem. Soc.* **2001**, *123*, 4803–4809.

139 A. Watts, I.J. Burnett, C. Glaubitz, G. Grobner, D.A. Middleton, P.J.R. Spooner, and P.T.F. Williamson, *Eur. Biophys. J. Biophys. Lett.* **1998**, *28*, 84–90.
140 M. Hong, *J. Biomol. NMR* **1999**, *15*, 114.
141 J. Pauli, M. Baldus, B. Rossum, H. Forster, H.J.M. deGroot, and H. Oschkinat, *ChemBioChem* **2001**, *2*, 272–281.
142 S.K. Straus, T. Bremi, and R.R. Ernst, *J. Biomol. NMR* **1998**, *12*, 39–50.
143 M. Baldus, D.G. Geurts, S. Hediger, and B.H. Meier, *J. Magn. Reson. A* **1996**, *118*, 140–144.
144 J. Egorova-Zachernyuk, Hollander, N. Fraser, P. Gast, A.J. Hoff, R. Cogdell, H.J.M. deGroot, and M. Baldus, *J. Biomol. NMR* **2001**, *19*, 243–253.
145 P.T.F. Williamson, G. Grobner, P.J.R. Spooner, K.W. Miller, and A. Watts, *Biochemistry* **1998**, *37*, 10854–10859.
146 P.T.F. Williamson, S. Bains, C. Chung, R. Cooke, and A. Watts, *FEBS Lett.* **2002**, *518*, 111–115.
147 S. Luca, D.V. Filippov, J.H. van Boom, H. Oschkinat, H.J.M. de Groot, and M. Baldus, *J. Biomol. NMR* **2001**, *20*, 325–331.
148 F. Creuzet, A. McDermott, R. Gebhard, K. van der Hoef, M.B. Spijker-Assink, J. Herzfeld, J. Lugtenburg, M.H. Levitt, and R.G. Griffin, *Science* **1991**, *251*, 783.
149 X. Feng, P.J.E. Verdegem, M. Eden, D. Sandstrom, Y.K. Lee, P.H.M. Bovee-Geurts, W.J. de Grip, J. Lugtenburg, H.J.M. de Groot, and M.H. Levitt, *J. Biomol. NMR* **2000**, *16*, 18.
150 C. Glaubitz, G. Grobner, and A. Watts, *Biochim. Biophys. Acta-Biomembr.* **2000**, *1463*, 151–161.
151 S.O. Smith, D. Song, S. Shekar, M. Groesbeek, M. Ziliox, and S. Aimoto, *Biochemistry* **2001**, *40*, 6553–6558.
152 J.R. Long, J.L. Dindot, H. Zerbroski, S. Kiihne, R.H. Clark, A.A. Campbell, P.S. Stayton, and G.P. Drobny, *Proc. Natl Acad. Sci. U.S.A.* **1998**, *95*, 12083–12087.
153 W.J. Shaw, J.R. Long, J.L. Dindot, A.A. Campbell, P.S. Stayton, and G.P. Drobny, *J. Am. Chem. Soc.* **2000**, *122*, 1709–1716.
154 P.T. Lansbury, P.R. Costa, J.M. Griffiths, E.J. Simon, M. Auger, K.J. Halverson, D.A. Kocisko, Z.S. Hendsch, T.T. Ashburn, R.G.S. Spencer, B. Tidor, and R.G. Griffin, *Nat. Struct. Biol.* **1995**, *2*, 990–998.
155 O.N. Antzutkin, J.J. Balbach, R.D. Leapman, N.W. Rizzo, J. Reed, and R. Tycko, *Proc. Natl. Acad. Sci. U. S. A.* **2000**, *97*, 13045–13050.
156 H.J.M. deGroot and S. Kiihne, editors. *The Future of Solid State NMR in Biology*, in chapter: Structural insight into the interaction of amyloid-protein with biological membranes by solid state NMR. Kluwer Academic Publishers, 2001.
157 J.M. Griffiths, T.T. Ashburn, M. Auger, P.R. Costa, R.G. Griffin, and P.T.J. Lansbury, *J. Am. Chem. Soc.* **1995**, *117*, 353–946.
158 K.A. Klug, L.A. Burzio, Waite J.H., and Schaefer J., *Arch. Biochem. Biophys.* **1996**, *333*, 221–224.
159 L.M. McDowell, L.A. Burzio, Waite J.H., and J. Schaefer, *J. Biol. Chem.* **1999**, *274*, 20293–20295.
160 L.M. McDowell, E.R. Cohen, and J. Schaefer, *J. Biol. Chem.* **1993**, *268*, 20768–20771.
161 K.L. Kramer, T.L. Hopkins, and J. Schaefer, *Insect Biochem. Molec.* **1995**, *25*, 1067–1080.
162 L.M. Mcdowell, D. Barkan, G.E. Wilson, and J. Schaefer, *Solid State NMR* **1996**, *7*, 203–210.
163 L.M. McDowell, C.A. Klug, D.D. Beusen, and Schaefer J., *Biochemistry* **1996**, *35*, 5395–5403.
164 D.D. Beusen, L.M. McDowell, U. Slomczynska, and Schaefer J., *J. Med. Chem.* **1995**, *38*, 2742–2747.
165 L.M. McDowell, M.A. McCarrick, D.R. Studelska, W.J. Guilford, D. Arnaiz, J.L. Dallas, D.R. Light, M. Whitlow, and J. Schaefer, *J. Med. Chem.* **1999**, *42*, 3910–3918.
166 W.C. Stallings, S.S. Abdel-Meguid, L.W. Lim, H.S. Shief, H.E. Dayringer, N.K. Leimgruber, R.A. Stegman, K.S. Anderson, J.A. Sikorski, S.R. Padgette, and G.M. Kishore, *Proc. Natl. Acad. Sci. U.S.A.* **1991**, *88*, 5046–5050.
167 L.M. McDowell, A. Schmidt, E.R. Cohen, D.R. Studelska, and J. Schaefer, *J. Mol. Biol.* **1996**, *256*, 160–171.
168 A.M. Christensen and J. Schaefer, *Biochemistry* **1993**, *32*, 2868–2873.
169 D.R. Studelska, C.A. Klug, D.D. Beusen, L.M. Mcdowell, and J. Schaefer, *J. Am. Chem. Soc.* **1996**, *118*, 5476–5477.

170 X. Feng, Y. K. Lee, D. Sandstrom, M. Eden, H. Maisel, A. Sebald, and M. H. Levitt, *Chem. Phys. Lett.* **1996**, *257*, 314–320.

171 A. S. Ulrich, A. Watts, I. Wallat, and M. P. Heyn, *Biochemistry* **1994**, *33*, 5370–5375.

172 J. C. Lansing, M. Hohwy, C. P. Janoniec, A. F. L. Creemers, J. Lugtenburg, J. Herzfeld, and R. G. Griffin, *Biochemistry* **2002**, *41*, 431–438.

173 P. R. Costa, J. D. Gross, M. Hong, and R. G. Griffin, *Chem. Phys. Lett.* **1997**, *280*, 95–103.

174 M. Hong, J. D. Gross, and R. G. Griffin, *J. Phys. Chem. B* **1997**, *101*, 5869–5874.

175 M. Hong, *J. Am. Chem. Soc.* **2000**, *122*, 3762–3770.

12
Determination of Protein Dynamics Using ^{15}N Relaxation Measurements

David Fushman

12.1
Introduction

One of the fundamental problems in understanding life at the molecular level is the relationship between structure, dynamics, and function in complex molecular systems like proteins. Proteins are molecular machines that perform most of the functions and control all key events in a living cell. Globular proteins are well packed and adopt ordered three-dimensional structures. Most importantly, they possess a variety of motions such as bond vibrations, side-chain rotations, segmental motions, domain movements etc. Although static three-dimensional structures alone provide extremely valuable information on the organization and interaction of protein molecules, it is motion that is required for most proteins to work. Dynamics are vital to protein function, which depends on alteration in three-dimensional structure in response to specific molecular interactions. The binding of a ligand to a protein (e. g., a substrate to an enzyme) generally results in changes in the protein's structure and dynamics; these changes can be local or also can involve other regions of the protein molecule. The relative orientation and motions of domains within many proteins are key to the control of multivalent recognition, or the assembly of protein-based cellular machines. In order to understand how proteins work, we need to know what motions are present in a protein and how they are related to the protein's biological function. Detailed knowledge of protein dynamics is, therefore, crucial to the understanding of the mechanisms of protein function, including such events as protein folding, ligand recognition, allostery, and catalysis.

Protein dynamics are extremely complex and difficult to analyze, because a variety of motions take place in the same molecule and at the same time. The key problem here is to determine which of these motions in a protein are essential for its biological function. In order to address this issue, we need detailed knowledge of protein dynamics.

Multidimensional NMR methods, combined with isotope labeling, can provide access to virtually every atom in a molecule, unique for protein structural studies. This not only allows characterization of the structure and interaction of proteins in their native milieu, but also provides unparalleled possibilities to obtain a complete atomic-level resolution picture of protein dynamics in a time range from picoseconds up to seconds, the range where most motions relevant to protein function take place. A significant number of ^{15}N and ^{13}C relaxation studies have been performed on a large number of proteins in the last decade, initiated by the pioneering work of Refs. [1] and [2]. In this chapter we review some experimental and analytical approaches to protein dynamics by ^{15}N relaxation measurements and outline the current picture of protein dynamics revealed by NMR.

12.2 Spectroscopic Techniques

Typical experiments involved in ^{15}N relaxation studies include measurements of the transverse ($R_2=1/T_2$) and longitudinal ($R_1=1/T_1$) relaxation rates of the ^{15}N nuclear spin and of the ^{1}H-^{15}N cross-relaxation rate, the latter being usually measured via the steady-state $^{15}N\{^{1}H\}$ nuclear Overhauser effect (NOE) (see e.g. Refs. [2–4]). These are HSQC-type experiments that are based on INEPT magnetization transfer in the ^{1}H-^{15}N pair and contain relaxation-specific pulse sequences inserted during the time period when magnetization is in the ^{15}N dimension. Typical NMR pulse sequences for these experiments can be found in Ref. [5]. Traditionally, all relaxation measurements are performed at a single resonance field (spectrometer), although more accurate approaches might require measurements of these parameters at multiple fields [6]. Some practical aspects of these measurements will be briefly mentioned here.

T_2 measurements usually employ either Carr-Purcell-Meiboom-Gill (CPMG) [7, 8] spin-echo pulse sequences or experiments that measure spin relaxation ($T_{1\rho}$) in the rotating frame. The time delay between successive 180° pulses in the CPMG pulse sequence is typically set to 1 ms or shorter to minimize the effects of evolution under the heteronuclear scalar coupling between ^{1}H and ^{15}N spins [3].

T_1 measurements are usually based on inversion-recovery-type experiments (see e.g. Ref. [9]). The longitudinal magnetization relaxes to its equilibrium value, which is not zero. The accuracy of T_1 determination then could depend on the duration of the longest relaxation delay used for the measurements. In order to circumvent this problem, these experiments are usually performed so that the contribution from the equilibrium magnetization is subtracted in the course of the experiment by alternating the receiver phase simultaneously with the sign of the initial longitudinal ^{15}N magnetization [2]. In a different approach, two sets of T_1 experiments, one with positive and one with negative initial longitudinal ^{15}N magnetization, are stored separately and then analyzed simultaneously [9]. As shown in Ref. [9], signal intensities measured in the heteronuclear NOE experiment can be used to assess the equilibrium magnetization for the T_1 experiment.

Cross-correlation effects between ^{15}N CSA and ^{1}H-^{15}N dipolar interactions [10] will result in different relaxation rates for the two components of the ^{15}N spin doublet, which could significantly complicate the analysis of the resulting bi-exponential decay of the decoupled signal in T_1 or T_2 experiments. To avoid this problem, 180° ^{1}H pulses are applied during the ^{15}N relaxation period [3, 4], which effectively averages the relaxation rates for the two components of the ^{15}N spin doublet.

The steady-state heteronuclear $^{15}N\{^{1}H\}$ NOEs are determined as a ratio of cross-peak intensities in two experiments, with and without presaturation of amide ^{1}H nuclear spins, usually referred to as NOE and NONOE experiments, respectively. Magnetization exchange between amide protons and water protons could affect the equilibrium ^{1}H magnetization in the NONOE experiment, and thereby increase the measured NOE values.

The effect could be considerable for solvent-exposed parts of the backbone and could render the NOE values inaccurate. These systematic errors could be minimized by using water flip-back pulses in order to avoid saturation of H_2O magnetization [11]. The NOE data are generally more susceptible to errors than R_1 and R_2 because (i) the NOE experiments start with the equilibrium ^{15}N magnetization that is ~ 10 fold lower than that of (^1H) in the R_1 and R_2 experiments, hence relatively low sensitivity, and (ii) the NOE values are derived from only two sets of measurements, whereas R_1 and R_2 data are obtained from fitting multiple sets of data; the latter is expected to result in a more efficient averaging of experimental errors.

12.3 Accuracy and Precision of the Method

Here we mention some practical aspects that could be useful for improving the accuracy and precision of the method.

12.3.1 Sampling Schemes

Typically, a series of several 2D spectra are recorded with various relaxation delays, ranging from very short to the longest delays, which usually correspond to 1.5–2 relaxation times. The delay values are usually selected so that they are uniformly distributed over this time interval or so that the signal values are uniformly spread. Another sparse sampling strategy proposed in [12] is based on an optimal sampling scheme for a monoexponential decay function; a five-point variant of this strategy uses one measurement at a very short relaxation delay and four measurements at 1.3 T_2 (or T_1).

12.3.2 Peak Integration

Relaxation rates for a particular ^{15}N nucleus are determined from a monoexponential fit of the corresponding signals in a series of 2D spectra. Strictly speaking, peak volume, not intensity, is a direct measure of the NMR signal. However, peak volume integration is usually not straightforward and could be prone to errors. Assuming that the measurements are performed on the same sample and under identical temperature conditions, peak widths are same for a given peak in all experiments of the series, so that the peak volumes are linearly related to intensities. This allows one to monitor NMR signal decay by measuring peak intensities, which proved to be a more accurate procedure than peak volume integration because of the computational errors associated with the latter.

12.3.3 Estimation of Experimental Errors

Since relaxation rates are derived from the measured peak intensities, the experimental errors in these parameters are typically assessed based on the level of noise in the NMR spectra. This noise represents experimental uncertainties in the measured peak intensities and can be measured by integrating areas in a 2D spectrum that are free from signals. Errors in peak intensities can also be obtained from duplicate measurements [13]. The uncertainties in R_1 or R_2 can then be derived using standard methods of error estimation for fitting parameters, see e.g. Ref. [14]. The experimental errors in NOE values can be assessed as follows: $\sigma_{NOE} = |NOE|\sqrt{(\delta_{NOE}/I_{NOE})^2 + (\delta_{NONOE}/I_{NONOE})^2}$ where I and δ denote the peak intensity and the level of experimental noise, respectively, in the NOE and NONOE experiments (recall that NOE = I_{NOE}/I_{NONOE}). Experimental errors in relaxation parameters could also be derived from the comparison of their values measured in multiple separate measurements [15].

The experimental precision of relaxation measurements depends on many factors: protein size and concentration, signal line width, number of transients per increment, etc. The higher the concentration and/or narrower the line width, the better is the signal-to-noise ratio. Typically, for a 1 mM sample of a relatively small protein (e.g. ubiquitin, M_w=8.6 kDa) at 25 °C (proton $T_2 \approx 50$ ms), relaxation measurements with 32 scans per t_1 increment could provide experimental uncertainty in R_2 or R_1 of about 1%. Heteronuclear $^{15}N\{^1H\}$ NOE measurements require a greater number of scans (>128) to reach reasonable levels of experimental precision (<2%). A complete set of R_1, R_2, and NOE measurements under these conditions typically takes less than 1 week of measurement time. Relaxation measurements at lower protein concentrations will require a significant increase in the number of scans, hence longer experimental time. For small to medium size proteins (up to ~ 15–20 kDa), relaxation data of reasonable precision could be obtained using conventional NMR probes for protein concentrations as low as 500 μM or even 250 μM [9]. Bigger proteins will require a greater number of scans, as the line-width scale is approximately linear with protein molecular weight. In addition, the decrease in T_2 (particularly ^1H) results in greater losses due to spin relaxation during the magnetization transfer (INEPT), which will result in the decrease in the signal. Additional gain in the signal-to-noise ratio could be obtained by reducing T_2s, which could be achieved, for example, by raising the temperature if the protein's stability permits, or by perdeuterating side chains. Increasing protein concentration usually helps to improve the sensitivity or reduce the required measurement time, although it could also promote protein aggregation (see below).

12.3.4
Noise Reduction

It proved helpful for the purpose of noise reduction to perform relaxation experiments in an interleaved fashion, as one pseudo-3D experiment, where the 2D planes in the F2 dimension correspond to various relaxation delays. The acquisition order (3-2-1) is selected so that cycling through various relaxation delays (in R_1 or R_2 experiments) or through NOE/NONOE 2D planes is performed prior to incrementing the evolution period in the indirect dimension (F1) (see e.g. Ref. [16]). The resulting pseudo-3D spectrum can be processed as a set of 2D spectra in $t1$ and $t3$ dimensions, and then analyzed in the usual way. This procedure reduces the noise arising from switching from one 2D experiment to the other and helps minimize temperature variations between the spectra acquired with different relaxation delays. It will also be helpful in those cases when protein degradation occurs in the course of measurements; a reduction in the signal due to sample instability will then affect the line shape in F1 but not the relaxation rate.

12.3.5 Temperature Control

Since both the overall and local protein dynamics depend on temperature, proper temperature control is essential for accurate interpretation of ^{15}N relaxation data [17]. Real temperature in the sample could be measured using calibration standards, e.g. methanol and/or ethyl glycol (e.g. Ref. [18]). However, even in this case, temperature consistency is a significant technical problem, as the actual sample temperature could vary from one experiment to the other. For example, in the T_2 experiment using CPMG, temperature variations between the 2D data sets could arise from the differences in the duration of relaxation delays and hence in the number of 180°-pulses applied. Temperature stability could be improved by using longer recycling delays between the transients, although at the cost of increased measurement time. Further temperature equalization could be obtained by inserting a compensating set of 180° pulses at the beginning of the recycling delay. The compensating pulses could be applied at a frequency far from the carrier position. This helps maintain the same temperature level between data sets corresponding to different relaxation delays, e.g. in the course of T_2 measurement, but, at the same time, can introduce another problem as it will amplify temperature differences between T_2 and T_1 and NOE experiments.

Temperature consistency between measurements performed on different spectrometers is particularly critical for accurate interpretation of the data (see Refs. [19, 20] for post-acquisition temperature consistency tests). However, temperature control and equalization are also important for the combined analysis of T_1, T_2, and NOE data measured on the same spectrometer, because of the possible temperature differences between these measurements. Fig. 12.1 illustrates the sensitivity of relaxation parameters to temperature variations. Accurate measurement of protein dynamics requires that all experiments be done at the same temperature. To improve temperature consistency between T_1, T_2, and

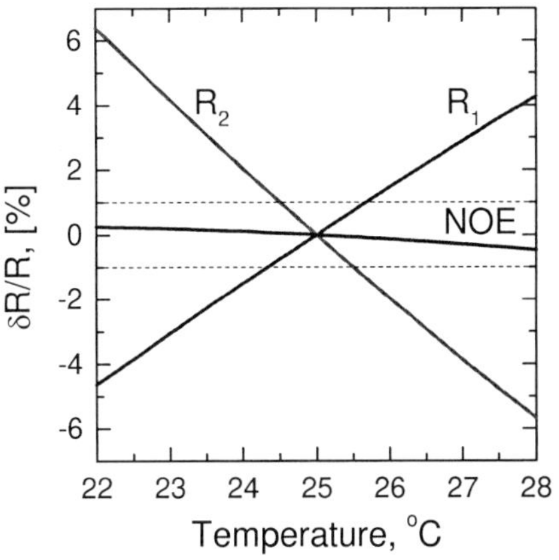

Fig. 12.1 Illustration of the temperature sensitivity of ^{15}N relaxation parameters, R_1, R_2, and NOE, as indicated. Shown are the relative deviations in these relaxation parameters from their values at 25 °C as a function of temperature in the range of ±3 °C. The expected variations in R_1 and R_2 due to temperature deviations of as little as ±1 °C are already greater than the typical level of experimental precision (±1%) of these measurements (indicated by the dashed horizontal lines). For simplicity, only temperature variation of the overall tumbling time of the molecule (due to temperature dependence of the viscosity of water) is taken into account; the effect of temperature variations on local dynamics is not considered here.

NOE measurements on the same spectrometer, these could be performed in an interleaved fashion, e.g. as one pseudo-3D experiment, similarly to that described above.

12.4 Basic Equations

The measured spin relaxation parameters (longitudinal and transverse relaxation rates, R_1 and R_2, and heteronuclear steady-state NOE) are directly related to power spectral densities (SD). These spectral densities, $J(\omega)$, are related via Fourier transformation with the corresponding correlation functions of reorientional motion. In the case of the backbone amide ^{15}N nucleus, where the major sources of relaxation are dipolar interaction with directly bonded ^{1}H and ^{15}N CSA, the standard equations read [21]:

$$R_1 = 3(d^2 + c^2)J(\omega_N) + d^2[3J(\omega_N) + J(\omega_H - \omega_N) + 6J(\omega_H + \omega_N)] \quad (1)$$

$$R_2 = \frac{1}{2}(d^2 + c^2)[4J(0) + 3J(\omega_N)] + \frac{1}{2}d^2[J(\omega_H - \omega_N) + 6J(\omega_H) + 6J(\omega_H + \omega_N)] + R_{ex} \quad (2)$$

$$NOE = 1 - |\gamma_H/\gamma_N| d^2[6J(\omega_H + \omega_N) - J(\omega_H - \omega_N)]/R_1 \,. \quad (3)$$

Here $d = -[\mu_0/(4\pi)]\gamma_H\gamma_N h/(4\pi r^3_{HN})$ and $c = -\omega_N\,\mathrm{CSA}/3$ represent contributions from the ^{15}N-^{1}H dipolar coupling and ^{15}N CSA, respectively; R_{ex} is the conformational exchange contribution (if any) to measured R_2. Equations (1–3) assume that the same autocorrelation function can be used to describe the spin relaxation-relevant modulation of both ^{1}H-^{15}N dipolar interaction and ^{15}N CSA. Corrections to these equations due to the noncollinearity of the two interactions are described in Ref. [22].

Equations (1–3) are widely used for protein dynamics analysis from relaxation measurements. The primary goals here are (A) to measure the spectral densities $J(\omega)$ and, most important, (B) to translate them into an adequate picture of protein dynamics. The latter goal requires adequate theoretical models of motion that could be obtained from comparison with molecular dynamics simulations (see for example Ref. [23]). However, accurate analysis of experimental data is an essential prerequisite for such a comparison.

In addition, cross-correlation effects between ^{15}N CSA and ^{1}H-^{15}N dipolar interaction could be measured, see e.g. Refs. [24–26]:

$$\eta_{xy} = dc[4J(0) + 3J(\omega_N)]P_2(\cos\beta) \quad (4)$$

$$\eta_z = 3dcJ(\omega_N)P_2(\cos\beta)\,. \quad (5)$$

Here β is the angle between the unique principal axis of the ^{15}N CSA tensor and the HN-bond, and $P_2(x)=½\,(3x^2-1)$ is the second Legendre polynomial. Although these measurements do not provide additional information (compared to Eqs. (1–3)) in terms of sampling the spectrum of protein motions, the η-terms could be used for the determination of the ^{15}N CSA tensor [19] and also for the identification of amide groups undergoing conformational exchange [26, 27] (see below). If the β angle was known, these parameters could also be used in addition to or in lieu of R_2 and R_1 to analyze protein dynamics.

As one can see from Eqs. (1–5), the number of experimentally measured parameters is less than the number of unknowns: independent characterization of the nine parameters in the right-hand side of Eqs. (1–5) from only five measurements is impossible. Therefore, several approaches were suggested in order to make possible characterization of protein dynamics from experimental data measured at a single resonance frequency. A "low-resolution" picture of protein dynamics can be obtained assuming constant values for r_{NH} (1.02A) and ^{15}N CSA (–160 or –170 ppm) and using various useful parametrizations for $J(\omega)$ [28–31].

The two major approaches, the model-free analysis and SD mapping will be briefly discussed here.

12.5 The Model-Free Approach

Recent progress in protein dynamics studies by NMR was greatly facilitated by the invention of the "model-free" formalism [28, 32]. In this approach, the local dynamics of a protein are characterized by an order parameter, S, measuring the amplitude of local motion on a scale from 0 to 1, and the correlation time of the motion, τ_{loc}. The model-free expression for the correlation function of local motion reads

$$C_{loc}(t) = S^2 + (1 - S^2)\exp(-t/\tau_{loc})\,. \tag{6}$$

In the extended model-free approach [29], the local dynamics are deconvolved into a fast and a slow motion $\tau_{fast} \ll \tau_{slow}$, $S^2 = S^2_{slow}S^2_{fast}$):

$$C_{loc}(t) = S^2 + (S^2_{fast} - S^2)\exp(-t/\tau_{slow}) + (1 - S^2_{fast})\exp(-t/\tau_{fast})\,. \tag{7}$$

Formally, S^2 represents a decrease in the autocorrelation function caused by the motion: $S^2=0$ corresponds to completely unrestricted motion of a bond (N–H in this case), while $S^2=0$ is expected if the bond reorientations are frozen. It was shown recently that the order parameter may be related to the statistical mechanical properties of a protein molecule [33–35]; hence, changes in the NMR-derived order parameters can indicate localized contributions to overall molecular entropy.

In the absence of a correlation between the local dynamics and the overall rotational diffusion of the protein, as assumed in the model-free approach, the total correlation function that determines the ^{15}N spin-relaxation properties (Eqs. (1–5)) can be deconvolved $(\tau_{fast}, \tau_{slow} \ll \tau_c)$:

$$C(t) = C_{ovrl}(t)C_{loc}(t)\,, \tag{8}$$

where C_{ovrl}(t) denotes an autocorrelation function of the ^{15}N-^{1}H dipolar vector interaction in the case of rigid body rotation, which is generally anisotropic (e.g. Ref. [36]). In the simplest case of isotropic overall rotation with a correlation time τ_c, $C_{ovrl}(t) = \frac{1}{5}e^{-t/\tau_c}$.

The model-free approach is essentially based on a parametrization of the spectral densities using a small number of fitting parameters, which then allows Eqs. (1–3) to be solved. The analysis of experimental data using this method will be discussed in a later section.

12.6 Reduced Spectral Densities Mapping

This approach is based on the idea of SD mapping [37, 38] that exploits the linear relationship (Eqs. 1–3) between the measured relaxation parameters and the spectral densities. The spectral density components, $J(0)$, $J(\omega_N)$, and $J(\omega_H)$ could be determined directly from the relaxation data, R_1, R_2, and NOE, using the reduced spectral density approach [30, 31] based on the observation that the spectral densities are slowly varying functions of the frequency around w_H:

$$J(0.87\omega_H) = \frac{1}{5d^2}\left|\frac{\gamma_N}{\gamma_H}\right|(1 - NOE)R_1 \quad (9\,a)$$

$$J(\omega_N) = \frac{R_1 - 7d^2 J(0.921\omega_H)}{3(d^2 + c^2)} \quad (9\,b)$$

$$J(0) = \frac{2R_2 - R_1 - 6d^2 J(\omega_H)}{4(d^2 + c^2)} \,. \quad (9\,c)$$

According to [31], the following two methods could be used in order to relate the high-frequency components of the spectral density in Eqs. (9 a–9 c):

(i) $J(\omega_H) \approx J(0.87\omega_H) \approx J(0.921\omega_H)$; or

(ii) $J(\omega) \propto 1/\omega^2$ for $\omega \approx \omega_H$, which leads to $J(0.921\omega_H) = 0.8923 J(0.87\omega_H)$ and $J(\omega_H) = 0.7569 J(0.87\omega_H)$ in Eqs. (9 b) and (9 c), respectively.

These two methods yield lower and upper bounds for $J(0)$ and $J(\omega_N)$, but have no effect on the values of $J(0.87\omega_H)$ directly determined from the experimental data. Note that the derivation of $J(0.87\omega_H)$ (Eq. 9 a) does not involve any assumption about the values of ^{15}N CSA.

This approach yields spectral densities. Although it does not require assumptions about the correlation function and therefore is not subjected to the limitations intrinsic to the model-free approach, obtaining information about protein dynamics by this method is no more straightforward, because it involves a similar problem of the physical (protein-relevant) interpretation of the information encoded in the form of SD, and is complicated by the lack of separation of overall and local motions. To characterize protein dynamics in terms of more palpable parameters, the spectral densities will then have to be analyzed in terms of model-free parameters or specific motional models derived e. g. from molecular dynamics simulations. The SD method can be extremely helpful in situations when no assumption about correlation function of the overall motion can be made (e. g. protein interaction and association, anisotropic overall motion, etc.; see e. g. Ref. [39] or, for the determination of the ^{15}N CSA tensor from relaxation data, Ref. [27]).

12.7
Multi-Field Approach

Both approaches outlined above provide a solution to Eqs. (1–3) by reducing the number of necessary dynamics-related parameters to three and by using certain constant values for the dipolar interaction and ^{15}N CSA, also assuming $R_{ex}=0$ (in the SD mapping). The dynamic picture derived from such analyses could be inaccurate if these parameters vary from one site to another. For example, the site-specific variations in ^{15}N CSA become important for the analysis of relaxation data at higher fields when the CSA contribution becomes comparable to that from the dipolar interaction, and for TROSY applications [6]. A more accurate approach to protein dynamics should allow independent derivation of protein spectral densities, site-specific ^{15}N CSA and dipolar interactions, as well as R_{ex} values. An approach has been suggested recently based on relaxation measurements at multiple fields (spectrometers) [6, 19, 20]. This method is a natural extension of previous approaches [37, 40] in that all major components in Eqs. (1–5), and not only SDs, are treated as fitting parameters. The approach is based on the fact that some terms in these equations are either field-independent [d, θ, and $J(0)$] or scale with field B_0 in a known way ($c \propto B_0$, $R_{ex} \propto B_0^2$). Therefore, when combining data measured at several fields, the number of experimental parameters increases faster than the number of unknowns, so that already a complete set of three-field data theoretically could allow determination of almost all unknowns in Eqs. (1–3) [6]. This could provide a useful tool for protein dynamics analysis free from the oversimplifying assumptions. The utility of this approach was demonstrated for the full characterization of the spectral densities, ^{15}N CSA tensor, and R_{ex} in ubiquitin, unbiased by the assumption of constant CSA [19, 20]. Similar approaches were recently used by other groups [41, 42]. The necessity to use multiple-field data for accurate analysis of protein motions in multidomain systems was recently demonstrated in Ref. [42].

12.8
Strategies for the Analysis of Protein Dynamics from ^{15}N Relaxation Data

The NMR-relevant motions that contribute to nuclear spin relaxation are a combination of local dynamics and the overall molecular tumbling in solution. Accurate analysis of NMR data requires a proper separation of these contributions. The strategies for such analysis are discussed below. As a practical example illustrating the methods described here, we use ^{15}N relaxation data for the Pleckstrin Homology (PH) domain from the β-adrenergic receptor kinase-1 (βARK). The structure of this protein (PDB access code 1bak) was solved by NMR [43], and its dynamics were characterized by ^{15}N relaxation measurements [43] as well as computer molecular dynamics simulations [23, 44]. The βARK PH domain structure comprises seven β-strands forming a β-sandwich flanked on one side by the C-terminal α-helix (Fig. 12.4b). Some of the loops connecting β-strands, as well as the extended N- and C-termini, are unstructured and flexible in solution. A summary of ^{15}N relaxation data collected at 500 MHz is presented in Fig. 12.2.

It is worth mentioning that the analytical approaches outlined here and currently used to treat relaxation data assume that the overall and local dynamics are not coupled. While this is a reasonable assumption for small, compact proteins, it might not be true for sys-

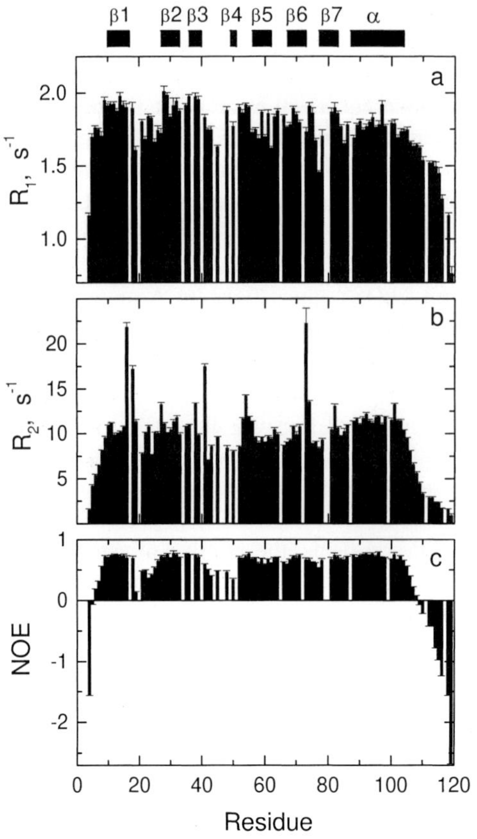

Fig. 12.2 Summary of the ^{15}N relaxation data for the βARK PH domain measured at 500 MHz (see Ref. [23]): **a** R_1, **b** R_2, and **c** heteronuclear NOE versus protein sequence.

tems possessing large-amplitude internal motions, e.g. multidomain systems and proteins with significant intersegmental flexibility. The approaches to these systems are starting to emerge [42, 45].

12.9
Overall Tumbling

The overall tumbling of a protein molecule in solution is the dominant source of NH-bond reorientations with respect to the laboratory frame, and hence is the major contribution to ^{15}N relaxation. Adequate treatment of this motion and its separation from the local motion is therefore critical for accurate analysis of protein dynamics in solution [46]. This task is not trivial because (i) the overall and internal dynamics could be coupled (e.g. in the presence of significant segmental motion), and (ii) the anisotropy of the overall rotational diffusion, reflecting the shape of the molecule, which in general case deviates from a perfect sphere, significantly complicates the analysis. Here we assume that the overall and local motions are independent of each other, and thus we will focus on the effect of the rotational overall anisotropy.

The anisotropy of the overall tumbling will result in the dependence of spin-relaxation properties of a given ^{15}N nucleus on the orientation of the NH-bond in the molecule. This orientational dependence is caused by differences in the apparent tumbling rates sensed by various internuclear vectors in an anisotropically tumbling molecule. Assume we have a molecule with the principal components of the overall rotational diffusion tensor D_x, D_y, and D_z (x, y, and z denote the principal axes of the diffusion tensor), and let $D_x < D_y < D_z$. For an internuclear vector parallel to the z-axis, its reorientations will be caused by molecular rotations around the x- and y-axes, but not z-axis, so that the apparent rotational diffusion rate for the vector will be determined by $D(z) = (D_x+D_y)/2$ (assuming small degree of anisotropy). On the contrary, reorientations of a vector perpendicular to the z-axis (e.g. along the x-axis) will be determined by $D(x) = (D_z+D_y)/2 > D(z)$ and hence will proceed faster than for the previous internuclear vector. These differences in the tumbling rates will lead to differences in the spin relaxation rates for the corresponding pairs of nuclei; the effect will increase with the anisotropy of the molecule. Theoretical expressions describing this phenomenon and relating spin relaxation rates to the overall rotational diffusion tensor can be found in Ref. [36]. These results are used to accurately analyze protein dynamics in the case of overall rotational anisotropy (e.g. Refs. [15, 23, 43, 47]). The structural information in the form of internuclear vector orientations can also be used in the structure refinement procedures, where the relaxation-derived orientational constraints for individual bond vectors could complement the "short-range" NOE information [48], or in the determination of interdomain orientation in multidomain systems [49, 50].

12.10 How Can We Derive the Rotational Diffusion Tensor of a Molecule from Spin-Relaxation Data?

12.10.1 Theoretical Background

Since the characterization of the overall rotational diffusion is a prerequisite for a proper analysis of protein dynamics from spin-relaxation data, we first focus on the theoretical basis of the method being used.

Several approaches to determination of the overall rotational diffusion tensor from ^{15}N relaxation data were suggested in the literature [15, 47, 49, 51–53]. The approach described here uses the orientational dependence of the ratio of spin-relaxation rates [49]

$$\rho \equiv \left(\frac{2R_2'}{R_1'} - 1\right)^{-1} = \frac{3J(\omega_N)}{4J(0)} = \frac{3/4}{1+(\omega_N\tau_x)^2}\left\{1 + \frac{(\omega_N\tau_x)^2}{(\omega_N\tau_x)^2 + (1+\frac{1}{6}\varepsilon)^2} \times \right.$$

$$\left. \frac{\varepsilon \sin^2\theta}{3+2\varepsilon+[1+\frac{1}{3}\varepsilon(2-3\sin^2\theta)]^2}\left[4+3\varepsilon+\frac{2}{9}\varepsilon^2 - \varepsilon\sin^2\theta\left(1+\frac{4+\frac{11}{3}\varepsilon+\frac{19}{18}\varepsilon^2+\frac{5}{54}\varepsilon^3}{(\omega_N\tau_x)^2+(1+\frac{2}{3}\varepsilon)^2}\right)\right]\right\}. \tag{10}$$

Here $\varepsilon \equiv D_\parallel/D_\perp - 1$, $\tau_x^{-1} \equiv 6\,D_\perp$, θ is the angle between an NH bond and the unique axis (associated with $D_\parallel$) of the rotational diffusion tensor, and ω_N is the ^{15}N resonance frequency. Given the orientation $\{\Theta, \Phi\}$ of the principal axes of the diffusion tensor with respect to the protein coordinate frame, the angle θ for each NH bond can be determined as

$$\theta = \cos^{-1}(\lambda_{ix} \cos \Phi \sin \Theta + \lambda_{iy} \sin \Phi \sin \Theta + \lambda_{iz} \cos \Theta) \tag{11}$$

where $\{\lambda_{ix}, \lambda_{iy}, \lambda_{iz}\}$ are the coordinates (in the protein coordinate frame) of a unit vector in the direction of the NH bond. These basic equations assume that the diffusion tensor is axially symmetric. In the general case of rotational anisotropy (presence of the rhombic component), the corresponding equations are more complex (see e.g. Ref. [36]), but the main idea that structural information is encoded in the relaxation rates still holds.

Here R_2' and R_1' are the transverse and longitudinal relaxation rates modified to subtract the contributions from high-frequency motions (see e.g. [20, 49]):

$$R_1' = R_1 - 7\left(\frac{0.87}{0.921}\right)^2 d^2 J(0.87\,\omega_H) = 3(c^2 + d^2)J(\omega_N) \tag{12 a}$$

$$R_2' = R_2 - \frac{13}{2}\left(\frac{0.87}{0.955}\right)^2 d^2 J(0.87\,\omega_H) = (c^2 + d^2)\left[2J(0) + \frac{3}{2}J(\omega_N)\right] \tag{12 b}$$

where $d^2 J(0.87\omega_H)$ is determined according to Eq. (9 a), and the conformational exchange contributions to R_2 are assumed to be negligible.

Equation (10) is exact in the absence of local motion. It provides a good approximation for the protein core residues, characterized by restricted mobility in the protein backbone: in this case both $J(0)$ and $J(\omega_N)$ scale as S^2, and therefore the contribution from local motion to ρ in Eq. (10) is canceled out. The approach described here is rather general and accommodates both large $(\omega_N \tau_x \gg 1)$ and small $(\omega_N \tau_x \leq 1)$ molecules and molecules with significant rotational anisotropy. Fig. 12.3 illustrates how well Eq.(10) describes the expected values for R_2'/R_1' in a wide range of anisotropies.

It is worth mentioning that parameter ρ is insensitive, to first order approximation, to modulation of the residue-specific ^{15}N chemical shift anisotropy tensor and/or dipolar interaction, as the (d^2+c^2) term in the R_2'/R_1' ratio is canceled out. The noncollinearity of the CSA and dipolar tensors will require corrections to Eqs. (10) and (12) for high degrees of rotational anisotropy $(D_\parallel/D_\perp > 1.5)$, as described in detail in Ref. [22].

Equation (10) allows the determination of the principal values of the diffusion tensor if the orientation of its principal axes frame is known. The latter information is included there in implicit form, via Eq. (11). The problem is that neither the principal axes nor principal values are known *a priori* and are to be determined simultaneously, as outlined below.

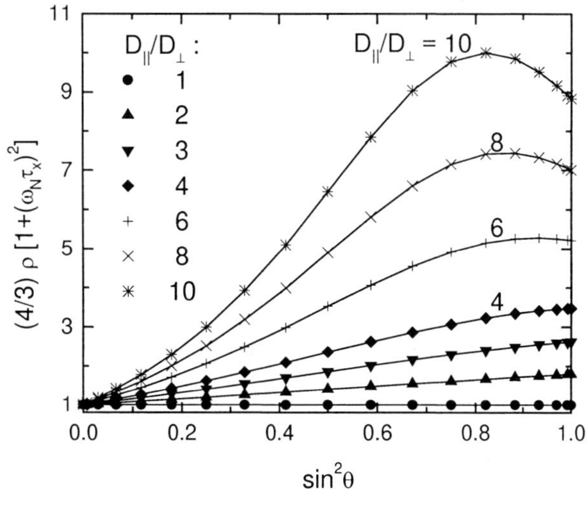

Fig. 12.3 Demonstration of the accuracy of Eq. (10) for a wide range of rotational anisotropies, $D_{\|}/D_{\perp}$, from 1 to 10. Symbols correspond to synthetic experimental data generated assuming overall tumbling with τ_c=5 ns and various degrees of anisotropy as indicated. Model-free parameters typical of restricted local backbone dynamics in protein core, S^2=0.87, τ_{loc} =20 ps, were used to describe the effect of local motions. The ^{1}H resonance frequency was set to 600 MHz. The solid lines correspond to the right-hand-side expression in Eq. (10).

12.10.2
Derivation of the Diffusion Tensor when Protein Structure is Known

When the structural information is available, then the components of the diffusion tensor can be derived from the minimization of the target function,

$$\chi^2 = \sum_i \left[\frac{\rho_i^{\text{exp}} - \rho_i^{\text{calc}}}{\sigma_{\rho i}} \right]^2 . \tag{13}$$

Here ρ^{exp} is the experimentally determined value of $(2R_2'/R_1' - 1)^{-1}$ derived directly from the experimental data according to Eqs. (12a, b) and (9a), and ρ^{calc} is its predicted value calculated using Eqs. (10) and (11) for a given set of the principal axes and principal values of the diffusion tensor; σ_ρ denotes the experimental errors in ρ^{exp}, and the sum runs over all available internuclear vectors. The minimization of the target function Eq. (13) for an axially symmetric tensor requires a search in a four-dimensional space of the principal values $\{D_{\|}, D_{\perp}\}$ and the Euler angles $\{\Phi, \Theta\}$. This method is implemented in our computer program R2R1. Its application for the determination of the rotational diffusion tensor for the βARK PH domain is illustrated in Fig. 12.4. The characteristics of the rotational diffusion tensor thus obtained can then be used as input parameters for the analysis of protein dynamics using computer program DYNAMICS, as described below.

In the most general case of a completely anisotropic diffusion tensor, six parameters have to be determined for the rotational diffusion tensor: three principal values and three Euler angles. This determination requires an optimization search in a six-dimensional space, which could be a significantly more CPU-demanding procedure than that for an axially symmetric tensor. Possible efficient approaches to this problem suggested recently include a simulated annealing procedure [54] and a two-step procedure [55].

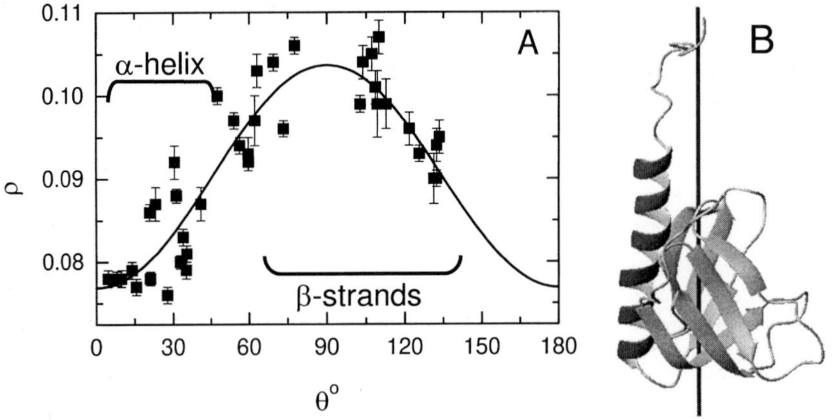

Fig. 12.4 The results of the determination of the rotational diffusion tensor of the βARK PH domain: **A** fit of the orientational dependence of the experimental values of ρ for the βARK PH domain and **B** a ribbon representation of the 3D structure of the protein, with the orientation of the diffusion axis indicated by a rod. Shown in **A** are the experimental (symbols) and the best fit (line) values of ρ (Eq. (10)) for those residues belonging to the well-defined secondary structure elements. The rotational diffusion tensor characteristics derived here are $\tau_x = 9.29 \pm 0.11$ ns and $D_\parallel/D_\perp = 1.39 \pm 0.05$ (resulting in $\tau_c = 8.23 \pm 0.15$ ns), $\Theta = 88 \pm 4°$ and $\Phi = 236 \pm 4.5°$. The probability that an improvement in the fit compared to the isotropic model could occur by chance is 2×10^{-13}. The rotational diffusion axis is oriented along the α-helix axis, and the angle between the two is about 20°. This plot also illustrates a clear separation between two sets of NH vectors, those belonging to the helix (hence oriented nearly parallel to the diffusion axis) and those in the β-strands. Because of the architecture of the PH domain fold, the latter NH vectors are almost all orthogonal to the helix and hence to the diffusion axis.

12.10.3
What Can We Do when Protein Structure is not Known? Preliminary Characterization of the Diffusion Tensor

In the absence of accurate structural information, the analysis based on anisotropic diffusion as discussed above cannot be applied. The use of the isotropic overall model is still possible (see below) because it does not require any structural knowledge. However, the isotropic model has to be validated, i.e. the degree of the overall rotational anisotropy has to be determined prior to such an analysis.

An estimate of the degree of rotational anisotropy can be obtained from the spread of the experimental values of ρ, assuming that the minimal and maximal observed values of ρ ($\rho_{\min}$ and $\rho_{\max}$) correspond to NH vectors oriented parallel ($\theta = 0$) and perpendicular ($\theta = 90°$) to the rotational diffusion axis. We obtain:

$$D_\perp = \frac{\omega_N}{3}\sqrt{\frac{\rho_{\min}}{3 - 4\rho_{\min}}} \quad \text{and} \quad D_\parallel/D_\perp = 1 + \frac{9}{4}\frac{\rho_{\max}/\rho_{\min} - 1}{3 - 4\rho_{\min}} \tag{14}$$

where the latter equation assumes $\varepsilon \ll 1$. Note that the assignment of $\rho_{\max}$ to the orthogonal orientation of NH vectors is not valid for strong anisotropies (e.g. $D_\parallel/D_\perp > 4$ in Fig. 12.3).

The diffusion tensor characteristics for the βARK PH domain derived using this approach are $D_{\|}/D_{\perp} = 1.45$, $\tau_x = 9.36$ ns, and $\tau_x = 8.15$ ns, in reasonable agreement with those derived using the actual fit (see Fig. 12.4).

These considerations assume that (i) we are dealing with a prolate molecule ($D_{\|} > D_{\perp}$ hence $\varepsilon > 0$), (ii) that the conformational exchange contribution is negligible, and (iii) that we have a uniform distribution of the orientations for the set of NH vectors available in the molecule. In the case of an oblate ellipsoid of revolution ($\varepsilon < 0$), ρ_{min} and ρ_{max} in these equations have to be interchanged, as the corresponding orientations will be flipped by 90°. Amide groups undergoing conformational exchange have to be excluded from this analysis: the presence of R_{ex} contributions to R_2 will lower the apparent values of ρ_{min} and hence could render this estimate inaccurate. The issue of incomplete orientational sampling is more complex and depends on the structure of the molecule (see Ref. [56]). In addition, it is worth mentioning that, even in the case of a uniform distribution, the number of vectors oriented parallel to any given direction is small, so that ρ_{max} could be underestimated. Here we assumed that the rotational diffusion tensor is axially symmetric. The presence of a rhombic component could be identified by the shape of the distribution of the values of ρ (see e.g. Refs. [55, 57]).

12.10.4 Isotropic Overall Model

The isotropic model is justified when the estimated degree of the overall rotational anisotropy is small. A $D_{\|}/D_{\perp}$ ratio of less than 1.1–1.2 could probably be considered as a reasonable value for the isotropic model, although an anisotropy as small as 1.17 can be reliably determined from ^{15}N relaxation measurements, as demonstrated in Ref. [15].

In the isotropic model, the overall rotational diffusion is characterized by a single parameter, the overall correlation time τ_c. The following steps could be used to determine τ_c.

(i) *Preliminary estimate of* τ_c. A preliminary estimate of τ_c is obtained as an average over the apparent correlation time values (τ_{ci}) for each NH bond; the latter could be calculated as [58]:

$$\tau_{ci} = \frac{1}{2\omega_N}\sqrt{\frac{6R_{2i}}{R_{1i}} - 7} \tag{15}$$

here subscript i assigns numbers to all NH-vectors. This value is more accurate if the R_2' and R_1' values, corrected for high-frequency components, are used (Eqs. (12 a, b)). A more refined estimate of τ_c is then obtained as an average over those residues that have their τ_{ci} values within 1–1.5 standard deviations from the mean.

(ii) *Optimization of* τ_c *via model selection.* This τ_c estimate can then be used as a starting value for the optimization procedure using computer program DYNAMICS, which is performed as follows. For a range of τ_c values (either automatically predetermined on a grid around the initial t_c estimate or input manually), the program performs model selection for all residues (see below). For each value of τ_c, the total residuals of fit and the number of degrees of freedom are obtained as a sum over all residues: $\chi^2_{tot} = \sum_i \chi^2_i$ and $df_{tot} = \sum_i df_i$. The optimal value of τ_c is determined as the one that yields the highest total number of degrees of freedom, df_{tot}. Usually, a relatively narrow range of τ_c values satisfies this criterion. The width of this range, typically about 0.1–0.2 ns, reflects the precision of τ_c estimation by this method. Further optimization is then possible based on the minimum of χ^2_{tot} in this τ_c range.

For the βARK PH domain example considered here, the τ_c estimate using Eq. (15) gave 8.22 ns, while the model-selection optimization yielded $\tau_c = 8.27$ ns, both values in good agreement with those derived using a more accurate fit (Fig. 12.3).

Note that, unlike other approaches [59], the optimization procedure implemented here performs model selection for each value of τ_c and is not restricted to the model selection determined for the starting value of τ_c.

12.11
Model Selection for NH Bond Dynamics

Here we describe the model selection algorithm that is used to derive microdynamic (model-free) parameters for each NH group from ^{15}N relaxation data. It is implemented in our program DYNAMICS [9]. Given the overall rotational diffusion tensor parameters (isotropic or anisotropic) derived as described above, this analysis is performed independently for each NH-group in order to characterize its local mobility.

The method is based on the model-free parameterization of the spectral density. Depending on the number of available experimental parameters, up to 7 different models for SD could be considered, listed in Tab. 12.1 [9].

These models are nested; the search starts with the simplest model and proceeds to the models of increasing degree of complexity (number of fitting parameters). An "Ockham's razor" principle is assumed here: if more than one model is consistent with the data, the simplest model is preferred. For each model of motion, all parameters are determined from fitting, based on the simplex algorithm, to minimize the following target function:

$$\chi_i^2 = \sum_k (R_{k_i}^{\text{exper}} - R_{k_i}^{\text{calc}})^2 / \sigma_{k_i}^2 \tag{16}$$

where the subscript k assigns numbers to all measured relaxation parameters available for residue i, σ_k is the corresponding experimental uncertainty in R_k, and the superscripts "ex-

Tab. 12.1 Parameters of the models of local motion for the relaxation data analysis using computer program DYNAMICS.

Model	S^2_{fast}	τ_{fast}	S^2_{slow}	τ_{slow}	R_{ex}	N_{par}
A	X	0	1		0	1
B	X	X	1		0	2
C	X	0	1		X	2
D	X	X	1		X	3
E	X	0	X	X	0	3
F	X	X	X	X	0	4
G	X	0	X	X	X	4
H	X	X	X	X	X	5

per" and "calc" denote experimental and calculated values, respectively. A reasonably good fit is expected to result in the χ_i^2 values about the number of "degrees of freedom", $df_i = N_i^{exper}$-N_i^{calc}, where the Ns indicate the numbers of experimental and fitting parameters, respectively. The goodness of fit test for each model and their comparison with one another are performed based on the χ_i^2 and df_i values using the F-statistics [60] as described in detail in Ref. [9]. Since no F-statistics comparison is possible when the number of degrees of freedom becomes zero (i.e. when $N_i^{exper} = N_i^{calc}$), the list of available models is limited. For example, for a typical set of relaxation measurements at a single field, $N_i^{exper} = 3$, so that only models A through C allow a statistically reliable assessment. Usually, models D and E are also included, and their acceptance conditions are somewhat arbitrarily set to a very low acceptance level for the χ_i^2, usually of the order of $\leq 10^{-2}$.

The summary of microdynamic parameters derived from ^{15}N relaxation data for the βARK PH domain is presented in Fig. 12.5.

Typically, residues from secondary structural elements such as α-helices or β-sheets may be fitted to a simple one-parameter (S^2) model with low residuals. Since increased mobility *reduces* R_2 rates in general, *increased* R_2 values well above those otherwise found

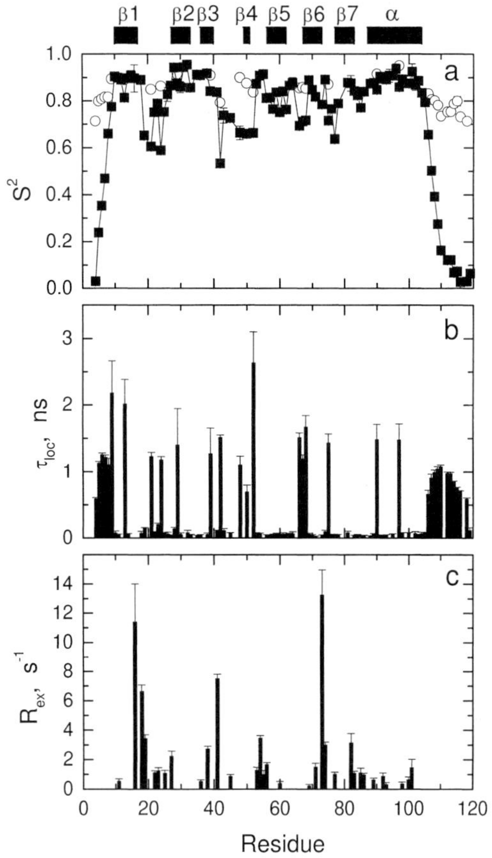

Fig. 12.5 Model-free characteristics of the backbone dynamics in the βARK PH domain: **a** squared order parameters, **b** local correlation time, and **c** conformational exchange contributions to R_2. These parameters were derived from ^{15}N relaxation data at 500 MHz (Fig. 12.2) using computer program DYNAMICS, as described in the text. Amides belonging to the well-defined secondary structure elements were treated assuming the anisotropic overall tumbling model with the axially symmetric diffusion tensor. The principal values and orientation of the latter are given in the legend to Fig. 12.4. The rest of the amides, comprising those in the flexible loops and termini, were treated assuming an isotropic model, with τ_c=8.27 ns. A constant value of –170 ppm for ^{15}N CSA was assumed here. In those cases when the "extended" model [29] (model E in Tab. 12.1) was selected, in addition to the total order parameter (in this case, $S^2 = S^2_{fast}\, S^2_{slow}$) represented by solid squares, the order parameter S^2_{fast} characterizing fast motion of the NH bond is also shown in **a** (open circles).

in residues from stable secondary structure indicate conformational exchange contributions. Residues from loops frequently require at least a two-parameter fit to take into account that there is a substantial contribution from internal motions that are not very fast. Residues displaying such motional behavior are often placed in sequence right before or after the termini of α-helices or β-sheets. Especially NH bonds with decreased values for the heteronuclear NOE quite often display motions on two different time scales and hence need the three-parameter fit S^2_{slow}, τ_{slow} and S^2_{fast}. This behavior is typical for residues at the unstructured N- and C-termini and in the loop regions.

12.12
Accuracy and Precision of the Model-Free Parameters

Let us briefly discuss possible limitations in the accuracy and precision of the dynamic picture of a protein derived using the model-free approach. Selection of a particular model of motion for each NH bond strongly depends on site-specific experimental values of relaxation parameters and is based on statistical tests that give preference to one model over another only with a certain probability. Experimental errors (systematic or random) in R_1, R_2, and NOE could therefore influence both the model selection and the actual values of the microdynamic parameters (S^2, τ_{loc}, R_{ex} etc.). The relationship between the errors in the relaxation rates and those in the model-free parameters is complex. For example, assume that the value of τ_c is fixed ($\tau_c = 5$ ns), the "true" values of the model-free parameters are $S^2 = 0.87$, $\tau_{loc} = 50$ ps, and there is no conformational exchange. A 10% increase in R_2 is likely to give rise to R_{ex} terms (0.7 s^{-1}); a similar increase in R_1 will bias model selection toward the "extended" model with reduced S^2 (0.81) and with τ_{slow} in the nanosecond range (1.9 ns), whereas a simultaneous increase in both R_1 and R_2 will result in increased S^2 (0.95) and τ_{loc} (270 ps). A 10% increase in NOE will cause only a slight change in S^2 (0.88). Note also that the values of τ_{loc} could be very sensitive to small variations in the relaxation rates [58].

Additional limitations in the accuracy of the derived dynamic parameters could be related to the limitations in the analytical approaches. For example, neglect of the overall rotational anisotropy could lead to considerable errors in the model-free parameters, as illustrated earlier [46]. As also shown in Ref. [6], the model-free parameters could be in error if the site-specific variations in ^{15}N CSA are not properly taken into account, particularly at higher fields (>600 MHz ^1H frequency).

It is worth mentioning that the "model-free" approach provides a semi-quantitative picture of protein dynamics, as it is practically impossible, with the means of a simple model having a limited number of fitting parameters, to fully and accurately describe the great variety of (often convoluted) motions which take place in a real protein. The order parameters, for example, allow a robust identification of mobile and rigid regions in a protein, but caution should be exercised when using order parameters to quantify changes in protein dynamics (e.g. upon ligand binding) or differences in backbone dynamics between various proteins.

12.13 Motional Models

The derivation of the SD or model-free parameters is just the beginning of the analysis of protein dynamics: what one finally wants to achieve is a picture of protein dynamics in terms of motional models.

The order parameter in the model-free approach allows a simple geometrical interpretation depending on a particular model of motion. For example, if a wobbling-in-cone model is used, then S is related to the semiangle α of the cone as $S = ½ \cos\alpha\,(\cos\alpha + 1)$. For the rotation on a cone model, $S = ½\,(3\cos^2\alpha - 1)$. A Gaussian angle fluctuation model [61] provides another way to look at NH-vector motion. This one is of particular interest because in some cases [62] the backbone motions observed in molecular dynamics simulations can be represented as a solid-body rocking of the peptide plane around the $C\alpha_{i-1}$–$C\alpha_i$ axis (crankshaft-type motions).

Which model provides the best representation for local mobility in a particular group remains unclear, as a detailed picture of protein dynamics is yet to be painted. This information is not directly available from NMR measurements that are necessarily limited by the number of experimentally available parameters. Additional knowledge is required in order to translate these experimental data into a reliable motional picture of a protein. At this stage, molecular dynamic simulations could prove extremely valuable, because they can provide complete characterization of atomic motions for all atoms in a molecule and at all instants of the simulated trajectory. This direction becomes particularly promising with the current progress in computational resources, when the length of a simulated trajectory approaches the NMR-relevant time scales [23, 63, 64].

12.14 Conformational Exchange

Conformational exchange in the intermediate time-range comprises those motions that modulate the isotropic contributions to spin Hamiltonian, already averaged by fast local and global motions. For example, conformational exchange could be produced by a two- or multiple-site interchange between two or multiple conformations characterized by different values of the isotropic chemical shift of the nucleus under observation. By their nature, these processes involve activation barriers. The conformational exchange leads to line-broadening effects, i.e. increased R_2; its contribution (R_{ex}, Eq. (2)) to R_2 can be assessed e.g. using the CPMG pulse sequence or rotating frame relaxation measurements. The processes of conformational exchange could take place over several time scales, but in practice their observation is limited to the range from µs to ms. The upper limit is determined by the experimental setup, principally by the repetition rate of the refocusing pulses in the CPMG pulse train. It can be varied (e.g. Refs. [9, 65]) in a limited range, which minimizes the effects of evolution due to heteronuclear scalar coupling (~ 94 Hz between ^1H and ^{15}N) [3, 4] on the one hand and avoids sample heating from the pulse train on the other. Recent approaches [66] allow an extension of the time window for R_{ex} observations beyond the 1 ms range. The lower limit of the accessible conformational exchange time scale depends on the lower limit of estimating R_{ex}, which is set by a level of

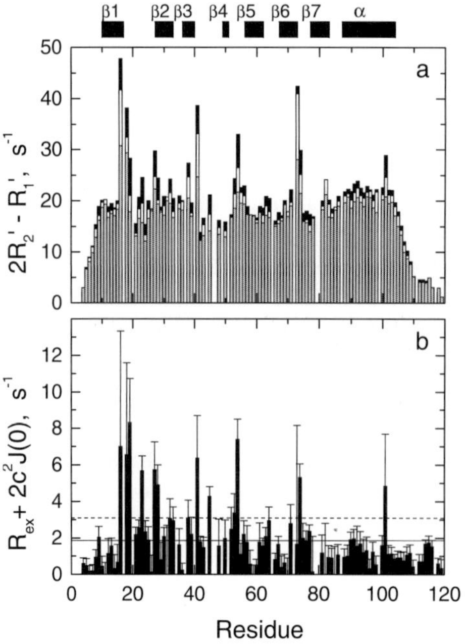

Fig. 12.6 Identification of the residues undergoing conformational exchange based on the field dependence of relaxation parameters: **a** the comparison of $2R_2'-R_1'$ values measured at three fields, 9.4, 11.7, and 14.1 T (shown as grey, white, and black bars, respectively); **b** the values of R_{ex} + $2c^2J(0)$ at 500 MHz directly derived from the difference between the $2R_2'-R_1'$ values (Eq. (17)) at 600 and 500 MHz (as described in Ref. [16]). This analysis does not require structural information. A strong increase in $2R_2'-R_1'$ with the field in panel **a** could be indicative of the conformational exchange, as further quantified in **b**. Although the contributions from CSA and R_{ex} cannot be separated based on these data alone, the expected level of $2c^2J(0)$ can be estimated based on ^{15}N CSA values. The horizontal lines in **b** indicate the expected level of $2c^2J(0)$ calculated for ^{15}N CSA values of –170 ppm (solid line) and –220 ppm (dashed line) and using model-free approach with $S^2=0.87$ and the overall (isotropic) and local correlation times of 8.27 ns and 50 ps, respectively. The latter CSA value corresponds to the maximum absolute ^{15}N CSA value observed in proteins [19]. Note the good correlation between the residues exhibiting elevated values in panel **b** and those having significant R_{ex} contributions as determined from the model-free analysis (Fig. 12.5c).

precision of the R_2 measurement: $R_{ex}>\delta R_2$, where δR_2 is the experimental error. Interestingly, this μs-ms range of motions is considered to be relevant to protein function.

It should be mentioned that rotational anisotropy of the molecule will result in an increase in the R_2 values for NH vectors having particular orientation with respect to the diffusion tensor frame [46]. This increase could be misinterpreted as conformational exchange contributions, and, *vice versa*, small values of R_{ex}, usually of the order or 1 s^{-1} or less, could be mistaken for the manifestation of the rotational anisotropy. Therefore, identification of residues subjected to conformational exchange is critical for accurate analysis of relaxation data. Additional approaches are necessary to distinguish between the two effects. As suggested earlier [27] (see also Ref. [26]), a comparison between R_2 and the cross-correlation rate η_{xy} could serve this purpose, as η_{xy} contains practically the same combination of spectral densities as R_2, but does not contain the R_{ex} term. The field dependence of R_{ex} ($R_{ex} \propto B_0^2$ in the fast exchange limit) can be used for these purposes [9, 16, 20], for example, by utilizing the following combination of relaxation rates (see Fig. 12.6)

$$2R_2' - R_1' = 4\,d^2J(0) + 2[R_{ex} + 2c^2J(0)]\,, \tag{17}$$

where the terms in the brackets scale as B_0^2, although a more accurate analysis has yet to separate the conformational exchange contribution from the CSA term having similar field dependence (see Ref. [20]).

Experimental approaches to direct characterization of the conformational exchange motions in proteins have been suggested earlier [67–69]. The most recent methods [66, 70–73] are based on a relaxation-compensated version of CPMG that alleviates the previous restriction on the duration of the refocusing delay due to evolution of magnetization from scalar couplings and dipole-dipole cross-correlations.

12.15 Effects of Self-Association

^{15}N relaxation studies of proteins are usually performed at protein concentrations in the millimolar range. Although the invention of the cryoprobe promises to eventually allow measurements at significantly lower protein concentrations, a millimolar-range level of protein concentration is currently required for measurements using conventional NMR probes to provide good sensitivity within a reasonable measurement time period. Protein aggregation that could occur under these conditions poses a problem for a robust data analysis. The presence of a protein in different aggregation states differing in their relaxation properties may severely complicate the analysis. The problem is at least two-fold. First, given an exchange process between the monomeric and dimerized/oligomerized forms, one has to determine the degree to which the measured relaxation rates sample the relaxation parameters for each of the states. Second, a complication arises from the overall anisotropy of the dimer, which must be taken into account even in those cases when the monomer is clearly isotropic. What could be done in this case?

As suggested in Ref. [9], in those cases when the dimerization is not apparently specific and the exchange is fast on the NMR-measurement time scale (R_2 from tens to hundreds of ms), this problem can be solved by using a "random-dimer" model. In this model, the resulting spectral density function is assumed to be a weighted sum of the monomeric and dimeric (anisotropic) SD, the latter being averaged over all possible (random) orientations sampled by each monomeric molecule in various dimers which it forms during the spin-relaxation time (illustrated in Fig. 12.7). The averaged spectral density function can be written in the following form [9]:

$$J(\omega) = J_{\text{mono}}(\omega)p_{\text{mono}} + J_{\text{dimer}}(\omega)p_{\text{dimer}}\,, \tag{18}$$

where p_{mono} and p_{dimer} denote the fraction of protein molecules in the monomeric and dimerized states, respectively. This approach accounts for both fast monomer-dimer exchange and anisotropy of the dimer rotation. It happens to be the model of choice in the case of the dynamin PH domain [9].

Fig. 12.7 Illusrration of the idea of the "random dimer" model. For each molecule the orientation of a selected NH vector is indicated; only three of an infinite number of possibilities of interdomain orientations in a dimer are shown.

12.16
Using ^{13}C Relaxation to Study Protein Dynamics

As demonstrated in this chapter, ^{15}N relaxation measurements provide access to backbone dynamics in proteins. The ^{15}N nuclei can also be used to gain information about side-chain motions [72], although their use as reporter groups here is inevitably limited to nitrogen-containing side chains (e.g., Trp, Asn, Gln, and Arg). Carbon is much more abundant in proteins, and therefore ^{13}C relaxation is potentially richer in the informational content about protein motions, both in the backbone and in the side chains. Both ^{15}N and ^{13}C relaxation data are necessary to fully characterize protein dynamics. Studying protein dynamics using ^{13}C relaxation is a real challenge. The ^{13}C relaxation measurements in uniformly labeled proteins are much more complicated than the ^{15}N measurements, primarily because of the magnetization transfer due to scalar coupling between adjacent ^{13}C nuclei, because of the contribution to ^{13}C relaxation from dipolar interactions between adjacent carbons, and because of the technical problems of achieving effective homonuclear decoupling of the ^{13}C nuclei. ^{13}C relaxation measurements in unlabeled protein samples (using ^{13}C natural abundance) [1, 74] help to avoid these problems, although these studies require highly soluble proteins and an extensive measurement time. An enhancement can be achieved here by random fractional ^{13}C labeling [75] or by site-specific ^{13}C-labeling [76, 77] at the cost of complex sample preparation. An approach that overcomes the multiple-spin problems in the CH_2 and CH_3 groups by fractional ^{2}H and uniform ^{13}C labeling was suggested in Ref. [78]. Another problem with ^{13}C relaxation is the influence of other, not directly attached, protons, so that the isolated spin-system approximation might not be valid. This problem could partially be addressed by fractional deuteration of side chains, although the extent of random incorporation of ^{2}H at various side-chain positions requires further analysis. ^{13}C relaxation measurements on the carbonyl and α-carbons are expected to complement our understanding of the backbone dynamics [79, 80]. While the experimental techniques for such measurements have been developed [81–86], the informational content of the first studies is limited, mostly because of the necessity to correctly assess contributions from various sources, predominantly the ^{13}C CSA and dipolar contributions from directly attached and through-space coupling to neighboring hydrogens and ^{13}C's. Site-specific fluctuations in the ^{13}C CSA tensors [87] and in the geometry of the local environment severely complicate this analysis.

Side-chain mobility is of particular interest because groups responsible for protein function are in many cases located in side chains rather than in the backbone. Gaining an insight into side-chain dynamics, therefore, could be necessary for understanding the relationship between protein dynamics and function. In contrast to protein backbone dynamics, the dynamics in the side chains have not been studied systematically, although promising experimental approaches are starting to emerge [72, 73, 78, 88–90].

12.17 What We Have Learned from Protein Dynamics Studies

A diverse picture of protein dynamics has emerged from ^{15}N NMR relaxation studies. MD simulations related to NMR relaxation of proteins performed in the last decade have primarily focused on motions in the protein backbone. Some earlier results may require reinterpretation using anisotropic models. Briefly, the backbone motions in the elements of well-defined secondary structure (β-strands, α-helices, tight turns) are largely restricted in their experimental or simulated amplitude ($0.8 < S^2 < 1$, $S^2 \approx 0.87$ on average), and mostly consist, in simulations, of ps time-range bond librations and restricted fluctuations of torsion angles involving rocking of the peptide plane [23, 91]. These correspond to oscillations within one potential well, and can be modeled either by diffusion within a cone or "wobbling in a cone" with the cone semiangle of $\sim 17^\circ$ or by an oscillating motion within a harmonic potential, e.g. a Gaussian axial fluctuation model [61]. Dynamics in the loops and in the termini, on the other hand, are largely influenced by fluctuations of the backbone dihedral angles (ϕ, ψ) that are less restricted in flexible segments than in the protein core. These motions are much slower, with a characteristic time in the ns range, and lead to greater amplitudes of bond reorientation. Not surprisingly, the complete picture of motion is more complex, and in many cases resembles infrequent jumps between various energetic wells. The picosecond time-scale bond librations are still present (as e.g. S_{fast}, τ_{fast} in Eq. (7)), and in some cases can be deconvolved from the slower motions [9, 23]. Protein termini, if they happen not to be fully folded, are virtually unrestricted, with the order parameter values asymptotically approaching zero toward a flexible terminus. Negative heteronuclear steady-state NOEs observed in these residues can be directly used as an indicator of unfolded parts of a protein construct [9].

12.18 Acknowledgements

I am grateful to David Cowburn for many inspiring and helpful discussions. Computer programs DYNAMICS and R2R1 described here are available from the author upon request.

12.19 References

1 Nirmala, N. and G. Wagner. *J. Am. Chem. Soc.*, 1988. **110**: p. 7556–7558.

2 Kay, L. E., D. A. Torchia, and A. Bax. *Biochemistry*, 1989. **28**(23): p. 8972–8979.

3 Kay, L.., L. K. Nicholson, F. Delaglio, A. Bax, and D. Torchia. *J. Magn. Reson.*, 1992. **97**: p. 359–375.

4 Palmer, A., N. Sketon, W. Chazin, P. Wright, and M. Rance. *Mol. Phys.*, 1992. **75**: p. 699–711.

5 Cavanagh, J., W. J. Fairbrother, A. J. P. III, and N. J. Skelton, *Protein NMR Spectroscopy*. 1996, San Diego: Academic Press.

6 Fushman, D. and D. Cowburn, *Nuclear magnetic resonance relaxation in determination of residue-specific ^{15}N chemical shift tensors in proteins in solution: protein dynamics, structure, and applications of transverse relaxation optimized spectroscopy*, in *Methods Enzymol.* T. James, U. Schmitz, and V. Doetsch, Editors. 2001. p. 109–126.

7 Carr, H. and E. M. Purcell. *Phys.Rev.*, 1954. **94**: p. 630.

8 Meiboom, S. and D. Gill. *Rev. Sci. Instrum.*, 1958. **29**: p. 688.

9 Fushman, D., S. Cahill, and D. Cowburn. *J. Mol. Biol.*, 1997. **266**: p. 173–194.

10 Goldman, M. *J. Magn. Reson.*, 1984. **60**: p. 437–452.

11 Grzesiek, S. and A. Bax. *J. Am. Chem. Soc.*, 1993. **115**: p. 12593–12594.

12 Jones, J. A. *J. Magn. Reson.*, 1997. **126**: p. 283–286.

13 Skelton, N., A. Palmer, M. Akke, J. Kordel, M. Rance, and W. Chazin. *J. Magn. Reson.*, 1993. **B102**: p. 253–264.

14 Press, W. H., S.A. Teukolsky, W. T. Vetterling, and B. P. Flannery, *Numerical Recipes in C.* 1992, NY: Cambridge University Press.

15 Tjandra, N., S. E. Feller, R. W. Pastor, and A. Bax. *J. Am. Chem. Soc.*, 1995. **117**: p. 12562–12566.

16 Camarero, J. A., D. Fushman, S. Sato, I. Giriat, D. Cowburn, D. P. Raleigh, and T. W. Muir. *J. Mol. Biol.*, 2001. **308**: p. 1045–1062.

17 Gagne, S., S. Tsuda, L. Spyracopoulos, L. Kay, and B. Sykes. *J. Mol. Biol.*, 1998. **278**: p. 667–686.

18 Braun, S., H.–O. Kalinowski, and S. Berger, *150 and more basic NMR experiments*. 1998, Weinheim: Wiley–VCH, 596 pp.

19 Fushman, D., N. Tjandra, and D. Cowburn. *J. Am . Chem. Soc.*, 1998. **120**(42): p. 10947–10952.

20 Fushman, D., N. Tjandra, and D. Cowburn. *J. Am. Chem. Soc.*, 1999. **121**: p. 8577–8582.

21 Abragam, A., *The Principles of Nuclear Magnetism*. 1961, Oxford: Clarendon Press.

22 Fushman, D. and D. Cowburn. *J. Biomol. NMR*, 1999. **13**: p. 139–147.

23 Pfeiffer, S., D. Fushman, and D. Cowburn. *J. Am. Chem. Soc.*, 2001. **123**: p. 3021–3036.

24 Tjandra, N. and A. Bax. *J. Am. Chem. Soc.*, 1997. **119**: p. 8076–8082.

25 Tjandra, N. and A. Bax. *J. Am. Chem. Soc.*, 1998. **119**: p. 9566–9567.

26 Kroenke, C. D., J. P. Loria, L. K. Lee, M. Rance, and A. G. I. Palmer. *J. Am. Chem. Soc.*, 1998. **120**: p. 7905–7915.

27 Fushman, D. and D. Cowburn. *J. Am. Chem. Soc.*, 1998. **120**: p. 7109–7110.

28 Lipari, G. and A. Szabo. *J. Am. Chem. Soc.*, 1982. **104**: p. 4546–4559.

29 Clore, G. M., A. Szabo, A. Bax, L. E. Kay, P.C. Driscoll, and A. M. Gronenborn. *J. Am. Chem. Soc*, 1990. **112**: p. 4989–4936.

30 Ishima, R. and K. Nagayama. *Biochemistry*, 1995. **34**: p. 3162–3171.

31 Farrow, N., O. Zhang, A. Szabo, D. Torchia, and L. Kay. *J. Biomol. NMR*, 1995. **6**: p. 153–162.

32 Lipari, G. and A. Szabo. *J. Am. Chem. Soc.*, 1982. **104**: p. 4559–4570.

33 Akke, M., R. Bruschweiler and A. Palmer. *J. Am. Chem. Soc.*, 1993. **115**: p. 9832–9833.

34 Li, Z., S. Raychaudhuri and A. J. Wand. *Protein Science*, 1996. **5**: p. 2647–2650.

35 Yang, D. and L. E. Kay. *J. Mol. Biol.*, 1996. **263**: p. 369–382.

36 Woessner, D. *J. Chem. Phys*, 1962. **37**: p. 647–654.

37 Peng, J. and G. Wagner. *J. Magn. Reson.*, 1992. **94**: p. 82–100.

38 Peng, J. and G. Wagner. *Biochemistry*, 1992. **31**: p. 8571–8586.

39 Farrow, N., O. Zhang, J. Forman-Kay, and L. Kay. *Biochemistry*, 1995. **34**: p. 868–878.

40 Vis, H., C.E. Vorgias, K.S. Wilson, R. Kaptein, and R. Boelens. *J. Biomol. NMR*, 1998. **11**(3): p. 265–277.

41 Kroenke, C.D., M. Rance, and A.G.I. Palmer. *J. Am. Chem. Soc.*, 1999. **121**(43): p. 10119–10125.

42 Baber, J.L., A. Szabo, and N. Tjandra. *J. Am. Chem. Soc.*, 2001. **123**(17): p. 3953–3959.

43 Fushman, D., T. Najmabadi-Haske, S. Cahill, J. Zheng, H. LeVine III and D. Cowburn. *J. Biol. Chem.*, 1998. **273**(5): p. 2835–2843.

44 Pfeiffer, S., D. Fushman, and D. Cowburn. *Proteins*, 1999. **34**: p. 206–217.

45 Tugarinov, V., Z. Liang, Y. Shapiro, J.H. Freed, and E. Meirovitch. *J. Am. Chem. Soc.*, 2001. **123**: p. 3055–3063.

46 Fushman, D. and D. Cowburn, *Studying protein dynamics with NMR relaxation*, in *Structure, Motion, Interaction and Expression of Biological Macromolecules*, R. Sarma and M. Sarma, Eds. 1998, Adenine Press: Albany, NY. p. 63–77.

47 Copie, V., Y. Tomita, S.K. Akiyama, S. Aota, K.M. Yamada, R.M. Venable, R.W. Pastor, S. Krueger, and D.A. Torchia. *J. Mol. Biol.*, 1998. **277**: p. 663–682.

48 Tjandra, N., D.S. Garrett, A.M. Gronenborn, A. Bax, and G.M. Clore. *Nat. Struct. Biol.*, 1997. **4**(6): p. 443–449.

49 Fushman, D., R. Xu, and D. Cowburn. *Biochemistry*, 1999. **38**: p. 10225–10230.

50 Hwang, P.M., N.R. Skrynnikov, and L.E. Kay. *J. Biomol. NMR*, 2001. **20**: p. 83–88.

51 Brüschweiler, R., X. Liao, and P.E. Wright. *Science*, 1995. **268**(5212): p. 886–889.

52 Lee, L.K., M. Rance, W.J. Chazin, and A.G. Palmer III. *J. Biomol. NMR*, 1997. **9**(3): p. 287–98.

53 Blackledge, M., F. Cordier, P. Dosset, and D. Marion. *J. Am. Chem. Soc.*, 1998. **120**: p. 4538–4539.

54 Hus, J.C., D. Marion, and M. Blackledge. *J. Mol. Biol.*, 2000. **298**(5): p. 927–936.

55 Ghose, R., D. Fushman, and D. Cowburn. *J. Magn. Reson.*, 2001. **149**: p. 214–217.

56 Fushman, D., R. Ghose, and D. Cowburn. *J. Am. Chem. Soc.*, 2000. **122**(43): p. 10640–10649.

57 Clore, G.M., A.M. Gronenborn, and A. Bax. *J. Magn. Reson.*, 1998. **133**(1): p. 216–221.

58 Fushman, D., R. Weisemann, H. Thuring, and H. Ruterjans. *J. Biomol. NMR*, 1994. **4**: p. 61–78.

59 Mandel, A.M., M. Akke, and A.G.I. Palmer. *J. Mol. Biol.*, 1995. **246**: p. 144–163.

60 Draper, N.R. and H. Smith, *Applied regression analysis*. 1981, New York: John Wiley & Sons. 709 pp.

61 Bruschweiler, R. and P. Wright. *J. Am. Chem. Soc.*, 1994. **116**: p. 8426–8427.

62 Lienin, S.F., T. Bremi, B. Brutscher, R. Brüschweiler, and R.R. Ernst. *J. Am. Chem. Soc.*, 1998. **120**(38): p. 9870–9799.

63 Chatfield, D.C., A. Szabo, and B.R. Brooks. *J. Am. Chem. Soc.*, 1998. **120**: p. 5301–11.

64 Duan, Y. and P.A. Kollman. *Science*, 1998. **282**: p. 740–744.

65 Orekhov, V., K.V. Pervushin, and A.S. Arseniev. *Eur. J. Biochem.*, 1994. **219**(3): p. 887–896.

66 Loria, J.P., M. Rance, and A.G.I. Palmer. *J. Am. Chem. Soc.*, 1999. **121**: p. 2331–2332.

67 Akke, M. and A.G. Palmer III,. *J. Am. Chem. Soc.*, 1996. **118**: p. 911–912.

68 Zinn–Justin, S., P. Berthault, M. Guenneugues, and H. Desvaux. *J. Biomol. NMR*, 1997. **10**: p. 363–372.

69 Ishima, R., P. Wingfield, S. Stahl, J. Kaufman, and D.A. Torchia. *J. Am. Chem. Soc.*, 1998. **120**: p. 10534–10542.

70 Millet, O., J. Loria, C.D. Kroenke, M. Pons, and A.G. Palmer. *J. Am. Chem. Soc.*, 2000. **122**: p. 2867–2877.

71 Wang, C., M.J. Grey, and A.G.I. Palmer. *J. Biomol. NMR*, 2001. **21**: p. 361–3666.

72 Mulder, F.A.A., N.R. Skrynnikov, B. Hon, F.W. Dahlquist, and L.E. Kay. *J. Am. Chem. Soc.*, 2001. **123**: p. 967–975.

73 Skrynnikov, N.R., F.A.A. Mulder, B. Hon, F.W. Dahlquist, and L.E. Kay. *J. Am. Chem. Soc.*, 2001. **123**: p. 4556–4566.

74 Palmer, A., M. Rance, and P. Wright. *J. Am. Chem. Soc*, 1991a. **113**: p. 4371–4380.

75 Wand, A., J. Urbauer, R. McEvoy, and R. Bieber. *Biochemistry*, 1996. **35**: p. 6166–6125.

76 Nicholson, L.K., L.E. Kay, D.M. Baldisseri, J. Arango, P.E. Young, A. Bax, and D.A. Torchia. *Biochemistry*, 1992. **31**(23): p. 5253–5263.

77 Siivari, K., M. Zhang, A. Palmer, and H. Vogel. *FEBS Lett.*, 1995. **366**: p. 104–109.

78 Yang, D. and L. Kay. *J. Magn. Reson.*, 1996. **B110**: p. 213–218.

79 Fischer, M. W. F., L. Zeng, Y. Pang, W. Hu, A. Majumdar, and E. R. P. Zuiderweg. *J. Am. Chem. Soc.*, 1997. **119**: p. 12629–12642.

80 Fischer, M. W. F., L. Zeng, A. Majumdar, and E. R. P. Zuiderweg. *Proc. Natl. Acad. Sci, USA*, 1998. **95**: p. 8016–8019.

81 Zheng, L., M. Fischer, and E. Zuiderweg. *J. Biomol. NMR*, 1996. **7**: p. 157–162.

82 Cordier, F., B. Brutcher, and D. Marion. *J. Biomol. NMR*, 1996. **7**: p. 163–168.

83 Allard, P. and T. Härd. *J. Magn. Reson.*, 1997. **126**: p. 48–57.

84 Dayie, K. T. and G. Wagner. *J. Am. Chem. Soc.*, 1997. **119**: p. 7797–7806.

85 Yamazaki, T., R. Muhandiram, and L. Kay. *J. Am. Chem. Soc*, 1994. **116**: p. 8266–8278.

86 Engelke, J. and H. Ruterjans. *J. Biomol. NMR*, 1995. **5**: p. 173–182.

87 Pang, Y. and E. R. Zuiderweg. *J. Am. Chem. Soc.*, 2000. **122**: p. 4841–4842.

88 Ishima, R., J. M. Louis, and D. A. Torchia. *J. Am. Chem. Soc.*, 1999. **121**: p. 11589–11590.

89 Ishima, R., A. P. Petkova, J. M. Louis, and D. A. Torchia. *J. Am. Chem. Soc.*, 2001. **123**: p. 6164–6171.

90 Mulder, F. A. A., B. Hon, A. Mittermaier, F. W. Dahlquist, and L. E. Kay. *J. Am. Chem. Soc.*, 2002. **124**: p. 1443–1451.

91 Fushman, D., O. Ohlenschläger, and H. Rüterjans. *J. Biomol. Struct. & Dyn.*, 1994. **4**: p. 61–78.

13
The Determination of Equilibrium Dissociation Constants of Protein-Ligand Complexes by NMR

Gordon C.K. Roberts

13.1
Introduction

In general terms, the essential requirement for measuring the dissociation constant of a protein-ligand complex spectroscopically is the identification of a signal (from either the protein or the ligand) which has a different value in the complex from that in the free protein or ligand. In this respect, NMR has real advantages, since of the many resonances from the protein and the ligand several are sure to change – in shift, relaxation and/or scalar coupling – on complex formation. Most NMR studies of complex formation are carried out in order to determine the magnitude and direction of these changes in NMR parameters and to use them to draw conclusions about the nature of the complex. Here we will consider only the more limited use of NMR to determine the equilibrium dissociation constant for complex formation. As we shall see, NMR is particularly useful for the measurement of relatively weak binding, typically dissociation constants (K_d) $\gtrsim 10^{-5}$ M. Accurate measurement of K_d requires the use of protein concentrations $\leq K_d$, and the low sensitivity of NMR in terms of the concentrations required obviously sets a practical limit to the range of dissociation constants which can be measured; this is discussed further below.

A key factor in the use of NMR for measuring dissociation constants is its sensitivity to the rate of "chemical exchange". Complex formation necessarily involves the exchange of the nuclei or molecules being observed between (at least) two states – the free ligand or protein and the complex. The fact that the appearance of the NMR spectrum is sensitive not only to the position of this equilibrium but also to the rates involved has a major influence on the design of NMR experiments for measuring dissociation constants and on the analysis of such data.

13.2
Chemical Exchange and NMR

In the context of NMR, "chemical exchange" refers to any process in which a nucleus exchanges between two or more environments in which its NMR parameters (chemical shift, scalar or dipolar coupling, relaxation) differ. The effect of this exchange process on the appearance of the NMR spectrum depends on the rate of exchange *relative to the mag-*

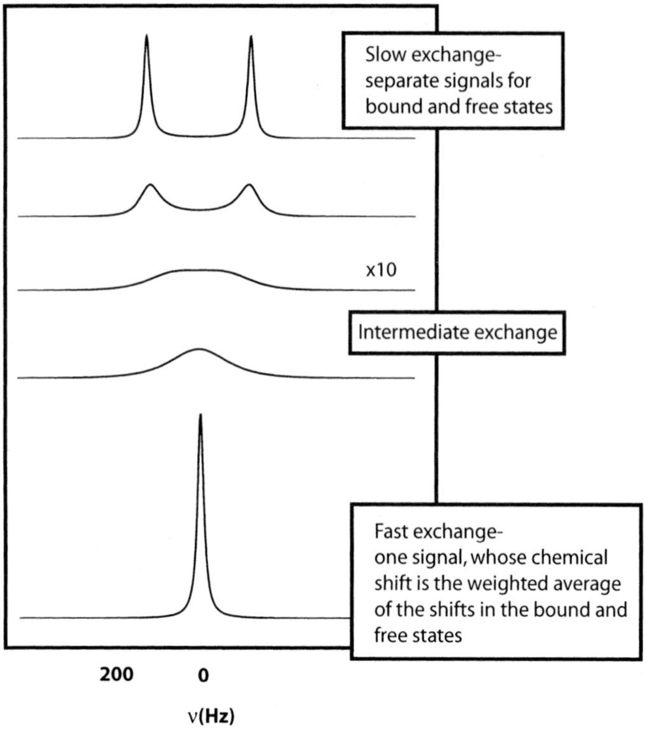

Fig. 13.1 Effects of the exchange rate on the appearance of the NMR spectrum from a system in which a nucleus is exchanging between two equally populated states in which it has a different chemical shift.

nitude of the difference in NMR parameters between the two states. The effects which exchange can have on the appearance of the spectra are illustrated in Fig. 13.1 for the simple case of exchange between two equally populated states. When the exchange is very slow on the "NMR timescale" – here relative to the magnitude of the chemical shift difference between the two states – two separate resonances are seen at the positions corresponding to the chemical shifts characteristic of the two states. At the other extreme, when the exchange is very fast relative to the chemical shift difference, a single resonance is seen, whose position is the average of the chemical shifts of the two states, weighted by their relative populations (see below). Between these extremes there are complex changes in line shape, which are very sensitive to the precise value of the rate of exchange. Detailed descriptions of these exchange effects can be found in a number of textbooks and reviews [1–7].

When attempting to use NMR to measure a dissociation constant, the basic experiment will be to vary the ligand concentration in the presence of a fixed concentration of protein. (The converse experiment, varying the protein concentration, may sometimes be carried out, but is generally less satisfactory because of problems with protein solubility and aggregation.) What one sees in this experiment will depend critically on the rate of

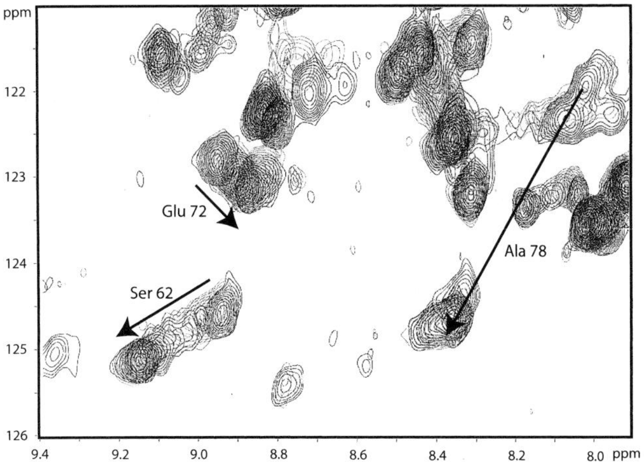

Fig. 13.2 Expanded region of the overlaid ^1H-^{15}N HSQC spectra from a titration of ^{15}N-labeled S100B with CapZ peptide. The protein concentration was 1 mM, and the spectra shown are for peptide concentrations of 0, 0.25, 0.5, 0.75, 1.0, 1.5, 2.5 and 3.0 mM. Chemical shift changes for the cross peaks of Ser62, Glu72 and Ala78 are indicated. From Ref. [8] with permission.

exchange of the molecules between the free state and the complex. If the exchange is slow on the "NMR timescale", the two resonances, corresponding to the bound and free states, will remain in the same positions, but their relative intensities will change. If the exchange is very fast, the position of the single resonance will change progressively as the relative amounts of free protein or ligand change.

It is important to recognise that the "NMR timescale" is a relative one. For a given equilibrium, different resonances will show slow, intermediate or fast exchange behavior, depending on how much their chemical shifts differ between the free and bound states. An example of this is shown in Fig. 13.2. This shows a small region of a superposition of a series of ^1H-^{15}N HSQC spectra of the calcium-binding regulatory protein S110B, alone and in the presence of increasing concentrations of a peptide corresponding to its binding site on the actin-capping protein CapZ [8]. Three residues whose ^1H and ^{15}N chemical shifts change on peptide binding are indicated. For Glu72, the chemical shift changes are small, and a progressive movement of the cross peak is seen, indicative of very fast exchange. This is also the case for Ser62, where the shift changes are larger, but here the progressive shift of the cross peak is accompanied by a marked broadening at intermediate concentrations, indicating that for this peak the exchange regime is between fast and intermediate exchange. Finally, for Ala78, which shows the largest changes in ^1H and ^{15}N shifts, the broadening at intermediate peptide concentrations is such that the peak disappears completely, characteristic of intermediate exchange rates (cf. Fig. 13.1). These line-broadening effects can be very valuable as a means of determining the dissociation rate constant of the complex; detailed procedures for doing this are described elsewhere [9]. As described below, they can, however, present complications in the accurate determination of dissociation constants.

Bearing in mind the relativity of the "NMR timescale", it is useful to estimate the range of rates which will produce line-broadening effects. For chemical shift differences [1)] in the range 10–1000 Hz, intermediate exchange behavior will be observed for dissociation rate constants in the range 10^2–10^4 s^{-1}; assuming that the association rate constant is diffusion-limited, ~10^7–10^8 $M^{-1}s^{-1}$, this will correspond to dissociation constants in the range 10^{-3}–10^{-6} M.

13.3 The Basic Equations

The following descriptions will be couched primarily in terms of chemical shift changes, which are the most widely used for the measurement of dissociation constants; the general principles apply equally to measurements based on other NMR parameters, and these will be touched on later.

$$\mathrm{E} + \mathrm{L} \underset{k_{-1}}{\overset{k_{+1}}{\rightleftharpoons}} \mathrm{EL}$$

$$K_d = \frac{k_{-1}}{k_{+1}} = \frac{[\mathrm{E}][\mathrm{L}]}{[\mathrm{EL}]} \tag{1}$$

In discussing the effects of exchange on NMR spectra, an important parameter is the lifetime of a particular state:

For a nucleus in the ligand molecule:

Lifetime in state EL, $\tau_{\mathrm{EL}} = 1/k_{-1}$ Lifetime in state L, $\tau_{\mathrm{L}} = 1/k_{+1}[\mathrm{E}]$

For a nucleus in the protein molecule:

Lifetime in state EL, $\tau_{\mathrm{EL}} = 1/k_{-1}$ Lifetime in state E, $\tau_{\mathrm{E}} = 1/k_{+1}[\mathrm{L}]$

Using Eq. (1), the lifetimes in the free state of the ligand and protein can be expressed more conveniently as:

Lifetime in state L, $\tau_{\mathrm{L}} = \frac{p_{\mathrm{L}}}{k_{-1} p_{\mathrm{EL}}}$ Lifetime in state E, $\tau_{\mathrm{E}} = \frac{p_{\mathrm{E}}}{k_{-1} p_{\mathrm{EL}}}$

Where p_X is the fractional population of species X, so that, for a ligand resonance,

$$p_{\mathrm{EL}} = \frac{[\mathrm{EL}]}{\mathrm{L_T}} \tag{2}$$

and

$$p_{\mathrm{L}} = \frac{[\mathrm{L}]}{\mathrm{L_T}} = 1 - p_{\mathrm{EL}}, \tag{3}$$

while for a protein resonance

1) Note that in the context of exchange effects on NMR spectra chemical shift differences must be measured in Hertz, not parts per million.

$$p_{EL} = \frac{[EL]}{E_T} \tag{4}$$

and

$$p_L = \frac{[E]}{E_T} = 1 - p_{EL}\,, \tag{5}$$

where L_T and E_T represent the total concentrations of ligand and protein, respectively. From Eq. (1), for a ligand resonance,

$$p_{EL} = \frac{[EL]}{L_T} = \frac{(E_T + L_T + K_d) - \sqrt{(E_T + L_T + K_d)^2 - 4E_T L_T}}{2L_T} \tag{6}$$

with an analogous equation for an enzyme resonance.

In discussions of exchange effects in NMR, a single lifetime is often used to characterize the exchange process

$$1/\tau = 1/\tau_{EL} + 1/\tau_L = k_{-1}\,(1 + p_{EL}/p_L) = k_{-1}/p_L \tag{7}$$

The spectrum in the presence of exchange in the absence of scalar coupling can be simulated by using McConnell's modification of the Bloch equations [10–12]. (When there is strong scalar coupling, or when the coupling changes because of the exchange process, the density matrix approach must be used [5]; a number of line-shape fitting programs based on this approach are available [5].) For the simple case, the line shape (amplitude as a function of frequency, ν) is given by the imaginary part of $G(\nu)$,

$$\frac{G(\nu) = iC[2P_L P_{EL}\tau - \tau^2(P_L a_{EL} + P_{EL} a_L)]}{P_L P_{EL} - \tau^2 a_L a_{EL}} \tag{8}$$

where C is a scaling factor and

$$a_L = 2\pi i(\nu_L - \nu) + 1/T_{2,L} + p_{EL}/\tau\,, \quad a_{EL} = 2\pi i(\nu_{EL} - \nu) + 1/T_{2,EL} + p_L/\tau$$

In general terms, the three exchange regimes are defined by:

Slow exchange	$1/\tau \ll 2\pi\lvert\nu_L - \nu_{EL}\rvert$
Intermediate exchange	$1/\tau \sim 2\pi\lvert\nu_L - \nu_{EL}\rvert$
Fast exchange	$1\tau \gg 2\pi\lvert\nu_L - \nu_{EL}\rvert$

Simpler expressions for the line shape are available for both the fast and slow exchange regimes (see Ref. [9]). The possibility of measuring the dissociation equilibrium constant in each of these three regimes will now be considered.

13.4
Slow Exchange

In this regime, separate peaks are observed for the free and bound states (cf. Fig. 13.1); when the relative amounts of these two states are altered by varying the ligand concentration, the positions of these two signals remain unaltered; only their relative intensity changes. In principle, this change in relative intensity can be used to estimate the dissociation constant; in practice this is only of any value under very restrictive conditions. First, the signals must be sufficiently well resolved to allow accurate intensity measurements; this can be achieved by appropriate isotope labeling – for example, using per-deuterated protein to study ligand resonances. Second, the signal-to-noise ratio must be sufficient to allow accurate intensity measurements at sub-stoichiometric ligand concentrations ([EL] $\sim 0.1 E_T$). This latter problem is compounded by the fact that accurate measurement of K_d requires that $E_T < K_d$. Slow exchange implies a slow rate of dissociation, which in turn implies a relatively low value of K_d and hence the need for a low protein concentration. For example, for a chemical shift difference of 100 Hz, observation of slow exchange would imply that $k_{-1} \leq 3 \times 10^2\ s^{-1}$, so that (assuming a diffusion-limited association rate) $K_d \leq 3 \times 10^{-5}$ M, and one would need to make accurate intensity measurements at protein concentrations of ~ 5 µM. Given the insensitivity of NMR, it is therefore very rarely practical to make reliable estimates of K_d from spectra in the slow exchange regime.

13.5
Intermediate Exchange

The intermediate exchange regime covers a relatively narrow range of exchange rates, no more than a factor of two either side of the "coalescence point" (Eq. (9)). Within this regime the full line-shape equation must be used in order to extract the relative populations of the bound and free states, their chemical shifts and the rate of exchange between them. In principle good estimates of these parameters can be obtained from fitting a series of spectra at different ligand concentrations, but in practice limitations of resolution and signal-to-noise ratio – the signals can be very broad at the coalescence point – mean that this is rarely achievable. In most circumstances a change in spectrometer frequency will be sufficient to shift the system into either slow exchange (at higher spectrometer frequency) or fast exchange (at lower spectrometer frequency), where the analysis will be easier.

13.6
Fast Exchange

The fast exchange regime is by far the most useful for the measurement of dissociation constants from changes in chemical shift, relaxation or other NMR parameters – a wide range of K_d values, from 10 µM (or perhaps a little less) to ≥ 10 mM, can be accessible.

13.6.1
Very Fast Exchange

When the exchange rate is very fast, the chemical shifts and relaxation rates of the nuclei of the ligand and the protein are completely averaged between the bound and free states, and a weighted average spectrum is observed. For ligand resonances

$$\delta_{obs,L} = p_L\delta_L + p_{EL}\delta_{EL} \tag{9}$$

$$R_{2,obs,L} = p_L R_{2,L} + p_{EL} R_{2,EL}\,, \tag{10}$$

where $\delta_{obs,L}$, δ_L and δ_{EL} are, respectively, the observed ligand chemical shift and the chemical shifts of the free ligand and of the ligand in the complex, with analogous definitions for R_2 $(=1/T_2)$, the transverse relaxation rate. Similarly for protein resonances

$$\delta_{obs,E} = p_E\delta_E + p_{EL}\delta_{EL} \tag{11}$$

$$R_{2,obs,E} = p_E R_{2,E} + p_{EL} R_{2,EL} \tag{12}$$

From Eq. (9), for example, for a ligand resonance

$$p_{EL} = \frac{\delta_{obs} - \delta_L}{\delta_{EL} - \delta_L} \tag{13}$$

and this can be combined with Eq. (6) to give

$$\frac{\delta_{obs} - \delta_L}{\delta_{EL} - \delta_L} = \frac{(E_T + L_T + K_d) - \sqrt{(E_T + L_t + K_d)^2 - 4E_tL_T}}{2L_T} \tag{14}$$

and fitting Eq. (14) (or the analogous equations derived from Eqs. (10)–(12)) to the data (δ_{obs} as a function of L_T) will yield estimates for K_d and δ_{EL}. This is the commonest approach used to determine K_d by NMR; a detailed protocol for experiments of this kind is given elsewhere [9].

13.6.2
The General Case of Fast Exchange

It is important to note that Eqs. (9)–(12) are valid only for *very* fast exchange. If the exchange is somewhat slower ("moderately fast" exchange), there will be an exchange contribution to the line width, and Eq. (10) for the ligand resonance[2)] will be replaced by

2) Eq. (15) and the following discussion are couched in terms of ligand resonances, but exactly the same considerations apply to protein resonances.

$$R_{2,\mathrm{obs}} = P_{\mathrm{L}} R_{2,\mathrm{L}} + P_{\mathrm{EL}} R_{2,\mathrm{EL}} + \frac{P_{\mathrm{EL}} P_{\mathrm{L}}^2 4\pi^2 (\delta_{\mathrm{EL}} - \delta_{\mathrm{L}})^2}{k_{-1}} \tag{15}$$

This equation predicts a different dependence of the line width on the ligand concentration from that predicted by Eq. (10). For large k_{-1}, Eq. (15) clearly reduces to Eq. (10), and, if changes in relaxation are being used to estimate K_d, it is generally sensible to fit the data using Eq. (15) rather than Eq. (10) (the only exception being paramagnetic relaxation, see below).

Most important, if the exchange contribution to the line width (the third term on the right hand side of Eq. (15)) is significant, Eqs. (9) and (14) will not be an accurate description of the change in chemical shift with ligand concentration. It has been shown by simulation experiments that the erroneous use of Eqs. (9) and (14) when exchange is not *very* fast can lead to errors of at least an order of magnitude in K_d [13]. For example, Fig. 13.3 A shows a series of simulated spectra of a ligand resonance at increasing values of L_T; the chemical shift difference between the bound and free states is 100 Hz and k_{-1} is 500 s^{-1}. There is clearly a progressive change in chemical shift as the ligand concentration is increased, as would be expected for fast exchange, and the signals appear to have Lorentzian shapes. However, broadening of the resonance line in the middle of the titration is also clearly evident; this arises from the exchange contribution in Eq. (15), which has a maximum when $p_{\mathrm{EL}} \approx 1/3$. While the chemical shifts obtained from these simulated spectra seem to be described reasonably satisfactorily by Eq. (14), as shown in Fig. 13.3 B, the K_d value estimated from the least-squares fitting of Eq. (14) to the data is in error by almost a factor of two, even when, as here, data is available for a wide range of ligand concentrations (from $0.1E_T$ to $30E_T$) [13]. When, more realistically, only ligand concentrations greater than the protein concentration are used in the analysis, the error becomes a factor of ten for the parameters used in Fig. 13.3 and can be as much as 100 for somewhat lower values of $k_{-1}/(\delta_{\mathrm{EL}} - \delta_{\mathrm{L}})$ [13].

This is a potentially serious source of error for combinations of $(\delta_{\mathrm{EL}} - \delta_{\mathrm{L}})$, k_{-1} and K_d which are very likely to be encountered in practice. It is thus essential to establish when it is safe to use Eq. (14) to analyze chemical shift changes. The simplest way to do this is to measure the line width of the resonance of interest as well as its position, and to analyze its dependence on ligand concentration (as described in Refs. [9, 13]) to estimate the magnitude of the exchange contribution (Eq. (15)); only when the exchange contribution is negligible – the very fast exchange condition is satisfied – is it safe to use Eq. (14). In the more general case of "moderately fast" exchange, where the simple Eq. (14) cannot be used, the alternative is either a full line-shape analysis or, more practically, a comparison of measured chemical shifts and line widths as a function of ligand concentration with those obtained from simulated spectra; detailed protocols for this approach are available [9]. Both in order to identify the exchange regime and to obtain accurate and precise values of K_d, it is important to make measurements at as low a ligand concentration as possible.

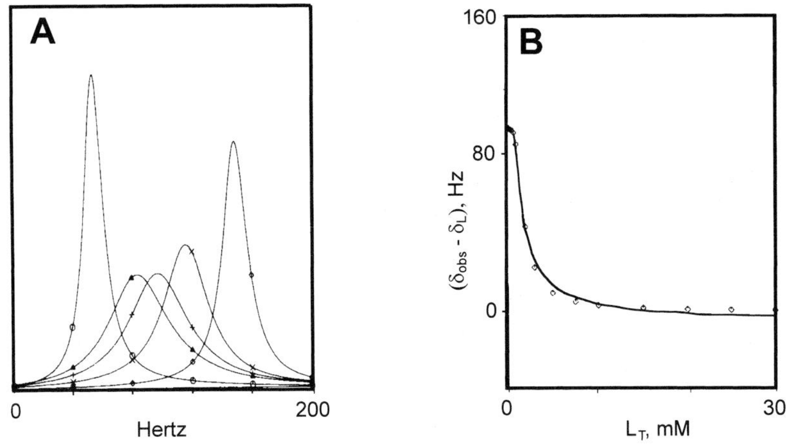

Fig. 13.3 Simulation of spectra of a ligand resonance in a system involving protein-ligand complex formation. The parameters used for the simulation were: $(\delta_{EL} - \delta_L)$ 100 Hz, k_{-1} 500 s^{-1}, K_d 10^{-5} M, E_T 10^{-3} M, L_T (0.1–30) 10^{-3} M, free line width 5 Hz, bound line width 20 Hz. **A** (left): Simulated line shapes for L_T 0.5 mM (○), 1.5 mM (△), 2.0 mM (+), 2.5 mM (×), 10 mM (□). **B** (right): Observed chemical shift $(\delta_{obs}-\delta_L)$ as a function of total ligand concentration, L_T. The points are values calculated from the exact simulation, and the curve is the best fit to Eq. (14), assuming no exchange contribution; the value of K_d estimated from the fit is 6.2×10^{-6} M. From Ref. [13] with permission.

13.6.3 Paramagnetic Relaxation

For those proteins which contain a paramagnetic center (e.g., heme iron, copper), usually at the active site, the bound ligand will experience a marked increase in relaxation rate due to dipolar interaction with the unpaired electron(s) of the paramagnetic center. While the paramagnetic relaxation effect is commonly used to obtain structural information (the distances between the paramagnetic center and ligand nuclei), it can also be used for the determination of K_d (see, for example, Refs. [14, 15]). It is mentioned separately here because, since the magnetic moment of the electron is 1800 times that of the proton, it is, when present, a dominant effect. It magnitude is such that it can only be measured under fast exchange conditions, with a large excess of ligand over protein. This can simplify the analysis (for $L_T \gg E_T$, $p_L \approx 1$), and it also allows one to use low enzyme concentrations, so that somewhat lower K_d values can be measured. On the other hand, these same experimental conditions mean that a long extrapolation to $R_{2,EL}$ or $R_{1,EL}$ is required, and this limits the precision of the measurement.

13.7 Conclusions

As discussed above, NMR can be a useful tool for the determination of dissociation constants provided that its strengths and limitations are recognized. In practice, it is only useful under fast exchange conditions, and thus for relatively weak binding: $K_d \gtrsim 10\ \mu M$. The upper limit to the dissociation constants which can be measured is usually set by the solubility of the ligand molecule, since concentrations $\gtrsim 10\ K_d$ are needed to obtain reliable estimates. Modest amounts of cosolvent (e.g., DMSO, dimethyl sulfoxide) are often used to ameliorate this problem, but most proteins will only tolerate small amounts <10% of such cosolvents, and for accurate data it is important to keep the cosolvent concentration constant during the titration. For dissociation constants in the 10 μM–10 mM range, NMR is certainly competitive with other available methods; it is not dependent on the presence of a chromophore and does not require the separation of bound and free ligand. Provided that simple precautions, described above, are taken to ensure that the appropriate analysis is used, K_d measurement by NMR is rapid, simple and accurate.

The emphasis in the discussion has been on the analysis of changes in chemical shift and relaxation rate because these are the parameters most commonly used for K_d measurements. Any protein-ligand binding process is bound to be accompanied by some change in chemical shift of a resonance or resonances from both the protein and the ligand, although of course the magnitude of such a shift change cannot in general be predicted. A number of resonances and indeed different spectrometer frequencies may need to be examined to find one with a shift large enough to be useful but small enough to be in fast exchange. Changes in relaxation rate are likely to be almost universal for ligand resonances, since the rotational correlation time of the ligand will increase substantially on binding to a macromolecule, though the magnitude of the effect can be complicated by internal motion in the complex. For the protein, on the other hand, changes in relaxation may occur but will be wholly dependent on local changes in mobility. The use of changes in relaxation to study ligand binding to proteins was one of the earliest biological applications of NMR [3, 16] and remains a valuable approach. In fact, of course, any NMR parameter which changes on complex formation and which shows fast exchange behavior (being described by an equation analogous to Eqs. (9)–(12)) can be used to measure K_d. Scalar coupling constants are likely to change if the conformation of the ligand or of protein side-chains changes on complex formation, and this can provide useful structural information; however, the magnitude of such changes is rarely large enough to be useful for K_d determination. On the other hand, NMR experiments which are used to screen for binding, such as pulsed-field-gradient measurements of translational diffusion or the measurement of the sign of intramolecular NOEs in the ligand (see, e.g., Ref. [17]), can also readily be adapted for quantitative determination of K_d – for example, see Ref. [18] for the use of translational diffusion measurements in this way.

13.8 References

1 Kaplan, J. I. and Fraenkel, G., *NMR of Chemically Exchanging Systems.* Academic Press, New York, **1980**.

2 Sandstrom, J., *Dynamic NMR Spectroscopy.* Academic Press, London, **1982**.

3 Jardetzky, O., and Roberts, G. C. K., *NMR in Molecular Biology.* Academic Press, New York, **1981**.

4 Sanders, J. K. M. and Hunter, B. K., *Modern NMR Spectroscopy.* Oxford University Press, Oxford, **1987**.

5 Nageswara Rao, B. D., *Methods Enzymol.* **1989**, *176*, 279–311.

6 Led, J. J., Gesmar, H., and Abildgaard, F., *Methods Enzymol.* **1989**, *176*, 311–329.

7 Berkowitz, B., and Balaban, R S., *Methods Enzymol.* **1989**, *176*, 330–341.

8 Kilby, P. M., Van Eldik, L. J., and Roberts, G. C. K., *Protein Sci.* **1997**, *6*, 2494–2503.

9 Lian, L.-Y., and Roberts, G. C. K., in *NMR of Macromolecules* (ed. Roberts, G.C.K.), Oxford University Press, Oxford, pp. 153–182, **1993**.

10 McConnell, H. M., *J. Chem. Phys.* **1958**, *28*, 430–435.

11 Leigh, J. S., jr., *J. Magn. Reson.* **1971**, *4*, 308–318.

12 McLaughlin, A. C. and Leigh, J. S., jr., *J. Magn. Reson.* **1973**, *9*, 296–305.

13 Feeney, J., Batchelor, J. G., Albrand, J. P. and Roberts, G. C. K., *J. Magn. Reson.* **1979**, *33*, 519–529.

14 Modi, S., Primrose, W. U., Boyle, J. M. B., Gibson, C. F., Lian, L.-Y., and Roberts, G. C. K. *Biochemistry* **1995**, *34*, 8982–8988.

15 Modi, S., Paine, M. J. I., Sutcliffe, M. J., Lian, L.-Y., Primrose, W. U., Wolf, C R. and Roberts, G. C. K., *Biochemistry* **1996**, *35*, 4540–4550.

16 Fischer, J. J. and Jardetzky, O., *J. Am. Chem. Soc.* **1965**, *87*, 3237–3281.

17 Roberts, G. C. K., *Drug Discov. Today* **2000**, *5*, 230–240.

18 Tillett, M. L., Horsfield, M. A., Lian. L.-Y. and Norwood, T. J., *J. Magn. Reson.* **1998**, *133*, 379–384.

14
Experiments in NMR-Based Screening

Carla Marchioro, Silvia Davalli, Stefano Provera, Markus Heller, Alfred Ross, and Hans Senn

14.1
Introduction

The drug research and development process has significantly changed over the past few years with the introduction of genomics/genetics approaches, high-throughput syntheses and biological screening. As a result of these developments, new high-throughput screening methods needed to be developed in order to select active compounds from the large number synthesized. Nuclear magnetic resonance (NMR) spectroscopy has experienced rapid growth and widespread application, thereby playing a crucial role in various aspects of the R & D process. Current attempts to improve the technique are based on increasing sensitivity, implementation of automation and development of hyphenated analysis methods [1]. Furthermore, the continuous developments in instrumentation (i.e. shielded magnets, cryogenic probes, etc.) (see Chapt. 3 and Ref. [2]) and in NMR methodology have expanded the use of NMR in different applications. As a result, NMR is now used as a screening technique for the identification of lead structures in addition to its classical use as a well-established tool for drug design and examination of structure-activity relationship (SAR) [3–8]. In fact, NMR techniques can now be directly used as a rapid and reliable method of characterizing compound mixtures with respect to their properties of binding to a target. This is especially helpful in identifying lead compounds and in helping them to progress to the candidate phase. Furthermore, NMR techniques do not require any specific set-up or knowledge of a protein's function, and hence the analysis is feasible for proteins for which no bio-assay is available. For many steps in the drug development process, NMR can reduce the time needed to find suitable drug candidates. Moreover, starting from a known binder, it may also increase success in modifying lead compounds to give better drug candidates.

Here, we will summarize the various NMR-based screening techniques. These will be subdivided into ligand-observe and target-observe techniques and classified according to a parameter that changes upon binding. Parameters sensitive to ligand-binding are chemical shifts, relaxation properties, and translational diffusion coefficients. Another set of experiments has been developed to facilitate observation of intermolecular cross-relaxation. Some of these parameters only indicate ligand binding, whereas others give site-specific information about the residues in the structure that are involved in binding. Site-specific information may be of prime importance in the drug design process, and NMR is uniquely able to deliver such information in comparison to biological assays or other spectroscopic methods. The design of libraries used for screening and the different strategies for lead development are described in more detail in Chapt. 17. Obviously, for a molecule to be useful as a lead compound it must bind to the target. This can be tested by all methods described in this chapter, and most of these do not require isotopically labeled compounds. However, some strategies require knowledge about the binding sites. For example, in the fragment-based approach [13], small molecules are sought which bind to the target. Once a ligand has been found by screening, the binding pocket is saturated by that ligand and another screening round is performed searching for a second binding pocket in spatial proximity to the first one (second-site screening). The process is summarized in Fig. 14.1 below.

The two different fragments are later linked chemically, resulting in a "biligand" with largely increased affinity. Importantly, this method only requires weakly binding fragments, which are likely to be found during NMR-based screening (and may not easily be found by other methods).

The fragment-based approach requires that binding to the target is identified by monitoring perturbation of chemical shifts in a ^{15}N-labeled protein or in an amino acid ^{13}C-selective labeled protein [9] (usually Ile, Leu or Val). Potential candidates can then be utilized for design of follow-up libraries or generally for larger libraries resulting from parallel synthesis. Another step is the optimization of the fragments themselves. Here, methods are useful that help to rank ligands binding to the same pocket according to their relative binding affinities. Such a step does not, in principle, require a labeled target, but may be done by diffusion or relaxation-based methods.

One limitation of NMR-based screening is the requirement that large protein quantities should yield concentrations giving reasonable durations of NMR experiments and reasonable throughput. Another limitation arises from low solubility (μM) of nondeveloped lead structures. Hydrophobic ligands are often required to enable binding to a hydrophobic

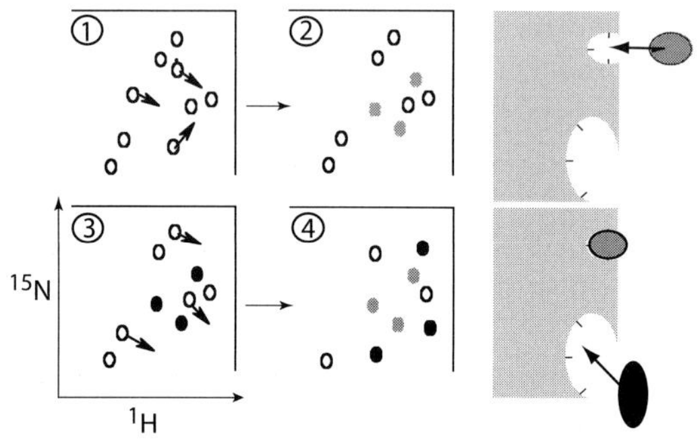

Fig. 14.1 Schematic drawing summarizing fragment screening by NMR. Spectra indicated on the left hand side correspond to [^{15}N,^{1}H]-HSQC spectra recorded on ^{15}N-labeled protein in the absence (spectrum 1) or presence (spectrum 2) of a first ligand. Cross peaks in the HSQC that have shifted upon addition of this ligand are indicated by gray circles (top row). Afterwards, second site-screening is performed (bottom row) while keeping the binding site for the first ligand saturated. Peaks which shift because of a second ligand binding at a different site are indicated by black circles.

pocket of the target. To achieve a reasonable throughput in a ligand-observe experiment, the concentration per substance in the mixture has to be around 100 μM or 0.5 mM in cryo- and conventional-probe technology, respectively. Experiments are typically performed at a ten-fold reduced protein concentration. Thus the amount of substance tested (MW 300 Da) and protein (30 kDa) per sample sums up to 15 μg (75 μg) and 150 μg (750 μg) respectively (numbers in brackets refer to noncryogenic probe equipment). All numbers have to be multiplied by roughly ten if a screening approach based on the observation of protein signals is employed. For comparison, the amount of ligand and protein needed in biological (U)HTS is given by 0.1–1 μg and 2.5–350 ng, respectively [10].

The required high concentration of ligands limits the number of compounds that can be tested simultaneously, since the total amount of ligands is normally limited for reasons of solubility and stability of the protein.

14.2 NMR-Based Screening

Generally, experiments used for screening can be categorized into those that observe the *target*, usually a protein, DNA, RNA or a complex of these, and those that aim at observing changes of the *ligand* properties upon binding. Usually, experiments that detect changes in the receptor molecule require isotope labeling of the target. For proteins, uniform ^{15}N labeling allows changes that occur upon binding to be detected very quickly and enables identification of the binding interface once the $^{15}N,^{1}H$ correlation map has been assigned. Alternatively, specific ^{13}C-methyl labeling may also be used, which is claimed to be more sensitive [11]. For methods of producing proteins in isotopically enriched form see Chapt. 1. It should be noted that amide moieties represent hydrophilic areas, while methyl groups are usually located in hydrophobic regions. The comparison of $[^{15}N,^{1}H]$-HSQC and $[^{13}C,^{1}H]$-HSQC screens may thus reveal additional information of interest.

Depending on the exchange kinetics between free ligand/protein and the complex, two different sets of signals (slow exchange) or a single set of averaged population-weighted signals (fast exchange) are observed [12]. Initial screening is very likely to yield weak binders, and hence spectra usually display fast-exchange characteristics. However, for highly optimized follow-up libraries, good binders may be detected that lead to the occurrence of a second set of peaks, which at reasonable sample concentration remain below the noise threshold in the experiment. Thus, for ligand-based methods performed without reference samples, high-affinity ligands (i.e. slow exchange kinetics) are hard to detect.

In the fast exchange case, an experimental parameter A such as the chemical shift, the scalar coupling or the relaxation rate is averaged according to

$$A_{\mathrm{eff}} = f_{\mathrm{b}} A_{\mathrm{b}} + (1 - f_{\mathrm{b}}) A_{\mathrm{f}} \tag{1}$$

in which f_{b} denotes the fraction of bound ligand. The fraction of bound ligand (or target) f_{b} depends on the dissociation constant K_{D} (see Chapt. 12) and can be calculated according to the formula:

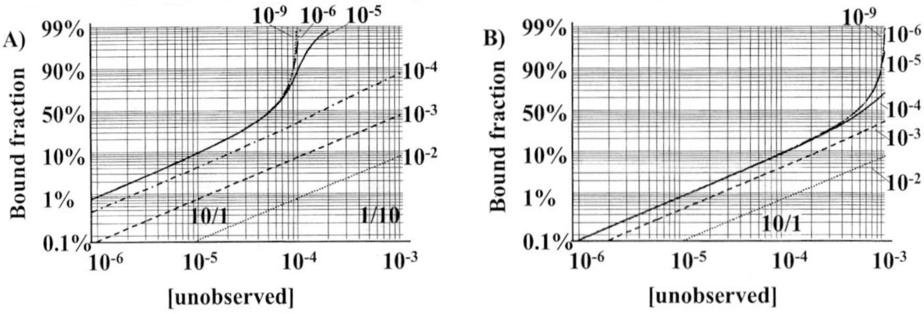

Fig. 14.2 Population of the bound state of a complex as a function of the (molar) concentration of the unobserved substance for different K_D values. The concentrations of the observed substance are 100 μM (cryo probe, panel A) and 1 mM (standard equipment, panel B), corresponding to typical experimental situations. The K_D areas shaded in gray correspond to the 10% bound-state situation. The K_D^{min} ranges shown are obtained for a 10:1 (ligand-observe) and 1/10 (protein-observe) situation respectively. The lines from bottom to top correspond to K_D values of 10 mM, 1 mM, 100 μM, 10 μM, 1 μM and 1 nM.

$$f_b = \frac{[L_0] + [P_0] + K_D - \sqrt{(-4)\cdot[L_0][P_0] + ([L_0] + [P_0] + K_d)^2}}{2[L_0]} \qquad (2)$$

Here $[L_0]$ and $[P_0]$ denote the total concentration of ligand and protein, respectively.

Fig. 14.2 displays the fraction of bound ligand as a function of the dissociation constant for two different ligand concentrations: (A) 100 μM, about the detection limit for 1D experiments by use of cryo-probes, and (B) 1 mM, a concentration for standard equipment. At these concentrations, typical measurement times are in the range of several minutes. From the figure the following conclusion can be drawn for ligand-observe techniques:

- Under the assumption that a given NMR method (see below) allows for the detection of 10% bound ligand, the upper limit of the K_D that will cause appreciable differences in the spectra equals the concentration of the molecule under observation (e.g. for a cryo-probe at a typical [substance]:[protein] ratio of 10:1, K_D values below 100 mM can be detected).
- K_D values one order of magnitude lower than the concentration of the observed substance cannot be discriminated by their effect on the spectra at the 10:1 substance:protein ratio (e.g. for a cryogenic-probe a 10 μM ligand will cause the same spectral differences as a 10 nM ligand, provided that both are in fast exchange).

Since protein-observe methods are usually performed at a substance:protein ratio of 0.1, they extend the minimum K_D detection limit by roughly two orders of magnitude. The price to be paid is a much higher consumption of both protein and substance to be tested.

14.2.1
Experiments Based on Chemical Shift Perturbations

Binding of ligands to targets leads to changes in the chemical shift. For example, formation of a hydrogen bond decreases the electron density at the acceptor atom and hence generally leads to a down-field shift. However, anisotropy effects, e.g. ring current effects, also need to be taken into account. The dissociation constant may be estimated from an NMR titration series by nonlinear curve fitting of the observed chemical shift vs concentration against the following equation:

$$\delta_{\mathrm{av}} = \delta_{\mathrm{f}} + (\delta_{\mathrm{b}} - \delta_{\mathrm{f}}) \cdot \left(\frac{[L_0] + [P_0] + K_D - \sqrt{([L_0] + [P_0] + K_D)^2 - 4 \cdot [L_0] \cdot K_D}}{2[L_0]} \right) \tag{3}$$

In addition to the definitions used in Eq. (2), δ_{f}, δ_{b} describe the chemical shift of the free and the bound state, respectively, while the observed chemical shift is given by δ_{av}. The interaction between a ligand and a target can be detected by monitoring the changes of the chemical shifts of both the ligand and the target signals, but more sensitive parameters (relaxation or diffusion rates) may be used for ligand-based NMR screening approaches (see below).

In contrast, motional properties of the target will not change significantly upon binding, and hence monitoring of induced chemical shift changes is the natural choice for this class of experiments. Because of the signal overlap in 1D spectra of proteins, changes in chemical shifts are most conveniently monitored in ^{15}N,^{1}H correlations of mixtures of one or more ligands with a ^{15}N-labeled protein. This method is referred to as chemical shift mapping [13]. Furthermore, specific binding can easily be distinguished from unspecific binding.

If the assignment is available or can be made in reasonable time (and preferably, but not necessarily, if the protein structure is approximately known), the location of the binding site may be easily derived. This method is currently implemented in the so-called SAR (Structure-Activity Relationship) by NMR technique [13].

Fig. 14.3 displays a superposition of two [^{15}N,^{1}H]-HSQC spectra of the target macromolecule gyrase B from *E. coli*, recorded in the absence (dark) and in the presence (light) of a binder. The comparison reveals resonances which have shifted upon addition of ligand and hence help to determine the binding site. For hundreds of data sets taken, the process of recognizing resonances whose positions have changed upon ligand binding may be automated (see Chapt. 18). If (nonreacting!) ligand mixtures are screened simultaneously, a subsequent deconvolution of data by additional experiments performed on the individual compounds of positively identified mixtures has to be performed. In contrast to other analytical methods (such as UV or mass spectrometry) or biological assays, the shift mapping applied to protein spectra is capable of finding weakly ($K_{\mathrm{D}} \approx 1$ mM) binding ligands that may serve as lead compounds for follow-up libraries (see Chapt. 17). In the SAR-by-NMR approach, two or three low-affinity ligands binding to different pockets in spatial proximity are synthetically linked. As the K_{D} values of subunits are multiplicative, highly affine bi- or tridentate ligands are derived. It should also be noted that

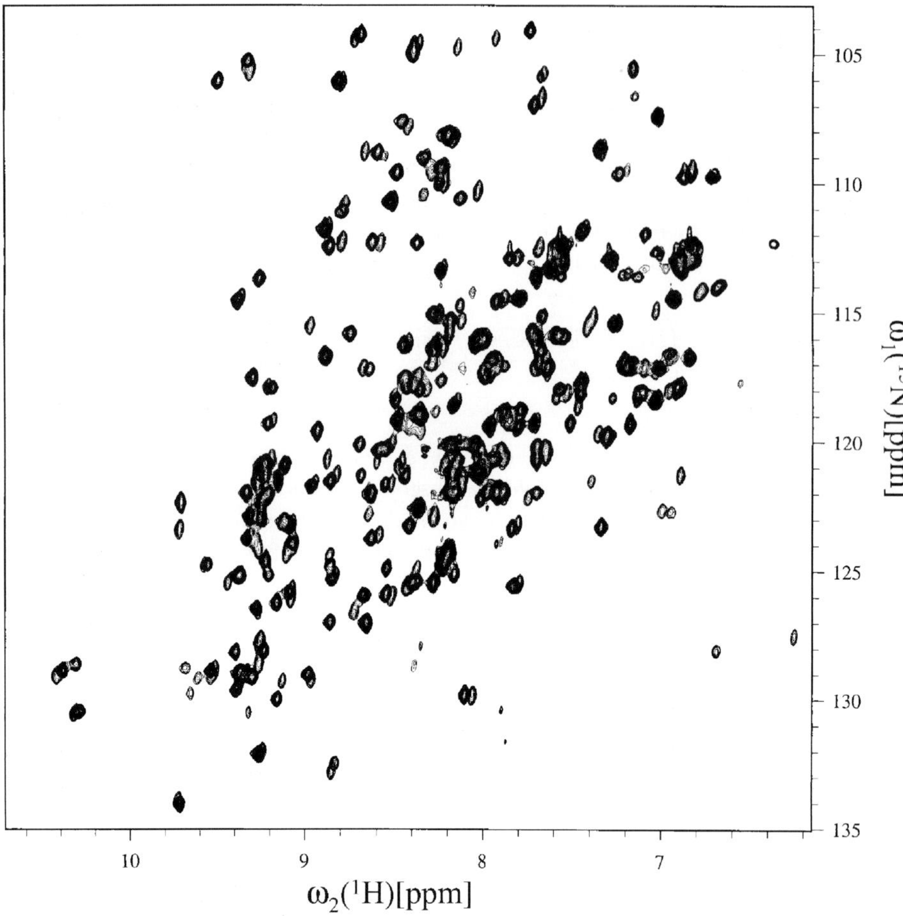

Fig. 14.3 Comparison of [$^{15}N/^{1}H$]-HSQC experiments before (dark) and after (light) addition of a ligand.

HSQC-based methods are presently limited to macromolecular complexes with a molecular weight below 30 or 100 kDa, depending on whether complete deuteration of the protein (in combination with TROSY) is used or not. Another limitation of the method is the need for large amounts of isotopically labeled protein, which can only be obtained if an effective expression system allowing growth in an isotope-labeled medium is available. Because of high costs, this has so far prevented large scale expression in nonprokaryotic systems. Ideally, all substances tested must be soluble in aqueous solution with conditions (pH, temperature etc.) optimized for protein stability.

The $^{15}N,^{1}H$ shift correlation maps are most conveniently recorded with a sensitivity-enhanced HSQC sequence with incorporated water flip-back pulses for reduced saturation transfer and pulsed-field gradients for coherence selection. The pulse sequence of the experiment is shown in Fig. 14.4A.

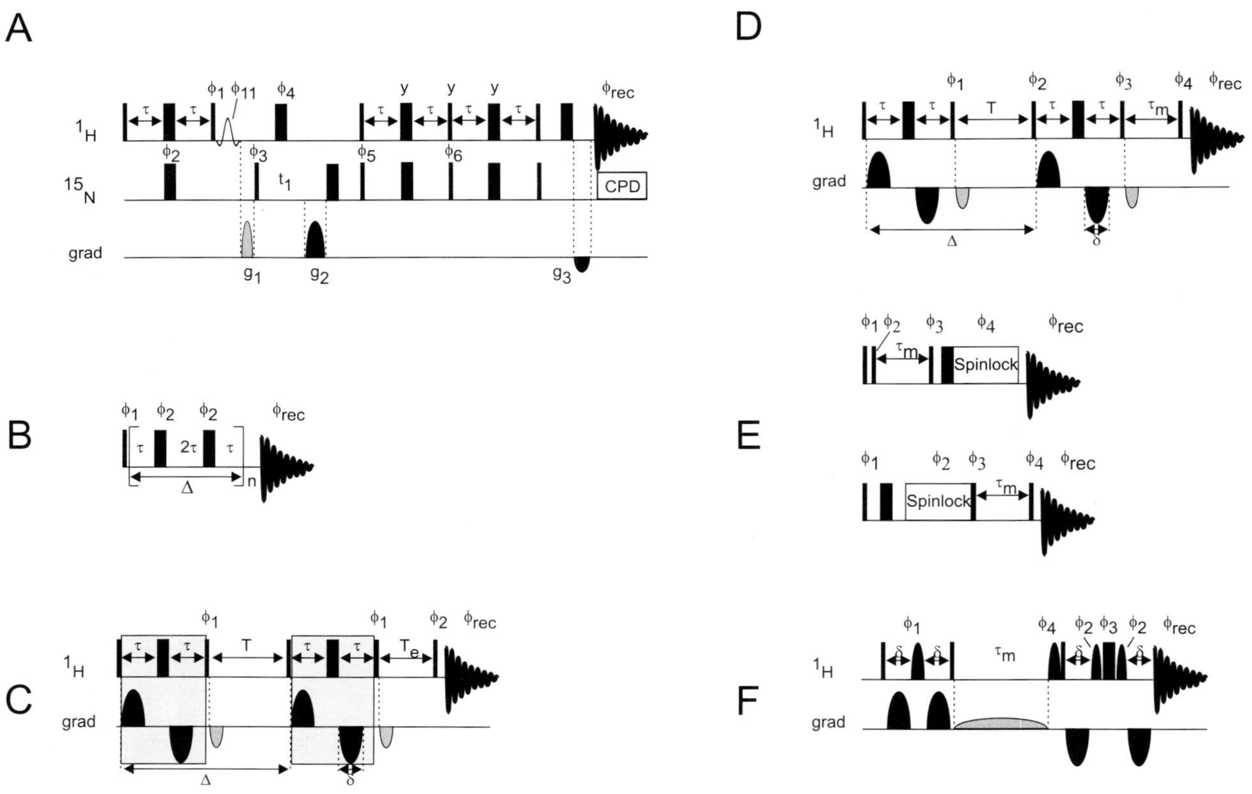

Fig. 14.4 Pulse sequences used for the experiments described in this chapter. **A** [^{15}N,^{1}H]-HSQC with water flip back and PFGs. The shaped pulse on the proton channel is a sine-shaped, 1.5 ms soft pulse; all other pulses are hard pulses. Gradients are applied as square or sine-shaped pulses. The sign of the last gradient is reversed for anti-echo selection together with the sign of phase 6. **B** CPMG sequence. **C** bpPFGLED sequence. The delay T denotes the diffusion delay. Typically, τ is set to 1 ms, T to 50–100 ms and T_e to 1.2 ms. **D** NOE-pumping experiment. **E** Reverse NOE pumping experiment (bottom) and reference experiment (top). **F** e-PHOGSY NOESY sequence. The water-selective 180° pulse sandwiched by the first two gradients has a gaussian shape and a duration of 40–50 ms . The mixing time is approx. 2 s. For further details, see Refs. [29, 30].

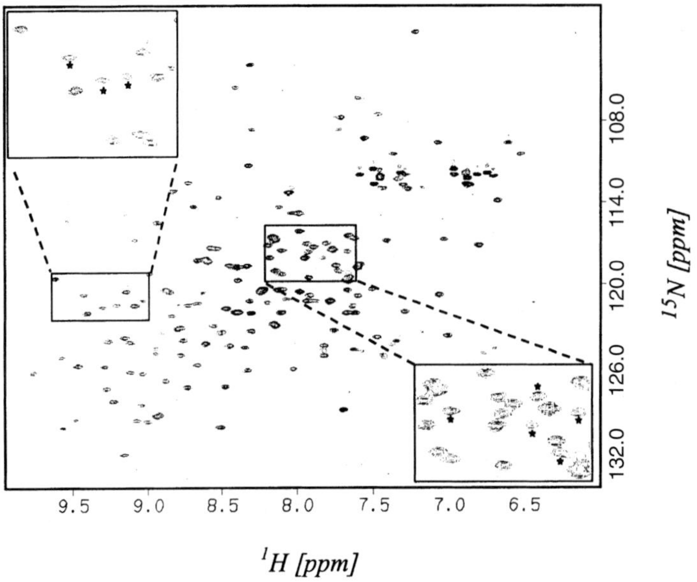

Fig. 14.5 Chemical shift changes due to a temperature change of 1° recorded on a sample of uniformly ^{15}N-labeled gyrase B from *E. coli.*

To exclude "false positive" hits, several test spectra for each biological target must be collected and analyzed to monitor all possible effects due to small changes in temperature, salt concentration and pH. Of prime importance is the perfect control of the experiment temperature, once the correct setting has been found in which the target is properly folded and biologically active. Differences in the NMR spectra, due to changes in resolution or amide exchange, need to be examined carefully in order to choose the best experimental set-up. Fig. 14.5 displays the remarkable effect that a slight difference in temperature may have on chemical shifts. Therein, peaks labeled with an asterisk experience frequency changes when the temperature changes by one degree.

In addition, since many potential drugs contain ionizable groups, changes in chemical shifts caused by shifts in pH will result in useful information for the design of the ligand, especially for deciding which functional groups should be attached. To avoid "false positive" hits, resulting for example from the addition of basic (or acidic) compounds that will lead to differences in the spectra even in the absence of interactions with a ligand, an NMR titration, monitoring changes in chemical shifts, will help to restrict responses to effects due to drug binding. In Fig. 14.6, two HSQC spectra on the same protein at two different values of the pH are displayed (at pH 6.5 (dark) and at pH 6.0 (light)).

14.2.2
Ligand-Observe Experiments

Another class of experiments is based on the analysis of changes in signal amplitudes as observed in ^{1}H spectra for ligand signals in the presence of the target. Provided that the

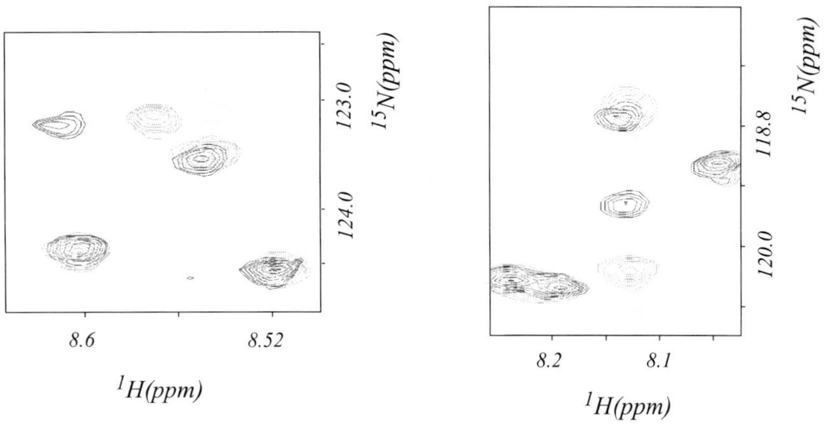

Fig. 14.6 Monitoring pH changes (6.0 vs 6.5) using the same sample as that in Fig. 14.5.

complex is in fast exchange on the NMR time scale, the observed parameters for the ligand are the population-weighted average of the free and bound form (see Eq. (1)). Non-interacting substances are not influenced. In principle, all experiments described in detail in the following part always give a response in the form of a reduction (disappearance) or increase (appearance) of signal amplitude. The intensity of the effect depends on the molecular weight of the target. For methods which rely on differences between a reference sample, recorded in the absence of the target, and the test sample, which contains the target molecule, the signal amplitude detected can be described by an exponential law

$$I(\tau) = I_0 \cdot e^{-k_{\mathrm{eff}} \cdot \tau} \tag{4}$$

Here k_{eff} describes the dynamic process (e.g. the transverse relaxation rate R_2, or the diffusion coefficient D) and τ describes a time constant typical for the experimental setup. By use of Eq.(1) the k_{eff} can be written as follows:

$$k_{\mathrm{eff}} = k_{\mathrm{f}} \cdot (p_{\mathrm{f}} + \frac{k_{\mathrm{b}}}{k_{\mathrm{f}}} \cdot p_{\mathrm{b}}) \tag{5}$$

Thus a binding event, increasing the population p_{b} of the bound state, will change the signal detected more drastically if the ratio of k_{b} to k_{f} is strongly dependent on the molecular weight of the target.

For experiments that involve intermolecular NOEs, the mathematical description is not straightforward. For the ranking presented at the end, we decided to show the steady-state NOE and the NOE build-up as a function of the MW of the target for a comparison. The different experiments in use are described in detail, together with examples, in the following sections. A ranking with respect to their practical applicability will be given at the end of this section.

14.2.3
Experiments Based upon Changes in Relaxation Properties of Ligands

Transverse and longitudinal relaxation of proton spins is mainly promoted by dipolar interactions between proximate spins. The relaxation efficiency depends on the distance separation (see Chapt. 11) of the interacting spins and on the reorientation time of the internuclear vector. The latter is determined by the motional properties of the molecule, which for rigid, isotropically tumbling molecules is described by an overall rotational correlation time. Binding of the ligand to the target leads to dramatic changes in overall tumbling for the bound fraction of the ligand, whereas the change for the target can be neglected. Usually, weak binding will lead to significant changes in linewidths of ligand signals. The overall correlation time can be estimated from MW_{target} using the Stokes-Einstein equation. Alternatively, it may be estimated from the following equation obtained by a fit of experimental data:

$$\tau_r[\mathrm{ns}] = \frac{9.18 \cdot 10^{-3}}{T[\mathrm{K}]} N^{0.93} \cdot \exp\left(\frac{2416}{T[\mathrm{K}]}\right) \quad (6)$$

in which N denotes the number of amino acids and T the temperature [14]. Additionally to the line broadening induced by changes in the motional properties, a further contribution R_2^{ex} may arise from the chemical exchange according to

$$R_2^{\mathrm{eff}} = \pi\Delta\nu_{1/2} = f_f R_{2,f} + f_b R_{2,b} + R_2^{\mathrm{ex}} \quad (7)$$

$$\text{with} \quad R_2^{\mathrm{ex}} = (\delta_b - \delta_f)^2 \frac{f_f^2 f_b}{k_{\mathrm{off}}}$$

Here $R_2^{eff} = (T_2^{eff})^{-1}$ is the observed relaxation rate at the given rates $R_{2,f}$, $R_{2,b}$ for the free and bound state respectively, and k_{off} is the off-rate of the complex. The result of this additional effect, which may dominate the linewidth, is displayed in Fig. 14.7 below.

As can be seen in Fig. 14.7, the effective relaxation rate is very sensitive to the chemical shift difference between the free and the bound state. Since protons usually have a different environment upon binding, the changes in relaxation rates are not uniform for all signals.

The effects due to enhanced transverse relaxation can be experimentally determined from a CPMG pulse experiment [15, 16], although the determination of $T_{1\rho}$ (by ROESY-type experiments) is more attractive from a practical point of view. Usually it suffices to record relaxation-weighted experiments using the methods described above to reveal binding.

14.2.4
Diffusion-Editing Experiments

Affinity NMR utilizes the changes in translational diffusion to probe an interaction of a ligand with a target [17–20]. The diffusion-editing experiments are based on the fact that upon binding of the ligand to the target the hydrodynamic radius r_H of the ligand increases

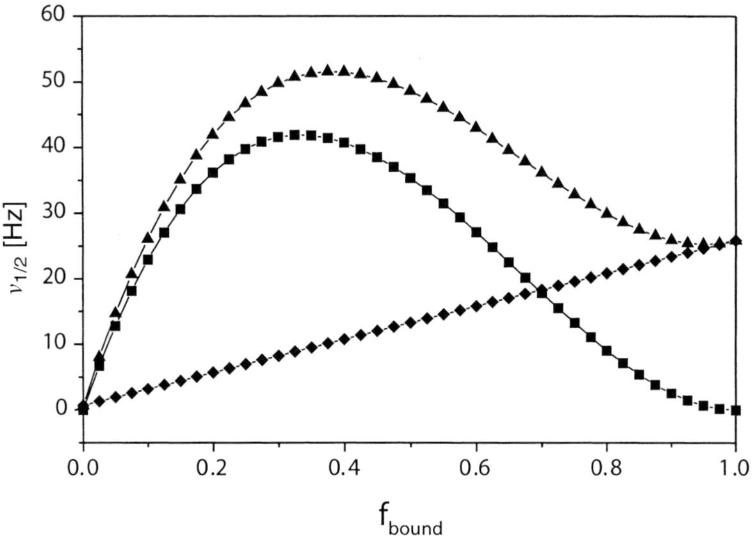

Fig. 14.7 Additional contributions to the line widths of signals in the presence (triangles) or absence (diamonds) of exchange. The middle curve (squares) corresponds to the exchange term in Eq. (7) (divided by π). For the calculations $k_{off}=1000\ s^{-1}$, $\Delta\nu=150$ Hz were used.

dramatically, and hence the diffusion rate decreases. This is a principle property of the molecule, and hence all signals from the ligand will be influenced in the same way.

In principle, the hydrodynamic radius may be calculated when the structure is known. However, such calculations contain large errors due to uncertainties in the extent of the hydration shell. One can use the following empirical formula to estimate hydration radii compatible with NMR results

$$r_H[\text{nm}] = \frac{\text{MW}[\text{kDa}]^{0.41}}{1.55} \tag{8}$$

which can be found under www.protein-solutions.com/calc.htm.

With the help of the Stokes-Einstein relation, the translational diffusion coefficient may be calculated according to

$$D = \frac{k_B T}{6\pi\eta_w r_h} \tag{9}$$

Fig. 14.8 illustrates the relative decrease of the diffusion coefficient upon binding to the target. It reveals that for large differences in mass the change in the diffusion coefficient is large.

Diffusion constants are most conveniently measured with the bpPFGLED sequence depicted in Fig. 14.4C [21]. Therein, magnetization is stored as polarization along the z-axis during the diffusion delay in order to reduce losses due to T_2 decay. Furthermore, an ad-

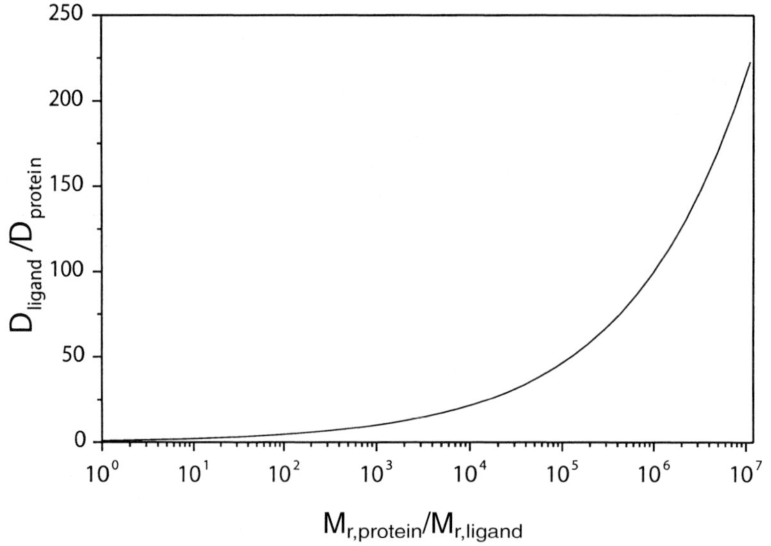

Fig. 14.8 D_{lig}/D_{prot} plotted against MW_{prot}/MW_{lig}. With increasing MW_{prot}/MW_{lig}, D_{lig}/D_{prot} increases as well, i.e. the effective diffusion coefficient of the ligand decreases dramatically.

ditional *z*-filter at the end of the sequence allows for the decay of eddy currents caused by the pulsed-field gradients. In order to yield optimal results, the lock circuit should be highly damped during the measurements.

The signal decay due to diffusion in a series of 1D spectra is given, in line with Eq. (4), by

$$S(2\tau_e + T + T_e) = \frac{1}{2} S(0) \exp\left[-\frac{2\tau}{T_2} - \frac{T + T_e}{T_1} - (g\delta\gamma)^2 D\left(\Delta - \frac{\delta}{3}\right)\right] \tag{10}$$

The experimental parameters (gradient strength (g), gradient length (δ) and diffusion delay (T)) will be selected to optimally suppress the signals of nonbinding compounds and to allow for the detection of resonances of the biological target and bound ligands. In the following, the use of diffusion editing is demonstrated for GV143253A (a GlaxoSmithKline compound) binding to human serum albumin (HSA). In Fig. 14.9, the 1D NMR spectrum of GV143253A is shown (lower left) together with the spectrum from the diffusion-weighted experiment (upper left). The right hand side displays spectra resulting from addition of the compound to the biological target. All experiments were recorded using the BPPSTE sequence (Bipolar Pulse Pair STimulated Echo; see also Ref. [17] for a general review) (Fig. 14.4 C). Whereas all ligand resonances disappear in the gradient-edited spectrum recorded on the ligand in the absence of the target, the ligand signals become visible in the presence of HSA. This clearly proves that GV143253A binds to HSA.

Fig. 14.10 displays spectra from an experiment recorded on the mixture of two compounds, GV143253A and carnitine, in the presence of HSA to show the ability of the

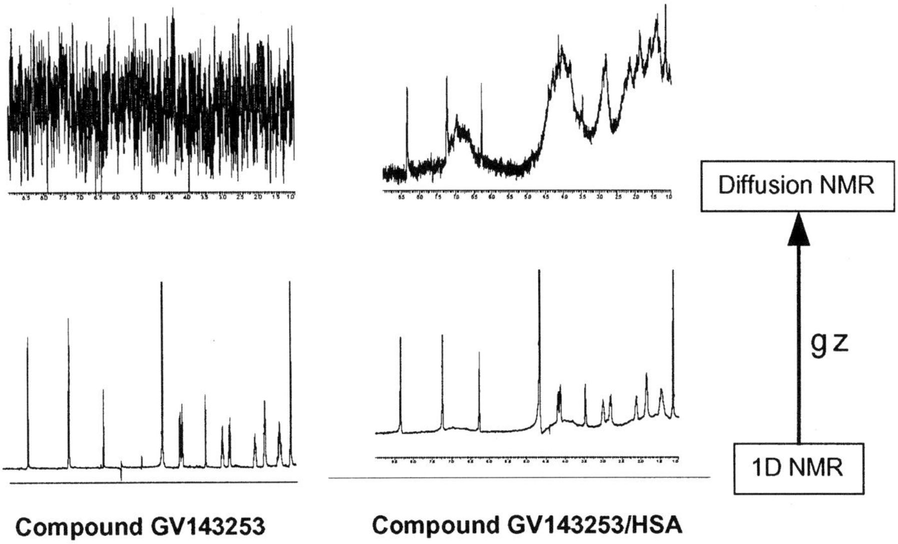

Fig. 14.9 Affinity NMR experiment on GV143253A/HAS (see text).

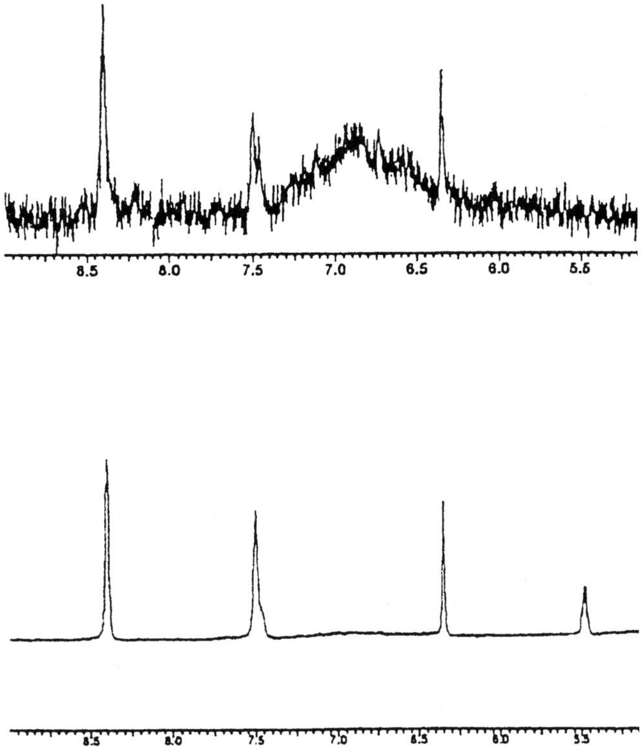

Fig. 14.10 Affinity NMR experiment on Carnitine+ GV143253A/HSA.

technique to discriminate between "binders" and "nonbinders": the peaks from carnitine disappear in the diffusion-weighted spectra while signals from GV143253A are retained. This observation suggests application of the method to mixtures of compounds in order to detect "hits" that interact with the biological target without need for deconvolution into binary mixtures [18, 20, 22]. In order to unambiguously define interacting ligands in the mixture, diffusion-encoded spectroscopy (DECODES) can be applied [20, 23].

Fig. 14.11 shows results from an experiment performed on a mixture of three compounds with the corresponding peaks a, b and c in the presence of HSA. By varying the ratio between compounds and the protein it is possible to rank binding ligands according to their relative binding affinities. In the presented case, at the chosen three different ratios of ligands to protein, the three compounds display similar signal intensities, indicating that their binding constants are similar. In contrast, should one compound bind much more tightly than the others, the signal decrease upon lowering the ligand-to-protein ratio would be more pronounced than for the more weakly binding ligands.

Moreover, by adding a reference compound (with known binding properties), a hit with a binding constant in the desired range of the reference compound can be found. An interesting consequence is the ability of the technique to detect competition between compounds. Since the diffusion method detects only signals from the free ligand, there is no upper limit to the size of the biological target and no requirement for isotope labeling.

The signal attenuation in the diffusion-weighted spectra does *not* depend on the binding mode and position but *only* on the effective diffusion coefficient and on the experimental parameters mentioned above.

One limitation of this method is the need to resolve at least one NMR signal per ligand from those of the biological target. If this is not possible, spectra subtraction using spectral differencing methods (see also Ref. [11]) or the use of isotope labeling may help. In the spectral differencing method, a diffusion-weighted spectrum recorded on the target alone is subtracted from data recorded on the mixture. However, the method is prone

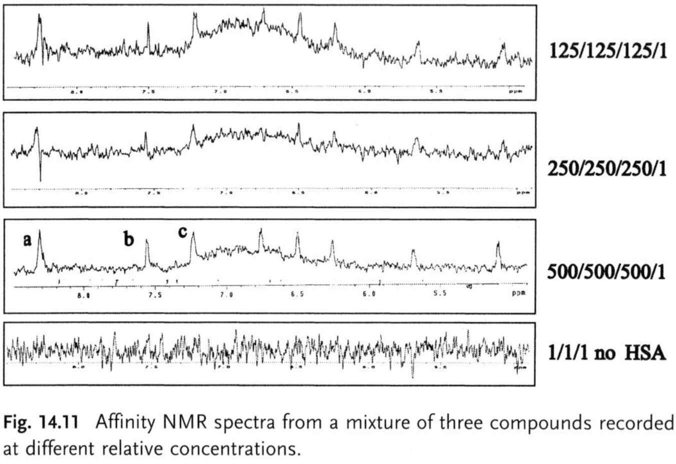

Fig. 14.11 Affinity NMR spectra from a mixture of three compounds recorded at different relative concentrations.

to subtraction artifacts. If the biological target is ^{13}C-labeled, its signals may be removed by isotope-filtering NMR techniques (see Chapt. 16).

In summary, the diffusion-based NMR screening strategy results in several advantages if compared to techniques based on differences in chemical shifts. It increases the throughput of the NMR-based screen and abolishes the need for assignment of the protein, similarly to other ligand-observe methods. Again, one limit might be the solubility in aqueous medium and problems arising from aggregation of ligands in solution.

14.2.5 NOE-Based Techniques

Recently, a number of NMR screening techniques based on the nuclear Overhauser effect (NOE) have been developed.

The saturation transfer difference (STD) technique [24–26] uses the difference of two spectra. The STD spectrum itself is recorded with saturation of target resonances (mostly the "methyl hump" of the protein is saturated). A second experiment, in which irradiation is performed far-off-resonance, is acquired as a reference. The "STD module" contains the following elements: *irradiation – excitation – relaxation filter* (optional). The relaxation filter can be used to efficiently suppress residual target magnetization prior to acquisition. It is well known that the steady-state NOE leads to highly efficient spin diffusion for large molecules or aggregates [24], thereby usually precluding its use for structure determination of high-molecular-weight molecules. Hence, upon saturation of target signals, efficient spin diffusion leads to almost complete saturation of all target resonances via *intramolecular* spin diffusion. In the case of binding, *intermolecular* spin diffusion leads to saturation of ligand resonances as well. In the reference experiment, all signal intensities remain unchanged, since no target signal is irradiated (off-resonance irradiation). Subtraction of the STD from the reference experiment yields only signals of binding compounds, because their signal intensities have changed. The subtraction can be done in several ways. The simplest approach is by subtraction after acquisition of both spectra, i.e. both data sets are stored separately and subtraction is done manually. In this approach, slight variations in sample temperature or magnetic field homogeneity may lead to serious subtraction artifacts. Preferably, subtraction is done in an interleaved manner by phase cycling the receiver phase concomitantly with switching between on- and off-resonance irradiation such that only the difference spectrum is recorded. Although this approach is very insensitive to slight variations of the temperature or the shim due to the interleaved sampling scheme, it may sometimes be preferable to acquire both spectra in an interleaved manner, but store the datasets in separate memory allocations for subsequent manual manipulations. STD can be used to screen single ligands for binding, but it may also be applied for the screening of compound mixtures. In this case, the "STD module" can be combined with other 2D experiments (e.g. STD-TOCSY) in order to unambiguously identify binding compounds without the need for further deconvolution of the mixture.

STD experiments allow for a mapping of the binding epitope (group epitope mapping, GEM) [25]. The degree of saturation of ligand resonances depends on the distance of the protons involved. Protons in close proximity to the target molecule are saturated to a higher degree, thus leading to a stronger STD signal. This information can be used to identify moieties that are important for binding. Recently, STD NMR has been used to determine the binding specificity of peptidic ligands to an integrin [26]. Most of the applications of this method found in the literature are used for the characterization of the binding of saccharide ligands [25, 27, 28]. Although STD is most commonly applied in solution state NMR, it can be used advantageously in HR-MAS NMR [29]. Here, the target is immobilized on controlled pore glass (CPG). Spin diffusion is much faster for immobilized targets, and therefore the STD effect is much more pronounced when compared to the solution state. The use of HR-MAS ensures enough resolution to identify binding compounds via their ^{1}H NMR spectra. Since the free ligand is observed, STD works best in the fast exchange regime, i.e. with K_d values in the millimolar or upper micromolar range. Although claims have been made [30] that the method works even for complexes displaying dissociation constants as low as 8×10^{-9} the authors believe that for these tight binders too little free ligand is available to enable successful observation of the free ligand with reasonable sensitivity. One drawback of the STD method is the need for a target resonance isolated from all irradiated signals. Thus, in principle, all ligand spectra have to be checked for signals residing in the irradiated spectral region.

In the transferred NOE (trNOE) experiment, a 1D or 2D NOESY experiment is recorded [31]. The intermolecular NOE build-up to be used as a parameter in Eq. (5) arises from the bound state but is observed via the free ligand, requiring rapid exchange between bound and free states. The technique is described in more detail in Chapt. 16.

The NOE pumping technique uses *intermolecular* cross relaxation [32]. The pulse sequence is depicted in Fig. 14.4 D and consists of a diffusion filter element for nonlabeled samples with a subsequent NOE mixing period. In the diffusion filter, ligand magnetization is effectively dephased while protein magnetization is largely preserved. During the NOE mixing time, both intra- and intermolecular cross relaxation occur. Ligand signals are easily distinguished from protein resonances because of their much narrower line widths and their different NOE build-up curves.

In the reverse NOE pumping (RNP) experiment [33], a relaxation ($T_{1\rho}/T_2$) filter is followed by the NOE mixing element (see Fig. 14.4 E). During the former, protein magnetization effectively decays because of rapid transverse relaxation. During the mixing time, polarization is transferred from a binding ligand to the protein. The signals of binding ligands are modulated by longitudinal relaxation as well as by intermolecular cross relaxation. A second (reference) experiment is performed using the second sequence in Fig. 14.4 E. Therein, the order of the two elements is interchanged and the resulting ligand signals are mainly modulated by T_1 relaxation during the mixing period. Subtracting the FIDs from the two experiments (the RNP and the reference spectrum) results in a spectrum containing only signals of binding ligands. Furthermore, the RNP should allow for group epitope mapping as described above for the STD method, although no group epitope mapping based on RNP data has been reported in the literature to date.

The waterLOGSY method [34, 35] uses the fact that for low-molecular-weight ligands the intermolecular water-ligand NOE will be negative only when the ligand is bound to the protein. This is because of the dramatic change in the correlation time of the ternary water-ligand-protein complex compared to that of the water-ligand complex. Bound water is quite often found at the interface between the protein and the ligand. A rough estimation of the

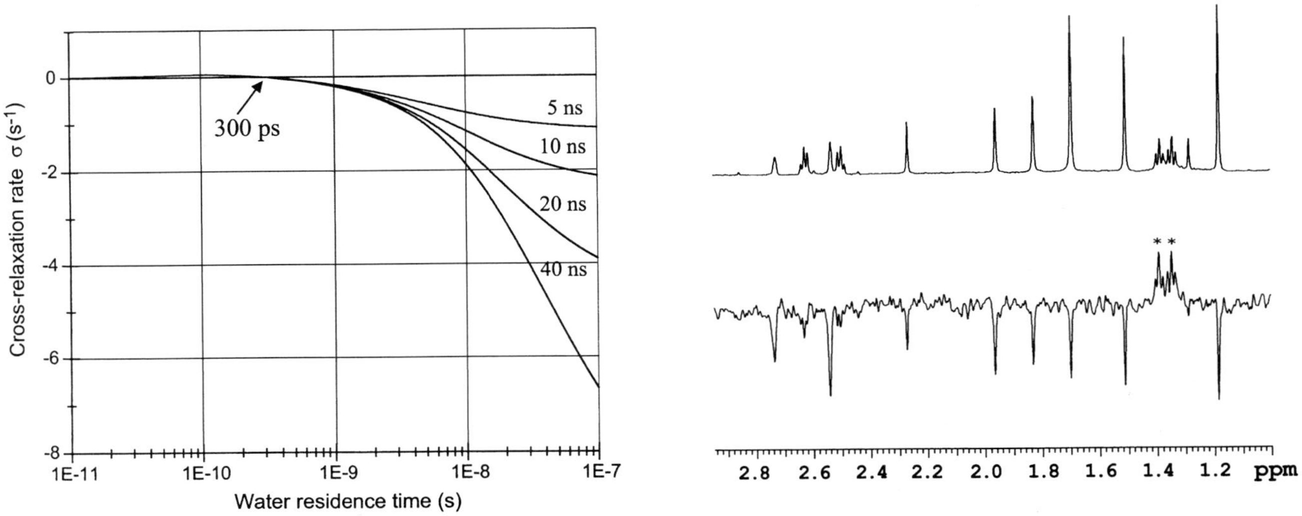

Fig. 14.12 Left: Intermolecular cross-relaxation rate at 600 MHz as a function of the water residence time (with permission from Ref. [39]) Right: waterLOGSY experiment recorded on a ten-compound mixture with CDK2 (bottom) and reference 1D spectrum (top). Spectra were collected at 19 °C and 600 MHz; the concentration of CDK2 was approx. 10 μM. The asterisks indicate signals due to a ligand binding to CDK2 (with permission from Ref. [35]).

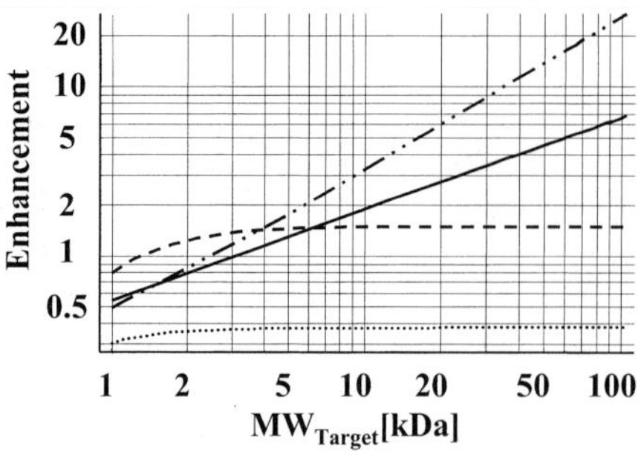

Fig. 14.13 Graphical representation of the effect of MW on T_2 (dashed-dotted), on the translational diffusion rate D (solid), on the steady state NOE (dashed) and on the build-up of the NOE (dotted). All values are normalized to a 300 Da molecular weight molecule. For the calculation of the parameters involving dipolar relaxation we used a formula that can be found in the literature [40]. For the dependence of the translational diffusion parameter we assumed a model of an unfolded polymer in a good solvent (upper limit) where $R_g \approx MW^{3/5}$. It should be noted that the figure should only be read qualitatively, as the results for the NOE-based parameters will be influenced to a large degree by spin diffusion.

entropic loss upon binding of water yields –9 kJ mol^{-1}, whereas the enthalpic gain can be as high as –16 kJ mol^{-1}, resulting, in the most favorable cases, in a gain of –7 kJ mol^{-1} for a singly bound water [36], corresponding to a tenfold improvement in the binding constant of the ligand at 298 K. The residence times of water in protein cavities lie in the range between a few ns to a few hundred μs [37], long compared to the effective correlation time of 0.3 ns, at which the intermolecular NOE changes its sign. The dependence of the intermolecular cross-relaxation rate on the water residence time is depicted in Fig. 14.12 (left). Since bound water exchanges with bulk water, the large quantity of bulk water magnetization increases the sensitivity of the experiment such that low protein concentrations may be used. The waterLOGSY technique is thus similar to the transfer NOE technique, but with the difference that the water-ligand NOE is not directly detected. Rather, the change in the sign of the water-ligand NOE indicates binding. To identify a "target-amplified" ligand-water NOE, the e-PHOGSY NOE experiment depicted in Fig. 14.4 F is performed. Therein, water magnetization is selectively restored at $-z$ and transferred onto the ligand via intermolecular cross relaxation. After the mixing period, residual water magnetization is destroyed by a watergate element [38].

Fig. 14.12 b displays results from a waterLOGSY experiment recorded on a ten-compound mixture in the presence of the cyclin-dependent kinase 2 (CDK2) (bottom) and reference 1D spectrum (top). Positive and negative signals in the lower spectrum identify CDK2 interacting and noninteracting molecules, respectively. The asterisk indicates the methyl group of the CDK2 ligand ethyl α-(ethoxycarbonyl)-3-indoleacrylate.

14.2.6 Comparison of Methods

One set of methods described above is based on a comparison of a reference spectrum (no target added) with a test spectrum. These methods have the clear disadvantage that two samples have to be used. Therefore, slight variations of sample (pH, salt concentration etc.) and spectrometer (shim) conditions will complicate the data analysis. Relaxation-based methods are more sensitive than diffusion-based methods due to an increase in the apparent molecular weight. Hence, these experiments generally yield more clearly visible effects than experiments exploiting differences in the diffusion rates. However, a clear advantage of the latter is that all signals of a substance experience the same change in signal intensity upon binding, thereby facilitating data evaluation.

The second set of experiments, where only one sample is required, always involves an intermolecular NOE transfer. The STD experiment exploits the spin-diffusion mediated steady-state NOE. In our experience, this method yields reliable results and can be easily implemented. Ligand signals are easily distinguished from protein signals by inserting a relaxation filter prior to acquisition. Care has to be taken to avoid direct saturation of ligand resonances.

Reverse NOE pumping works robustly and reliably. Here, the huge difference of $T_{1\rho}/T_2$ for test substance and target ensures a high difference in the initial z-magnetization at the beginning of the NOE mixing, thus resulting in a high sensitivity. Such a huge difference is also realized in the waterLOGSY method by frequency selective pulses.

A little different is the situation when looking at the NOE pumping experiment. Here, the signals of the test substance are suppressed more than those of the target. The less pronounced MW dependence (compared to $T_{1\rho}/T_2$) of the translational diffusion constant reduces the difference in z-magnetization and hence the size of the NOE difference. Here, a reference experiment will ideally only contain amplitude-reduced signals from the target. If a ligand binds, NMR signals of this substance will appear in the spectrum. In our hands it turned out that upcoming signals remained close to the noise threshold of the data, thus complicating data analysis.

To summarize, for methods using two samples, relaxation-based experiments are more favorable because of the strong dependence of the relaxation rates on the apparent molecular weight. For single-sample techniques, both the STD and the RNP represent suitable methods. The NOE pumping experiment suffers from the weaker dependence of the apparent diffusion coefficient on the molecular weight. One should keep in mind that the much higher sensitivity of chemical shift changes induced on the protein if tenfold excess ligand concentration is employed makes this type of experiment the best choice if a large quantity of isotopically labeled target can be obtained. In addition, the analysis of 2D spectra obtained can be automated more easily (see Chapt. 19).

Further methods utilizing spin-labels are introduced in Chapt. 15.

14.3 References

1 Ross, A.; Schlotterbeck, G.; Klaus, W.; Senn, H. *J. Biomol. NMR* **2000**, *16*, 139–146.
2 Keifer, P.A. *Curr. Opin. Biotechnol.* **1999**, *10*, 34–41.
3 Moore, J.M. *Curr. Opin. Biotechnol.* **1999**, *10*, 54–58.
4 Moore, J.M. *Biopolymers* **1999**, *51*, 221–243.
5 Roberts, G.C.K. *Curr. Opin. Biotechnol.* **1999**, *10*, 42–47.
6 Roberts, G.C.K. *Drug Discov. Today* **2000**, *5*, 230–240.
7 Stockman, B.J. *Prog. NMR Spectrosc.* **1998**, *33*, 109–151.
8 Diercks, T.; Coles, M.; Kessler, H. *Curr. Opin. Chem. Biol.* **2001**, *5*, 285–291.
9 Hajduk, P.J.; Augeri, D.J.; Mack, J.; Mendoza, R.; Yang, J.; Betz, S.F.; Fesik, S.W. *J. Am. Chem. Soc.* **2000**, *122*, 7898–7904.
10 Apfel, C., Hoffman LaRoche, personal communication
11 Hajduk, P.J.; Olejniczak, E.T.; Fesik, S.W. *J. Am. Chem. Soc.* **1997**, *119*, 12257–12261.
12 Harris, R. K. *Nuclear Magnetic Resonance Spectroscopy, A Physiochemical View*; Longman Scientific & Technical: New York, 1983.
13 Shuker, S.B.; Hadjuk, P. J.; Meadows, R.P.; Fesik, S. W. *Science* **1996**, *274*, 1531–1534.
14 Daragan, V.A.; Mayo, K.H. *Prog. NMR Spectrosc.* **1997**, *31*, 63–105.
15 Meiboom, S.; Gill, D. *Rev. Sci. Instrum.* **1958**, *29*, 688–691.
16 Carr, H.Y.; Purcell, E.M. *Phys. Rev.* **1954**, *94*, 630–638.
17 Johnson, C.S. *Prog. NMR Spectrosc.* **1999**, *34*, 203–256.
18 Lin, M.; Shapiro, M.J.; Wareing, J.R. *J. Am. Chem. Soc.* **1997**, *119*, 5249–5250.
19 Ma, Y.; Liu, M.; Mao, X.; Nicholson, J. K.; Lindon, J.C. *Magn. Reson. Chem.* **1999**, *37*, 269–273.
20 Lin, M.; Shapiro, M.J.; Wareing, J.R. *J. Org. Chem.* **1997**, *62*, 8930–8931.
21 Wu, D.; Chen, A.; Jr., C.S.J. *J. Magn. Reson. A* **1995**, *115*, 260–264.
22 Bleicher, K.; Lin, M.; Shapiro, M. J.; Wareing, J.R. *J. Org. Chem.* **1998**, *63*, 8486–8490.
23 Lin, M.; Shapiro, M.J. *J. Org. Chem.* **1996**, *61*, 7617–7619.
24 Neuhaus, D.; Williamson, M. *The Nuclear Overhauser Effect in Structural and Conformational Analysis*; VCH: Weinheim, 1989.
25 Mayer, M.; Meyer, B. *J. Am. Chem. Soc.* **2001**, *123*, 6108–6617.
26 Meinecke, R.; Meyer, B. *J. Med. Chem.* **2001**, *44*, 3059–3065.
27 Haselhorst, T.; Weimar, T.; Peters, T. *J. Am. Chem. Soc.* **2001**, *123*, 10705–10714.
28 Kooistra, O.; Herfurth, L.; Lüneberg, E.; Frosch, M.; Peters, T.; Zähringer, U. *Eur. J. Biochem.* **2002**, *269*, 573–582.
29 Klein, J.; Meinecke, R.; Mayer, M.; Meyer, B. *J. Am. Chem. Soc.* **1999**, *121*, 5336–5337.
30 Mayer, M.; Meyer, B. *Angew. Chem. Int. Ed.* **1999**, *38*, 1784–1788.
31 Ni, F. *Prog. NMR Spectrosc.* **1994**, *26*, 517–606.
32 Chen, A.; Shapiro, M.J. *J. Am . Chem. Soc.* **1998**, *120*, 10258–10259.
33 Chen, A.; Shapiro, M.J. *J. Am. Chem. Soc.* **2000**, *122*, 414–415.
34 Dalvit, C.; Floersheim, P.; Zurini, M.; Widmer, A. *J. Biomol. NMR* **1999**, *14*, 23–32.
35 Dalvit, C.; Pevarello, P.; Tatò, M.; Veronesi, M.; Vulpetti, A.; Sundström, M. *J. Biomol. NMR* **2000**, *18*, 65–68.
36 Connely, P.R. In *Structure-Based Drug Design Thermodynamics, Modeling and Strategy*, Ladbury, J. E., Connely, P.R., Eds.; Springer Verlag: Berlin, 1997; pp 143–159.
37 Otting, G. *Prog. NMR Spectrosc.* **1997**, *31*, 259–285.
38 Piotto, M.; Saudek, V.; Sklenar, V. *J. Biomol. NMR* **1992**, *2*, 661–665.
39 Dalvit, C.; Fogliatto, G.; Stewart, A.; Veronesi, M.; Stockman, B. *J. Biomol. NMR* **2001**, *21*, 349–359.
40 Sudmeier, J.L.; Anderson, S.E.; Frye, J.S. *Concept. Magn. Reson.* **1990**, *2*, 197–212.

15
The Use of Spin Labels in NMR-Supported Lead Finding and Optimization

Wolfgang Jahnke

15.1
Introduction

Pharmaceutical research strives to discover compounds that cure diseases. Although biopharmaceuticals (such as antibodies) have been applied in the treatment of illnesses, small organic molecules are still by far the most important class of pharmaceuticals today. Almost all drug discovery programs are target-directed, in that they have a defined target molecule – often a protein, but sometimes a nucleic acid – which is thought to cause or convey harmful effects. Modulation of the target molecule's function, e.g. inhibition of an enzyme, can therefore cure or alleviate diseases or pathological situations.

Inhibition of a target protein by a small organic molecule requires identification of a class of compounds with some of the desired activity ("lead finding") and subsequent optimization ("lead optimization") of this compound class to yield a drug candidate with suitable potency, toxicity, and pharmacokinetic properties. Since the target is generally an identified and isolated protein, the identification of a potential new medicine boils down to the identification of ligands that bind and inhibit the target protein.

Among several other techniques, NMR spectroscopy has high potential to support the drug discovery process. NMR can be used both in lead finding and lead optimization, by addressing the following questions:

- Does a particular compound bind to the target protein? Where and how tightly does it bind?
- How can the compound be optimized?

The first question is typically asked in lead finding, where NMR is used either to screen compounds for binding to the target molecule or to validate hits identified in another screen. A variety of NMR techniques are available for this purpose, and most of them are outlined in this book (see Chapt. 14). Binding can for example be detected by chemical shift perturbations, changes in relaxation or diffusion properties, protein-ligand NOEs, or water-ligand NOEs. The principles, advantages and disadvantages of NMR screening will not be discussed here in detail, since they are extensively described in this book and in several reviews [1–3].

The second question is of prime importance in lead optimization, and structural information can greatly assist this process. X-ray crystallography and NMR can equally and in parallel contribute structural information, and some of the techniques to obtain structural information, such as isotope editing (see Chapt. 17) or transfer NOE (see Chapt. 16), are presented in this book.

Recently, we have demonstrated that organic nitroxide radicals, so-called spin labels, can be efficiently used by NMR spectrocopy in drug discovery research [4, 5]. Spin labels have dramatic effects on the relaxation properties of neighboring protons, and can be employed both for lead finding and lead optimization. In this chapter, the theory of action of spin labels is outlined, together with practical aspects of working with spin labels. Then examples are given for spin labels applied in first-site and second-site NMR screening. The chapter ends with a section on linker design, and an outlook.

15.2
Basic Theory of Spin Labels

Spin labels have a long history in magnetic resonance spectroscopy [6, 7]. Discovered and synthesized in the early 1960s, they were first applied to biological systems in the late 1960s and 1970s. Most of the applications involved electron spin resonance (ESR), which can give information on dynamic processes in the millisecond to nanosecond range [8]. Systems under investigation were often enzymes, but also nucleic acids. The application of spin labels to NMR was based on the increase of relaxation rates of neighboring protons caused by the paramagnetic center, so-called paramagnetic relaxation enhancement (PRE). It can be used to measure long distances up to 20 Å, which are too long to be measured by NOE, or to study the dynamics of peptides or proteins by sampling all conformations that lead to short proton-radical distances.

The magnitude of the paramagnetic relaxation enhancement (PRE) caused by dipolar interactions depends on the square of the gyromagnetic ratios of both involved spins, the inverse sixth power of the inter-spin distance, and the correlation time τ_c of the vector connecting the two spins. For the transverse relaxation rate enhancement, $R2_{para,}$ of a spin I,

$$R2_{para} = \frac{1}{15} S(S+1) \frac{\gamma_I^2 g^2 \beta^2}{r^6} \left(4\tau_c + \frac{3\tau_c}{1+\omega_i^2 \tau_c^2} \right) \qquad (1)$$

where S is the electron spin, γ_I is the gyromagnetic ratio of the proton, g the electronic g factor, β the Bohr magneton, and r the distance between the electron spin and the nuclear spin [6, 9]. A similar equation holds for the longitudinal relaxation rate enhancement, $R1_{para}$. Since the electron gyromagnetic ratio is 658 times that of the proton, proton relaxation rates in the vicinity of a paramagnetic center are drastically larger than the corresponding diamagnetic (proton-proton) relaxation rates. It is essentially this factor of $(658)^2$ which we aim to exploit to enhance the detection of ligand binding.

In a protein-ligand complex the correlation time τ_c of the vector connecting the electron and nuclear spins, depends on the rotational correlation time of the protein-ligand complex, τ_r, on the electron relaxation time, τ_s, and on the lifetime of the complex, τ_m [6, 9]:

$$\frac{1}{\tau_c} = \frac{1}{\tau_r} + \frac{1}{\tau_s} + \frac{1}{\tau_m} \qquad (2)$$

The electronic relaxation time, τ_s, plays a key role in paramagnetic systems. For organic nitroxide radicals, τ_s is typically in the order of 100 ns, and therefore much longer than the rotational correlation time, τ_r, which is typically in the order of a few nanoseconds. Since τ_m is usually even longer than τ_s, τ_c is dominated by τ_r, the same correlation time that governs relaxation in diamagnetic systems. Given the high gyromagnetic ratio and the long effective correlation time, organic nitroxide radicals drastically increase relaxation rates of neighboring protons, while having little effect on chemical shifts. Organic nitroxide radicals or spin labels are therefore potent relaxation reagents.

Some paramagnetic transition metals, for example Co(II), Ni(II), low-spin Fe(III), or most lanthanides [with the exception of Gd(III)], on the other hand, have very short electron relaxation times τ_s in the order of 10^{-13}–10^{-12} s. The electron relaxation times therefore dominate the effective correlation time of the electron-proton vector. Since the relaxation enhancements depend on τ_s, relaxation rates for nuclei with very short (small) τ_s are only slightly affected. Since the magnetic field produced by the paramagnetic transition metals is anisotropic, it does not average out and leads to significant chemical shift changes of neighboring protons. These paramagnetic transition metals are therefore shift reagents. The effect on chemical shifts depends on the third power of the distance between electron and proton, rather than on the sixth power like the relaxation enhancement due to the relaxation reagents [10].

An additional consequence of paramagnetism is partial alignment of the paramagnetic molecule in the magnetic field, because of its anisotropic magnetic susceptibility [11]. As a consequence, dipolar couplings do not average out completely, and the residual dipolar couplings can be measured using appropriate NMR experiments. These dipolar couplings have become powerful long-range constraints for NMR structure determination [12, 13].

Ligands displaying low-affinity binding exchange rapidly between the bound and unbound states and the relaxation properties are time and population averaged. In principle, paramagnetic relaxation rates can be as much as $658^2 = 430{,}000$ times larger than diamagnetic relaxation rates, and this is the theoretical factor of possible reduction in protein concentration for such weakly binding ligands that still allows the detection of protein-ligand interactions. In practice, however, the distance between spin label and proton is considerably larger than that between two protons. Since paramagnetic relaxation enhancement effects decay with the inverse sixth power of the distance, a large part of this theoretical enhancement factor vanishes. In the example given below, the detection of ligand binding to FKBP, the closest spin label is at a distance of approximately 12 Å from the ligand, about 4.5 times the distance from neighboring protons. Considering the inverse sixth power distance dependence and the square dependence on gyromagnetic ratios, the effective relaxation enhancement is then $658^2/4.5^6$, about 50. This value corresponds in fact qualitatively to the experimentally observed relaxation enhancement, and is equivalent to the possible reduction in protein concentration.

Clearly, the potential to use spin labels as a means to reduce protein concentration for detection of protein-ligand interactions, given by the factor of 658^2, is tremendous. The sixth-power dependence on electron-proton distance underlines the need to carefully design the residue type which is to be spin labeled. Residues that can be spin labeled include lysine, tyrosine, cysteine, histidine, and methionine [7, 14]. At least one residue of this type should be as close as possible to the binding site, but must not interfere with ligand binding. Availability of the three-dimensional structure or a reliable homology model is obviously highly advantageous for the design of a spin label experiment. In our experience, it is almost always possible to identify residues within 10–12 Å from the ligand binding site that can be spin labeled and which are not expected to interfere with ligand binding.

15.2.1
Some Practical Aspects of Work with Spin Labels

Organic nitroxide spin labels are commercially available with a variety of functional groups, and the synthesis of an even larger number of spin labels is described in the literature [15]. Residues to be spin labeled include lysine, tyrosine, cysteine, histidine, and methionine [7, 14]. Spin labeling of protein side chains is a chemical and post-translational modification of proteins, which is achieved after the target protein has been expressed in a functional and well-folded form. The labeling is in most cases an easy undertaking which requires only half a working day. For the case of lysine modification using 1-oxyl-2,2,5,5-pyrroline-3-carboxylate-*N*-hydroxy-succinimide ester as the spin label reagent, protein and spin label reagent are mixed for 2 h at room temperature in aqueous solution at slightly basic pH. Excess spin label is removed by gel filtration on a Superdex G25 column. The reaction can be followed by HPLC, and the extent of spin labeling is analyzed by mass spectrometry.

Rarely, labeling will be site-directed due to the lack of regiochemical control of the labeling reaction. Therefore it is important to emphasize that separation of the differently spin-labeled species is generally not necessary. Although it can be advantageous to work with singly spin-labeled protein (see below), for general screening a heterogenous protein preparation is sufficient, as long as a significant fraction of protein is spin labeled within a distance of 10–15 Å from the active site.

Spin labels are generally very stable and chemically inert. The four methyl groups flanking the paramagnetic center (see Fig. 15.1 C) serve two purposes: first, they sterically hinder dimerization of the nitroxide, and second, they prevent disproportionation [15]. Spin labels can be treated with most organic solvents and are stable in aqueous solutions in the pH range 3–10 [15]. When working with proteins, however, it should be noted that reducing agents such as DTT or *β*-mercaptoethanol have to be avoided since they reduce the paramagnetic nitroxide to the diamagnetic *N*-hydroxylamine. In our experience, it is also advantageous to add small amounts of EDTA in order to complex any catalytic amounts of metals, to bubble nitrogen through the solution in order to remove dissolved oxygen, and to store the spin-labeled protein at 4 °C. Then the spin-labeled protein is chemically stable for several months.

15.3
Applications of Spin Labels in NMR Screening

15.3.1
Primary NMR Screening Using Spin Labels: SLAPSTIC

As discussed in Chapt. 14 and in several reviews, two distinct strategies can be pursued to detect protein-ligand interactions by NMR: observation of the protein resonances or observation of the ligand resonances [1, 3]. Observation of the protein resonances is usually done by means of [^{15}N,^{1}H]- or [^{13}C,^{1}H]-HSQC-type experiments, while observation of the ligand resonances can be achieved e.g. by $T_{1\rho}$ experiments, which measure the transverse relaxation rates of the ligand in the presence and absence of protein (Fig. 15.1). An unbound compound is a small molecule, and therefore transverse relaxation is slow. If it binds to the target protein, however, it behaves like a large molecule, and transverse relaxation becomes very efficient. In an exchanging system, one can observe the resonances of the free ligand, but with a relaxation rate that is a weighted average of the relaxation rates in the free and bound states (cf. Eq. (3)).

In primary NMR screening, the following reasons make observation of the ligand resonances often preferable:

1. Protein demands are significantly lower, and proteins do not need isotope labeling.
2. The molecular size of the protein is not an obstacle. In fact, most detection methods observing ligand resonances work better for larger proteins.
3. The method is less prone to artefacts caused by slight pH changes [16].
4. If compound mixtures are screened, mixtures containing a hit need not be deconvoluted, but the hit can be identified directly.
5. Limited compound solubility is not an obstacle as long as it is not below 20 μM (see below).

The drawback of methods which are based on observation of ligand resonances is that they rely on a significant exchange of ligand between bound state and free state during

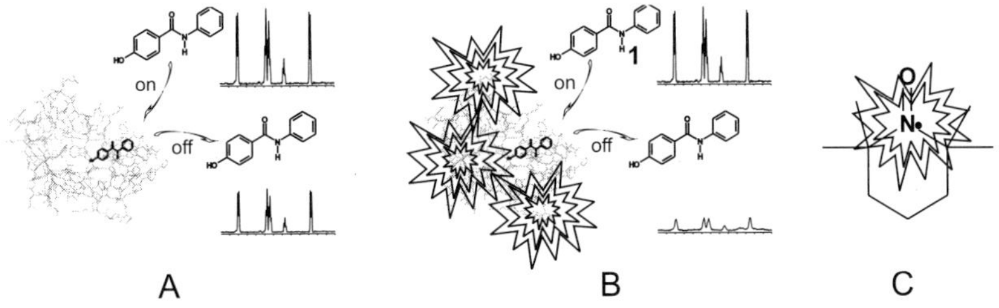

Fig. 15.1 Principle of the $T_{1\rho}$ experiment (**A**) and of the SLAPSTIC experiment (**B**). The $T_{1\rho}$ experiment makes use of the increased transverse relaxation rate of the ligand in the bound state, which leads to slightly reduced signal intensity (**A**). Signal intensity is drastically reduced or completely quenched by paramagnetic relaxation enhancement from the spin-labeled protein in the case of the SLAPSTIC technique (B). A common spin label is TEMPO (2,2,6,6-tetramethyl-1-piperidine-*N*-oxyl, C).

the time scale of the experiment. If the dissociation constant is so low (and binding concomitantly strong) that the ligand does not significantly dissociate during the experiment, the properties of the free ligand are not averaged with the properties of the bound ligand, and it appears as if the ligand does not bind the target protein. The bound resonances of the ligand cannot usually be observed since the protein concentration is typically much lower than the ligand concentration.

In order to test for strong ligand binding using ligand-detected methods, one can add to the compound under investigation a known weak "reporter" ligand and observe the relaxation properties of this reporter ligand. If the compound under investigation indeed binds strongly to the protein, it will block all protein binding sites, so that the reporter ligand cannot compete with binding, and relaxes like a nonbinding compound. This is an indirect proof of strong binding of the compound under investigation [24, 25].

In the previous section, it was shown that relaxation of a proton by a paramagnetic moiety is orders of magnitude stronger than relaxation by another proton. In $T_{1\rho}$ experiments, this paramagnetic relaxation enhancement can be constructively used to further enhance the relaxation rate of the ligand in the bound state. In an exchanging system, the observed transverse relaxation rate, $R2_{\text{obs}}$, with a paramagnetic protein target is

$$R2_{\text{obs}} = (1 - p_{\text{b}})R2_{\text{free}} + p_{\text{b}}R2_{\text{bound}} + p_{\text{b}}R2_{\text{para}} + R2_{\text{ex}} \tag{3}$$

Where p_{b} is the fraction of bound protein, $R2_{\text{free}}$ is the (small) transverse relaxation rate in the unbound state, $R2_{\text{bound}}$ is the (larger) transverse relaxation rate in the bound state, due to the increase in correlation time and spin density, $R2_{\text{para}}$ is the (much larger) paramagnetic relaxation enhancement (see Eq. (1)]) and $R2_{\text{ex}}$ accounts for exchange broadening in the intermediate exchange regime, which can often be neglected for weak (high micromolar) binding affinities.

The potential of the SLAPSTIC method [5] (**s**pin **la**bels attached to **p**rotein **s**ide chains as a **t**ool to identify **i**nteracting **c**ompounds) is demonstrated with the FK506 binding protein, FKBP, as a model system. A variety of FKBP ligands are known in the literature, among them *para*-hydroxy-benzanilide (**1**) (Fig. 15.1), which binds to FKBP with a dissociation constant K_{D}=1.1 mM.[1] The structure of FKBP is well known, and visual inspection showed that several lysine residues reside within a distance of 12–15 Å from the active site without being involved in ligand binding. We therefore chose to spin label lysine residues, which was readily achieved using published protocols [5].

The benefit of using spin labels for primary NMR screening stems from the occurrence of $p_{\text{b}}R2_{\text{para}}$ in the equation for the observed relaxation rate. Figure 15.2 illustrates the effect of the spin label on the relaxation rates of **1**. This figure shows $T_{1\rho}$ experiments of the mixture of FKBP with ligand **1** and with a mixture of four noninteracting aromatic compounds. The upper row shows one-dimensional proton spectra after a short (10 ms) spin-lock period, and the lower row shows one-dimensional proton spectra after a long (200 ms) spin-lock period.[2] In the absence of protein (left column), the signal de-

1) This K_{D} value was determined by NMR titration of FKBP with **1**
2) The spin lock period is used to translate broad lines into less intense lines

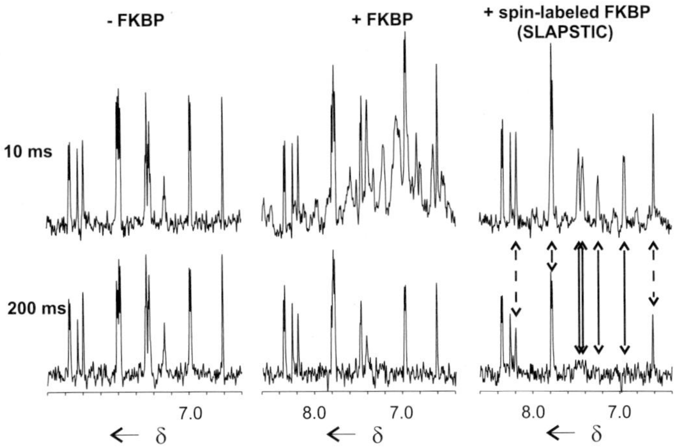

Fig. 15.2 $T_{1\rho}$ experiments performed on a mixture of *para*-hydroxy-benzanilide (**1**) and five other compounds with FKBP. Spectra in the upper and lower row correspond to spin-lock periods of 10 ms and 200 ms, respectively. Higher attenuation at 200 ms, as visible in the SLAPSTIC spectra (right) means faster relaxation in the bound state and easier detection of binding. Resonances of **1** are marked with black arrows, resonances of methyl-4-methoxythiophene-3-carboxylate, a compound that binds to FKBP very weakly, are marked with dashed arrows. The remaining signal at 7.8 ppm stems from another compound.

cay is very small since the transverse relaxation rate of unbound **1** is small. In the presence of 60 μM FKBP (middle column), signals of the interacting compound **1** are partially attenuated because of the term $p_b R2_{bound}$ in Eq. (3). In the presence of 20 μM spin-labeled FKBP (SLAPSTIC), however, paramagnetic effects described by $p_b R2_{para}$ in Eq. (3) are so large that the resonances of ligand **1** are completely quenched, and even the resonances of another compound, methyl-4-methoxythiophene-3-carboxylate, which was not known to bind to FKBP and subsequently shown to bind with $K_D = (9 \pm 2)$ mM, are partially attenuated. This demonstrates the high sensitivity of the SLAPSTIC technique.

Knowing the exact concentrations of protein and ligand, as well as the dissociation constant of the complex, the transverse relaxation rates can be calculated as $R2_{free} = 0.6\ s^{-1}$, $R2_{bound} = 15\ s^{-1}$, and $R2_{para} = 700\ s^{-1}$. Paramagnetic relaxation enhancement therefore magnifies the transverse relaxation in the bound state by a factor of about 50, which is in line with our theoretical estimates (see above). It should be emphasized that the potential of the SLAPSTIC method is even higher than that: because of the inverse sixth-power distance dependence of paramagnetic relaxation enhancement (Eq. (16.1)), paramagnetic effects will be another order of magnitude larger if the average distance of the closest spin label to the binding site is reduced by 30% in comparison to FKBP.

15.3.2
Protein Amounts Needed for SLAPSTIC Screening

The drastic paramagnetic effect on transverse relaxation rates can be used to make the distinction between binding and nonbinding compounds so clear that analysis of SLAPS-

TIC experiments can easily be automated. Figure 15.2 shows an additional advantage of SLAPSTIC spectra compared to conventional $T_{1\rho}$ experiments: since the protein resonances themselves are also quenched by the spin label and since only small protein amounts are employed, there is virtually no background signal from the protein even with short spin-lock periods. Protein background signals are often strongly disturbing when accurately measuring $T_{1\rho}$ relaxation rates (Fig. 15.2 upper middle).

Alternatively, the drastic paramagnetic effect on transverse relaxation rates can be used to decrease the protein concentration, while maintaining clear discrimination in transverse relaxation rate. How low can protein concentrations become while significant paramagnetic relaxation enhancement is still observed? The crucial parameter in Eq. (3) is p_b, the fraction of bound ligand. If p_b is too low, the observed transverse relaxtion rate, $R2_{obs}$, is dominated by $R2_{free}$, and no sufficient distinction is possible between binding and nonbinding compounds. However, the larger $R2_{para}$, the smaller p_b may become while still allowing the relaxation enhancement effect to be observed.

Fig. 15.3 shows calculated values for the fraction of bound ligand as a function of ligand concentration for three different dissociation constants and two different protein concentrations. It is important to realize that the fraction of bound ligand depends almost linearly on protein concentration and also strongly on binding affinity. The depen-

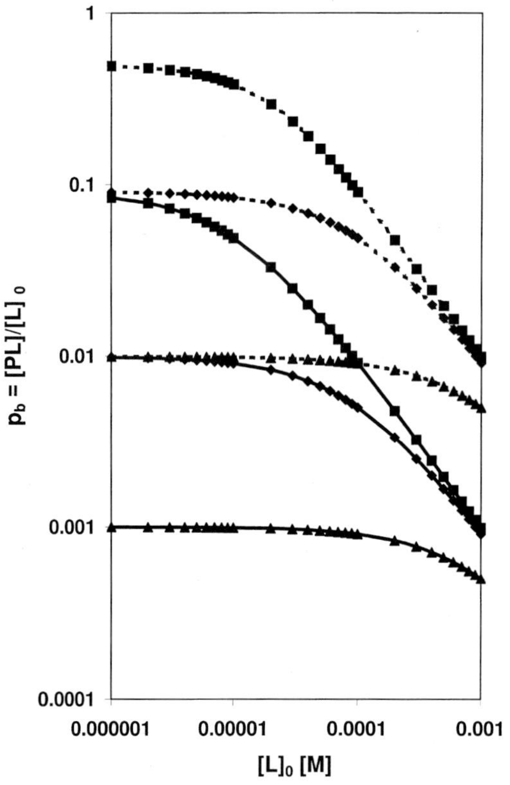

Fig. 15.3 Calculated percentage of bound ligand (p_b=[PL]/[L]$_0$, where [PL] is the concentration of the protein–ligand complex, and [L]$_0$ is the total ligand concentration) as a function of total ligand concentration. The graph is in double logarithmic scale. The dissociation constants, K_D, are 10 μM (squares), 100 μM (diamonds), and 1 μM (triangles). Protein concentrations are 1 μM (solid curves) and 10 μM (dashed curves).

dence on total ligand concentration is relatively small for weak affinities and larger for strong affinities. The contribution of paramagnetic relaxation enhancement to the observed transverse relaxation rate is $p_b R2_{para}$. If $R2_{para}$ is 10-fold higher because of a well-placed spin label, p_b can be 10 times lower with equal observed transverse relaxation rate, $R2_{obs}$. Because of the almost linear dependence of p_b on protein concentration, this allows a 10-fold reduction in protein concentration in a $T_{1\rho}$-weighted experiment while maintaining the observed transverse relaxation rate reduction.

Alternatively, the protein concentration, and thereby p_b, can be adjusted so that $p_b R2_{para}$ is sufficiently large that a threshold binding affinity can just be detected using SLAPSTIC. Thus, the protein concentration can be tuned for the desired K_D sensitivity of the experiment, so that only moderately strong ligands, but not very weak ligands, are detected in the experiment.

It should be emphasized that NMR screening is not the only method of obtaining mildly potent first inhibitors. Other biochemical assays can yield such ligands as well, and often do in our experience, so that the first ligand can simply be an optimized ligand from high-throughput screening, a natural or a published compound, or a generic ligand for specific protein classes, e.g. ATP (or ATP analogs) for kinases, or NADP (or NADP analogs) for oxido-reductases [17]. The power of NMR comes in by characterizing the structure of the complex with this first ligand and by identifying second-site ligands that bind to the target in the vicinity and at the same time. In fact, screening for second-site ligands can essentially only be done by NMR, which is a very powerful technique for this task.

15.3.3
Validation and Preliminary Optimization of Primary NMR Screening Hits

Once a primary ligand has been identified by the SLAPSTIC method, another NMR-based screening method, or some other screening technique, it will usually be validated by NMR methods. If isotope-labeled protein is available, one can run ^{15}N-^{1}H or ^{13}C-^{1}H correlation experiments. These will show the approximate binding site of the ligand, if resonance assignments are available, or will at least show whether the set of resonances affected by the ligand is overlapping with, or is even a subset of, the resonances affected by binding of a known ligand. In the latter case, it can be assumed that the binding sites of the identified and the known ligands are overlapping or even identical. If isotope-labeled protein is not available, NOEs between protein and ligand can still be detected. They can usually not be assigned to specific protein protons, but can certainly be classified as belonging to aromatic, amide, aliphatic, or methyl groups, which can be important information if the structure of the target or a reliable homology model are available. Observation of protein-ligand NOEs, STD effects [19], or observation of selective line broadening of ligand resonances indicate the parts of the ligand in contact with the protein, which is important information for chemistry to optimize, derivatize or link the ligands.

From interesting new ligands, a substructure or similarity search will reveal compounds that can be tested by the same NMR methods, so that some knowledge about the structure-activity relationship is built up. This can lead to optimization of the ligand and to increased binding affinity, although improvement by more than an order of magnitude is rare. More pronounced improvement of the binding affinity can occur if two ligands or ligand classes are identified that bind to distinct binding sites of the protein. If these binding sites are partially overlapping, both compounds cannot bind simultaneously, and a merged compound may be designed that exploits both individual binding sites.

If the binding sites of the two ligands are nonoverlapping and if both ligands bind to the target at the same time, an attractive option is to link the two ligands in order to obtain a single linked ligand that occupies both individual binding sites. Theoretically, the binding energy of the individual ligands should add in the linked compound, so that the binding affinities should multiply. Even if the linker between the two ligands is not optimal and binding affinity is thereby lost, this strategy is the most attractive to yield nanomolar inhibitors out of initially micromolar or millimolar ligands. The next section describes how second-site ligands can be identified by using spin labels.

15.3.4
Second-Site NMR Screening Using Spin Labels

Identification of a second-site ligand, that is a ligand which binds simultaneously to the first ligand at a second, neighboring binding site, requires saturation of the first binding site by the first ligand to prevent the second ligand from binding to the first binding site. Unfortunately, this is often not possible because of the weak affinity and poor water solubility of the first ligand. For example, if K_D for the first ligand is 200 μM and its aqueous solubility is 100 μM, only about 33% of all first binding sites are saturated. Any test compound in a second-site screen can then bind to the first binding site instead of the desired second binding site and thus produce a false positive response. These false positive hits are difficult and time-consuming to identify, but can be completely removed by employing spin labels, as described in the following [4].

A true and desired second-site ligand binds to the target at the same time (and therefore at nonoverlapping binding sites) and in the vicinity of the first-site ligand. It can be detected by using a spin-labeled analog of the first-site ligand. Then a quenching effect of the spin label on other compounds is observed if and only if both compounds bind at the same time and in the vicinity (Fig. 15.4). If the putative second ligand actually binds to the first binding site, it will never bind at the same time as the spin-labeled first ligand, and will therefore not produce a false positive result. If a compound does not bind the target at all, it will accordingly never be in the close vicinity of the spin-labeled first ligand. The average distance between a test compound and the spin-labeled first ligand in the absence of target protein is too large for any quenching effects to be observed on the test compound. Only simultaneous binding to the target protein brings the spin-labeled first ligand and the second-site ligand close enough together for a long enough time to allow observation of paramagnetic relaxation enhancement on the second-site ligand caused by the spin-labeled first ligand.

As an example, the identification of a second-site ligand to the anti-apoptotic protein Bcl-xL is shown in Fig. 15.4. Compound **2**, a weak ligand for Bcl-xL (K_D=140 μM), was identified by an ELISA-based high-throughput screening assay, and second-site ligands were sought by NMR screening to improve its potency after linking. The observation of small specific chemical shift differences of **2** in the presence and absence of Bcl-xL,

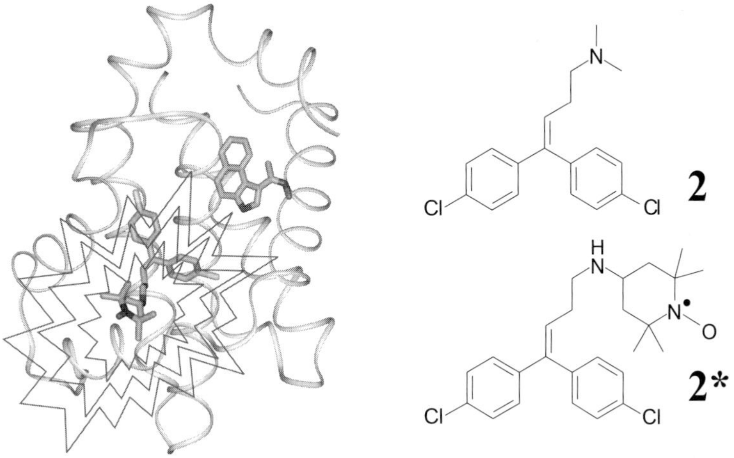

Fig. 15.4 Principle of second-site screening using a spin-labeled first ligand. First-site ligand **2** was spin-labeled to yield **2***. The quenching effects of **2*** on the resonances of any second-site ligand are observed.

knowledge about its structure-activity relationship, and modeling studies, suggested that the aromatic groups are primarily interacting with Bcl-xL, while the substituents of the tertiary amine are noncritical for binding. Therefore, **2** was derivatized with TEMPO at the tertiary amine to yield spin-labeled **2***. The ELISA-based assay confirmed that the binding affinity of **2*** and **2** are not significantly different.

For second-site NMR screening, Bcl-xL and spin-labeled ligand **2*** were incubated with a mixture of eight aromatic compounds, and paramagnetic relaxation enhancement effects from **2*** to any of the compounds were investigated by $T_{1\rho}$ relaxation weighted experiments [4]. Fig. 15.5 shows the corresponding $T_{1\rho}$ weighted spectra, with short and long mixing times in the upper and lower row, respectively. The left column was recorded in the absence of Bcl-xL. It serves as a control inasmuch as none of the compounds experiences paramagnetic relaxation enhancement, since the average distance in diluted solution between noninteracting ligands is large. The right column shows the same spectra in the presence of Bcl-xL. While most compounds still do not experience any paramagnetic relaxation enhancement, the resonances of one compound, marked by the arrows, are completely quenched. This compound is therefore unambigously detected as second-site ligand. It should be stressed that there are essentially no sources of false positive detection in this spin-label experiment. If a compound experiences paramagnetic relaxation enhancement, it can be safely assumed that the compound binds simultaneously and in the vicinity of the spin-labeled first ligand, and that it therefore is a true second-site ligand.

Another example of successful second-site screening is presented in Chapt. 20 of this book.

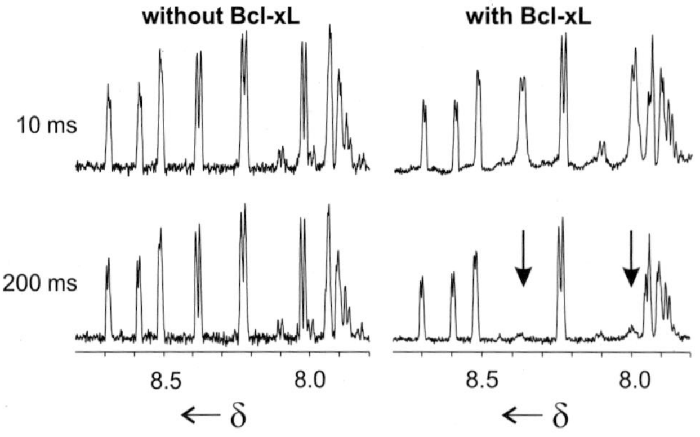

Fig. 15.5 Identification of a second-site ligand for Bcl-xL using spin-labeled first ligand **2***. The figure shows $T_{1\rho}$ spectra as in Fig. 15.2, in the absence (left) and presence (right) of Bcl-xL.

15.4 Linker Design

Both the first ligand and the second-site ligand are generally low-affinity ligands with dissociation constants in the micromolar range. They need to be chemically linked in order to obtain a high-affinity, nanomolar ligand. Favorable placement of the linker is crucial for the potency of the linked compound. The optimal linker should satisfy three criteria. First, it should exert little strain to the two individual components, and should allow them to occupy exactly the same binding site in the same orientation in the linked compound as in the unlinked fragments. Second, while allowing the two components to have sufficient flexibility to adopt optimal binding orientations, the linker should be as rigid as possible in order to preform the bioactive conformation of the compound and to reduce the entropic cost of binding [18]. Thirdly, the linker must not have unfavorable interactions with the protein; if it has interactions at all, they should be optimized to affect binding positively.

In the ideal case, the structure of the ternary complex is known by X-ray crystallography or NMR. This would be of great help in designing the linker by molecular modeling. Unfortunately, this ideal case is rather rare. Fortunately, there are several other clues to designing a successful linkage with the help of NMR. The atoms on both fragments to which the linker should be attached can be identified by a combination of various techniques:

- Observation of chemical shift changes or selective line broadening of the ligand resonances upon addition of protein indicate which parts of the ligand are in contact with the protein. Residues that are unaffected in position and line width are probably not in contact with the protein, and are therefore candidates for linker attachment. Alternatively, the binding epitopes can be mapped by STD-type experiments [19].
- In the case of second-site screening with a spin-labeled first ligand, the dependence of the paramagnetic relaxation enhancement on the inverse sixth power of the distance (Eq. (1)) leads to differential quenching effects on the second ligand, depending on its

proximity to the paramagnetic center. Those resonances of the second ligand that are most strongly affected are located nearest to the first ligand and are therefore primary candidates for linker attachment. The absolute distance between those protons and the first ligand can be estimated using Eq. (1).
- If site-specific spin-labeled protein is available [20], quantification of quenching effects on both first-site and second-site ligands can additionally aid in constructing a model of the ternary complex.
- In any case, a transfer NOE experiment should be recorded on the ternary mixture between protein, first and second ligand. Since both first and second ligand bind weakly to the protein, they are likely to exhibit transferred NOEs that reflect their bound conformation. If both ligands bind in close proximity, i.e. if there are protons on both ligands that are within 4 Å in the ternary complex, then these protons can show an interligand NOE effect [21]. In that case, the respective atoms for linker attachment are easily identified, and the required length of the linker can be estimated from the strength of the NOE [22]. If interligand NOEs cannot be identified although transferred NOEs are present, the distance between both ligands may be too large, and it may be worthwhile to investigate ligand analogs.

Once the atoms are defined to which the linker is attached and the length of the linker has been estimated, the exact nature of the linker will be designed on the basis of structure-activity relationships of the individual ligands or the structure or a homology model of the protein.

15.5 Conclusions and Outlook

The cases presented so far have mostly concerned the identification of first-site and second-site ligands for lead generation. However, spin labels can also be used to refine and characterize the structure of a ligand bound to a protein. In cases of weakly binding ligands, it is often possible to determine the conformation of the bound ligand using transfer NOE experiments. However, it is very difficult to generate a reliable model for the complex, particularly when dealing with large proteins. Protein-ligand NOEs may then not be detectable, and, if they are, it may not be possible to assign them if the protein is not isotopically labeled.

In the absence of assigned NOEs, spin labels can be very helpful to generate a docking model of the protein-ligand complex. It is often possible to introduce a paramagnetic center at a defined position. This can be a spin label attached to a cysteine side chain if a single unpaired cysteine is in the vicinity of the binding site. It can be a spin-labeled version of a cofactor, or it can even be a paramagnetic metal to substitute a diamagnetic metal in metal-binding proteins. Introduction of this paramagnetic center then permits distances between different protons in the weakly binding ligand to be measured with respect to a fixed center of reference, namely the paramagnetic center. This strategy would be analogous to a recently presented method of refining protein structures by tagging a protein with an N-terminal Cu(II) or Ni(II) binding (ATCUN) motif and measuring distances between protein protons and the paramagnetic metal [23].

In conclusion, spin labels can be used to identify and characterize interactions between proteins and ligands. The main advantages are reduced protein consumption and increased robustness of the experiment against detection of false positives. Spin labels can aid in the identification of ligands by NMR screening, identifying both first-site ligands by the SLAPSTIC method, and second-site ligands by using spin-labeled first ligands. Both methods are now extensively and successfully used in our laboratory. After the identification of first-site and second-site ligands, suitable linkers can be designed by the protocols outlined above, using information from interligand NOEs, differential line broadening, and quantification of paramagnetic relaxation enhancement. This procedure can result in the NMR-based discovery of nanomolar ligands for a protein target of essentially any molecular size.

15.6 References

1 J.W. Peng, C.A. Lepre, J. Fejzo, N. Abdul-Manan, J.M. Moore, *Methods Enzymol.* **2001**, *338*, 202.

2 T. Diercks, M. Coles, H. Kessler, *Curr. Opin. Chem. Biol.* **2001**, *5*, 285.

3 M.J. Shapiro, J.R. Wareing, *Curr. Opin. Drug Discovery Dev.* **1999**, *2*, 396.

4 W. Jahnke, L.B. Perez, C.G. Paris, A. Strauss, G. Fendrich, C.M. Nalin, *J. Am. Chem. Soc.* **2000**, *122*, 7394.

5 W. Jahnke, S. Ruedisser, M. Zurini, *J. Am. Chem. Soc.* **2001**, *123*, 3149.

6 R.A. Dwek, *Monographs on Physical Biochemistry: Nuclear Magnetic Resonance (N.M.R) in the Biochemistry. Applications to enzyme systems.* Oxford Univ. Press, New York, **1973**.

7 P.A. Kosen, *Methods Enzymol.* **1989**, *177*, 86.

8 P. Jost, O.H. Griffith, *Methods Pharmacol.* **1972**, *2*, 223.

9 T.R. Krugh, *Spin Labeling* **1976**, 339.

10 J.A. Peters, J. Huskens, D.J. Raber, *Prog. Nucl. Magn. Reson. Spectrosc.* **1996**, *28*, 283.

11 J.R. Tolman, J.M. Flanagan, M.A. Kennedy, J. H. Prestegard, *Proc. Natl. Acad. Sci. U. S. A.* **1995**, *92*, 9279.

12 J.H. Prestegard, *Nat. Struct. Biol.* **1998**, *5*, 517.

13 N. Tjandra, J.G. Omichinski, A. M. Gronenborn, G.M. Clore, A. Bax, *Nat. Struct. Biol.* **1997**, *4*, 732.

14 R.L. Lundblad, Editor, *Techniques in Protein Modification*, CRC, Boca Raton, Fla., **1995**.

15 L.J. Berliner, *Prog. Bioorg. Chem.* **1974**, *3*, 1.

16 A. Ross, H. Senn, *Drug Discov. Today* **2001**, *6*, 583.

17 D.S. Sem, M. Pellecchia, *Curr. Opin. Drug Discov. Dev.* **2001**, *4*, 479.

18 G. Thoma, J.L. Magnani, J. T. Patton, B. Ernst, W. Jahnke, *Angew. Chem. Int. Ed. Engl.* **2001**, *40*, 1941.

19 M. Mayer, B. Meyer, *J. Am. Chem. Soc.* **2001**, *123*, 6108.

20 V.W. Cornish, D.R. Benson, C.A. Altenbach, K. Hideg, W.L. Hubbell, P.G. Schultz, *Proc. Natl. Acad. Sci. U.S.A.* **1994**, *91*, 2910.

21 D. Li, E.F. DeRose, R. E. London, *J. Biomol. NMR* **1999**, *15*, 71.

22 R.E. London, *J. Magn. Reson.* **1999**, *141*, 301.

23 L.W. Donaldson, N.R. Skrynnikov, W.-Y. Choy, D.R. Muhandiram, B. Sarkar, J.D. Forman-Kay, L.E. Kay, *J. Am. Chem. Soc.* **2001**, *123*, 9843.

24 C. Dalvit, M. Flocco, S. Knapp, M. Mostardini, R. Perego, B.J. Stockmann, M. Veronesi, M. Varasi, *J. Am. Chem. Soc.* **2002**, *124*, 7702.

24 W. Jahnke, P. Floersheim, C. Ostermeier, X. Zhang, R. Hemmig, K. Hurth, D. Uzunov, *Angew. Chem.* **2002**, *114*, 3570; *Angew. Chem. Ind. Ed.* **2002**, *41*, 3420.

16
NMR of Weakly Binding Ligands

Marcel J. J. Blommers, and Simon Rüdisser

16.1
Introduction

Key processes in molecular cell biology are regulated by interactions between biomolecules. Protein-protein and protein-ligand interactions, e.g. in signal transduction pathways, rely on the subtle interaction between atoms at the binding interface of the involved molecules. Because biomolecules often have many interacting partners, the interaction is not necessarily strong but mostly specific. The study of molecular recognition gives insight into the complex network of signaling in life. This is not only of interest for structural biology, but also for rational drug design, since if the precise mode of molecular recognition is understood, one may develop compounds that inhibit or activate disease related biological signalling specifically. An understanding of the molecular details of these interactions is therefore a prerequisite for rational ligand design. Rational ligand design does not only aim at higher affinities but also at increased specificity to ultimately reduce side effects. A prerequisite for rational ligand design is to account for all effects of protein-ligand interactions. It is still a major difficulty to describe protein-ligand interactions theoretically and predict binding affinities and structures of complexes. One major reason for these problems lies in the complexity of the physics which needs to account not only for structural parameters but also for thermodynamics and kinetics. Whereas thermodynamic and kinetic data cannot be deduced from X-ray structures, NMR offers a whole range of methods of studying these effects. In the situation where the interaction is weak, one of the traditional methods that can be applied to obtain structural information of the bound ligand is the so-called transferred NOE (trNOE) method [1–3].

16.2
The Dynamic Equilibrium

In a mixture of protein and ligand there are three species present in equilibrium: the free ligand (L), the free protein (P) and the protein-ligand complex (PL). Assuming one binding site, the equilibrium concentrations [P], [L] and [PL] are described by:

$$K_D = \frac{[P][L]}{[PL]} \tag{1}$$

K_D is the dissociation constant of the complex (for a more general treatment see Chapt. 13). In a dynamic equilibrium K_D is given by $K_D = k_{off}/k_{on}$ where k_{off} is the rate constant for the dissociation of the protein-ligand and k_{on} is the association rate constant. The upper limit of k_{on} is diffusion controlled and in the order of 10^8 M^{-1} s^{-1}. However, k_{on} can be smaller than the diffusion limit if there are activation barriers to overcome upon association. Activation barriers may arise because of conformational reorientation of the ligand and/or protein required for binding, rearrangements of the hydration shells of the ligand or the binding pocket, etc. Although k_{off} is governed by the same process, the activation barrier for dissociation are higher since "good binders" have strong interactions with the protein, i.e. hydrogen bonds, hydrophobic interactions, van der Waals forces, etc., which need to be overcome to remove the ligand from the binding pocket.

16.3
Transferred NOE (trNOE)

Exchange of magnetization due to cross-relaxation (NOE, nuclear Overhauser effect) does lead to intensity changes of individual resonances which provide valuable information about spatial and motional characteristics of the spins involved [4, 5]. It is currently mostly measured in two-dimensional NMR, where the NOE is measured as cross-peak intensity. Cross-relaxation is caused by mutual spin flips in dipolar coupled spin pairs.

Cross-relaxation and auto-relaxation for a two-spin system is described by the Solomon [6] equation:

$$\frac{d}{dt}\begin{pmatrix}\langle \hat{I}_{1z}\rangle \\ \langle \hat{I}_{2z}\rangle\end{pmatrix} = \begin{pmatrix}-R_{auto} & R_{cross} \\ R_{cross} & -R_{auto}\end{pmatrix}\begin{pmatrix}\langle \hat{I}_{1z}\rangle - \langle \hat{I}_{1z}\rangle^{eq} \\ \langle \hat{I}_{2z}\rangle - \langle \hat{I}_{2z}\rangle^{eq}\end{pmatrix} \tag{2}$$

The symbols R_{auto} and R_{cross} within the relaxation matrix are the auto- and cross-relaxation rates, respectively. $\langle \hat{I}_{1z}\rangle$ and $\langle \hat{I}_{2z}\rangle$ are the longitudinal magnetizations of spin 1 and 2, respectively, and the brackets indicate averaging over the whole ensemble of spins. R_{cross} in terms of the spectral densities is given by

$$R_{cross} = -\frac{1}{10}\left(\frac{\gamma^2\hbar}{r^3}\right)^2 \{J(0) - 6J(2\omega_0)\} \tag{3}$$

where the constants γ and $\hbar$ have their usual meaning and $J(0)$ and $J(2\omega_0)$ are the spectral densities at zero frequency and twice the Larmor frequency, respectively. For isotropic tumbling, we have an exponentially decaying autocorrelation function, and in the absence of internal motion the spectral density depends on the correlation time τ_c in the following way:

$$J(\omega) = \frac{\tau_c}{(1 + \omega^2\tau_c^2)} \tag{4}$$

Hence R_{cross} depends heavily on the correlation time of the molecule. R_{cross} is negative for short correlation times (small molecules in nonviscous solution) and positive for long

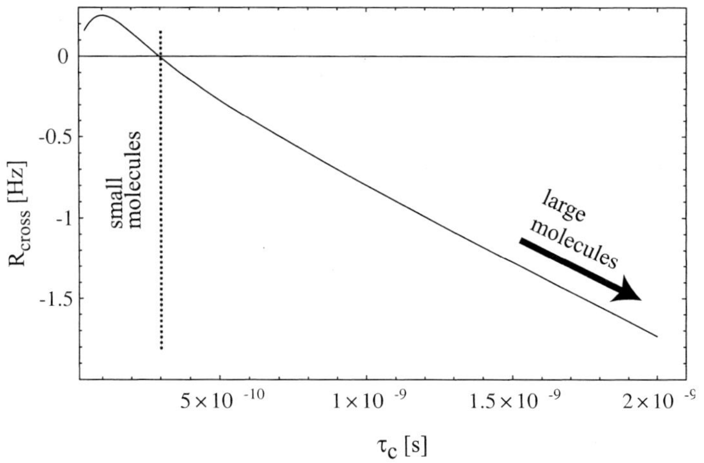

Fig. 16.1 The dependence of R_{cross} for a pair of protons on the rotational correlation time τ_c. The curve starts at the left with a τ_c corresponding to molecules with about 200 g mol^{-1} and ends on the right hand side with a τ_c corresponding to a molar mass of 5000 g mol^{-1}. For the calculation a proton–proton distance of 0.2 nm and a ^{1}H frequency of 600 MHz have been used.

correlation times (large molecules or molecules in viscous solutions). For sufficiently slow molecular motions the condition $\omega^2\tau_c^2 \gg 1$ is satisfied.

Thus, in the fast tumbling limit $R_{cross} = \frac{5}{10}\left(\frac{\gamma^2\hbar}{r^3}\right)^2\tau_c$ and in the slow tumbling limit

$$R_{cross} = -\frac{1}{10}\left(\frac{\gamma^2\hbar}{r^3}\right)^2\tau_c\,.$$

Up to this point only overall motion of the molecule has been considered, but often there is internal motion, in addition to overall molecular tumbling, which needs to be considered to obtain a correct expression for the spectral density function. Here we apply the model-free approach to treat internal motion where the unique information is specified by a generalized order parameter S, which is a measure of the spatial restriction of internal motion, and the effective correlation time τ_e, which is a measure of the rate of internal motion [7, 8]. The model-free approach only holds if internal motion is an order of magnitude (<0.3 ns) faster than overall reorientation and can therefore be separated from overall molecular tumbling. The spectral density has the following simple expression in the model-free formalism:

$$J(\omega) = \frac{2}{5}\left[\frac{S^2\tau_c}{(1+\omega^2\tau_c^2)} + \frac{(1-S^2)\tau}{(1+\omega^2\tau^2)}\right] \tag{5}$$

$$\tau^{-1} = \tau_c^{-1} + \tau_e^{-1} \tag{6}$$

In the transferred NOE experiment, R_{cross} is the cross-relaxation rate of the ligand weighted average over the free and bound state (Eq. 16.7). The experiment works best when the rates in the two states are very different, which is usually the case for small ligands binding to protein targets. In a transferred NOE experiment, spatial and motional information of the ligand in the bound state is transferred to the free state (dynamic equilibrium) whose resonances are observed [1–3, 9–11]. It should also be noted that ligands interacting with slow tumbling proteins reach their trNOE maximum build-up quickly, while nonbinding molecules show a much slower build-up. However, a prerequisite to deduce the bound ligand conformation by studying the free state is that k_{off} is comparable to or larger than the relaxation rates in the bound form. For a tightly binding ligand with a K_{D} of approx. 10^{-9} M k_{off} is not sufficiently large and one has to resort to the direct study of the protein-ligand complex. If the exchange rate k_{off} is much faster than the relaxation rates, the effective relaxation rate is simply the population weighted average of bound and free state (neglecting the possibility of having different conformations in the bound and/or free state and neglecting intermolecular cross-relaxation):

$$R_{\mathrm{cross}} = aR_{\mathrm{cross}}^{\mathrm{bound}} + (1-a)R_{\mathrm{cross}}^{\mathrm{free}} \tag{7}$$

a is the fraction of bound ligand, and $R_{\mathrm{cross}}^{\mathrm{bound}}$ and $R_{\mathrm{cross}}^{\mathrm{free}}$ are the corresponding relaxation matrix elements of the ligand for the bound and free state, respectively. If the ligand is a small molecule, the NOE or $R_{\mathrm{cross}}^{\mathrm{free}}$ is close to zero and can be neglected. This is one of the key aspects that makes the trNOE experiment attractive, since the information obtained is almost exclusively related to the bound conformation (see Fig. 16.2).

In general, a relatively direct and straightforward means of analysis may be performed in the case of slow exchange on the chemical shift time-scale by combining the relaxation matrices of the free and bound state with the kinetic matrix to describe the effect of exchange [12]. For the two spin systems described above the expanded relaxation matrix $\underline{R}$ can be written as:

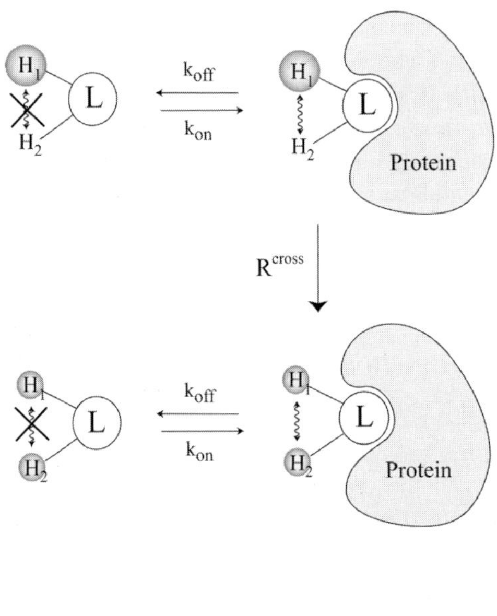

Fig. 16.2 A simplified scheme of the trNOE concept. The ligand L in the free state has negligible cross-relaxation between protons H_1 and H_2 because of its rapid tumbling motion. Upon binding to the much slower tumbling protein R^{cross} becomes effective and leads to a transfer of magnetization from H_1 to H_2. Because of the dynamic equilibrium the ligand is released back into solution where it is still in the magnetization state corresponding to the bound form. The same concept is also applicable to trCCR and trRDC (see Sects. 16.4 and 16.5).

$$\underline{R} = \begin{bmatrix} -R_{\text{auto}}^{\text{free}} & R_{\text{cross}}^{\text{free}} & 0 & 0 \\ R_{\text{cross}}^{\text{free}} & -R_{\text{auto}}^{\text{free}} & 0 & 0 \\ 0 & 0 & -R_{\text{auto}}^{\text{bound}} & R_{\text{auto}}^{\text{bound}} \\ 0 & 0 & R_{\text{cross}}^{\text{bound}} & -R_{\text{auto}}^{\text{bound}} \end{bmatrix} \tag{8}$$

The exchange process is then described by the kinetic matrix $\underline{K}$ expressed in terms of k_{on} and k_{off}:

$$\underline{K} = \begin{bmatrix} -K_{\text{on}} & 0 & k_{\text{off}} & 0 \\ 0 & -k_{\text{on}} & 0 & k_{\text{off}} \\ k_{\text{on}} & 0 & -k_{\text{off}} & 0 \\ 0 & k_{\text{on}} & 0 & -k_{\text{off}} \end{bmatrix} \tag{9}$$

The kinetic matrix $\underline{K}$ reflects the fact that only corresponding spins in the free and bound state are connected by exchange. The complete matrix governing relaxation is simply the sum of $\underline{R}$ and $\underline{K}$:

$$\frac{d}{dt}\begin{pmatrix} \langle \hat{I}_{1z}^{\text{free}} \rangle \\ \langle \hat{I}_{2z}^{\text{free}} \rangle \\ \langle \hat{I}_{1z}^{\text{bound}} \rangle \\ \langle \hat{I}_{2z}^{\text{bound}} \rangle \end{pmatrix} = (\underline{R} + \underline{K}) \begin{pmatrix} \langle \hat{I}_{1z}^{\text{free}} \rangle & - & \langle \hat{I}_{1z}^{\text{free}} \rangle^{\text{eq}} \\ \langle \hat{I}_{2z}^{\text{free}} \rangle & - & \langle \hat{I}_{2z}^{\text{free}} \rangle^{\text{eq}} \\ \langle \hat{I}_{1z}^{\text{bound}} \rangle & - & \langle \hat{I}_{1z}^{\text{bound}} \rangle^{\text{eq}} \\ \langle \hat{I}_{2z}^{\text{bound}} \rangle & - & \langle \hat{I}_{2z}^{\text{bound}} \rangle^{\text{eq}} \end{pmatrix} \tag{10}$$

Alternatively, the much more common situation in trNOE studies involves fast exchange on the chemical shift time scale where the observed resonance shifts are weighted averages of the corresponding shift in the free and bound state [13]. A full account of the complete relaxation matrix and conformational exchange effects for n spins has been performed by London et al. [13], and a similar treatment was later incorporated into the programme CORCEMA [14].

Intermolecular cross-relaxation between ligand and protein cannot be neglected and can lead to serious errors in the structure of the bound conformation of the ligand if derived from trNOE intensities, because of spin diffusion effects. In the simulations shown in Fig. 16.3 the trNOE is influenced by the presence of a third proton of the protein. If the distance between the protons within the ligand is small, the protein may function as a relaxation sink, and as a result the measured NOE is smaller than expected. In contrast, if the distance is large, e.g. 4 Å, the NOE transfer will increase by relaxation via the additional proton, and as a result the measured trNOE is larger than expected on the basis of the ligand-bound conformation. Hence, the analysis of trNOE intensities is greatly simplified if fully deuterated proteins are used.

Indirect NOE effects can also be identified by means of a transferred ROE experiment [16]. Disadvantages of the trROE over the trNOE experiments are their lower sensitivity and the fact that cross-relaxation in the free ligand cannot be neglected. Several techniques have been developed to suppress spin diffusion in two-dimensional NOE spectroscopy. QUIET-BAND-NOESY and QUIET-ET-NOESY [17, 18] eliminate migration of magnetization through consecutive steps by the application of a doubly-selective inversion pulse in the middle of the mixing time. Similar concepts date back to work by Massefski and Red-

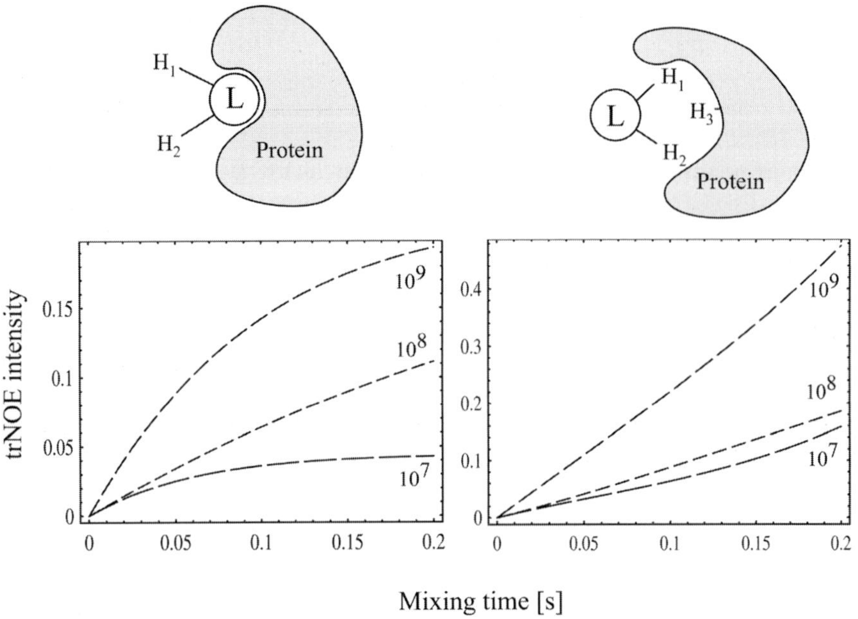

Fig. 16.3 The calculated trNOE intensity as a function of mixing time without (left part) and with (right part) spin diffusion through intermolecular cross-relaxation. For the calculations the distance between the protons was assumed to be 2 Å, and the rotational correlation time used was 0.1 ns for the free ligand and 10 ns for the protein. k_{on} was assumed to be 10^8 M^{-1} s^{-1}; trNOE build-up curves are shown for three different values of k_{off}. The corresponding k_{off} is indicated next to the corresponding curve. The influence of k_{off} on trNOE intensity has previously been studied by Lippens et al. [15]. Note the different scales on the left and right hand sides.

field [19], where weak irradiation is applied during the mixing time in a 2D NOESY experiment. Spin diffusion pathways involving H^{CH}/H^{CH3} or H^{CH2} can be eliminated in an isotope-edited NOE experiment. However, a ^{13}C-labeled protein is required to selectively invert either H^{CH2} or H^{CH}/H^{CH3} magnetization in the middle of the mixing time [20]. It is further desirable to remove cross peaks stemming from intramolecular interactions of the protein in order to eliminate overlap with trNOE cross peaks. Filtering methods based on the difference between the $T_{1\rho}$ values of the ligand and protein can be used to obtain a 2D trNOE spectrum in which the background cross peaks from the protein are removed [21].

Intermolecular trNOEs between protein and ligand are sometimes desirable and can be put in good use to detect binding and to validate the binding model. These may be observed for short mixing times in 2D trNOEs acquired with many scans at the expense of resolution in t_1. The so-called saturation transfer experiment is accomplished by irradiating the protein resonances and observing signals of the free ligand [22].

Another complication of the trNOE method may be the possibility that the ligand interacts nonspecifically with the protein. This has been discussed by Rao and coworkers [23]. Specificity of the interaction can best be demonstrated by a competition experiment with a strong binder.

The usefulness of trNOE methods to derive the bound conformation of a ligand has been shown in practical examples where trNOE had a direct impact on ligand optimization. A recent application of trNOE shows how to increase the potency of a highly flexible Farnesyltransferase (FTase) ligand by designing conformational restraints to reduce the entropic barrier for complex formation [24]. In the described example the calculated FTase bound conformation, which was consistent with trNOE intensities, was used to design a macrocyclic ring closure. It was thus possible to synthesize an inhibitor with enhanced FTase and reduced GGT-ase-I inhibitory effects. Studies of natural ligand and peptides by trNOE are often facilitated by incorporating a trNOE step into 3D heteronuclear experiments to reduce spectral overlap [25–27].

A nice example of an indirect structure determination using the trNOE method is the study of the conformation of a loop of the membrane protein bacteriorhodopsin (BR) [28]. Antibodies were raised against BR, and subsequently the complex of a heptapetide derived from BR was studied in complex with the antibody by trNOE. The bound conformation is a reasonably good representation of the conformation of the peptide in its native state in BR.

The interligand trNOE offers an avenue to study ternary complexes of protein and two ligands, thus giving information on the relative orientation of the two ligands [29]. This can be especially useful in linker design in the context of second-site screening by NMR [30]: if two low-affinity ligands bind in close proximity to each other and are synthetically linked they might combine to a higher affinity ligand. Designing an appropriate linker is often crucial, and any input from trNOE can reduce the synthetic effort and the number of compounds to be tested considerably.

Screening of compound mixtures in large excess is possible in trNOE studies even if a small amount of protein is present (for more screening techniques see Chapt. 14). This has been demonstrated by the example of angiotensin-converting enzyme and different furylacryloyl (fa) amino acid derivatives [23]. In this approach a description of the active site of the protein has been achieved by scanning the binding pocket with a set of weakly binding ligands by trNOE methods. This has been proven to be especially useful since no X-ray structure of the protein was available. In the emerging field of NMR-based screening in drug discovery, trNOE of mixtures of compounds are easily screened by acquiring 2D NOE spectra, and the ligand interacting with the protein-target is identified by the sign of the cross peaks [31, 32]. Also, the trNOE experiment may contribute to the characterization of ligands by the determination of dissociation constants [33].

If the ligand contains exchangeable protons, magnetization may be transferred from bulk water to the exchangeable ligand protons, from where magnetization is further relayed to nonexchangeable ligand protons. This offers an approach to selectively perturbing hydroxyl protons, for example, in a carbohydrate ligand by means of a manipulation of the water magnetization. A careful analysis of build-up rates following water inversion and ligand-based ^{13}C filtering experiments made it possible to place the hydrogen bonding protons of α-methyl-D-mannopyranoside in the binding pocket of the mannose binding protein [34]. Likewise, bulk water magnetization is utilized to screen mixtures of compounds by NMR. In the so-called WaterLOGSY experiment [35], magnetization is transferred from bulk water through chemical exchange to exchangeable ligand and protein protons as well as to hydration water of the protein. Magnetization is further relayed during the long mixing time period through exchangeable and nonexchangeable protons to the ligand. The signal of the unbound ligand is observed. Binding and nonbinding ligands can easily be distinguished by the sign of the signals.

The method works best for ligands displaying dissociation constants in the μM to mM (10^{-2} to 10^{-5} M) range. Typically, 10- to 20-fold excess of ligand with respect to the protein concentration is used. This translates into concentrations of approx. 50 μM protein (conventional probe, 600 MHz). The pH should be set such that the protein is active and the ligand binds. The trNOE experiment may not work if the k_{off} of the ligand is too small. In this case, optimization might be obtained by, e.g., raising the temperature, adding small amounts of organic solvent, or by a change of the ionic strength.

16.4
Transferred Cross-Correlated Relaxation (trCCR)

Since the discovery of the nuclear Overhauser effect (NOE, see previous section) [4, 5] and scalar coupling constants [36, 37] decades ago, NMR-derived structure calculations of biomolecules largely depended on the measurement of these two parameters [38]. Recently it became possible to use cross-correlated relaxation (CCR) to directly measure angles between bond vectors [39] (see also Chapt. 7). In addition, residual dipolar couplings of weakly aligned molecules were discovered to measure the orientation of bond vectors relative to the alignment tensor (see Sect. 16.5). Measurement of cross-correlated relaxation was described experimentally earlier for homonuclear cases [40, 41] and is widely used in solid-state NMR [42–44].

CCR is easily measured by heteronuclear NMR experiments of isotopically labeled molecules. The information extracted from these experiments will significantly improve the resolution of NMR structures, especially bound conformations of weakly bound ligands, since *J*-couplings cannot be used in this case. The reason is the fact that the nonbound conformation significantly contributes to the averaged values of the coupling constant.

Cross-correlated dipolar relaxation can easily be explained in a simplified form. All coherences between nuclear spins, which finally give rise to NMR signals, decay with a certain rate and eventually disappear. In dipolar relaxation, the relaxation of a spin is mediated by the fluctuating magnetic field caused by adjacent spins. Cross-correlated dipolar relaxation indicates that the dipolar coupled spin pair is not isolated, but experiences fluctuating magnetic fields from other spin pairs which influence its relaxation rate. Since both dipolar interactions are modulated by the same overall tumbling they display a very similar time-dependence, giving rise to cross-correlation effects. This particular dipolar interaction between these magnetic moments depends on the angle between them [45]:

$$\Gamma_{\text{dipole-dipole}} = \frac{2}{5}\frac{\gamma_i\gamma_j}{r_{ij}^3}\frac{\gamma_k\gamma_l}{r_{kl}^3}\hbar^2\frac{1}{2}(3\cos^2\theta - 1)\tau_c \qquad (11)$$

θ is the angle between the internuclear vectors *i-j* and *k-l* (Fig. 16.4). r denotes bond distances which are well known, and τ_c denotes the rotational correlation time, which can be measured independently. All other factors are constants. Note that the equation is valid only for isotropic reorientation. The angle θ between bond vectors is then the sole unknown

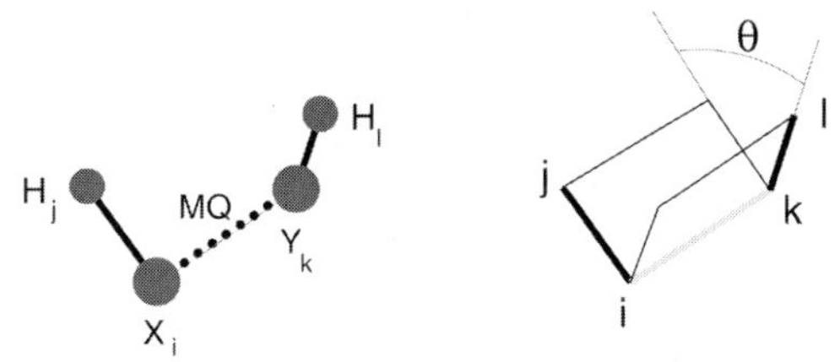

Fig. 16.4 Schematic representation of a pair of bond vectors. The relaxation rate of the multiple quantum coherence (MQ) is dependent on the angle θ measured between the bond vectors.

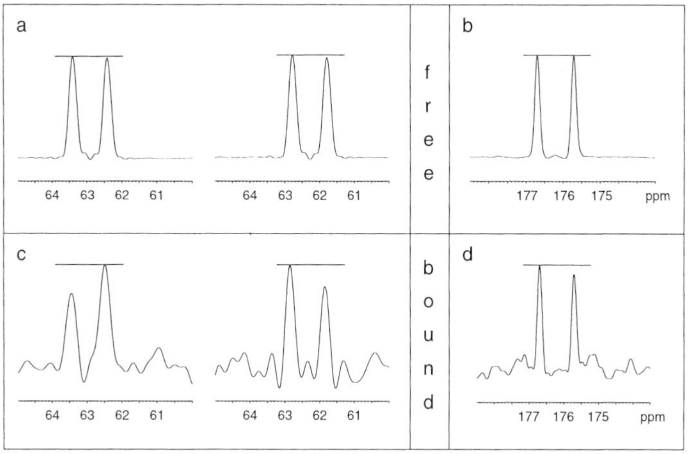

Fig. 16.5 Cross sections through cross peaks taken from the HN(CO)CA derived experiments at the ^1H_N frequency. The cross peaks in the dipole-dipole CCR experiment (**a**, **c**) are observed at the $^{13}C\alpha$ frequency. Two spectra with different parameters are recorded [52]. The cross peaks are splitted by $^1J_{C\alpha H\alpha}$. The cross peaks in the two spectra are shifted relatively by $^1J_{NHH\alpha}$. The cross peaks in the dipole-CSA CCR experiment (**b**, **d**) are observed at the $^{13}C'$ frequency. They are splitted by the $^1J_{C\alpha H\alpha}$ coupling. Data are shown for the free tetrapeptide (**a**, **b**) and for a mixture of 0.5 mM tetrapeptide and 40 μM STAT-6 (**c**, **d**) [51]. The horizontal line indicates the intensity of the highest peak. In absence of CCR, each pair of peaks would have equal intensity.

in the equation, and can be readily and precisely determined by measurement of the CCR rate. It should be emphasized that θ is measured *directly,* without the need of experimental calibration, in contrast to, e.g., the Karplus curve for scalar J-coupling constants.

Cross-correlated dipolar relaxation can conveniently be measured using triple-resonance experiments. These utilize the ^1H, ^{13}C, and ^{15}N nuclei in isotopically enriched biomolecules. In triple-resonance NMR experiments designed for that purpose, e.g. a nondecoupled heteronuclear multiple quantum experiment, each signal is split by the (large) heteronuclear one-bond coupling. The individual lines relax with different rates because of the phenomenon of CCR (differential line broadening, Fig. 16.5). Thus, the relaxation rate can be directly extracted from the intensity of individual lines. The equations for the CCR rates given above are approximations for rigid isotropic tumbling macromolecules.

If one likes to include molecular motional effects of the ligand when bound to the macromolecule, one can introduce an order parameter using the model-free approach, as has been applied for the interpretation of CCR rates by Carlomagno et al. [46].

Cross-correlated dipolar relaxation can be measured between a variety of nuclei. The measurement requires two central nuclear spins, each of which is directly attached to a remote nuclear spin (Fig. 16.4). The central spin and its attached remote spin must be connected via a large scalar coupling, and the remote spin must be the primary source of dipolar relaxation for the central spin. The two central spins do not need to be scalar coupled, although the necessity to create multiple quantum coherence between them requires them to be close together in a scalar or dipolar coupled network. In practice, the central spins will be heteroatoms (e.g. ^{13}C or ^{15}N in isotopically enriched biomolecules), and the remote spins will be their directly attached protons.

A practical example is the peptide backbone where the amide ^{15}N nuclei and the preceding $^{13}C\alpha$ nuclei are the central nuclei and their directly attached protons are the remote nuclei [47]. This choice of atoms allows the measurement of the angle θ between the H_N-N bond vector and the $H\alpha$-$C\alpha$ bond vector of the preceding residue in a peptide. Assuming the planarity of the *trans* peptide bond between the two amino acid residues, the angle θ depends only on the protein backbone angle ψ. Measurement of θ is thus an indirect measurement of ψ, which is otherwise challenging to determine using conventional NMR methodology.

Apart from the relaxation mechanism described here, other mechanisms such as relaxation involving cross-correlation between dipole-dipole coupling and chemical shift anisotropy (CSA) can also provide structural information [48, 49]. The expression for this relaxation rate in case of axial symmetric CSA tensors is

$$\Gamma_{\text{dipole-CSA}} = \frac{4}{15}\varpi_j \frac{\gamma_k \gamma_l}{r_{kl}^3} \hbar \vec{\sigma} \frac{1}{2}(3\cos^2\theta - 1)\tau_c \tag{12}$$

in which $\vec{\sigma}$ denotes the difference $\sigma_{||} - \sigma_{\perp}$. This relaxation mechanism can also be explored with the aim of extracting structural information.

In the case of a ligand that is weakly bound to a macromolecule in the fast exchange regime, the CCR rates are averaged with the population of free and bound conformation in analogy to the previously described trNOE (Sect. 16.3) [50, 51].

$$\Gamma = k(a)\frac{1}{2}(3\cos^2\theta^{\text{bound}} - 1)\tau_c^{\text{bound}} + k(1-a)\frac{1}{2}(3\cos^2\theta^{\text{free}} - 1)\tau_c^{\text{free}} \tag{13}$$

in which the constant k is different for the two types of CCR mechanisms (see equations 11 and 12) and a is the fraction bound. In the case of weak binding and if the ligand is sufficiently small, the last term in the equation can be neglected, and the CCR rate is directly related to the angle θ^{bound} (and $a\tau_c^{\text{bound}}$).

That the second term indeed can be neglected is illustrated in the following example. The CCR rates were measured in a small peptide weakly bound to STAT-6, a 100 kDa protein. The signals that are split by the heteronuclear ^1J coupling are of equal intensity for the free peptide. After addition of protein, the signal intensities of the individual peaks change significantly. They are all significantly less intense, which is caused by an increased rate of auto relaxation. Additionally, they relax differently, which is caused by CCR. The CCR rate can be extracted [52, 53] by measuring the signal intensities of the peaks. In this case, when the signals in Fig. 16.5 are labelled *a–f* (from the left to the right)

$$\Gamma_{\text{dipole-dipole}} = \frac{1}{4t} \ln\left(\frac{I_a\, I_d}{I_b\, I_c}\right) \tag{14}$$

$$\Gamma_{\text{dipole-CSA}} = \frac{1}{2t} \ln\left(\frac{I_e}{I_f}\right) \tag{15}$$

Where t is the total of the constant time delays in the CCR experiment and I represents the intensities of the peaks.

Alternatively CCR rates can be measured from a so-called cross and reference experiment [54].

In this example, the projection angle involving the H_N-N bond and Cα-Hα bond is measured. Because the peptide bond is fixed, the relation between CCR rate and the peptide backbone can be directly described by converting the angle θ *to* ψ. To describe the angular dependance for the dipole-dipole CCR we use the function $\Gamma = k a \tau_c f(\psi)$ with

$$f_{\text{dipole-dipole}}(\psi) = \frac{1}{2}[3\cos^2\theta_1 - 1] \tag{16}$$

$$\cos\theta_1 = 0.1628 + 0.8188\cos(\psi - 120^\circ) \tag{17}$$

and for the dipole-CSA CCR rate we use the function

$$f_{\text{dipole-CAS}}(\psi) = \frac{1}{2}[3\cos^2\theta_{2x} - 1]\sigma_x + \frac{1}{2}[3\cos^2\theta_{2y} - 1]\sigma_y + \frac{1}{2}[3\cos\theta_{2z} - 1]\sigma_z \tag{18}$$

$$\cos\theta_{2x} = -0.3095 + 0.3531\cos(\psi - 120^\circ) \tag{19}$$

$$\cos\theta_{2y} = -0.1250 - 0.8740\cos(\psi - 120^\circ) \tag{20}$$

$$\cos\theta_{2z} = -0.9426\cos(\psi - 120^\circ) \tag{19}$$

in which the angle θ denotes the projection angle between the bond vectors HN and CH (θ_1) or the projection angle between the bond vector H_N and the Cα CSA tensor (θ_2) [49]. The CSA tensor may be derived from solid-state experiments or calculated. In case of data redundancy, the CSA tensor may also be refined explicitly.

This results in typical curves as depicted in Fig. 16.6.

From the CCR rate one can easily discriminate between an α-helical conformation ($\psi \approx -60^\circ$) and a β-sheet ($\psi \approx 120^\circ$) conformation, as indicated in Fig. 16.6. The curves in the figure are scaled by $a\tau_c$. A more precise determination of the torsion angle is also possible in the case that both CCR rates are measured. By combining the two rates, the term $a\tau_c$ as well as the angle ψ can be determined:

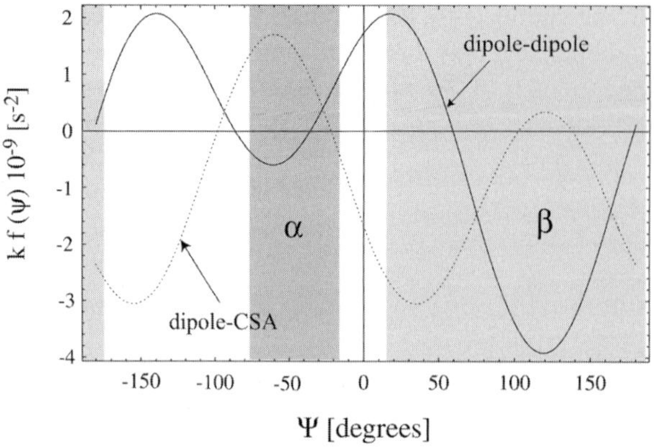

Fig. 16.6 Theoretical curves of the dipole-dipole CCR rate and the dipole-CSA CCR rate as a function of the peptide backbone torsion angle ψ. The sterically allowed regions are indicated by a gray background.

$$\frac{\Gamma_{\text{dipole-dipole}}}{\Gamma_{\text{dipole-CSA}}} = \frac{k_1\, a\tau_c^{\text{bound}}\, f_1(\psi^{\text{bound}})}{k_2\, a\tau_c^{\text{bound}}\, f_2(\psi^{\text{bound}})} \tag{22}$$

the subscripts 1 and 2 denote the different constants and trigonometric functions that correspond with dipole-dipole and dipole-CSA CCR, respectively. When the value for ψ is derived, automatically the scaling of the curves depicted in Fig. 16.6 is known and can be applied for the determination of other torsion angles in the molecule. If more torsion angles are determined using transferred CCR, the value of $a\tau_c$ can be refined against all torsion angles. It is obvious from Fig. 16.6 that for one value of the CCR rate more than one solution is possible. By combining two rates, the number of possible solutions is reduced, especially if one also considers energetically unfavorable regions in the plot. Perhaps the most appropriate structure determination protocol refines the experimental CCR rates together with trNOE-derived distances against calculated CCR rates during structural refinement, as has been done previously for *J*-couplings [55].

The transferred CCR finds applications for all ligands for which CCR effects can be measured. In most cases this would require specific or uniform labeling of the ligand with NMR sensitive nuclei. A second example is given for the measurement of the sugar conformation of a weakly bound nucleotide. The nucleotide has been labeled with ^{13}C. The sugar conformation is best described by the so-called pseudorotation phase angle P and pucker amplitude ψ_m. Analysis of these two parameters as functions of the endocyclic torsion angles reveals that the measurement of only two projection angles by CCR would be enough to determine the conformation of the ribose ring in the bound state. Thus, the measurement of $\Gamma_{H1'C1'C2'H2'}$ and $\Gamma_{H3'C3'C4'H4'}$ would reveal the ribose conformation. This is analogous to the observation that the combination of the ^3J coupling constants $J_{H1'H2'}$ and $J_{H3'H4'}$ describe the sugar conformation [56].

The relation between the CCR rates and the pseudorotation phase angle can be calculated using the general formula for $\Gamma_{\text{dipole-dipole}}$ by converting the values of the projection angles into P and ψ_m [47] according to

$$\theta_{1'2'} = 121.4^\circ + 1.03\psi_m \cos(P - 144^\circ) \tag{23}$$

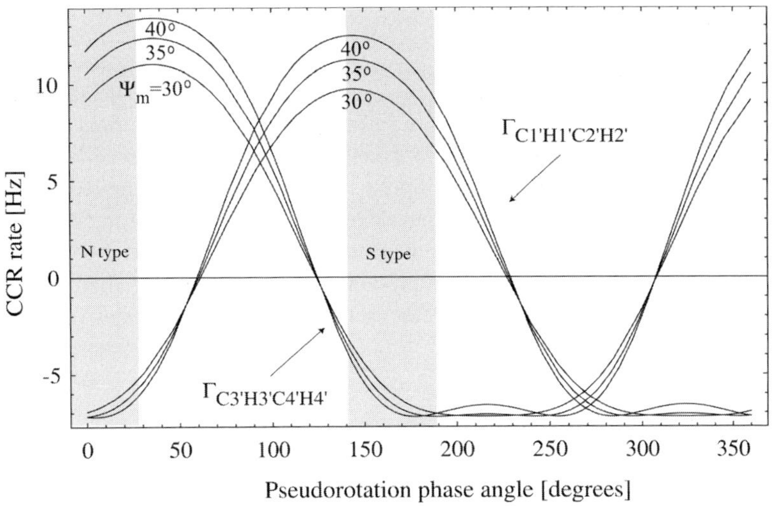

Fig. 16.7 Theoretical curves of the CCR rate as a function of the ribose pseudorotation phase angle P for three different pucker amplitudes ψ_m=30, 35 and 40°. A rotational correlation time of 1.5 ns and a C-H distance of 1.07 Å has been used for the calculation.

$$\theta_{3'4'} = -124.0^\circ + 1.09\psi_m \cos(P + 144^\circ) \tag{24}$$

θ is the projection angle. The function is shown in Fig. 16.7 [54]. The two pucker modes N-type (~C3′-endo) and S-type (~C2′-endo) can be discriminated from the sign of these CCR rates.

From the examples presented above it is clear that trCCR can be measured for many ligands which allow to determine the CCR rates. However, methods for isotope labeling of the ligand need to be available. The other limitation of the method is the fact that the exchange between free and bound state must be fast on the time scale of the NMR experiment. Thus, considering that the CCR is measured during a period of 10–20 ms, the formed complex should dissociate much faster to make the CCR of the bound state measurable on the resonances of the free state. The restriction on k_{off} is therefore more critical than in the trNOE experiment. In the latter experiment, the mixing time is usually 10-fold longer.

16.5 Transferred Residual Dipolar Couplings (trRDC)

At high magnetic fields, molecules with anisotropic magnetic susceptibility experience a preferred orientation relative to the magnetic field. For these cases residual dipolar coupling can be measured, which provides information on the orientation of internuclear vectors relative to the alignment frame [57, 58] (see also Chapt. 8). This method is interesting since the structural information obtained provides long-range structural restraints. The residual dipolar coupling (RDC, Δ^1J) can be expressed as

$$\Delta^1 J_{ij} = -\frac{\gamma_i \gamma_j h}{2\pi^2 r_{ij}^3} x \frac{1}{2} (3\cos^2\theta - 1) \tag{25}$$

θ is the angle between the *ij* vector and the external magnetic field. In analogy with NOE and CCR, internal motion can be taken into account by inserting the order parameter into the equation. The residual dipolar coupling is also a function of the fraction, x, of the protein that is aligned with the magnetic field. In contrast to NOE and CCR, the residual dipolar coupling (RDC) is not dependent on the rotational correlation time τ_c. Provided that the fraction of aligned protein is large enough, the method can be applied for the weakly bound ligands in the fast exchange regime. The residual dipolar coupling is

$$\Delta^1 J = a\Delta^1 J^{\text{bound}} + (1-a)\Delta^1 J^{\text{free}} \tag{26}$$

The first term in the equation should be significantly larger than the second term if the macromolecule is partly aligned ($\Delta J^{\text{bound}} \gg \Delta J^{\text{free}}$), because the fraction x in Eq. (25) increases because of the much larger anisotropic magnetic susceptibility of the complex. It should be noted that for most cases of weakly aligned molecules, the effect might be too small, at commonly used protein concentrations (ca. 50 μM), for successfully applying the method of transferred RDC. Similarly to the case of scalar couplings, the observed values are linearly averaged over bound and free state, leading to the same limitations. Since the experiments are usually carried out in an excess of free ligand, the RDC for the bound state should be significantly larger. If in the favorite cases the method can be applied to weakly bound ligands, it has the following two features. First, local angular restraints can be derived that could be used to complement other experimental data like trNOE and trCCR. Second, the orientation of the ligand relative to the alignment tensor can be obtained. If the protein can be uniformly labeled with ^{15}N, the distribution of RDCs gives the appearance of a powder spectrum from which the values corresponding to the principal axes can be deduced [59]. When the orientation of ligand and macromolecule relative to the alignment tensor are combined, the bound conformation of the ligand can be docked onto the macromolecule.

This concept has been worked out for a peptide weakly bound to oriented rhodopsin [60]. In this case rhodopsin was anchored in a membrane bilayer. From the ^{31}P chemical shift of 23 ppm it was concluded that the bicelles are almost fully aligned with the membrane-normal parallel to the field. This alignment would result in a dipolar coupling in the order of 20 kHz. The high alignment is likely caused by the presence of transmembrane helices of the protein and many double bonds of the fatty acids in the membrane. Another way to increase the alignment of the macromolecule with the magnetic field is the use of a paramagnetic center [57].

Pioneering studies have also been published for weakly aligned protein complexes [61, 62]. Note that the ratio of bound ligand to free ligand must be optimized in order to achieve the desired accuracy when extrapolating the RDCs of the bound state. Since the protein does not interfere with the measurement of RDCs of the ligand, it is recommended to use a high concentration of the former.

16.6 Summary

Since the recent reviews on transferred NOE in 1993 and 1994 [1–3], much progress has been made, and new techniques have been added to the area of transferred methods. Drawbacks of the trNOE-based structure elucidation have been greatly alleviated by eliminating spin diffusion by experimental means and a more rigorous treatment of spin diffusion and exchange on a theoretical basis. These spin diffusion effects still form a major problem which is not yet solved. Future research may focus on the treatment of internal dynamics, especially in cooperation with other transferred methods, i.e. transferred CCR and transferred RDC. Transferred CCR and transferred RDC open new avenues for the study of weakly binding ligands, and it will be especially interesting to see how these different transferred methods will contribute synergistically to our understanding of structure and dynamics of weakly bound ligands. The problem of protein-mediated spin diffusion is not present in the latter two techniques. In addition to the methods described here, the development of novel transferred methods can also be foreseen. Especially, the further exploration of paramagnetic effects seems to be very promising in this respect. The new methods can potentially provide complementary information to the NOE.

Screening by NMR, which recently became increasingly important in drug discovery, also benefits from transferred methods. For NMR screening purposes, observation of the ligand signal is superior to direct detection of protein signals for the identification of binding molecules. In this field transferred methods are most promising, and one can expect rapid progress in the development of transferred NMR screening methods.

16.7 References

1 F. Ni, *Progr. NMR Spectrosc.* **1994**, 26, 517–606.

2 L.Y. Lian, I.L. Barsukov, M.J. Sutcliffe, K.-H. Sze, G. C. K. Roberts, *Methods Enzymol.* **1994**, 239, 657–700.

3 A. P. Campbell, B. D. Sykes, *Annu. Rev. Biophys. Biomol. Struct.* **1993**, 22, 99–122.

4 S. Macura, R. R. Ernst, *Mol. Phys.* **1980**, 41, 95–117.

5 D. Neuhaus, M. Williamson, *The Nuclear Overhauser Effect in Structural and Conformational Analysis*, **1989**, VCH, New York.

6 I. Solomon, *Phys. Rev.* **1955**, 99, 559.

7 G. Lipari, A. Szabo, *J. Am. Chem. Soc.* **1982**, 104, 4546–4559.

8 G. Lipari, A. Szabo, *J. Am. Chem. Soc.* **1982**, 104, 4559–4570.

9 J. P. Albrand, B. Birdsall, J. Feeney, C.K. Roberts, A.S.V. Burgen, *Int. J. Biol. Macromol.* **1979**, 1, 37–41.

10 G.M. Clore, A.M. Gronenborn, *J. Magn. Reson.* **1982**, 48, 402–417.

11 G.M. Glore, A.M. Gronenborn, *J. Magn. Reson.* **1983**, 53, 423–442.

12 J. Jeener, B.H. Meier, P. Bachmann, R.R. Ernst, *J. Chem. Phys.* **1979**, 71, 4546.

13 R.E. London, M.E. Perlman, D.G. Davis, *J. Magn. Reson.* **1992**, 97, 79–98.

14 H.N.B. Moseley, E.V. Curto, N.R. Krishna, *J. Magn. Res. B* **1995**, 108, 243–261.

15 G.M. Lippens, C. Cerf, K. Hallenga, *J. Magn. Reson.* **1992**, 99, 268–281.

16 S.R. Arepalli, C.P. Glaudemans, G.D. Jr. Daves, P. Kovac, A. Bax, *J. Magn. Reson. B.* **1995**, 106, 195–198.

17 S.J.F. Vincent, C. Zwahlen, G. Bodenhausen, *J. Biomol. NMR* **1996**, 7, 169–172.

18 S.J.F. Vincent, C. Zwahlen, C.B. Post, J.W. Burgner, G. Bodenhausen, *Proc. Natl. Acad. Sci. USA* **1997**, 94, 4383–4388.

19 W. Massefski Jr., A.G. Redfield, *J. Magn. Reson.* **1988**, 78, 150–155.

20 N. Jurani, Z. Zolnai, S. Macura, *J. Biomol. NMR* **1997**, 9, 317–322.

21 T. Scherf, J. Anglister, *Biophys. J.* **1993**, 64(3), 754–761.

22 A. Heerschap, J. A. L. I. Walters, J. R. Mellema, C. W. Hilbers, *Biochemistry* **1986**, 25(9), 2707–2713.

23 N. Murali, G. K. Jalori, S. B. Landy, B. D. Rao, *Biochemistry* **1993**, 32, 12941–12948.

24 C. J. Dinsmore, M. J. Bogusky, J. C. Culberson, J. M. Bergman, C. F. Homnick, C. B. Zartman, S. D. Mosser, M. D. Schaber, R. G. Robinson, K. S. Koblan, H. E. Huber, S. L. Graham, G. D. Hartman, J. R. Huff, T. M. Williams, *J. Am. Chem. Soc.* **2001**, 123(9), 2107–2108.

25 V. S. Pashkov, T. A. Balashova, V. Zhemaeva, N. N. Sikilinda, M. A. Kutuzov, N. G. Abdulaev, A. S. Arseniev, *FEBS Lett.* **1996**, 381, 119–122.

26 H. Kusunoki, K. Wakamatsu, K. Sato, T. Miyazawa, T. Kohno, *Biochemistry* **1998**, 37, 4782–4790.

27 K. Scheffler, J.–R. Brisson, R. Weisemann, J. L. Magnani, W. T. Wong, B. Ernst, T. Peters, *J. Biomol. NMR* **1997**, 9, 423–436.

28 V. S. Pashkov, T. A. Balashova, L. V. Zhemaeva, N. N. Sikilinda, M. A. Kutuzov, N. G. Abdulaev, A. S. Arseniev, *FEBS Lett.* **1995**, 381, 119–122.

29 R. E. London, *J. Magn. Reson.* **1999**, 141, 301–311.

30 W. Jahnke, L. B. Perez, C. G. Paris, A. Strauss, G. Fendrich, C. M. Nalin, *J. Am. Chem. Soc.* **2000**, 122, 7394–7395.

31 J. Fejzo, C. A. Lepre, J. W. Peng, G. W. Bemis, Ajay, M. A. Murcko, J. A. Moore, *Chem. Biol.* **1999**, 6, 755–769.

32 B. Meyer, T. Weimar, T. Peters, *Eur. J. Biochem.* **1997**, 246(3), 705–709.

33 I. L. Barsukov, L. Y. Lian, J. Ellis, K. H. Sze, W. V. Shaw, G. C. Roberts, *J. Mol. Biol.* **1996**, 262(4), 543–558.

34 E. W. Sayers, J. L. Weaver, J. H. Prestegard, *J. Biomol. NMR* **1998**, 12, 209–222.

35 C. Dalvit, G. Fogliatto, A. Stewart, M. Veronesi, B. Stockman, *J. Biomol. NMR* **2001**, 21, 349–359.

36 M. Karplus, *J. Chem. Phys.* **1959**, *30*, 11–15.

37 V. F. Bystrov, *Prog. NMR Spectrosc.* **1976**, *10*, 41–81.

38 K. Wüthrich, *NMR of Proteins and Nucleic Acids* **1986**, Wiley, New York.

39 B. Reif, M. Hennig, C. Griesinger *Science* **1997**, *276*,1230–1233.

40 C. Dalvit, G. Bodenhausen *J. Am. Chem. Soc.* **1988**, 110, 7924–2926.

41 C. Dalvit, G. Bodenhausen *Adv. Magn. Res.* **1990**, 14, 1–33.

42 X. Feng, Y. K. Lee, S. Sandström, M. Eden, H. Maisel, A. Sebald, M. H. Levitt, *Chem. Phys. Lett.* **1996**, *257*, 314–320.

43 X. Feng, P. J. E. Verdegem, Y. K. Lee, D. Sandström, M. Eden, P. Bovee-Geurts, W. J. de Grip, J. Lugtenburg, H. . M. de Groot, M. H. Levitt, *J. Am. Chem. Soc.* **1997**, *119*, 6853–6857.

44 K. Schmidt-Rohr, *J. Am. Chem. Soc.* **1996**, *118*, 7601–7603.

45 B. Reif, A. Diener, M. Henning, M. Maurer, C. Griesinger, *J. Magn. Reson.* **2000**, 143, 45–68.

46 T. Carlomagno, M. J. J. Blommers, J. Meiler, B. Cuenoud, C. Griesinger, *J. Am. Chem. Soc.* **2001**, 123, 7364–7370.

47 J. L. Markley, A. Bax, Y. Arata, C. W. Hilbers, R. Kaptein, B. D. Sykes, P. E. Wright, K. Wüthrich, *J. Biomol. NMR* **1998**, 12, 1–23.

48 M. Tessari, H. Vis, R. Boelens, R. Kaptein, G. Vuister, *J. Am. Chem. Soc.* **1997**, 119, 8985–8990.

49 D. Yang, R. Konrat, L. E. Kay, *J. Am. Chem. Soc.* **1997**, 119, 11938–11940.

50 T. Carlomagno, I. C. Felli, M. Czech, R. Fischer, M. Sprinzl, C. Griesinger, *J. Am. Chem. Soc.* **1999**, 121, 1945–1948.

51 M. J. J. Blommers, W. Stark, C. E. Jones, D. Head, C. E. Owen, W. Jahnke, *J. Am. Chem. Soc.* **1999**, 1949–1953.

52 D. Yang, K. H. Gardner, L. E. Kay, *J. Biomol. NMR* **1998**, 11, 213–220.

53 D. Yang, L. E. Kay, *J. Am. Chem. Soc.* **1998**, 120, 9880–9887.

54 I. C. Felli, C. Richter, C. Griesinger, H. Schwalbe, *J. Am. Chem. Soc.* **1999**, 121, 1956–1957.

55 M. Eberstadt, D. F. Mierke, M. Köck, H. Kessler, *Helv. Chim. Act.* **1992**, 75, 2583–2592.

56 S. S. Wijmenga, M. M. W. Mooren, C. W. Hilbers in: NMR of macromolecules, A practical approach, Ed. G. C. K. Roberts, Oxford University Press, 1993.

57 J. R. Tolman, J. M. Flanagan, M. A. Kennedy, J. H. Prestegard, *Proc. Natl. Acad. Sci. USA* **1995**, 92, 9297–9283.

58 N. Tjandra, A. Bax, *Science* **1997**, 278, 1111–1114.

59 G. M. Clore, A. M. Gronenborn, A. Bax, *J. Magn. Reson.* **1998**, 133, 216–221.

60 B. W. Koenig, D. C. Mitchell, S. König, S. Grzesiek, B. J. Litman, A. Bax, *J. Biomol. NMR* **2000**, 16, 121–125.

61 P. J. Bolon, H. M. Al-Hashimi, J. H. Prestegard, *J. Mol. Biol.* **1999**, 293, 107–115.

62 H. Shimizu, A. Donohue-Rolfe, S. Homans *J. Am. Chem. Soc.* **1999**, 121, 5815–5816.

17
Isotope Filter and Editing Techniques

Gerd Gemmecker

17.1
General Concept

Isotope filtering/editing NMR techniques make use of differential isotopic labeling to simplify spectra and thus more easily extract information from complex systems. In the case of biomolecular NMR these will generally be intermolecular complexes between one biomacromolecule (for example, a protein) and a second species: either another protein, a nucleic acid, or a ligand (generally a small organic molecule).

The basic idea of isotope filtering/editing techniques is to selectively observe the subspectra of the labeled or unlabeled components only. A prerequisite for the application is of course the different isotopic composition of the different components. In an ideal case, we will have one compound which is completely labeled with the NMR active isotope(s) of one (or several) elements, and the other compound completely devoid of these isotopes (see Fig. 17.1).

In the field of biomolecular NMR and drug research, there are only a limited number of pairs of NMR active and inactive stable isotopes as would be required for differential labeling techniques (Tab. 17.1). For the case of carbon, the roles are clear: ^{13}C acts as the NMR active isotope, while ^{12}C is the corresponding inactive isotope. In the cases of hydrogen and nitrogen there are two NMR active isotopes each: one spin-½ isotope with reasonably high γ/resonance frequency (^{1}H, ^{15}N) and one quadrupolar isotope with lower resonance frequency (^{2}H, ^{14}N) and small coupling constants to other nuclei. These latter ones can – in good approximation – be treated as "NMR inactive", that is, they are not used for detection and there is generally no need for ^{2}H- or ^{14}N-decoupling when detecting other isotopes.

The situation is again different for ^{19}F and ^{31}P, which are the only naturally occurring isotopes of these two elements, both being spin-½ nuclei with a high gyromagnetic ratio γ. In addition, these elements occur only rarely in biomacromolecules (with the exception of ^{31}P in nuclei acids!) or organic ligands. Therefore, they can be employed as NMR probes in a very straightforward way: by just acquiring a ^{19}F or ^{31}P 1D spectrum. In the following, these isotopes will not be considered any further.

In general, the concentrations of biomolecular NMR samples are in the 0.1–1 mM range, so that the direct detection of isotope-edited spectra of low γ nuclei (^{13}C, ^{15}N) would suffer from very low sensitivity. In addition, the observation of the unlabeled moi-

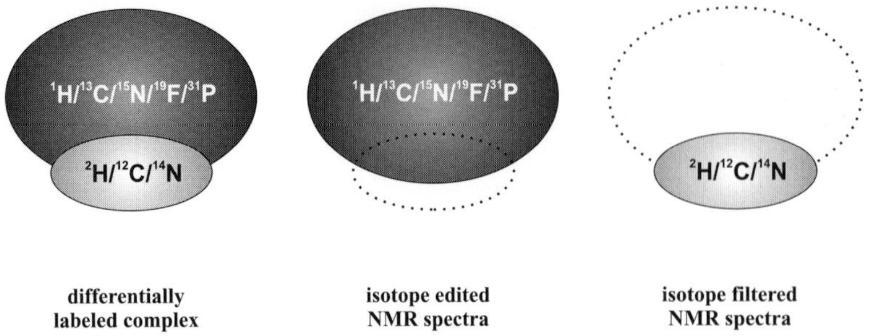

Fig. 17.1 The concept of isotope filtering and isotope editing, in an idealized way (for a more realistic picture, see Fig. 17.2!). In a complex consisting of two (or more) differentially labeled compounds (left), the NMR spectra could be considerably simplified by either selecting signals only from the moiety enriched with NMR active isotopes (isotope editing, center), or by selectively observing signals from the unlabeled compound via isotope filtering (right).

Tab. 17.1 NMR properties of stable isotope pairs in biomolecular NMR

NMR active isotope	*Natural abundance*	*Gyromagnetic ratio γ (^1H = 100)*	*Corresponding "NMR inactive" isotope*	*Natural abundance*	*Gyromagnetic ratio γ (^1H=100)*
^{1}H	99.985%	100.00	^{2}H *	0.015%	15.35
^{13}C	1.11%	25.14	^{12}C	–	–
^{15}N	0.37%	–10.13	^{14}N *	99.63%	7.22
^{19}F	100.0%	94.08	–	–	–
^{31}P	100.0%	40.48	–	–	–

* quadrupolar nuclei

ety would be completely impossible by direct detection (^{12}C: no magnetic moment; ^{2}H, ^{14}N: low γ quadrupolar nuclei).

Therefore all biomolecular NMR spectra are usually ^{1}H-detected, and the following definitions will be used throughout this chapter, although their use in the literature is not always consistent (Fig. 17.2):

- *Isotope editing* means selecting protons which are directly bound to an NMR-active heteronucleus (^{1}H-^{13}C, ^{1}H-^{15}N).
- *Isotope filtering* suppresses these signals, leaving only the signals of protons bound to the NMR inactive carbon and nitrogen isotopes (^{1}H–^{12}C, ^{1}H–^{14}N), as well as protons not bound to carbon or nitrogen at all (mainly –SH, –OH).

These methods generally rely on $^1J_{H,X}$ couplings (X=^{13}C, ^{15}N), since these are (i) relatively large (requiring only short coupling evolution delays) and (ii) fairly uniform (allowing exact tuning of the delays). In the case of ^{31}P, isotope editing is still possible via long-range ^{1}H, ^{31}P couplings (but less efficient because of their small size), while isotope filtering is seriously impeded by their nonuniformity.

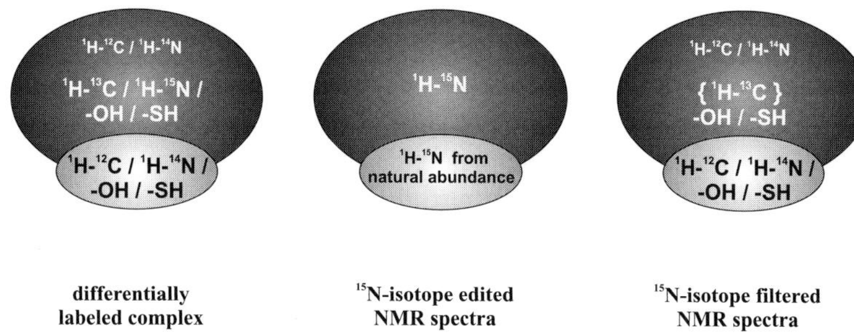

Fig. 17.2 A realistic picture of the possibilities and limitations of isotope filtering and isotope editing, shown using the example of ^{15}N. The selection of the isotope-labeled moiety is not significantly perturbed by the low level of ^{15}N natural abundance (center). However, in the ^{15}N-filtered case (right), only the ^{15}N-bound protons of the labeled moiety are suppressed, while other protons will interfere with the observation of signals from the unlabeled component (right). If the labeled component is also ^{13}C enriched, then an additional ^{13}C filter can also remove its ^{1}H-^{13}C signals (indicated by the brackets in the picture on the right).

The general detection of ^{1}H signals suggests another, easy way for removing signals from specific components: if a molecule can be prepared in perdeuterated form, then there are no ^{1}H signals left to suppress, and only the non-perdeuterated components will be observed. This approach requires only standard pulse sequences and will not be discussed here any further; examples with either perdeuterated protein [1–3] or ligand [4] can be found in the literature.

Of course, isotope filtering is not restricted to such binary systems, but can also be applied to multicomponent systems such as multiprotein complexes, or one or more ligands bound to proteins or protein complexes. It is also conceivable to construct single molecules from sections with different isotopic labeling in order to selectively observe one part by isotopic filtering. However, in these cases a specific synthetic approach has to be designed to allow for efficient incorporation of the isotope labels into the appropriate parts only (for example, by fragment condensation or inteins see also Chapt. 1) [5].

17.2 Sample Requirements

17.2.1 Complex Size and Concentration

In the isotope edited / filtered spectra of a protein-ligand complex, the species actually observed is generally the complex itself. This is an important difference from transferred NOE or saturation difference techniques, where the existence of an equilibrium between free and bound species – and a certain rate of exchange between them – is essential (Chapts. 13 and 16). The general conditions for isotope filtering/editing are therefore identical to those required for standard protein NMR: sample concentrations are usually limited by availability and solubility of the components to the order of 1 mM. Considerably lower concentrations will reduce the sensitivity of the experiments to unacceptable levels, although concentrations in the 100 μM range are already made accessible by some of the newest advances in NMR technology, such as high-field magnets and cryoprobes [6].

Generally the sensitivity of the isotope filtered/edited version of an NMR experiment will be comparable to that of the corresponding standard experiment. However, some reduction in signal intensity will occur caused by the additional pulses (due to pulse imperfections and B_1 inhomogeneity) and delays (due to relaxation) of the filter elements. These losses can become significant in the case of large molecular weight complexes.

Size restrictions are again similar to that encountered in protein NMR: up to ca. 20–30 kDa complex size, line widths will usually be acceptable to allow detection of the ^{1}H resonances of the labeled or unlabeled component, including NOESY transfer steps. Beyond that, deuteration of the protein component becomes essential to reduce excessive line broadening. For the observation of the (^{15}N-) labeled component (i.e., isotope editing), the use of TROSY techniques will further extend the size limit (Chapt. 10); however, this approach does not work for unlabeled components.

When combining isotope filtering/editing with coherence transfer steps to multidimensional experiments, then further size restrictions apply. For example, isotope edited / filtered ^{1}H TOCSY or COSY experiments are generally limited to systems of < 10 kDa, because of their sensitivity to T_2 relaxation. In larger systems, heteronuclear correlation spectroscopy can be used for the correspondingly labeled component, while structural information about both the labeled and unlabeled moiety can be extracted from isotope edited/ filtered NOESY spectra, respectively.

17.2.2
Complex Affinity

From the general concept of isotope filtering/editing (see Sect. 17.1), it becomes clear that the existence of additional species other than the complex itself will lead to a complication of the obtained spectra and should be avoided. Specifically, to prevent additional signals from significant concentrations of the free components, the systems under study should normally be in the tight binding regime (requiring sub-μM binding affinities for concentrations in the order of 1 mM, cf. Fig. 17.3), exhibiting slow exchange kinetics. These conditions differ clearly from the intermediate affinity situation preferred for transferred NOEs (cf. Chapt. 16).

In addition, the complex partners should be present in near-stochiometric ratio. In any case, the component to be studied must not be in excess: if the protein component is of interest, the numerous signals from any additional free protein will heavily interfere with the already complicated spectra of the complexed protein. When observing the ligand, the very sharp and intense lines of any free ligand will most probably obscure the much broader lines of the bound ligand.

However, it is possible to add the non-observed component (usually the ligand) in excess (solubility permitting) in order to cope with weak (i.e., μM–mM) binding affinities. Naturally, this requires an even higher performance of the isotope filtering technique used to suppress the (excess) component.

The preparation of these complexes is not always trivial, especially in the case of low solubility of one of the free components (usually the ligand). The transfer of the ligand

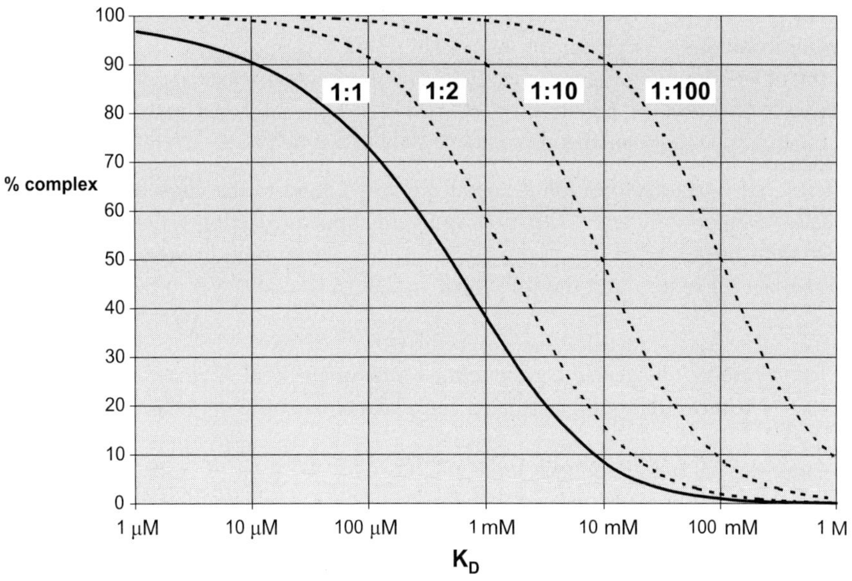

Fig. 17.3 Percentage of complex formation as a function of binding affinity K_D. The curves shown are valid for a 1:1 complex stochiometry and a 1 mM concentration of the component to be observed, with the other component either in equimolar ratio or in 2-fold, 10-fold or 100-fold excess. To achieve quantitative complexation with an equimolar complex partner, sub-μM affinities are required. However, complete complex formation is even possible for mM affinities if the other component can be added in excess.

from the solid to the complexed state – via a dissolved form of very low concentration – can take considerable time and require additional tricks, such as sonication, lyophilization or the use of small amounts of water-miscible organic solvents (for example, dimethylsulfoxide). Some practical hints on this and other aspects of isotope filtering can be found in a recent review [7].

In the case of a symmetric (protein-protein) homodimer, the preparation of molecules with differently labeled monomers is often far from trivial, and special approaches have been described in the literature [8, 9]. However, when successful, differential isotopic labeling of a symmetric homodimer, in combination with isotope editing/filtering techniques, offers a unique access to the NMR investigation of the monomer interfaces.

17.2.3 Labeling Pattern

In reality, the ideal isotope labeling pattern (one component 100% labeled, the other one 0%) is almost never fulfilled: first, a 100% isotopic enrichment will be generally very hard to reach (and very costly!) for chemically synthesized as well as for overexpressed compounds. On the other hand, even the nonenriched components will always contain NMR-active isotopes at natural abundance unless they are derived from specially isotope-depleted (and again very expensive) starting material.

For practical purposes, it suffices that at least one of the two conditions is fulfilled:

1. One component is practically *completely devoid* of a specific NMR active isotope, while the other component is enriched with this isotope to a reasonable level, or
2. One component is practically *100% enriched* with an NMR active isotope, while the other component contains this isotope only to a much lower degree.

In case (1), *isotope editing* will select all protons bound to the chosen NMR-active heteronuclear isotope, ^{15}N or ^{13}C. In first approximation (neglecting different line widths and relaxation rates for the two components), the intensity ratio between the desired signals of the labeled moiety and the unwanted signals of the other component can be described by

$$\frac{\text{(concentration of labeled component)} \times \text{(enrichment level of this component)}}{\text{(concentration of unlabeled component)} \times \text{(natural abundance)}} \tag{1}$$

Since the natural abundance levels of ^{13}C and ^{15}N are on the order of 1%, interference from residual signals of the unlabeled moiety can be neglected as long as it is not present in large excess. Furthermore, nonoptimized editing (from imperfections in the pulse sequence etc.) will just lower the overall sensitivity, but not affect the selectivity.

In case (2), *isotope filtering* will (ideally!) remove all protons bound to the chosen heteronuclear isotope(s). In the case of ^{13}C and/or ^{15}N filtering (requiring ^{13}C and/or ^{15}N labeling of one component), there will be approximately the following selectivity for the unlabeled component:

$$\frac{\text{concentration of unlabeled component}}{\text{(conc. of labeled component)} \times \text{(fraction of protons not bound to labeled heteroatoms)}} \tag{2}$$

Now the second factor of the denominator comprises

- for ^{13}C filtering: all non^{13}C-bound protons, that is, the fraction of ^{1}H-^{12}C pairs from nonquantitative ^{13}C enrichment, *plus* all protons bound to N, O and S
- for ^{15}N filtering: all non^{15}N-bound protons, that is, the fraction of ^{1}H-^{14}N pairs from nonquantitative ^{15}N enrichment, *plus* all protons bound to C, O and S
- for ^{13}C and ^{15}N filtering: all protons not bound to either ^{13}C or ^{15}N, that is, all ^{1}H-^{12}C *and* ^{1}H-^{14}N pairs from nonquantitative ^{13}C, ^{15}N enrichment, *plus* all –OH and –SH protons

In general, there will be considerable interference (so-called *cross-talk*) from signals of the labeled component, when the enrichment levels in overexpressed proteins are only in the usual 95% range. For better results, special precautions have to be taken to reach incorporation levels of 98–99% (use of highly enriched media, isotope-labeled inoculum, etc.) [7]. Even then, the residual signals can interfere with the observation of weak cross peaks of the unlabeled component, especially NOE signals, which are themselves only in the 1% range. In contrast to, e.g., the more uniform TOCSY cross peaks, NOE signal intensities usually span a range of more than two orders of magnitude, because of the r^{-6} dependence on the interproton distance *r*. Hence, NOESY spectra are quite sensitive to even weak cross-talk from the suppressed component. As an example, in a ^{13}C-edited NOESY experiment, a short-distance (ca. 2.2 Å) NOE cross peak even from the only 1.1% natural abundance ^{13}C content of the *unlabeled* moiety will show up with an intensity similar to that of a long-distance (ca. 4.5 Å) NOE of the selected >90% ^{13}C-labeled complex partner.

The situation is worse in the case of only ^{13}C- *or* ^{15}N-filtering: now all ^{14}N- and ^{15}N-bound protons will be detected (for ^{13}C filtering), and all ^{12}C- and ^{13}C-bound protons in the case of ^{15}N filtering. This will especially cause problems in the case of aromatic ^{12}C-bound protons, which will be obscured by ^{14}N-bound amide protons in a ^{13}C-filtered spectrum, and vice versa in ^{15}N-filtered spectra.

Also, imperfections of the isotope filters (e.g., from compromises in the delay settings, see Sect. 17.3.2) will lead to additional leakage of signals from the (unwanted) labeled component, thus further lowering the selectivity.

From all this it can be concluded that, in general, isotope editing is much easier to use and interpret than isotope filtering. However, it requires that the component of interest should be available in isotope-labeled form (with reasonable enrichment level). This is very often not the case, especially with many ligands that result from a multi-step laboratory synthesis. In this case isotope filtering is the only – although sometimes arduous – way to obtain structural information by NMR about the ligand part in receptor-ligand complexes.

It should be mentioned that – for isotope editing – the labeling can be limited to specific sites of interest within one component, making the labeling (especially of synthetic compounds) less costly (for examples see Sect. 17.4) [10, 11]. An interesting application of "reverse" amino acid type-specific labeling, in conjunction with isotope filtering, is, e.g., the (notoriously difficult) identification and assignment of important aromatic resonances, by introducing unlabeled phenylalanine into a uniformly ^{13}C-labeled protein [12].

17.3 NMR Techniques

17.3.1 Heteronuclear Shift Correlations

The most straightforward isotope-editing method for selecting protons bound to a heteronucleus and suppressing all others is the simple acquisition of a spectrum with an indirect heteronuclear dimension (in the literature the term *isotope editing* is often used as a synonym for these techniques). This can be accomplished by a simple 2D HMQC or HSQC shift correlation, or a more elaborate 3D technique including an additional NOESY or TOCSY step (3D X-edited NOESY/TOCSY etc.), or even 4D experiments with a second heteronuclear shift dimension [13, 14].

All standard pulse sequences for these experiments include some mechanism for selecting ^{1}H-X pairs and suppressing the other ^{1}H signals (phase cycling and/or pulsed field gradients, see below). As a further advantage, residual signal intensity of protons not bound to X (because of imperfect suppression) will not lead to a cross peak in the ^{1}H,X plane after Fourier transformation, but merely contribute to axial peaks at the spectral edge. Therefore, these techniques can be considered to be the first choice for the observation of the isotope-labeled component (see Sect.17.4).

Obviously, this approach cannot be used for selecting the nonisotope-labeled components. In the following we will consider isotope filtering/editing techniques that do not use heteronuclear chemical shift evolution.

17.3.2
Filtering and Editing Techniques

The basis for the discrimination between protons directly bound to a specific NMR-active heteronucleus, ^{1}H-X, and all others lies in the development of antiphase coherence of these protons with respect to the heteronucleus, under the influence of the scalar coupling $^1J_{\mathrm{H,X}}$:

$$^1\text{H-X:}\ \ \mathrm{I_x} \longrightarrow \mathrm{I_x}\ \cos(\pi\,^1J_{\mathrm{H,X}}t) + 2\,\mathrm{I_y}\,\mathrm{S_z}\ \sin(\pi\,^1J_{\mathrm{H,X}}t) \tag{3}$$

(using cartesian product operator formalism [15], with I standing for proton, S for the heteronucleus; starting with I_x phase coherence, which could be generated by a $90°_y$ proton pulse, for example).

After a time $\tau = {}^1/_{2J}$, the cosine term will be zero, and the sine term unity, so that inphase ^{1}H coherence is completely converted into heteronuclear antiphase coherence, $2\,I_y\,S_z$. For protons not bound to an I_x spin, nothing happens (neglecting chemical shift evolution, other couplings, relaxation etc.), and they will stay at I_x coherence:

$$^1\text{H:}\ \ \mathrm{I_x} \xrightarrow{\ \tau = {}^1/_{2J}\ } \mathrm{I_x} \tag{4}$$

Long-range $^nJ_{\mathrm{H,X}}$ couplings (with n>1) can be neglected in this discussion, since they are at least one order of magnitude smaller than the large $^1J_{\mathrm{H,X}}$ couplings, which for ^{15}N (amides) amount to ca. 90–92 Hz and for ^{13}C range from ca. 125 Hz (methyl) to over 160 Hz (aromatics).

According to Eqs. (3) and (4), after a time of $\tau = {}^1/_{2J}$ the two terms for ^{1}H-X and ^{1}H differ in two important aspects:

(a) Only the term for ^{1}H-X contains the X spin (as S_z factor) and hence will be affected by phase cycling rf pulses at the X spin frequency.

(b) The proton components of the two terms are orthogonal and can be separately affected by ^{1}H pulses with appropriate pulse phase.

Most filtering techniques rely on the application of X spin rf pulses (a) for differentiation, a scheme which can be readily understood considering the action of a 90° (X) pulse at the time $\tau = {}^1/_{2J}$:

$$^1\text{H-X: } 2\,\mathrm{I_y}\,\mathrm{S_z} \xrightarrow{\ 90^\circ_{\pm x}(\mathrm{X})\ } \mp 2\,\mathrm{I_y}\,\mathrm{S_y} \tag{5}$$

$$^1\text{H: }\mathrm{I_x} \xrightarrow{\ 90^\circ_{\pm x}(\mathrm{X})\ } \mathrm{I_x} \tag{6}$$

Thus, the phase of the X pulse (± x) directly influences the phase of the term generated from the ^{1}H,X antiphase term (Eq. (5)), but does not affect the phase of the in-phase term at all (Eq. (6)) Simple additive combination of two scans with a $90^\circ_{\pm x}$ (X) pulse will therefore lead to a selective cancellation of the signals from ^{1}H-X pairs, while a subtractive combination will suppress all signals from protons not bound to X.

This scheme is applied in the so-called X-half-filter technique (Fig. 17.4 a, c), with the only difference of an additional 90°_x (X) pulse with constant phase [16, 17]. Instead of generating heteronuclear multiple quantum coherence $2\,I_y\,S_y$, which cannot readily be detected (in case of *selecting* the ^{1}H-X pairs), one now always ends up with proton antiphase coherence, but with the same phase alternation:

$$^1\text{H-X: } 2\,I_y\,S_z \xrightarrow{90^\circ_{+x}(\text{X}),\,90^\circ_{\pm x}(\text{X})} \mp 2\,I_y\,S_z \tag{7}$$

$$^1\text{H: } I_x \xrightarrow{90^\circ_{+x}(\text{X}),\,90^\circ_{\pm x}(\text{X})} I_x \tag{8}$$

Again, addition or subtraction of two scans with the two phase combinations leads to suppression or selection of the ^{1}H-X pairs. In the case of *selecting* ^{1}H-X, another delay of length $\tau = ½_J$ is usually added to refocus this term back to detectable in-phase magnetization (Fig. 17.4 c).

An alternative way of realizing an isotope filter is shown in Fig. 17.4b, where the 90° phase difference between the two proton magnetizations is exploited [18]. A second 90°_y (^{1}H) pulse (of same phase as the excitation pulse) at the end of the period $\tau = {}^1/_{2J}$ leaves the heteronuclear antiphase magnetization of the X-bound protons unaffected, while the other protons are converted to z magnetization:

$$^1\text{H-X: } 2\,I_y\,S_z \xrightarrow{90^\circ_y(^1\text{H})} 2\,I_y\,S_z \tag{9}$$

$$^1\text{H: } I_x \xrightarrow{90^\circ_y(^1\text{H})} -I_z \tag{10}$$

In the pulse sequence of Fig. 17.4 b, any coherence is then rapidly dephased by a pulsed field gradient, while the z magnetization is not sensitive to these gradients and can be restored to in-phase coherence by another 90° proton pulse. Thus, this technique can only be employed to destroy magnetization of the labeled component and select the unlabeled signals. This gradient purging approach has also been elegantly combined with the X pulse phase cycling method described above [19].

The single filter elements (as shown in Figs. 17.4 a–c) are often called X-*half filters* since each of them acts only in one dimension of a 2D experiment, to be distinguished from X-*filters* that select (or suppress) ^{1}H-X pairs in both dimensions of a 2D experiment [17, 20, 21]. Of course, X-half filters can be employed twice in a 2D experiment, to yield isotope selection in both dimensions (see Sect. 17.3.3).

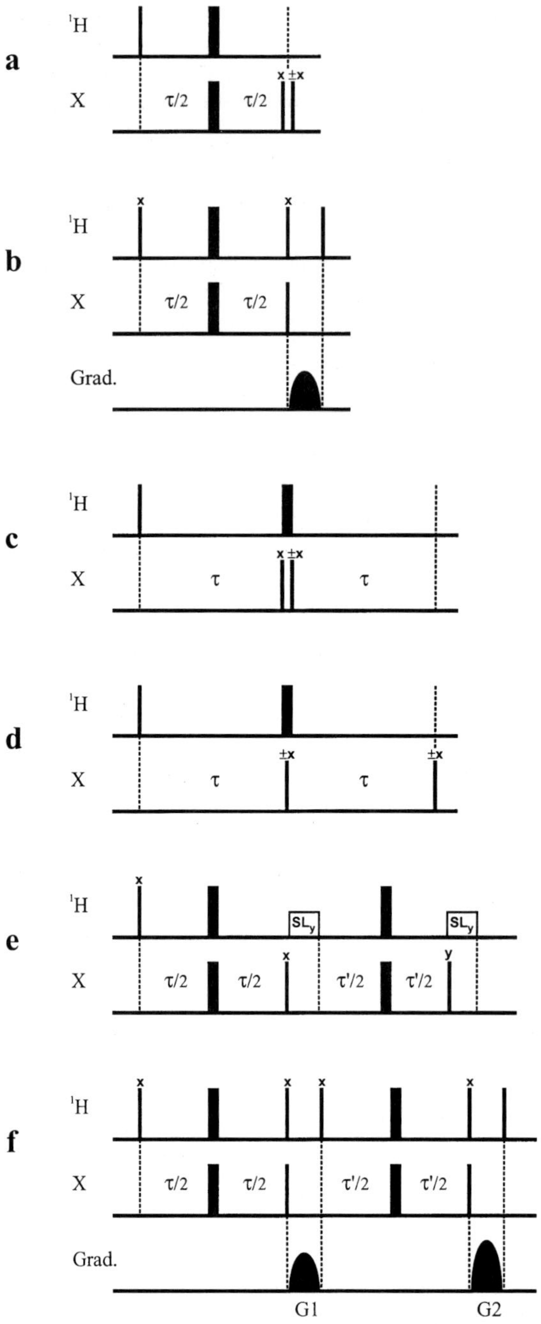

Fig. 17.4 Common filter elements: **a** X-half filter based on X pulse phase cycling [16, 17], **b** X-half filter with purge gradient [18], **c** X-half filter as in **a**, but with refocusing period for the heteronuclear antiphase magnetization [16, 17]. Sequences **d** [22], **e** [23] and **f** [18] show double filters based on single filter elements; the delays τ and τ' can be set to slightly different values to cover a broader range of ^1J coupling constants (see text for a more detailed description).

In the case of *suppressing* the ^{1}H-X signals, it turns out that the performance of a single filtering step is often not sufficient. From Eq. (3) it follows that the ^{1}H-X term is only *completely* converted into antiphase (and subsequently eliminated), if the delay τ *exactly* equals $^1/_{2J}$. Otherwise a fraction cos $(\pi {}^1J_{H,X}\,\tau)$ stays in phase and will *not* be removed by the filter. Especially in the case of ^{13}C with its wide range of $^1J_{H,X}$ values (see above) it is virtually impossible to match this condition simultaneously for all ^{1}H-^{13}C pairs in a molecule, and the leakage occurring from these signals can easily obscure the desired signals of the unlabeled ligand.

This was the reason for the development of the double filter [22, 23], which serves a double purpose: first, by combining two independent filtering steps the filtering efficiency is greatly increased [this requires that the two 90°(X) pulses be independently phase cycled or the two purge gradients do not interfere, respectively]. Second, there is now the possibility to tune both filtering steps to slightly different $^1J_{H,X}$ values, with different settings for τ and τ'. Both aspects will further reduce leakage, especially in the case of ^{13}C [23].

The simplest double tuned filter can be constructed by a concatenation of two X-half filters and removal of redundant 180° pulse pairs (Fig. 17.4 d) [22]. Alternatively, it can also be realized by keeping the 180° pulse pairs and adding short spin-lock periods (to dephase the ^{1}H-^{13}C magnetization which is orthogonal to the spin-lock axis, Fig. 17.4 e) [23], or it is based on the gradient-purging scheme of Fig. 17.4 b, resulting in the double filter shown in Fig. 17.4 f [18].

An interesting approach to the problem of nonuniform $^1J^1$H,^{13}C values was proposed by Zwahlen et al. [24]. They realized that there is an almost linear relationship between the ^{13}C chemical shifts in a protein and its $^1J^1$H,^{13}C coupling constant (Fig. 17.5, a similar relationship holds for nucleic acids). In the X-filter element (Fig. 17.4 a) the optimum timing for the 180° (X) pulse is after a time $^\tau/_2 = ^1/_{4J}$, for complete conversion of ^{1}H-^{13}C magnetization into the antiphase term. Because of the different ^1J values, this would be after ca. 1.5 ms for aromatic carbons, ca. 1.7 ms for C$^\alpha$, and ca. 2.0 ms for methyl groups. This problem can be solved elegantly by employing a so-called adiabatic inversion pulse instead of a normal 180° (X) pulse. The adiabatic pulse does not perform a practically simultaneous inversion on all carbons, but "sweeps" through the carbon resonance frequencies in a sequential manner. For the filtering element, this sweep has to be adjusted so that it starts at the low-field edge of the carbon spectrum (the aromatic region) and then proceeds toward the high field (methyl) edge in exactly the time difference needed for optimum filtering (ca. 0.5 ms, see Ref. [24] for further details).

As expected, the performance of this filter over the whole ^{13}C shift range surpasses that of a conventional single or even double filter. The authors have successfully applied this technique to a 2D NOESY spectrum with ^{13}C-filtering in both dimensions, acquired on the 26 kDA complex between a [U-^{13}C,^{15}N]-labeled DNA-binding protein domain and a 15-mer DNA ligand (unlabeled) [24].

In many investigations of protein-ligand complexes, the labeled component (usually the protein) is available in [U-^{13}C, ^{15}N] double-labeled form. In this case, both the ^{13}C- and ^{15}N-bound protons can be suppressed simultaneously, leaving ideally a ligand spectrum that is essentially free from all protein signals [22, 23, 25]. This double filtering is especially useful for the observation of ligand signals in the range of ca. 5–10 ppm, where they could be obscured by either ^{15}N-bound amide protons or ^{13}C-bound aromatic protons. Since both filters

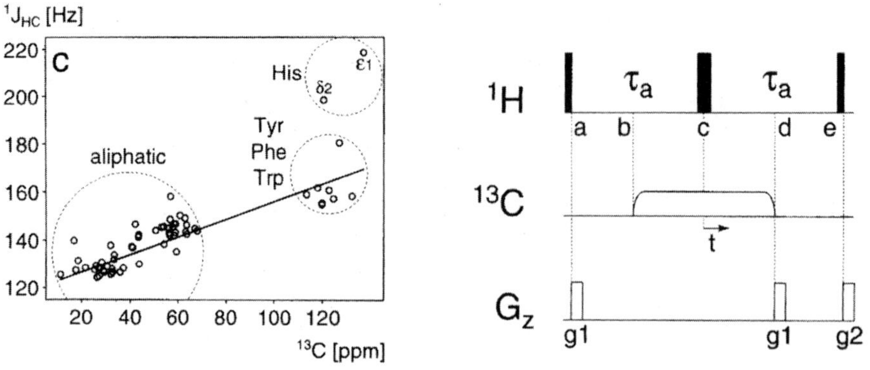

Fig. 17.5 Use of an adiabatic inversion pulse in a ^{13}C filter. An almost linear relationship exists between a ^{13}C chemical shift and the corresponding ^{1}J^{1}H,^{13}C coupling constant (left panel). With an adiabatic 180° pulse on ^{13}C, sweeping from high to low carbon frequencies, an optimum evolution of heteronuclear antiphase can be achieved for a wide range of ^{1}J values (right panel). Reprinted with permission from Ref. [24]. Copyright 1997 American Chemical Society.

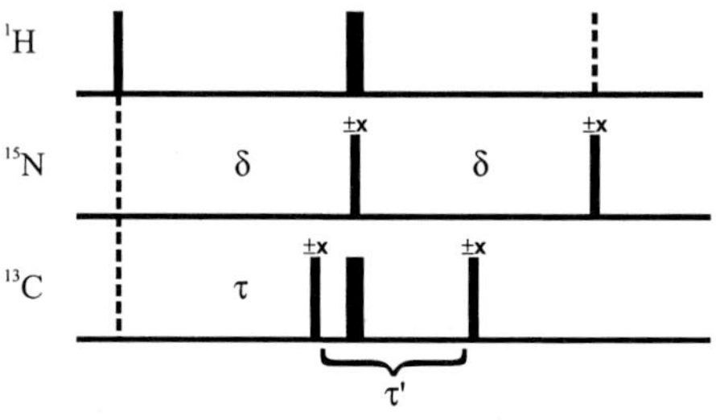

Fig. 17.6 Simultaneous double filter on ^{13}C and ^{15}N; different settings for the delays τ and τ' can be used for double tuning for ^{13}C [22, 23, 25].

work completely independently on mutually exclusive sets of (protein) protons, they can be implemented in a parallel fashion (Fig. 17.6) that does not extend the total duration beyond that needed for the longer of the two filters (that is, the one on ^{15}N).

Again, a combined ^{13}C, ^{15}N-filter can also be constructed from other filtering elements, for example, a gradient-purging scheme [19].

The performance of the isotope filter is a critical step for the acquisition of interpretable NMR spectra of mixed labeled/unlabeled complexes. Therefore several additional variations with promising results have been proposed in the literature, but cannot be discussed within the scope of this chapter [26–28].

17.3.3
Selection of Intra-/Intermolecular NOEs

If we consider a standard homonuclear ^{1}H, ^{1}H 2D experiment like NOESY, TOCSY or COSY, then all these different filter elements can be applied either before or after the mixing step, or both. Correspondingly this will lead to a selection in F1 only, F2 only or both F1 and F2, of the 2D spectrum.

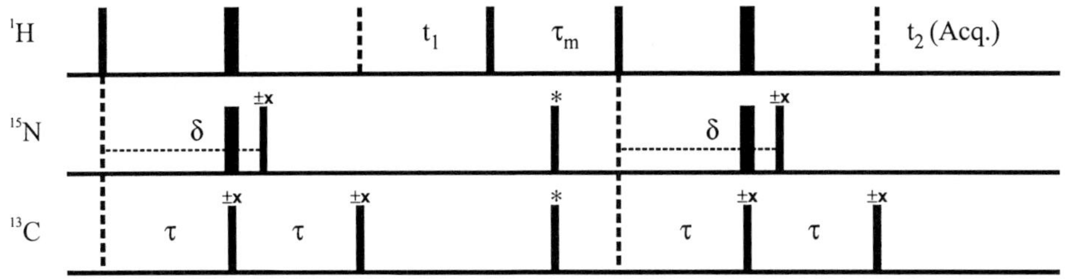

Fig. 17.7 2D NOESY experiment with ^{13}C, ^{15}N filter in both ^{1}H dimensions, selecting intramolecular NOEs within the unlabeled components [22]. The sequence uses double filters for ^{13}C, single filters for ^{15}N, and additional ^{13}C, ^{15}N purge pulses during the NOE mixing time (*).

In NMR experiments where coherence transfer is only realized through scalar couplings (COSY, TOCSY), only intramolecular signals within one component will occur, so that in principle one filtering element (that is, a *half-filter*) is enough to distinguish between signals within the labeled and unlabeled moiety (in TOCSY spectra intermolecular signals can arise mainly from ROE contributions, so that filtering in both dimensions is to be preferred).

NOE transfer, however, clearly can occur in an intramolecular as well as in an intermolecular fashion. Here the different possible combinations of isotope filtering and editing (selecting/suppressing X-bound protons in F1 only, F2 only, or both in F1 and F2) allow one to select certain classes of NOE signals. Isotope editing in F1 *and* F2 will select only intramolecular NOEs in the labeled component, isotope filtering in F1 *and* F2 intramolecular NOEs in the unlabeled component (one possible version is shown in Fig. 17.7). In addition, editing in one dimension and filtering in the other dimension will allow one to observe exclusively intermolecular NOEs between the labeled and unlabeled component. Numerous applications of these different versions can be found in the literature (see also Sect. 17.1.4).

In reality, some complications arise from the fact that, for example, the ^{15}N-bound amide protons of a protein component are *not* suppressed by a ^{13}C filter, and an *intramolecular* NOE between such a ^{1}H-^{15}N protein signal (in the ^{13}C-filtered dimension) and a ^{1}H-^{13}C protein signal (in the ^{13}C-edited dimension) could be confused with a real *intermolecular* NOE between ^{1}H-^{12}C of the ligand and a ^{1}H-^{13}C of the protein. There are two ways to solve this ambiguity: without ^{15}N decoupling only the ^{15}N bound amide signals will split into a characteristic doublet due to the $^{1}J^{1}H,^{15}N$ coupling. More elegantly, the use of the above-mentioned combined ^{13}C, ^{15}N filter (Figs. 17.3 and 17.6) will virtually suppress *all* protein signals (except for OH and SH groups).

17.4 Applications

If a ligand is available in isotope-labeled form, then the use of standard isotope-edited techniques (preferentially with at least one heteronuclear dimension) will allow straightforward access to its structure in the bound state without having to solve the much more complex problem of the protein structure.

A classical example for this approach is the structure determination of the cyclophilin-cyclosporin A complex. In the beginning, selectively [^{13}C-MeLeu^{9}, MeLeu^{10}]-labeled cyclosporin A bound to unlabeled cyclophilin was used to prove the existence of a *trans*-9, 10 amide bond by isotope-edited NOESY [10]. Later on, both components, the protein cyclophilin as well as the ligand cyclosporin A, were accessible in isotope-labeled form. This made it possible to employ isotope editing techniques (generally including a heteronuclear dimension) for the selective observation of the [U-^{13}C]- or [U-^{15}N]-labeled ligand [29, 30] as well as for the binding protein cyclophilin [31], and finally to solve the complete structure of the complex [32, 33].

With another immunophilin, FK binding protein (FKBP), experiments were performed using isotope editing of the [U-^{13}C]-labeled inhibitor ascomycin (bound to unlabeled FKBP) [34], as well as by isotope filtering with unlabeled ascomycin derivatives (bound to labeled FKBP) [35].

More recent applications comprise, for example, the identification of the binding site of 18 kDa human cardiac troponin C for the drug bepridil [36]. For this study, the unlabeled ligands were bound to selectively [$^{13}CH_3$-Met, Phe-d_8]-labeled protein (Fig. 17.8). First, the $^{13}CH_3$-Met signals of troponin C were easily identified from ^{13}C-HSQC spectra. In a 2D NOESY spectrum with ^{13}C-editing in one dimension, intermolecular NOEs could

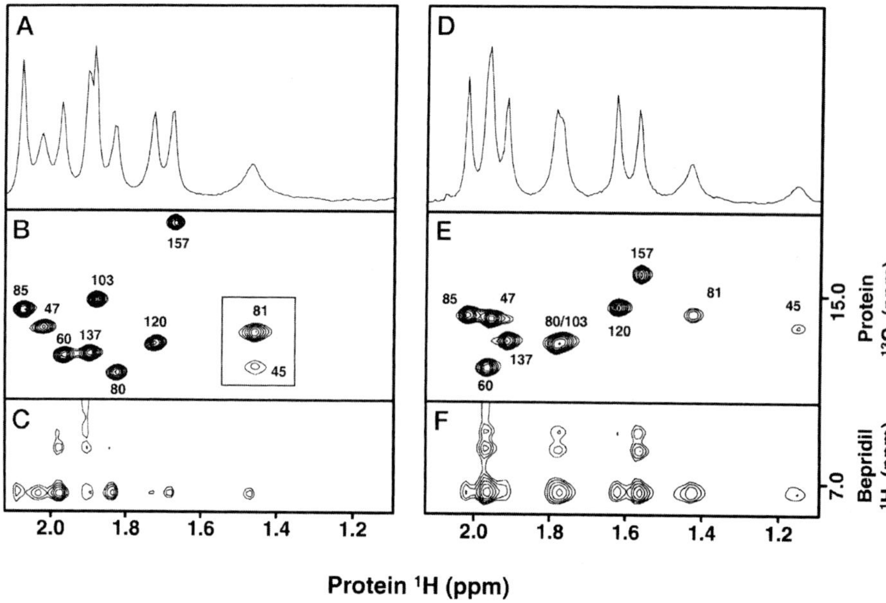

Fig. 17.8 Intermolecular NOEs between [$^{13}CH_3$-Met, Phe-d_8]-labeled cardiac troponin C and the drug bepridil (left panels, drug:protein 1.5:1; right panels, 3.5:1). **A, D**: methyl region of the ^{1}H spectrum; **B, E**: HSQC spectra showing the $^{13}CH_3$-Met signals of the protein; **C, F**: section from the NOESY spectrum with ^{13}C-editing in one dimension, showing intermolecular NOEs between the ^{1}H signals of the protein $^{13}CH_3$ groups (horizontal dimension) and the aromatic signals of bepridil (vertical dimension). Reproduced with permission from [36]. Copyright 1998 American Society of Biochemistry and Molecular Biology, Inc.

be detected from the now known methionin methyl ^{1}H shifts to the aromatic signals of the ligand bepridil [21]. This region was not obscured by any protein signals due to the deuteration of all aromatic and amide resonances, through incorporation of Phe-d_8 and use of D_2O as solvent.

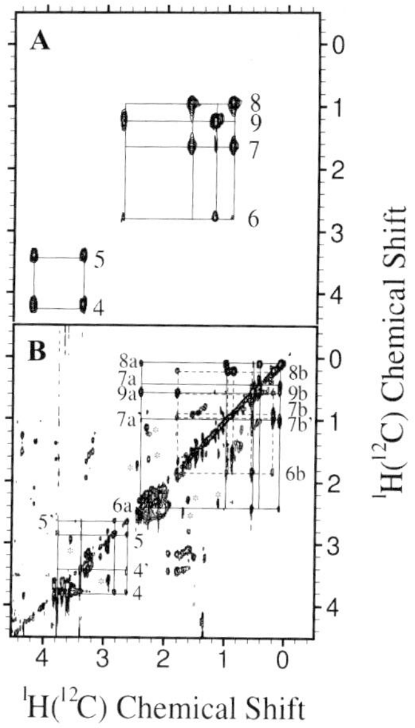

Fig. 17.9 The ^{1}H assignment of the unlabeled pheromone 2-*sec*-butyl-4,5-dihydrothiazole, bound to [U-^{13}C, ^{15}N]-labeled mouse major urinary protein I, was accomplished by comparing the 2D TOCSY spectrum of free ligand (**A**) to the ^{13}C-filtered TOCSY spectrum of the complex (**B**). Reproduced with permission from Ref. [37]. Copyright 1999 American Chemical Society.

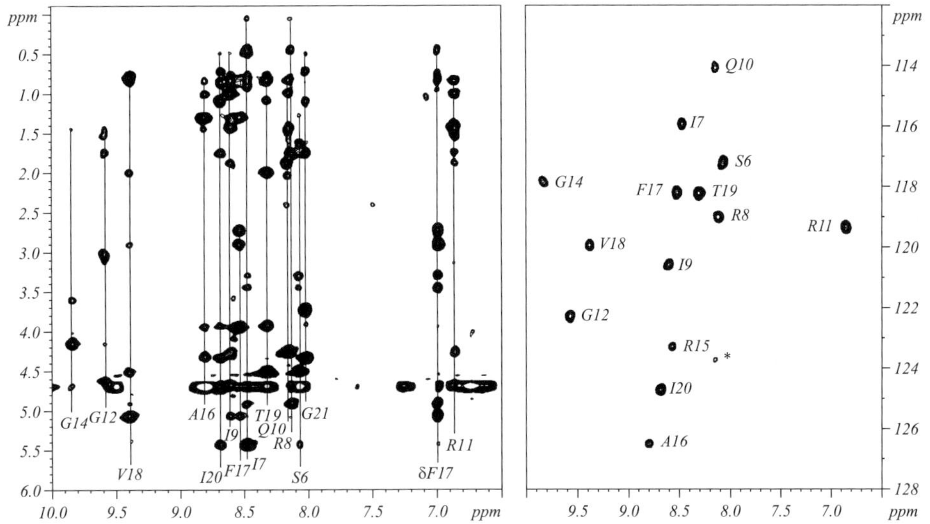

Fig. 17.10 F1,F2 ^{13}C, ^{15}N-filtered NOESY spectrum of a complex of an unlabeled HIV peptide with an [U-^{13}C, ^{15}N]-labeled antibody fragment (left). The chemical shifts of the peptide amide protons were assigned from a ^{15}N-HSQC spectrum of ^{15}N-labeled peptide bound to unlabeled protein (right). Reproduced with permission from Ref. [11].

Isotope filtering has also been successfully applied to the study of the unlabeled pheromone 2-sec-butyl-4,5-dihydrothiazole, bound to [U-^{13}C, ^{15}N]- or [U-^{15}N]-labeled mouse major urinary protein I (19 kDa) [37]. The ^{1}H signals of the ligand could be assigned by comparison of the TOCSY signals of the free ligand to the signals in a ^{13}C-filtered (in F1) 2D TOCSY spectrum of the complex (Fig. 17.9). Intermolecular NOEs could then be extracted from a 3D NOESY-^{13}C-HSQC spectrum, with ^{13}C-filter in the indirect ^{1}H dimension (based on the adiabatic filter shown in Fig. 17.5) and ^{13}C-editing (in the form of an explicit ^{13}C dimension) in the direct ^{1}H dimension. Together with the 3D structure of the protein (from NMR studies of the labeled protein), this led to the determination of the complete structure of the complex.

The nature of antigen-antibody recognition is of prime interest for the understanding of the immune system, and isotope-filtering techniques can also contribute to this area. An unlabeled 18-mer HIV-1_{IIIB} V3 peptide (P1053) was bound to a 25 kDa [U-^{13}C,^{15}N]-labeled Fv anti-gp120 antibody fragment. Assignments of the P1053 peptide were confirmed from a ^{13}C-filtered TOCSY employing ^{13}C-heteronuclear dephasing during the ^{1}H spin-lock (not shown) [38]. An ^{13}C, ^{15}N-filtered NOESY (in F1 and F2) selected intra-peptide NOEs, revealing a β-hairpin conformation with a type VIb turn of the peptide 10 (Fig. 17.10) [11]. In addition, a ^{13}C-edited 2D NOESY on [^{13}C,^{15}N-Gly^{12}, Pro^{13}, Gly^{14}]-la-

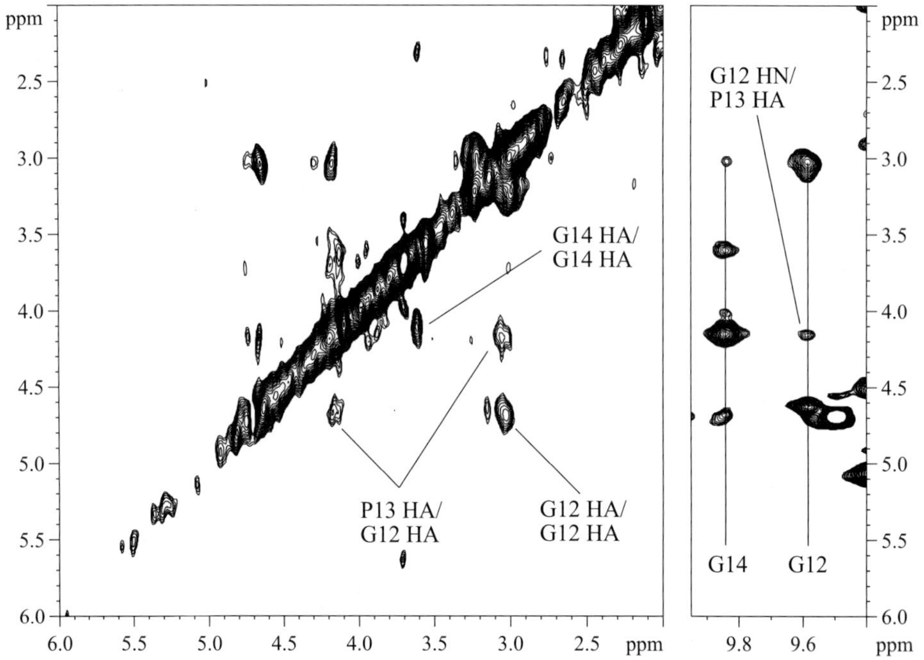

Fig. 17.11 Region from a ^{13}C-edited NOESY spectrum of an HIV peptide ^{13}C,^{15}N- labeled at positions Gly12, Pro13 and Gly14. The characteristic NOEs for a *cis* configuration of the Gly-Pro peptide bond are indicated (left). Similarly, the expected cross peak between Gly12-H^{N} and Pro13-H$^{\alpha}$ can be found in a NOESY spectrum with ^{13}C,^{15}N filter in both dimensions. Reproduced with permission from Ref. [11].

beled P1053 complexed by unlabeled protein unambiguously confirmed the existence of the sequential H_i^{α}-H_{i+1}^{α} NOEs characteristic for a *cis* peptide bond (Fig. 17.11). Finally, another sample of [^{15}N-Ser^6-Ile^{20}]-labeled P1053 bound to the unlabeled Fv fragment yielded amide proton NOEs in a ^{15}N-edited 3D NOESY-HSQC spectrum (not shown).

An even larger complex (ca. 38 kDa) between two calmodulin molecules and a dimeric SEF2-1 mimicking peptide (SEF2-1mp) has been investigated recently, using [U-^{15}N]- and [U-^{13}C,^{15}N]-labeled calmodulin [39]. ^{13}C,^{15}N-filtered 2D TOCSY and NOESY spectra yielded the ^{1}H assignment and conformation of SeF2-1mp when bound to calmodulin. Intermolecular NOEs were detected in a 3D NOESY dataset with ^{13}C,^{15}N-filtering in one ^{1}H dimension and ^{13}C,^{15}N-editing in the other ^{1}H dimension, including an explicit heteronuclear ^{13}C,^{15}N dimension according to [24].

These examples show the wide region of possible applications for isotope editing and filtering, with respect to the type of complexes as well as the range of molecular weights involved. It can be assumed that, especially in the field of drug design, the study of protein-ligand complexes by these techniques will be of continuing and growing interest in the future.

17.5 References

1 S.H. Seeholzer, M. Cohn, J.A. Putkey, A.R. Means, H. L. Crespi, *Proc. Natl. Acad. Sci. U.S.A.* **1986**, *83*, 3634.

2 S.W. Fesik, *J. Med. Chem.* **1991**, *34*, 2937–2945.

3 A.M. Petros, S.W. Fesik, in: Methods Enzymol., vol. 239, eds. T.L. James, N. J. Oppenheimer, Academic Press, San Diego CA 1994, p. 717–739.

4 S.W. Fesik, E.R.P. Zuiderweg, *J. Am. Chem. Soc.* **1989**, *111*, 5013.

5 T. Otomo, N. Ito, Y. Kyogoku, T. Yamazaki, *Biochemistry* **1999**, *38* (49), 16040–16044.

6 P.J. Hajduk, T. Gerfin, J.M. Boehlen, M. Haberli, D. Marek, S.W. Fesik, *J. Med. Chem.* **1999**, *42* (13), 2315–2317.

7 A.L. Breeze, *Prog. NMR Spectrosc.* **2000**, *36*, 323–372.

8 M.G.M. Burgering, R. Boelens, M. Caffrey, J.N. Berg, R. Kaptein, *FEBS Lett.* **1993**, *330*, 105–109.

9 T.M. Handel, P.J. Domaille, *Biochemistry* **1996**, *35*, 6569–6584.

10 S.W. Fesik, R.T. Gampe Jr., T.F. Holzman, D.A. Egan, R. Edalji, J.R. Luly, R. Simmer, R. Helfrich, V. Kishore, D.H. Rich, *Science* **1990**, *250*, 1406–1409.

11 V. Tugarinov, A. Zvi, R. Levy, J. Anglister, *Nature Struct. Biol.* **1999**, *6*, 331–335.

12 G.W. Vuister, S.-J. Kim, C. Wu, A. Bax, *J. Am. Chem. Soc.* **1994**, *116*, 9206–9210.

13 L.E. Kay, G.M. Clore, A. Bax, A.M. Gronenborn, *Science* **1990**, *249* (4967), 411–414.

14 E.R.P. Zuiderweg, A.M. Petros, S.W. Fesik, E.T. Olejniczak, *J. Am. Chem. Soc.* **1991**, *113*, 370–372.

15 O.W. Sørensen, G.W. Eich, M.H. Levitt, G. Bodenhausen, R.R. Ernst, Prog. *NMR Spectrosc.* **1999**, *34*, 163.

16 G. Otting, H. Senn, G. Wagner, K. Wüthrich, *J. Magn. Reson.* **1986**, *70*, 500–505.

17 G. Otting, K. Wüthrich, *Q. Rev. Biophys.* **1990**, *23*, 39–96.

18 K. Ogura, H. Terasawa, F. Inagaki, *J. Biomol. NMR* **1996**, *8*, *492*–498.

19 C. Dalvit, S. Cottens, P. Ramage, U. Hommel, *J. Biomol. NMR* **1999**, *13*, 43–50.

20 E. Wörgötter, G. Wagner, K. Wüthrich, *J. Am. Chem. Soc.* **1986**, *108*, 6162–6167.

21 E. Wörgötter, G. Wagner, M. Vašák, J.H. R. Kägi, K. Wüthrich, *J. Am. Chem. Soc.* **1988**, *110*, 2388–2393.

22 M. Ikura, A. Bax, *J. Am. Chem. Soc.* **1992**, *114*, 2433–2440.

23 G. Gemmecker, E.T. Olejniczak, S.W. Fesik, *J. Magn. Reson.* **1992**, *96*, 199–204.

24. C. Zwahlen, P. Legault, S.J.F. Vincent, J. Greenblatt, R. Konrat, L.E. Kay, *J. Am. Chem. Soc.* **1997**, *119*, 6711–6721.
25 G. Gemmecker, W. Jahnke, H. Kessler, *J. Am. Chem. Soc.* **1993**, *115*, 11610–11621.
26 A.C. Stuart, K.A. Borzilleri, J.M. Withka, A.G. Palmer III, *J. Am. Chem. Soc.* **1999**, *121*, 5346–5347.
27 P. Andersson, G. Otting, *J. Magn. Reson.* **2000**, *144* (1), 168–70.
28 J. Iwahara, J.M. Wojciak, R.T. Clubb, *J. Biomol. NMR* **2001**, *19*, 231–141.
29 S.W. Fesik, R.T. Gampe Jr., H.L. Eaton, G. Gemmecker, E.T. Olejniczak, P. Neri, T.F. Holzman, D.A. Egan, R. Edalji, R. Simmer, et al., *Biochemistry* **1991**, 30 (26), 6574–6583.
30 C. Weber, G. Wider, B. von Freyberg, R. Traber, W. Braun, H. Widmer, K. Wüthrich, *Biochemistry* **1991**, *30* (26), 6563–6574.
31 L. Kallen, C. Spitzfaden, M.G. Zurini, G. Wider, H. Widmer, K. Wüthrich, M.D. Walkinshaw, *Nature* **1991**, *353* (6341), 276–279.
32 Y. Theriault, T.M. Logan, R. Meadows, L. Yu, E.T. Olejniczak, T.F. Holzman, R.L. Simmer, S.W. Fesik, *Nature* **1993**, *361* (6407), 88–91.
33 C. Spitzfaden, W. Braun, G. Wider, H. Widmer, K. Wüthrich, *J. Biomol. NMR* **1994**, *4* (4), 463–482.
34 A.M. Petros, G. Gemmecker, P. Neri, E.T. Olejniczak, D. Nettesheim, R.X. Xu, E.G. Gubbins, H. Smith, S.W. Fesik, *J. Med. Chem.* **1992**, *35* (13), 2467–2473.
35 A.M. Petros, M. Kawai, J.R. Luly, S.W. Fesik, *FEBS Lett.* **1992**, *308* (3), 309–314.
36. Q. Kleerekoper, W. Liu, D. Choi, J.A. Putkey, *J. Biol. Chem.* **1998**, *273*, 8153–8160.
37 L. Idek, M.J. Stone, S.M. Lato, M.D. Pagel, Z. Miao, A. Ellington, M. V. Novotny, *Biochemistry* **1999**, *38*, 9850–9861.
38 A. Bax, S. Grzesiek, A.M. Gronenborn, G.M. Clore, *J. Magn. Reson. A* **1994**, *106*, 269–273.

18
Strategies for NMR Screening and Library Design

Christopher A. Lepre

18.1
Introduction

In recent years, the practicality and productivity of screening small-molecule libraries by NMR has improved dramatically as a result of advances in hardware, experimental methods, and automation. It is now relatively straightforward to launch an NMR screening program using commercially available compounds, published methods, and off-the-shelf instrumentation. The question now facing many researchers is how best to apply NMR screening to drug discovery. The essential product of any screen is information. The problem is how to generate information that will be useful to a drug discovery program, particularly from NMR screening, which is not as widely accepted as conventional, activity-based high-throughput screening (HTS) and tends to produce weaker hits.

This review focuses upon NMR screening strategies and discusses how compound libraries can be designed to provide information pertinent to the specific strategy being pursued. The principles and tools of library construction are described and illustrated by the example of the approach used in our laboratory. The technical details of experimental methods will not be discussed here since they are reviewed in other chapters of this volume and elsewhere [1].

18.2
Choosing a Screening Strategy

18.2.1
Strategy Directs Library Design

When planning an NMR screening program, it is essential to tailor the screening strategy to the needs of the discovery program, taking into account the type of information that is desired and the availability of resources to follow up hits. Early-stage discovery programs typically require novel leads, while late-stage programs need to optimize lead compounds and broaden the scope of lead classes. Thus, screening a small library of high-molecular weight, lipophilic, proprietary scaffolds is unlikely to benefit a young project that needs novel scaffolds with good pharmacokinetic (PK) characteristics. In a mature program, where the focus is on optimizing a few selected potent compounds, there is little to be gained by identifying small, weakly binding scaffolds, and it will be difficult to persuade busy chemists to try linking them together. Program needs dictate the strategy, and the strategy in turn dictates the choice of compounds for the library.

18.2.2
Types of NMR Screening Strategies

A screening strategy is defined here as a process for using screening information to discover compounds with more desirable characteristics, most often higher potency but possibly greater novelty or solubility. Though many different approaches have been published, all may be categorized according to the manner in which the primary screening hits are followed up: screening hits may be connected together (combined), used to direct the search for more complex compounds (elaborated), or changed one part at a time (varied).

In the *combination* strategy (Fig. 18.1), simple molecular fragments are screened, and those that bind to the target are then connected together to construct more potent compounds. Assuming that the dissociation constants of the compounds are multiplicative, linking even very weak binders can theoretically produce a potent inhibitor; e.g. linking three

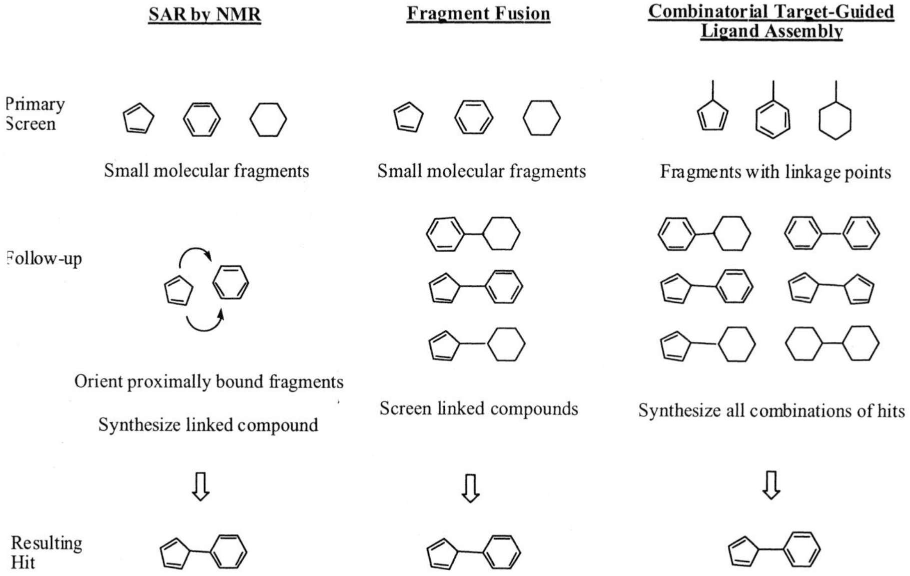

Fig. 18.1 Fragment combination strategies. *Left:* In SAR by NMR [5], small molecular fragments that bind to the target are identified using protein-detected experiments. In the follow-up step, compounds bound in close proximity to one another are identified by chemical shift perturbation and intermolecular NOE experiments, and distance information is used to design and synthesize a linked compound. *Center:* In fragment fusion [3], binding fragments are identified using ligand-detected experiments. Then, compounds containing various combinations of active fragments are tested for activity. *Right:* In combinatorial target-guided ligand assembly [4] a library of molecular fragments containing combinatorial linking groups is screened. Then, all possible combinations of active fragments are synthesized and tested for activity.

fragments with millimolar affinities could create a compound with mM×mM×mM=nM affinity [2]. In practice, this may not hold true because of entropic factors and changes in the binding orientations of the fragments when linked.

The combination approach requires that the screening library consist of small molecular fragments, possibly containing chemically accessible linkage sites but with few other functional groups (since these may limit the possible binding orientations or hinder linking). Because the compounds are so simple, a library of only a few hundred to a few thousand compounds is often sufficient to sample the desired scaffold classes. Small fragments have limited potential for making favorable interactions with the target, so nearly all of the resulting hits exhibit very low (e.g. millimolar) binding affinities.

Various approaches have been suggested for combining molecular fragments. Fragment fusion [3] is the testing of compounds (usually purchased from commercial sources) that contain two or more substructures known to bind to the target. Combinatorial target-guided ligand assembly [4] uses reactive groups built into synthetic fragments to make all possible combinations of binding fragments. Unfortunately, unless the proximity and relative orientations of the individual bound fragments are known, with the latter methods there is no way to know that the blindly linked compounds will have the correct conformation to bind to the protein. To address this problem, SAR (structure-activity relationship) by NMR [5] guides the synthesis of linked compounds using NMR-derived structural information: close contacts between bound fragments found using NOE experiments [6] and binding sites modeled using chemical shift perturbation data [7, 8]. Alternatively, it is possible to determine the structures of fragments soaked into protein crystals using X-ray crystallography [3, 9, 10].

The *elaboration* strategy (Fig. 18.2) consists of starting from the primary screening hits and systematically building upon them to create more complex molecules. The goal is to add functional groups and ring systems that will form new contacts with the target and increase potency without disrupting binding. An iterative series of screens is carried out with successive generations of compounds, each using information gained from the preceding screen. Follow-up screens can be carried out using HTS assays and usually produce much higher hit rates than HTS screens of random libraries, indicating that the follow-up libraries are enriched in active compounds and that the process converges toward more potent inhibitors. As with the combination strategy, it is very useful to know the structure of the bound scaffolds when deciding how to elaborate them.

The elaboration strategy requires that the compounds screened be simple enough that further elaboration is possible without creating molecules too complex to serve as good leads (MW >500). It is also desirable to screen molecules that are amenable to synthetic modification or for which many analogs are commercially available, since this facilitates the construction of follow-up libraries. An example is the SHAPES strategy [3], which uses compounds containing scaffolds from known drugs or proprietary classes to bias the selection of compounds for further HTS screening. Further details of this method are described in other sections of this paper. Needle screening uses small compounds (MW <300) intended to serve as molecular probes of the active site [10]. These compounds are based on known inhibitors and contain only the minimal structural elements predicted to make essential contacts with the protein by molecular modeling. Follow-up compounds are synthesized using information from SAR and the X-ray structures of bound hits.

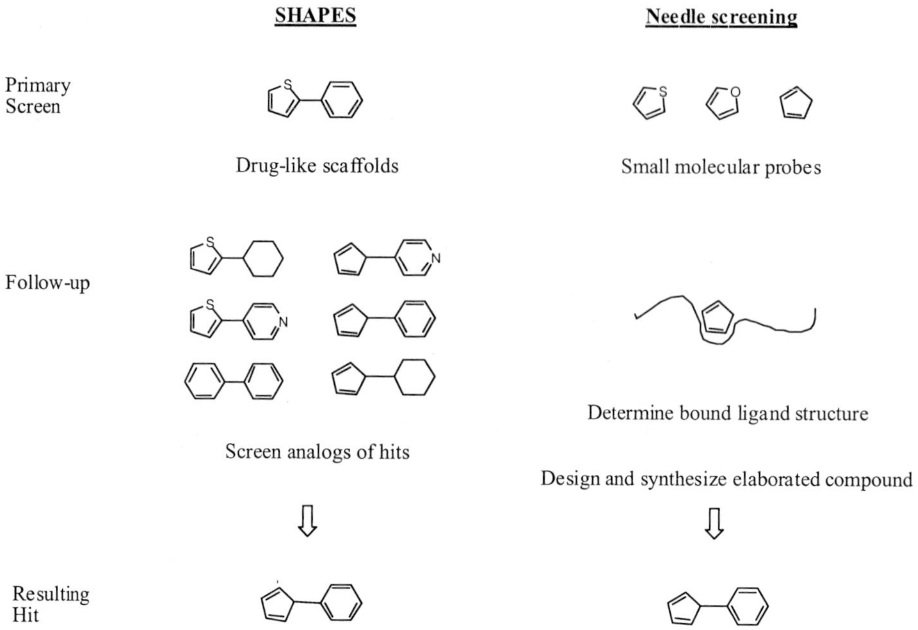

Fig. 18.2 Elaboration-based strategies. *Left:* In SHAPES screening [3], simple drug-like scaffolds are screened using ligand-detected experiments. In the follow-up step, information about binding scaffolds is used to design a follow-up library comprised of analogous compounds. Successive rounds of screening lead to increasingly complex and potent inhibitors. *Right:* In needle screening [10], the primary screening library consists of small molecular fragments (needles) chosen to probe specific interactions in the protein active site. Follow-up compounds are synthesized by attaching additional groups to the needles, guided by SAR and structural information.

In the *variation* strategy (Fig. 18.3), systematic changes are made to permute specific portions of the binding molecules. When applied as a primary screening tool, this strategy requires a library of compounds containing multiple rings or functional groups that are synthetically amenable to being independently varied. An example of this is the SHAPES Linking Library [11], which consists of compounds containing drug-like scaffolds connected by linkers that are accessible to combinatorial chemistry. Hits may be followed up by individually permuting the scaffolds and synthesizing the combinations. Alternatively, one may screen individual combinatorial scaffolds that contain multiple positions for substitutions, such as triazine [12], and then systematically explore the SAR at each position. An advantage of these approaches over methods involving the linking of individual fragments is that the original screening hit is already a linked compound, so it is much easier to synthesize linked follow-up compounds with fragments correctly oriented for binding. Unfortunately, it is also more difficult to test all possible combinations of scaffolds, since the number of linked compounds that must be screened increases geometrically with the number of groups to be varied.

Systematic variation is also useful for optimizing certain properties (e.g., solubility, novelty, toxicity, or metabolic stability) of a lead molecule. The fragment optimization

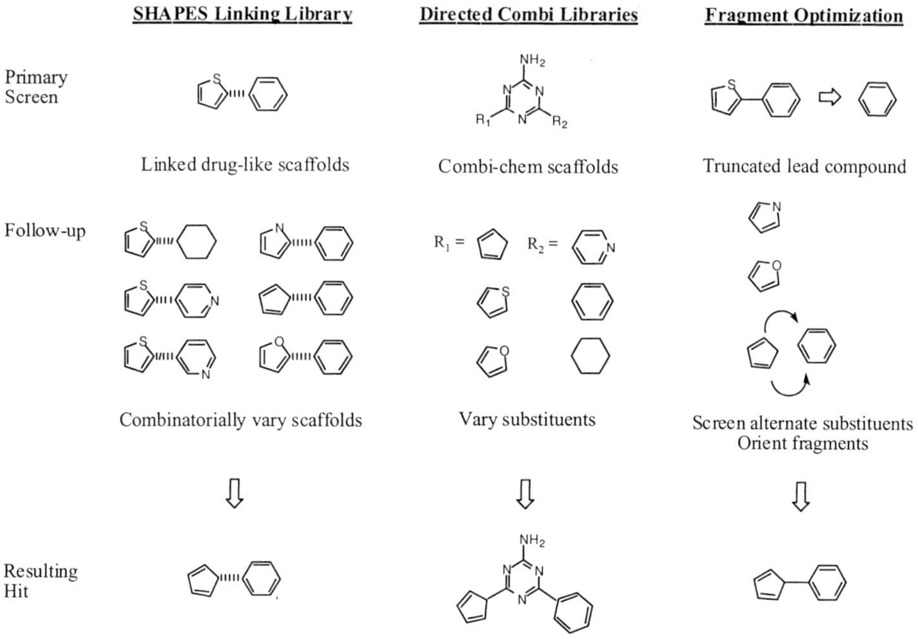

Fig. 18.3 Variation-based strategies. *Left:* The SHAPES Linking Library [11] consists of drug-like scaffolds connected by linkers that are amenable to combinatorial chemistry (dashed line). Hits are followed up by synthesizing a combinatorial library in which the scaffolds are systematically varied. *Center:* Directed Combinatorial Libraries [12] are comprised of scaffolds containing multiple sites for substituents. Hits are followed up by combinatorially varying the substituents. *Right:* Fragment Optimization [13] begins by removing the substituents from the binding core of a lead compound, then screening a library of small molecular fragments to identify replacement substituents. These new fragments are oriented relative to the binding core using structural methods, and new compounds are synthesized.

approach [13] consists of screening a small fragment library against the target in the presence of a bound core compound that has been truncated to remove the piece to be optimized. When new binding fragments are identified, their positions are determined relative to the truncated core and the fragments are synthetically linked.

The choice of strategy for any particular project depends on the type of information sought and the availability of structural data and chemistry resources. In practice, elements of several strategies are often combined. For example, the SHAPES [3] and Structure Based Screening [14] strategies use primary screening hits both to bias the search for analogs and to generate crystal structures of bound hits that are then elaborated or linked together. If the goal is to find novel lead scaffolds, then any of the three principal strategies can be applied. On the other hand, proprietary scaffolds are best screened using elaboration-based strategies, since the goal will be to start from an existing scaffold rather than to assemble a completely new one from small fragments. If the proprietary compounds contain linked scaffolds or are amenable to combinatorial chemistry, then variation-based strategies may also be used. Many methods require synthetic chemistry (e.g., combinatorial target-guided ligand assembly, needle screening), and some fragment linking methods require structural information as well (SAR by NMR, fragment optimization). If these resources are unavailable, then methods based on a biased search using commercially available compounds (e.g. SHAPES) are the best option.

18.2.3
Types of NMR Screening Libraries

The selection of compounds for an NMR screening library is dictated by its intended purpose. There are four basic types of libraries: general screening, targeted, focused, and follow-up.

General screening libraries are the most common and are used in primary screens to identify novel leads. They typically consist of compounds that are as chemically diverse as possible while still satisfying the requirements of the NMR method (such as being highly soluble, chemically stable, and perhaps drug-like; this is described further in Sect. 18.3). In many cases, the hits from such a library can be readily pursued by purchasing commercially available analogs. Analogs are usually purchased rather than synthesized because the hits are relatively weak, and unless a very large library is screened it is unlikely that there will be enough hits within a given compound class to generate meaningful SAR information.

Targeted libraries consist of compounds likely to bind to a family of related proteins, such as analogs of substrates, cofactors, or known inhibitors. Similarities between the active sites of homologous proteins lead to higher hit rates and potencies for targeted compounds than are obtained from a general screening library. The targeted library must represent the principal classes of known binders while conforming to experimental requirements, and the compounds satisfying these criteria are limited in number and novelty. Since analogs of the hits are usually not commercially available, synthetic chemistry is required for follow-up. If the compounds represent proprietary classes of scaffolds, then analogs may be readily available from the corporate archive; however, such compounds are relatively expensive if synthesized individually (estimated at $ 6000 per compound in 1993 [15]). Proprietary molecules are attractive because their synthetic chemistry has been worked out and their pharmacokinetics and toxicology are sometimes understood, but compounds based on drugs tend to be lipophilic and often exhibit aqueous solubilities too low for NMR.

Combinatorial chemistry offers a route to large targeted libraries at a reduced cost per compound. Combinatorial libraries have been targeted against enzyme classes by incorporating recognition elements (such as transition state mimics) with known affinity for the target class into the molecules. Examples of targeted libraries and their associated recognition elements appear in Tab. 18.1. When screening combinatorial libraries by NMR, it is important to ensure that they can be made in sufficient quantities as discrete, purified compounds and that they are sufficiently soluble. Furthermore, if mixtures are to be screened using ligand detection, there must be sufficient variety around the common core that the individual components can be resolved in the NMR spectra. Finally, although highly diverse substituents may be available, the use of a common scaffold and chemistry inherently limits the diversity and novelty of many combinatorial libraries. For example, peptides can

Tab. 18.1 Targeted combinatorial libraries and their associated recognition elements

Target class	*Type of inhibitor*	*Recognition element*	*Reference*
Asp proteases	Transition state isostere	Statine, hydroxyethylamine	139, 140
Ser proteases	P1 pocket binder	Benzamide, α-keto amide	19, 141
	Proline analog	Tetrahydroisoquinoline	142
	Beta turn mimic	Urazole	143
Cys proteases	Covalent binder	Michael acceptor	144
Metalloproteases	Metal ligand	Thiol, hydroxamate	145, 146
Phosphatases	Substrate mimic	Cinnamate, phosphonate	147, 148
Kinases	Inhibitor analog	Purine, dibenzylaniline	149, 150
Calcium channels	Inhibitor analog	Dialkyldipeptidyl amine	151

Tab. 18.2 Focused combinatorial libraries and their associated recognition elements

Target class	*Type of inhibitor*	*Recognition element*	*Reference*
Plasmepsin II	Transition state mimic	Hydroxyethylamine	20
Factor XA	P1 pocket binder	Benzamidine	19
IL1β-converting enzyme	Covalent binder	α-activated ketone	18
Cyclin-dependent kinase	Inhibitor analog	Purine	22
MAP kinase p38	Inhibitor analog	Bis-aryl urea	152
Delta opioid receptor	Agonist analog	Diarylmethyl piperazine	23
Delta opioid receptor	Antagonist analog	Hydroxyphenyl piperidine	153

be difficult to optimize into novel, nonpeptidic leads and can also be difficult to screen by NMR because of low aqueous solubility.

Focused libraries are directed against a specific protein. The specific binding elements of substrates or known inhibitors for that target can be used to design libraries; combinatorial examples [16–19] are listed in Tab. 18.2. If the structure of the target is known, then the structures of compounds likely to bind to the target may be predicted by using molecular modeling methods such as protein-based virtual screening and structure-based drug design [20–23]. Virtual screening typically refers to computational docking of molecules in the active site of the protein in order to identify the most promising candidates [24–31]. In one published example, comparison of docking and NMR screening results for 3300 commercially available compounds against FKBP-12 [32] showed that virtual screening could reduce the number of compounds to be screened by two- to three-fold without eliminating any that actually bound. More impressive results have been obtained from enzymatic assays of libraries designed by virtual screening of commercially available compounds [33, 34] and combinatorial reagents [20, 21, 35], or compounds constructed from molecular fragments computationally fitted into the target active site [36–38]. Virtual screening must significantly enhance hit rates in order to justify the expense of constructing new focused libraries for every primary screening target. In practice, they are rarely used for primary NMR screens and are more commonly created to follow up primary screening hits.

It has been proposed that certain molecular fragments have an inherently higher propensity to bind to protein active sites, and that a library enriched in these universal "privileged scaffolds" will be more likely to produce leads [156]. A retrospective analysis of NMR screening results for a wide variety of proteins in our own laboratory (unpublished results) revealed that certain scaffolds bound more frequently, but they were different scaffolds from those previously published and they varied according to target class. So, although we do not discount the possibility that privileged scaffolds exist, our experience indicates that their identification from historical screening data may be highly biased by the composition of the screening library and the nature of the screening targets.

The construction of *follow-up* libraries is a critical element of elaboration and variation strategies. A primary NMR screen of a general screening library usually only yields a few hits within a given scaffold class and no detailed structural information, so it is not immediately apparent how best to exploit them. Thus, the first step in following up primary screening hits is typically to search databases of readily available compounds for analogs to screen, with the intent of discerning a relationship between chemical structure and activity. Functional groups are added or modified to increase the number of interactions between the molecule and the target, thereby increasing the potency, or replaced with isosteres [39–41] to broaden the lead class. The molecular core is permuted in order to discover equipotent alternative scaffolds with significantly different structures (referred to as a scaffold hop [42]).

This approach is predicated on the assumption that compounds with similar molecular structures possess similar biological properties. Although counter-examples exist (cases where very small changes in molecular structure result in dramatic changes in biological activity), this assumption is invariably taken for granted since in its absence there is essentially no other rationale by which to follow up hits! Unfortunately, NMR hits tend to suffer from a local minimum problem: testing compounds similar to weak hits usually results in more weak hits. This is because weak binders presumably bind in a nonoptimal fashion, and a search of similar compounds simply discovers more analogs with the same binding mode, and thus the same weak binding. To move out of the local minimum requires testing more disparate compounds, which dramatically increases the size of the follow-up library. The challenge is to strike the right balance between compounds that are similar enough to the original hits to be active while being different enough to be significantly more potent.

For this reason, it is important to use an intelligent search strategy when seeking follow-up compounds to elaborate upon simple scaffolds. Analogs of primary screening hits are easily found by searching databases of compounds using commercially available similarity matching [43] or nearest-neighbor searching [44] tools. However, a conventional global similarity search, in which the entire molecule is matched to the target structure, may fall prey to the local minimum problem because the compounds matched cannot be significantly larger than the original hit. A substructure similarity search [45], which matches only the target substructure, is preferable because it finds compounds that contain scaffolds resembling the primary screening hit while allowing free variation elsewhere in the molecule.

In addition to the commonly used MACCS structural fragment keys (MDL Information Systems, Inc., San Leandro CA) and Daylight or UNITY fingerprints (Daylight

Chemical Information, Irvine CA; Tripos, Inc. St. Louis MO), similarity searching can be carried out around a well-defined compound class using local descriptors such as atom pairs [46, 47] or topomeric shape [48, 49]. Also, ligand-based pharmacophore searches are able to identify follow-up compounds that are less obvious and more diverse than similarity searches [30, 50–54]. The problem with the latter methods is defining the molecular shape or pharmacophore specifically enough to be useful when there are few hits within a compound class and they cannot be reliably aligned (as is often the case for NMR hits in the absence of detailed structural information).

18.3 Designing NMR Screening Libraries

18.3.1 Dealing with Diversity

Molecular diversity is defined by the distribution of the values of descriptors (e.g. MW, ClogP, molecular fingerprints) for a given set of molecules. Molecular diversity is not an inherent property; rather, it is completely dependent upon the particular descriptors used. Care should be taken to use only descriptors that are correlated with biological activity, because irrelevant descriptors can overwhelm relevant ones and skew the compound selection in favor of unimportant factors [55]. In particular, traditional QSAR descriptors are poorly suited for diverse searching around weak hits because they were developed for sets of relatively similar compounds [55, 56]. The best-validated descriptors for selecting diverse libraries from large compound databases are structural keys and 2D fingerprints [55–58]. In addition, pharmacophores and molecular steric fields can be used to design analog libraries for well-defined series of compounds [58]. Detailed comparisons of molecular diversity methods may be found in several recent reviews [26, 59].

The term "neighborhood behavior" was coined to describe the correlation of molecular descriptors with biological activity [55]. As seen in Fig. 18.4, a neighborhood is the region immediately surrounding each compound when plotted in descriptor space. Compounds within the same neighborhood have similar biological activities: it has been shown that 85% of all compounds with better than 85% similarity to any active compound will typically be active themselves [55] (in this case, similarity was measured using the well-known Tanimoto index and MACCS keys as descriptors). It is inherently inefficient to test multiple compounds from the same neighborhood in a primary screen. A uniformly distributed (maximally diverse) set of compounds avoids overlap between neighborhoods and so is more efficient for discovering active compounds than a random set. When hits are found, the likelihood of finding other hits is highest for compounds within the neighborhood of the original hit. Since this neighborhood may represent a local minimum, it may also be necessary to search nearby neighborhoods to find significantly more potent compounds. In real-life libraries, compounds tend to cluster in descriptor space [55] (Fig. 18.5). Where the neighborhoods of these compounds overlap, it is most efficient to test representative compounds from the centroids of the clusters.

It is worth noting that when relatively few compounds are screened and the descriptor space is very large or the neighborhoods very small, each compound covers very little

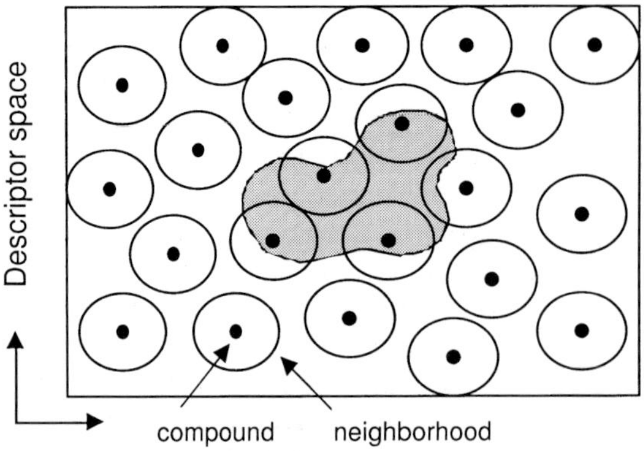

Fig. 18.4 An ideal compound library. Compounds (black dots) are uniformly distributed in 2D descriptor space, a plane defined by two molecular descriptors that correlate with biological activity but not with one another. The compounds are surrounded by nonoverlapping neighborhoods (circles) of >85% similarity. Active compounds (in the gray region) are found most efficiently by testing one compound from each neighborhood. When following up hits, active analogs are most likely to be found within the neighborhood surrounding each hit. Adapted from Patterson, et al. 1996 [55].

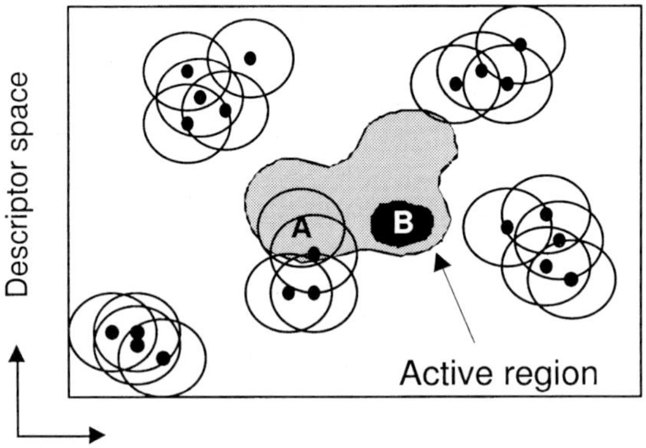

Fig. 18.5 A typical compound library. Compounds are clustered in descriptor space (for example, around synthetically accessible scaffolds). For clusters with overlapping neighborhoods, efficiency may be increased by testing representative compounds from each cluster; analogs of hits may be chosen from the same cluster. Where the cluster is located in a local minimum of activity (point A) it may be necessary to search well outside the neighborhood to find a significantly more potent compound (point B). Adapted from Patterson, et al. 1996 [55].

space and overlaps are rare. In this case, approximately the same amount of descriptor space is covered by either maximally diverse or randomly selected sets of compounds.

A number of different approaches have been used for selecting diverse sets of compounds. One of the simplest is maximum dissimilarity, in which each new compound is chosen to be as dissimilar as possible to those already selected [60]. This can drastically reduce the library size without significantly reducing the likelihood of discovering classes of active compounds. For example, when an 85% similarity cutoff was used to select 487 compounds (38%) from a set of 1283 biologically active compounds, all 55 biologically active classes of compounds were included in the selection (compared to an average of 7 for random selection) [58].

A possible drawback to using maximum dissimilarity to construct primary screening libraries is that selection is biased towards unique compounds, so close analogs are unavailable for follow-up. Advanced dissimilarity algorithms have been developed to address this problem [61, 62]. Alternatively, commercially available clustering software can be used to identify groups of similar compounds so that representative compounds may be chosen from each, and hits may be followed up easily by choosing additional compounds from the active clusters. Nonhierarchical Jarvis-Patrick clustering [63] is computationally the fastest, but may produce a large number of singletons (unclustered compounds) or a few large, heterogeneous clusters [64]. Hierarchical clustering by Ward's method [65] using MACCS keys as descriptors is reportedly the best for separating active and inactive compounds [64]. When one compound is chosen from each cluster, the contribution of small clusters may dominate the selection. To avoid this bias, multiple compounds may be chosen from large clusters, or small clusters and singletons may be reclustered in a cascaded manner [66].

Dissimilarity and clustering methods only describe the compounds that are in the input set; voids in diversity space are not obvious, and if compounds are added then the set must be re-analyzed. Cell-based partitioning methods address these problems by dividing descriptor space into cells, and then populating those cells with compounds [67, 68]. The library is chosen to contain representatives from each cell. The use of a partition-based method with BCUT descriptors [69] to design an NMR screening library has recently been described [70].

Other factors besides diversity should be considered when designing a library. Libraries designed purely to maximize diversity are systematically biased away from ideal drug properties, with many compounds that are large, overly flexible, and too lipophilic to be readily soluble in water [71]. The following sections describe additional criteria that have been used to design libraries containing molecules that are more "drug-like" or "lead-like" in character.

18.3.2
Optimizing Molecular Complexity

It is important to design a screening library with an appropriate level of molecular complexity. Small, simple compounds are more likely to bind than complex molecules, but they have lower activities and are less selective [56, 72]. As molecular complexity increases, more interactions are possible between the molecule and sites on the protein, increasing the likelihood of potent binding. However, models suggest that the overall probability of finding a complementary match between molecule and receptor declines exponentially with increasing molecular complexity [72]. Consequently, there is an optimum window in which molecules are complex enough to give detectable inhibition but not so complex as to make hits too rare in the first place (Fig. 18.6).

NMR screening methods can detect very weak binders, which shifts the "measurable inhibition" curve (Fig. 18.6) toward less complex molecules. So, it is theoretically more

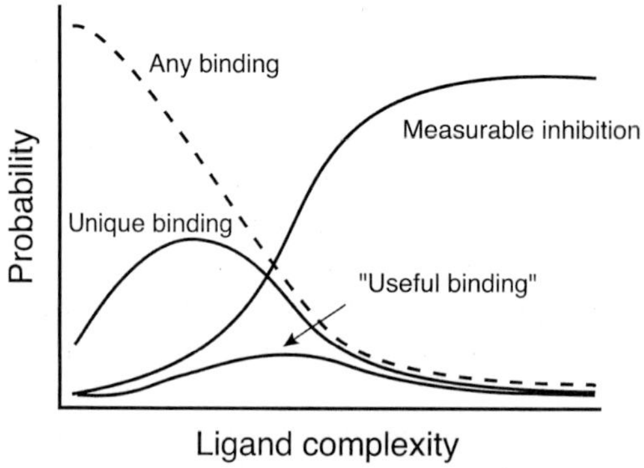

Fig. 18.6 Ligand binding probability as a function of molecular complexity. The total probability of any ligand binding (dashed line) is highest for small, simple molecules and drops off rapidly with increasing complexity. The probability of unique binding (a single orientation) peaks at intermediate levels of complexity, then drops off as the total probability declines. The likelihood of observing inhibition above the threshold of detection (measurable inhibition) increases with complexity and then levels off. For binding to be "useful", it must be measurable and have a unique binding mode; the probability of useful binding is thus the product of the probabilities of measurable and unique binding, and it peaks at moderate levels of complexity. This peak defines the optimal region for screening. Adapted from Hann, et al. 2001 [72].

likely that an NMR screen will detect nonspecific binders (compounds with multiple binding modes). Although nonspecific binding is a concern, in our experience it is seldom a serious problem. In fact, we find that NMR screens consistently produce lower rates of false positives than HTS screens, probably because the latter are much more susceptible to compounds interfering with the assay system.

So, how large should screening compounds be? They should be as large as necessary for detectable binding but no larger. Since NMR can detect millimolar binders, this size can be very small indeed. For molecules with up to 10 heavy atoms, the binding free energy contribution per heavy atom can be as high as –1.5 kcal/mol, so even very small compounds can, in principle, bind very tightly [73]. In practice, this is extremely rare. Typical results are exemplified by a library of combinatorial monomers used at GlaxoSmithKline, in which compounds of 100–250 Da yield activities of 0.2–10 mM [72]. By comparison, most drugs have molecular weights between 150 and 500 [74–77]. In order to balance the need to find a reasonable number of hits against the need for specificity, we find it is advisable to screen compounds across the intermediate range of ca. 100–350 Da.

In addition to molecular weight, percentage saturation of the standard Daylight fingerprint has been used to describe the degree of molecular functionality [78]. Compounds with saturation outside a certain range (e.g., $<10\%$ and $>60\%$ saturation) can be excluded in order to remove overly simple or complicated molecules that would otherwise pass a simple molecular weight cutoff.

18.3.3
Selecting for Drug-like Character

In recent years, it has become common practice to bias screening libraries in favor of molecules possessing characteristics similar to those of known drugs. Drug-like compounds are presumed to possess more attractive toxicological and ADME properties, and this advantage is thought to outweigh the reduced likelihood of finding novel, as yet undiscovered classes of drugs. Three approaches have been used to predict drug-likeness: selection based on physicochemical properties, functional group filtering, and topology-based classification.

Physicochemical profiling of drug databases produced the popular "rules of five" [74], which state that most orally available compounds possess a molecular weight $\leq$ 500, $\leq$5 hydrogen bond donors (OH and NH), $\leq$10 hydrogen bond acceptors (O and N), and log $P \leq 5$. Other studies have further extended and refined the descriptor set. Table 18.3 summarizes some of the molecular properties that have been used to define drug-like character and their preferred ranges. These descriptors are readily implemented as filters that remove molecules with undesirable properties from lists of candidates. Alternatively, physicochemical descriptors may be used as the basis for machine learning programs that separate drugs from non-drugs [76, 79, 80] and predict ADME and toxicity properties [81–85].

Chemists and biologists have long known that certain chemical moieties are likely to produce false positives in biochemical assays because of their chemical reactivity [86]. Software filters are now routinely used at Vertex [24, 87] and other companies to flag compounds containing functional groups known empirically to contribute to reactivity, insolubility, toxicity, or poor ADME. Table 18.4 lists examples of such undesirable functional groups.

Topologically based methods compare the molecular structures (atom type and connectivity) of drugs and nondrugs to identify distinguishing features. The molecular features used for classification range from small to large, including atom types [88–90], molecular fragments and functional groups [75, 91–94], scaffolds and building blocks [95, 96], and whole molecules [97]. The classification methods include scoring functions [88, 98], decision trees [93] and neural networks [80, 90, 97, 99, 100]. These methods correctly classify 80–90% of known drugs, while classifying 10–30% of nondrugs as drugs [79, 89, 90, 93]. Although this performance is impressive (claimed to be comparable to the intuition of a medicinal chemist), nonspecialists often prefer easy-to-use filtering methods.

It has been pointed out recently that compounds selected on the basis of drug-like properties may actually be sub-optimal for use as drug discovery leads [101]. When drug-like compounds are screened, the resulting hits typically have molecular weights (350–450) and log P values (3–4.5) at the high end of the preferred "rules of five" ranges [101]. In the process of optimizing the potency of these drug-like leads, the molecules become more complex and lipophilic (median molecular weight increase of 69 Da, log P increase of 0.43, 1 ring and 2 rotatable bonds typically added) [102], adversely affecting their pharmacokinetic properties. Compounds with molecular weight (100–350) and log P (1–3) at the low end of the preferred range offer better starting points, since addition of lipophilic groups promises to improve both potency and oral bioavailability [101]. Libraries consist-

Tab. 18.3 Descriptors used to select drug-like molecules [74–77, 98, 154]

Descriptor	*Values*
Molecular weight	150 to 450
Number of atoms	10 to 70
H-bond donors	0 to 3
H-bond acceptors	2 to 9
Rotatable bonds	2 to 8
Polar bonds	1 to 8
Number of rings	1 to 4
Ring size (number of atoms)	3 to 6
Unsubstituted cyclic atoms	1 to 8
Log P	–0.5 to 5
Molar refractivity	40 to 130
$^2K\alpha$ shape index	2 to 12

Tab. 18.4 Functional groups used to filter out undesirable compounds [24, 78, 86, 87, 155]

Acid, sulfonyl, or phosphoryl halides	Michael acceptors
Acetals	N or S mustards
Aldehydes	Nitro groups
Alpha halo ketones	Peroxides
Anhydrides	Phosphoranes
Beta carbonyl quaternary nitrogen	Phosphorus or sulfur halides
Beta lactams	Polyenes
Chloramidines	Primary alkyl halides
Cyanophosphonates	Quaternary C, Cl, I, P or S
Diimides, isocyanates, isothiocyanates	Quinones
Dicarbonyls	Silanes, silates, and silyl halides
Disulfides	Thio-, sulfonate- or phosphonate esters
Epoxides, thiepoxides, and aziridines	Thiols
Halogen bound to N or O	Triacyloximes
Halopyrimidines	Trihydroxyl phenyl
Imines	Triflates

ing of small, polar, lead-like molecules (particularly those with a single charge at physiological pH) are thus better than drug-like libraries for producing useful leads [101]. Such compounds also tend to be highly soluble, which is beneficial for NMR experiments.

18.3.4
Solubility Requirements

Aqueous solubility of compounds is a critical issue, since NMR screens must be run at relatively high compound concentrations, anywhere from millimolar for NOE, chemical shift perturbation or affinity NMR methods [3, 5, 103–108] to ca. 50 μM for saturation-transfer methods [109–111] or screens run using a cryoprobe [112]. The ability to identify potentially insoluble compounds would obviously be very useful when designing screening libraries.

It is difficult to accurately predict aqueous solubility from chemical structure, because it involves disruption of the crystal lattice as well as solvation of the compound. Simple methods based on log P and melting temperature have been widely used [113, 114]. Recently, various prediction methods have been reported [115–125] that are able to predict aqueous solubility to within ca. 0.5 log units (roughly a factor of 3 in concentration). Although these predictors may not be precise or robust enough to select final compounds, they can be used as rough filters for narrowing the list of candidates.

There is currently no substitute for experimentally testing the solubility of all compounds before screening. Even this simple measurement is not trivial. The common practice of diluting DMSO stock solutions into buffer and measuring solubility based on apparent absorption (A) or, more precisely, extinction (ε) at 650 nm is very unreliable. In our experience, A_{650} (or ε_{650}) measurements grossly overestimate solubility, since solutions that appear optically clear often contain high molecular weight aggregates (detectable by more sensitive light-scattering instruments and by nephelometry). Furthermore, the precipitation of compounds is often kinetically controlled [74], so that a compound diluted from DMSO into buffer will precipitate slowly over many hours. The result can be disastrous for the NMR screen, since aggregates can produce both false negatives and false positives regardless of screening method. Our preference is to test all compounds for aggregation prior to screening using the sign of cross peaks in NOESY spectra to assess the rotational correlation time.

18.3.5
Designing Mixtures

Screening compounds in mixtures rather than singly has obvious advantages of higher throughput and reduced protein consumption. There are some drawbacks, however, which can be partly alleviated by careful mixture design:

Deconvolution of mixtures to identify individual hits is necessary for large, spectrally unresolved mixtures and for any screens using protein detection. The total number of data points required to screen and deconvolute mixtures in one step is minimized when the number of compounds per mixture equals $1/(\text{hit rate})^{1/2}$ [126]. When the hit rate is relatively low, so that mixtures are unlikely to contain more than one hit, it is possible to avoid deconvolution by assaying each compound in duplicate, with each replicate in an orthogonal mixture [127]. When the hit rate is relatively high, the optimal mixture size becomes so small (e.g., mixtures of 3 for a 10% hit rate) that it may be more convenient to screen individual compounds [128]. Otherwise, for mixtures of more than four compounds, it is more efficient to carry out a two-step deconvolution, in which active mixtures of N compounds are divided into smaller mixtures of n compounds and re-tested; active n-compound mixtures are then tested singly [126]. The efficiency is highest when n is the square root of N [126]. When a ligand-detected screening method is used, deconvolution can be avoided by designing the mixtures so that there are resolved NMR resonances for every compound in the mixture [3].

Competition between components of a mixture becomes more likely as the hit rate and pool size increase. For a 10% hit rate, competing hits will be found in 26% of mixtures of ten compounds, but only 2.7% in mixtures of three [11]. Unless mixtures are deconvoluted, competition can lead to false negatives when the competing compounds have different affinities. For screens using protein detection, competition can also produce false positives due to the additive effect of multiple weak binders. Competition can be controlled by limiting the size of mixtures and pooling dissimilar compounds to reduce the likelihood of two compounds binding at the same site.

Reactions between compounds in mixtures are more common than is generally recognized. Mixtures of 10 compounds stored in DMSO show evidence of chemical reactions between components in 25% of random mixtures, but this is reduced to 9% in mixtures strategically designed to separate acids from bases and electrophiles from nucleophiles [78]. Care must be taken in designing strategic mixtures, first to designate the compound types correctly (especially challenging for compounds of unknown pK_a) [78], and then to adjust for the additive effect of combining multiple acids or bases into one mixture (requiring high buffer concentrations to control pH) [1, 129]. Decreasing mixture size can dramatically reduce the likelihood of inter-component reactions. For example, decreasing mixture size from 10 to 5 reduces the number of potential pairwise interactions by 4.5-fold [11]. Although the screening of random 100-compound mixtures has been reported [112], the high likelihood of interactions between components (>90% of mixtures) makes the practicality of such large mixtures rather questionable. In practice, most laboratories screen pools of 4–10 compounds.

18.4 Implementing a Strategy

18.4.1 Choosing an Experimental Method

The choice of experimental method used for the screen is based primarily on the strategy and the nature of the target. SAR by NMR [5] and fragment optimization [3] employ protein-detected heteronuclear methods that require isotopically labeled protein, but problems with spectral complexity and line broadening have in the past limited these methods to proteins of less than approximately 30 kDa. Recent methods propose to increase this limit to 100 kDa by using ^{13}C-labeled methyls in combination with deuterated protein [130], selective observation of solvent-exposed amides in a TROSY experiment using deuterated protein [131], or protein containing a unique pair of ^{13}C and ^{15}N labels in adjacent amino acids [14]. Despite these advances, protein-based methods are still not applicable to very large or membrane-bound proteins, and because of the current cost and difficulty of expressing isotopically labeled proteins in insect cells or yeast, they are generally limited to proteins that express well in bacteria.

Ligand-detected screening methods are preferred for targets that have high molecular weights, are membrane-bound, cannot be labeled, have low solubility, are expressed in nonbacterial systems, or cannot be obtained in large quantities. Although these methods do not identify the ligand binding site, this information can be obtained by a variety of other methods (competition experiments, enzyme assays, chemical shift perturbation, crystallography, etc.). The advantages of using unlabeled protein, low protein consumption, and unlimited molecular weight typically outweigh the disadvantage of having to determine the binding site in a separate experiment.

Saturation transfer difference [103, 109] is by far the most sensitive ligand-detected method, capable of screening at protein concentrations as low as 100 nM. This is 500-fold less protein than required by protein-detected methods, and is even below the concentration used in some enzyme assays; 1000 compounds may be screened (in mixtures of five) against a 50 kDa target using 0.5 mg of protein. It is possible to screen relatively insoluble receptor proteins by immobilizing [110] or stabilizing them in synthetic membranes [136]. For example, the *trans*-membrane integrin $\alpha_{IIb}\beta_3$ has been successfully screened while embedded in liposomes [132].

Another advantage of the STD method is that the threshold of detection may be set at the desired level by adjusting the protein concentration [1]. If a targeted library is being screened, then it may be advisable to screen using line-broadening experiments [3, 133] at low ligand:protein ratio in order to detect relatively tight binding (< 1 μM) ligands. A detailed comparison of experimental methods and their protein requirements may be found in Ref. [1].

18.4.2
NMR Screening at Vertex

The NMR screening program at Vertex uses the SHAPES strategy, which is based on ligand-detected screening of drug-like scaffolds [1, 3, 11]. At various times, we have used transferred NOE, line broadening, relaxation difference, diffusion, and saturation transfer difference methods to detect binding, and currently favor STD because of its high sensitivity [1, 3]. Information about binding scaffolds is used to drive combination, elaboration, and variation strategies. NMR screens are carried out using the original drug-like SHAPES library as well as a combi-chem-friendly SHAPES Linking Library, gene-family-specific libraries, and follow-up libraries based on previous hits.

The original SHAPES library consists of commercially available compounds that represent scaffolds commonly found in known drugs. Two Vertex modeling studies using shape descriptors [91, 96] found that a surprisingly small number of molecular scaffolds and side chains were sufficient to describe a large percentage of known drugs covering a wide variety of indications. We constructed a library from commercially available compounds that contained a selection of the most common of these structures, shown in Fig. 18.7. Compounds with undesirable physicochemical properties and functional groups were excluded by using our REOS ("Rapid Elimination Of Swill") software filter [24, 87]. This filter routinely rejects approximately 60% of candidate molecules, though care must be taken to avoid over-filtering. For example, with overly conservative settings the REOS filter rejects 73% of known drugs from the Comprehensive Medicinal Chemistry database. Similar problems have been reported for other classification methods [27, 88]. A combination of filtering and diversity ranking is very effective for narrowing a list of candidate compounds. As seen in Fig. 18.8, applying an 85% similarity threshold to a REOS-filtered set of commercially available compounds removes 42% of the compounds; the combined pass rate for the two filters is only 23%.

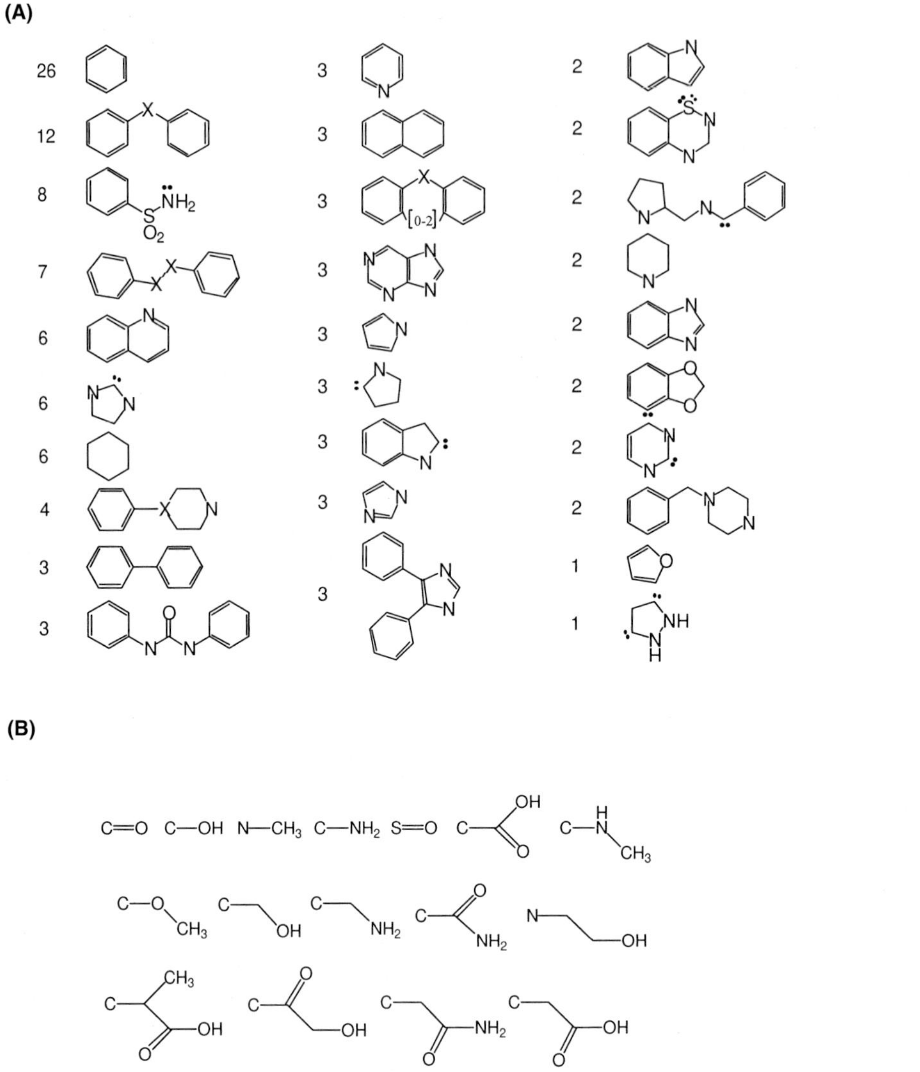

Fig. 18.7 Molecular structures used in the SHAPES library. (**A**) Common drug frameworks used in the SHAPES library. The numbers denote the frequency of occurrence of each framework, and lone pairs indicate points of side-chain attachment X represents a C, N, O or S atom. (**B**) Common drug side chains used in the SHAPES library. The left-most atom indicates the point of attachment to the framework. Reproduced from Peng, et al. 2001[1]. © 2001 Academic Press.

The SHAPES Linking Library was designed to facilitate the use of combinatorial chemistry to follow up screening hits [11]. This library consists primarily of commercially available compounds containing two drug-like scaffolds connected by a linkage that is synthetically accessible. To construct this library, a database of commercially available compounds was filtered to select for drug-likeness and the presence of the desired molecular

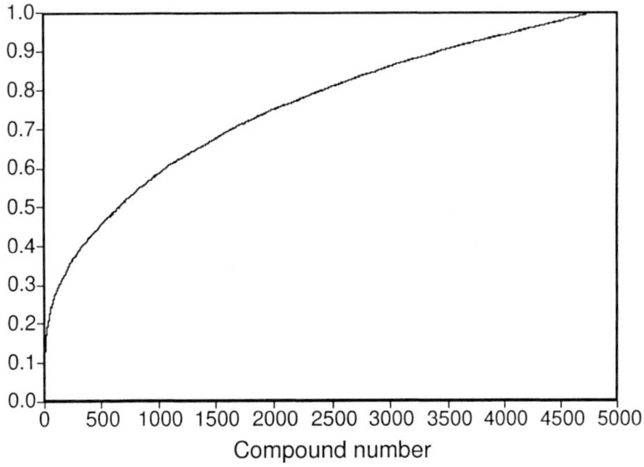

Fig. 18.8 Similarity profile for filtered set of commercially available compounds. 5000 randomly selected compounds from the Available Chemicals Directory that pass the REOS filter were ranked according to their Tanimoto similarity scores (vertical axis) using Daylight fingerprints. 2886 compounds (58%) had similarity scores below 0.85.

features, then edited based on predicted solubility [11]. This reduced set was subjected to Jarvis-Patrick clustering [63], and compounds were manually selected from the centroids of the clusters. In addition to facilitating the purchase of follow-up compounds, centroid picking has the advantage of selecting less complex compounds; the latter is an unintended artifact of clustering based on molecular fingerprints [56]. Cascaded clustering [66] was used to reduce the number of small clusters and singletons. Selected compounds were prepared as stock solutions in DMSO using sonication to aid dissolution. In addition to LC-MS analysis to check compound purity and identity, NMR was used to test the solubility of individual compounds and mixtures of five compounds. By the end of this process, a database of 1.2 million compounds was narrowed to a final library of ca. 500 compounds [11].

Our targeted libraries are comprised of compounds that inhibit members of a target gene family. Most of these are proprietary compounds, such as favored leads from discovery projects, representatives from in-house combinatorial libraries, novel hits from HTS screens, and analogs of known substrates or inhibitors. All of these compounds were selected manually with input from a variety of chemists and modelers, and emphasis was placed on compounds with high solubility, good PK characteristics, available analogs, and activity against multiple members of the gene family.

Follow-up libraries consist of compounds purchased or synthesized around hits from a primary SHAPES screen. The most desirable hits have known chemistry, a clear relationship between structure and binding affinity, and no competing patents. Experience has shown that the best results are obtained when the search for analogs is as broad as possible. This is partly because of the inherent limitations of similarity searches using bit strings, which tend to exaggerate the diversity of small molecule libraries and produce counter-intuitive similarities (for example, failing to recognize topologically similar molecules with different heteroatom substitution patterns) [134, 135]. For this reason, it is always wise to manually edit the results of similarity searches. In addition, broad searches are more likely to avoid the previously described local minimum problem. Also, the number of compounds must be large enough to give a reasonable likelihood of finding hits above the detection limit of the follow-up assay. Given an HTS hit rate around 0.3%, a 500-compound general screening library has a 76% chance of yielding one or two hits. In our experience, a 500-compound follow-up library generally produces HTS-detectable hits, since HTS screens of SHAPES follow-up libraries usually produce hit rates around tenfold higher than random compound libraries. Large improvements in potency demand large follow-up libraries, since linear improvements in potency require geometrically more compounds [15].

If a model is available for the active site of the protein, then virtual screening methods may be used to prioritize the choice of follow-up compounds. For example, hits from the SHAPES Linking Library may be followed up by conceptually breaking the compounds apart at the linker and docking commercially available reagents to replace each portion of the molecule [11]. Docking programs are currently limited in their ability to predict binding modes and relative affinities *a priori* because of difficulty accounting for flexibility and water bound in the active site. The addition of knowledge about which scaffolds bind and proximity restraints from chemical shift perturbation or NOE experiments promises to improve the predictive ability of these docking models.

SHAPES screening has been implemented in several different ways in order to contribute to projects at various stages of development. We find that NMR screening has little impact if run concurrently with HTS, since the resources available for follow-up are invariably directed toward the more potent HTS hits. Because SHAPES screening needs no assay development and requires only a few milligrams of protein, new targets can be screened long before HTS. We have found that even with follow-up screens of only a few hundred compounds it is commonplace to find inhibitors with affinities ranging from micromolar to nanomolar. When a high-throughput assay is unavailable, such inhibitors can be also used as potential probe molecules for use in assay development.

SHAPES screening can also be run in "target rescue" mode to find new leads after HTS screens have failed to generate sufficient viable leads or to identify scaffolds that are clear of intellectual property conflicts. We have found that SHAPES screening often identifies classes of compounds that were tested in the original HTS screen, but were not identified as hits at 30 μM. Why are these classes missed by HTS? One possible explanation is that compounds in the HTS general screening library are typically more complex, which reduces the likelihood of binding (see Fig. 18.6). The simpler SHAPES screening compounds are more likely to bind, and this binding can be observed by NMR even when too weak to be detected under standard HTS screening conditions.

Finally, we have recently implemented SHAPES screening as part of a "chemogenomics"-based approach [137] by screening proprietary scaffolds against multiple members of a gene family. This permits us to find new indications for existing scaffolds, determine intrinsic selectivity patterns, and assess the "drugability" of new targets.

18.5 Conclusion

As instrumentation and experimental methods improve, the role of NMR screening in drug discovery continues to evolve. Of the various criteria for selecting compounds, it is fair to say that solubility remains the most stringent, but the use of probes with cryogenically cooled preamplifiers [138] may mitigate this problem by allowing screens to be run at lower concentrations. Given that advances continue to be made in competing methods for detecting ligand binding (such as mass spectroscopy, surface plasmon resonance, capillary electrophoresis, calorimetry, etc.), such improvements may be essential if NMR is to remain an attractive method for screening.

Like other screening methods, NMR screening is subject to the "garbage in, garbage out" rule; the results are only as good as the compounds that are tested. It continues to be challenging to design screening libraries that will provide meaningful information, and in particular to design follow-up libraries that bridge the gap between weakly binding NMR screening hits and leads with affinities high enough to be detected by conventional enzymatic assays. The increased use of combinatorial libraries should increase the speed with which follow-up libraries can be synthesized. At the same time, the increased use of genomic information will make gene-family-targeted libraries more useful. Regardless of future advances, thoughtful selection of a screening strategy and intelligent design of primary screening and follow-up libraries will continue to be essential.

18.6 Acknowledgements

The author is grateful to many colleagues at Vertex who have contributed to the success of SHAPES screening, in the NMR, modeling, protein biochemistry, enzymology, crystallography and chemistry groups. In addition, Jon Moore, Jeff Peng and John Thomson are thanked for their insightful comments and critical reading of the manuscript.

18.7 References

1 Peng, J.W., Lepre, C.A., Fejzo, J., Abdul-Manan, N., Moore, J. *Methods Enzymol.* **2001**, *339*, 202–230.
2 Hajduk, P.J., Meadows, R.P., Fesik, S.J. *Q. Rev. Biophys.* **1999** *32* (3) 211–240.
3 Fejzo, J., Lepre, C.A., Peng, J.W., Bemis, G.W., Ajay, Murcko, M. A., Moore, J. M. *Chem. Biol.* **1999** 6 (10) 755–69.
4 Maly, D.J., Choong, I.C., Ellman, J.A. *Proc. Natl. Acad. Sci. U S A* **2000** 97 (6) 2419–24.
5 Shuker, S.B., Hajduk, P.J., Meadows, R.P., Fesik, S.W. *Science* **1996** *274* (5292) 1531–4.
6 Hajduk, P.J., Sheppard, G., Nettesheim, D. G., Olejniczak, E.T., Shuker, S.B., Meadows, R.P., Steinman, D.H., Carrera, J.G.M., Marcotte, P.A., Severin, J., Walter, K., Smith, H., Gubbins, E., Simmer, R., Holzman, T.F., Morgan, D.W., Davidsen, S.K., Summers, J.B., Fesik, S.W. *J. Am. Chem. Soc.* **1997** *119* 5818–27.
7 Medek, A., Hajduk, P.J., Mack, J., Fesik, S. W. *J. Am. Chem. Soc.* **2000** *122* 1241–1242.
8 Ross, A. *Automation of NMR measurements and data evaluation for systematically screening interactions of small molecules with target proteins.* in *NMR: Drug Discovery and Design – Post-Genomic Analysis.* **2000.** McLean, Virginia: Cambridge Healthtech Institute.
9 Hajduk, P.J., Boyd, S., Nettesheim, D., Nienaber, V., Severin, J., Smith, R., Davidson, D., Rockway, T., Fesik, S. W. *J. Med. Chem.* **2000** *43* 3862–3866.
10 Boehm, H.-J., Boehringer, M., Bur, D., Gmuender, H., Huber, W., Klaus, W., Kostrewa, D., Kuehne, H., Luebbers, T., Meunier-Keller, N., Mueller, F. *J. Med. Chem.* **2000** *43* 2664–2674.
11 Lepre, C.A. *Drug Discov. Today* **2001** 6 (3) 133–140.

12 Hajduk, P.J., Dinges, J., Schkeryantz, J.M., Janowick, D., Kaminski, M., Tufano, M., Augeri, D.J., Petros, A., Nienaber, V., Zhong, P., Hammond, R., Coen, M., Beutel, B., Katz, L., Fesik, S.W. *J. Med. Chem.* **1999** *42* (3852–3859).

13 Hajduk, P.J., Gomtsyan, A., Didomenico, S., Cowart, M., Bayburt, E.K., Solomon, L., Severin, J., Smith, R., Walter, K., Holzman, T.F., Stewart, A., McGaraughty, S., Jarvis, M.F., Kowaluk, E.A., Fesik, S.J. *J. Med. Chem.* **2000** *43* (25) 4781–4786.

14 Weigelt, J., van Dongen, M., Uppenberg, J., Schultz, J., Wikström, M. *J. Am. Chem. Soc.* **2002**, *124* (11) 2446–2447.

15 Young, S.S., Sheffield, C.F., Farmen, M. *J. Chem. Inf. Comp. Sci.* **1997** *37* 892–899.

16 Dankwardt, S.M., Billideau, R.J., Lawley, LK., Abbott, S.C., Martin, R.L., Chan, C.S., VanWart, H.E., Walker, K.A.M. *Bioorg. & Med. Chem. Lett.* **2000** *10* 2513–2516.

17 Szardenings, A.K., Antonenko, V., Campbell, D.A., DeFrancisco, N., Ida, S., Shi, L., Sharkov, N., Tien, D., Wang, Y., Navre, M. *J. Med. Chem.* **1999** *42* 1348–1357.

18 Warmus, J.S., Ryder, T.R., Hodges, J.C., Kennedy, R.M., Brady, K.D. *Bioorg. Med. Chem. Lett.* **1998** *8* 2309–2314.

19 Mohan, R., Yun, W., Buckman, B.O., Liang, A., Trinh, L., Morrissey, M.M. *Bioorg. Med. Chem. Lett.* **1998** *8* 1877–1882.

20 Haque, T.S., Skillman, G., Lee, C.E., Habashita, H., Gluzman, I.Y., Ewing, T.J.A., Goldberg, D.E., Kuntz, I.D., Ellman, J.A. *J. Med. Chem.* **1999** *42* 1428–1440.

21 Reich, S.H., Johnson, T., Wallace, M.B., Kephart, S.E., Fuhrman, S.A., Worland, S. T., Matthews, D.A., Hendrickson, T.F., Chan, F., Meador, J., Ferre, R.A., Brown, E.L., DeLisle, D.M., Patick, A.K., Binford, S.L., Ford, C.E. *J. Med. Chem.* **2000** *43* 1670–1683.

22 Chang, Y.-T., Gray, N.S., Rosania, G.R., Sutherlin, D.P., Kwon, S., Norman, T.C., Sarohia, R., Leost, M., Meijer, L., Schultz, P.G. *Chem. Biol.* **1999** *6* 361–375.

23 Cottney, J., Rankovic, Z., Morphy, J.R. *Bioorg. Med. Chem. Lett.* **1999** 9 1323–1328.

24 Walters, W.P., Stahl, M.T., Murcko, M.A. *Drug Discov. Today* **1998** *3* (4) 160–178.

25 Böhm, H.-J.,Stahl, M. *Curr. Opin. Chem. Biol.* **2000** 4 283–286.

26 Drewry, D.H.,Young, S.S. *Chemometr. Intell. Lab.* **1999** *48* 1–20.

27 Leach, A.R.,Hann, M.M. *Drug Discov. Today* **2000** 5 (8) 326–336.

28 Van Drie, J.H., Lajiness, M.S. *Drug Discov. Today* **1998** *3* (6) 274–283.

29 Higgs, R.E., Bemis, K.G., Watson, I.A., Wikel, J. H. *J. Chem. Inf. Comput. Sci.* **1997** *37* 861–870.

30 Good, A.C., Lewis, R.A. *J. Med. Chem.* **1997** *40* 3926–3936.

31 Cramer, R.D., Patterson, D.E., Clark, R.D., Soltanshahi, F., Lawless, M. S. *J. Chem. Inf. Comput. Sci.* **1998** *38* 1010–1023.

32 Muegge, I., Martin, Y.C., Hajduk, P.J., Fesik, S. W. *J. Med. Chem.* **1999** *42* 2498–2503.

33 Perola, E., Xu, K., Kollmeyer, T. M., Kaufmann, S.H., Predergast, F.G., Pang, Y.-P. *J. Med. Chem.* **2000** *43* 401–408.

34 Schapira, M., Raaka, B.M., Samuels, H.H., Abagyan, R. *Proc. Natl. Acad. Sci. USA* **2000** *97* (3) 1008–1013.

35 Brown, R.D.,Martin, Y.C. *J. Med. Chem.* **1997** *40* 2304–2313.

36 Böhm, H.-J., Banner, D., Weber, L. *J. Comput.-Aid. Mol. Des.* **1999** *13* 51–56.

37 Julián-Ortiz, J.V., Gálvez, J., Muoz-Collado, C., García-Domenech, R., Gimeno-Cardona, C. *J. Med. Chem.* **1999** *42* 3308–3314.

38 Makino, S., Kayahara, T., Tashiro, K., Takahashi, M., Tsuji, T., Shoji, M. *J.Comput.-Aid. Mol. Des.* **2001** *15* 553–559.

39 Kubinyi, H. *Perspect. Drug Discov. Des.* **1998** *9/10/11* 225–252.

40 Lipinski, C.A., ed. *Bioisosterism in drug design.* Annu. Rep. Med. Chem., ed. R.C. Allen. Vol. 21., Academic Press, Inc. **1986**.

41 Patani, G.A.,LaVoie, E. J. *Chem. Rev.* **1996** *96* 3147–3176.

42 Schneider, G., Neidhart, W., Giller, T., Schmid, G. *Angew. Chem. Int. Ed. Engl.* **1999** *38* (19) 2894–2896.

43 Willet, P., Barnard, J. M., Downs, G. M. *J. Chem. Inf. Comput. Sci.* **1998** *38* 983–996.

44 Stanton, D.T., Morris, T. W., Roychoudhury, S., Parker, C.N. *J. Chem. Inf. Comput. Sci.* **1999** *39* 21–27.

45 Hagadone, T.R. *J. Chem. Inf. Comput. Sci.* **1992** *32* 515–521.

46 Carhart, R.E., Smith, D.H., Venkataraghavan, R. *J. Chem. Inf. Comput. Sci.* **1985** *25* 64–73.

47 Sheridan, R.P., Miller, M.D., Underwood, D.J., Kearsly, S. K. *J. Chem. Inf. Comput. Sci.* **1996** *36* 128–136.

48 Cramer, R.D., Poss, M.A., Hermsmeier, M.A., Caulfield, T.J., Kowala, M.C., Valentine, M.T. *J. Med. Chem.* **1999** *42* 3919–3933.

49 Andrews, K.M.,Cramer, R.D. *J. Med. Chem.* **2000** *43* 1723–1740.

50 Pickett, S.D., Mason, J.S., Mclay, I.M. *J. Chem. Inf. Comput. Sci.* **1996** *36* 1214–1223.

51 McGregor, M.J., Muskal, S.M. *J. Chem. Inf. Comput. Sci.* **1999** *39* 569–574.

52 Marriott, D.P., Dougall, I.G., Meghani, P., Liu, Y.-J., Flower, D.R. *J. Med. Chem.* **1999** *42* 3210–3216.

53 Pickett, S.D., McLay, I.M., Clark, D.E. *J. Chem. Inf. Comput. Sci.* **2000** *40* 263–272.

54 Mason, J.S.,Hermsmeier, M.A. *Curr. Op. Chem. Biol.* **1999** *3* 342–349.

55 Patterson, D.E., Cramer, R.D., Ferguson, A.M., Clark, RD., Weinberger, L.E. *J. Med. Chem.* **1996** *39* 3049–3059.

56 Bayada, D.M., Hamersma, H., Van Geerestein, V.J. *J. Chem. Inf. Comput. Sci.* **1999** *39* 1–10.

57 Brown, R.D.,Martin, Y.C. *J. Chem. Inf. Comput. Sci.* **1997** *37* 1–9.

58 Matter, H. *J. Med. Chem.* **1997** *40* 1219–1229.

59 Gorse, D., Rees, A., Kaczorek, M., Lahana, R. *Drug Discov. Today* **1999** *4* (6) 257–264.

60 Holliday, J.D., Ranade, S.S., Willett, P. *Quant. Struct.–Act. Relat.* **1995** *14* 501–506.

61 Clark, R. D. *J. Chem. Inf. Comput. Sci.* **1997** *37* 1181–1188.

62 Clark, R.D., Langton, W.J. *J. Chem. Inf. Comput. Sci.* **1998** *38* 1079–1086.

63 Jarvis, R.A., Patrick, E.A. *IEEE Trans. Comput.* **1973** *C-22* (11) 1025–1034.

64 Brown, R.D., Martin, Y.C. *J. Chem. Inf. Comput. Sci.* **1996** *36* 572–584.

65 Ward, J.H. *J. Am. Stat. Assoc.* **1963** *58* 236–245.

66 Menard, P.R., Lewis, R.A., Mason, J.S. *J. Chem. Inf. Comput. Sci.* **1998** *38* 497–505.

67 Mason, J.S., Pickett, S.D. *Perspect. Drug Discov. Des.* **1997** *7/8* 85–114.

68 Bayley, M.J., Willett, P. *J. Mol. Graph. Mod.* **1999** *17* 10–18.

69 Pearlman, R.S., Smith, K.M. *Perspect. Drug Discov. Des.* **1998** *9/10/11* 339–353.

70 Stockman, B.J., Farley, K.A., Angwin, D.T. *Methods Enzymol.* **2001** *338* (A) 230–231.

71 Martin, E.J., Critchlow, RE. *J. Comb. Chem.* **1999** *1* 32–45.

72 Hann, M.M., Leach, A.R., Harper, G. *J. Chem. Inf. Comput. Sci.* **2001** *41* 856–864.

73 Kuntz, I.D., Chen, K., Sharp, K.A., Kollman, P.A. *Proc. Natl. Acad. Sci. USA* **1999** *96* 9997–10002.

74 Lipinski, C.A., Lombardo, F., Dominy, B.W., Feeny, P. J. *Adv. Drug Deliv. Rev.* **1997** *23* 3–25.

75 Ghose, A.K., Viswanadhan, V.N., Wendoloski, J.J. *J. Comb. Chem.* **1999** *1* (1) 55–68.

76 Gillet, V.J., Willet, P., Bradshaw, J. *J. Chem. Inf. Comput. Sci.* **1998** *38* 165–179.

77 Oprea, T.I. *J. Comput.–Aid. Mol. Des.* **2000** *14* 251–264.

78 Hann, M., Hudson, B., Lewell, X., Lifely, R., Miller, L., Ramsden, N. *J. Chem. Inf. Comput. Sci.* **1999** *39* 897–902.

79 Ajay, Walters, W.P., Murcko, M.A. *J. Med. Chem.* **1998** *41* (18) 3314–3324.

80 Ajay, Bemis, G.W., Murcko, M.A. *J. Med. Chem.* **1999** *42* 4942–4951.

81 Clark, D.E., Pickett, S.D. *Drug Discov. Today* **2000** *5* (2) 49–58.

82 Mitchell, T., Showell, G. A. *Curr. Opin. Drug Discov. Develop.* **2001** *4* (3) 314–318.

83 Blake, J.F. *Curr. Opin. Biotechnol.* **2000** *11* (104–107).

84 Cronin, M.T.D. *Curr. Opin. Drug Discov. Develop.* **2000** *3* (3) 292–297.

85 Stenberg, P., Luthman, K., Artursson, P. *J. Contr. Rel.* **2000** 231–243.

86 Rishton, G.M. *Drug Discov. Today* **1997** *2* (9) 382–384.

87 Walters, W.P., Murcko, M.A., *Library filtering systems and prediction of drug-like properties*, in *Virtual screening for bioactive molecules*, H.-J. Böhm,G. Schneider, Editors. **2000**, Wiley-VCH: New York. p. 15–32.

88 Wang, J., Ramnarayan, K. *J. Comb. Chem.* **1999** *1* 524–533.

89 Sadowski, J., Kubinyi, H. *J. Med. Chem.* **1998** *41* 3325–3329.

90 Frimurer, T.M., Bywater, R., Naerum, L., Lauritsen, L.N., Brunak, S. *J. Chem. Inf. Comput. Sci.* **2000** *40* 1315–1324.

91 Bemis, G.W., Murcko, M.A. *J. Med. Chem.* **1999** *42* (25) 5095–9.

92 Lewell, X.Q., Judd, D.B., Watson, S.P., Hann, M.M. *J. Chem. Inf. Comput. Sci.* **1998** *38* 511–522.

93 Wagener, M., van Geerestein, V.J. *J. Chem. Inf. Comput. Sci.* **2000** *40* 280–292.

94 Muegge, I., Heald, S L., Brittelli, D. *J. Med. Chem.* **2001** *44* (12) 1841–1846.

95 Xu, J.,Stevenson, J. *J. Chem. Inf. Comput. Sci.* **2000** *40* 1177–1187.

96 Bemis, G.W., Murcko, M.A. *J. Med. Chem.* **1996** *39* (15) 2887–93.

97 Burden, F.R., Winkler, D.A. *J. Chem. Inf. Comput. Sci.* **1999** *39* 236–242.

98 Xu, J., Stevenson, J. *J. Chem. Inf. Comput. Sci.* **2000** *40* 1177–1187.

99 Sadowski, J. *Curr. Opin. Chem. Biol.* **2000** *4* 280–282.

100 Sadowski, J., *Database profiling by neural networks*, in *Virtual screening for bioactive molecules*, H.-J. Böhm,G. Schneider, Eds. **2000**, Wiley-VCH: New York. p. 117–129.

101 Teague, S.J., Davis, A.M., Leeson, P.D., Oprea, T. *Angew. Chem. Int. Ed. Engl.* **1999** *38* (24) 3743–3747.

102 Oprea, T.I., Davis, A.M., Teague, S.J., Leeson, P. D. *J. Chem. Inf. Comput. Sci.* **2001** *41* 1308–1315.

103 Meyer, B., Weimar, T., Peters, T. *Eur. J. Biochem.* **1997** *246* (3) 705–709 .

104 Chen, A., Shapiro, M.J. *J. Am. Chem. Soc.* **1998** *120* 10258–10259.

105 Lin, M., Shapiro, M.J. *J. Org. Chem.* **1996** *61* 7617–9.

106 Lin, M.L., Shapiro, M.J., Wareing, J.R. *J. Am. Chem. Soc.* **1997** *119* 5349–5250.

107 Lin, M., Shapiro, M.J., Wareing, J.R. *J. Org. Chem.* **1997** *62* (25) 8930–8931.

108 Gonnella, N., Lin, M., Shapiro, M. J., Wareing, J.R., Zhang, X. *J. Magn. Reson.* **1998** *131* (2) 336–8.

109 Mayer, M., Meyer, B. *Angew. Chem. Int. Ed. Engl.* **1999** *38* (12) 1784–1788.

110 Klein, J., Meinecke, R., Mayer, M., Meyer, B. *J. Am. Chem. Soc.* **1999** .

111 Dalvit, C., Pevarello, P., Tato, M., Vulpetti, A., Sundstrom, M. *J. Biomol. NMR* **2000** *18* 65–68.

112 Hajduk, P.J., Gerfin, T., Boehlen, J.-M., Haberli, M., Marek, D., Fesik, S.W. *J. Med. Chem.* **1999** *42* 2525–2517.

113 Meylan, W.M., Howard, P.H. *J. Pharm. Sci.* **1995** *84* (1) 83–92.

114 Meylan, W.M., Howard, P.H., Boethling, R. S. *Environ. Toxicol. Chem.* **1996** *15* (2) 100–106.

115 McElroy, N.R., Jurs, P.C. *J. Chem. Inf. Comput. Sci.* **2001** *41* 1237–1247.

116 Jorgensen, W.L., Duffy, E.M. *Bioorg. & Med. Chem. Lett.* **2000** *10* 1155–1158.

117 Abraham, M.H., Le, J. *J. Pharm. Sci.* **1999** *88* (9) 868–880.

118 Huuskonen, J., Salo, M., Taskinen, J. *J. Pharm. Sci* **1997** *86* (4) 450–454.

119 Huuskonen, J., Salo, M., Taskinen, J. *J. Chem. Inf. Comput. Sci.* **1998** *38* 450–456.

120 Huuskonen, J. *J. Chem. Inf. Comput. Sci.* **2000** *40* 773–777.

121 Huuskonen, J. *Comb. Chem. High Tech. Scr.* **2001** *4* 311–316.

122 Klopman, G., Zhu, H. *J. Chem. Inf. Comput. Sci.* **2001** *41* 439–445.

123 Mitchell, B.E., Jurs, P.C. *J. Chem. Inf. Comput. Sci.* **1998** *38* 489–496.

124 Jain, N.,Yalkowsky, S.H. *J. Pharm. Sci.* **2001** *90* (2) 234–252.

125 Ran, Y., Jain, N., Yalkowsky, S.H. *J. Chem. Inf. Comput. Sci.* **2001** *41* 1208–1217.

126 Teixido, J., Michelotti, E.L., Tice, C.M. *J. Comb. Chem.* **2000** *2* 658–674.

127 Snider, M. *J. Biomol. Screen.* **1998** *3* (3) 169–170.

128 Ross, A., Schlotterbeck, G., Klaus, W., Senn, H. *J. Biomol. NMR* **2000** *16* 139–146.

129 Ross, A., Senn, H. *Drug Discov. Today* **2001** *6* (11) 583–593.

130 Hajduk, P.J., Augeri, D. J., Mack, J., Mendoza, R., Yang, J., Betz, S.F., Fesik, S.W. *J. Am. Chem. Soc.* **2000** *122* 7898–7904.

131 Pellecchia, M., Meininger, D., Shen, A. L., Jack, R., Kasper, C.B., Sem, D.S. *J. Am. Chem. Soc.* **2001** *123* 4633–4634.

132 Meinecke, R., Meyer, B. *J. Med. Chem.* **2001** *44* 3059–3065.

133 Hajduk, P.J., Olejniczak, E.T., Fesik, S.W. *J. Am. Chem. Soc.* **1997** *119* 12257–61.

134 Dixon, S.L., Koehler, R.T. *J. Med. Chem.* **1999** *42* 2887–2900.

135 Flower, D.R. *J. Chem. Inf. Comput. Sci.* **1998** *38* 379–386.

136 Meyer, B. *STD NMR to screen libraries and characterize binding.* in *NMR: drug discovery and design – Post-genomic analysis.* **2000.** McLean, Virginia: Cambridge Healthtech Institute.

137 Caron, P.R., Mullican, M.D., Mashal, R.D., Wilson, K.P., Su, M.S., Murcko, M.A. *Curr. Op. Chem. Biol.* **2001** *5* 464–470.

138 Busse-Grawitz, M. E., Roeck, W., *NMR Probe Head With Cryogenically Cooled Preamplifiers*, in *United States Patent 5,814,992.* **1998**, Spectrospin AG, Fallenden, Switzerland: USA.

139 Lee, C. E., Kick, E. K., Ellman, J. A. *J. Am. Chem. Soc.* **1998** *120* 9735–9747.

140 Dolle, R. E., Guo, J., O'Brien, L., Jin, Y., Piznik, M., Bowman, K. J., Li, W., Egan, W. J., Cavallaro, C. L., Roughton, A. L., Zhao, Q., Reader, J. C., Orlowski, M., Jacob–Samuel, B., Carroll, C. D. *J. Comb. Chem.* **2000** *2* 716–731.

141 Bastos, M., Maeji, N. J., Abeles, R. H. *Proc. Natl. Acad. Sci. USA* **1995** *92* 6738–6742.

142 Vendeville, S., Bourel, L., Davioud-Charvet, E., Grellier, P., Deprez, B., Sergheraert, C. *Bioorg. Med. Chem. Lett.* **1999** *9* 437–442.

143 Ogbu, C. O., Qabar, M. N., Boatman, D., Urban, J., Meara, J. P., Ferguson, M. D., Tulinsky, J., Lum, C., Babu, S., Blaskovich, M. A., Nakanishi, H., Ruan, F., Cao, B., Minarik, R., Little, T., Nelson, S., Nguyen, M., Gall, A., Kahn, M. *Bioorg. Med. Chem. Lett.* **1998** *8* 2321–2326.

144 Matthews, D. A., Dragovich, P. S., Webber, S. E., Fuhrman, S. A., Patick, A. K., Zalman, L. S., Hendrickson, T. F., Love, R. A., Prins, T. J., Marakovits, J. T., Zhou, R., Tikhe, J., Ford, C. E., Meador, J. W., Ferre, R. A., Brown, E. L., Binford, S. L., Brothers, M. A., DeLisle, D. M., Worland, S. T. *Proc. Natl. Acac. Sci. USA* **1999** 96 11000–11007.

145 Burns, C. J., Groneberg, R. D., Salvino, J. M., McGeehan, G., Condon, S. M., Morris, R., Morrissette, M., Mathew, R., Darnbrough, S., Neuenschwander, K., Scotese, A., Djuric, S. W., Ullrich, J., Labaudiniere, R. *Angew. Chem. Int. Ed. Engl.* **1998** *37* (20) 2848–2850.

146 Szardenings, A. K., Harris, D., Lam, S., Shi, L., Tien, D., Wang, Y., Patel, D. V., Navre, M., Campbell, D. A. *J. Med. Chem.* **1998** *41* 2194–2200.

147 Cao, X., Moran, E. J., Siev, D., Lio, A., Ohashi, C., Mjalli, A. M. M. *Bioorg. Med. Chem. Lett.* **1995** 5 (24) 2953–2958.

148 Li, Z., Yeo, S. L., Pallen, C. J., Ganesan, A. *Bioorg. Med. Chem. Lett.* **1998** *8* 2443–2446.

149 Green, J. *J. Org. Chem.* **1995** *60* 4287–4290.

150 Gray, N. S., Wodicka, L., Thunnissen, A.-M. W. H., Norman, T. C., Kwon, S., Espinoza, F. H., Morgan, D. O., Barnes, G., LeClerc, S., Meijer, L., Kim, S.-H., Lockhart, D. J., Schultz, P. G. *Science* **1998** *281* 533–538.

151 Ryder, T. R., Hu, L.-Y., Rafferty, M. F., Millerman, E., Szoke, B. G., Tarczy-Hornocii, K. *Bioorg. Med. Chem. Lett.* **1999** 9 1813–1818.

152 Dumas, J., Sibley, R., Riedl, B., Monahan, M. K., Lee, W., Lowinger, T. B., Redman, A. M., Johnson, J. S., Kingery-Wood, J., Scott, W. J., Smith, R. A., Bobko, M., Schoenleber, R., Ranges, G. E., Housley, T. J., Bhargava, A., Wilhelm, S. M., Shrikhande, A. *Bioorg. Med. Chem. Lett.* **2000** *10* 2047–2050.

153 Thomas, J. B., Fall, M. J., Cooper, J. B., Rothman, R. B., Mascarella, S. W., Xu, H., Partilla, J. S., Dersch, C. M., McCollough, K. B., Cantrell, B. E., Zimmerman, D. M., Carroll, F. I. *J. Med. Chem.* **1998** *41* 5188–5197.

154 Brown, R. D., Hassan, M., Waldman, M. *J. Mol. Graph. Mod.* **2000** *18* 427–437.

155 Lewis, R. A., Mason, J. S., McLay, I. M. *J. Chem. Inf. Comput. Sci.* **1997** *37* 599–614.

156 Hajduk, P. J., Bures, M., Praestgaard, J., Fesik, S. W. *J. Med. Chem.* **2000** *43*, 3443–3447.

19
Strategies for Hit Finding Using NMR

Werner Klaus and Hans Senn

19.1
Introduction

The present chapter describes the developments made and our experience in the area of biomolecular screening by the NMR group at Roche Basle. It is not intended to provide a comprehensive review of the subject. Rather we give an overview of the concepts and the currently available tools and show how they are applied by our group. We also try to illustrate some practical aspects of NMR screening.

The development of a pharmaceutically relevant drug is a multi-step process. It starts with the attempt to identify the mechanism(s) of a disease and to reduce its complexity to a system which can be studied by biological, chemical, and physical methods. Once the function (or malfunction) of a biological macromolecule (usually a protein) has been implicated in contributing to a disease state, the search begins for small molecules that modulate its activity in the test tube in such a way that the desired beneficial effect results. These molecules, called "hits", do not necessarily possess the attributes of a drug that can be administered to humans. They serve as a starting point for the conversion into "leads" which are amenable to chemical modifications. Once a lead has been optimized it can be developed further into a drug candidate, which is subjected to clinical trials.

One of the established methods for finding hits is "high-throughput screening" (HTS). This is a powerful method enabling several thousands of compounds per day to be tested. However, there are drawbacks. In many cases the assay for the detection of a hit comprises coupled enzymatic reactions, and a test compound may falsely lead to positive results because of its interference with components of the enzyme cascade. Problems may arise further from the chemical substances that make up the compound depository which are tested in HTS. Especially for companies with historically grown collections of compounds, the quality of the chemicals is an uncontrolled parameter in the sense that it is not always known whether the compounds are still unmodified: they may have decomposed, precipitated or formed aggregates (polymers).

NMR spectroscopy can be used to help in validating the hits resulting from such a biological screen by confirming the supposed interaction. Herein NMR offers the advantage that it studies a system which has been reduced to its basic components, namely the drug target – in the majority of cases a protein – and a putative ligand. Binding effects are readily observed either for the NMR signals of the small molecule employing one-dimensional NMR techniques or by monitoring changes in the 2D (two-dimensional) spectrum of the target molecule. For the later approach, isotopic labeling of the macromolecule by ^{15}N is almost always mandatory. Another application of NMR might be called *de novo* screening as it works in a way similar to its biological equivalent. Its purpose is to find low molecular weight chemical moieties that bind to a drug target and which are readily amenable to chemical modifications transforming them into lead structures. Here, the target molecule is exposed to either one compound or simultaneously to a mixture of compounds, and their interaction is observed via changes in the spectrum of the drug target. This approach, called "SAR (structure-activity relationship) by NMR", has been pioneered by the NMR group at Abbott Laboratories [1]. Although claims have been made that more than 200,000 compounds could be tested in less than one month [2], thus making the method comparable to HTS, doubts remain whether this number will be achieved in practice.

19.2
Hit Finding by NMR: the "Needle" Concept for *de novo* Screening

Needle screening aims at the detection of low molecular weight compounds that are reduced to the minimal structural elements while still fitting into an active site and fulfilling the essential binding requirements. This approach has several advantages over a "standard screening" of a library of chemicals: (i) a needle lacks unwanted structural elements which, for example, might result in toxicity or metabolic instability; (ii) a later introduction of tailored side chains during a subsequent optimization process does not *per se* lead to oversized molecules having, for example, bioavailability problems; (iii) a needle and its derivatives are likely to be more easily and cheaply synthesized than complex natural products [3]. The concept hinges critically on the detection of weak interactions, as a needle is not likely to show strong binding to a target protein. NMR is perfectly suited for this purpose as affinities in the millimolar to micromolar range can be detected readily by various methods, either as effects in the spectrum of a ligand when the protein is added ("ligand observe") or in that of a protein when the ligand is added ("protein observe").

Usually, only weakly binding molecules are found during the initial screening using the needle approach. These molecules have to be improved in their binding affinity by, for example, the addition of further fragments that interact with the target at adjacent sites. One way of identifying them might be "subsite screening". This method can be used in two cases: (i) when a given compound occupies only a part of an extended binding site (at a subsite A), to identify one or more additional ligands that would fill (at a subsite B) in a complementary manner the total binding site; or (ii) when the active site of a protein is made up of more than one binding pocket, to saturate a first pocket with a given compound and to search for ligands binding at an adjacent second pocket. Fig. 19.1 illustrates the two situations. The given compound and the molecule interacting with low affinity at subsite B or at the second pocket may then be chemically linked. Although the untethered molecules might only bind in the micromolar to millimolar range, the binding affinity of a linked compound is, in principle, the product of the binding constants of the individual fragments plus a term that accounts for the changes in binding affinity that are due to linking [1].

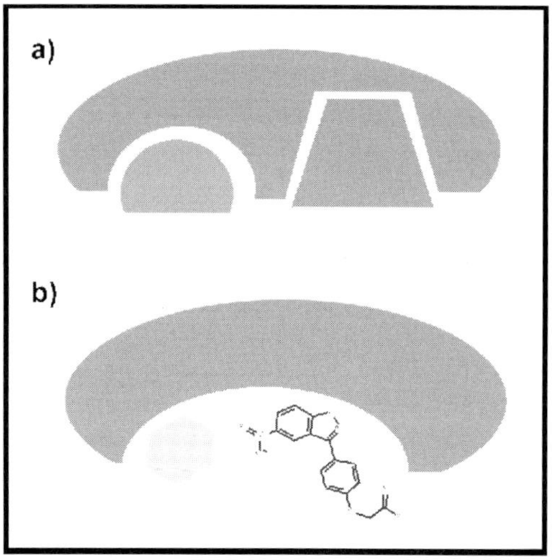

Fig. 19.1 Schematic views of the ligand pocket of an enzyme. **a** Two adjacent pockets, each occupied by a needle fragment. **b** One large cavity that may be filled by two needle fragments.

19.3 Requirements for NMR-based Screening

For the application of biomolecular NMR in the process of hit identification and hit validation, four important components have to be optimized: (i) supply of the target molecule, (ii) selection of ligands, (iii) hardware for automated sample preparation and high-sensitivity measurements, and (iv) software tools for automated data evaluation.

19.3.1 The Sample

Certainly of highest importance is the supply of the target molecule, normally a protein, in a water-soluble or membrane-mimicking form in order to investigate the system of drug-target and putative ligand by using high-resolution NMR methods. Although the concentration of protein per NMR sample can be decreased to 10–50 μM, the complexity (and hence the molecular weight) of target proteins has grown. Accordingly there is still the need for 1–10 mg for the recording of 1D spectra or ~100 mg protein (preferably ^{15}N-labeled for heteronuclear 2D measurements) if one wants to characterize a larger number of ligands. This requirement excludes many drug targets from important disease areas such as the central nervous system, because here proteins often belong to the class of membrane-bound receptors and are therefore currently not amenable to NMR screening. Although, in principle, any protein which can be obtained in reasonable quantities could be subjected to screening, in practice an *E. coli* expression system is mandatory for the large scale fermentation of biomass if one wants to do 2D heteronuclear NMR spectroscopy which necessitates ^{15}N-labeling. Another important factor is a dedicated biochemistry laboratory for the purification and reconstitution of natively folded and active proteins. The effort to be spent in this area can be considerable, as it is essential to be able to keep the proteins in solution for a long period of time without aggregation, precipitation or proteolytic degradation. To achieve this one can add stabilizing agents such as, for example, 3-[(3-cholamidopropyl)dimethylammonio]-1-propanesulfonate hydrate (CHAPS) or glycerol to prevent the formation of aggregates or dithiothreitol (DTT) as an anti-oxidative, protective compound [4, 5]. Sometimes one has to go back to the level of molecular biology and modify the clones to introduce point mutations or to create gene constructs of different lengths that may finally result in the expression of a more "robust" protein. In a search for optimal conditions, the stability with respect to temperature and pH can be monitored conveniently both via circular dichroism (CD) and NMR spectroscopy. Ideally, the macromolecular target would have its optimum of stability at room temperature and at a neutral pH, i.e. under conditions comparable to those normally found in conventional, biological screening efforts. Small deviations can be accepted, but for an enzyme they are normally limited by the range of the pH-activity profile. In any case, the effects of the variation of the pH of the solution on the chemical shifts of the NMR signals of the target molecule have to be investigated. In the NMR screen, the differentiation between binding and nonbinding compounds relies on the observation of shifted resonance lines. Therefore, it is mandatory to identify those that are intrinsically pH sensitive. Otherwise, potential pH changes induced by the addition of the testing molecules may result in false positive hits. In principle a stronger buffer system can be used, but pH changes will not be prevented completely [6]. A minor effect that has to be taken into account is that shifts can be induced by the solvent used to dissolve the library compounds. The proportion of such a solvent in the screening solution is between 1 and 5%, which for DMSO-d_6 amounts to concentrations between 142 and 710 mM. For most proteins this is high enough to lead to detectable specific interactions of the solvent, allowing their binding site(s) on the surface of the target molecule to be identified. For example, Dalvit and colleagues were able to detect the presence of one molecule of DMSO bound in the substrate binding site of the protein FKBP12 [7]. A substructure search of small molecules similar to DMSO resulted in the identification of molecules with improved binding affinities.

19.3.2
The Ligands

The selection, build-up, and maintenance of a collection of small molecules (designated as shapes, needles, binding motifs...) with specific features is another factor of importance (see Chapt. 18). There are at least two possible ways of creating such libraries. On the one hand, one could perform an unbiased screen by the testing of randomly chosen compounds that constitute a universal library. On the other hand, one could construct a dedicated library that exploits specific properties of a given target molecule (three-dimensional structure, similarity to other members of a family of proteins, etc). This requires input from modelers, cheminformatics experts, medicinal chemists and NMR spectroscopists, to include more NMR-specific aspects such as solubility and easy handling of substances (toxicity, storage properties). The compounds that make up the libraries are stored as DMSO-d_6 stock solutions in 96- or 384-well plates, a format that many robotic systems can handle. Routinely the libraries are checked by optical methods for solubility in aqueous buffer at two pH values (6.5; 8.5) at a concentration of 5 mM in the presence of 2.5% DMSO-d_6. Another check is for the tendency of the compounds to change the pH of the protein buffer when they are added. As this may induce shifts in the resonances of the protein target (see above), it is helpful to have this information at hand and to take it into account during the evaluation of the results. Both tests are performed in dedicated laboratories in an automated manner using optical methods for the detection of changes [8].

Another point to consider in the context of libraries of compounds for NMR-based screening is "cocktails versus single compounds". Obviously the testing of mixtures of 10–100 individual compounds in a single NMR measurement is very attractive as it will increase the number of substances to be tested for a given amount of the target molecule. If no changes in the NMR screen are observed, then the components of the mixture can be regarded as inactive. If such a cocktail shows effects, it has to be broken down into clusters of smaller size that have to be re-tested until the active compound is identified. How often this "deconvolution" has to be performed depends on the initial hit rate and the number of components of the cocktail [9]. Only when both are favorable the full advantage of using mixtures is exploited. A high hit rate may require many cycles of deconvolution, thus reducing the benefits of using a cocktail. An important factor in the building of mixtures is the selection of compounds to be combined. Obviously one has to prevent substances from modifying each other's physico-chemical properties, e.g., one must avoid pooling together strong acids and bases. When more than ten compounds are to be combined, then the initial concentration of the stock solutions in, for example, DMSO has to be high in order to keep the final ratio of that solvent at an acceptable level below 5%. Therefore the solubility in DMSO may impose a restriction. Otherwise the concentration of the individual compounds in the mixture would be low, which in turn has effects on the detection threshold of the screen. Because the concentration of a weakly binding molecule has to be in the range of its putative K_D value, a minimum value of 100 μM should be maintained in order to detect ligands of a correspondingly low affinity [2]. The ligand has to be present in an at least tenfold excess over the target molecule, and therefore 2D HSQC (heteronuclear single quantum correlation) measurements on ^{15}N-labeled protein would have to be performed at a protein concentration of about 10 μM, which is time-consuming even with the most sophisticated NMR equipment. For "ligand-observe" techniques, the requirements regarding concentration of the compounds are less severe, but the problem of the mutual "compatibility" of the constituents of the mixture still remains.

19.3.4
Automation of NMR Experiments and Hardware Improvements

The automation of sample preparation is clearly a requirement if one wants to measure a reasonably high number of test compounds. While sample changers have been used for a long time in analytical NMR laboratories, sample preparation robots have been introduced only recently, especially in the field of biomolecular NMR. They offer the great advantage that the samples are freshly prepared under identical conditions and delivered to the NMR instrument in a "just-in-time" fashion. Thus, they circumvent the disadvan-

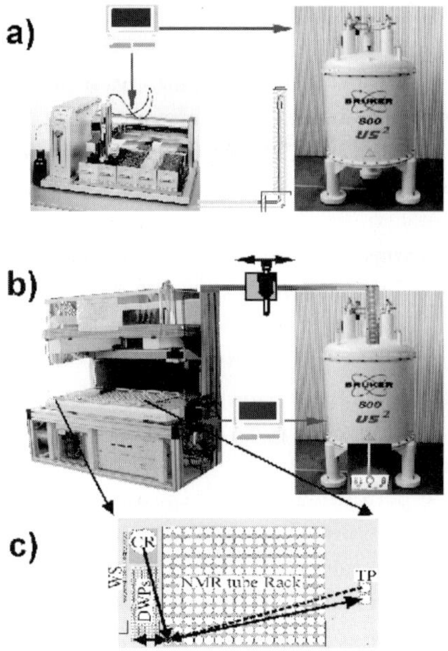

Fig. 19.2 Two set-ups for automated "just-in-time" sample preparation for NMR screening. The communication between the preparation robot and the spectrometer is handled and controlled by PC software. **a** A Gilson 215 liquid handling robot is connected via a capillary to a flow-through NMR probe head. **b** A Tecan Genesis robot for the preparation of samples in discrete NMR tubes. The system is connected via the Bruker SampleRail to the NMR spectrometer. **c** Close-up view on the worktable of the Tecan rack. The arrows indicate liquid transfer steps by the pipetting arm and the forth and back (dotted line) transfer of the NMR samples to the TP by the robot arm, respectively. CR: cooling rack for storage of the macromolecules target; DWPs: deep well plates for storage of libraries; TP: sample transfer port to SampleRail; WS: washing station.

tage of a traditional sample changer, in which samples would spend different lengths of time between preparation and measurement. For this purpose, two systems have been used in our laboratory (Fig. 19.2). One consists of a Gilson liquid-handling pipetting robot (system 215; Gilson, Middleton, WI) connected via a capillary line with a flow-through probehead. This combination allows the automated acquisition of NMR spectra in a "continuous" fashion. Manual interference is limited to the supply of protein solution and the screening compounds, which are stored in a 96 deep-well plate format. Based on the results of a preceding determination of the solubility of the library substances, those compounds are excluded that were found to precipitate upon transfer from the DMSO stock solution into the aqueous buffer system in which the protein was kept. This reduces the risk of a clogging of the capillaries of the liquid handling system or of the flow-through probehead. For 2D HSQC measurements an overall throughput of 20 samples per day including measurements over the weekend is achieved with the current instrumentation. The number of samples could be increased if a cryo-cooled flow-through probe head were to become available.

An alternative set-up has been developed in our laboratory for the preparation of discrete samples. It consists of a Genesis sample preparation robot (Tecan, Hombrechtikon, Switzerland) connected to the so-called Bruker SampleRail (Bruker Biospin, Karlsruhe, Germany) [8]. The Genesis system accommodates a container in which the target molecule is stored, two deep well plates with the stock solutions of the compounds to be tested, and racks in which up to 192 premounted disposable NMR tubes can be placed. The pipetting arm of the Genesis mixes a small aliquot (10–20 μL) of the testing substance to 500 μL of the target molecule directly in a standard 5 mm NMR tube. A second moveable arm takes up the sample and delivers it to the SampleRail system, which transfers it to the magnet for insertion. After the acquisition of the NMR spectrum, the sample is transported back and placed at its original position within the Genesis rack. This combination of hardware can be used with any conventional high-sensitivity probe head. Obviously the highest throughput in terms of number of samples can be obtained when the spectrometer is equipped with a cryo-cooled probe head.

The field of NMR instrumentation has witnessed a continuous improvement in the magnetic field strength of the superconducting magnets, the stability of the console electronics and the performance of probe heads. The commercial introduction of the cryogenic NMR probe technology, in which the preamplifier and the radio frequency coils of the probe are cooled to low temperatures, has significantly increased the signal-to-noise ratio in NMR spectra [2]. Another area in which large improvements were achieved is the miniaturization of probe heads. By the reduction of the size of the coils the signal-to-noise ratio is improved, the requirements for the volume of a sample are reduced, and concomitantly the absolute amount of protein is reduced [G. Schlotterbeck, manuscript in preparation].

19.3.5
Automation in Spectral Analysis

The software is available for handling the data associated with sample preparation, measurement, spectral evaluation and data interpretation [8]. The statistical method of Principal Component Analysis (PCA) is very useful for the interpretation of 2D spectra in order to spot compounds that interact with the screening target. Its application is to a large extent automated, and its implicit data reduction conveniently allows for the simultaneous visualization of a large number of data sets [10]. PCA uses a matrix of data with rows of observations (=spectral regions) and columns corresponding to different samples (=spectra, one spectrum per screened compound) to represent a multivariate data table. It can be considered as the projection of the data on a low-dimensional space, usually comprising two to five dimensions. The projected values are the "scores" along the principal components (PC) of the corresponding sample. Outliers with respect to the majority of the data manifest themselves by their large scores along one or more axes. The visual inspection of 2D displays of PC^i versus PC^j allows one to find combinations that show a clear separation of spectra from the "average" [11].

Opportunities exist for the improvement of software tools for the docking of a ligand to a target molecule based on NMR-derived chemical shift perturbations. These secondary shift changes – readily deduced from 2D HSQC-spectra – reflect the binding location and orientation of the small molecule within the three-dimensional structure of the target [12]. Any information gained from them would provide a useful input for the process of the rational drug design of more elaborate chemical structures with improved binding capacities.

In the following, some examples will be given that illustrate the concepts and applications of the various methods mentioned above.

19.4
Examples

19.4.1
Gyrase

DNA gyrase is a well-established target for the development of anti-bacterial drugs [13]. It catalyzes the ATP-dependent introduction of negative supercoils into bacterial DNA as well as the decatenation and unknotting of DNA. Based on detailed 3D structural information on the ATP-binding site, rational drug design was used to generate new lead structures which are expected to bind at that site [3]. At the early stage of the design process it is not very likely that compounds with high affinity will be found. Therefore it is essential to have a method at hand for the validation of hits that interact weakly with the target protein. NMR is especially suited to this, as it offers a rapid way to confirm binding and to localize the interaction site of a ligand on the protein. The effects of binding are monitored via the observation of changes in chemical shifts of resonance lines of the ^{15}N-labeled protein in 2D ^{1}H-^{15}N HSQC spectra [14, 15]. Provided that the assignment of the macromolecule is known, one can identify the interaction site of a ligand in a sequence-specific manner and map it onto the structure of the protein. Alternatively, if an assignment is not available, the shifts induced by a compound with known binding properties may serve as a reference against which the shift pattern of a putative hit is compared. In the case of gyrase, the sequence-specific NMR assignment of the 24 kDa fragment of subunit B had been obtained in-house [16]. NMR spectroscopy supported this project in two ways: (i) as an efficient biophysical method to validate hits found in a "needle screening" and (ii) for the identification of small compounds supposed to bind to a part of the active site ("subsite screening").

In order to find novel ligands that interact at the ATP-site of gyrase, a "3D-structure-based needle screening" combining several biophysical methods was performed [3]. A flow chart of the various steps is given in Fig. 19.3. To detect binding by NMR, 2D ^{1}H-^{15}N HSQC spectra of ^{15}N-labeled protein were recorded. This technique offers the additional and very important advantage that the binding site of a ligand can be detected in a sequence-specific manner and mapped onto the three-dimensional structure of the target molecule (Fig. 19.4). This helped to exclude several classes of needles which had been tested as positives in the biological assay but did not show binding in the NMR measure-

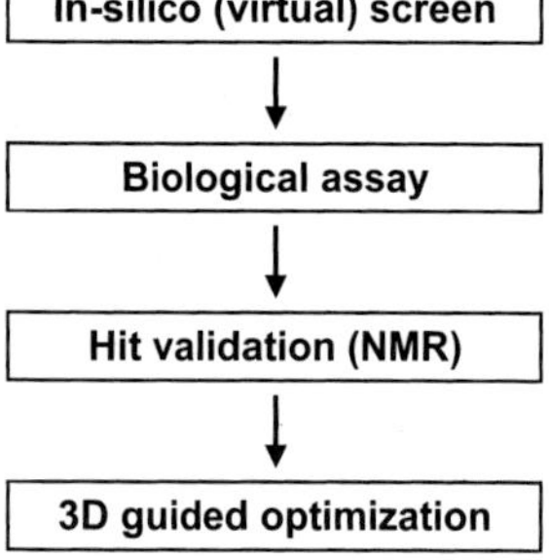

Fig. 19.3 Flow chart of 3D structure-based biased needle screening as applied to the bacterial enzyme, gyrase.

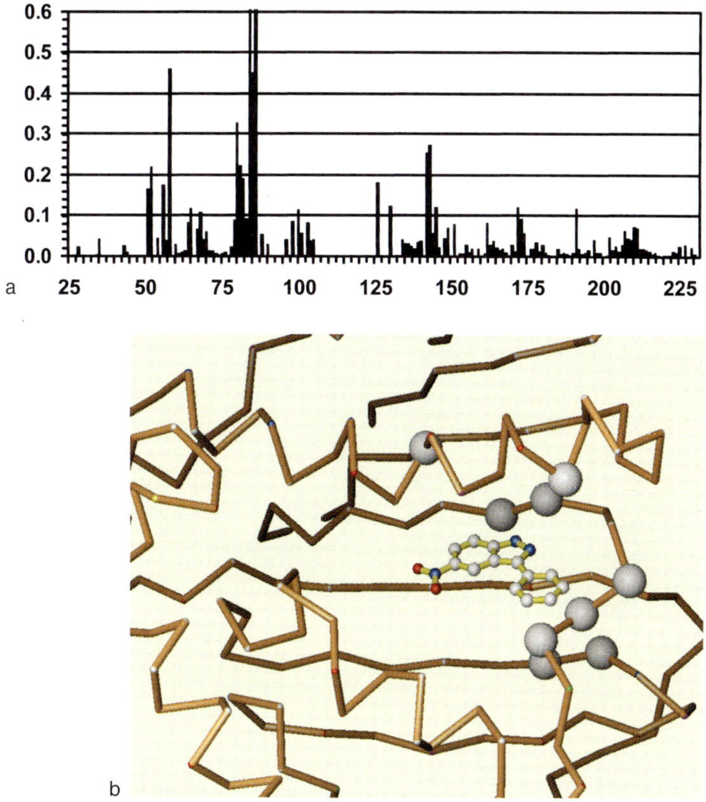

Fig. 19.4 **a** Plot of the amino acid sequence versus differences in chemical shift of the ^{1}H-^{15}N cross peaks in the 2D HSQC spectrum of the 24 kD fragment of GyrB (23-234) upon complexation with the compound Ro 05-8733. **b** X-ray structure of the complex of the GyrB fragment and Ro 05-8733. The ligand is shown in a ball-and-stick representation; the protein backbone is displayed by tubes connecting the C_{α} atoms. White balls along the protein backbone indicate residues for which a difference in chemical shift greater than 0.2 ppm was observed.

ments at all or which led to precipitation of the protein. In some cases the ligands were found to be insoluble under the conditions used for NMR. As the requirements regarding the concentrations of protein and ligand and their solubility properties are similar for NMR and X-ray crystallography, it was possible to predict that attempts to obtain crystals of a "complex" of these (because of their insolubility) false-positive ligands and the gyrase fragment would not lead to the desired complexes but to crystals of the "empty" apo-structure. Table 19.1 shows some needles that were found, together with their K_D-values as determined by NMR titrations [17].

Another contribution of NMR to the gyrase project was based on an approach termed "subsite screening by NMR". For gyrase, the situation corresponded to case (1) as defined in section 19.2. Fig. 19.5 (see p. 428) provides a space-filling model of the active site of GyrB in the complex with a phenylnitroindazole as determined by X-ray crystallography

Tab. 19.1 Chemical structures of needle molecules as found in the gyrase project. The K_D values were determined by NMR titrations [17]

Needle	*K_D from NMR*	*Name*
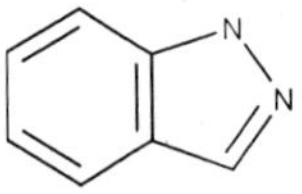	10 mM	Indazole
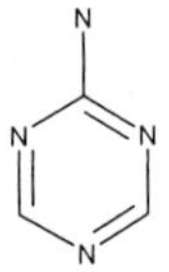	50 mM	2-Amino-1,3,5-triazine
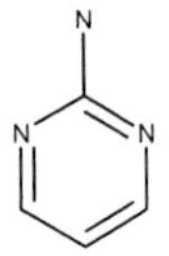	100 mM	2-Aminopyrimidine
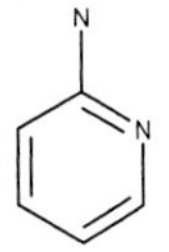	200 mM	2-Aminopyridine
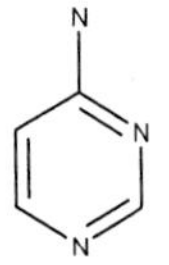	200 mM	4-Aminopyrimidine

(D. Kostrewa, unpublished results). It is evident that the cavity is not completely filled by the ligand and that in addition another small molecule might be accommodated. In the actual NMR experiments, the protein was saturated in a first step with a twofold molar excess of a more soluble derivative of the ligand shown in Fig. 19.5, which partially covers the ATP binding site and shows moderately high affinity. In a second step, an excess amount of a (presumably) weakly binding "needle" molecule was added to the solution, and the effects on the binary complexes were observed in 2D HSQC spectra. The needles were taken from a library that comprises about 520 molecules of MW <150. About 20% of the 250 soluble compounds which were measured led to clear changes in the 2D spectrum of the protein. All the differences mapped unambiguously to a cluster of residues forming a contiguous epitope on the surface of GyrB. However, the binding interface that they constitute resides on the back of the central β-sheet of the protein, while the ATP-site – the actual target of the screening efforts – is located on the front face. Therefore, the NMR results point to a second binding site which is separated from the first site by the body of the protein and hence cannot be exploited by chemistry in this case.

19.4.2
Peptide Deformylase

Peptide Deformylase (PDF) removes the formyl group from the methionine residue at the N-terminus of a nascent polypeptide chain during the bacterial protein synthesis [18, 19]. PDF is essential for this process, for which there is no counterpart in mammals, and therefore it has been considered to be a valid target for the development of antibiotic drugs [20]. The protein is an iron-containing enzyme with some structural relationship to the zinc metalloproteinases [21]. The iron atom can be replaced by other divalent cations like Ni^{2+}, Co^{2+}, or Zn^{2+} with decreasing enzymatic activities. Most inhibitors bind to the active site via a hydroxamic acid group that chelates the metal ion. A *de novo* NMR screening was performed aimed at finding novel binding motifs that occupy the hydrophobic S1′ pocket of the enzyme, in which normally the side chain of the methionine residue of the substrate is anchored. Such a moiety, when chemically linked to hydroxamic acid, could serve as a starting point for a subsequent rational drug design [22]. Because of its diamagnetic nature the Zn^{2+}-form of the enzyme was used for NMR spectroscopy.

For the screen, the compounds used were from the same library as that used in the gyrase B project. The statistical method of "principal component analysis" (PCA) was applied to the evaluation of the 2D ^{1}H-^{15}N HSQC spectra of PDF [10]. The result of the PCA when applied to the data set of about 300 spectra is shown in Fig. 19.6. The combination of PC^2 versus PC^4 was found as the one which gave the best separation. Evidently, there are two classes of spectral features: class A with large scores along PC^2 and small ones along PC^4, and class B with the reversed situation. An overlay of the spectra of the two most outlying samples with a reference spectrum revealed that there is a subset of shifted peaks for ligands of class A which is clearly disjunct from those signals that disappear for class B. To get insight into the physico-chemical origin that caused the change of spectral patterns of both classes, the corresponding ligand structures were checked and compared. All class B compounds contained a hydroxyl moiety as a functional group, and it was therefore plausible that the observed changes were due to a small change of the pH of the sample caused by the acidic properties of these ligands. This was verified by recording 2D spectra of samples of the apo-protein at pH values changed by +/− 0.5 pH units relative to the standard screening conditions. Inclusion of these spectra in the PCA clearly showed the data set with the lower pH as a member of class B, while that with the higher pH did not show up as an outlier. Furthermore, visual inspection of the spectral changes revealed that those associated with class B were different from those in class A. As the NMR assignment of PDF was known from literature [23] and confirmed in-house, it could be verified that the protein resonances perturbed in class B were either Glu- or Asp-residues or in spatial proximity to these amino acids. Therefore, the observed effects could be attributed to a partial protonation of the titrable side chains. Changes grouped in class A could be mapped to residues that make up the active site of the protein.

The binding constants of selected ligands of class A were determined to obtain their ranking with respect to their affinity to the protein and to select a chemically attractive compound for further detailed characterization of its mode of binding. The compound HUB093 was chosen by medicinal chemists as interesting. Using uniformly ^{13}C-labeled

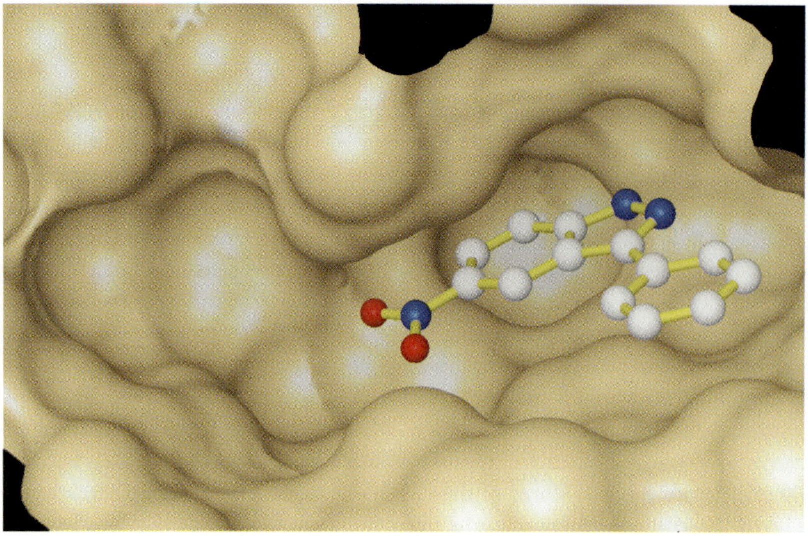

Fig. 19.5 Surface representation of the ATP-binding site of the 24 kD fragment of GyrB (23-234) filled with the compound Ro 05-8733. The figure was prepared using the program MOLOC [24] with the coordinates of the X-ray structure of the protein-ligand complex (D. Kostrewa, unpublished results). Ro 05-8733 is shown in a ball-and-stick representation.

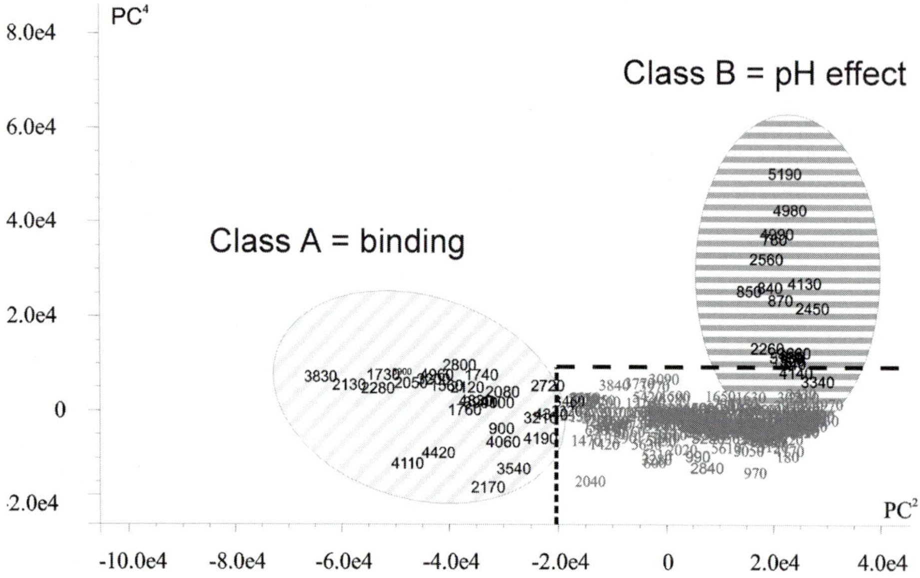

Fig. 19.6 Statistical data analysis using Principal Component Analysis (PCA). The plot of PC^2 versus PC^4 is shown. Ligands of "Class A" were shown to interact with the protein at its binding pocket, while those of "Class B" caused changes in the pH of the solution.

PDF-protein, the spectral changes induced by a titration of this ligand into the protein solution were monitored in a series of ^{1}H-^{13}C-HSQC experiments. Because of the hydrophobic nature of the binding site, several methyl signals of PDF were affected. In conjunction with distance information derived from a NOESY-spectrum of this sample, a model of the structure of a complex of the protein with HUB093 was generated. To support this task a software tool was incorporated into the program MOLOC [24] that exploits the effects of the ring current associated with aromatic systems like HUB093 on the chemical shifts of neighboring atoms of the binding protein. For perturbations caused by aromatic ring currents, the sign of the chemical shift changes as a function of ring orientation, while the intensity varies as a function of the distance from the ring [25]. Affected atoms with an orientation out of plane relative to the aromatic ring show negative upfield shifts, whereas an in-plane orientation leads to positive downfield shifts. Thus, the spatial distribution of positive and negative perturbation effects allows one to place the plane of an aromatic ring system within the three-dimensional structure of the target molecule [26]. The localization of an aromatic ligand can be refined by an automated minimization of the difference between experimentally determined shift perturbations and those calculated based on the Johnson-Bovey model for ring current effects [25]. As this optimization problem is invariant with respect to rotations of the ring around an axis perpendicular to its plane, the identification of NOEs is mandatory for a more detailed answer when modeling the protein-ligand complex. Fig. 19.7 shows the result of a MOLOC-docking of ligand HUB093 into the structure of

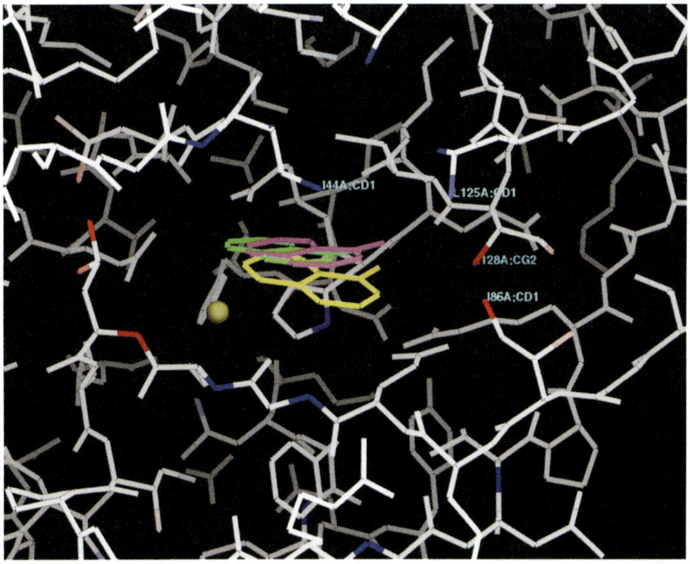

Fig. 19.7 Three possible orientations of the ligand HUB093 within the S1′ pocket of the enzyme peptide deformylase. Heavy atoms of the protein are coloured according to secondary chemical shift effects caused by the ring current of the aromatic ligand. *Blue*: negative secondary shift for N or C; *red*, positive shift difference. The four atoms labelled in *cyan* show intermolecular NOEs between the protein and the methyl group of the compound.

Tab. 19.2 Chemical structures of needle molecules as found in the project on peptide deformylase. The K_D values were determined by NMR titrations [17]

Needle	*K_D from NMR*	*Needle*	*K_D from NMR*
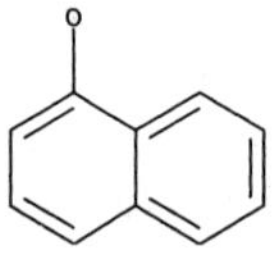	0.5 mM	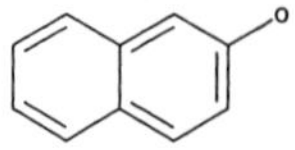	0.9 mM
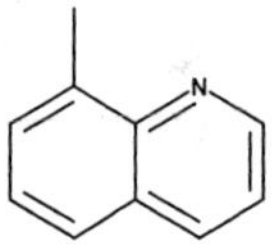	0.6 mM	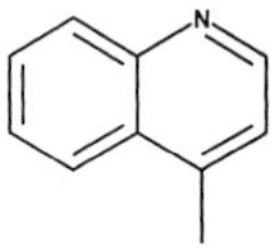	1.0 mM
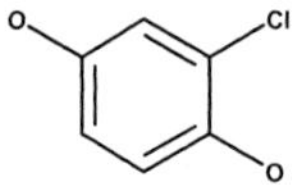	0.7 mM	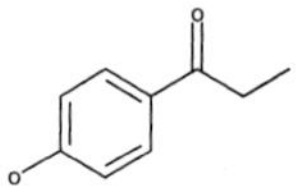	1.1 mm
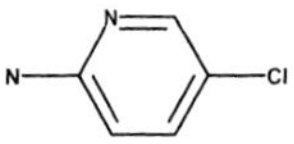	0.7 mM	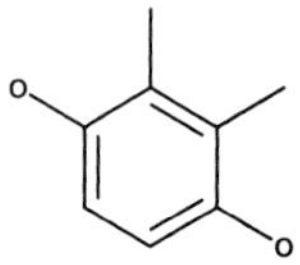	1.1 mM

PDF based on ring current shifts and NOEs. Table 19.2 shows the chemical structures of needle molecules as found in this project.

19.4.3
MMP-1

Matrix metalloproteinases (MMPs) are a class of zinc- and calcium-dependent enzymes that are responsible for the metabolism of extracellular matrix proteins [27]. Increased activity of MMPs has been associated with pathological diseases such as arthritis, cancer, multiple sclerosis and Alzheimer's disease [28–31]. Therefore, they constitute an important group of drug targets. Their inhibition is accomplished by blocking the active site of the catalytic domain with ligands that contain hydroxamic or carboxylic acids to chelate the Zn metal. The identification of low molecular weight compounds that contain different scaffolds may lead to the development of a new class of specific inhibitors.

Knowledge of the three-dimensional structure of metalloproteases and their inhibitor complexes together with structural information extracted from the Cambridge Structure Database regarding pharmacologically relevant Zn-ligand complexes was employed by the Roche Modeling Group to select small molecules for a target-biased library termed the "Zn-needle" library. As the one-dimensional NMR screen was to be performed with a flow-through probe head, it was important to test the compounds for solubility in aqueous buffer. It turned out that about a quarter of the molecules were soluble enough to be measured under the conditions of the NMR screening.

The effects of the compounds on metalloproteases were detected in one-dimensional ^1H spectra of the catalytic domain of human fibroblast collagenase-1 (MMP-1) by observing the well-resolved lines of His-residues located in the active site of this enzyme. Based on the experience gained in the project on deformylase, an internal marker was included to identify those compounds that caused a change of the pH of the aqueous solution which otherwise might be mistaken as a binding effect. To this end, imidazole was used as a buffer system, as it offers besides its buffer capacity the advantage that the chemical shifts of its resonance lines can be taken as indicators of the pH. Fig. 19.8 shows an expansion of a representative collection of 1D spectra of MMP-1 in the absence (top trace) and in the presence of needle compounds (other traces). In the spectral range from 13.5 to 11.0 ppm, four of the five resonance lines belong to atoms of the active site and are shifted to various extents. However, as evidenced by the unchanged position of the imidazole signal at 8.35 ppm, only the compounds of traces 2 and 3 had no influence on the pH of the solution and may therefore be considered as binders to the enzyme. The interactions caused by these molecules were further evaluated in 2D HSQC spectra, for which an in-house assignment had been obtained based on shift data available from the BioMagResBank (entry 4064). If the effects could be mapped to residues that make up the active site, this was taken as unambiguous confirmation of binding of the ligand to the target site. Interestingly, two compounds that had been identified in the 1D spectra as binders led to protein aggregation during the time required for recording the 2D spectra and had to be discarded. For the other ligands, binding to the active site was confirmed, and the K_D-values of their complexes with the protein were determined in a series of NMR titrations.

In order to learn more about the Zn-binding properties of those molecules that had been classified as "poorly soluble" in the solubility assay, the well plate with the highest proportion of them was selected to represent the different structural motifs for metal binding that make up the Zn-needle library. The compounds were investigated as mixtures of protein and one ligand at a time in a ratio of 1 : 20 in standard NMR tubes. For each molecule a 1D and a 2D spectrum was recorded to monitor both the degree of solubility and the interaction with the target protein. The 2D data were evaluated using an improved version of the nmrPipe software, which combines the automated processing of 2D spectra and their subsequent analysis via PCA in an integrated fashion without the need for user interference [32]. The result is displayed in Fig. 19.9. In the left hand part, the score plot shows the result of the PCA of about 100 2D spectra. In the representation of PC^1 versus PC^2 it becomes obvious that the majority of the spectra are very similar to each other (coloured in yellow) as evidenced by their clustering. However, there are outliers (coloured in cyan) originating from those spectra in which the addition of a ligand has caused shift perturbations. In the right hand part of Fig. 19.9, a superposition is given of a spectrum representing the majority population of a noninteracting ligand and one of the outliers in the PCA plot. Clear differences between the two spectra were mapped to residues that constitute the known binding epitope of MMP-1, thus confirming the classification as Zn chelators. Two interesting observations were made in the course of this part of the project. First, although the compounds were taken from the

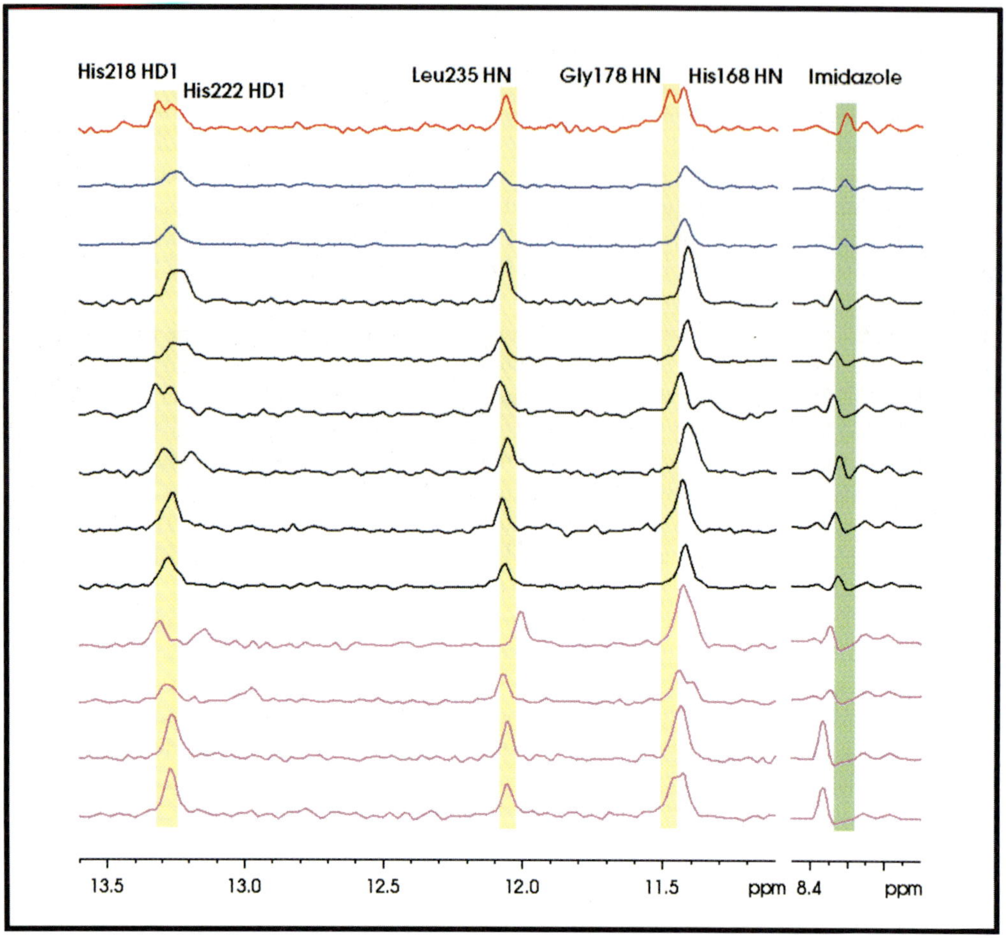

Fig. 19.8 Series of the low-field region of the one-dimensional ^{1}H spectra of unlabeled MMP-1 without and with various ligands. The assignment of the signals of the protein is indicated. In the region between 13.6 ppm and 11.1 ppm only His168 does not belong to the active site. The top trace corresponds to the apo-form of the enzyme. The lines in blue show shifts at the active site residues while the imidazole signal is not shifted. The effects on the NMR signals in black (medium effects) and magenta (strong effects) are caused by changes in the pH (as indicated by the shifted imidazole line) induced by the ligands.

subset of poorly dissolving molecules, their residual solubility was high enough for them to interact with the protein target, leading to shift changes. Second, no binding motif was found in addition to the ones that had shown up in the previous screen of the soluble compounds. Instead, it turned out that the substances were mainly aromatic derivatives of the basic scaffolds identified before. Fig. 19.10 provides an overview of the sequence of events in this project. As an interesting side aspect, it should be mentioned that the work described here inspired a member of the modeling group to postulate that

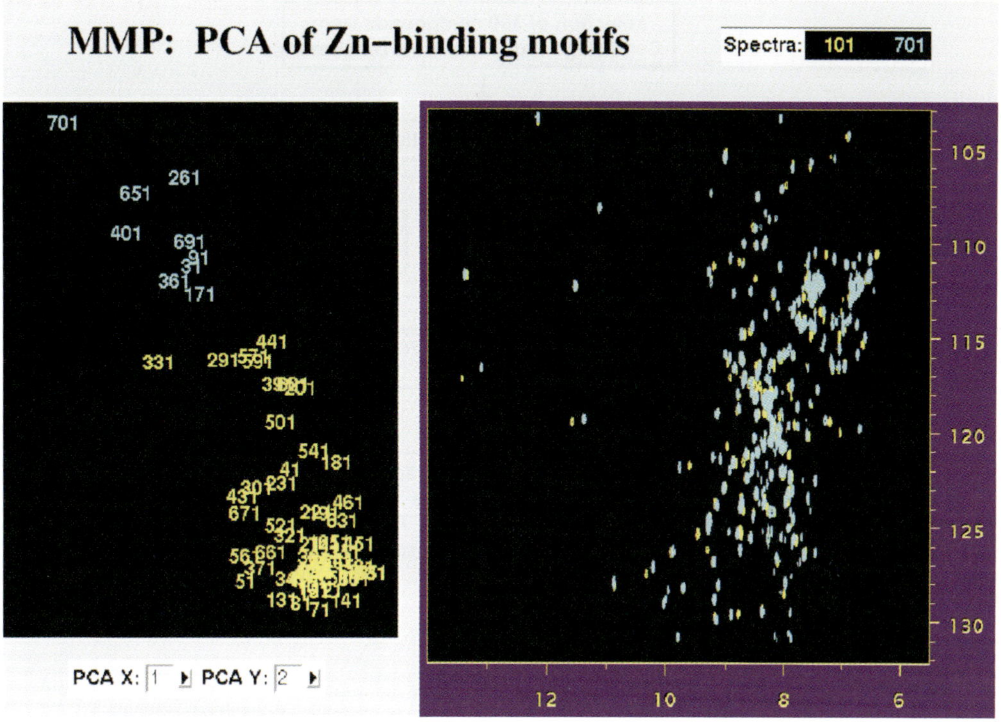

Fig. 19.9 Left: nmrDraw-integrated display of the result of a "Principal Component Analysis" (PCA) for about 100 2D spectra. Numbers in yellow correspond to spectra in which a testing compound does not lead to shift changes of the protein. Numbers in cyan represent spectra whose PCA scores differ from the majority population of spectra. Right: Overlay of two 2D-HSQC-spectra of MMP-1. Differences in chemical shift are observed between spectrum 101 (yellow) and spectrum 701 (cyan).

a certain other molecule might be a good chelator of Zn in MMPs. Subsequently this compound was tested under the same conditions as all the zinc needles and the hypothesis was confirmed. The dissociation constant was determined as 15 mM, a value that compares favorably with the affinity of the well-known hydroxamic acid moiety which serves as a "standard" building block for Zn chelation.

19.5 Summary

The examples described in this chapter are meant to illustrate the actual application of NMR for the study of protein-ligand interactions in the pharmaceutical industry. In our group, the focus has been mainly on "protein observe" techniques, as we have been fortunate enough to have a dedicated laboratory for providing both unlabeled and labeled proteins in large quantities. At present, two-dimensional HSQC spectra offer the highest content of information which is not accessible that rapidly by other methods. 2D NMR spectroscopy provides

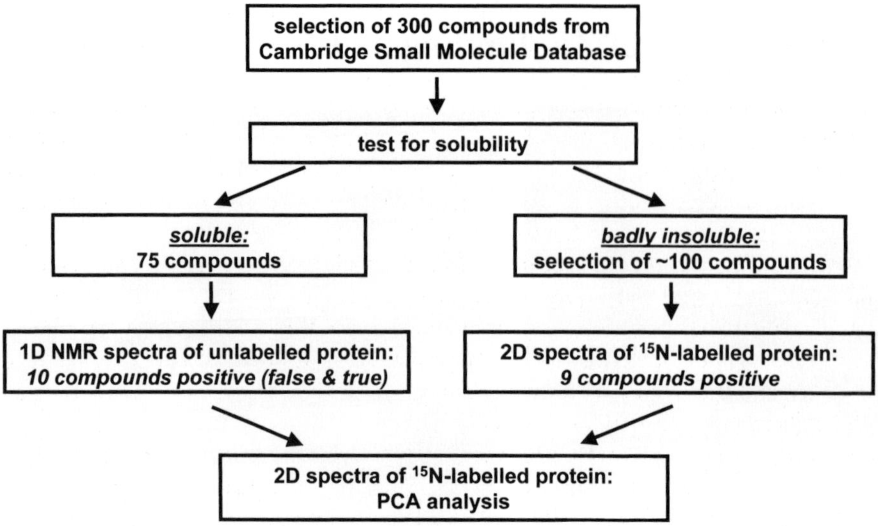

Fig. 19.10 Flow chart for the NMR identification of novel scaffolds for Zn binding.

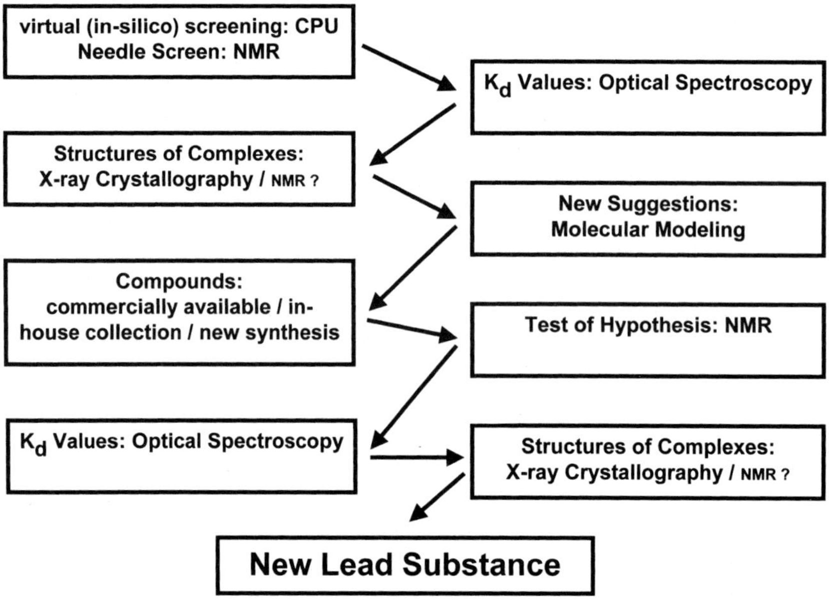

Fig. 19.11 Scheme for a structure-based biophysical screening for lead finding.

answers to the question whether a given ligand binds at the active site or somewhere else, but the price one has to pay is the greater effort required for obtaining the labeled samples. In addition, there is an upper limit to the size of the proteins which are currently amenable to this kind of NMR investigations. For the "ligand-observe" one-dimensional methods there is no such limitation, but the information one gains is solely on the low molecular weight ligand, enabling one to determine whether it binds and which parts of it interact with the target. It is not difficult to predict that techniques like saturation transfer difference (STD) or cross saturation will find their application for those drug targets that are enzyme complexes, membrane-associated proteins or other high molecular weight entities [33–36]. Other biophysical techniques like surface plasmon resonance [37, 38] and analytical ultracentrifugation [39] or traditional biochemical methods do not suffer from size limitiations. They have their strengths in the determination of affinity constants and binding stoichiometry, and they require only small amounts of protein. However, it is not straightforward for them to localize the binding site.

With respect to the structure determination of protein-ligand complexes, the method of choice is, in general, not NMR spectroscopy but X-ray crystallography. This provides detailed information on the binding site, and the picture of a protein structure or complex readily inspires modelers or medicinal chemists to speculate what could be done chemically to arrive at a new "lead" substance. However, the information on flexibility or crystallographic uncertainties as given by B-factors is not always taken into account. NMR spectroscopy, on the other hand, offers an immediate access to these parameters, although their visualization as a bundle of structures, for example, is more demanding on a conceptual basis. Considering the power of X-ray crystallography, NMR presently has only a role as a niche player in the field of the determination of 3D structures. In terms of time and effort it can only compete when the size of the protein under investigation is not too large and when it can be (doubly-) labeled in quantities of 10 mg or more.

Without doubt, the validation of hits as found in a biological screen is a powerful application of NMR in pharmaceutical research. Problems caused by false-positive hits which are related to the physico-chemical properties can be recognized quite readily, and the question whether a given substance binds to a target can be answered definitely. NMR may also serve as a "guiding" technique for X-ray crystallography to select ligands with increased chances of crystallization, as the two methods have similar requirements with respect to solubility and concentration of compounds to be investigated. Sometimes ligands of low solubility are readily measured by those techniques that work at very low concentrations but turn out to be problematic for both NMR and X-ray spectroscopy. These compounds can be immediately identified by simple 1D NMR measurements, and, in our experience, they will fail to produce complexes that would give useful crystals. On the other hand, substances that behave well in NMR have a high probability of resulting in good crystals.

The other area in which NMR has considerable impact is the search for low molecular weight compounds that bind specifically to a given target. Quite often there is structural information on the binding site for those biomolecular targets that are amenable to NMR investigations, or such information can be obtained. Knowledge of the spatial location of hydrogen-bond donors or acceptors and the distribution of hydrophobic or hydrophilic patches can be converted into a pharmacophore hypothesis. This can be used for a virtual ("in-silico") computer-search in, for example, the "Available Chemicals Directory (ACD)" for the extraction of possible "hits" which are subsequently validated by NMR. Optical spectroscopy or surface plasmon resonance may be used for the determination of dissociation constants, and, for the most interesting compounds, X-ray crystallography or NMR may determine their three-dimensional structures in complexes with the target. Based on this information, molecular modelers might suggest further candidate molecules to be tested in another round of a hit-finding cycle. In Fig. 19.11, a general scheme of such a structure-based lead discovery process is given.

Certainly biomolecular NMR is not "the" single method which is important for hit identification in pharmaceutical research. It is always a combination of techniques and a team effort that leads to a successful drug. This can involve biologists (basic understanding, assay development, bio-informatics), chemists (both bench chemists and modelers), screening specialists (HTS/natural products) and spectroscopists (X-ray, optical methods, surface plasmon resonance, NMR).

19.6 Acknowledgements

We gratefully acknowledge the contributions of and discussions with our colleagues at Hoffmann-LaRoche, Basel: B. Gsell, A. Ross and G. Schlotterbeck. We also thank F. Delaglio for making available his software.

19.7 References

1 S.B. Shuker, P.J. Hajduk, R.P. Meadows, S.W. Fesik, *Science* **1996**, *274*, 1531–1534.
2 P.J. Hajduk, T. Gerfin, J.-M. Boehlen, M. Häberli, D. Marek, S.W. Fesik, *J. Med. Chem.* **1999**, *42*, 2315–2317.
3 H.-J. Böhm, M. Böhringer, D. Bur, H. Gmünder, W, Huber, W. Klaus, D. Kostrewa, H. Kühne, T. Lübbers, N. Meunier-Keller, F. Müller, *J. Med. Chem.* **2000**, *43*, 2664–2674.
4 W.F. Simonds, G. Koski, R.A. Streaty, L.M. Hjelmeland, W.A. Klee, *Proc. Natl. Acad. Sci. U.S.A.* **1980**, *77*, 4623–4627.
5 R.K. Scopes, *Protein Purification (2nd edition)*, Springer-Verlag, New York, USA, **1987**.
6 R.J. Beynon, J.S. Easterby, *Buffer Solutions: The Basics*, Chapter 3, BIOS Scientific Publishers, **1996**.
7 C. Dalvit, P. Floersheim, M. Zurini, A. Widmer, *J. Biomol NMR* **1999**, *14*, 23–32.
8 A. Ross, H. Senn, *Drug Discov. Today* **2001**, *6*, 583–593.
9 D.C. Schriemer, O. Hindsgaul, *Comb. Chem. High Throughput Screen* **1998**, *1*, 155–170.
10 A. Ross, G. Schlotterbeck, W. Klaus, H. Senn, *J. Biomol NMR* **2000**, *16*, 139–146.
11 J. E. Jackson, *A User's Guide to Principal Components*, John Wiley, New York, **1991**.
12 A. Medek, P. J. Hajduk, J. Mack, S. W. Fesik, *J. Am. Chem. Soc.* **2000**, *122*, 1241–1242.
13 R.J. Reece, A. Maxwell, *CRC Crit. Rev. Biochem. Mol. Biol.* **1991**, *26*, 335–375.
14 G. Bodenhausen, D. J. Ruben, *Chem. Phys. Lett.* **1980**, *69*, 185–188.
15 S. Grzesiek, A. Bax, *J. Am. Chem. Soc.* **1993**, *115*, 12593–12594.
16 W. Klaus, A. Ross, B. Gsell, H. Senn, *J. Biomol. NMR* **2000**, *16*, 357–358.
17 L.-Y. Lian, G.C.K. Roberts, in: *NMR of Macromolecules–A Practical Approach*, Oxford University Press, Oxford, UK, **1993**.
18 J.M. Adams, *J. Mol. Biol.* **1968**, *33*, 571–589.
19 L.A. Ball, P. Kaesberg, *J. Mol. Biol.* **1973**, *79*, 531–537.

20 D. Mazel, S. Pochet, P. Marliere, *EMBO J.* **1994**, *13*, 914–923.

21 A. Becker, I. Schlichting, W. Kabsch, D. Groche, S. Schultz, A. F. V. Wagner, *Nat. Struct. Biol.* **1998**, *5*, 1053–1058.

22 Ch. Apfel, D. W. Banner, D. Bur, M. Dietz, T. Hirata, Ch. Hubschwerlen, H. Locher, M. G. P. Page, W. Pirson, G. Rossé, J.–L. Specklin, *J. Med. Chem.* **2000**, *43*, 2324–2331.

23 F. Dardel, S. Ragusa, Ch. Lazennac, S. Blanquet, T. Meinnel, *J. Mol. Biol.* **1998**, 280, 501–513.

24 Gerber Molecular Design Inc. *MOLOC Manual*, **2000**.

25 C. E. Johnson, F. A. Bovey, *J. Chem. Phys.* **1958**, *29*, 1012–1014.

26 M. A. McCoy, D. F. Wyss, *J. Biomol. NMR*, **2000**, *18*, 189–198.

27 J. F. Woessner, *FASEB J.* **1991**, *5*, 2145–2154.

28 A. J. P. Docherty, J. O'Connell, T. Crabbe, S. Angal, G. Murphy, *Trends Biotech.* **1992**, *10*, 200–207.

29 G. J. P. Murphy, R. M. Hembry, *J. Rheumatol.* **1992**, *19*, 61–64.

30 N. Peress, E. Perillo, S. Zucker, *J. Neuropathol. Exp. Neurol.* **1995**, *54*, 16–22.

31 D. Wernicke, C. Seyfert, B. Hinzmann, E. Gromnica-Ihle, *J. Rheumatol.* **1996**, *23*, 590–595.

32 F. Delaglio, *NMRPipe Vers. 1.7.156*, **1999**.

33 M. Mayer, B. Meyer, *J. Am. Chem. Soc.* **2001**, *123*, 6108–6117.

34 H. Takahashi, T. Nakanishi, K. Kami, Y. Arata, I. Shimada, *Nat. Struct. Biol.* **2000**, *7*, 220–223.

35 A. Ramos, G. Kelly, D. Hollingworth, A. Pastore, T. Frenkiel, *J. Am. Chem. Soc.* **2000**, *122*, 11311–11314.

36 A. N. Lane, G. Kelly, A. Ramos, T. A. Frenkiel, *J. Biomol. NMR* **2001**, *21*, 127–139.

37 A. Szabo, L. Stolz, R. Granzow, *Curr. Opin. Struct. Biol.* **1995**, 5, 699–705.

38 R. L. Rich, D. G. Myszka, *Curr. Opin. Biotechnol.* **2000**, 11, 54–61.

39 T. M. Schuster, J. M. Toedt, *Curr. Opin. Struct. Biol.* **1996**, 6, 650–658.

20
Strategies for Drug Discovery Using NMR

Marcel J. J. Blommers, Andreas Flörsheimer, and Wolfgang Jahnke

20.1
Introduction

High-throughput screening and structure-based drug design have become extremely important and are successful components in the discovery and development of new therapeutics [1]. There are, however, many interesting protein targets for which it has apparently been very difficult to find low molecular weight molecules that are functional and fulfil all the properties of a desired drug candidate, such as bioavailability, lack of toxicity, ease of formulation, and functionality in man (see Chapt. 18). Drug discovery is a high-risk process if one considers that the universe of all possible drug molecules with a molecular weight less than 400 Da has been estimated to contain 10^{10}–10^{50} molecules [2], in contrast to the very small proprietary libraries (10^6) used for high-throughput screening.

20.1.1
Many Drugs are Modular

A new screening strategy by NMR may significantly contribute to an improved lead finding process. Such a strategy has already been successfully implemented in several companies including Abbott [3] and Vertex [4]. The screening concept is based on the recognition that most drugs are built up of two or more fragments that are linked together (Fig. 20.1) [5].

NMR screening is ideal to find such fragments of drugs that bind to sub-pockets of the binding site. If such fragments can be found and later be linked together, drug candidate molecules may be discovered using this method. The potential drug-binding site is considered to be divided into sub-sites (Fig. 20.2). Small fragments (also often called building blocks) of drug-like molecules can be tested for binding to one of these sub-sites. If the fragments to be tested are drug-like, diverse, and chemically stable compounds, the many possibilities to link them would result in a large number of virtual molecules. Consider the hypothetical case of a ligand comprising three fragments which bind to three sub-sites. In the context of drug discovery, this is a significant combinatorial problem. If in a combinatorial-chemistry approach only 100 fragments were to be tested for each of the subsites, this would require the synthesis and testing of at least $100 \times 100 \times 100 = 10^6$ molecules, not considering the number of possible linkers. NMR screening is a fragment-based approach and would only require the screening of 3×100 compounds in this hypothetical case.

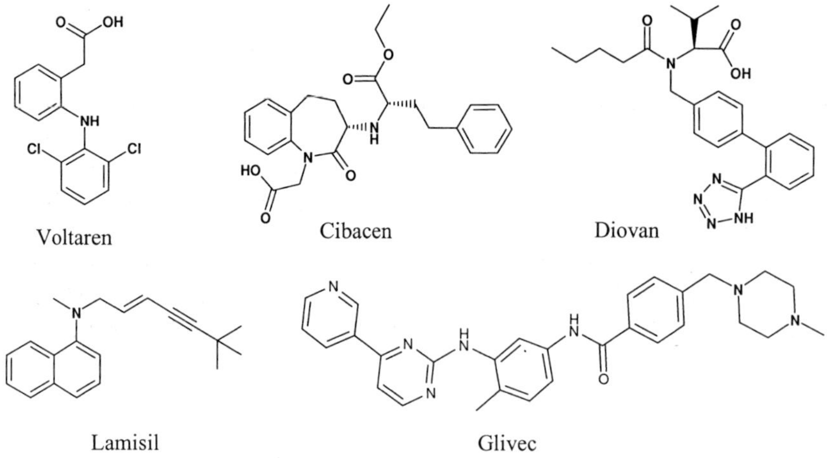

Fig. 20.1 Drugs are often composed of two or more fragments or building blocks. Thus, the discovery of new drugs may well be achieved by finding the building blocks of the new drug and linking them together.

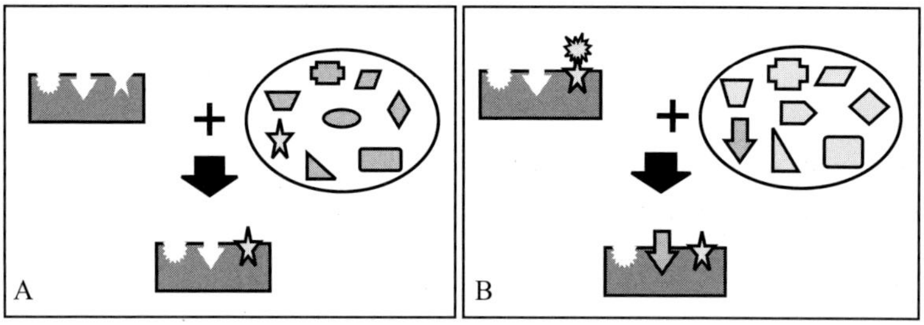

Fig. 20.2 Concept of NMR screening of fragments. **A** First-site screening. A compound library of small fragments or scaffolds is tested for binding to a target protein. The compound library and binding site are drawn schematically. The binding site contains three sub-sites. Only compounds that are complementary in structure bind to a sub-site: the star would bind to one of the sub-sites. **B** Second-site screening. The ligand that was found in first-site screening is spin-labeled. The spin label is used to select only binders to proximate binding sites, i.e., to one of the two sub-sites. Correct covalent linkage of ligands that bind to the sub-binding sites would boost affinity and specifcity.

20.1.2
Fragments can be Assembled Piece by Piece

An important aspect of this approach is the discovery of fragments that bind in spatial proximity, thereby offering the possibility to link them chemically. These fragments can be found by attaching a spin label to one of the ligands identified by first-site screening. The spin label affects the relaxation of a second ligand if it binds simultaneously and in close proximity to the spin-labeled ligand [6]. Linking fragments that are small and therefore likely to bind only very weakly ($K_D > 10\ \mu M$) would boost the affinity (and specificity) of the resulting compounds. NMR is an excellent method to detect such weakly binding ligands [7, 8].

A variant of fragment-based screening is reverse screening. In this approach, a known ligand is decomposed into smaller fragments. Second site screening is performed in the presence of one of these (spin-labeled) component fragments. In order to circumvent e.g. patent issues or toxicity problems, one could use the strategy of reverse screening to discover alternative lead structures.

In terms of free binding energies, the linkage of weakly binding ligands provides not only the sum of the binding energies of the fragments and linkers but also an additional (favorable) entropic contribution for fixing only one ligand instead of binding the individual components simultaneously:

$$\Delta G = \Delta G^{A} + \Delta G^{B} + \Delta G^{c} + \Delta G^{\text{linker}} \qquad (1)$$

or

$$K_D = K_D^A K_D^b K_D^c c_{\text{linker}} \qquad (2)$$

Linking of the fragments is thus boosting the affinity (and specificity) of the resulting compounds [9].

20.1.3
Three Alternative Approaches for the Discovery of Linked Fragments: In Silico Screening, SAR-by-NMR and Second-Site Screening

Once a ligand is discovered by screening methods there are opportunities to improve its binding affinity and target specificity. This has been recognized in structure-based drug design [1]. The binding pocket is usually much larger than the ligand discovered by screening methods. There are several computational approaches to including the unoccupied proximate sub-pockets of the binding site into the design of improved ligands. For example, one could connect fragments from a fragment library to the first-site ligand [10], or alternatively one could screen in silico for the binding of fragments that bind to proximate binding sites [11] (see Fig. 20.3). This results in the design of compounds that would have better binding properties.

Since NMR can readily detect the weak binding of fragments, it has emerged as an experimental approach to screen for weakly binding ligands. Although it has long been known that [^{15}N,^{1}H]-HSQC spectra are excellent tools for studying the complexation of proteins [7, 12], they were first used systematically for screening compound libraries by Fesik and coworkers using ^{15}N-labeled FKBP (FK506-binding protein) (SAR-by-NMR, [8]). HSQC spectra of ^{15}N-labeled FKBP were recorded with mixtures of ten compounds each. Second-site screening can be performed by saturating the first binding site and monitoring changes in chemical shifts of amino acids in proximate sites. The method requires isotope labeling of the target protein, and combines results of assignment and structure determination with the screening interpretation. This approach has led to the discovery

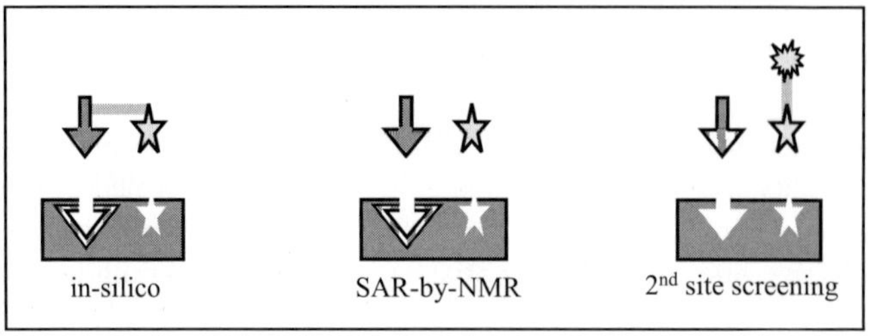

Fig. 20.3 Schematic representation of fragment-based screening strategies. Left panel: in silico docking and *de novo* design, middle panel: SAR-by-NMR, and right panel: second-site screening. In docking and *de novo* design methods, ligands are identified which have high complementarity to the binding site of the protein [11]. In SAR-by-NMR, the identification of fragments binding in mutual proximity is performed by saturating a first binding site and screening for fragments that bind close to the first one. This is done by monitoring changes of specific resonances of amino acids close to the binding site of the first fragment [8]. Second-site screening relies on the spin labeling of the first site ligand. Binding of a ligand to the second site is detected on the resonances of the fragment [6]. In all panels the region in which the measured perturbation occurs is indicated schematically by a hatched area.

of high-affinity ligands for FKBP. The assignment may be partly circumvented if the binding site is well characterized, e.g., by X-ray crystallography, and if amino acids in the binding site can be selectively labeled with ^{15}N and/or ^{13}C.

Second-site screening can also be performed in an alternative way, in which a paramagnetic moiety is attached to the first-site ligand [6] (see also Chapt. 15). The paramagnetic center, e.g. TEMPO, increases the relaxation properties of ligands that bind simultaneously in proximity to the first binding site. In contrast to SAR-by-NMR, perturbation of the NMR signals of the ligands rather than the signals of the protein itself are monitored. The application of second-site screening using spin labels does not require the production of isotope-labeled target protein. It is also less critical with respect to the size of the target protein. The big advantage of the former method is the availability of structural information about the relative binding of the fragments in the binding pockets, which is an advantage for the linker design.

NMR thus has potential for lead finding using this new screening strategy. This will broaden the application of NMR spectroscopy in drug discovery.

20.1.4
Hit Validation by NMR

It is important to recognize that, apart from its role in screening, NMR can be also be used to validate binding of hits from conventional high-throughput screening (HTS). It is important to validate these hits, since false-positive hits are often found. The detection of false-positive hits may obviously lead to a waste of chemistry resources if these are followed up. Thus, the use of NMR to test the binding of ligands from HTS is a useful complementary binding assay in order to validate the results from HTS.

20.2
Detection of Ligand Binding by NMR for Hit Validation or NMR Screening

For the validation of hits from HTS and for the identification of weakly binding ligands in the NMR library, a variety of NMR experiments capable of detecting binding exist. Sometimes it is very useful to have different methods, because a particular experiment may work better in one case than another. This depends on the solubility of the ligand, the strength of binding, the availability of ^{15}N and/or ^{13}C labeled protein, physico-chemical properties of the protein, etc.

20.2.1
Change in Relaxation Properties

Small ligands (molar mass <400 Da) tumble rapidly in solution and have small relaxation rates. Large molecules tumble slowly and relax fast. Hence, a protein as well as a ligand bound to a protein relaxes faster than the nonbinders.

Also, weakly binding ligands that are in exchange between the free and the bound state exhibit relaxation properties that present the population-weighted average of the free and the bound state (see Chapt. 13). Thus, binding of ligands to macromolecules may be detected by increased relaxation rates upon adding protein. If the relaxation rate is drastically increased, this can even be observed by an increase in line width in a simple 1-dimensional ^{1}H spectrum (see Fig. 20.4).

If the effects are not as drastic, $T_{1\rho}$ spectra must be measured in order to enhance the effect on the signal intensities and hence facilitate detection of the effect (Fig. 20.5). Firstly, reference $T_{1\rho}$ spectra are measured on compound mixtures in the absence of protein. The $T_{1\rho}$ experiment contains a spin-lock period with a variable delay of between 0 and 200 ms and an acquisition pulse that enables detection of NMR signals in H_2O samples without saturation of water, such as WATERGATE [13] or excitation sculpting [14]. Subsequently, the experiment is repeated after addition of protein. Typical concentrations of protein and compounds are 20 μM and 200 μM, respectively. Hits are found by estimating the $T_{1\rho}$ relaxation rate from relative line intensities of $T_{1\rho}$ filtered spectra recorded for at least two different settings for the length of the spin-lock period, and by comparing the values for compounds with and without protein. In some cases, differential $T_{1\rho}$ times may be used to identify protons that are in direct contact with the protein [15].

The increase in relaxation rate can be enhanced tenfold or more by employing spin labels coupled to the protein [16]. This allows a concomitant reduction in protein concentration.

20.2.2
Changes in Chemical Shift

The chemical shift is very sensitive to the local environment [17]. Thus, if a ligand binds, changes in chemical shift of the protein are induced. Changes in chemical shift upon binding may also occur because of conformational changes within the protein. It is furthermore important to check that the changes which are observed in the chemical

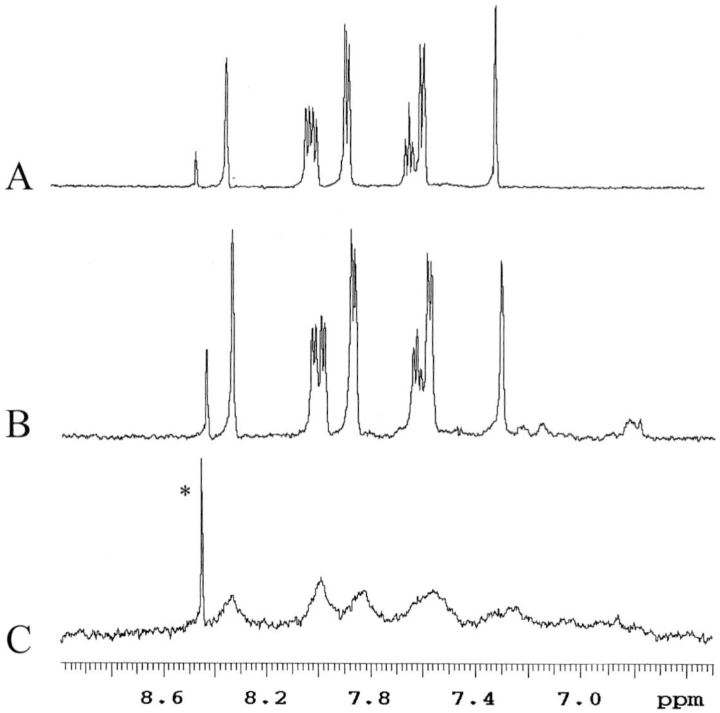

Fig. 20.4 In this example, binding of a ligand is detected by a change in line width of the ligand resonances which is drastic enough to be detected in a regular ^{1}H experiment. The spectral region displayed contains the aromatic resonances of a small compound (panel **A**). No ligand binding was detected for protein I (panel **B**), whereas binding to protein II was shown by line broadening of the resonances of the ligand (panel **C**). The concentration of ligand and protein were 100 μM and 10 μM, respectively. The narrow resonance indicated with an asterisk (*) corresponds to a nonbinding compound. Thus, the line broadening is caused by binding and not by field inhomogeneity effects.

shift are not caused by changes in pH, ionic strength or DMSO (compounds are usually added as DMSO solution). Changes in chemical shifts can (in favorable cases) be observed in 1-dimensional spectra, e.g. in cases when labeled protein is not available (Fig. 20.6) or in 2-dimensional spectra of a ^{15}N-labeled protein. [^{15}N,^{1}H]-HSQC spectra are often used since the amino acids in direct contact with the ligand can be identified by comparison of the assigned fingerprint of the unligated protein with that of the protein-ligand complex. If the structure is known, the binding pocket can be localized relatively easily. Binding constants can be determined by titration of the compound to the protein and following the chemical shifts of protein resonances (see Chapt. 13). Although the HSQC screening method is very robust, it requires labeling of the protein. Furthermore, if mixtures are measured, additional experiments are required to find out which compound(s) from the mixture actually bind(s).

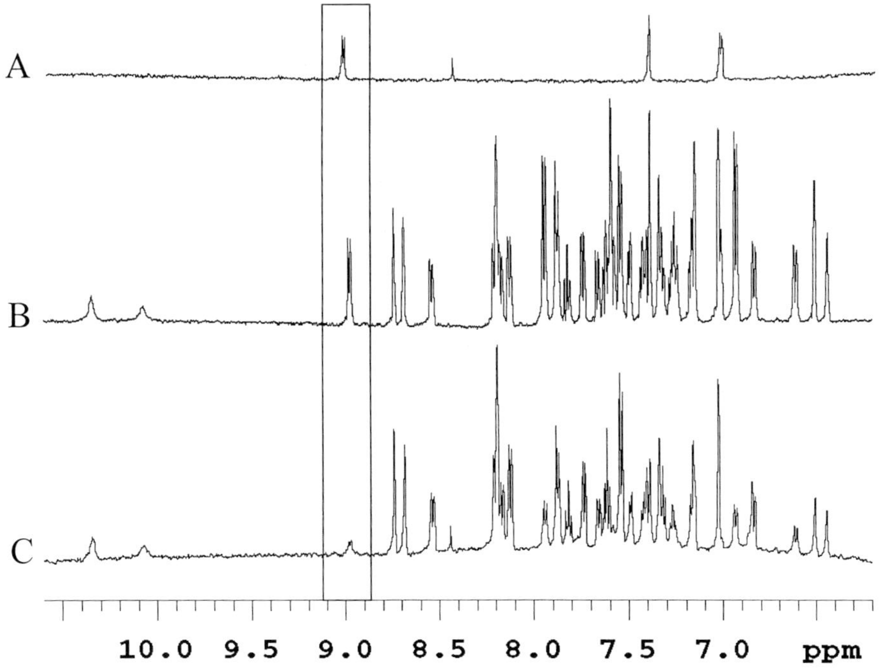

Fig. 20.5 Example of first-site screening using $T_{1\rho}$ experiments. A spin-lock time of 200 ms is used. Mixtures of ten compounds are measured. By comparison of the $T_{1\rho}$ spectra obtained before (**B**) and after (**C**) adding protein in a 10:1 ratio of compound:protein, ligands are readily detected. A reduction in signal intensity of at least 50% is considered to present a hit in first-site screening. In the example shown, a compound to bind the protein is identified. Compound identification is obtained without additional experiments, because the spectra of all compounds that comprise the mixtures are known (**A**).

20.2.3
Saturation Transfer

Another experiment that can be helpful for detecting binding is the so-called water-LOGSY experiment [18] (see also Chapt. 14). In this experiment the water resonance is selectively excited. During the mixing time, its magnetization is transferred to the protein and the mixture of compounds via intermolecular cross-relaxation. The effect is measured on the free ligand provided that it is in fast exchange with the bound ligand. For a bound ligand, the water NOE signals in the ^1H spectrum have an opposite sign to that of compounds that do not interact with the protein because of the much longer overall correlation time of the ligand in the protein-ligand complex (Fig. 20.7).

One could also saturate isolated signals of the protein and measure transfer of magnetization to the bound ligand (STD experiment) [19]. Alternatively, one could use diffusion experiments to detect the binding of compounds by their reduced translational diffusion caused by binding to a macromolecule [20]. NOE pumping [21] combines the strategies of diffusion editing and saturation transfer.

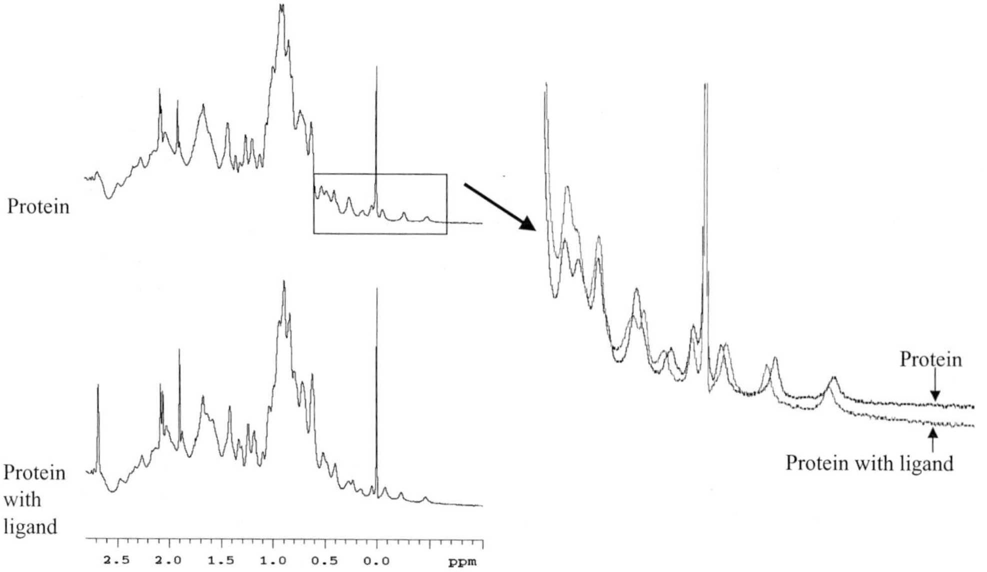

Fig. 20.6 Ligand binding is observed via changes of chemical shifts of the protein. In this example, changes in the high-field shifted resonances of the aliphatic ^{1}H resonances are observed upon ligand binding. The signal at 0 ppm is 2,2,-dimethyl-2-silapentane-5-sulfonate (DSS), which is added as a chemical shift reference. The spectra were measured at 200 μM protein and 1 mM ligand concentrations at 25 °C. Changes in chemical shifts are caused by direct effects of the ligand binding to protons in the binding pocket and/or by conformational changes of the protein upon binding.

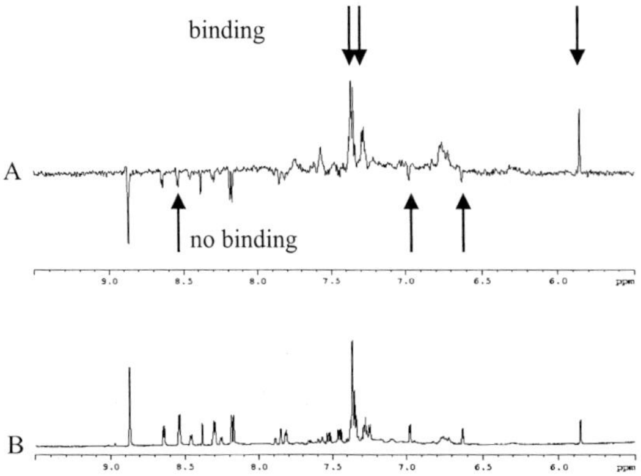

Fig. 20.7 Detection of ligand binding using water-LOGSY. The water LOGSY spectrum (**A**) is obtained after saturation of water and subsequent transfer of the magnetization to the resonances of the compounds in the compound mixture (300 μM) in the presence of 50 μM protein. Only the positive peaks correspond to resonances of bound compounds. The other compounds of the mixture do not interact with the protein. This spectrum can be compared with the regular one-dimensional spectrum (**B**).

20.2.4
Transferred NOE

Another valuable method that can be used for the validation of ligand binding is the transferred NOE experiment [22] (see also Chapt. 16). Small molecules tumble rapidly in solution, and the cross peaks of small molecules therefore have opposite sign to that of the diagonal peaks. If the ligand is weakly bound (10 mM > K_D > 10 μM), NOESY cross peaks are observed with the same sign. Observation of these transferred NOEs is a validation of binding.

For weak binding affinities (and high k_{off} rates), the NOE intensity is in first approximation a function of intraligand proton distance, r, and the fraction of bound ligand, a, [22]:

$$NOE = ka\frac{1}{r^6} \tag{3}$$

The NOE intensity can therefore be used to rank the compounds according to their binding affinity (Fig. 20.8). Thus, if in two ligands a cross peak corresponds to the same interproton distance (usually a fixed distance such as the distance between geminal methylene protons or aromatic protons), the higher NOE intensity indicates the better binder.

The distance information derived from the trNOE experiment may be relevant to determining the relative orientation of the bound fragments (Figs. 20.12 and 20.14).

Finally, in some cases NOEs between ligand and protein are detected. This identifies which protons are in contact with the protein.

In the following example, hit validation was done by measuring a trNOE spectrum of a mixture of seven compounds in the presence of the protein. For five compounds binding was confirmed by the appearance of NOESY cross peaks that had the same sign as the signals on the diagonal. Two compounds resulted in cross peaks of opposite sign. Thus, binding of these two compounds was not confirmed. It is likely, however, that these compounds bind weakly and compete for binding with one of the other compounds in the mixture. Among the other ligands, nmr006 and nmr176 bind best, according to the NOE intensity of the cross peaks.

20.3
"Free" Check for Compound Solubility and Integrity

20.3.1
Solubility

In the approach of second-site screening by NMR, a dedicated library of fragments should be available. The library ideally contains a diverse set of small drug-like molecules with a molecular weight of 100–300 Da [23, 24]. Our compound library was designed based on the application of several filters using the ACD database [25].

Before adding compounds to the library, their solubility was tested by NMR (Fig. 20.9). If the solubility (in DMSO/water ~ 1:99) was below 50 μM, the compound was excluded from the library. Thus, many compounds have a solubility that is less than 1 mM. Since it is important to measure always the same compound:protein ratio, the solubility data are useful for the preparation of the compound/protein mixtures.

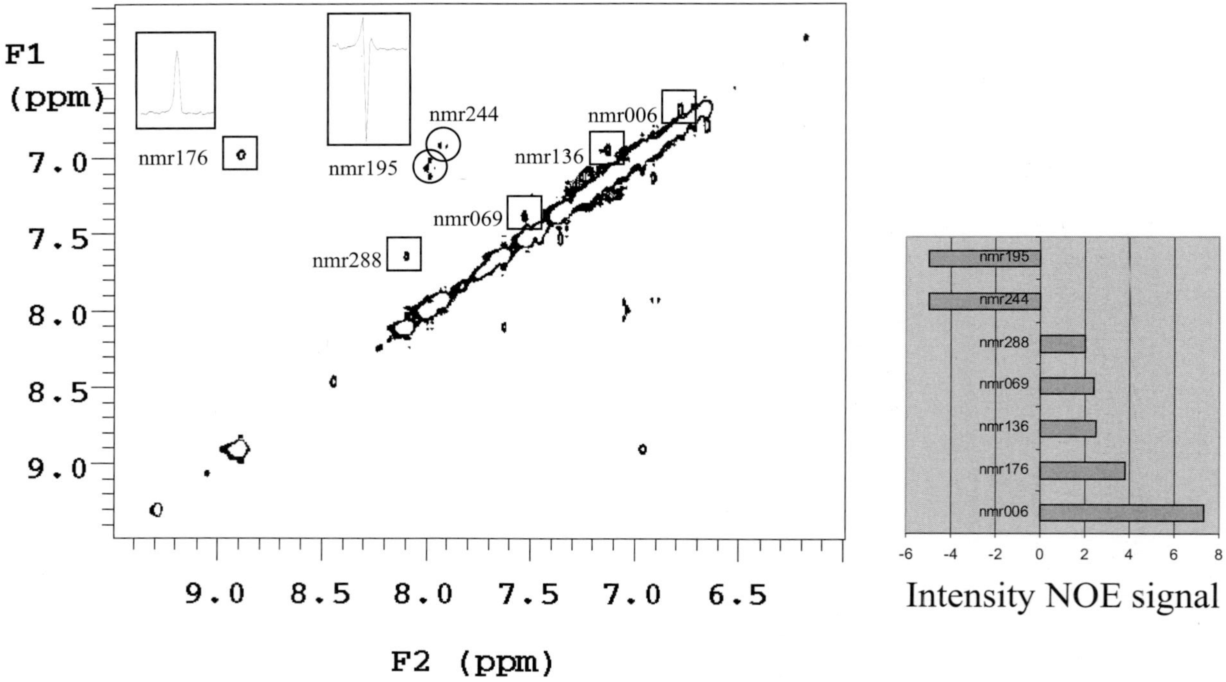

Fig. 20.8 Hit validation of a mixture of hits. A transferred NOE recorded with a mixing time of 100 ms is shown. The concentration of ligands and protein was 500 μM and 50 μM, respectively. The NOEs with the same sign as the diagonal (i.e., positive peaks – boxes, see cross section of the peak corresponding to nmr176) confirm binding. The negative peaks (circles, see cross section of the peak corresponding to nmr195) do not confirm binding. Because the intra-ligand proton distance of the corresponding NOE is about the same for all compounds of this mixture, the NOE intensity is a measure of the affinity. It is concluded that nmr176 and nmr006 are the best-binding ligands.

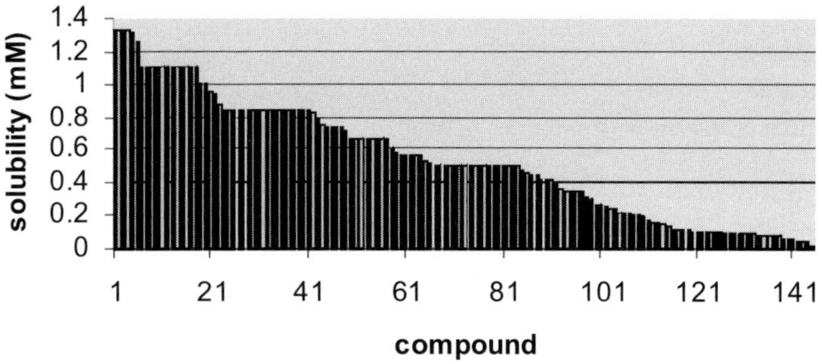

Fig. 20.9 Solubility of the compounds taken from the NOVARTIS NMR library for second-site screening. The solubility was determined at 25 °C by measuring the signal intensities of the compound and of DSS that was added as internal reference. The compound was dissolved in DMSO up to 100 mM. 5 µL of this solution was taken and diluted to 500 µL with 50 mM phosphate buffer, pH 7.0, in water. In the case of lower solubility in DMSO, 5 µL of the saturated solution was taken after removal of undissolved material by centrifugation. Precipitates in the final NMR samples were removed by centrifugation before measuring. Only results for the finally selected compounds are shown.

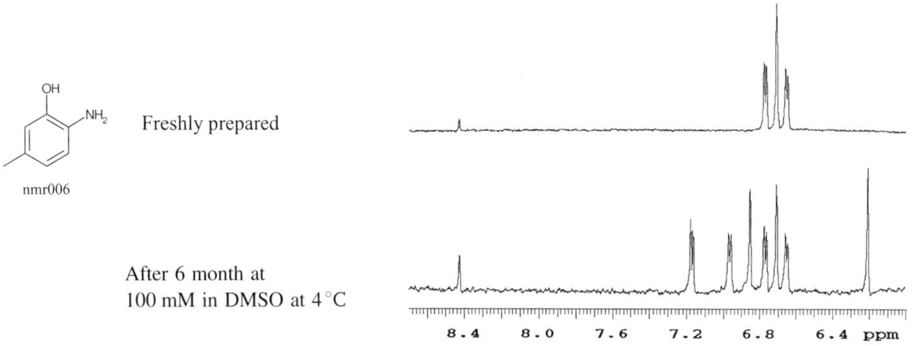

Fig. 20.10 Example of stability problems of 5-amino-*m*-cresol (nmr006), which is a compound of the NOVARTIS NMR compound library. The compound shown at the left shows the expected NMR spectrum (top) for the aromatic ^{1}H resonances. The compound is stable for weeks, but after storage for 6 months at 4 °C about 50% has reacted (see spectrum at bottom). The product is likely a biphenyl.

20.3.2
Compound Integrity

The compound stock solutions are stored in d_6-DMSO/D_2O 9:1 at 4 °C. The addition of D_2O lowers the melting temperature of DMSO and thus prevents the solvent from becoming solid at the storage temperature. Stability problems of the compounds have to be taken into account. To this end, a spectrum of the compound is recorded before and after adding the target protein. The need for such measures is shown by the example depicted in Fig. 20.10. In this figure, the NMR spectrum of one of the compounds, nmr006, is shown for a freshly prepared sample and a sample after 3 month storage at 4 °C. Such stability problems are not always foreseen, and therefore a quality control for each mixture is useful immediately before the mixture is used for screening.

20.4
Spin Labeling for Second-Site Screening

Ligands that bind to sites that are in proximity to binding sites of a previously identified ligand can be detected by covalently attaching a paramagnetic center to the first ligand [6, 26]. We often refer to such compounds as a spin-labeled ligand. In our approach used so far, the spin labels were based on TEMPO (2,2,6,6,-tetramethyl-1-piperidinyloxy), which contains an unpaired electron in the nitroxide moiety (Fig. 20.13). The incorporation of TEMPO is relatively straightforward (Fig. 20.11).

In many cases, a general principle for attaching the paramagnetic spin label TEMPO to ligands can be used: ligands and spin labels can be connected via amide bond formation using standard procedures of current peptide chemistry. For practical reasons, ligands functionalized with a carboxylic acid group and TEMPO functionalized with an amino group are commercially available (Fig. 20.11).

The spin label has an effect on the relaxation of protons that is measurable up to a distance of 20 Å [27]. The effect of the presence of an unpaired electron on the T_2 relaxation time is

$$\Delta\left(\frac{1}{T_2}\right) = \frac{1}{15} S(S+1) \frac{\gamma^2 g^2 \beta^2}{r^6} \left(4\tau_c + \frac{3\tau_c}{1+\omega_H^2 \tau_c^3}\right) \tag{4}$$

S is electron spin, γ is the gyromagnetic ratio for protons, g is the electronic g factor, β the Bohr magneton, r the distance between electron spin and proton, and ω the resonance frequency of protons. τ_c is the rotational correlation time. This increase in T_2 relaxation caused by the paramagnetic center is observed as an increase in the line width of the proton resonances. A similar equation is valid for $T_{1\rho}$ relaxation. The effect is strongly distance dependent, and therefore gives information on the relative orientation of the first and second ligand.

Second-site screening can be applied using the $T_{1\rho}$-filtered experiments described above. To each of the freshly prepared mixtures of compounds and protein, 20 μM spin label was added. The $T_{1\rho}$ spectra recorded with a 200 ms spin-lock are compared, and those compounds that show increased relaxation after adding the spin label are identified as second site ligands. An example of second-site screening using a spin-labeled first-site ligand is given in Fig. 20.13.

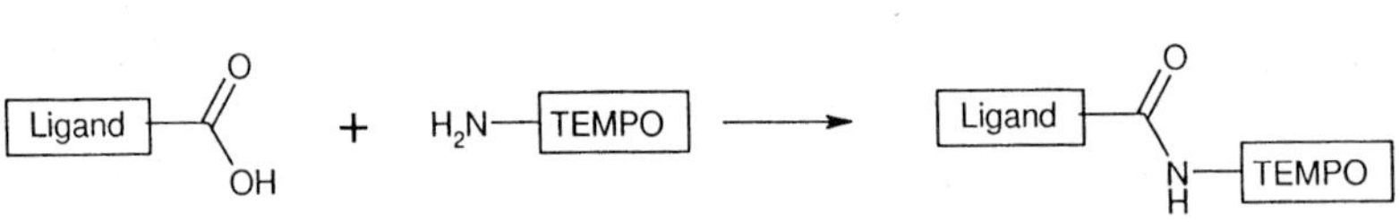

Fig. 20.11 Connection of carboxylic acid functionalized ligand with amino-TEMPO.

20.5 Intermolecular trNOE for Linker Design

If two bound ligands possess protons within 5 Å, transferred NOEs may be observed between resonances of the two ligands, as is illustrated in Fig. 20.12. This is very valuable information about which protons come close in the ternary complex, and therefore may guide the construction of an appropriate linkage of the two ligands.

20.6 A Case Study: Application of Second-Site Screening to Tubulin

Tubulin is an important target for the treatment of cancer. Despite the relatively long use of cytostatica in cancer chemotherapy, the spectrum of compounds with the same mode of action has been quite small until recently. It was mainly the discovery of paclitaxel (Taxol®) that stimulated the search for novel substances of natural or synthetic origin that interact with tubulin and microtubules [28]. Since the structure of paclitaxel and other natural compounds allowed only minor alterations, the discovery of new relatively simple molecules with the same mode of action would give the tubulin-targeted pharma research an enormous impulse.

The starting point for second-site screening was a previously discovered ligand, 3-methylpyridine, which was found to bind tubulin using the reverse screening approach. Using the recently developed second-site screening method applying spin-labeled ligands [6], compounds were tested for binding at a proximate binding site. For this purpose, a TEMPO spin label was covalently attached to 3-methylpyridine.

A typical example of second-site screening using $T_{1\rho}$ experiments is shown in Fig. 20.13.

After identification of seven second-site ligands, a transferred NOE experiment was carried out for these 7 compounds together with 3-methylpyridine and tubulin. This experiment was aimed at observing possible intermolecular NOEs between 3-methylpyridine and at least one of the second-site ligands. Such NOEs were not observed. However, intermolecular NOEs involving resonances of two of the seven second-site ligands in the mixture were observed. These two fragments are shown in Fig. 20.14. Thus, if these two ligands were "correctly" linked together, an increase of the binding affinity would be expected. The synthesis of four molecules was proposed; however, there was no suitable commercial building block available to derivatize one of the fragments. After discussing alternatives, it was decided to use an analog of the fragment. This analog was tested, and also showed binding. Three linked molecules were synthesized. Their chemical structures are presented in Fig. 20.15.

The three linked compounds were tested for tubulin binding by $T_{1\rho}$ experiments. For comparison, the binding of Epothilone A was measured as well. Epothilone A is a known natural ligand that binds to tubulin at the paclitaxel binding site. The results are depicted in Fig. 20.16. Two of the three compounds bind significantly better than Epothilone A. The binding of compound AB1 is also demonstrated in the spectra in Fig. 20.16, in which the aromatic region of the spectrum of the compound is compared before and after the addition of tubulin. Binding to the target is indicated by the observed line broadening.

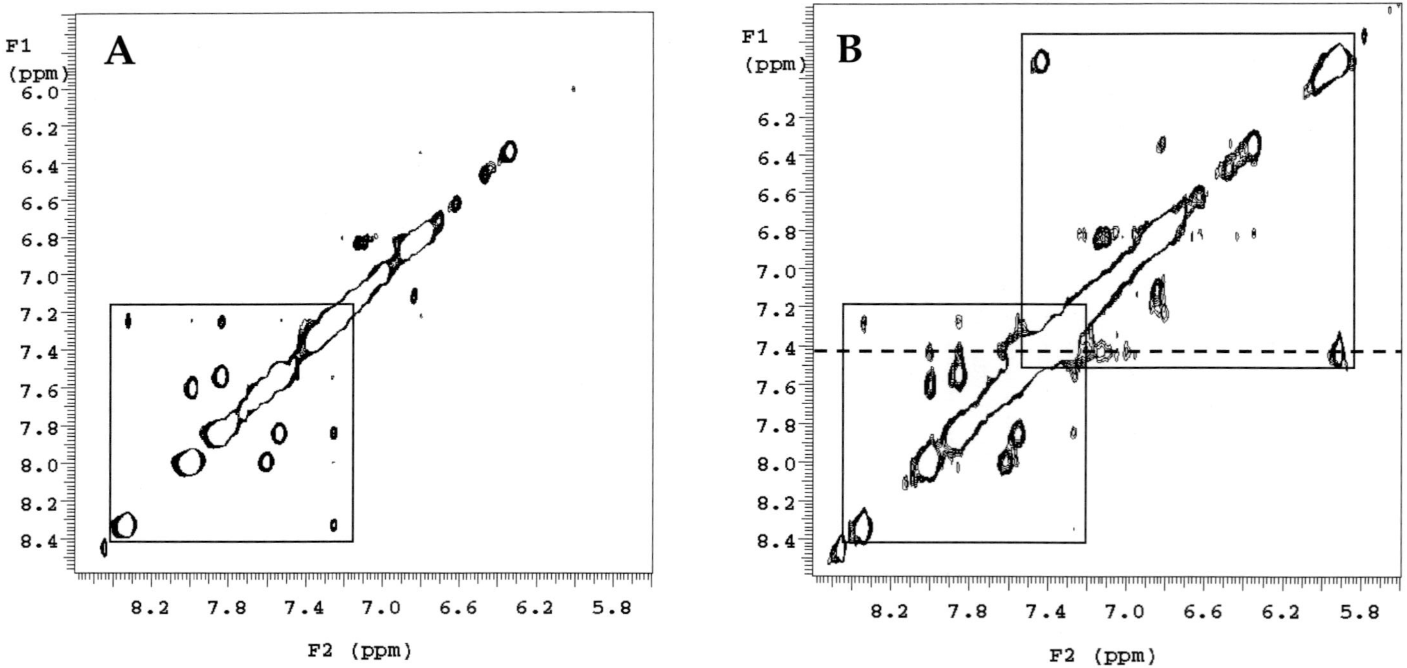

Fig. 20.12 Ligands in proximate binding pockets can result in inter-ligand transferred NOEs, provided the distance between some protons in the different ligands is within 5 Å. In spectrum **A**, trNOEs validate the binding of the first ligand to protein III. The concentration of ligand and protein was 500 μM and 50 μM, respectively. Resonances of the ligand are only found in the boxed region. Spectrum **B** was obtained after adding 500 μM of a second ligand that was found by second-site screening. The resonances of the second ligand are located within the second box. Thus, the binding of both ligands to the target protein was confirmed by the sign of the cross peaks. In addition, new cross peaks are observed, indicated by the dashed line. These cross peaks are inter-ligand NOEs, and can be of use in linker design.

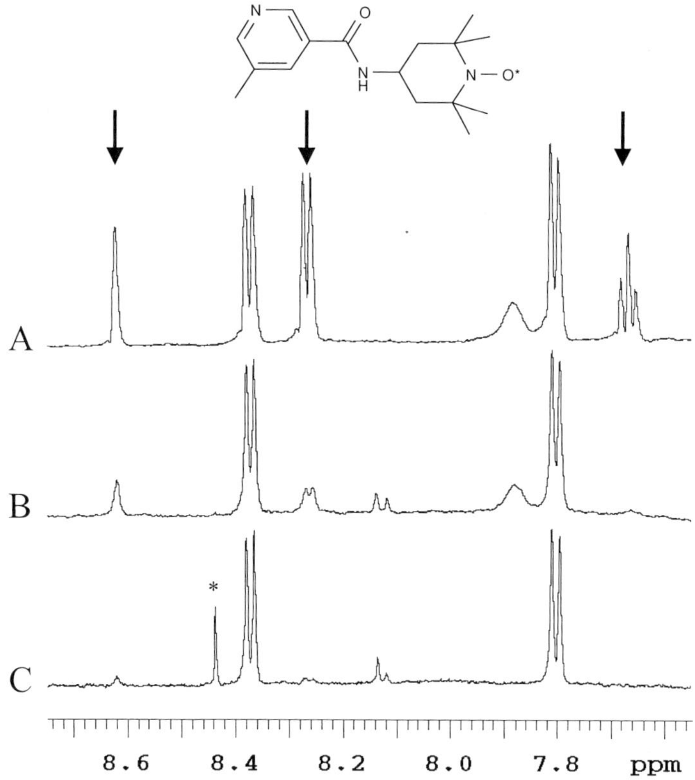

Fig. 20.13 Second-site screening by NMR makes use of a spin-labeled ligand. A small section of the $T_{1\rho}$ spectrum of a mixture of ten compounds is shown (**A**). The spin-lock time is 200 ms. The spectra displayed in **B** and **C** correspond to the same experiment but after adding tubulin (**B**) and after adding tubulin and the spin-labeled compound displayed at the top (**C**). The resonances indicated with arrows correspond to a ligand, identified to bind tubulin because of increased transverse relaxation (compare **A** and **B**). This ligand is identified to also bind close to the spin-labeled ligand, because additional relaxation caused by the paramagnetic ligand is observed (compare **B** and **C**). The signal marked with an asterisk (*) is due to an impurity.

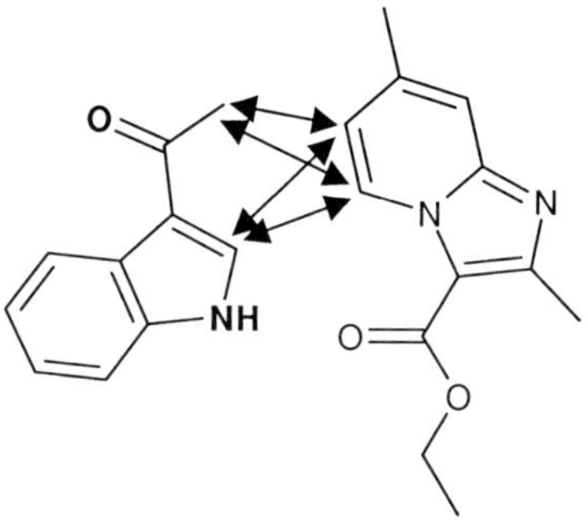

Fig. 20.14 Linker design of a molecule that connects two tubulin-binding ligands was based on the intermolecular trNOEs observed as indicated by arrows. It is noted that the model allows for possible hydrogen bond formation between the two ligands. Modeling showed that about six atoms are required to bring the fragments into the same orientation as that observed in the trNOE experiment.

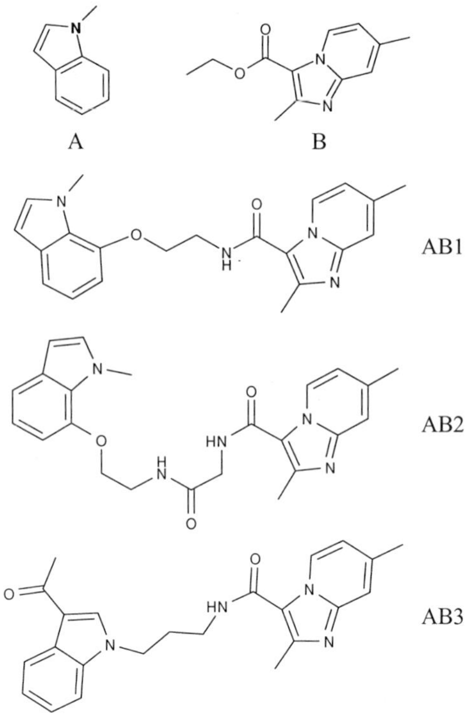

Fig. 20.15 Chemical structures of three compounds that were synthesized, together with the two fragments that were originally found to bind tubulin in proximate binding sites.

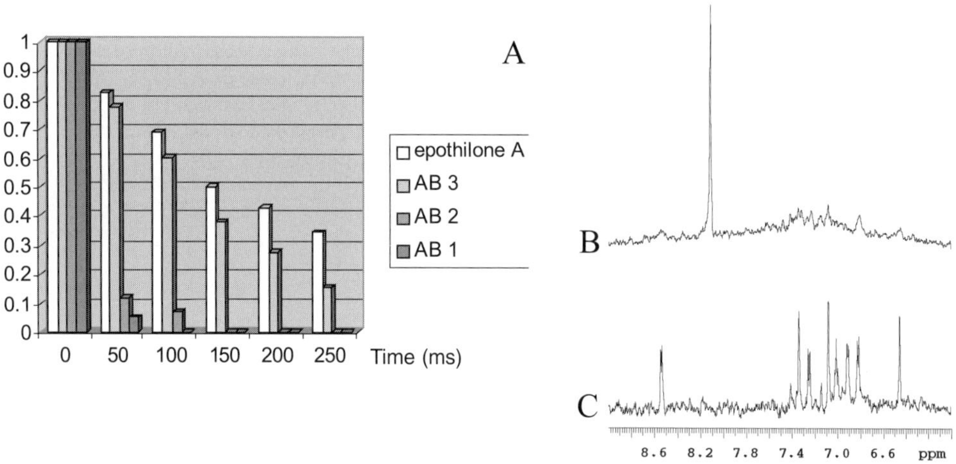

Fig. 20.16 Comparison of the binding of three compounds based on NMR screening with epothilone A. The compounds are in tenfold excess with respect to tubulin. The intensity of the NMR signal is measured as a function of the $T_{1\rho}$ filter delay (**A**). Compounds AB1 and AB2 relax significantly faster than epothilone A. The binding of these compounds is also obvious on the basis of the observed line broadening. This is shown in the $T_{1\rho}$-filtered one-dimensional spectra recorded for AB1 in the presence (**B**) and absence (**C**) of tubulin. A filter delay of 10 ms was used.

In the ideal case, a reliable model of the complex between the two ligands and the protein should be available for the linker design. Such a model can be obtained using X-ray crystallography or NMR spectroscopy. If such a model is not available, the design of a suitable linker is difficult. However, the previous example shows that even in the absence of such a model, alternative information such as the presence of intermolecular NOEs involving the two ligands can be successfully used for linker design.

20.7 An Integrated NMR Approach

The use of HTS [29, 30] in drug design has a long track record in the discovery of lead compounds. However, suitable leads may not always be found. This is not surprising if one considers that the currently available libraries used for HTS are only a small fraction of the estimated universe of possible compounds [2]. Therefore, it is very important to integrate new approaches for lead finding into the drug discovery process. Fig. 20.17 shows a flow chart that integrates NMR screening methodology into this process.

Once a protein target is selected and is produced in the mg amounts required for NMR screening, first-site screening can be started. For screening using 1-dimensional $T_{1\rho}$ relaxation filtered experiments, as described in this manuscript, one requires about 5 mg of a 25 kDa protein for the screening of 100 compounds, provided that screening is done in mixtures of 10 compounds on a 600 MHz spectrometer equipped with a regular probe. The use of SLAPSTIC [16] reduces this amount about tenfold. It is an advantage that the protein can be screened as soon as a properly folded protein or protein domain is available. Since the NMR binding assay is similar for each protein, no special assay needs to be developed. Thus, NMR screening can be done in an early phase of the drug discovery process, prior to or in parallel with HTS. However, screening is also useful after the HTS has been completed. Using the so-called reverse screening approach, leads from HTS can be redesigned. The lead molecule, which often comprises three to four fragments is split up into separate fragments. Some of these are used in conjunction with second-site screening, with the aim of replacing one of the fragments by a better one. In this way, a high affinity lead may acquire more desired properties. This approach is thus also useful if a putative drug candidate lacks important characteristics (e.g., bioavailablility, solubility, or patentability). If high-affinity ligands are available, fragments that bind to the desired binding pocket can be selected by competition experiments with these known ligands.

Apart from the discovery of ligands, NMR screening could potentially be used to validate a target. Well-defined binding pockets characterize a protein that may be successfully targeted by a drug. Analysis of fragments that have the property of binding to almost any protein that may be used as a drug target have shown that, e.g., biphenyls are very unselective, and bind to many different targets [31]. Thus, we propose to create a designed sub-library with such compounds to quickly validate the presence of such binding pockets, to be used in the early target validation phase.

It is perhaps too early to apply the screening to membrane-bound proteins, such as G-protein coupled receptors (GPCRs). However, it should be noted that the screening method does not exclude the screening of membrane-bound proteins. If the protein of interest

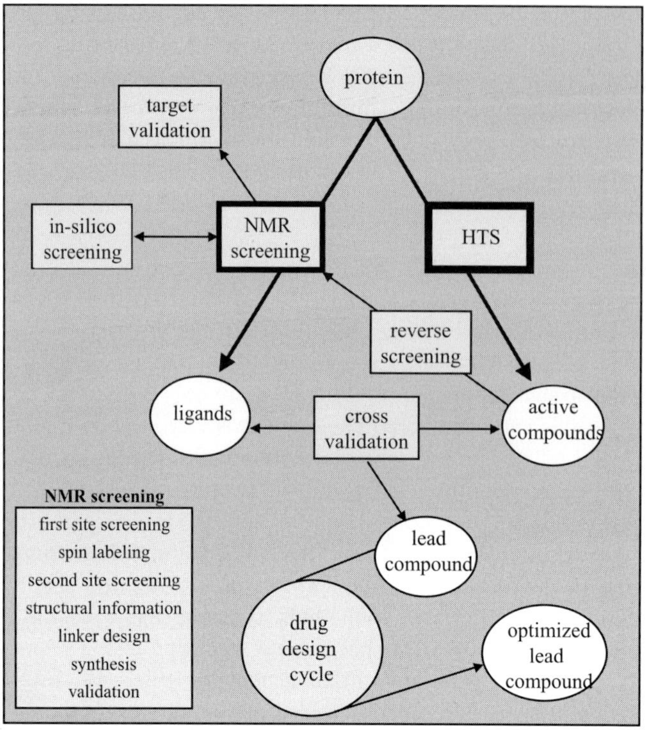

Fig. 20.17 Concept of the role of NMR screening in drug discovery. NMR screening may play a complementary role to HTS. NMR screening encompasses the following steps: first-site screening, spin labeling of a hit, second-site screening, linker design and synthesis (if possible using structural information), and validation of binding. HTS screening includes assay development, screening of compound libraries, plant extracts, microbial growths and combinatorial libraries. Also compound profiling of hits forms an integrated part of an HTS. The drug design cycle is a multidisciplinary approach that includes structure determination by protein crystallography or protein NMR. This often requires additional efforts in cloning, expression and purification of proteins or protein domains. Also chemistry, biology and molecular modeling play a critical role.

can be functionally reconstituted in a membrane system (e.g. vesicles), NMR can in principle detect weak binding of small molecules to such systems. Of course, the screening experiment requires a dedicated NMR library, which excludes compounds that bind to membranes. The latter can be tested by $T_{1\rho}$ experiments.

As has already been pointed out, ligands developed by linking compounds discovered by NMR screening may bind selectively and with high affinity to the target protein, but may not necessarily inhibit or stimulate the protein. Therefore, it is very important to cross-validate the result with a biochemical *in vitro* assay, and/or an *in vivo* model. Validation of results from HTS by NMR is also very important in order to detect possible false positives. The NMR data will not verify the biological activity, but will at least validate the interaction by detection of binding of the hit to the target protein. Only good validated compounds should be declared lead compounds.

Lead compounds can be optimized using standard medicinal chemistry, but if the protein is produced in amounts required for NMR screening it is also likely that a crystal structure will be obtained. The combined application of X-ray crystallography and NMR spectroscopy is especially fruitful. NMR can play a role in hit validation and screening. In addition, NMR may give structural information in various ways, e.g., the bound conformation of a ligand or its docking mode. In the meantime, a protein crystallographer could be successful in finding the conditions under which the protein-ligand complex crystallizes. If this turns out to be difficult, the full three-dimensional structure of the protein-ligand complex could be determined by NMR. The three-dimensional structure combined with molecular modeling efforts gives the medicinal chemist the opportunity to improve the affinity of the compound by rational structure-based design [1].

20.8 References

1 Böhm, H.-J., Klebe, G., Kubinyi, H. Wirkstoffdesign. Der Weg zum Arzneimittel. 1996 Spektrum Akademischer Verlag, Heidelberg.
2 Villar, H. O., Koehler, R. T. *Mol. Divers.* **2000**, *5*, 13–24.
3 Hajduk, P. J., Meadows, R. P., Fesik, S. W. *Q. Rev. Biophys.* **1999**, *32*, 211–240.
4 Moore, J. M. *Biopolymers.* **1999**, *51*, 221–243.
5 Bemis G. W., Murcko M. A. *J. Med. Chem.* **1996**, *39*, 2887–2893.
6 Jahnke, W., Perez, L. B., Paris, C. G., Strauss, A., Fendrich, G., Nalin, C. M. *J. Am. Chem. Soc.* **2000**, *122*, 7394–7395.
7 Thewes, T., Constantine, K., Byeon, I.–J. L., Llinas, M. *J. Biol. Chem.* **1989**, *265*, 3906–3915.
8 Shuker, S. B., Hajduk, P. J., Meadows, R. P., Fesik, S. W. *Science* **1996**, *274*, 1531–1534.
9 Jencks, W. P. *Proc. Natl. Acad. Sci USA* **1981**, *78*, 4046–4050.
10 Rotstein, S. H., Murcko, M. A. *J. Med. Chem.* **1993**, *36*, 1700–1710.
11 Böhm, H.-J. *J. Comput. Aided Mol. Des.* **1992**, *6*, 61–78.
12 Burkhard, P. Protein ligand interactions. PhD Thesis Biozentrum, Basel 1995.
13 Piotto, M., Saudek, V., Sklenar, V. *J. Biomol. NMR* **1992**, *2*, 661–665.
14 Callighan, D., West, J., Kumar, S., Schweitzer, B. I., Logan, T. M *J. Magn. Reson. B* **1996**, *112*, 82–85.
15 LaPlante, S. R., Aubry, N., Deziel, R., Ni, F., Xu, P. *J. Am. Chem. Soc.* **2000**, *122*, 12530–12535.
16 Jahnke, W., Rüdisser, S., Zurini, M. *J. Am. Chem. Soc.* **2001**, *123*, 3149–3150.
17 Szilagyi, L. *Prog. Nucl. Magn. Reson. Spectrosc.* **1995**, *27*, 325–443.
18 Dalvit, C., Pevarello, P., Tat, M., Veronesi, M., Vulpetti, A., Sundström, M. *J. Biomol. NMR* **2000**, *18*, 65–68.
19 Mayer, M., Meyer, B. *Angew. Chem.* **1999**, *111*, 1902–1906.
20 Hajduk, P. J., Olejniczak, E. T., Fesik, S. W. *J. Am. Chem. Soc.* **1997**, *119*, 12257–12261.
21 Chen, A., Shapiro, M. J. *J. Am. Chem. Soc.* **1998**, *120*, 10258–10259.
22 Ni, F. *Prog. NMR Spectrosc.* **1994**, *26*, 517–606.
23 Fejzo, J., Lepre, C. A., Peng, J. W., Bemis, G. W., Murcko, M., Moore, J. M. *Chem. Biol.* **1999**, 6, 755–769.
24 Lepre, C. A. *Drug Discov. Today* **2001**, *6*, 133–140.
25 Blommers, M. J. J., Davies, J., Jacoby, E. *Curr. Top. Med. Chem.* (in press)
26 Jahnke, W. *ChemBioChem* **2002** (in press)
27 Solomon, I., Bloembergen, N. *J. Chem. Phys.* **1956**, *25*, 261–266.
28 Hamel, E. *Med. Res. Rev.* **1996**, *16*, 207–231.
29. Fernandes, P. B. *Curr. Opin. Chem. Biol.* **1998**, *2*, 597–603.
30. Silverman, L., Campbell, R., Broach, J. R. *Curr. Opin. Chem. Biol.* **1998**, *2*, 397–403.
31 Hajduk, P. J., Bures, M., Prestegaard, J., Fesik, S. W. *J. Med. Chem.* **2000**, *43*, 3443–3447.

21
NMR-Based Drug Design: Approaches for Very Large Proteins

Maurizio Pellecchia, Xuemei Huang, David Meininger, and Daniel S. Sem

21.1
Introduction

While genomics efforts are producing unprecedented volumes of data that are being used to identify potential new drug targets [1], we are now left with the daunting challenge of translating targets to drug leads. In fact, with the overwhelming amount of information arising from structural proteomics efforts [2, 3], there is a greater need in pharmaceutical research for novel techniques capable of exploiting this information rapidly. Although structure-based drug design is certainly a powerful tool in general, the techniques used are somewhat limited by the long time required and the unpredictable nature of the structure determination process. In this chapter we will review two very recent NMR-based approaches to accelerate drug discovery, which can be applied to different general problems: NMR-DOC (Nuclear Magnetic Resonance DOcking of Compounds) is a technique that enables the rapid docking of compounds into their binding site once a model of the three-dimensional structure of the target protein is known, and NMR-SOLVE (Nuclear Magnetic Resonance Structurally Oriented Library Valency Engineering) is a method that enables the rapid structural characterization of ligand binding mode to enzymes with two adjacent binding pockets. NMR-SOLVE works in the complete absence of any 3D-structural data on the protein targets [4].

A major limitation in using protein NMR spectroscopy in drug discovery has been the molecular weight limitation imposed by nuclear spin relaxation (line broadening) and increased spectral complexity associated with macromolecules larger than ~35 kDa [5]. The most recent developments in NMR spectroscopy aimed at overcoming these problems will be briefly reviewed in Sect. 21.2.

21.2
NMR with Very Large Proteins

NMR spectroscopy has long been known for its ability to provide structural and dynamics information at atomic resolution, and it is increasingly being used in pharmaceutical drug discovery and development [6]. However, its use in structure-based drug design [7] has been limited because of the difficulty in obtaining structural information on protein-ligand complexes of molecular weight exceeding ~35 kDa, as well as the lengthy and potentially ambiguous process of resonance assignments [8]. In this section we will discuss the most recent solutions to the two major problems associated with obtaining useful NMR spectra of very large proteins: spectral complexity and line broadening.

21.2.1
Protein Perdeuteration and SEA-TROSY (Solvent Exposed Amides with Transverse Relaxation Optimized Spectroscopy)

Chemical shift mapping by monitoring the perturbation of the amide chemical shift of ^{15}N-labeled proteins upon the addition of ligands is a simple and straightforward method of detecting ligand binding. It is used to characterize the target's binding site, as well as to measure dissociation constants and binding kinetics [7], (see Chapt. 14). The recent advances of ^{2}H labeling, in combination with the revolutionary spectral improvement provided by the development of TROSY-type experiments (see Chapt. 10) [9], reduces the problems of nuclear spin relaxation, particularly in ^{15}N,^{1}H correlation experiments. These advances have allowed complexes of molecular weight >50 kDa to be studied, where only one of the components (with $\sim$ 200 amino acid residues) was ^{15}N/^{2}H labeled [10]. However, for even larger proteins (>300 amino acid residues), the problem of resonance overlap still represents a major obstacle for resonance assignment and chemical shift perturbation studies.

SEA-TROSY [11] was developed to reduce the problem of resonance overlap in ^{15}N, ^{1}H correlation spectra of very large proteins (>70 kDa) with a large number of amino acid residues (>300). The idea builds on the fact that binding sites are largely comprised of solvent-exposed loops, whereas amides from residues buried in the interior of the protein are not likely to be involved in intermolecular interactions. The selection of solvent-exposed amide protons in a perdeuterated ^{15}N-labeled sample dissolved in H_2O is obtained with the pulse scheme of Fig. 21.1. The sequence starts with a ^{15}N double filter (see Chapt. 17) [12], which serves to eliminate all the magnetization coming from amide protons. Water magnetization is not affected by the ^{15}N filter and subsequently returned to the z-axis at the last 90° ^{1}H pulse prior to τ_m (Fig. 21.1). At this time, water z magnetization is allowed to exchange with amide protons during a variable mixing time, τ_m. Backbone amides that are exposed to the solvent will acquire magnetization from the solvent that can be subsequently detected with a TROSY-type experiment [9, 13]. The resulting spectrum looks like a ^{15}N,^{1}H TROSY spectrum, but contains only backbone amides that are solvent exposed, and therefore the number of resonances is greatly reduced. Dramatic spectral simplification is achieved by removing those cross peaks that correspond to buried amides and are therefore of little interest in the characterization of protein-ligand interactions. Thus we observe only the signals that are relevant for the drug design process. The signal intensity (I) of a given amide proton in the resulting ^{15}N,^{1}H TROSY spectrum at a given mixing time, τ_m, is related to the exchange rate, k_{ex}, according to [14]:

$$I = I_\infty(1 - e^{-k_{ex}\cdot\tau_m}) \qquad (1)$$

where I_∞ is the intensity at infinite mixing time (complete exchange). The individual exchange rates, k_{ex}, depend on the amino acid type, the sequence, the pH of the solution

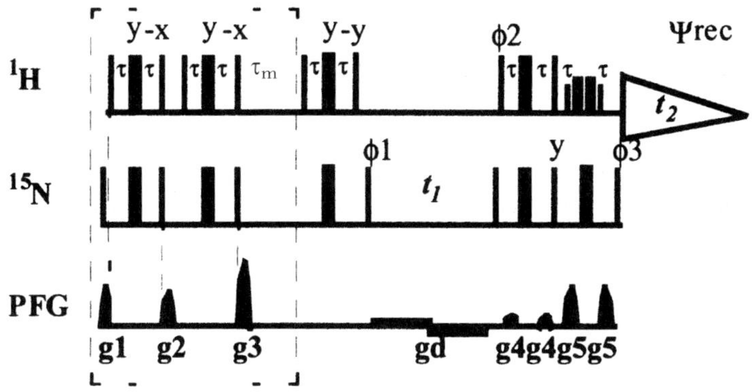

Fig. 21.1 Pulse sequence to selectively observe solvent-exposed amide protons with TROSY (SEA-TROSY). Narrow and thin bars represent 90 and 180° rf pulses, respectively. Unless specified otherwise, pulse phases are along the x axis. The pulsed field gradients are of 500 ms duration with strengths of $g_1 = 20$ G cm^{-1}, $g_2 = 30$ G cm^{-1}, $g_3 = 40$ G cm^{-1}, $g_4 = 15$ G cm^{-1}, $g_5 = 55$ G cm^{-1}. The bipolar gradient g_d is 0.5 G cm^{-1} and is used to avoid radiation damping effects during t_1 [29]. The delay τ was set to 2.7 ms. The phase cycle is: $\phi_1 = y, -y, -x, x$; $\phi_2 = y$; $\phi_3 = x$; $\Psi_{rec} = x, -x, -y, y$. A phase sensitive spectrum in the ^{15}N dimension is obtained by recording a second FID for each t_1 value, with $\phi_1 = -y, y, -x, x$, $\phi_2 = -y$ and $\phi_3 = -x$, and the data processed as for echo/anti-echo data. The SEA element is outlined by the dashed rectangle. (Taken from Ref. [11]).

and the degree of protection [15]. Generally, backbone amides located in loop regions exchange very fast, whereas amides involved in creating secondary structure elements exchange slowly and have high protection factors.

The SEA element can be combined with many triple resonance TROSY-type experiments (see Chapt. 10) [16] on triply ^{13}C/^{2}H/^{15}N labeled samples. Examples include SEA-HNCA-TROSY, SEA-HNCACB-TROSY, SEA-HN(CO)CA-TROSY etc. (for the strategy of resonance assignment in large proteins see Chapt. 4), which could be used to determine backbone resonance assignments for solvent-exposed loop regions. The SEA-TROSY spectrum of the uniformly ^{15}N/^{2}H-labeled catalytic portion (residues 57–678, 71 kDa molecular weight) of rat cytochrome P450 oxidoreductase [17] is shown in Fig. 21.2 (right). The reference ^{15}N,^{1}H TROSY spectrum of cytochrome P450 oxidoreductase is shown in Fig. 21.2 (left). Here deuteration and TROSY dramatically improved the quality of the spectrum, which contained virtually no signals when recorded with a nondeuterated sample and without TROSY (data not shown). However, only ~ 300 out of ~ 600 expected cross-peaks can be resolved, since most of the spectrum is characterized by severe resonance overlap that makes resonance assignments an improbable task and chemical shift mapping studies practically impossible.

The spectrum simplification in the SEA-TROSY (Fig. 21.2) is dramatic, as is also evident from the 1D traces taken from a crowded region of the spectra. The SEA-TROSY spectrum (Fig. 21.2) exhibits a loss in signal-to-noise ratio by a factor of only 0.7 on average. This small loss, however, is largely compensated for by the spectral simplification achieved. The cross-peak intensity is modulated by Eq. (1), so that a best compromise between spectral simplification and signal-to-noise ratio can be empirically derived by com-

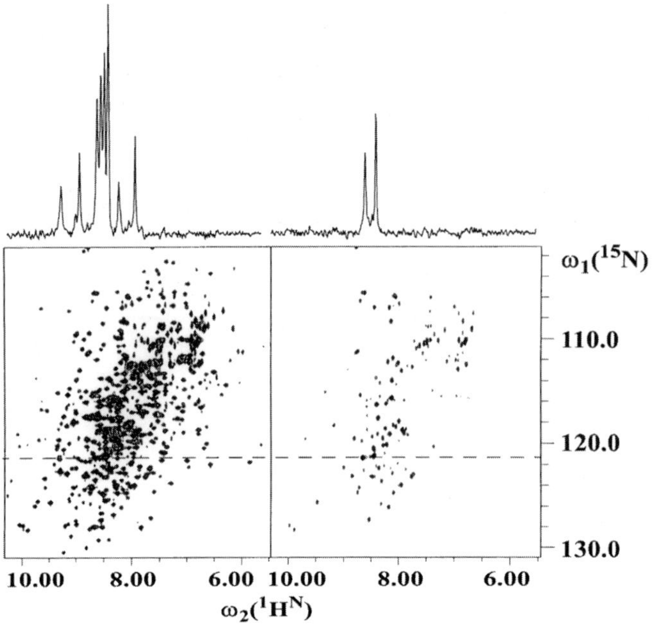

Fig. 21.2 TROSY (left) and SEA-TROSY (right) $^{15}N,^{1}H$ correlation spectra of 0.5 mM $^{2}H/^{15}N$-labeled cytochrome P450 oxidoreductase from rat liver (95% H_2O, 5% D_2O; $T=303$ K, pH=7.5). U-$^{2}H/^{15}N$ and U-$^{2}H/^{15}N/^{13}C$-labeled cytochrome P450 oxidoreductase from rat liver, lacking the N-terminal 56 residues and the C-terminal WS sequence. The TROSY spectrum was obtained with the same pulse scheme but without the SEA element (dashed box in Fig. 21.1). The SEA-TROSY spectrum was obtained with the pulse scheme reported in Fig. 21.1 with a mixing time τ_m of 100 ms. The spectra were recorded on a Bruker DRX700 spectrometer operating at 700 MHz ^{1}H frequency. 128 (t_1)×1024 (t_2) complex points were recorded. Traces along ω_2 are taken at the position indicated by the dashed line. (Taken from Ref. [11]).

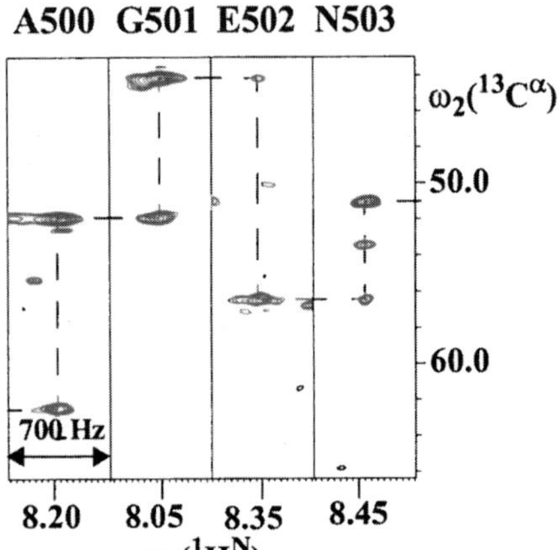

Fig. 21.3 Representative $\omega_2(^{13}C^{\alpha}) - \omega_3(^{1}H^{N})$ strips from a 3D SEA-HNCA-TROSY spectrum recorded on a 0.5 mM solution of $^{15}N/^{2}H/^{13}C$-labeled sample of cytochrome P450 oxidoreductase. The assignment of the loop of residues Ala 500 to Asn 503 has been obtained via the sequential $^{13}C^{\alpha}(i)/^{13}C^{\alpha}(i\text{–}1)$ connectivities indicated by the dashed line. (Taken from Ref. [11]).

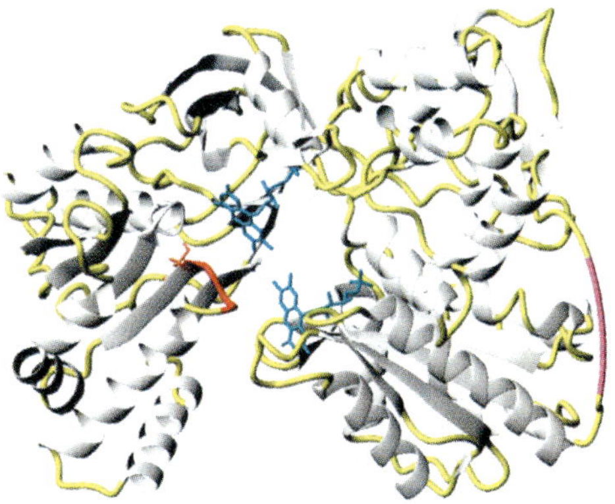

Fig. 21.4 Three-dimensional structure of the enzyme cytochrome P450 oxidoreductase. The cofactors FAD and FMN are depicted in light blue. Loop regions are represented by cylindrical rods (yellow); α-helices and β-sheets (in white) are represented by ribbons and arrows, respectively. Resonance assignments for the residues located in the key loops regions [30] highlighted in purple and red were obtained with a 3D SEA-HNCA-TROSY experiment.

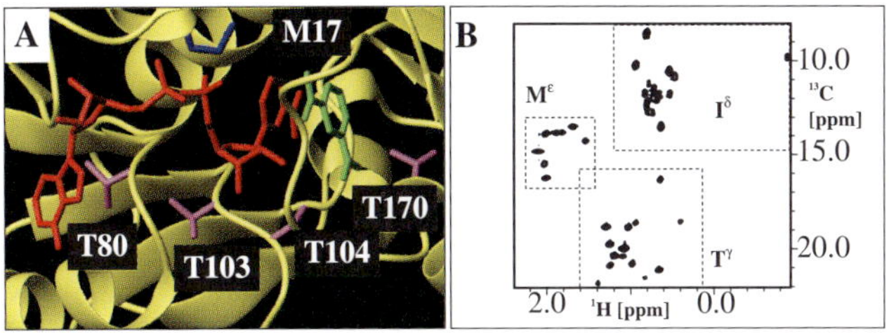

Fig. 21.5 **A** Ribbon drawing representing the binding site region of DHPR (PDB code 1arz) in complex with NADH and 2,6-pyridinedicarboxylate (PDC). The cofactor NADH is depicted in red and the substrate analog PDC in green. Heavy atom side chains of Thr and Met residues close to either ligand are displayed and numbered. All figures displaying three-dimensional structures were generated with MOLMOL [31]. **B** 2D [^{13}C, ^{1}H] HMQC spectra recorded with a sample of U-^{2}H, ^{13}C/^{1}H Thr, ^{13}C$^{\delta}$/^{1}H Ile, ^{13}C$^{\varepsilon}$/^{1}H Met labeled DHPR ([MIT]-DHPR). Met, Ile and Thr regions are enclosed in dashed rectangles. DHPR uniformly enriched in ^{2}H and ^{15}N and containing ^{1}H/^{15}N/^{13}C-labeled threonine, ^{1}H/^{13}C(δ-methyl)-labeled isoleucine, and ^{1}H/^{13}C(ε-methyl)-labeled methionine was produced in a BioFlo 3000 fermentor (New Brunswick Scientific) using a modified version of the supplemented M9 minimal media dissolved in 99.9% deuterium oxide (Aldrich). The fermentation protocol utilized is a modified version of that found in Meininger et al. [36] with 100 mg L^{-1} ^{1}H/^{15}N/^{13}C-labeled threonine and 200 mg L^{-1} ^{1}H/^{13}C (ε-methyl)-labeled methionine (Cambridge Isotope Laboratories) added to the culture 10 min prior to addition of the feed solution.
All NMR experiments were performed on a Bruker DRX700 spectrometer operating at 700 MHz ^{1}H frequency and equipped with a triple resonance probe and a triple axis gradient coil. DHPR concentration was $\sim$ 75 μM (300 mM monomer) in 25 μM d$_{11}$-Tris buffer, pH=7.8 and T=303 K. (Taken from Ref. [4]).

paring SEA-TROSY spectra recorded at different mixing times. In practice, a range of mixing times from 10 to $\sim$ 400 ms needs to be tried. Loop regions can be easily assigned with a 3D SEA-HNCA-TROSY sub-spectrum, since the number of possible ^{13}C$^{\alpha}$ connectivities is largely reduced (Fig. 21.3). The assignment process can then be focused on loop regions of the protein of interest, which are more likely to be involved in protein function (Fig. 21.4). The process of mapping the sequentially connected stretches of amino acids onto the sequence of the protein can also be facilitated by the knowledge of the three-dimensional structure of the enzyme, which allows identification of the sequences of solvent-exposed loop regions. However, in most cases, additional triple resonance experiments such as 3D SEA-HNCACB-TROSY may also be needed.

In conclusion, the selection of solvent-exposed amide protons with the SEA-TROSY-type experiments represents an attractive solution to the problem of resonance overlap encountered with very large proteins with a very large number of amino acid residues. These can provide partial backbone assignments without requiring selective labeling, thus enabling traditional chemical shift mapping and other NMR-based drug design methods with very large proteins to be performed. It is a spectral simplification strategy that is ideally suited to NMR-based drug design, since it removes cross peaks from residues that are not likely to be involved in protein-ligand interactions.

21.2.2
Protein Perdeuteration and Selective Amino Acid Labeling

In theory, a structural description of protein-ligand binding interactions requires that only key interactions occurring in the binding site of the target protein be characterized. The NMR-DOC and NMR-SOLVE (Sect. 21.3) methods employ several protocols to rapidly obtain such structural information. These methods rely on the selective labeling and/or probing key amino acids in the binding site. In this section we illustrate experiments involving the enzyme dihydrodipicolinate reductase (DHPR), a homotetramer of MW 120 kDa involved in lysine and bacterial cell wall biosynthesis [18]. Since this enzyme is well beyond the current molecular weight limit for traditional protein NMR spectroscopy, we will review the selective labeling schemes and deuteration that enables high-resolution NMR data to be acquired.

The three-dimensional X-ray structure of the enzyme [19] reveals that several Thr residues occur in both the NADH cofactor and substrate binding sites (Fig. 21.5 A see p. 463). A Met residue (Met17) is also present at the interface between the cofactor NADH and a substrate analog pyridine-2,6-dicarboxylate (PDC) (Fig. 21.5 A). Therefore, we prepared a sample of DHPR that was selectively labeled in these amino acid residues as follows: $^{13}C^{\varepsilon}/^{1}H$ Met, $^{13}C^{\varepsilon}/^{1}H$ Ile, $^{13}C/^{1}H$ Thr and uniformly ^{2}H-labeled elsewhere ([MIT]-DHPR). This labeling can be achieved by supplementing the media with appropriate commercially available labeled amino acids, $^{12}C/^{2}H$-labeled glucose and D_2O [20] (see also the caption to Fig. 21.5 for details).

The high level of deuteration enormously reduces the line broadening due to the ^{1}H-^{1}H dipole-dipole relaxation of the methyl protons of Met, Ile and Thr. Furthermore, the selective ^{13}C labeling of these methyl groups can be used to resolve the methyl proton resonance overlap.

However, in order to observe all the expected cross peaks, one for each Thr, Ile and Met methyl in DHPR in a $^{13}C,^{1}H$ correlation spectrum, the NMR parameters need to be optimized for maximum sensitivity and resolution as follows. First, based on the large chemical shift difference of Thr $^{13}C^{\gamma}$ ($\sim$ 18 ppm) and $^{13}C^{\beta}$ ($\sim$ 70 ppm), selective WURST adiabatic decoupling [21] is used during the ^{13}C evolution to decouple $^{13}C^{\gamma}$ from $^{13}C^{\beta}$, resulting in much narrower lines in the Thr $^{13}C^{\gamma}$ dimension. This line narrowing dramatically reduces the overlap among the 14 $^{13}C^{\gamma}/^{1}H^{\gamma}$ resonances in DHPR (Fig. 21.5 B), wherein Thr $^{13}C^{\gamma}/^{1}H^{\gamma}$ cross peaks are much narrower than those corresponding to Ile $^{13}C^{\delta}/^{1}H^{\delta}$. Second, instead of the HSQC (heteronuclear single quantum correlation) scheme, an HMQC (heteronuclear multiple quantum correlation) magnetization transfer has been adopted, since ^{1}H-^{1}H dipole-dipole relaxation contributing to fast decay of ^{13}C,^{1}H multiple-quantum coherences will be largely attenuated [8]. In highly protonated ^{13}C-labeled protein samples, HSQC sequences may exhibit better relaxation properties than HMQC experiments for certain resonances because of strong dipole-dipole ^{1}H-^{1}H relaxation during the heteronuclear evolution time between the given ^{13}C-attached proton and all the surrounding protons. Dephasing due to proton-proton scalar couplings and, to a minor extent, ^{1}H chemical shift anisotropy are other sources leading to line broadening in the ^{13}C dimension. Proton-proton dipole-dipole relaxation is especially efficient for methylene and methyl sites that contain very short proton-proton distances. The selectively labeled samples, however, will be mostly deuterated, and proton-proton dipole-dipole interactions can occur (in this particular case) only between Met, Thr and Ile residues. As Thr and Met residues of interest are not in close proximity to each other (>6 Å), these dipole-dipole interactions are very small, so that the HMQC experiment is superior. Indeed, with these modifications, all expected cross peaks are clearly observed in the 2D ^{13}C,^{1}H correlation spectrum of [MIT]-DHPR, even at a relatively low enzyme concentration, measured in about 30 min with a DRX700 Bruker spectrometer (Fig. 21.5 B see p. 463). Such sensitivity and resolution are unprecedented for such a large protein and suggest the feasibility of the application of NMR in the study of even larger protein systems [22].

21.3 NMR-Based Drug Design Techniques

The most recent NMR-based drug design techniques described in the following sections build on the recent advances, reported in Sect. 21.2, that make the study of very large proteins possible. Two different strategies will be illustrated, each strategy being applicable in different cases. The NMR-DOC method is applicable when a model of the three-dimensional structure of the target protein is known. This method will find more applications as the structural genomics efforts produce large volumes of structural information. The second method, NMR-SOLVE, is a semi-rational method to guide the construction of combinatorial bi-ligand libraries, and is applicable to protein targets that possess two adjacent binding sites. In this case it is not necessary to know the three-dimensional structure of the target protein.

21.3.1 NMR-DOC (Nuclear Magnetic Resonance DOcking of Compounds)

Although the labeling scheme reported in Sect. 21.2 largely eliminates the problems of spectral complexity and line broadening that characterize the NMR spectra of large proteins, the traditional methods for obtaining resonance assignments [8] are no longer applicable. However, in the example reported in Sect. 21.2, the resonance assignments for the key residues Thr104 and Met17 can be easily obtained by expressing and purifying

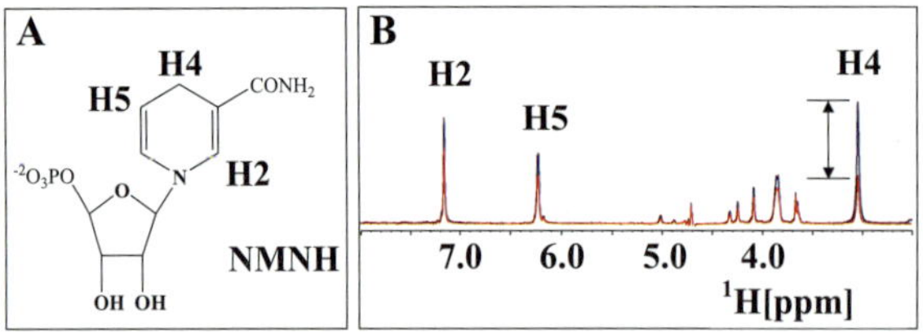

Fig. 21.6 Establishing ligand binding with a cross-saturation experiment. **A** Chemical structure of NMNH. **B** Reference spectrum (blue) and binding site-saturated spectrum (red) of NMNH (500 μM) in the presence of [MIT]-DHPR at 2.5 μM (10 μM monomer). Cross-saturation was performed with a train of IBURP pulses [32] of 8 ms duration spaced by 10 ms and positioned at 1 ppm (cross-saturated spectrum) and at –2.5 ppm (reference spectrum). (Taken from Ref. [4]).

[MIT]-labeled single point mutants, T104S and M17I. In fact, the assignments can then be easily obtained by comparing the 2D $^{13}C,^{1}H$ correlation spectra of [MIT]-T104S-DHPR and [MIT]-M17I-DHPR with that of wild-type [MIT]-labeled enzyme.

Once the binding site residues are assigned, compound screening and identification can be performed with simple experiments that provide information, in a relatively short time, not only on binding affinity (via titration) but also on inhibitor binding mode. The following paragraphs will illustrate the results of NMR-DOC with [MIT]-DHPR using nicotinamide mononucleotide (NMNH, Fig. 21.6 A) as a test and validation cofactor mimic, since its binding mode is easily extrapolated from the X-ray structure of NADH bound to DHPR [19].

Ligand binding experiments are performed with a low ($\sim$ 10 μM) concentration of labeled protein and about 1 mM (or less) of the given ligand. Binding of NMNH to [MIT]-DHPR is quickly verified using a saturation transfer experiment (see Chapts. 11 and 14) [23–25]. The saturation of binding site nuclei can be achieved through selective saturation of the aliphatic region of the spectrum that contains the resonances of the labeled Thr and Met residues. The saturation is transferred to the ligand only if it is bound to the protein, as evidenced by a difference spectrum (Fig. 21.6 A). Cross-saturation experiments with unlabeled protein are discussed in greater detail in Chapts. 14 and 16. Using a selectively methyl-labeled protein sample makes these experiments particularly informative, since most of the protonation is expected to be in its binding site (Fig. 21.5). Sparse methyl ^{1}H labeling minimizes spin-diffusion, and therefore observation of a cross-saturation effect is a likely indicator of specific binding in the binding pocket of the protein. Once a "structurally screened" hit is discovered, its orientation in the binding site can be more accurately determined by measuring intermolecular NOEs with previously assigned resonances in the binding site of the enzyme. In DHPR, intramolecular NOEs are observed, which can be used to determine the bioactive conformation of the ligand as well as very strong intermolecular NOEs between the nicotinamide ring protons and the methyl groups of residues Met17 and Thr104 (Fig. 21.7). The observation of very strong intermolecular NOEs is a consequence of the large molecular weight of the complex (fast cross-relaxation) on the one hand, and low proton density (due to the labeling scheme used) on the other, which results in a very efficient intermolecular proton-proton magnetization transfer via cross relaxation. In addition, inter-ligand (PDC to NADH in this ex-

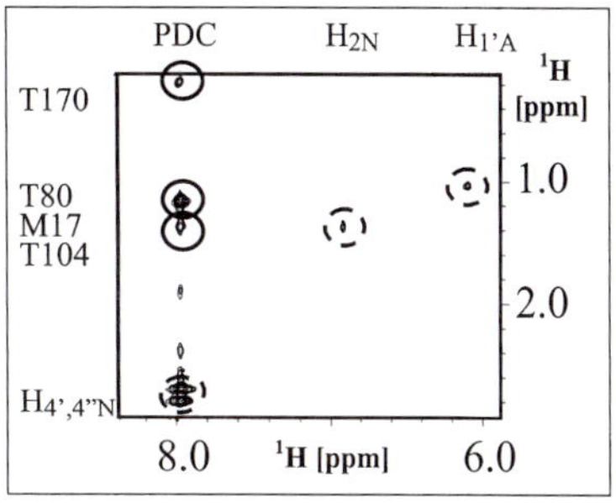

Fig. 21.7 **A** Portion of a 2D [^{1}H, ^{1}H]-NOESY spectrum of NADH (500 μM) in complex with [MIT]-DHPR (75 μM) and PDC (500 μM). Protein-NADH NOEs are labeled with dashed circles. PDC-protein NOEs are labeled with solid circles. Inter-ligand PDC-NADH NOEs are labeled with dotted circles. Typical 2D [^{1}H, ^{1}H]-NOESY [33] spectra were acquired with 256×2048 complex points and with mixing times between 50 ms and 500 ms. The measuring time for a 2D [^{1}H, ^{1}H]-NOESY varied from ~ 12 h to 48 h, depending on the ligand concentration (between 500 μM and 2 mM). Occasional ambiguities due to proton overlap among Thr and Met residues were resolved by recording a 3D [^{13}C, ^{1}H]-resolved [^{1}H, ^{1}H]-NOESY experiment. QUIET-NOESY (Quenching Undesirable Indirect External Trouble in NOESY) experiments [34] were also performed to avoid artificial NOE cross peaks arising from spin diffusion. (Taken from Ref. [4]).

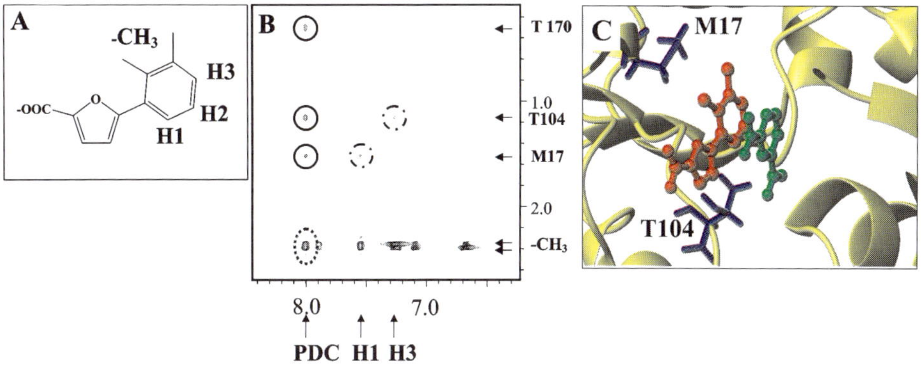

Fig. 21.8 **A** Structure of the NADH mimic TTM0.029.A85. **B** Portion of a 2D [^{1}H, ^{1}H]-NOESY spectrum of TTM0.029.A85 (500 μM) in complex with [MIT]-DHPR (75 μM) and PDC (500 μM). The NOEs are labeled according to the use in Fig. 21.7. **C** Structure of TTM0.029.A85 (red) docked into the three-dimensional structure of DHPR (yellow) bound to PDC (green). The structure was calculated with DYANA [35] based on the NOE data obtained with the ternary complex of TTM0.029.A85, [MIT]-DHPR and PDC, and the X-ray structure of DHPR bound to PDC. The position of PDC and the coordinates of the enzyme were kept fixed during the structure calculation. The energy-minimized coordinates of the inhibitor were obtained using InsightII (Molecular Simulation Inc.). This model was subsequently linked by a dummy linker of ~ 100 Å and encompassing 56 dummy torsion angles to the model of DHPR bound to PDC. Random torsion angles were assigned to the linker to generate a model of the complex with random initial positioning of the ligand, while the coordinates of DHPR and PDC were kept fixed. Subsequently, a variable target function was minimized in the linker torsion angle space versus the NOE distance constraints between the ligand and both protein and substrate analog (PDC). Twenty structures were calculated with 5000 iterations per structure. The best seven structures with the lower target function converged into the final structure shown. (Taken from Ref. [4]).

ample) NOEs also provide extremely useful information on the orientation and the distance of two binders relative to each other [26]. These inter-ligand NOEs appear in the 2D [^{1}H, ^{1}H]-NOESY spectrum of the ternary complex (Fig. 21.7) along with the protein-ligand NOEs. Protein deuteration limits spin diffusion and, most importantly, transverse nuclear spin relaxation, and therefore permits the use of more concentrated protein samples (300 μM in the examples reported). This in turn provides increased signal-to-noise ratio for both protein-ligand and inter-ligand NOEs. Observation of bound-state inter-ligand NOEs provides an enormous advantage over the transferred inter-ligand NOE effect, which is limited to rapidly dissociating molecules. The latter effect becomes rather small when one or both ligands dissociates slowly, as for tight binders [26].

These NOE cross peaks for the DHPR-NMNH-PDC ternary complex are consistent with the X-ray structure of the DHPR-NADH-PDC ternary complex (Fig. 21.5A), and can therefore be used as constraints for docking the NMNH ligand into the binding site of DHPR. As a more novel example, an NMR DOC model was created for the complex of a weak-binding inhibitor ($K_D \approx 500$ μM) (Fig. 21.8A, see p. 467) with the enzyme DHPR when complexed to PDC. This model was constructed based on DHPR-ligand and PDC-ligand NOE derived constraints, which were obtained in a few days (Fig. 21.8 B, see p. 467). An attractive application of the NMR-DOC experiment is the design of bi-ligand inhibitors of enzymes through linking fragment compounds such as those shown in Fig. 21.8 C in order to obtain higher affinity ligands (for general considerations in this strategy see Chapt. 18). In fact, we designed several inhibitors with nanomolar affinity for DHPR using this strategy.

The NMR-DOC method can be summarized in the following steps. First, based on the three-dimensional structure of the enzyme, a set of amino acids to be selectively labeled is identified. Methyl-containing amino acid residues seem particularly well suited for NMR-DOC because of their unique chemical shifts and their favorable relaxation properties. Aromatic residues could also be used in conjunction with TROSY-type experiments (see Chapt. 10) [27]. Second, selectively labeled wild-type protein and the appropriate single point mutants are expressed and purified to obtain the resonance assignments for key binding site residues. Third, NOEs between a ligand and the assigned binding site residues are collected. Finally, these NOEs are converted into distance constraints used to dock the ligand into the three-dimensional structure of the target protein. Once the first two steps are performed to identify binding site residues, NMR data are gathered and ligand docking is performed in a few days for a given ligand. Thus, structural information is collected for large protein-ligand complexes with unprecedented speed. It should be noted that assignment of interface residues "Thr104" and "Met17" (step 2 of NMR-DOC) could also have been done in the complete absence of any three-dimensional structure of DHPR and without doing site-directed mutagenesis. The method developed to accomplish the assignment of interface residues in this scenario and to provide information useful in the design of bi-ligand compounds and libraries is called NMR-SOLVE. It is discussed in the next section, with application to an enzyme of unknown structure.

21.3.2
NMR-SOLVE
(Nuclear Magnetic Resonance Structurally Oriented Library Valency Engineering)

In the previous section, we discussed how ligand-protein and ligand-ligand NOEs are used to obtain a model of the three-dimensional structure of a ternary complex, which can be used to design a linker between two ligands to produce bi-ligand compounds with increased potency. Alternatively, a linker reaching from a first ligand towards the second binding site can be designed using the NMR data. A combinatorial library can then be placed at the end of this linker (see Chapt. 18 for general aspects) aimed at targeting the second binding pocket, thus producing bi-ligands in a combinatorial manner. This strategy, called NMR-SOLVE, is very attractive, since it works also for protein targets for which no three-dimensional structure is available. The method has been applied to DHPR and is illustrated here with the enzyme 1-deoxy-D-xylulose-5-phosphate reductoisomerase (DOXPR) [28], a homo-tetrameric enzyme of MW 174 kDa for which there is no protein with >20% homology in the Protein Data Bank. We uniformly $^{2}H/^{12}C$ labeled DOXPR, with the exception of the methyl groups of Met, Thr and Ile that were ^{13}C- and ^{1}H-labeled to obtain [MIT]-DOXPR. The MIT amino acid labeling scheme was chosen based on our survey of oxidoreductase three-dimensional structures, which revealed an average of four to five of these residues in the NAD-binding sites. However, similar experiments could also be performed with most other amino acid types. In particular, selective side-chain $^{13}C/^{1}H$ labeling of Val, Tyr, Phe, Trp and His has also been reported [20].

The [^{13}C, ^{1}H]-HMQC spectrum for [MIT]-DOXPR and the 2D [^{1}H, ^{1}H]-NOESY spectrum for DOXPR in complex with its cofactor NADPH are shown in Fig. 21.9. From their unique ^{1}H chemical shifts, the observed NOEs clearly indicate that there is an Ile as well as a Met proximal to the nicotinamide ring portion of NADPH (Fig. 21.9 A). Proximity of another Ile residue to the adenosine ring is also observed (Fig. 21.9 B). In a related application of NMR-SOLVE, interface residues have also been assigned by comparing spectra of protein complexed with a chemically altered version of the cofactor, provided that there is a rather good dispersion of the signals in the 2D spectra. Rather than NOEs, chemical shift perturbations are now being monitored. Similar experiments have been performed with active-site paramagnetic probes [4].

Although it is not possible to assign these residues to a specific location in the primary sequence of the enzyme, the detection of NOEs between a novel cofactor mimic and these residues provides the necessary information on the orientation of inhibitors with respect to the cofactor. For example, the H_{2N} proton of NADPH shows NOEs to a Met in DOXPR (Fig. 21.9 B) and a Thr in DHPR (Fig. 21.7). Since the H_{2N} of the nicotinamide ring must be proximal to the substrate binding site, because the nicotinamide ring participates in a chemical reaction with the substrate, these Met and Thr residues must be at the interface of the cofactor and substrate sites in both DOXPR and DHPR. If any novel ligand shows NOEs to a Thr in DHPR with this chemical shift (Fig. 21.8 B), it is known to be proximal to the substrate site. No three-dimensional structure is needed for this analysis, although in the case of DHPR a structure allowed us to assign the interface residue to Thr 104. But such assignment is clearly not needed for NMR-SOLVE. As such, it is possible to identify ligands and portions of these ligands that reside in the catalytic part of the binding site

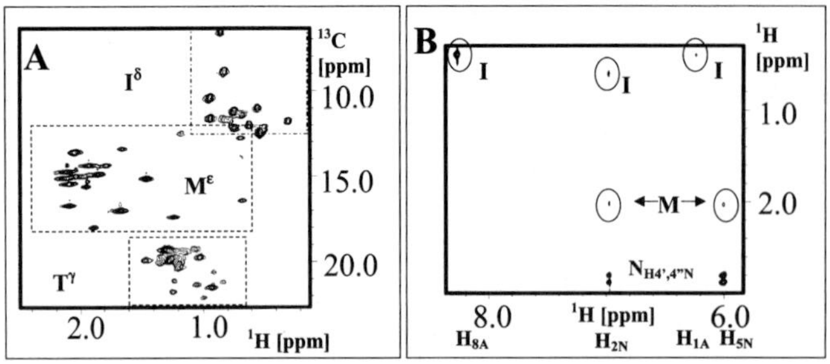

Fig. 21.9 NMR-SOLVE with DOXPR. **A** 2D [^{13}C, ^{1}H]-HMQC spectrum recorded with a 75 μM (300 μM monomer) sample of U-^{2}H, ^{13}C/^{1}H Thr, $^{13}C^{\delta}/^{1}$H Ile, $^{13}C^{\varepsilon}/^{1}$H Met labeled DOXPR ([MIT]-DOXPR). Met, Ile and Thr regions are enclosed in dashed rectangles. **B** Portion of a 2D [^{1}H, ^{1}H]-NOESY spectrum of the complex between $NADP^{+}$ and [MIT]-DOXPR with intermolecular NOEs circled. (Taken from Ref. [4]).

close to the nicotinamide ring of NADH. This is, by necessity, at the interface of the cofactor and substrate sites. This in turn provides the structural information needed to optimally place the linker on the cofactor mimic for creating a bi-ligand or bi-ligand library that targets the substrate binding site, as a source of specificity. In conclusion, NMR-SOLVE can be used to guide the construction of large combinatorial libraries with the goal of finding specific bi-ligand inhibitors, while NMR-DOC deals with the design of individual bi-ligands from specific pairs of compounds. At Triad Therapeutics we use these strategies to design gene-family focused bi-ligand combinatorial libraries (NMR-SOLVE) and to design specific bi-ligand inhibitors for a given enzyme (NMR-DOC). Several nM bi-ligand inhibitors have been found against both DOXPR and DHPR using both strategies.

21.4
Conclusions

Genomic research is driving the identification and validation of many new drug targets, frequently of unknown three-dimensional structures. There is thus a great need for translating this target information into inhibitors and eventually drugs. Appropriate structure-based methods need to be developed that are much faster than present approaches. The techniques discussed in this chapter illustrate that with the appropriate protein isotope-labeling schemes and novel pulse sequences, solution NMR spectroscopy represents a rapid and efficient way to provide descriptive structural information on protein-ligand complexes with unprecedented speed. Traditional ligand binding studies via chemical shift mapping can be performed even for proteins with >600 amino acid residues using SEA-TROSY-type experiments. The NMR-DOC method is applicable to even larger proteins and can rapidly provide very useful structural information on protein ligand complexes if a reference three-dimensional structure of a given enzyme is known. For enzymes with unknown three-dimensional structure with two adjacent binding sites, as in the dehydrogenase gene family, the NMR-SOLVE method provides structural information on ligand-binding mode. This information is gathered in days and is used to guide medicinal chemistry in the design of bi-ligands and bi-ligand libraries by focusing on the catalytic region of the binding site. The unprecedented speed with which this information can be obtained provides a much needed tool to dramatically increase our understanding of protein-ligand interactions and to guide drug design. It will help to fulfill the promise of structural genomics in basic and applied research and in systems-based approaches to drug discovery.

21.5 References

1 J. Drews, *In Search of Tomorrow's Medicines*, Springer, New York, USA, **1998**.
2 G.T. Montelione, D. Zheng, Y.J. Huang, K.C. Gunsalus, T. Szyperski, *Nat. Struct. Biol.* **2000**, *7*, (Suppl): 982–985.
3 S.K. Burley, *Nat. Struct. Biol.* **2000**, *7*, (Suppl): 932–934.
4 M. Pellecchia, D. Meininger, Q. Dong, E. Chang, R. Jack, D. Sem, *J. Biomol. NMR* **2002**, *22*, 165–173.
5 L. E. Kay, *Methods Enzymol.* **2001**, *339*, 174–203.
6 D.S. Sem, M. Pellecchia, *Curr. Opin. Drug Disc.* **2001**, *4*, 479–493.
7 P.J. Hajduk, R.P. Meadows, S.W. Fesik, *Science* **1997**, *278*, 497–499.
8 J. Cavanagh, W.J. Fairbrother, A.G. Palmer III, N.J. Skelton, *Protein NMR Spectroscopy, Principles and Practice*, Academic Press, **1996**.
9 K. Pervushin, R. Riek, G. Wider, K. Wüthrich, *Proc. Natl. Acad. Sci. USA* **1997**, *94*, 12366–12371.
10 M. Pellecchia, P. Sebbel, U. Hermanns, K. Wüthrich, R. Glockshuber, *Nat. Struct. Biol.* **1999**, *6*, 336–339.
11 M. Pellecchia, D. Meininger, A.L. Shen, R. Jack, C.B. Kasper, D.S. Sem, *J. Am. Chem. Soc.*, **2001**, *123*, 4633–4634.
12 G. Otting, K. Wüthrich, *Q. Rev. Biophys.* **1990**, *23*, 39–96.
13 K. Pervushin, G. Wider, K. Wüthrich, *J. Biomol. NMR* **1998**, *12*, 345–348.
14 G. Gemmeker, W. Jahnke, H. Kessler, *J. Am. Chem. Soc.* **1993**, *115*, 11620–11621.
15 Y. Bai, J. S. Milne, L. Mayne, S. W. Englander *Prot. Struct. Func. Gen.* **1993**, *17*, 75–86.
16 M. Salzmann, K. Pervushin, G. Wider, H. Senn, K. Wüthrich, *Proc. Natl. Acad. Sci. USA* **1998**, *95*, 13585–13590.
17 A.L. Shen, L.H. Porter, T.E. Wilson, C.B. Kasper, *J. Biol. Chem.* **1989**, *264*, 7584–7589.
18 W. Farkas, C. Gilvarg, *J. Biol. Chem.* **1965**, *240*, 4717–4722.
19 G. Scapin, S. G. Reddy, R. Zheng, J.S. Blanchard, *Biochemistry* **1997**, *36*, 15081–15088.
20 N.K Goto, L.E. Kay, *Curr. Opin. Struct. Biol.* **2000**, *10*, 585–592.
21 E. Kupce, R. Freeman, *J. Magn. Reson.* B, **1995**, *109*, 329–333.
22 P.J. Hajduk, D.J. Augeri, J. Mack, R. Mendoza, J. Yang, S.F. Betz, S.W. Fesik, *J. Am. Chem. Soc.* **2000**, *122*, 7898–7904.
23 D. Neuhaus, M.P. Williamson, *The Nuclear Overhauser Effect in Structural and Conformational Analysis*, Wiley–VCH, New York, **2000**.
24 J. Klein, R. Meinecke, M. Mayer, B. Meyer, *J. Am. Chem. Soc.* **1999**, *121*, 5336–5337.
25 C. Dalvit, P. Pevarello, M. Tato', M. Veronesi, A. Vulpetti, M. Sundstrom, *J. Biomol. NMR* **2000**, *18*, 65–68.
26 D. Li, E.F. DeRose, R.E. London, *J. Biomol. NMR* **1999**, *15*, 71–76.
27 K. Pervushin, R. Riek, G. Wider, K. Wüthrich, *J. Am. Chem. Soc.* **1998**, *120*, 6394–6400.
28 T. Kuzuyama, S. Takahashi, M. Takagi, H. Seto, *J. Biol. Chem.* **2000**, *275*, 19928–19932.
29 V. Sklenar, *J. Magn. Reson. A.* **1995**, *114*, 132–135.
30 M. Wang, D.L. Roberts, R. Paschke, T.M. Shea, B.S. S. Masters, J.-J.P. Kim *Proc. Natl. Acad. Sci.* **1997**, *94*, 8411–8416.
31 R. Koradi, M. Billeter, K. Wüthrich, *J. Mol. Graph.* **1996**, *14*, 51–55.
32 H. Geen, R. Freeman, *J. Magn. Reson.* **1991**, *93*, 93–141.
33 Anil-Kumar, R.R. Ernst, K. Wüthrich, *Biochem. Biophys. Res. Comm.* **1980**, *95*, 1–6.
34 S.J.F. Vincent, C. Zwahlen, P.H. Bolton, T.M. Logan, G. Bodenhausen, *J. Am. Chem. Soc.* **1996**, *118*, 3531–3532.
35 P. Güntert, C. Mumenthaler, K. Wüthrich, *J. Mol. Biol.* **1997**, *273*, 283–298.
36 D.P. Meininger, M. Rance, M.A. Starovasnik, W.J. Fairbrother, N.J. Skelton *Biochemistry* **2000**, *39*, 26–36.

Subject Index